D0580935

ENGLISH
DICTIONARY

Published by Collins
An imprint of HarperCollins Publishers
Westerhill Road
Bishopbriggs
Glasgow G64 2QT

UK Edition:
Seventeenth edition 2016, 2018
ISBN 978-0-00-814167-7

Australian Edition:
Eleventh edition 2016, 2018
ISBN 978-0-00-814170-7

New Zealand Edition:
First Edition 2018
978-0-00-830130-9

South African Edition:
First Edition 2018
978-0-00-814643-6

10 9 8 7 6 5

www.collinsdictionary.com
www.collins.co.uk/languagesupport

Typeset by Davidson Publishing
Solutions, Glasgow

Printed in Italy by Grafica Veneta S.p.A.

A catalogue record for this book is available from the British Library.

If you would like to comment on any aspect of this book, please contact us at the given address or online.
E-mail: dictionaries@harpercollins.co.uk
 facebook.com/collinsdictionary
 @collinsdict

Acknowledgements
We would like to thank those authors and publishers who kindly gave permission for copyright material to be used in the Collins Corpus. We would also like to thank Times Newspapers Ltd for providing valuable data.

Foreword

In 1902 the first edition of the *Collins Gem Dictionary* was published with the aim of providing a comprehensive picture of current English that was clear and accessible to all—something that would be expansive in its coverage of the language yet small enough to fit in the pocket. Over a hundred years later on and now entering its 17th edition, the goal remains the same.

This edition, like all its predecessors, has been compiled with ease of use in mind. The layout is clear and clutter-free, taking you to the information you want with the minimum of fuss.

All main entry words, including abbreviations and combining forms, are listed in strict alphabetical sequence. Phrases and idioms are included in the meanings of the main entry word. Related words are shown in the same paragraph as the main entry word after an arrow, for example *absurdity* at *absurd*. Compounds are shown in alphabetical order at the end of the paragraph after an arrow, for example *air bag* at *air*.

The stripped back design is matched by definitions that are straightforward and direct. Where a word has more than one sense, the most common in today's language is given first. Other senses of a word—for example technical or historical senses—are explained after its primary present-day meaning. We have included simple pronunciations for certain words that may be unfamiliar or whose spelling is misleading.

We work hard at Collins to retain our long-held place at the forefront of research into new words and changes in the way English is used. As such we have watched attentively as new forms of electronic communication have blossomed over the past few years. With them have come new meanings of words like *meme* and *viral* as well as new questions of style and etiquette. Reflecting this we have included a Practical Writing Guide supplement that addresses how to write well across all mediums, whether it's in a tweet or in a letter.

This dictionary has been put together with reference to a powerful language tool developed over the years by our lexicographers: the Collins Corpus.

The corpus is a 4.5-billion-word database of lexical information and was the first of its kind. A constant flow of text is fed into it from sources all over the world – books, newspapers, websites, and even transcripts of radio and TV shows. By analysing this data Collins editors can build up a razor-sharp picture of today's language. But it's in the discovery of new words that this way of working really comes into its own – our 'monitor corpus' compares each new batch of text with what has gone before, automatically alerting our lexicographers to new coinages as soon as they begin to establish themselves in the common vocabulary. The result is a dictionary like this one, one which registers the newest arrivals to the English language, be it *emoji* or *vape*.

This *Collins Gem Dictionary* is a worthy addition to the Gem's venerable history as the most up-to-date and user-friendly little dictionary in the world.

Collins Gem – The world's bestselling little dictionary.

Contents

Editorial Staff

Editors

Ian Brookes

Mary O'Neill

Supplement

Martin and Simon Toseland

For the publisher

Gerry Breslin

Hannah Dove

Kerry Ferguson

Michelle Fullerton

Abbreviations

AD	anno Domini	*meteorol*	meteorology
adj	adjective	*mil*	military
adv	adverb	*n*	noun
anat	anatomy	*N*	North
archit	architecture	*naut*	nautical
astrol	astrology	*NZ*	New Zealand
Austral	Australia(n)	*obs*	obsolete
BC	before Christ	*offens*	offensive
biol	biology	*orig*	originally
Brit	British	*photog*	photography
chem	chemistry	*pl*	plural
C of E	Church of England	*prep*	preposition
conj	conjuction	*pron*	pronoun
E	East	*psychol*	psychology
e.g.	for example	*®*	trademark
esp.	especially	*RC*	Roman Catholic
etc.	et cetera	*S*	South
fem	feminine	*S Afr*	South Africa(n)
foll	followed	*Scot*	Scottish
geom	geometry	*sing*	singular
hist	history	*US*	United States
interj	interjection	*usu.*	usually
lit	literary	*vb*	verb
masc	masculine	*W*	West
med	medicine	*zool*	zoology

Aa

a *adj* indefinite article, used before a noun being mentioned for the first time

AA Alcoholics Anonymous; Automobile Association

aardvark *n* S African anteater with long ears and snout

AB able-bodied seaman

aback *adv* **taken aback** startled or disconcerted

abacus *n* beads on a wire frame, used for doing calculations

abalone [ab-a-**lone**-ee] *n* edible sea creature with a shell lined with mother of pearl

abandon *v* desert or leave (one's wife, children, etc.); give up (hope etc.) altogether ▷ *n* lack of inhibition ▶ **abandoned** *adj* deserted; uninhibited ▶ **abandonment** *n*

abase *v* humiliate or degrade (oneself) ▶ **abasement** *n*

abashed *adj* embarrassed and ashamed

abate *v* make or become less strong ▶ **abatement** *n*

abattoir [ab-a-**twahr**] *n* place where animals are killed for food

abbess *n* nun in charge of a convent

abbey *n* dwelling place of, or a church belonging to, a community of monks or nuns

abbot *n* head of an abbey of monks

abbreviate *v* shorten (a word) by leaving out some letters ▶ **abbreviation** *n* shortened form of a word or words

ABC[1] *n* alphabet; basics of a subject

ABC[2] Australian Broadcasting Corporation

abdicate *v* give up (the throne or a responsibility) ▶ **abdication** *n*

abdomen *n* part of the body containing the stomach and intestines ▶ **abdominal** *adj*

abduct *v* carry off, kidnap ▶ **abduction** *n* ▶ **abductor** *n*

aberration *n* sudden change from what is normal, accurate, or correct; brief lapse in control of one's thoughts or feelings ▶ **aberrant** *adj* showing aberration

abet *v* **abetting, abetted** help or encourage in wrongdoing ▶ **abettor** *n*

abeyance *n* **in abeyance** not in use

abhor *v* **-horring, -horred** detest utterly ▶ **abhorrent** *adj* hateful, loathsome ▶ **abhorrence** *n*

abide *v* endure, put up with; (*obs*) stay or dwell e.g. *abide with me* ▶ **abide by** *v* obey (the law, rules, etc.) ▶ **abiding** *adj* lasting

ability *n*, *pl* **-ties** competence, power; talent

abject *adj* utterly miserable; lacking all self-respect ▶ **abjectly** *adv*

abjure *v* deny or renounce on oath

ablative *n* case of nouns in Latin and other languages, indicating source, agent, or instrument of action

ablaze *adj* burning fiercely

able *adj* capable, competent ▶ **ably** *adv* ▶ **able-bodied** *adj* strong and healthy

ablutions *pl n* act of washing

abnormal *adj* not normal or usual ▶ **abnormally** *adv* ▶ **abnormality** *n*

aboard *adv*, *prep* on, in, onto, or into (a ship, train, or plane)

abode *n* home, dwelling

abolish v do away with ▸ **abolition** n ▸ **abolitionist** n person who wishes to do away with something, esp. slavery

abominable adj detestable, very bad ▸ **abominable snowman** large apelike creature said to live in the Himalayas ▸ **abominably** adv

abomination n someone or something that is detestable

aborigine [ab-or-**rij**-in-ee], **aboriginal** n original inhabitant of a country or region; (**A-**) original inhabitant of Australia ▸ **aboriginal** adj

abort v have an abortion or perform an abortion on; have a miscarriage; end a plan or process before completion ▸ **abortive** adj unsuccessful

abortion n operation to end a pregnancy; (informal) something grotesque ▸ **abortionist** n person who performs abortions, esp. illegally

abound v be plentiful ▸ **abounding** adj

about prep concerning, on the subject of; in or near (a place) ▷ adv nearly, approximately; nearby **about to** shortly going to **not about to** determined not to ▸ **about-turn** n complete change of attitude

above adv, prep over or higher (than); greater (than); superior (to) ▸ **above board** in the open, without dishonesty

abracadabra n supposedly magic word

abrasion n scraped area on the skin

abrasive adj harsh and unpleasant in manner; tending to rub or scrape ▷ n substance for cleaning or polishing by rubbing

abreast adv, adj side by side **abreast of** up to date with

abridge v shorten by using fewer words ▸ **abridgment**, **abridgement** n

abroad adv to or in a foreign country; at large

abrogate v cancel (a law or agreement) formally ▸ **abrogation** n

abrupt adj sudden, unexpected; blunt and rude ▸ **abruptly** adv ▸ **abruptness** n

abs pl n (informal) abdominal muscles

abscess n inflamed swelling containing pus

> **SPELLING TIP**
> There is a silent c in the middle of **abscess**

abscond v leave secretly

abseil [ab-**sale**] v go down a steep drop by a rope fastened at the top and tied around one's body

absent adj not present; lacking; inattentive ▷ v stay away ▸ **absently** adv ▸ **absence** n being away; lack ▸ **absentee** n person who should be present but is not ▸ **absenteeism** n persistent absence from work or school ▸ **absent-minded** adj inattentive or forgetful ▸ **absent-mindedly** adv

absinthe n strong green aniseed-flavoured liqueur

absolute adj complete, perfect; not limited, unconditional; pure e.g. absolute alcohol ▸ **absolutely** adv completely ▷ interj certainly, yes ▸ **absolutism** n government by a ruler with unrestricted power

absolve v declare to be free from blame or sin ▸ **absolution** n

absorb v soak up (a liquid); take in; engage the interest of (someone) ▸ **absorption** n ▸ **absorbent** adj able to absorb liquid ▸ **absorbency** n

abstain v choose not to do something; choose not to vote ▶ **abstainer** n ▶ **abstention** n abstaining, esp. from voting ▶ **abstinence** n abstaining, esp. from drinking alcohol ▶ **abstinent** adj

abstemious adj taking very little alcohol or food ▶ **abstemiousness** n

abstract adj existing as a quality or idea rather than an actual object; theoretical; (of art) using patterns of shapes and colours rather than realistic likenesses ▷ n summary; abstract work of art; abstract word or idea ▷ v summarize; remove ▶ **abstracted** adj lost in thought ▶ **abstraction** n

abstruse adj not easy to understand

absurd adj incongruous or ridiculous ▶ **absurdly** adv ▶ **absurdity** n

abundant adj plentiful ▶ **abundantly** adv ▶ **abundance** n

abuse v use wrongly; ill-treat violently; speak harshly and rudely to ▷ n prolonged ill-treatment; harsh and vulgar comments; wrong use ▶ **abuser** n ▶ **abusive** adj ▶ **abusively** adv ▶ **abusiveness** n

abut v abutting, abutted be next to or touching

abysmal adj (informal) extremely bad, awful ▶ **abysmally** adv

abyss n very deep hole or chasm

AC alternating current

a/c account

acacia [a-kay-sha] n tree or shrub with yellow or white flowers

academy n, pl -mies society to advance arts or sciences; institution for training in a particular skill; (Scot) secondary school ▶ **academic** adj of an academy or university; of theoretical interest only ▷ n lecturer or researcher at a university ▶ **academically** adv ▶ **academician** n member of an academy

acanthus n prickly plant

ACAS (in Britain) Advisory Conciliation and Arbitration Service

ACC (in New Zealand) Accident Compensation Corporation

accede v consent or agree (to); take up (an office or position)

accelerate v (cause to) move faster ▶ **acceleration** n ▶ **accelerator** n pedal in a motor vehicle to increase speed

> **SPELLING TIP**
> There is a double c in accelerate, but only one l

accent n distinctive style of pronunciation of a local, national, or social group; mark over a letter to show how it is pronounced; stress on a syllable or musical note ▷ v place emphasis on

accentuate v stress, emphasize ▶ **accentuation** n

accept v receive willingly; agree to; consider to be true ▶ **acceptance** n ▶ **acceptable** adj tolerable; satisfactory ▶ **acceptably** adv ▶ **acceptability** n

access n means of or right to approach or enter ▷ v obtain (data) from a computer ▶ **accessible** adj easy to reach ▶ **accessibility** n

accession n taking up of an office or position

accessory n, pl -ries supplementary part or object; person involved in a crime although not present when it is committed

accident n mishap, often causing injury; event happening by chance

a

▸ **accidental** adj happening by chance or unintentionally ▸ n (Music) symbol indicating that a sharp, flat, or natural note is not a part of the key signature ▸ **accidentally** adv

acclaim v applaud, praise ▸ n enthusiastic approval ▸ **acclamation** n

acclimatize v adapt to a new climate or environment ▸ **acclimatization** n

accolade n award, honour, or praise; award of knighthood

accommodate v provide with lodgings; have room for; oblige, do a favour for; adapt or adjust (to something) ▸ **accommodation** n house or room for living in ▸ **accommodating** adj obliging

> **SPELLING TIP**
> **Accommodation** and **accommodate** have two *c*s, but remember they have two *m*s as well

accompany v **-nying, -nied** go along with; occur with; provide a musical accompaniment ▸ **accompaniment** n something that accompanies; (Music) supporting part that goes with a solo ▸ **accompanist** n

accomplice n person who helps another to commit a crime

accomplish v manage to do; finish ▸ **accomplishment** n completion; personal ability or skill ▸ **accomplished** adj expert, proficient

accord n agreement, harmony ▸ v fit in with **of one's own accord** voluntarily

> **USAGE NOTE**
> Do not confuse *of one's own accord* with *on one's own account* 'for one's own benefit'

accordance n **in accordance with** conforming to or according to

according adv **according to** as stated by; in conformity with ▸ **accordingly** adv in an appropriate manner; consequently

accordion n portable musical instrument played by moving the two sides apart and together, and pressing a keyboard or buttons to produce the notes ▸ **accordionist** n

accost v approach and speak to, often aggressively

account n report, description; business arrangement making credit available; record of money received and paid out with the resulting balance; person's money held in a bank; importance, value ▸ v judge to be **on account of** because of ▸ **accountable** adj responsible to someone or for something ▸ **accountability** n

accounting n skill or practice of maintaining and auditing business accounts ▸ **accountant** n person who maintains and audits business accounts ▸ **accountancy** n

accoutrements pl n clothing and equipment for a particular activity

accredited adj authorized, officially recognized

accretion [ak-kree-shun] n gradual growth; something added

accrue v **-cruing, -crued** increase gradually ▸ **accrual** n

accumulate v gather together in increasing quantity ▸ **accumulation** n ▸ **accumulative** adj ▸ **accumulator** n (Brit & Aust) rechargeable electric battery

accurate adj exact, correct ▸ **accurately** adv ▸ **accuracy** n

accursed adj under a curse; detestable

accusative n grammatical case indicating the direct object

accuse v charge with wrongdoing
▶ **accused** n ▶ **accuser** n
▶ **accusing** adj ▶ **accusation** n
▶ **accusatory** adj

accustom v make used to
▶ **accustomed** adj usual; used (to);
in the habit (of)

ace n playing card with one symbol
on it; (informal) expert; (Tennis)
unreturnable serve ▷ adj (informal)
excellent

acerbic [ass-**sir**-bik] adj harsh or
bitter ▶ **acerbity** n

acetate [ass-it-tate] n (Chem)
salt or ester of acetic acid; (also
acetate rayon) synthetic textile
fibre

acetic [ass-**see**-tik] adj of or
involving vinegar ▶ **acetic acid**
colourless liquid used to make
vinegar

acetone [ass-it-tone] n colourless
liquid used as a solvent

acetylene [ass-**set**-ill-een] n
colourless flammable gas used in
welding metals

ache n dull continuous pain ▷ v be in
or cause continuous dull pain

achieve v gain by hard work
or ability ▶ **achievement** n
something accomplished

Achilles heel [a-**kill**-eez] n small
but fatal weakness

Achilles tendon n cord
connecting the calf muscle to the
heel bone

achromatic adj colourless; (Music)
with no sharps or flats

acid n (Chem) one of a class of
compounds, corrosive and sour
when dissolved in water, that
combine with a base to form a
salt; (slang) LSD ▷ adj containing
acid; sour-tasting; sharp or sour
in manner ▶ **acidic** adj ▶ **acidify**
v ▶ **acidity** n ▶ **Acid (House)** n
type of funk-based electronically

edited disco music with hypnotic
sound effects ▶ **acid rain** rain
containing acid from atmospheric
pollution ▶ **acid test** conclusive
test of value

acknowledge v admit,
recognize; indicate recognition
of (a person); say one has
received ▶ **acknowledgment,
acknowledgement** n

acme [ak-mee] n highest point of
achievement or excellence

acne [ak-nee] n pimply skin disease

acolyte n follower or attendant;
(Christianity) person who assists
a priest

aconite n poisonous plant with
hoodlike flowers; poison obtained
from this plant

acorn n nut of the oak tree

acoustic adj of sound and
hearing; (of a musical instrument)
not electronically amplified
▶ **acoustics** n science of sounds
▷ pl features of a room or building
determining how sound is heard
within it ▶ **acoustically** adv

acquaint v make familiar, inform
▶ **acquainted** adj ▶ **acquaintance** n
person known; personal
knowledge

acquiesce [ak-wee-**ess**] v
agree to what someone
wants ▶ **acquiescence** n
▶ **acquiescent** adj

> **USAGE NOTE**
> The use of to after acquiesce was
> formerly regarded as incorrect,
> with in being preferred, but
> either is now acceptable

acquire v gain, get ▶ **acquisition** n
thing acquired; act of getting

acquisitive adj eager to
gain material possessions
▶ **acquisitiveness** n

acquit v -quitting, -quitted
pronounce (someone) innocent;

a

behave in a particular way ► **acquittal** n

acre n measure of land, 4840 square yards (4046.86 square metres) ► **acreage** [ake-er-rij] n land area in acres

acrid [ak-rid] adj pungent, bitter

acrimonious adj bitter in speech or manner ► **acrimony** n

acrobat n person skilled in gymnastic feats requiring agility and balance ► **acrobatic** adj ► **acrobatics** pl n acrobatic feats

acronym n word formed from the initial letters of other words, such as NASA

across adv, prep from side to side (of); on or to the other side (of) ► **across the board** applying equally to all

acrostic n lines of writing in which the first or last letters of each line spell a word or saying

acrylic n, adj (synthetic fibre, paint, etc.) made from acrylic acid ► **acrylic acid** strong-smelling corrosive liquid

ACT Australian Capital Territory

act n thing done; law or decree; section of a play or opera; one of several short performances in a show; pretended attitude ▷ v do something; behave in a particular way; perform in a play, film, etc. ► **act of God** unpredictable natural event ► **acting** n art of an actor ▷ adj temporarily performing the duties of ► **actor, actress** n person who acts in a play, film, etc.

actinium n (Chem) radioactive chemical element

action n process of doing something; thing done; lawsuit; operating mechanism; minor battle ► **actionable** adj giving grounds for a lawsuit ► **action**

replay rerun of an event on a television recording

active adj moving, working; busy, energetic; (Grammar) (of a verb) in a form indicating that the subject is performing the action, e.g. threw in Kim threw the ball ► **actively** adv ► **activity** n state of being active; (pl -**ties**) leisure pursuit ► **activate** v make active ► **activation** n ► **activator** n ► **activist** n person who works energetically to achieve political or social goals ► **activism** n

actual adj existing in reality ► **actually** adv really, indeed ► **actuality** n

actuary n, pl -**aries** statistician who calculates insurance risks ► **actuarial** adj

actuate v start up (a device)

acuity [ak-kew-it-ee] n keenness of vision or thought

acumen [ak-yew-men] n ability to make good judgments

acupuncture n medical treatment involving the insertion of needles at various points on the body ► **acupuncturist** n

acute adj severe; keen, shrewd; sharp, sensitive; (of an angle) less than 90° ▷ n accent (´) over a letter to indicate the quality or length of its sound, as in café ► **acutely** adv ► **acuteness** n

AD anno Domini

ad n (informal) advertisement

adage n wise saying, proverb

adagio adv, n, pl -**gios** (Music) (piece to be played) slowly and gracefully

adamant adj unshakable in determination or purpose ► **adamantly** adv

Adam's apple n projecting lump of thyroid cartilage at the front of the throat

adapt v alter for new use or new conditions ▶ **adaptable** adj ▶ **adaptability** n ▶ **adaptation** n thing produced by adapting something; adapting ▶ **adaptor, adapter** n device for connecting several electrical appliances to a single socket

add v combine (further quantities); join (to something); say or write further

addendum n, pl -**da** addition; appendix to a book etc.

adder n small poisonous snake

addict n person who is unable to stop taking drugs; (informal) person devoted to something ▶ **addicted** adj ▶ **addiction** n ▶ **addictive** adj causing addiction

addition n adding; thing added **in addition** besides, as well ▶ **additional** adj ▶ **additionally** adv ▶ **additive** n something added, esp. to a foodstuff, to improve it or prevent deterioration

addled adj confused or unable to think clearly

address n place where a person lives; direction on a letter; location; formal public speech ▷ v mark the destination, as on an envelope; make a speech; give attention to (a problem, task, etc.) ▶ **addressee** n person addressed

SPELLING TIP
Remember to double the d and the s in **address**

adduce v mention something as evidence or proof

adenoids [ad-in-oidz] pl n mass of tissue at the back of the throat ▶ **adenoidal** adj having a nasal voice caused by swollen adenoids

adept adj, n very skilful (person)

adequate adj sufficient, enough; not outstanding ▶ **adequately** adv ▶ **adequacy** n

adhere v stick (to); be devoted (to) ▶ **adherence** n ▶ **adherent** n devotee, follower ▶ **adhesion** n sticking (to); joining together of parts of the body that are normally separate, as after surgery

adhesive n substance used to stick things together ▷ adj able to stick to things

ad hoc adj, adv (Latin) for a particular purpose only

adieu [a-dew] interj (lit) farewell, goodbye

ad infinitum [ad in-fin-eye-tum] adv (Latin) endlessly

adipose adj of or containing fat

adj adjective

adjacent adj near or next (to); having a common boundary; (Geom) (of a side in a right-angled triangle) lying between a specified angle and the right angle

adjective n word that adds information about a noun or pronoun ▶ **adjectival** adj

adjoin v be next to ▶ **adjoining** adj

adjourn v close (a court) at the end of a session; postpone temporarily; (informal) go elsewhere ▶ **adjournment** n

adjudge v declare (to be)

adjudicate v give a formal decision on (a dispute); judge (a competition) ▶ **adjudication** n ▶ **adjudicator** n

adjunct n subordinate or additional person or thing

adjure v command (to do); appeal earnestly

adjust v adapt to new conditions; alter slightly so as to be suitable ▶ **adjustable** adj ▶ **adjuster** n ▶ **adjustment** n

adjutant [aj-oo-tant] n army officer in charge of routine administration

ad-lib v -**libbing, -libbed** improvise a speech etc. without preparation ▷ *n* improvised remark

admin *n* (*informal*) administration

administer *v* manage (business affairs); organize and put into practice; give (medicine or treatment)

administrate *v* manage (an organization) ▶ **administrator** *n*

administration *n* management of an organization; people who manage an organization; government e.g. *the Bush administration*

administrative *adj* of the management of an organization

admiral *n* highest naval rank ▶ **Admiralty** *n* (in Britain) former government department in charge of the Royal Navy

admire *v* regard with esteem and approval ▶ **admirable** *adj* ▶ **admirably** *adv* ▶ **admiration** *n* ▶ **admirer** *n* ▶ **admiring** *adj* ▶ **admiringly** *adv*

admissible *adj* allowed to be brought in as evidence in court ▶ **admissibility** *n*

admission *n* permission to enter; entrance fee; confession

admit *v* -**mitting**, -**mitted** confess, acknowledge; concede the truth of; allow in ▶ **admittance** *n* permission to enter ▶ **admittedly** *adv* it must be agreed

admixture *v* mixture; ingredient

admonish *v* reprove sternly ▶ **admonition** *n*

ad nauseam [ad naw-zee-am] *adv* (*Latin*) to a boring or sickening extent

ado *n* (*lit*) fuss, trouble

adobe [ad-**oh**-bee] *n* sun-dried brick

adolescence *n* period between puberty and adulthood ▶ **adolescent** *n*, *adj* (person) between puberty and adulthood

adopt *v* take (someone else's child) as one's own; take up (a plan or principle) ▶ **adopter**, **adoptor** *n* ▶ **adoption** *n* ▶ **adoptive** *adj* related by adoption

adore *v* love intensely; worship ▶ **adorable** *adj* ▶ **adoration** *n* ▶ **adoring** *adj* ▶ **adoringly** *adv*

adorn *v* decorate, embellish ▶ **adornment** *n*

adrenal [ad-**reen**-al] *adj* near the kidneys ▶ **adrenal glands** glands covering the top of the kidneys

adrenalin, adrenaline *n* hormone secreted by the adrenal glands in response to stress

adrift *adj*, *adv* drifting; without a clear purpose

adroit *adj* quick and skilful ▶ **adroitly** *adv* ▶ **adroitness** *n*

adsorb *v* (of a gas or vapour) condense and form a thin film on a surface ▶ **adsorption** *n*

adulation *n* uncritical admiration

adult *adj* fully grown, mature ▷ *n* adult person or animal ▶ **adulthood** *n*

adulterate *v* spoil something by adding inferior material ▶ **adulteration** *n*

adultery *n*, *pl* -**teries** sexual unfaithfulness of a husband or wife ▶ **adulterer**, **adulteress** *n* ▶ **adulterous** *adj*

adv adverb

advance *v* go or bring forward; further (a cause); propose (an idea); lend (a sum of money) ▷ *n* forward movement; improvement; loan ▷ *pl* approaches to a person with the hope of starting a romantic or sexual relationship ▷ *adj* done or happening before an event **in advance** ahead ▶ **advanced** *adj* at a late stage in development; not elementary ▶ **advancement** *n* promotion

advantage n more favourable position or state; benefit or profit; (Tennis) point scored after deuce **take advantage of** use (a person) unfairly; use (an opportunity) ▶ **advantageous** adj ▶ **advantageously** adv

advent n arrival; (**A-**) season of four weeks before Christmas ▶ **Adventist** n member of a Christian sect that believes in the imminent return of Christ (also **Seventh Day Adventist**)

adventitious adj added or appearing accidentally

adventure n exciting and risky undertaking or exploit ▶ **adventurer**, **adventuress** n person who unscrupulously seeks money or power; person who seeks adventures ▶ **adventurous** adj

adverb n word that adds information about a verb, adjective, or other adverb ▶ **adverbial** adj

adversary [ad-verse-er-ree] n, pl **-saries** opponent or enemy

adverse adj unfavourable; antagonistic or hostile ▶ **adversely** adv ▶ **adversity** n very difficult or hard circumstances

advert n (informal) advertisement

advertise v present or praise (goods or services) to the public in order to encourage sales; make (a vacancy, event, etc.) known publicly ▶ **advertisement** n public announcement to sell goods or publicize an event ▶ **advertiser** n ▶ **advertising** adj, n

SPELLING TIP
Some verbs can be spelt ending in either -ise or -ize, but **advertise** and **advise** always have an s

advice n recommendation as to what to do ▶ **advise** v offer advice to; notify (someone) ▶ **advisable** adj prudent, sensible ▶ **advisability** n ▶ **advisory** adj giving advice ▶ **advised** adj considered, thought-out e.g. ill-advised ▶ **advisedly** adv deliberately

SPELLING TIP
The verb is **advise** with an s and the noun is **advice** with a c

adviser, **advisor** n person who offers advice, e.g. on careers to students or school pupils

advocaat n liqueur with a raw egg base

advocate v propose or recommend ▷ n person who publicly supports a cause; (Scot & S Afr) barrister ▶ **advocacy** n

adze n tool with an arched blade at right angles to the handle

aegis [ee-jiss] n sponsorship, protection

aeolian harp [ee-oh-lee-an] n musical instrument that produces sounds when the wind passes over its strings

aeon [ee-on] n immeasurably long period of time

aerate v put gas into (a liquid), as when making a fizzy drink ▶ **aeration** n

aerial adj in, from, or operating in the air; relating to aircraft ▷ n metal pole, wire, etc., for receiving or transmitting radio or TV signals ▶ **aerial top dressing** spreading of fertilizer from an aeroplane onto remote areas

aerobatics pl n stunt flying ▶ **aerobatic** adj

aerobics n exercises designed to increase the amount of oxygen in the blood ▶ **aerobic** adj

aerodrome n small airport

aerodynamics n study of how air flows around moving solid objects ▶ **aerodynamic** adj

aerofoil n part of an aircraft, such as the wing, designed to give lift

aerogram n airmail letter on a single sheet of paper that seals to form an envelope

aeronautics n study or practice of aircraft flight ▶ **aeronautical** adj

aeroplane n powered flying vehicle with fixed wings

aerosol n pressurized can from which a substance can be dispensed as a fine spray

> **SPELLING TIP**
> Be careful not to confuse the spelling aer- with air-

aerospace n earth's atmosphere and space beyond

aesthetic [iss-thet-ik] adj relating to the appreciation of art and beauty ▶ **aesthetics** n study of art, beauty, and good taste ▶ **aesthetically** adv ▶ **aesthete** [eess-theet] n person who has or affects an extravagant love of art ▶ **aestheticism** n

aether n same as **ether**

aetiology [ee-tee-ol-a-jee] n same as **etiology**

afar adv from afar from or at a great distance

affable adj friendly and easy to talk to ▶ **affably** adv ▶ **affability** n

affair n event or happening; sexual relationship outside marriage; thing to be done or attended to ▶ pl personal or business interests; matters of public interest

affect[1] v act on, influence; move (someone) emotionally

> **USAGE NOTE**
> Do not confuse affect meaning 'influence' with effect meaning 'accomplish'

affect[2] v put on a show of; wear or use by preference ▶ **affectation** n attitude or manner put on to impress ▶ **affected** adj displaying affectation; pretended

affection n fondness or love ▶ **affectionate** adj loving ▶ **affectionately** adv

affianced [af-fie-anst] adj (old-fashioned) engaged to be married

affidavit [af-fid-**dave**-it] n written statement made on oath

affiliate v (of a group) link up with a larger group ▶ **affiliation** n

affinity n, pl -**ties** close connection or liking; close resemblance; chemical attraction

affirm v declare to be true; uphold or confirm (an idea or belief) ▶ **affirmation** n ▶ **affirmative** n, adj (word or phrase) indicating agreement

affix v attach or fasten ▶ n word or syllable added to a word to change its meaning

afflict v give pain or grief to ▶ **affliction** n

affluent adj having plenty of money ▶ **affluence** n wealth

afford v have enough money to buy; be able to spare (the time etc.); give or supply ▶ **affordable** adj

afforest v plant trees on ▶ **afforestation** n

affray n (Brit, Aust & NZ Law) noisy fight, brawl

affront v, n insult

Afghan adj of Afghanistan or its language ▶ **Afghan hound** large slim dog with long silky hair

aficionado [af-fish-yo-**nah**-do] n, pl -**dos** enthusiastic fan of something or someone

afield adv far afield far away

aflame adj burning

afloat adv, adj floating; at sea

afoot adv, adj happening, in operation

aforesaid, aforementioned adj referred to previously

aforethought adj premeditated e.g. with malice aforethought

Afr Africa(n)

afraid adj frightened; regretful

afresh adv again, anew

African adj of Africa ▷ n person from Africa ▶ **African violet** house plant with pink or purple flowers and hairy leaves

Afrikaans n language used in S Africa, descended from Dutch

Afrikaner n White S African whose mother tongue is Afrikaans

Afro- combining form African e.g. Afro-Caribbean

aft adv at or towards the rear of a ship or aircraft

after prep following in time or place; in pursuit of; in imitation of ▷ adv at a later time than ▷ adv at a later time ▶ **afters** pl n (Brit informal) dessert

afterbirth n material expelled from the womb after childbirth

aftercare n support given to a person discharged from a hospital or prison; regular care required to keep something in good condition

aftereffect n result occurring some time after its cause

afterglow n glow left after a source of light has gone; pleasant feeling left after an enjoyable experience

afterlife n life after death

aftermath n results of an event considered together

afternoon n time between noon and evening

aftershave n lotion applied to the face after shaving

afterthought n idea occurring later; something added later

afterwards, afterward adv later

Ag (Chem) silver

again adv once more; in addition

against prep in opposition to or contrast with; in contact with; as a protection from

agape adj (of the mouth) wide open; (of a person) very surprised

agaric n fungus with gills on the underside of the cap, such as a mushroom

agate [ag-git] n semiprecious form of quartz with striped colouring

age n length of time a person or thing has existed; time of life; latter part of human life; period of history; long time ▷ v **ageing** or **aging, aged** make or grow old ▶ **aged** adj [ay-jid] old; [rhymes with raged] being at the age of ▶ **ageing, aging** n, adj ▶ **ageless** adj apparently never growing old; seeming to have existed for ever ▶ **age-old** adj very old

agency n, pl **-cies** organization providing a service; business or function of an agent; (old-fashioned) power or action by which something happens

agenda n list of things to be dealt with, esp. at a meeting

agent n person acting on behalf of another; person or thing producing an effect

agent provocateur [azh-on prov-vok-at-tur] n, pl **agents provocateurs** [azh-on prov-vok-at-tur] person employed by the authorities to tempt people to commit illegal acts and so be discredited or punished

agglomeration v make confused mass or cluster

aggrandize v make greater in size, power, or rank ▶ **aggrandizement** n

aggravate v make worse; (chiefly informal) annoy ▶ **aggravating** adj ▶ **aggravation** n

a

SPELLING TIP
Remember that there is an *a* (not an *e*) in the middle of **aggravate**

aggregate *n* total; rock consisting of a mixture of minerals; sand or gravel used to make concrete ▷ *adj* gathered into a mass; total or final ▷ *v* combine into a whole ▶ **aggregation** *n*

aggression *n* hostile behaviour; unprovoked attack ▶ **aggressive** *adj* showing aggression; forceful ▶ **aggressively** *adv* ▶ **aggressiveness** *n* ▶ **aggressor** *n*

SPELLING TIP
Note the double *g* and double *s* in **aggressive**

aggrieved *adj* upset and angry

aggro (*slang*) *n* (*Brit, Aust & NZ*) aggressive behaviour ▷ *adj* (*Aust & NZ*) aggressive

aghast *adj* overcome with amazement or horror

agile *adj* nimble, quick-moving; mentally quick ▶ **agility** *n*

agitate *v* disturb or excite; stir or shake (a liquid); stir up public opinion for or against something ▶ **agitation** *n* ▶ **agitator** *n*

aglow *adj* glowing

AGM annual general meeting

agnostic *n* person who believes that it is impossible to know whether God exists ▷ *adj* of agnostics ▶ **agnosticism** *n*

ago *adv* in the past

agog *adj* eager or curious

agony *n*, *pl* **-nies** extreme physical or mental pain ▶ **agonize** *v* worry greatly; (cause to) suffer agony ▶ **agonizing** *adj* ▶ **agony aunt** journalist who gives advice in an agony column ▶ **agony column** newspaper or magazine feature offering advice on personal problems

agoraphobia *n* fear of open spaces ▶ **agoraphobic** *n*, *adj*

agrarian *adj* of land or agriculture

agree *v* **agreeing, agreed** be of the same opinion; consent; reach a joint decision; be consistent; (foll by *with*) be suitable to (one's health or digestion) ▶ **agreeable** *adj* pleasant and enjoyable; prepared to consent ▶ **agreeably** *adv* ▶ **agreement** *n* agreeing; contract

agriculture *n* raising of crops and livestock ▶ **agricultural** *adj* ▶ **agriculturalist** *n*

agronomy [ag-ron-om-mee] *n* science of soil management and crop production ▶ **agronomist** *n*

aground *adv* onto the bottom of shallow water

ague [aig-yew] *n* (*old-fashioned*) periodic fever with shivering

ahead *adv* in front; forwards

ahoy *interj* shout used at sea to attract attention

AI artificial insemination; artificial intelligence

aid *v*, *n* (give) assistance or support

aide *n* assistant

aide-de-camp [aid-de-kom] *n*, *pl* **aides-de-camp** [aid-de-kom] military officer serving as personal assistant to a senior

AIDS acquired immunodeficiency syndrome: a viral disease that destroys the body's ability to fight infection

AIH artificial insemination by husband

ail *v* trouble, afflict; be ill ▶ **ailing** *adj* sickly ▶ **ailment** *n* illness

aileron *n* movable flap on an aircraft wing which controls rolling

aim *v* point (a weapon or missile) or direct (a blow or remark) at a target; propose or intend ▷ *n* aiming; intention, purpose ▶ **aimless** *adj* having no purpose ▶ **aimlessly** *adv*

ain't (*not standard*) am not; is not; are not; has not; have not

air *n* mixture of gases forming the earth's atmosphere; space above the ground; sky; breeze; quality or manner; tune ▷ *pl* affected manners ▷ *v* make known publicly; expose to air to dry or ventilate **on the air** in the act of broadcasting on radio or television ▶ **airless** *adj* stuffy ▶ **air bag** vehicle safety device which inflates automatically in a crash to protect the driver or passenger when he or she is thrown forward ▶ **airborne** *adj* carried by air; (of aircraft) flying ▶ **airbrush** *n* atomizer spraying paint by compressed air ▶ **airfield** *n* place where aircraft can land and take off ▶ **air force** branch of the armed forces responsible for air warfare ▶ **air gun** gun fired by compressed air ▶ **air hostess** female flight attendant ▶ **airlift** *n* transport of troops or cargo by aircraft when other routes are blocked ▷ *v* transport by airlift ▶ **airlock** *n* air bubble blocking the flow of liquid in a pipe; airtight chamber ▶ **airmail** *n* system of sending mail by aircraft; mail sent in this way ▶ **airman** *n* member of the air force ▶ **air miles** miles of free air travel that can be earned by buying airline tickets and various other products ▶ **airplay** *n* broadcast performances of a record on radio ▶ **airport** *n* airfield for civilian aircraft, with facilities for aircraft maintenance and passengers ▶ **air raid** attack by aircraft ▶ **airship** *n* lighter-than-air self-propelled aircraft ▶ **airspace** *n* atmosphere above a country, regarded as its territory ▶ **airstrip** *n* cleared area where aircraft can take off and

land ▶ **airtight** *adj* sealed so that air cannot enter

air conditioning *n* system that controls the temperature and humidity of the air in a building ▶ **air conditioner**

aircraft *n* any machine that flies, such as an aeroplane ▶ **aircraft carrier** warship for the launching and landing of aircraft

airing *n* exposure to air for drying or ventilation; exposure to public debate

airline *n* company providing scheduled flights for passengers and cargo ▶ **airliner** *n* large passenger aircraft

airworthy *adj* (of aircraft) fit to fly ▶ **airworthiness** *n*

airy *adj* **airier, airiest** well-ventilated; light-hearted and casual ▶ **airily** *adv*

aisle [rhymes with **mile**] *n* passageway separating seating areas in a church, theatre, etc., or row of shelves in a supermarket

ajar *adj, adv* (of a door) partly open

akimbo *adj* **with arms akimbo** with hands on hips and elbows outwards

akin *adj* **akin to** similar, related

alabaster *n* soft white translucent stone

à la carte *adj, adv* (of a menu) having dishes individually priced

alacrity *n* speed, eagerness

à la mode *adj* fashionable

alarm *n* sudden fear caused by awareness of danger; warning sound; device that gives this; alarm clock ▷ *v* fill with fear ▶ **alarming** *adj* ▶ **alarmist** *n* person who alarms others needlessly ▶ **alarm clock** clock which sounds at a set time to wake someone up

alas *adv* unfortunately, regrettably

albatross n large sea bird with very long wings

albeit conj even though

albino n, pl **-nos** person or animal with white skin and hair and pink eyes

album n book with blank pages for keeping photographs or stamps in; long-playing record

albumen n egg white

albumin, albumen n protein found in blood plasma, egg white, milk, and muscle

alchemy n medieval form of chemistry concerned with trying to turn base metals into gold and to find the elixir of life ▶ **alchemist** n

alcohol n colourless flammable liquid present in intoxicating drinks; intoxicating drinks generally ▶ **alcoholic** adj of alcohol ▷ n person addicted to alcohol ▶ **alcoholism** n addiction to alcohol

alcopop n (Brit, Aust & S Afr informal) alcoholic drink that tastes like a soft drink

alcove n recess in the wall of a room

aldehyde n one of a group of chemical compounds derived from alcohol by oxidation

alder n tree related to the birch

alderman n formerly, senior member of a local council

ale n kind of beer

alert adj watchful, attentive ▷ n warning of danger ▷ v warn of danger; make (someone) aware of (a fact) **on the alert** watchful ▶ **alertness** n

A level n (Brit) (pass in a subject at) the advanced level of the GCE

alfalfa n kind of plant used to feed livestock

alfresco adv, adj in the open air

algae [al-jee] pl n plants which live in or near water and have no true stems, leaves, or roots

algebra n branch of mathematics using symbols to represent numbers ▶ **algebraic** adj

ALGOL n (Computing) programming language for mathematical and scientific purposes

algorithm n logical arithmetical or computational procedure for solving a problem

alias adv also known as ▷ n false name

alibi n plea of being somewhere else when a crime was committed; (informal) excuse

alien adj foreign; repugnant (to); from another world ▷ n foreigner; being from another world ▶ **alienate** v cause to become hostile ▶ **alienation** n

alight¹ v step out of (a vehicle); land

alight² adj on fire; lit up

align [a-line] v bring (a person or group) into agreement with the policy of another; place in a line ▶ **alignment** n

alike adj like, similar ▷ adv in the same way

alimentary adj of nutrition ▶ **alimentary canal** food passage in the body

alimony n allowance paid under a court order to a separated or divorced spouse

A-line adj (of a skirt) slightly flared

aliquot (Maths) adj of or denoting an exact divisor of a number ▷ n exact divisor

alive adj living, in existence; lively **alive to** aware of **alive with** swarming with

alkali [alk-a-lie] n substance which combines with acid and neutralizes it to form a salt ▶ **alkaline** adj ▶ **alkalinity** n ▶ **alkaloid** n any of a group of organic compounds containing nitrogen

all adj whole quantity or number (of) ▷ adv wholly, entirely; (in the score of games) each **give one's all** make the greatest possible effort ▶ **all in** adj exhausted; (of wrestling) with no style forbidden ▷ adv with all expenses included ▶ **all right** adj adequate, satisfactory; unharmed ▷ interj expression of approval or agreement ▶ **all-rounder** n person with ability in many fields

Allah n name of God in Islam

allay v reduce (fear or anger)

allege v state without proof ▶ **alleged** adj ▶ **allegedly** adv ▶ **allegation** n unproved accusation

allegiance n loyalty to a person, country, or cause

allegory n, pl -ries story with an underlying meaning as well as the literal one ▶ **allegorical** adj

allegretto adv, n, pl -tos (Music) (piece to be played) fairly quickly or briskly

allegro adv, n, pl -gros (Music) (piece to be played) in a brisk lively manner

alleluia interj same as **hallelujah**

allergy n, pl -gies extreme sensitivity to a substance, which causes the body to react to it ▶ **allergic** adj having or caused by an allergy ▶ **allergen** n substance capable of causing an allergic reaction

alleviate v lessen (pain or suffering) ▶ **alleviation** n

alley n narrow street or path; long narrow enclosure in which tenpin bowling or skittles is played

alliance n state of being allied; formal relationship between countries or groups for a shared purpose

alligator n reptile of the crocodile family, found in the southern US and China

alliteration n use of the same sound at the start of words occurring together, e.g. moody music ▶ **alliterative** adj

allocate v assign to someone or for a particular purpose ▶ **allocation** n

allot v -lotting, -lotted assign as a share or for a particular purpose ▶ **allotment** n distribution; portion allotted; small piece of public land rented to grow vegetables on

allotrope n any of two or more physical forms in which an element can exist

allow v permit; set aside; acknowledge (a point or claim) ▶ **allow for** v take into account ▶ **allowable** adj ▶ **allowance** n amount of money given at regular intervals; amount permitted **make allowances for** treat or judge (someone) less severely because he or she has special problems; take into account

alloy n mixture of two or more metals ▷ v mix (metals)

allspice n spice made from the berries of a tropical American tree

allude v (foll by to) refer indirectly to ▶ **allusion** n indirect reference ▶ **allusive** adj

USAGE NOTE
Allude is followed by to. Be careful not to confuse this word with elude meaning 'escape'. Also be careful not to confuse allusion with illusion meaning 'fallacy' or 'fantasy'.

allure n attractiveness ▷ v entice or attract ▶ **alluring** adj

alluvium n fertile soil deposited by flowing water ▶ **alluvial** adj

ally n, pl -lies country, person, or group with an agreement to support another ▷ v -lying, -lied ▶ **ally oneself with** join as an ally ▶ **allied** adj

alma mater n school, university, or college that one attended

almanac n yearly calendar with detailed information on anniversaries, phases of the moon, etc.

almighty adj having absolute power; (informal) very great ▸ n **the Almighty** God

almond n edible oval-shaped nut which grows on a small tree

almoner n (Brit) formerly, a hospital social worker

almost adv very nearly

alms [ahmz] pl n (old-fashioned) gifts to the poor

aloe n plant with fleshy spiny leaves ▸ pl bitter drug made from aloe leaves

aloft adv in the air; in a ship's rigging

alone adj, adv without anyone or anything else

along prep over part or all the length of ▸ adv forward; in company with others ▸ **alongside** prep, adv beside (something)

> **USAGE NOTE**
> Avoid following alongside with of. Its use is unnecessary

aloof adj distant or haughty in manner ▸ **aloofness** n

alopecia [al-loh-pee-sha] n loss of hair

aloud adv in an audible voice

alpaca n Peruvian llama; wool or cloth made from its hair

alpenstock n iron-tipped stick used by climbers

alpha n first letter in the Greek alphabet ▸ **alpha male** dominant male animal or person in a group

alphabet n set of letters used in writing a language ▸ **alphabetical** adj in the conventional order of the letters of an alphabet ▸ **alphabetically** adv ▸ **alphabetize** v put in alphabetical order

alpine adj of high mountains; (A-) of the Alps ▸ n mountain plant

already adv before the present time; sooner than expected

alright adj, interj all right

> **USAGE NOTE**
> The form alright, though very common, is still considered by many people to be wrong or less acceptable than all right

Alsatian n large wolflike dog

also adv in addition, too ▸ **also-ran** n loser in a race, competition, or election

alt. combining form (informal) alternative e.g. alt.rock

altar n table used for Communion in Christian churches; raised structure on which sacrifices are offered and religious rites are performed ▸ **altarpiece** n work of art above and behind the altar in some Christian churches

alter v make or become different ▸ **alteration** n

altercation n heated argument

alter ego n second self; very close friend

alternate v (cause to) occur by turns ▸ adj occurring by turns; every second (one) of a series ▸ **alternately** adv ▸ **alternation** n ▸ **alternator** n electric generator for producing alternating current ▸ **alternating current** electric current that reverses direction at frequent regular intervals

alternative n one of two choices ▸ adj able to be done or used instead of something else; (of medicine, lifestyle, etc.) not conventional ▸ **alternatively** adv

although conj despite the fact that

altimeter [al-tim-it-er] n instrument that measures altitude

altitude *n* height above sea level

alto *n, pl* -**tos** (*Music*) short for **contralto**; (singer with) the highest adult male voice; instrument with the second-highest pitch in its group

altogether *adv* entirely; on the whole; in total

altruism *n* unselfish concern for the welfare of others ▸ **altruistic** *adj* ▸ **altruistically** *adv*

aluminium *n* (*Chem*) light silvery-white metal that does not rust

alumnus [al-**lumm**-nuss] *n, pl* -**ni** [-nie] graduate of a college ▸ **alumna** [al-**lumm**-na] *n fem, pl* -**nae** [-nee]

always *adv* at all times; for ever

alyssum *n* garden plant with small yellow or white flowers

AM amplitude modulation; (in Britain) Member of the National Assembly for Wales

am *v* see **be**

a.m. ante meridiem: before noon

amakhosi *n* plural of **inkhosi**

amalgam *n* blend or combination; alloy of mercury and another metal

amalgamate *v* combine or unite ▸ **amalgamation** *n*

amandla [ah-**mand**-lah] *n* (*S Afr*) political slogan calling for power to the Black population

amanuensis [am-man-yew-en-siss] *n, pl* -**ses** [-seez] person who writes from dictation

amaranth *n* imaginary flower that never fades; lily-like plant with red, green, or purple flowers

amaryllis *n* lily-like plant with large red, pink, or white flowers

amass *v* collect or accumulate

amateur *n* person who engages in a sport or activity as a pastime rather than as a profession; person unskilled in something ▸ *adj* not professional ▸ **amateurish** *adj* lacking skill ▸ **amateurishly** *adv*

amatory *adj* relating to romantic or sexual love

amaze *v* surprise greatly, astound ▸ **amazing** *adj* ▸ **amazingly** *adv* ▸ **amazement** *n*

Amazon *n* strong and powerful woman; legendary female warrior ▸ **Amazonian** *adj*

ambassador *n* senior diplomat who represents his or her country in another country ▸ **ambassadorial** *adj*

amber *n* clear yellowish fossil resin ▸ *adj* brownish-yellow

ambergris [**am**-ber-greece] *n* waxy substance secreted by the sperm whale, used in making perfumes

ambidextrous *adj* able to use both hands with equal ease

ambience *n* atmosphere of a place

ambient *adj* surrounding

ambiguous *adj* having more than one possible meaning ▸ **ambiguously** *adv* **ambiguity** *n*

ambit *n* limits or boundary

ambition *n* desire for success; something desired, goal ▸ **ambitious** *adj* ▸ **ambitiously** *adv*

ambivalence *n* state of feeling two conflicting emotions at the same time ▸ **ambivalent** *adj* ▸ **ambivalently** *adv*

amble *v* walk at a leisurely pace ▸ *n* leisurely walk or pace

ambrosia *n* (*Myth*) food of the gods ▸ **ambrosial** *adj*

ambulance *n* motor vehicle designed to carry sick or injured people

ambush *n* act of waiting in a concealed position to make a surprise attack; attack from a concealed position ▸ *v* attack from a concealed position

ameliorate [am-**meal**-yor-rate] *v* make (something) better ▸ **amelioration** *n*

a

amen interj so be it: used at the end of a prayer

amenable adj likely or willing to cooperate

amend v make small changes to correct or improve (something) ▸ **amendment** n

amends pl n **make amends for** compensate for

amenity n, pl -**ties** useful or enjoyable feature

American adj of the United States of America or the American continent ▸ n person from America or the American continent ▸ **Americanism** n expression or custom characteristic of Americans

Americano n, pl -**nos** espresso coffee diluted with hot water

amethyst [am-myth-ist] n a bluish-violet variety of quartz used as a gemstone

amiable adj friendly, pleasant-natured ▸ **amiably** adv ▸ **amiability** n

amicable adj friendly ▸ **amicably** adv

amid, amidst prep in the middle of, among ▸ **amidships** adv at or towards the middle of a ship

> **USAGE NOTE**
> The form amidst is old-fashioned. Nowadays use amid or among.

amino acid [am-**mean**-oh] n organic compound found in protein

amiss adv wrongly, badly ▸ adj wrong, faulty **take something amiss** be offended by something

amity n friendship

ammeter n instrument for measuring electric current

ammonia n strong-smelling alkaline gas containing hydrogen and nitrogen; solution of this in water

ammonite n fossilized spiral shell of an extinct sea creature

ammunition n bullets, bombs, and shells that can be fired from or as a weapon; facts that can be used in an argument

amnesia n loss of memory ▸ **amnesiac** adj, n

amnesty n, pl -**ties** general pardon for offences against a government

amniocentesis n, pl -**ses** removal of some amniotic fluid to test for possible abnormalities in a fetus

amniotic fluid n fluid surrounding a fetus in the womb

amoeba [am-**mee**-ba] n, pl -**bae**, -**bas** microscopic single-celled animal able to change its shape

amok [a-**muck**, a-**mock**] adv **run amok** run about in a violent frenzy

among, amongst prep in the midst of; in the group or number of; to each of e.g. divide it among yourselves

> **USAGE NOTE**
> Among is generally used when more than two things are mentioned: It's hard to choose among friends. The form amongst is rather old-fashioned.

amoral [aim-**mor**-al] adj without moral standards ▸ **amorality** n

amorous adj feeling, showing, or relating to sexual love ▸ **amorously** adv

amorphous adj without distinct shape

amortize v pay off (a debt) gradually by periodic transfers to a sinking fund

amount n extent or quantity ▸ v (foll by to) be equal or add up to

amour n (secret) love affair

amp n ampere; (informal) amplifier

ampere [am-**pair**] n basic unit of electric current

ampersand n the character (&), meaning and

amphetamine [am-fet-am-mean] *n* drug used as a stimulant

amphibian *n* animal that lives on land but breeds in water; vehicle that can travel on both land and water ▶ **amphibious** *adj* living or operating both on land and water

amphitheatre *n* open oval or circular building with tiers of seats rising round an arena

amphora [am-for-a] *n, pl* **-phorae** two-handled ancient Greek or Roman jar

ample *adj* more than sufficient; large ▶ **amply** *adv*

amplifier *n* device used to amplify a current or sound signal

amplify *v* **-fying, -fied** increase the strength of (a current or sound signal); explain in more detail; increase the size or effect of ▶ **amplification** *n*

amplitude *n* greatness of extent

ampoule *n* small sealed glass vessel containing liquid for injection

amputate *v* cut off (a limb or part of a limb) for medical reasons ▶ **amputation** *n*

amuck *adv* same as **amok**

amulet *n* something carried or worn as a protection against evil

amuse *v* cause to laugh or smile; entertain or divert ▶ **amusing** *adj* ▶ **amusement** *n* state of being amused; something that amuses

an *adj* form of **a** used before vowels, and sometimes before **h**

> **USAGE NOTE**
> This form of the indefinite article is also used with abbreviations which have a vowel sound when read aloud: *an MA*. It is old-fashioned to use *an* before words like *hotel* or *historic*.

anabolic steroid *n* synthetic steroid hormone used to stimulate muscle and bone growth

anachronism [an-nak-kron-iz-zum] *n* person or thing placed in the wrong historical period or seeming to belong to another time ▶ **anachronistic** *adj*

anaconda *n* large S American snake which kills by constriction

anaemia [an-neem-ea-a] *n* deficiency in the number of red blood cells ▶ **anaemic** *adj* having anaemia; pale and sickly; lacking vitality

anaesthetic [an-niss-thet-ik] *n, adj* (substance) causing loss of bodily feeling ▶ **anaesthesia** [an-niss-theez-ea-a] *n* loss of bodily feeling ▶ **anaesthetist** [an-neess-thet-ist] *n* doctor trained to administer anaesthetics ▶ **anaesthetize** *v*

anagram *n* word or phrase made by rearranging the letters of another word or phrase

anal [ain-al] *adj* of the anus

analgesic [an-nal-jeez-ik] *n, adj* (drug) relieving pain ▶ **analgesia** *n* absence of pain

analogous *adj* similar in some respects

analogue *n* something that is similar in some respects to something else ▶ *adj* displaying information by means of a dial

analogy *n, pl* **-gies** similarity in some respects; comparison made to show such a similarity ▶ **analogical** *adj*

analysis *n, pl* **-ses** separation of a whole into its parts for study and interpretation; psychoanalysis ▶ **analyse** *v* make an analysis of (something); psychoanalyse ▶ **analyst** *n* person skilled in analysis ▶ **analytical, analytic** *adj* ▶ **analytically** *adv*

anarchism *n* doctrine advocating the abolition of government

anarchist n person who advocates the abolition of government; person who causes disorder ▶ **anarchistic** adj

anarchy [an-ark-ee] n lawlessness and disorder; lack of government in a state ▶ **anarchic** adj

anathema [an-nath-im-a] n detested person or thing

anatomy n, pl -**mies** science of the structure of the body; physical structure; person's body; detailed analysis ▶ **anatomical** adj ▶ **anatomically** adv ▶ **anatomist** n expert in anatomy

ANC African National Congress

ancestor n person from whom one is descended; forerunner ▶ **ancestral** adj ▶ **ancestry** n lineage or descent

anchor n heavy hooked device attached to a boat by a cable and dropped overboard to fasten the ship to the sea bottom ▶ v fasten with or as if with an anchor ▶ **anchorage** n place where boats can be anchored ▶ **anchorman, anchorwoman** n broadcaster in a central studio who links up and presents items from outside camera units and other studios; last person to compete in a relay team

anchorite n religious recluse

anchovy [an-chov-ee] n, pl -**vies** small strong-tasting fish

ancient adj dating from very long ago; very old ▶ **ancients** pl n people who lived very long ago

ancillary adj supporting the main work of an organization; used as an extra or supplement

and conj in addition to; as a consequence; then, afterwards

andante [an-dan-tay] n, adv (Music) (piece to be played) moderately slowly

andiron n iron stand for supporting logs in a fireplace

androgynous adj having both male and female characteristics

android n robot resembling a human; Android® software package for smartphones; smartphone that uses this software

anecdote n short amusing account of an incident ▶ **anecdotal** adj

anemometer n instrument for recording wind speed

anemone [an-nem-on-ee] n plant with white, purple, or red flowers

aneroid barometer n device for measuring air pressure, consisting of a partially evacuated chamber in which variations in pressure cause a pointer on the lid to move

aneurysm, aneurism [an-new-riz-zum] n permanent swelling of a blood vessel

anew adv once more; in a different way

angel n spiritual being believed to be an attendant or messenger of God; person who is kind, pure, or beautiful ▶ **angelic** adj ▶ **angelically** adv

angelica n aromatic plant; its candied stalks, used in cookery

Angelus [an-jell-uss] n (in the Roman Catholic Church) prayers recited in the morning, at midday, and in the evening; bell signalling the times of these prayers

anger n fierce displeasure or extreme annoyance ▶ v make (someone) angry

angina [an-jine-a] n heart disorder causing sudden severe chest pains (also **angina pectoris**)

angle¹ n space between or shape formed by two lines or surfaces that meet; divergence between these, measured in degrees;

corner; point of view ▷ v bend or place (something) at an angle

angle² v fish with a hook and line; (foll by for) try to get by hinting ▶ **angling** n

angler n person who fishes with a hook and line

Anglican n, adj (member) of the Church of England ▶ **Anglicanism** n

anglicize v make or become English in outlook, form, etc.

Anglo- combining form English e.g. Anglo-Scottish; British e.g. Anglo-American

Anglo-Saxon n member of any of the W Germanic tribes that settled in England from the fifth century AD; language of the Anglo-Saxons ▷ adj of the Anglo-Saxons or their language

angophora n Australian tree related to the eucalyptus

angora n variety of goat, cat, or rabbit with long silky hair; hair of the angora goat or rabbit; cloth made from this hair

Angostura Bitters® pl n bitter tonic, used as a flavouring in alcoholic drinks

angry adj **-grier, -griest** full of anger; inflamed e.g. an angry wound ▶ **angrily** adv

angst n feeling of anxiety

angstrom n unit of length used to measure wavelengths

anguish n great mental pain ▶ **anguished** adj

angular adj (of a person) lean and bony; having angles; measured by an angle ▶ **angularity** n

anhydrous adj (Chem) containing no water

aniline n colourless oily liquid obtained from coal tar and used for making dyes, plastics, and explosives

animal n living creature with specialized sense organs and capable of voluntary motion, esp. one other than a human being; quadruped ▷ adj of animals; sensual, physical

animate v give life to; make lively; make a cartoon film of ▷ adj having life ▶ **animated** adj ▶ **animation** n technique of making cartoon films; liveliness and enthusiasm ▶ **animator** n

animism n belief that natural objects possess souls ▶ **animist** n, adj ▶ **animistic** adj

animosity n, pl **-ties** hostility, hatred

animus n hatred, animosity

anion [an-eye-on] n ion with negative charge

anise [an-niss] n plant with liquorice-flavoured seeds

aniseed n liquorice-flavoured seeds of the anise plant

ankle n joint between the foot and leg ▶ **anklet** n ornamental chain worn round the ankle

annals pl n yearly records of events

anneal v toughen (metal or glass) by heating and then slow cooling

annelid n worm with a segmented body, such as an earthworm

annex v seize (territory); take (something) without permission; join or add (something) to something larger ▶ **annexation** n

annexe n extension to a building; nearby building used as an extension

annihilate v destroy utterly ▶ **annihilation** n

anniversary n, pl **-ries** date on which something occurred in a previous year; celebration of this

anno Domini [an-no dom-in-eye] adv (Latin) (indicating years numbered from the supposed year

a

of the birth of Christ) in the year of our Lord

annotate v add notes to (a written work) ▸ **annotation** n

announce v make known publicly; proclaim ▸ **announcement** n ▸ **announcer** n person who introduces radio or television programmes

annoy v irritate or displease ▸ **annoyance** n

annual adj happening once a year; lasting for a year ▸ n plant that completes its life cycle in a year; book published once every year ▸ **annually** adv

annuity n, pl -**ties** fixed sum paid every year

annul v -**nulling**, -**nulled** declare (something, esp. a marriage) invalid ▸ **annulment** n

annular [an-new-lar] adj ring-shaped

Annunciation n (Christianity) angel Gabriel's announcement to the Virgin Mary of her conception of Christ

anode n (Electricity) positive electrode in a battery, valve, etc. ▸ **anodize** v coat (metal) with a protective oxide film by electrolysis

anodyne n something that relieves pain or distress ▸ adj relieving pain or distress

anoint v smear with oil as a sign of consecration

anomaly [an-nom-a-lee] n, pl -**lies** something that deviates from the normal, irregularity ▸ **anomalous** adj

anon adv (obs) in a short time, soon

anon. anonymous

anonymous adj by someone whose name is unknown or withheld; having no known name ▸ **anonymously** adv ▸ **anonymity** n

anorak n light waterproof hooded jacket

anorexia n psychological disorder characterized by fear of becoming fat and refusal to eat (also **anorexia nervosa**) ▸ **anorexic** adj, n

another adj, pron one more; different (one)

answer n reply to a question, request, letter, etc.; solution to a problem; reaction or response ▸ v give an answer (to); be responsible to (a person); respond or react ▸ **answerable** adj (foll by for or to) responsible for or accountable to ▸ **answering machine** device for answering a telephone automatically and recording messages

ant n small insect living in highly organized colonies ▸ **anteater** n mammal which feeds on ants by means of a long snout; same as **echidna**; same as **numbat** ▸ **ant hill** mound built by ants around their nest

antacid n substance that counteracts acidity, esp. in the stomach

antagonist n opponent or adversary ▸ **antagonism** n open opposition or hostility ▸ **antagonistic** adj ▸ **antagonize** v arouse hostility in, annoy

Antarctic n the Antarctic area around the South Pole ▸ adj of this region

> **SPELLING TIP**
> Note there is a c after the r in **Antarctic**

ante n player's stake in poker ▸ v -**teing**, -**ted** or -**teed** place (one's stake) in poker

ante- prefix before in time or position e.g. antedate; antechamber

antecedent n event or circumstance happening or

existing before another ▷ *adj* preceding, prior

antedate *v* precede in time

antediluvian *adj* of the time before the biblical Flood; old-fashioned

antelope *n* deerlike mammal with long legs and horns

antenatal *adj* during pregnancy, before birth

antenna *n* (*pl* **-nae**) insect's feeler; (*pl* **-nas**) aerial

anterior *adj* to the front; earlier

anteroom *n* small room leading into a larger one, often used as a waiting room

anthem *n* song of loyalty, esp. to a country; piece of choral music, usu. set to words from the Bible

anther *n* part of a flower's stamen containing pollen

anthology *n*, *pl* **-gies** collection of poems or other literary pieces by various authors ▶ **anthologist** *n*

anthracite *n* hard coal burning slowly with little smoke or flame but intense heat

anthrax *n* dangerous disease of cattle and sheep, communicable to humans

anthropoid *adj* like a human ▷ *n* ape, such as a chimpanzee, that resembles a human

anthropology *n* study of human origins, institutions, and beliefs ▶ **anthropological** *adj* ▶ **anthropologist** *n*

anthropomorphic *adj* attributing human form or personality to a god, animal, or object ▶ **anthropomorphism** *n*

anti- *prefix* against, opposed to e.g. *anti-war*; counteracting e.g. *antifreeze*

anti-aircraft *adj* for defence against aircraft attack

antibiotic *n* chemical substance capable of destroying bacteria ▷ *adj* of antibiotics

antibody *n*, *pl* **-bodies** protein produced in the blood, which destroys bacteria

anticipate *v* foresee and act in advance of; look forward to ▶ **anticipation** *n* ▶ **anticipatory** *adj*

anticlimax *n* disappointing conclusion to a series of events

anticlockwise *adv, adj* in the opposite direction to the rotation of the hands of a clock

antics *pl n* absurd acts or postures

anticyclone *n* area of moving air of high pressure in which the winds rotate outwards

antidote *n* substance that counteracts a poison

antifreeze *n* liquid added to water to lower its freezing point, used esp. in car radiators

antigen [an-tee-jen] *n* substance, usu. a toxin, causing the blood to produce antibodies

anti-globalization *n* opposition to globalization

antihero *n*, *pl* **-roes** central character in a book, film, etc., who lacks the traditional heroic virtues

antihistamine *n* drug used to treat allergies

antimacassar *n* cloth put over a chair-back to prevent soiling

antimony *n* (*Chem*) brittle silvery-white metallic element

antipasto *n*, *pl* **-tos** appetizer in an Italian meal

antipathy [an-tip-a-thee] *n* dislike, hostility ▶ **antipathetic** *adj*

antiperspirant *n* substance used to reduce or prevent sweating

antiphon *n* hymn sung in alternate parts by two groups of singers ▶ **antiphonal** *adj*

a

antipodes [an-tip-pod-deez] *pl n* any two places diametrically opposite one another on the earth's surface **the Antipodes** Australia and New Zealand ▶ **antipodean** *adj*

antipyretic *adj* reducing fever ▷ *n* drug that reduces fever

antiquary *n, pl* **-quaries** student or collector of antiques or ancient works of art ▶ **antiquarian** *adj* of or relating to antiquities or rare books ▷ *n* antiquary

antiquated *adj* out-of-date

antique *n* object of an earlier period, valued for its beauty, workmanship, or age ▷ *adj* made in an earlier period; old-fashioned

antiquity *n* great age; ancient times ▶ **antiquities** *pl n* objects dating from ancient times

antiracism *n* policy of challenging racism and promoting racial tolerance

antirrhinum *n* two-lipped flower of various colours

anti-Semitism *n* discrimination against Jews ▶ **anti-Semitic** *adj*

antiseptic *adj* preventing infection by killing germs ▷ *n* antiseptic substance

anti-social *adj* avoiding the company of other people; (of behaviour) harmful to society

antistatic *adj* reducing the effects of static electricity

antithesis [an-tith-iss-iss] *n, pl* **-ses** [-seez] exact opposite; placing together of contrasting ideas or words to produce an effect of balance ▶ **antithetical** *adj*

antitoxin *n* (serum containing) an antibody that acts against a toxin

antitrust *adj* (Aust & S Afr) (of laws) opposing business monopolies

antivirus *n, adj* (software) protecting computers from viruses

antler *n* branched horn of male deer

antonym *n* word that means the opposite of another

anus [ain-uss] *n* opening at the end of the alimentary canal, through which faeces are discharged

anvil *n* heavy iron block on which metals are hammered into particular shapes

anxiety *n, pl* **-ties** state of being anxious

anxious *adj* worried and tense; intensely desiring ▶ **anxiously** *adv*

any *adj, pron* one or some, no matter which ▷ *adv* at all e.g. *it isn't any worse* ▶ **anybody** *pron* anyone ▶ **anyhow** *adv* anyway ▶ **anyone** *pron* any person; person of any importance ▶ **anything** *pron* ▶ **anyway** *adv* at any rate, nevertheless; in any manner ▶ **anywhere** *adv* in, at, or to any place

Anzac *n* (in World War 1) a soldier serving with the Australian and New Zealand Army Corps ▶ **Anzac Day** 25 April, a public holiday in Australia and New Zealand commemorating the Anzac landing at Gallipoli in 1915

AOB (on the agenda for a meeting) any other business

aorta [eh-or-ta] *n* main artery of the body, carrying oxygen-rich blood from the heart

apace *adv* (lit) swiftly

apart *adv* to or in pieces; to or at a distance; individual, distinct

apartheid *n* former official government policy of racial segregation in S Africa

apartment *n* room in a building; flat

apathy *n* lack of interest or enthusiasm ▶ **apathetic** *adj*

ape *n* tailless monkey such as the chimpanzee or gorilla; stupid, clumsy, or ugly man ▷ *v* imitate

aperient [ap-**peer**-ee-ent] *adj* having a mild laxative effect ▷ *n* mild laxative

aperitif [ap-per-rit-**teef**] *n* alcoholic drink taken before a meal

aperture *n* opening or hole

APEX (*Brit, NZ & S Afr*) Advance Purchase Excursion: reduced fare for journeys booked a specified period in advance

apex *n* highest point

aphasia *n* disorder of the central nervous system that affects the ability to speak and understand words

aphid [eh-fid], **aphis** [eh-fiss] *n* small insect which sucks the sap from plants

aphorism *n* short clever saying expressing a general truth

aphrodisiac [af-roh-**diz**-zee-ak] *n* substance that arouses sexual desire ▷ *adj* arousing sexual desire

apiary *n, pl* -**ries** place where bees are kept

apiculture *n* breeding and care of bees

apiece *adv* each

aplomb *n* calm self-possession

apocalypse *n* end of the world; event of great destruction **the Apocalypse** book of Revelation, the last book of the New Testament ▶ **apocalyptic** *adj*

Apocrypha [ap-**pok**-rif-fa] *pl n* **the Apocrypha** collective name for the 14 books of the Old Testament which are not accepted as part of the Hebrew scriptures

apocryphal [ap-**pok**-rif-al] *adj* (of a story) of questionable authenticity

apogee [ap-oh-jee] *n* point of the moon's or a satellite's orbit that is farthest from the earth; highest point

apology *n, pl* -**gies** expression of regret for wrongdoing; (foll by *for*)

poor example (of); ▶ **apologetic** *adj* showing or expressing regret ▶ **apologetically** *adv* ▶ **apologetics** *n* branch of theology concerned with the reasoned defence of Christianity ▶ **apologist** *n* person who formally defends a cause ▶ **apologize** *v* make an apology

> **SPELLING TIP**
> Remember that there is one *p* and one *l* in **apology**.

apoplexy *n* (*Med*) stroke ▶ **apoplectic** *adj* of apoplexy; (*informal*) furious

apostasy [ap-**poss**-stass-ee] *n, pl* -**sies** abandonment of one's religious faith or other belief ▶ **apostate** *n, adj*

a posteriori [eh poss-steer-ee-or-rye] *adj* involving reasoning from effect to cause

Apostle *n* one of the twelve disciples chosen by Christ to preach his gospel; (**a-**) ardent supporter of a cause or movement ▶ **apostolic** *adj*

apostrophe [ap-**poss**-trof-fee] *n* punctuation mark (') showing the omission of a letter or letters in a word, e.g. *don't*; or forming the possessive, e.g. *Jill's car*; digression from a speech to someone an imaginary or absent person or thing

apothecary *n, pl* -**caries** (*obs*) chemist

apotheosis [ap-poth-ee-**oh**-siss] *n, pl* -**ses** [-seez] perfect example; elevation to the rank of a god

app *n* computer program designed for a particular purpose

appal *v* -**palling**, -**palled** dismay, terrify ▶ **appalling** *adj* dreadful, terrible

> **SPELLING TIP**
> The verb **appal** has two *p*s, but only one *l*. If you extend it with an ending beginning with a

a

vowel, you must add another *l*, as in **appalling**

apparatus *n* equipment for a particular purpose

apparel *n* (*old-fashioned*) clothing

apparent *adj* readily seen, obvious; seeming as opposed to real
▸ **apparently** *adv*

> **SPELLING TIP**
> **Apparently** is spelt with two *a*s – and then an *e*

apparition *n* ghost or ghostlike figure

appeal *v* make an earnest request; attract, please, or interest; request a review of a lower court's decision by a higher court ▸ *n* earnest request; attractiveness; request for a review of a lower court's decision by a higher court
▸ **appealing** *adj*

appear *v* become visible or present; seem; be seen in public
▸ **appearance** *n* appearing; outward aspect

appease *v* pacify (a person) by yielding to his or her demands; satisfy or relieve (a feeling)
▸ **appeasement** *n*

appellant *n* person who makes an appeal to a higher court

appellation *n* (*formal*) name, title

append *v* join on, add
▸ **appendage** *n* thing joined on or added

appendicitis *n* inflammation of the appendix

appendix *n*, *pl* **-dices** or **-dixes** separate additional material at the end of a book; (*Anat*) short closed tube attached to the large intestine

> **USAGE NOTE**
> Extra sections at the end of a book are *appendices*. The plural *appendixes* is used in medicine

appertain *v* (foll by *to*) belong to; be connected with

appetite *n* desire for food or drink; liking or willingness
▸ **appetizer** *n* thing eaten or drunk to stimulate the appetite
▸ **appetizing** *adj* stimulating the appetite

applaud *v* show approval of by clapping one's hands; approve strongly ▸ **applause** *n* approval shown by clapping one's hands

apple *n* round firm fleshy fruit that grows on trees in **apple-pie order** (*informal*) very tidy

appliance *n* device with a specific function

applicable *adj* relevant, appropriate ▸ **applicability** *n*

applicant *n* person who applies for something

application *n* formal request; act of applying something to a particular use; diligent effort; act of putting something onto a surface

appliqué [ap-**plee**-kay] *n* kind of decoration in which one material is cut out and attached to another

apply *v* **-plying, -plied** make a formal request; put to practical use; put onto a surface; be relevant or appropriate **apply oneself** concentrate one's efforts
▸ **applied** *adj* (of a skill, science, etc.) put to practical use

appoint *v* assign to a job or position; fix or decide *e.g.* *appoint a time*; equip or furnish
▸ **appointment** *n* arrangement to meet a person; act of placing someone in a job; the job itself ▸ *pl* fixtures or fittings

apportion *v* divide out in shares

apposite *adj* suitable, apt
▸ **apposition** *n* grammatical construction in which two nouns or phrases referring to the same thing are placed one after another

without a conjunction e.g. *my son the doctor*

appraise v estimate the value or quality of ▶ **appraisal** n

appreciate v value highly; be aware of and understand; be grateful for; rise in value ▶ **appreciable** adj enough to be noticed ▶ **appreciably** adv ▶ **appreciation** n ▶ **appreciative** adj feeling or showing appreciation

apprehend v arrest and take into custody; grasp (something) mentally ▶ **apprehension** n dread, anxiety; arrest; understanding ▶ **apprehensive** adj fearful or anxious

apprentice n someone working for a skilled person for a fixed period in order to learn his or her trade ▷ v take or place (someone) as an apprentice ▶ **apprenticeship** n

apprise v make aware (of)

appro n **on appro** (*Brit, Aust, NZ & S Afr informal*) on approval

approach v come near or nearer (to); make a proposal or suggestion to; begin to deal with (a matter) ▷ n approaching or means of approaching; approximation ▶ **approachable** adj ▶ **approach road** smaller road leading into a major road

approbation n approval

appropriate adj suitable, fitting ▷ v take for oneself; put aside for a particular purpose ▶ **appropriately** adv ▶ **appropriateness** n ▶ **appropriation** n

approve v consider good or right; authorize, agree to ▶ **approval** n consent; favourable opinion ▶ **on approval** (of goods) with an option to be returned without payment if unsatisfactory

approx. approximate(ly)

approximate adj almost but not quite exact ▷ v (foll by to) come close to; be almost the same as ▶ **approximately** adv ▶ **approximation** n

appurtenances pl n minor or additional features

Apr April

après-ski [ap-ray-skee] n social activities after a day's skiing

apricot n yellowish-orange juicy fruit like a small peach ▷ adj yellowish-orange

April n fourth month of the year ▶ **April fool** victim of a practical joke played on April 1 (**April Fools' Day**)

a priori [eh pry-or-rye] adj involving reasoning from cause to effect

apron n garment worn over the front of the body to protect the clothes; area at an airport or hangar for manoeuvring and loading aircraft; part of a stage in front of the curtain

apropos [ap-prop-poh] adj, adv appropriate(ly) **apropos of** with regard to

apse n arched or domed recess, esp. in a church

apt adj having a specified tendency; suitable; quick to learn ▶ **aptly** adv ▶ **aptness** n ▶ **aptitude** n natural ability

aqualung n mouthpiece attached to air cylinders, worn for underwater swimming

aquamarine n greenish-blue gemstone ▷ adj greenish-blue

aquaplane n board on which a person stands to be towed by a motorboat ▷ v ride on an aquaplane; (of a motor vehicle) skim uncontrollably on a thin film of water

aquarium n, pl **aquariums** or **aquaria** tank in which fish and other underwater creatures are kept; building containing such tanks

aquatic *adj* living in or near water; done in or on water ▸ **aquatics** *pl n* water sports

aquatint *n* print like a watercolour, produced by etching copper

aqua vitae [ak-wa vee-tie] *n (obs)* brandy

aqueduct *n* structure carrying water across a valley or river

aqueous *adj* of, like, or containing water

aquiline *adj* (of a nose) curved like an eagle's beak; of or like an eagle

Arab *n* member of a Semitic people originally from Arabia ▷ *adj* of the Arabs ▸ **Arabian** *adj* of Arabia or the Arabs ▸ **Arabic** *n* language of the Arabs ▷ *adj* of Arabic, Arabs, or Arabia

arabesque [ar-ab-besk] *n* ballet position in which one leg is raised behind and the arms are extended; elaborate ornamental design

arable *adj* suitable for growing crops on

arachnid [ar-rak-nid] *n* eight-legged invertebrate, such as a spider, scorpion, tick, or mite

Aran *adj* (of sweaters etc.) knitted in a complicated pattern traditional to the Aran Islands, usu. with natural unbleached wool

arbiter *n* person empowered to judge in a dispute; person with influential opinions about something

arbitrary *adj* based on personal choice or chance, rather than reason ▸ **arbitrarily** *adv*

> **SPELLING TIP**
> Don't forget the second of the three **r**s in **arbitrary**.

arbitration *n* hearing and settling of a dispute by an impartial referee

chosen by both sides ▸ **arbitrate** *v* ▸ **arbitrator** *n*

arboreal *adj* of or living in trees

arboretum [ahr-bore-ee-tum] *n, pl* **-ta** place where rare trees or shrubs are cultivated

arboriculture *n* cultivation of trees or shrubs

arbour *n* glade sheltered by trees

arc *n* part of a circle or other curve; luminous discharge of electricity across a small gap between two electrodes ▷ *v* form an arc

arcade *n* covered passageway lined with shops; set of arches and their supporting columns

arcane *adj* mysterious and secret

arch¹ *n* curved structure supporting a bridge or roof; something curved; curved lower part of the foot ▷ *v* (cause to) form an arch ▸ **archway** *n* passageway under an arch

arch² *adj* superior, knowing; coyly playful ▸ **archly** *adv* ▸ **archness** *n*

arch- combining form chief, principal e.g. **archenemy**

archaeology *n* study of ancient cultures from their physical remains ▸ **archaeological** *adj* ▸ **archaeologist** *n*

archaic [ark-kay-ik] *adj* ancient; out-of-date ▸ **archaism** [ark-kay-iz-zum] *n* archaic word or phrase

archangel [ark-ain-jell] *n* chief angel

archbishop *n* chief bishop

archdeacon *n* priest ranking just below a bishop

archdiocese *n* diocese of an archbishop

archer *n* person who shoots with a bow and arrow ▸ **archery** *n*

archetype [ark-ee-type] *n* perfect specimen; original model ▸ **archetypal** *adj*

archipelago [ark-ee-pèl-a-go] *n, pl* **-gos** group of islands; sea full of small islands

architect n person qualified to design and supervise the construction of buildings ▶ **architecture** n style in which a building is designed and built; designing and construction of buildings ▶ **architectural** adj

architrave n (Archit) beam that rests on columns; moulding round a doorway or window

archive [ark-ive] n (often pl) collection of records or documents; place where these are kept ▶ **archival** adj ▶ **archivist** [ark-iv-ist] n person in charge of archives

Arctic n the Arctic area around the North Pole ▷ adj of this region; (a-) (informal) very cold

ardent adj passionate; eager, zealous ▶ **ardently** adv ▶ **ardour** n passion; enthusiasm, zeal

arduous adj hard to accomplish, strenuous ▶ **arduously** adv

are¹ v see **be**

are² n unit of measure, 100 square metres

area n part or region; size of a two-dimensional surface; subject field

arena n seated enclosure for sports events; area of a Roman amphitheatre where gladiators fought; sphere of intense activity

aren't are not

areola n, pl -lae or -las small circular area, such as the coloured ring around the human nipple

argon n (Chem) inert gas found in the air

argot [ahr-go] n slang or jargon

argue v -guing, -gued try to prove by giving reasons; debate; quarrel, dispute ▶ **arguable** adj ▶ **arguably** adv ▶ **argument** n quarrel; discussion; point presented for or against something ▶ **argumentation** n process of reasoning methodically

▶ **argumentative** adj given to arguing

SPELLING TIP

There is an **e** at the end of **argue**, but you should leave it out when you write **argument**

argy-bargy n, pl -bargies (informal) squabbling argument

aria [ah-ree-a] n elaborate song for solo voice, esp. one from an opera

arid adj parched, dry; uninteresting ▶ **aridity** n

aright adv rightly

arise v arising, arose, arisen come about; come into notice; get up

aristocracy n, pl -cies highest social class ▶ **aristocrat** n member of the aristocracy ▶ **aristocratic** adj

arithmetic n calculation by or of numbers ▷ adj of arithmetic ▶ **arithmetical** adj ▶ **arithmetically** adv

ark n (Old Testament) boat built by Noah, which survived the Flood; (A-) (Judaism) chest containing the writings of Jewish Law

arm¹ n either of the upper limbs from the shoulder to the wrist; sleeve of a garment; side of a chair ▶ **armful** n as much as can be held in the arms ▶ **armchair** n upholstered chair with side supports for the arms ▶ **armhole** n opening in a garment through which the arm passes ▶ **armpit** n hollow under the arm at the shoulder

arm² v supply with weapons; prepare (a bomb etc.) for use ▶ **arms** pl n weapons; military exploits; heraldic emblem

armada n large number of warships

armadillo n, pl -los small S American mammal covered in strong bony plates

a

Armageddon n (New Testament) final battle between good and evil at the end of the world; catastrophic conflict

armament n military weapons; preparation for war

armature n revolving structure in an electric motor or generator, wound with coils carrying the current

armistice [arm-miss-stiss] n agreed suspension of fighting

armour n metal clothing formerly worn to protect the body in battle; metal plating of tanks, warships, etc. ▸ **armourer** n maker, repairer, or keeper of arms or armour ▸ **armoury** n place where weapons are stored

army n, pl **armies** military land forces of a nation; great number

aroma n pleasant smell ▸ **aromatic** adj ▸ **aromatherapy** n massage with fragrant oils to relieve tension

arose v past tense of **arise**

around prep, adv on all sides (of); from place to place (in); somewhere in or near; approximately

arouse v stimulate, make active; awaken

arpeggio [arp-pej-ee-oh] n, pl **-gios** (Music) notes of a chord played or sung in quick succession

arr. arranged (by); arrival; arrive(d)

arraign [ar-rain] v bring (a prisoner) before a court to answer a charge; accuse ▸ **arraignment** n

arrange v plan; agree; put in order; adapt (music) for performance in a certain way ▸ **arrangement** n

arrant adj utter, downright

arras n tapestry wall-hanging

array n impressive display or collection; orderly arrangement, esp. of troops; (poetic) rich clothing ▸ v arrange in order; dress in rich clothing

arrears pl n money owed **in arrears** late in paying a debt

arrest v take (a person) into custody; stop the movement or development of; catch and hold (the attention) ▸ n act of taking a person into custody; slowing or stopping ▸ **arresting** adj attracting attention, striking

arrive v reach a place or destination; happen; come; (informal) attain success ▸ **arrival** n arriving; person or thing that has just arrived

arrogant adj proud and overbearing ▸ **arrogantly** adv ▸ **arrogance** n

arrogate v claim or seize without justification

arrow n pointed shaft shot from a bow; arrow-shaped sign or symbol used to show direction ▸ **arrowhead** n pointed tip of an arrow

arrowroot n nutritious starch obtained from the root of a W Indian plant

arse n (vulgar slang) buttocks or anus ▸ **arsehole** n (vulgar slang) anus; stupid or annoying person

arsenal n place where arms and ammunition are made or stored

arsenic n toxic grey element; highly poisonous compound of this ▸ **arsenical** adj

arson n crime of intentionally setting property on fire ▸ **arsonist** n

art n creation of works of beauty, esp. paintings or sculpture; works of art collectively; skill ▸ pl nonscientific branches of knowledge ▸ **artist** n person who produces works of art, esp. paintings or sculpture; person skilled at something; artiste ▶ **artiste** n professional entertainer such as a singer or dancer ▶ **artistic** adj ▶ **artistically** adv ▶ **artistry** n artistic skill ▶ **arty** adj (informal) having an affected interest in art

artefact n something made by human beings

arteriosclerosis [art-ear-ee-oh-skler-**oh**-siss] n hardening of the arteries

artery n, pl **-teries** one of the tubes carrying blood from the heart; major road or means of communication ▶ **arterial** adj of an artery; (of a route) major

artesian well [art-**teez**-yan] n well bored vertically so that the water is forced to the surface by natural pressure

Artex® n (Brit) textured covering for ceilings and walls

artful adj cunning, wily ▶ **artfully** adv ▶ **artfulness** n

arthritis n painful inflammation of a joint or joints ▶ **arthritic** adj, n

arthropod n animal, such as a spider or insect, with jointed limbs and a segmented body

artichoke n flower head of a thistle-like plant, cooked as a vegetable

article n written piece in a magazine or newspaper; item or object; clause in a document; (Grammar) any of the words the, a, or an

articled adj bound (as an apprentice) by a written contract

articulate adj able to express oneself clearly and coherently; (of speech) clear, distinct; (Zool) having joints ▷ v speak or say clearly and coherently ▶ **articulately** adv ▶ **articulated** adj jointed

▶ **articulated vehicle** large vehicle in two separate sections joined by a pivoted bar ▶ **articulation** n

artifice n clever trick; cleverness, skill ▶ **artificer** [art-**tiff**-iss-er] n craftsman

artificial adj man-made, not occurring naturally; made in imitation of something natural; not sincere ▶ **artificial insemination** introduction of semen into the womb by means other than sexual intercourse ▶ **artificial intelligence** branch of computer science aiming to produce machines which can imitate intelligent human behaviour ▶ **artificial respiration** method of restarting a person's breathing after it has stopped ▶ **artificially** adv ▶ **artificiality** n

artillery n large-calibre guns; branch of the army who use these

artisan n skilled worker, craftsman

artless adj free from deceit or cunning; natural, unpretentious ▶ **artlessly** adv

arum lily [**air**-rum] n plant with a white funnel-shaped leaf surrounding a spike of flowers

arvie n (S Afr informal) afternoon

as conj while, when; in the way that; that which e.g. do as you are told; since, seeing that; for instance ▷ adv, conj used to indicate amount or extent in comparisons e.g. he is as tall as you ▷ prep in the role of, being e.g. as a mother, I am concerned

asafoetida n strong-smelling plant resin used as a spice in Eastern cookery

a.s.a.p. as soon as possible

asbestos n fibrous mineral which does not burn ▶ **asbestosis** n lung disease caused by inhalation of asbestos fibre

a

ASBO (Brit) anti-social behaviour order: a civil order made against a persistently anti-social person

ascend v go or move up ▸ **ascent** n ascending; upward slope ▸ **ascendant** adj dominant or influential ▸ **in the ascendant** increasing in power or influence ▸ **ascendancy** n condition of being dominant **the Ascension** (Christianity) passing of Jesus Christ from earth into heaven

ascertain v find out definitely ▸ **ascertainable** adj ▸ **ascertainment** n

ascetic [ass-set-tik] n, adj (person) abstaining from worldly pleasures and comforts ▸ **asceticism** n

ascorbic acid [ass-core-bik] n vitamin C

ascribe v attribute, as to a particular origin ▸ **ascription** n

aseptic [eh-sep-tik] adj free from harmful bacteria

asexual [eh-sex-yew-al] adj without sex ▸ **asexually** adv

ash¹ n powdery substance left when something is burnt ▸ pl remains after burning, esp. of a human body after cremation **the Ashes** cricket trophy competed for in test matches by England and Australia ▸ **ashen** adj pale with shock ▸ **ashtray** n receptacle for tobacco ash and cigarette butts ▸ **Ash Wednesday** first day of Lent

ash² n tree with grey bark

ashamed adj feeling shame

ashlar n square block of hewn stone used in building

ashore adv towards or on land

ashram n religious retreat where a Hindu holy man lives

Asian adj of the continent of Asia or any of its peoples or languages ▸ n person from Asia or a descendant of one; person from the Indian subcontinent or a descendant of one ▸ **Asian pear** apple-shaped pear with crisp flesh

aside adv on one side; out of other people's hearing e.g. he took me aside to tell me his plans ▸ n remark not meant to be heard by everyone present

asinine adj stupid, idiotic

ask v say or write (something) in a form that requires an answer; make a request or demand; invite

askance [ass-kanss] adv **look askance at** look at with an oblique glance; regard with suspicion

askew adv, adj to one side, crooked

aslant adv, prep at a slant (to), slanting (across)

asleep adj sleeping; (of limbs) numb

asp n small poisonous snake

asparagus n plant whose shoots are cooked as a vegetable

aspect n feature or element; position facing a particular direction; appearance or look

aspen n kind of poplar tree

asperity n roughness of temper

aspersion n **cast aspersions on** make derogatory remarks about

asphalt n black hard tarlike substance used for road surfaces etc.

asphodel n plant with clusters of yellow or white flowers

asphyxia [ass-fix-ee-a] n suffocation ▸ **asphyxiate** v suffocate ▸ **asphyxiation** n

aspic n savoury jelly used to coat meat, eggs, fish, etc.

aspidistra n plant with long tapered leaves

aspirate (Phonetics) v pronounce with an h sound ▸ n h sound

aspire v (foll by to) aspire (for), hope (to do or be) ▸ **aspirant** n person who aspires ▸ **aspiration** n strong desire or aim

aspirin n drug used to relieve pain and fever; tablet of this

ass n donkey; stupid person

assagai n same as **assegai**

assail v attack violently
▶ **assailant** n

assassin n person who murders a prominent person ▶ **assassinate** v murder (a prominent person)
▶ **assassination** n

assault n violent attack ▷ v attack violently ▶ **assault course** series of obstacles used in military training

assay n analysis of a substance, esp. a metal, to ascertain its purity ▷ v make such an analysis

assegai [**ass**-a-guy] n slender spear used in S Africa

assemble v collect or congregate; put together the parts of (a machine) ▶ **assemblage** n collection or group; assembling ▶ **assembly** n (pl **-blies**) assembled group; assembling ▶ **assembly line** sequence of machines and workers in a factory assembling a product

assent n agreement or consent ▷ v agree or consent

assert v declare forcefully; insist upon (one's rights etc.) **assert oneself** put oneself forward forcefully ▶ **assertion** n ▶ **assertive** adj ▶ **assertively** adv

assess v judge the worth or importance of; estimate the value of (income or property) for taxation purposes ▶ **assessment** n ▶ **assessor** n

asset n valuable or useful person or thing ▷ pl property that a person or firm can sell, esp. to pay debts

asseverate v declare solemnly

assiduous adj hard-working ▶ **assiduously** adv ▶ **assiduity** n

assign v appoint (someone) to a job or task; allot (a task); attribute

▶ **assignation** n assigning; secret arrangement to meet

▶ **assignment** n task assigned; assigning

assimilate v learn and understand (information); absorb or be absorbed or incorporated
▶ **assimilable** adj ▶ **assimilation** n

assist v give help or support ▶ **assistance** n ▶ **assistant** n helper ▷ adj junior or deputy

assizes pl n (Brit) court sessions formerly held in each county of England and Wales

associate v connect in the mind; mix socially ▷ n partner in business; friend or companion ▷ adj having partial rights or subordinate status e.g. *associate member* ▶ **association** n society or club; associating

assonance n rhyming of vowel sounds but not consonants, as in *time* and *light*

assorted adj consisting of various types mixed together ▶ **assortment** n assorted mixture

assuage [ass-**wage**] v relieve (pain, grief, thirst, etc.)

assume v take to be true without proof; take upon oneself e.g. *he assumed command*; pretend e.g. *I assumed indifference* ▶ **assumption** n thing assumed; assuming

assure v promise or guarantee; convince; make (something) certain; insure against loss of life ▶ **assured** adj confident; certain to happen ▶ **assuredly** adv definitely ▶ **assurance** n assuring or being assured

USAGE NOTE
When used in the context of business, *assurance* and *insurance* have the same meaning

astatine n (Chem) radioactive nonmetallic element

a

aster n plant with daisy-like flowers

asterisk n star-shaped symbol (*) used in printing or writing to indicate a footnote etc. ▷ v mark with an asterisk

astern adv at or towards the stern of a ship; backwards

asteroid n any of the small planets that orbit the sun between Mars and Jupiter

asthma [ass-ma] n illness causing difficulty in breathing ▶ **asthmatic** adj, n

astigmatism [eh-stig-mat-tizzum] n inability of a lens, esp. of the eye, to focus properly

astir adj (old-fashioned) out of bed; in motion

astonish v surprise greatly ▶ **astonishment** n

astound v overwhelm with amazement ▶ **astounding** adj

astrakhan n dark curly fleece of lambs from Astrakhan in Russia; fabric resembling this

astral adj of stars; of the spirit world

astray adv off the right path

astride adv, prep with a leg on either side (of)

astringent adj causing contraction of body tissue; checking the flow of blood from a cut; severe or harsh ▷ n astringent substance ▶ **astringency** n

astrolabe n instrument formerly used to measure the altitude of stars and planets

astrology n study of the alleged influence of the stars, planets, and moon on human affairs ▶ **astrologer** n ▶ **astrological** adj

astronaut n person trained for travelling in space

astronautics n science and technology of space flight ▶ **astronautical** adj

astronomy n scientific study of heavenly bodies ▶ **astronomer** n ▶ **astronomical** adj very large; of astronomy ▶ **astronomically** adv

astrophysics n science of the physical and chemical properties of stars, planets, etc. ▶ **astrophysical** adj ▶ **astrophysicist** n

Astroturf® n artificial grass

astute adj perceptive or shrewd ▶ **astutely** adv ▶ **astuteness** n

asunder adv (obs or poetic) into parts or pieces

asylum n refuge or sanctuary; old name for a psychiatric hospital

asymmetry n lack of symmetry ▶ **asymmetrical, asymmetric** adj

asymptote [ass-im-tote] n straight line closely approached but never met by a curve

at prep indicating position in space or time, movement towards an object, etc. e.g. at midnight; throwing stones at windows

atavism [at-a-viz-zum] n recurrence of a trait present in distant ancestors ▶ **atavistic** adj

ate v past tense of eat

atheism [aith-ee-iz-zum] n belief that there is no God ▶ **atheist** n ▶ **atheistic** adj

atherosclerosis n, pl -**ses** disease in which deposits of fat cause the walls of the arteries to thicken

athlete n person trained in or good at athletics ▶ **athletic** adj physically fit or strong; of an athlete or athletics ▶ **athletics** pl n track-and-field sports such as running, jumping, throwing, etc. ▶ **athletically** adv ▶ **athleticism** n

athwart prep across ▷ adv transversely

atlas n book of maps

atmosphere n mass of gases surrounding a heavenly body,

esp. the earth; prevailing tone or mood (of a place etc.); unit of pressure ▶ **atmospheric** *adj* ▶ **atmospherics** *pl n* radio interference due to electrical disturbance in the atmosphere

atoll *n* ring-shaped coral reef enclosing a lagoon

atom *n* smallest unit of matter which can take part in a chemical reaction; very small amount ▶ **atom bomb** same as **atomic bomb**

atomic *adj* of or using atomic bombs or atomic energy; of atoms ▶ **atomic bomb** bomb in which the energy is provided by nuclear fission ▶ **atomic energy** nuclear energy ▶ **atomic number** number of protons in the nucleus of an atom ▶ **atomic weight** ratio of the mass per atom of an element to one twelfth of the mass of a carbon atom

atomize *v* reduce to atoms or small particles

atomizer *n* device for discharging a liquid in a fine spray

atonal [eh-tone-al] *adj* (of music) not written in an established key

atone *v* make amends (for sin or wrongdoing) ▶ **atonement** *n*

atop *prep* (lit) on top of

atrium *n, pl* **atria** upper chamber of either half of the heart; central hall extending through several storeys of a modern building; main courtyard of an ancient Roman house

atrocious *adj* extremely cruel or wicked; horrifying or shocking; (informal) very bad ▶ **atrociously** *adv* ▶ **atrocity** *n* wickedness; (pl **-ties**) act of cruelty

atrophy [at-trof-fee] *n, pl* **-phies** wasting away of an organ or part ▷ *v* **-phying, -phied** (cause to) waste away

attach *v* join, fasten, or connect; attribute or ascribe ▶ **attached** *adj* (foll by *to*) fond of ▶ **attachment** *n* affection; accessory that can be fitted to a device; computer file sent with an e-mail

attaché [at-tash-shay] *n* specialist attached to a diplomatic mission ▶ **attaché case** flat rectangular briefcase for papers

attack *v* launch a physical assault (against); criticize; set about (a job or problem) with vigour; affect adversely ▷ *n* act of attacking; sudden bout of illness ▶ **attacker** *n*

attain *v* achieve or accomplish (a task or aim); reach ▶ **attainable** *adj* ▶ **attainment** *n* accomplishment

attar *n* fragrant oil made from roses

attempt *v* try, make an effort ▷ *n* effort or endeavour

attend *v* be present at; go regularly to a school, college, etc.; look after; pay attention; apply oneself (to) ▶ **attendance** *n* attending; number attending ▶ **attendant** *n* person who assists, guides, or provides a service ▷ *adj* accompanying ▶ **attention** *n* concentrated direction of the mind; consideration; care; alert position in military drill ▶ **attentive** *adj* giving attention; considerately helpful ▶ **attentively** *adv* ▶ **attentiveness** *n*

attenuated *adj* weakened; thin and extended ▶ **attenuation** *n*

attest *v* affirm the truth of, be proof of ▶ **attestation** *n*

attic *n* space or room within the roof of a house

attire *n* (formal) fine or formal clothes

attired *adj* dressed in a specified way

attitude *n* way of thinking and behaving; posture of the body

attorney *n* person legally appointed to act for another; (US & SAfr) lawyer

a

attract v arouse the interest or admiration of; draw (something) closer by exerting a force on it ▸ **attraction** n power to attract; something that attracts ▸ **attractive** adj ▸ **attractively** adv ▸ **attractiveness** n

attribute v (usu. foll by to) regard as belonging to or produced by ▷ n quality or feature representative of a person or thing ▸ **attributable** adj ▸ **attribution** n ▸ **attributive** adj (Grammar) (of an adjective) preceding the noun modified

attrition n constant wearing down to weaken or destroy

attune v adjust or accustom (a person or thing)

atypical [eh-**tip**-ik-al] adj not typical

Au (Chem) gold

aubergine [oh-bur-zheen] n (Brit) dark purple tropical fruit, cooked and eaten as a vegetable

aubrietia [aw-bree-sha] n trailing plant with purple flowers

auburn adj (of hair) reddish-brown

auction n public sale in which articles are sold to the highest bidder ▷ v sell by auction ▸ **auctioneer** n person who conducts an auction

audacious adj recklessly bold or daring; impudent ▸ **audaciously** adv ▸ **audacity** n

audible adj loud enough to be heard ▸ **audibly** adv ▸ **audibility** n

audience n group of spectators or listeners; formal interview

audio adj of sound or hearing; of or for the transmission or reproduction of sound ▸ **audio typist** typist trained to type from a dictating machine ▸ **audiovisual** adj (esp. of teaching aids) involving both sight and hearing

audit n official examination of business accounts ▷ v **auditing,**

audited examine (business accounts) officially ▸ **auditor** n

audition n test of a performer's ability for a particular role or job ▷ v test or be tested in an audition

auditorium n, pl -toriums or -toria area of a concert hall or theatre where the audience sits

auditory adj of or relating to hearing

au fait [oh fay] adj (French) fully informed; expert

Aug August

auger n tool for boring holes

aught pron (obs) anything whatever

augment v increase or enlarge ▸ **augmentation** n

au gratin [oh grat-tan] adj covered and cooked with breadcrumbs and sometimes cheese

augur v be a sign of (future events) ▸ **augury** n foretelling of the future; (pl -ries) omen

August n eighth month of the year

august [aw-gust] adj dignified and imposing

auk n northern sea bird with short wings and black-and-white plumage

aunt n father's or mother's sister; wife of parent's sibling ▸ **auntie, aunty** n, pl **aunties** (informal) aunt ▸ **Aunt Sally** (Brit, NZ & S Afr) figure used in fairgrounds as a target; target of abuse or criticism

au pair n young foreign woman who does housework in return for board and lodging

aura n distinctive air or quality of a person or thing

aural adj of or using the ears or hearing

aureole, aureola n halo

au revoir [oh riv-vwahr] interj (French) goodbye

auricle n upper chamber of the heart; outer part of the ear ▸ **auricular** adj

aurochs n, pl **aurochs** recently extinct European wild ox

aurora n, pl **-ras** or **-rae** bands of light sometimes seen in the sky in polar regions ▶ **aurora australis** aurora seen near the South Pole ▶ **aurora borealis** aurora seen near the North Pole

auscultation n listening to the internal sounds of the body, usu. with a stethoscope, to help with diagnosis

auspices [aw-spiss-siz] pl n **under the auspices of** with the support and approval of

auspicious adj showing signs of future success, favourable ▶ **auspiciously** adv

Aussie n, adj (informal) Australian

Aust Australia(n)

austere adj stern or severe; ascetic or self-disciplined; severely simple or plain ▶ **austerely** adv ▶ **austerity** n

Australasian n, adj (person) from Australia, New Zealand, and neighbouring islands

Australia Day n (Aust) public holiday on 26 January

Australian n, adj (person) from Australia **Indigenous Australian** same as **aborigine**

autarchy [aw-tar-kee] n absolute power or autocracy

autarky [aw-tar-kee] n policy of economic self-sufficiency

authentic adj known to be real, genuine ▶ **authentically** adv ▶ **authenticity** n ▶ **authenticate** v establish as genuine ▶ **authentication** n

author n writer of a book etc.; originator or creator ▶ **authorship** n

authority n, pl **-ties** power to command or control others; (often pl) person or group having this power; expert in a particular field ▶ **authoritarian** n, adj (person) insisting on strict obedience to authority ▶ **authoritative** adj recognized as being reliable; possessing authority ▶ **authoritatively** adv ▶ **authorize** v give authority to; give permission for ▶ **authorization** n

autism n (Psychiatry) developmental condition characterized by difficulties with social communication and interaction ▶ **autistic** adj

auto- combining form self- e.g. autobiography

autobiography n, pl **-phies** account of a person's life written by that person ▶ **autobiographical** adj ▶ **autobiographically** adv

autocrat n ruler with absolute authority; dictatorial person ▶ **autocratic** adj ▶ **autocratically** adv ▶ **autocracy** n government by an autocrat

autocross n motor-racing over a rough course

Autocue® n electronic television prompting device displaying a speaker's script, unseen by the audience

autogiro, autogyro n, pl **-ros** self-propelled aircraft resembling a helicopter but with an unpowered rotor

autograph n handwritten signature of a (famous) person ▷ v write one's signature on or in

automat n (US) vending machine

automate v make (a manufacturing process) automatic ▶ **automation** n

automatic adj (of a device) operating mechanically by itself; (of a process) performed by automatic equipment; done without conscious thought; (of a firearm) self-loading ▷ n self-loading firearm; vehicle with automatic transmission ▶ **automatically** adv

automaton n robot; person who acts mechanically

automobile n (US) motor car

autonomy n self-government ▶ **autonomous** adj

autopsy n, pl -sies examination of a corpse to determine the cause of death

autosuggestion n process in which a person unconsciously influences his or her own behaviour or beliefs

autumn n season between summer and winter ▶ **autumnal** adj

auxiliary adj secondary or supplementary; supporting ▷ n, pl -ries person or thing that supplements or supports ▶ **auxiliary verb** verb used to form the tense, voice, or mood of another, such as will in I will go

avail v be of use or advantage (to) ▷ n use or advantage e.g. to no avail **avail oneself of** make use of

available adj obtainable or accessible ▶ **availability** n

avalanche n mass of snow or ice falling down a mountain; sudden overwhelming quantity of anything

avant-garde [av-ong-**gard**] n group of innovators, esp. in the arts ▷ adj innovative and progressive

avarice [av-a-riss] n greed for wealth ▶ **avaricious** adj

avast interj (Naut) stop

avatar n (Hinduism) appearance of a god in animal or human form; (Computing) graphical representation of a person in a virtual environment, such as an online role-playing game

Ave Avenue

avenge v take revenge in retaliation for (harm done) or on behalf of (a person harmed) ▶ **avenger** n

avenue n wide street; road between two rows of trees; way of approach

aver [av-**vur**] v **averring, averred** state to be true

average n typical or normal amount or quality; result obtained by adding quantities together and dividing the total by the number of quantities ▷ adj usual or typical; calculated as an average ▷ v calculate the average of; amount to as an average

averse adj (usu. foll by to) disinclined or unwilling ▶ **aversion** n strong dislike; person or thing disliked

avert v turn away; ward off

avian adj relating to birds ▶ **avian flu** same as **bird flu**

aviary n, pl **aviaries** large cage or enclosure for birds

aviation n art of flying aircraft ▶ **aviator** n

avid adj keen or enthusiastic; greedy (for) ▶ **avidly** adv ▶ **avidity** n

avocado n, pl -dos pear-shaped tropical fruit with a leathery green skin and yellowish-green flesh

avocation n (old-fashioned) occupation; hobby

avocet n long-legged wading bird with a long slender upward-curving bill

avoid v prevent from happening; refrain from; keep away from ▶ **avoidable** adj ▶ **avoidance** n

avoirdupois [av-er-de-**poise**] n system of weights based on pounds and ounces

avow v state or affirm; admit openly ▸ **avowal** n ▸ **avowed** adj ▸ **avowedly** adv

avuncular adj (of a man) friendly, helpful, and caring towards someone younger

await v wait for; be in store for

awake v **waking, awoke, awoken** emerge or rouse from sleep; (cause to) become alert ▸ adj not sleeping; alert

awaken v awake

award v give (something, such as a prize) formally ▸ n something awarded, such as a prize

aware adj having knowledge, informed ▸ **awareness** n

awash adv washed over by water

away adv from a place e.g. go away; to another place e.g. put that gun away; out of existence e.g. fade away; continuously e.g. laughing away b adj not present; distant e.g. two miles away; (Sport) played on an opponent's ground

awe n wonder and respect mixed with dread ▸ v fill with awe ▸ **awesome** adj inspiring awe; (slang) excellent or outstanding ▸ **awestruck** adj filled with awe

awful adj very bad or unpleasant; (informal) very great; (obs) inspiring awe ▸ **awfully** adv in an unpleasant way; (informal) very

awhile adv for a short time

awkward adj clumsy or ungainly; embarrassed; difficult to use or handle; inconvenient ▸ **awkwardly** adv ▸ **awkwardness** n

awl n pointed tool for piercing wood, leather, etc.

awning n canvas roof supported by a frame to give protection against the weather

awoke v past tense of **awake** ▸ **awoken** v past participle of **awake**

AWOL adj (Mil) absent without leave

awry [a-rye] adv, adj with a twist to one side, askew; amiss

axe n tool with a sharp blade for felling trees or chopping wood; (informal) dismissal from employment etc. ▸ v (informal) dismiss (employees), restrict (expenditure), or terminate (a project)

axil n angle where the stalk of a leaf joins a stem

axiom n generally accepted principle; self-evident statement ▸ **axiomatic** adj self-evident

axis n, pl **axes** (imaginary) line round which a body can rotate or about which an object or geometrical figure is symmetrical; one of two fixed lines on a graph, against which quantities or positions are measured ▸ **axial** adj

axle n shaft on which a wheel or pair of wheels turns

axolotl n aquatic salamander of central America

ayatollah n Islamic religious leader in Iran

aye, ay interj yes ▸ n affirmative vote or voter

azalea [az-zale-ya] n garden shrub grown for its showy flowers

azimuth n arc of the sky between the zenith and the horizon; horizontal angle of a bearing measured clockwise from the north

Aztec n, adj (person) of the race ruling Mexico before the Spanish conquest in the 16th century

azure adj, n (of) the colour of a clear blue sky

Bb

BA Bachelor of Arts

baa v **baaing, baaed** make the characteristic bleating sound of a sheep ▷ n cry made by a sheep

baas n (S Afr) boss

babble v talk excitedly or foolishly; (of streams) make a low murmuring sound ▷ n muddled or foolish speech

babe n baby

babel n confused mixture of noises or voices

baboon n large monkey with a pointed face and a long tail

baby n, pl **-bies** very young child or animal; (slang) sweetheart ▷ adj comparatively small of its type ▶ **babyish** adj ▶ **baby-sit** v take care of a child while the parents are out ▶ **baby-sitter** n

baccarat [back-a-rah] n card game involving gambling

bacchanalia [back-a-nail-ee-a] n wild drunken party or orgy

bach [batch] (NZ) n small holiday cottage ▷ v look after oneself when one's spouse is away

bachelor n unmarried man; person who holds the lowest university or college degree

> **SPELLING TIP**
> Not **batchelor**, but **bachelor**: the correct spelling has no t

bacillus [bass-ill-luss] n, pl **-li** [-lie] rod-shaped bacterium

back n rear part of the human body, from the neck to the pelvis; part or side of an object opposite the front; part of anything less often seen or used; (Ball games) defensive player or position ▷ v (cause to) move backwards; provide money for (a person or enterprise); bet on the success of; (foll by onto) have the back facing towards ▷ adj situated behind; owing from an earlier date ▷ adv at, to, or towards the rear; to or towards the original starting point or condition ▶ **backer** n person who gives financial support ▶ **backing** n support; musical accompaniment for a pop singer ▶ **backward** adj directed towards the rear; physically or materially underdeveloped ▶ **backwardness** n ▶ **backwards** adv towards the rear; with the back foremost; in the reverse of the usual direction ▶ **back up** v support ▶ **backup** n support or reinforcement; reserve or substitute

backbencher n Member of Parliament who does not hold office in the government or opposition

backbiting n spiteful talk about an absent person

backbone n spinal column; strength of character

backchat n (informal) impudent replies

backcloth, backdrop n painted curtain at the back of a stage set

backdate v make (a document) effective from a date earlier than its completion

backfire v (of a plan) fail to have the desired effect; (of an engine) make a loud noise like an explosion

backgammon n game played with counters and dice

background n events or circumstances that help to explain something; person's social class, education, or experience; part of a scene or picture furthest from the viewer

backhand n (Tennis etc.) stroke played with the back of the hand facing the direction of the stroke ▸ **backhanded** adj ambiguous or implying criticism e.g. a backhanded compliment ▸ **backhander** n (slang) bribe

backlash n sudden and adverse reaction

backlog n accumulation of things to be dealt with

backpack n large pack carried on the back ▷ v go travelling or hiking with a backpack ▸ **backpacker** n

backside n (informal) buttocks

backslide v relapse into former bad habits ▸ **backslider** n

backstage adv, adj behind the stage in a theatre

backstroke n swimming stroke performed on the back

backtrack v return by the same route by which one has come; retract or reverse one's opinion or policy

backwash n water washed backwards by the motion of a boat; repercussion

backwater n isolated or backward place or condition

backwoods pl n remote sparsely populated area

bacon n salted or smoked pig meat

bacteria pl n, sing -**rium** large group of microorganisms, many of which cause disease ▸ **bacterial** adj
▸ **bacteriology** n study of bacteria
▸ **bacteriologist** n

> **USAGE NOTE**
> Note that the word bacteria is already plural and does not need an '-s'. The singular form is bacterium

bad adj **worse**, **worst** of poor quality; lacking skill or talent;

harmful; immoral or evil; naughty or mischievous; rotten or decayed; unpleasant ▸ **badly** adv ▸ **badness** n ▸ **bad bank** financial institution set up to limit or regulate underperforming assets owned by other banks
▸ **bad debt** debt that is not collectible and therefore worthless to the creditor

bade v a past tense of **bid**

badge n emblem worn to show membership, rank, etc.

badger n nocturnal burrowing mammal of Europe, Asia, and N America with a black and white head ▷ v pester or harass

badinage [bad-in-nahzh] n playful and witty conversation

badminton n game played with rackets and a shuttlecock, which is hit back and forth over a high net

Bafana Bafana [bah-fan-na] pl n (S Afr) South African national soccer team

baffle v perplex or puzzle ▷ n device to limit or regulate the flow of fluid, light, or sound ▸ **bafflement** n

bag n flexible container with an opening at one end; handbag or piece of luggage; (offens) ugly or bad-tempered woman ▷ v **bagging**, **bagged** put into a bag; succeed in capturing, killing or scoring ▸ **baggy** adj (of clothes) hanging loosely

bagatelle n something of little value; board game in which balls are struck into holes

bagel n hard ring-shaped bread roll

baggage n suitcases packed for a journey

bagpipes pl n musical wind instrument with reed pipes and an inflatable bag

bail n (Law) money deposited with a court as security for a person's reappearance in court ▷ v pay bail for (a person)

bail², **bale** v (foll by out) remove (water) from (a boat); (informal) help (a person or organization) out of a predicament; make an emergency parachute jump from an aircraft

bail³ n (Cricket) either of two wooden bars across the tops of the stumps

bailey n outermost wall or court of a castle

bailiff n sheriff's officer who serves writs and summonses; landlord's agent

bairn n (Scot) child

bait n piece of food on a hook or in a trap to attract fish or animals ▷ v put a piece of food on or in (a hook or trap); persecute or tease

> **SPELLING TIP**
>
> The phrase with bated breath is sometimes wrongly spelled with baited breath

baize n woollen fabric used to cover billiard and card tables

bake v cook by dry heat as in an oven; make or become hardened by heat ▸ **bakeoff** n baking competition ▸ **baking powder** powdered mixture containing sodium bicarbonate, used as a raising agent in baking

baker n person whose business is to make or sell bread, cakes, etc. ▸ **baker's dozen** thirteen ▸ **bakery** n, pl **-eries** place where bread, cakes, etc. are baked or sold

bakkie n (SAfr) small truck

Balaclava, Balaclava helmet n close-fitting woollen hood that covers the ears and neck

balalaika n guitar-like musical instrument with a triangular body

balance n state in which a weight or amount is evenly distributed; amount that remains e.g. the balance of what you owe; weighing device; difference between the credits and debits of an account ▷ v weigh in a balance; make or remain steady; consider or compare; compare or equalize the money going into or coming out of an account

balcony n, pl **-nies** platform on the outside of a building with a rail along the outer edge; upper tier of seats in a theatre or cinema

bald adj having little or no hair on the scalp; plain or blunt; (of a tyre) having a worn tread ▸ **balding** adj becoming bald ▸ **baldness** n

balderdash n stupid talk

bale¹ n large bundle of hay or goods tightly bound together ▷ v make or put into bales

bale² v same as **bail²**

baleful adj vindictive or menacing ▸ **balefully** adv

balk, baulk v be reluctant to (do something); thwart or hinder

Balkan adj of any of the countries of the Balkan Peninsula: Romania, Bulgaria, Albania, Greece, the former Yugoslavia, and the European part of Turkey

ball¹ n round or nearly round object, esp. one used in games; single delivery of the ball in a game ▷ pl (vulgar slang) testicles; nonsense ▷ v form into a ball ▸ **ball bearings** steel balls between moving parts of a machine to reduce friction ▸ **ball cock** device with a floating ball and a valve for regulating the flow of water ▸ **ballpoint, ballpoint pen** n pen with a tiny ball bearing as a writing point

ball² n formal social function for dancing ▸ **ballroom** n

ballad n narrative poem or song; slow sentimental song

ballast n substance, such as sand, used to stabilize a ship when it is not carrying cargo

ballet n classical style of expressive dancing based on conventional steps; theatrical performance of this ▷ **ballerina** n female ballet dancer

ballistics n study of the flight of projectiles, such as bullets ▷ **ballistic missile** missile guided automatically in flight but which falls freely at its target

balloon n inflatable rubber bag used as a plaything or decoration; large bag inflated with air or gas, designed to float in the atmosphere with passengers in a basket underneath ▷ v fly in a balloon; swell or increase rapidly in size ▷ **balloonist** n

ballot n method of voting; actual vote or paper indicating a person's choice ▷ v -**loting**, -**loted** vote or ask for a vote from

ballyhoo n exaggerated fuss

balm n aromatic substance used for healing and soothing; anything that comforts or soothes

Balmain bug n edible Australian shellfish

balmy adj **balmier**, **balmiest** (of weather) mild and pleasant

baloney n (informal) nonsense

balsa [bawl-sa] n very light wood from a tropical American tree

balsam n soothing ointment; flowering plant

baluster n set of posts supporting a rail

balustrade n ornamental rail supported by balusters

bamboo n tall treelike tropical grass with hollow stems

bamboozle v (informal) cheat or mislead; confuse, puzzle

ban v **banning**, **banned** prohibit or forbid officially ▷ n official prohibition

banal [ban-nahl] adj ordinary and unoriginal ▷ **banality** n

banana n yellow crescent-shaped fruit

band¹ n group of musicians playing together; group of people having a common purpose ▷ **bandsman** n ▷ **bandstand** n roofed outdoor platform for a band ▷ **band together** v unite

band² n strip of some material, used to hold objects; (Physics) range of frequencies or wavelengths between two limits

bandage n piece of material used to cover a wound or wrap an injured limb ▷ v cover with a bandage

bandanna, bandana n large brightly coloured handkerchief or neckerchief

B & B bed and breakfast

bandicoot n ratlike Australian marsupial

bandit n robber, esp. a member of an armed gang ▷ **banditry** n

bandolier n shoulder belt for holding cartridges

bandwagon n **jump, climb on the bandwagon** join a party or movement that seems assured of success

bandwidth n the range of frequencies used in a telecommunications signal

bandy adj -**dier**, -**diest** (also **bandy-legged**) having legs curved outwards at the knees ▷ v -**dying**, -**died** exchange (words) in a heated manner; use (a name, term, etc.) frequently

bane n person or thing that causes misery or distress ▷ **baneful** adj

bang n short loud explosive noise; hard blow or loud knock ▷ v hit or knock, esp. with a loud noise; close (a door) noisily ▷ *adv* precisely; with a sudden impact

banger *n* (Brit & Aust informal) old decrepit car; (slang) sausage; firework that explodes loudly

bangle *n* bracelet worn round the arm or the ankle

banish *v* send (someone) into exile; drive away ▶ **banishment** *n*

banisters *pl n* railing supported by posts on a staircase

banjo *n*, *pl* **-jos** or **-joes** guitar-like musical instrument with a circular body

bank¹ *n* institution offering services such as the safekeeping and lending of money; any supply, store, or reserve ▷ *v* deposit (cash or cheques) in a bank ▶ **banking** *n* ▶ **banknote** *n* piece of paper money ▶ **bank on** *v* rely on

bank² *n* raised mass, esp. of earth; sloping ground at the side of a river ▷ *v* form into a bank; cause (an aircraft) or (of an aircraft) to tip to one side on turning

bank³ *n* arrangement of switches, keys, oars, etc. in a row or in tiers

banker *n* manager or owner of a bank

bankrupt *n* person declared by a court to be unable to pay his or her debts ▷ *adj* financially ruined ▷ *v* make bankrupt ▶ **bankruptcy** *n*

banksia *n* Australian evergreen tree or shrub

banner *n* long strip of cloth displaying a slogan, advertisement, etc.; placard carried in a demonstration or procession; advert that extends across the top of a web page

bannisters *pl n* same as **banisters**

banns *pl n* public declaration, esp. in a church, of an intended marriage

banquet *n* elaborate formal dinner

banshee *n* (in Irish folklore) female spirit whose wailing warns of a coming death

bantam *n* small breed of chicken ▶ **bantamweight** *n* boxer weighing up to 118lb (professional) or 54kg (amateur)

banter *v* tease jokingly ▷ *n* teasing or joking conversation

Bantu *n* group of languages of Africa; (offens) Black speaker of a Bantu language

baobab [bay-oh-bab] *n* African tree with a thick trunk and angular branches

baptism *n* Christian religious ceremony in which a person is immersed in or sprinkled with water as a sign of being cleansed from sin and accepted into the Church ▶ **baptismal** *adj* ▶ **baptize** *v* perform baptism on

Baptist *n* member of a Protestant denomination that believes in adult baptism by immersion

bar¹ *n* rigid length of metal, wood, etc.; solid, usu. rectangular block, of any material; anything that obstructs or prevents; counter or room where drinks are served; heating element in an electric fire; (Music) group of beats repeated throughout a piece of music ▷ *v* **barring, barred** secure with a bar; obstruct; ban or forbid ▷ *prep* (also **barring**) except for **the Bar** barristers collectively ▶ **barman, barmaid** *n*

bar² *n* unit of atmospheric pressure

barb *n* cutting remark; point facing in the opposite direction to the main point of a fish-hook etc. ▶ **barbed** *adj* ▶ **barbed wire** strong wire with protruding sharp points

barbarian *n* member of a primitive or uncivilized people ▶ **barbaric**

adj cruel or brutal ▶ **barbarism** *n* condition of being backward or ignorant ▶ **barbarity** *n* state of being barbaric or barbarous; (*pl* **-ties**) vicious act ▶ **barbarous** *adj* uncivilized; brutal or cruel

barbecue *n* grill on which food is cooked over hot charcoal, usu. outdoors; outdoor party at which barbecued food is served ▷ *v* cook (food) on a barbecue

barber *n* person who cuts men's hair and shaves beards

barbiturate *n* drug used as a sedative

bar code *n* arrangement of numbers and parallel lines on a package, which can be electronically scanned at a checkout to give the price of the goods

bard *n* (*lit*) poet

bare *adj* unclothed, naked; without the natural or usual covering; unembellished, simple; just sufficient ▷ *v* uncover ▶ **barely** *adv* only just ▶ **bareness** *n* ▶ **barebones** *adj* basic, containing just the essential elements e.g. *a barebones summary*

bareback *adj*, *adv* (of horse-riding) without a saddle

barefaced *adj* shameless or obvious

barefoot, barefooted *adj*, *adv* with the feet uncovered

bargain *n* agreement establishing what each party will give, receive, or perform in a transaction; something bought or offered at a low price ▷ *v* negotiate the terms of an agreement ▶ **bargain for** *v* anticipate or take into account

barge *n* flat-bottomed boat used to transport freight ▷ *v* (*informal*) push violently ▶ **barge in, barge into** *v* interrupt rudely

barista [bar-**ee**-sta] *n* person who makes and sells coffee in a coffee bar

baritone *n* (singer with) the second lowest adult male voice

barium *n* (*Chem*) soft white metallic element

bark[1] *n* loud harsh cry of a dog ▷ *v* (of a dog) make its typical cry; shout in an angry tone

bark[2] *n* tough outer layer of a tree

barley *n* tall grasslike plant cultivated for grain

barmy *adj* **-mier, -miest** (*slang*) insane

barn *n* large building on a farm used for storing grain

barnacle *n* shellfish that lives attached to rocks, ship bottoms, etc.

barney *n* (*informal*) noisy fight or argument

barometer *n* instrument for measuring atmospheric pressure ▶ **barometric** *adj*

baron *n* member of the lowest rank of nobility; powerful businessman ▶ **baroness** *n* ▶ **baronial** *adj*

baronet *n* commoner who holds the lowest hereditary British title

baroque [bar-**rock**] *n* highly ornate style of art, architecture, or music from the late 16th to the early 18th century ▷ *adj* ornate in style

barque [bark] *n* sailing ship, esp. one with three masts

barra *n* (*Aust informal*) short for **barramundi**

barrack *v* criticize loudly or shout against (a team or speaker)

barracks *pl n* building used to accommodate military personnel

barracouta *n* large Pacific fish with a protruding lower jaw and strong teeth

barracuda *n* tropical sea fish

barrage [bar-**rahzh**] *n* continuous delivery of questions, complaints, etc.; continuous artillery fire; artificial barrier across a river to control the water level

barramundi n edible Australian fish

barrel n cylindrical container with rounded sides and flat ends; tube in a firearm through which the bullet is fired ▶ **barrel organ** musical instrument played by turning a handle

barren adj (of a woman or female animal) incapable of producing offspring; (of land) unable to support the growth of crops, fruit, etc. ▶ **barrenness** n

barricade n barrier, esp. one erected hastily for defence ▶ v erect a barricade across (an entrance)

barrier n anything that prevents access, progress, or union

barrister n (Brit, Aust & NZ) lawyer qualified to plead in a higher court

barrow¹ n wheelbarrow; movable stall used by street traders

barrow² n mound of earth over a prehistoric tomb

barter v trade (goods) in exchange for other goods ▶ n trade by the exchange of goods

basalt [bass-awlt] n dark volcanic rock ▶ **basaltic** adj

base¹ n bottom or supporting part of anything; fundamental part; centre of operations, organization, or supply; starting point ▶ v (foll by on or upon) use as a basis (for); (foll by at or in) to station or place ▶ **baseless** adj

base² adj dishonourable or immoral; of inferior quality or value ▶ **baseness** n

baseball n team game in which runs are scored by hitting a ball with a bat then running round four bases; ball used for this

basement n partly or wholly underground storey of a building

bash (informal) v hit violently or forcefully ▶ n heavy blow; party

bashful adj shy or modest ▶ **bashfully** adv ▶ **bashfulness** n

BASIC n computer programming language that uses common English words

basic adj of or forming a base or basis; elementary or simple ▶ **basics** pl n fundamental principles, facts, etc. ▶ **basically** adv

basil n aromatic herb used in cooking

basilica n rectangular church with a rounded end and two aisles

basilisk n legendary serpent said to kill by its breath or glance

basin n round open container; sink for washing the hands and face; sheltered area of water where boats may be moored; catchment area of a particular river

basis n, pl -ses fundamental principles etc. from which something is started or developed

bask v lie in or be exposed to something, esp. pleasant warmth

basket n container made of interwoven strips of wood or cane ▶ **basketwork** n

basketball n team game in which points are scored by throwing the ball through a high horizontal hoop; ball used for this

Basque n, adj (member or language) of a people living in the W Pyrenees in France and Spain

basque n tight-fitting bodice for women

bas-relief n sculpture in which the figures project slightly from the background

bass¹ [base] n (singer with) the lowest adult male voice ▶ adj of the lowest range of musical notes

bass² *n* edible sea fish

basset hound *n* smooth-haired dog with short legs and long ears

bassoon *n* low-pitched woodwind instrument

bastard *n* (*offens*) obnoxious or despicable person; person born of parents not married to each other

baste¹ *v* moisten (meat) during cooking with hot fat

baste² *v* sew with loose temporary stitches

bastion *n* projecting part of a fortification; thing or person regarded as defending a principle

bat¹ *n* any of various types of club used to hit the ball in certain sports ▷ *v* **batting, batted** strike with or as if with a bat ▶ **batsman** *n* (*Cricket*) person who bats or specializes in batting

bat² *n* nocturnal mouse-like flying animal

batch *n* group of people or things dealt with at the same time

bated *adj* **with bated breath** in suspense or fear

bath *n* large container in which to wash the body; act of washing in such a container ▷ *pl* public swimming pool ▷ *v* wash in a bath ▶ **bathroom** *n* room with a bath, sink, and usu. a toilet

Bath chair *n* wheelchair for an invalid

bathe *v* swim in open water for pleasure; apply liquid to (the skin or a wound) in order to cleanse or soothe; (foll by *in*) fill (with) e.g. *bathed in sunlight* ▶ **bather** *n*

bathos [bay-thoss] *n* sudden ludicrous change in speech or writing from a serious subject to a trivial one

batik [bat-**teek**] *n* process of printing fabric using wax to cover areas not to be dyed; fabric printed in this way

batman *n* officer's servant in the armed forces

baton *n* thin stick used by the conductor of an orchestra; short bar transferred in a relay race; police officer's truncheon

battalion *n* army unit consisting of three or more companies

batten *n* strip of wood fixed to something, esp. to hold it in place ▶ **batten down** *v* secure with battens

batter¹ *v* hit repeatedly ▶ **battering ram** large beam used to break down fortifications

batter² *n* mixture of flour, eggs, and milk, used in cooking

battery *n*, *pl* **-teries** device that produces electricity in a torch, radio, etc.; group of heavy guns operating as a single unit ▷ *adj* kept in series of cages for intensive rearing

battle *n* fight between large armed forces; conflict or struggle ▷ *v* struggle

battle-axe *n* (*informal*) domineering woman; (formerly) large heavy axe

battlement *n* wall with gaps along the top for firing through

battleship *n* large heavily armoured warship

batty *adj* **-tier, -tiest** (*slang*) eccentric or crazy

bauble *n* trinket of little value

bauera *n* small evergreen Australian shrub

baulk *v* same as **balk**

bauxite *n* claylike substance that is the chief source of aluminium

bawdy *adj* **bawdier, bawdiest** (of writing etc.) containing humorous references to sex

bawl *v* shout or weep noisily

bay¹ *n* stretch of coastline that curves inwards

b

bay² n recess in a wall; area set aside for a particular purpose e.g. *loading bay*

bay³ v howl in deep prolonged tones

bay⁴ n Mediterranean laurel tree ▷ **bay leaf** its dried leaf, used in cooking

bay⁵ adj, n reddish-brown (horse)

bayonet n sharp blade that can be fixed to the end of a rifle ▷ v **-neting, -neted** stab with a bayonet

bazaar n sale in aid of charity; market area, esp. in Eastern countries

bazooka n portable rocket launcher that fires an armour-piercing projectile

BBC British Broadcasting Corporation

BC before Christ

BCE before Common Era

BCG® antituberculosis vaccine

be v, present sing 1st person **am**, 2nd person **are**, 3rd person **is**, present pl **are**, past sing 1st person **was**, 2nd person **were**, 3rd person **was**, past pl **were**, present participle **being**, past participle **been** exist or live; used as a linking between the subject of a sentence and its complement e.g. *John is a musician*; forms the progressive present tense e.g. *the man is running*; forms the passive voice of all transitive verbs e.g. *a good film is being shown on television tonight*

beach n area of sand or pebbles on a shore ▷ v run or haul (a boat) onto a beach ▷ **beachhead** n beach captured by an attacking army on which troops can be landed

beacon n fire or light on a hill or tower, used as a warning

bead n small piece of plastic, wood, etc., pierced for threading on a string to form a necklace etc.; small drop of moisture ▷ **beaded**

adj ▷ **beading** n strip of moulding used for edging furniture ▷ **beady** adj small, round, and glittering e.g. *beady eyes*

beagle n small hound with short legs and drooping ears

beak¹ n projecting horny jaws of a bird; (slang) nose ▷ **beaky** adj

beak² n (Brit, Aust & NZ slang) judge, magistrate, or headmaster

beaker n large drinking cup; lipped glass container used in laboratories

beam n broad smile; ray of light; narrow flow of electromagnetic radiation or particles; long thick piece of wood, metal, etc., used in building ▷ v smile broadly; divert or aim (a radio signal, light, etc.) in a certain direction

bean n seed or pod of various plants, eaten as a vegetable or used to make coffee etc.

beanie n close-fitting woollen hat

bear¹ v **bearing, bore, borne** support or hold up; bring e.g. *to bear gifts*; (past participle (passive) **born**) give birth to; tolerate or endure; hold in the mind ▷ **bearable** adj ▷ **bear out** v show to be truthful

bear² n large heavy mammal with a shaggy coat ▷ **bearskin** n tall fur helmet worn by some British soldiers

beard n hair growing on the lower parts of a man's face ▷ **bearded** adj

bearer n person who carries, presents, or upholds something

bearing n relevance (to); person's general social conduct; part of a machine that supports another part, esp. one that reduces friction ▷ pl sense of one's own relative position

beast n large wild animal; brutal or uncivilized person ▷ **beastly** adj unpleasant or disagreeable

beat v **beating, beat, beaten** or **beat** hit hard and repeatedly; move (wings) up and down; throb rhythmically; stir or mix vigorously; overcome or defeat ▷ n regular throb; assigned route, as of a policeman; basic rhythmic unit in a piece of music ▶ **beatbox** n machine that produces musical beats ▶ **beat up** v injure (someone) by repeated blows or kicks

beatify [bee-**at**-if-fie] v **-fying, -fied** (RC Church) declare (a dead person) to be among the blessed in heaven: the first step towards canonization ▶ **beatific** adj displaying great happiness ▶ **beatification** n ▶ **beatitude** n (Christianity) any of the blessings on the poor, meek, etc., in the Sermon on the Mount

beau [boh] n, pl **beaux** or **beaus** boyfriend or admirer; man greatly concerned with his appearance

Beaufort scale n scale for measuring wind speeds

beautician n person who gives beauty treatments professionally

beautiful adj very attractive to look at; very pleasant ▶ **beautifully** adv

beautify v **-fying, -fied** make beautiful ▶ **beautification** n

beauty n, pl **-ties** combination of all the qualities of a person or thing that delight the senses and mind; very attractive woman; (informal) something outstanding of its kind

beaver n amphibious rodent with a big flat tail ▶ **beaver away** v work industriously

becalmed adj (of a sailing ship) motionless through lack of wind

became v past tense of **become**

because conj on account of the fact that ▶ **because of** on account of

beck¹ n at someone's **beck and call** having to be constantly available to do as someone asks

beck² n (N English) stream

beckon v summon with a gesture

become v **-coming, -came, -come** come to be; (foll by of) happen to; suit ▶ **becoming** adj attractive or pleasing; appropriate or proper

bed n piece of furniture on which to sleep; garden plot; bottom of a river, lake, or sea; layer of rock ▶ **go to bed with** have sexual intercourse with ▶ **bed down** v go to or put into a place to sleep or rest ▶ **bedpan** n shallow bowl used as a toilet by bedridden people ▶ **bedridden** adj confined to bed because of illness or old age ▶ **bedrock** n solid rock beneath the surface soil; basic facts or principles ▶ **bedroom** n ▶ **bedsit**, **bedsitter** n furnished sitting room with a bed

bedding n sheets and covers that are used on a bed

bedevil v **-illing, -illed** harass, confuse, or torment

bedlam n noisy confused situation

bedraggled adj untidy, wet, or dirty

bee¹ n insect that makes wax and honey ▶ **beehive** n structure in which bees live ▶ **beeswax** n wax secreted by bees, used in polishes etc.

bee² n social gathering to carry out a communal task e.g. quilting bee ▶ **spelling bee** contest in which players are required to spell words ▶ **working bee** voluntary group doing a job for charity

beech n tree with a smooth greyish bark

beef n flesh of a cow, bull, or ox ▶ **beefy** adj like beef; (informal) strong and muscular ▶ **beefburger** n flat grilled or fried cake of minced beef ▶ **beefeater** n yeoman warder at the Tower of London

b

been v past participle of **be**

beep n high-pitched sound, like that of a car horn ▷ v (cause to) make this noise

beer n alcoholic drink brewed from malt and hops ▶ **beery** adj

beet n plant with an edible root and leaves ▶ **beetroot** n type of beet plant with a dark red root

beetle n insect with a hard wing cover on its back

befall v (old-fashioned) happen to (someone)

befit v be appropriate or suitable for ▶ **befitting** adj

before conj, prep, adv indicating something earlier in time, in front of, or preferred to e.g. before the war; brought before a judge; death before dishonour ▶ **beforehand** adv in advance

befriend v become friends with

beg v **begging, begged** solicit (for money or food), esp. in the street; ask formally or humbly

began v past tense of **begin**

beget v **-getting, -got** or **-gat, -gotten** or **-got** (old-fashioned) cause or create; father

beggar n person who lives by begging ▶ **beggarly** adj

begin v **-ginning, -gan, -gun** start; bring or come into being ▶ **beginner** n person who has just started learning to do something ▶ **beginning** n

begonia n tropical plant with waxy flowers

begrudge v envy (someone) the possession of something; give or allow unwillingly

beguile [big-**gile**] v cheat or mislead; charm or amuse ▶ **beguiling** adj

begun v past participle of **begin**

behalf n **on behalf of** in the interest of or for the benefit of

behave v act or function in a particular way; conduct (oneself) properly

behaviour n manner of behaving

behead v remove the head from

beheld v past of **behold**

behest n order or earnest request

behind prep, adv indicating position to the rear, lateness, responsibility, etc. e.g. behind the wall; behind schedule; the reasons behind her departure ▷ n (informal) buttocks

behold v **-holding, -held** (old-fashioned) look (at) ▶ **beholder** n

beholden adj indebted or obliged

behove v (old-fashioned) be necessary or fitting for

beige adj pale brown

being n state or fact of existing; something that exists or is thought to exist; human being ▷ v present participle of **be**

belabour v attack verbally or physically

belated adj later or too late ▶ **belatedly** adv

belch v expel wind from the stomach noisily through the mouth; expel or be expelled forcefully e.g. smoke belched from the factory ▷ n act of belching

beleaguered adj struggling against difficulties or criticism; besieged by an enemy

belfry n, pl **-fries** part of a tower where bells are hung

Belgium sausage n (NZ) large smooth bland sausage

belie v show to be untrue

belief n faith or confidence; opinion; principle accepted as true, often without proof

believe v accept as true or real; think, assume, or suppose **believe in** be convinced of the truth or existence of ▶ **believable** adj ▶ **believer** n

Belisha beacon [bill-**lee**-sha] n (Brit) flashing orange globe mounted on a post, marking a pedestrian crossing

belittle v treat as having little value or importance

bell n hollow, usu. metal, cup-shaped instrument that emits a ringing sound when struck; device that rings or buzzes as a signal

belladonna n (drug obtained from) deadly nightshade

bellbird n Australasian bird with bell-like call

belle n beautiful woman, esp. the most attractive woman at a function

bellicose adj warlike and aggressive

belligerent adj hostile and aggressive; engaged in war ▷ n person or country engaged in war ▸ **belligerence** n

bellow v make a low deep cry like that of a bull; shout in anger ▷ n loud deep roar

bellows pl n instrument for pumping a stream of air into something

belly n, pl -**lies** part of the body of a vertebrate which contains the intestines; stomach; front, lower, or inner part of something ▷ v -**lying, -lied** (cause to) swell out ▸ **bellyful** n (slang) more than one can tolerate

belong v (foll by to) be the property of; (foll by to) be a part or member of ▸ **belongings** pl n personal possessions

beloved adj dearly loved ▷ n person dearly loved

below prep, adv at or to a position lower than, under

belt n band of cloth, leather, etc., worn usu. around the waist; long narrow area e.g. a belt of trees;

circular strip of rubber that drives moving parts in a machine ▷ v fasten with a belt; (slang) hit very hard; (slang) move very fast

bemoan v express sorrow or dissatisfaction about

bemused adj puzzled or confused

bench n long seat; long narrow work table **the bench** judge or magistrate sitting in court, or judges and magistrates collectively ▸ **benchmark** n criterion by which to measure something

bend v **bending, bent** (cause to) form a curve; (often foll by down) incline the body ▷ n curved part **the bends** (informal) decompression sickness ▸ **bendy** adj

beneath adv, prep below; not worthy of

Benedictine adj of an order of Christian monks and nuns founded by Saint Benedict

benediction n prayer for divine blessing

benefactor, benefactress n someone who supports a person or institution by giving money ▸ **benefaction** n

beneficent [bin-**eff**-iss-ent] adj charitable or generous ▸ **beneficence** n

beneficial adj helpful or advantageous

beneficiary n, pl -**ciaries** person who gains or benefits

benefit n something that improves or promotes; advantage or sake e.g. I'm doing this for your benefit; payment made by a government to a poor, ill, or unemployed person ▷ v -**fiting, -fited** do or receive good

benevolence n inclination to do good; act of

kindness ▶ **benevolent** adj
▶ **benevolently** adv

benighted adj ignorant or
uncultured

benign [bin-**nine**] adj showing
kindness; (of a tumour)
not threatening to life
▶ **benignly** adv

bent v past of **bend** ▷ adj curved;
(slang) dishonest or corrupt; (Brit &
Aust offens slang) homosexual ▷ n
personal inclination or aptitude
bent on determined to pursue (a
course of action)

bento, bento box n thin
lightweight box divided into
compartments, which contain
small separate dishes comprising a
Japanese meal

benzene n flammable poisonous
liquid used as a solvent, insecticide,
etc.

bequeath v dispose of (property)
as in a will ▶ **bequest** n legal gift
of money or property to someone
who has died

berate v scold harshly

bereaved adj having recently lost
a close friend or relative through
death ▶ **bereavement** n

bereft adj (foll by of) deprived

beret [ber-**ray**] n round flat close-
fitting brimless cap

berg¹ n iceberg

berg² n (SAfr) mountain

bergamot n small Asian tree, the
fruit of which yields an oil used in
perfumery

beri-beri n disease caused by
vitamin B deficiency

berk n (Brit, Aust & NZ slang) stupid
person

berm n (NZ) narrow grass strip
between the road and the footpath
in a residential area

berry n, pl -**ries** small soft
stoneless fruit

berserk adj **go berserk** become
violent or destructive

berth n bunk in a ship or train; place
assigned to a ship at a mooring ▷ v
dock (a ship)

beryl n hard transparent mineral

beryllium n (Chem) toxic silvery-
white metallic element

beseech v -**seeching**, -**sought** or
-**seeched** ask earnestly; beg

beset v trouble or harass constantly

beside prep at, by, or to the side of;
as compared with **beside oneself**
overwhelmed or overwrought
▶ **besides** adv, prep in addition

besiege v surround with military
forces; overwhelm, as with
requests

besotted adj infatuated

besought v a past of **beseech**

bespeak v indicate or suggest
▶ **bespoke** adj (esp. of a suit)
made to the customer's
specifications

best adj most excellent of a
particular group etc. ▷ adv
in a manner surpassing all
others ▷ n most outstanding
or excellent person, thing, or
group in a category **best man**
groom's attendant at a wedding
▶ **bestseller** n book or other
product that has sold in great
numbers

bestial adj brutal or savage; of or
like a beast ▶ **bestiality** n

bestir v cause (oneself) to become
active

bestow v present (a gift) or confer
(an honour) ▶ **bestowal** n

bestride v have or put a leg on
either side of

bet n the act of staking a sum
of money or other stake on the
outcome of an event; stake risked
▷ v **betting**, **bet** or **betted** make or
place (a bet); (informal) predict

betel [bee-tl] *n* Asian climbing plant, the leaves and nuts of which can be chewed

bête noire [bet nwahr] *n, pl* **bêtes noires** person or thing that one particularly dislikes

betide *v* happen (to)

betoken *v* indicate or signify

betray *v* hand over or expose (one's nation, friend, etc.) treacherously to an enemy; disclose (a secret or confidence) treacherously; reveal unintentionally ▶ **betrayal** *n* ▶ **betrayer** *n*

betrothed *adj* engaged to be married ▶ **betrothal** *n*

better *adj* more excellent than others; improved or fully recovered in health ▷ *adv* in or to a more excellent manner; in or to a greater degree ▷ *pl n* one's superiors ▷ *v* improve upon

bettong *n* short-nosed rat kangaroo

between *prep, adv* indicating position in the middle, alternatives, etc.

> **USAGE NOTE**
> *Between* is used when two people or things are mentioned: *the trade war between Europe and America.* Otherwise use *among*.

betwixt *prep, adv* (*old-fashioned*) between

bevel *n* slanting edge ▷ *v* **-elling, -elled** cut a bevel on (a piece of timber etc.)

beverage *n* drink

bevy *n, pl* **bevies** flock or group

bewail *v* express great sorrow over

beware *v* be on one's guard (against)

bewilder *v* confuse utterly ▶ **bewildering** *adj* ▶ **bewilderment** *n*

bewitch *v* attract and fascinate; cast a spell over ▶ **bewitching** *adj*

beyond *prep* at or to a point on the other side of; outside the limits or scope of ▷ *adv* at or to the far side of something

bi- *combining form* two or twice e.g. *bifocal; biweekly*

biannual *adj* occurring twice a year ▶ **biannually** *adv*

bias *n* mental tendency, esp. prejudice; diagonal cut across the weave of a fabric; (*Bowls*) bulge or weight on one side of a bowl that causes it to roll in a curve ▷ *v* **-asing, -ased** or **-assing, -assed** cause to have a bias ▶ **biased, biassed** *adj*

bib *n* piece of cloth or plastic worn to protect a young child's clothes when eating; upper front part of dungarees etc.

Bible *n* sacred writings of the Christian religion; (**b-**) book regarded as authoritative ▶ **biblical** *adj*

bibliography *n, pl* **-phies** list of books on a subject; list of sources used in a book etc. ▶ **bibliographer** *n*

bibliophile *n* person who collects or is fond of books

bibulous *adj* addicted to alcohol

bicarbonate *n* salt of carbonic acid ▶ **bicarbonate of soda** powder used in baking or as medicine

bicentenary *n, pl* **-naries** 200th anniversary

biceps *n* muscle with two origins, esp. the muscle that flexes the forearm

bicker *v* argue over petty matters

bicycle *n* vehicle with two wheels, one behind the other, pedalled by the rider

bid *v* **bidding, bade, bidden** say (a greeting); command; (*past tense* **bid**) offer (an amount) in an attempt to buy something ▷ *n* offer of a specified amount; attempt ▸ **bidder** n ▸ **biddable** adj obedient ▸ **bidding** n command

biddy-bid, biddy-biddy n, pl **-bids** or **-biddies** (NZ) low-growing plant with hooked burrs

bide v **bide one's time** wait patiently for an opportunity

bidet [bee-day] n low basin for washing the genital area

biennial adj occurring every two years ▸ n plant that completes its life cycle in two years

bier n stand on which a corpse or coffin rests before burial

bifocals pl n spectacles with lenses permitting near and distant vision

big adj **bigger, biggest** of considerable size, height, number, or capacity; important through having power, wealth, etc.; elder; generous ▸ adv on a grand scale ▸ **bighead** n (informal) conceited person ▸ **big-headed** adj ▸ **big shot, bigwig** n (informal) important person

bigamy n crime of marrying a person while still legally married to someone else ▸ **bigamist** n ▸ **bigamous** adj

bigot n person who is intolerant, esp. regarding religion or race ▸ **bigoted** adj ▸ **bigotry** n

bijou [bee-zhoo] adj (of a house) small but elegant

bike n (informal) bicycle or motorcycle

bikini n woman's brief two-piece swimming costume

bilateral adj affecting or undertaken by two parties

bilberry n bluish-black edible berry

bilby n, pl **-bies** Australian marsupial with long pointed ears and grey fur

bile n bitter yellow fluid secreted by the liver

bilge n (informal) nonsense; ship's bottom

bilingual adj involving or using two languages

bilious adj sick, nauseous

bill[1] n statement of money owed for goods or services supplied; draft of a proposed new law; poster; (chiefly US & Canadian) piece of paper money; list of events, such as a theatre programme ▸ v send or present a bill to; advertise by posters

bill[2] n bird's beak

billabong n (Aust) stagnant pool in an intermittent stream

billboard n large outdoor board for displaying advertisements

billet v-**leting, -leted** assign a lodging to (a soldier) ▸ n accommodation for a soldier in civil lodgings

billet-doux [bill-ee-doo] n, pl **billets-doux** love letter

billhook n tool with a hooked blade, used for chopping etc.

billiards n game played on a table with balls and a cue

billion n one thousand million; formerly, one million million ▸ **billionth** adj

billow n large sea wave ▸ v rise up or swell out ▸ **billowy, billowing** adj

billy, billycan n, pl **-lies** or **-lycans** metal can or pot for cooking on a campfire

biltong n (SAfr) strips of dried meat

bimbo n (slang) attractive but empty-headed young person, esp. a woman

bin n container for rubbish or for storing grain, coal, etc.

binary adj composed of two parts; (Maths, Computing) of or in a counting system with only two digits, 0 and 1

bind v **binding, bound** make secure with or as if with a rope; place (someone) under obligation; enclose and fasten (the pages of a book) between covers ▶ n (*informal*) annoying situation ▶ **binder** n firm cover for holding loose sheets of paper together ▶ **binding** n anything that binds or fastens; book cover

bindi-eye n small Australian plant with burlike fruit

bindweed n plant that twines around a support

binge n (*informal*) bout of excessive indulgence, esp. in drink

bingo n gambling game in which numbers are called out and covered by the players on their individual cards

binoculars pl n optical instrument consisting of two small telescopes joined together

binomial n, adj (mathematical expression) consisting of two terms

bio- combining form life or living organisms e.g. biology

bioactive adj having an effect on living tissue

biochemistry n study of the chemistry of living things ▶ **biochemist** n

biodegradable adj capable of being decomposed by natural means

biodiesel n biofuel for use in diesel engines

biodiversity n existence of a wide variety of species in their natural environment

biofuel n fuel derived from renewable biological resources

biographer n person who writes an account of another person's life

biography n, pl -**phies** account of a person's life by another person ▶ **biographical** adj

biological adj of or relating to biology

biology n study of living organisms ▶ **biologist** n

biometric adj of any automated system using physiological or behavioural traits as a means of identification

bionic adj having a part of the body that is operated electronically

biopsy n, pl -**sies** examination of tissue from a living body

BIOS n (*Computing*) built-in software that controls main functions of a computer

biosphere n part of the earth's surface and atmosphere inhabited by living things

biotechnology n use of microorganisms, such as cells or bacteria, in industry and technology

bioterrorism n use of viruses, bacteria, etc., by terrorists ▶ **bioterrorist** n

biped [bye-ped] n animal with two feet

biplane n aeroplane with two sets of wings, one above the other

bipolar adj having two poles; having two extremes; (*Psychiatry*) (of a person) manic-depressive

birch n tree with thin peeling bark; birch rod or twigs used, esp. formerly, for flogging offenders

bird n creature with feathers and wings, most types of which can fly; (*slang*) young woman ▶ **bird flu** form of flu occurring in poultry in SE Asia, capable of spreading to humans (also **avian flu**)

birdie n (*Golf*) score of one stroke under par for a hole

biretta n stiff square cap worn by the Catholic clergy

Biro® n ballpoint pen

birth n process of bearing young; childbirth; act of being born; ancestry **give birth to** bear (offspring) ▶ **birth control**

b

any method of contraception
▶ **birthday** n anniversary of the
day of one's birth ▶ **birthmark** n
blemish on the skin formed before
birth ▶ **birthright** n privileges
or possessions that someone is
entitled to at birth

biscuit n small flat dry sweet or
plain cake

bisect v divide into two equal parts

bisexual adj sexually attracted
to both men and women
▶ **bisexuality** n

bishop n clergyman who governs a
diocese; chessman which is moved
diagonally ▶ **bishopric** n diocese or
office of a bishop

bismuth n (Chem) pinkish-white
metallic element

bison n, pl -**son** large hairy animal
of the cattle family, native to N
America and Europe

bisque n thick rich soup made from
shellfish

bistro n, pl -**tros** small restaurant

bit[1] n small piece, portion, or
quantity **a bit** rather, somewhat
bit by bit gradually

bit[2] n metal mouthpiece on a bridle;
drilling part of a tool

bit[3] v past tense of **bite**

bit[4] n (Maths, Computing) single digit
of binary notation, either 0 or 1

bitch n female dog, fox, or wolf;
(offens) spiteful woman ▷ v
(informal) complain or grumble
▶ **bitchy** adj ▶ **bitchiness** n

bite v biting, bit, bitten grip, tear,
or puncture the skin, as with the
teeth or jaws; take firm hold of
or act effectively upon ▷ n act of
biting; wound or sting inflicted by
biting; snack ▶ **biter** n ▶ **biting** adj
piercing or keen; sarcastic

bitter adj having a sharp
unpleasant taste; showing or
caused by hostility or resentment;

extremely cold ▷ n beer with a
slightly bitter taste ▷ pl bitter-
tasting alcoholic drink ▶ **bitterly**
adv ▶ **bitterness** n

bittern n black marsh bird with a
booming call

bitumen n black sticky substance
obtained from tar or petrol

bivalve n, adj (marine mollusc)
with two hinged segments to
its shell

bivouac n temporary camp in the
open air ▷ v -**acking, -acked** camp
in a bivouac

bizarre adj odd or unusual

blab v blabbing, blabbed reveal
(secrets) indiscreetly

black adj of the darkest colour,
like coal; (**B-**) (of a person)
dark-skinned; without hope;
angry or resentful e.g. black
looks; unpleasant in a macabre
manner e.g. black comedy ▷ n
darkest colour; (**B-**) member of
a dark-skinned race; complete
darkness ▷ v make black; (of trade
unionists) boycott (goods or
people) ▶ **blackness** n ▶ **blacken** v
make or become black; defame or
slander ▶ **black magic** magic used
for evil purposes ▶ **black market**
illegal trade in goods or currencies
▶ **black sheep** person who is
regarded as a disgrace by his or her
family ▶ **black spot** place on a road
where accidents frequently occur

blackball v exclude from a group
▷ n (NZ) hard boiled sweet with
black-and-white stripes

BlackBerry® n hand-held wireless
device incorporating e-mail,
browser and mobile-phone
functions

blackberry n small blackish
edible fruit

blackbird n common European
thrush

blackboard n hard black surface used for writing on with chalk

blackboy n Australian plant with grasslike leaves and a spike of small white flowers

blackbutt n Australian eucalyptus tree with hard wood used as timber

blackcurrant n very small blackish edible fruit that grows in bunches

blackfish n small dark Australian estuary fish

blackguard [blag-gard] n unprincipled person

blackhead n black-tipped plug of fatty matter clogging a skin pore

blackleg n person who continues to work during a strike

blacklist n list of people or organizations considered untrustworthy etc.

blackmail n act of attempting to extort money by threats ▷ v (attempt to) obtain money by blackmail

blackout n extinguishing of all light as a precaution against an air attack; momentary loss of consciousness or memory ▶ **black out** v extinguish (lights); lose consciousness or memory temporarily

blacksmith n person who works iron with a furnace, anvil, etc.

black snake n venomous Australian snake

black swan n black Australian swan with a red beak

bladder n sac in the body where urine is held; hollow bag which may be filled with air or liquid

blade n cutting edge of a weapon or tool; thin flattish part of a propeller, oar, etc.; leaf of grass

blame v consider (someone) responsible for ▷ n responsibility for something that is wrong ▶ **blameless** adj ▶ **blameworthy** adj deserving blame

blanch v become white or pale; prepare (vegetables etc.) by plunging them in boiling water

blancmange [blam-monzh] n jelly-like dessert made with milk

bland adj dull and uninteresting ▶ **blandly** adv

blandishments pl n flattery intended to coax or persuade

blank adj not written on; showing no interest or expression ▷ n empty space; cartridge containing no bullet ▶ **blankly** adv ▶ **blank verse** unrhymed verse

blanket n large thick cloth used as covering for a bed; concealing cover, as of snow ▷ v cover as with a blanket

blare v sound loudly and harshly ▷ n loud harsh noise

blarney n flattering talk

blasé [blah-zay] adj indifferent or bored through familiarity

blaspheme v speak disrespectfully of (God or sacred things) ▶ **blasphemy** n ▶ **blasphemous** adj ▶ **blasphemer** n

blast n explosion; sudden strong gust of air or wind; sudden loud sound, as of a trumpet ▷ v blow up (a rock etc.) with explosives ▶ **blastoff** n launching of a rocket

blatant adj glaringly obvious ▶ **blatantly** adv

blaze¹ n strong fire or flame; very bright light ▷ v burn or shine brightly

blaze² n mark made on a tree to indicate a route

blazer n lightweight jacket, often in the colours of a school etc.

blazon v proclaim publicly

bleach v make or become white or colourless ▷ n bleaching agent

bleak adj exposed and barren; offering little hope

b

bleary adj **-rier, -riest** with eyes dimmed, as by tears or tiredness ▸ **blearily** adv

bleat v (of a sheep, goat, or calf) utter its plaintive cry ▸ n cry of sheep, goats, and calves

bleed v **bleeding, bled** lose or emit blood; draw blood from (a person or animal); (informal) obtain money by extortion

bleep n short high-pitched sound made by an electrical device ▸ v make a bleeping sound ▸ **bleeper** n small portable radio receiver that makes a bleeping signal

blemish n defect or stain ▸ v spoil or tarnish

blench v shy away, as in fear

blend v mix or mingle (components or ingredients); look good together ▸ n mixture ▸ **blender** n electrical appliance for puréeing vegetables etc.

bless v make holy by means of a religious rite; call upon God to protect; endow with health, talent, etc. ▸ **blessed** adj holy ▸ **blessing** n invoking of divine aid; approval; happy event

blether (Scot) v talk, esp. foolishly or at length ▸ n conversation

blew v past tense of **blow¹**

blight n person or thing that spoils or prevents growth; withering plant disease ▸ v frustrate or disappoint

blighter n (informal) irritating person

blimp n small airship

blind adj unable to see; unable or unwilling to understand; not determined by reason e.g. blind hatred ▸ v deprive of sight; deprive of good sense, reason, or judgment ▸ n covering for a window; something that serves to conceal the truth ▸ **blindly** adv ▸ **blindness** n

blindfold v prevent (a person) from seeing by covering the eyes ▸ n piece of cloth used to cover the eyes

blink v close and immediately reopen (the eyes); shine intermittently ▸ n act of blinking **on the blink** (slang) not working properly

blinkers pl n leather flaps on a horse's bridle to prevent sideways vision

blip n spot of light on a radar screen indicating the position of an object

bliss n perfect happiness ▸ **blissful** adj ▸ **blissfully** adv

blister n small bubble on the skin; swelling, as on a painted surface ▸ v (cause to) have blisters ▸ **blistering** adj (of weather) very hot; (of criticism) extremely harsh

blithe adj casual and indifferent ▸ **blithely** adv

blitz n violent and sustained attack by aircraft; intensive attack or concerted effort ▸ v attack suddenly and intensively

blizzard n blinding storm of wind and snow

bloat v cause to swell, as with liquid or air

bloater n (Brit) salted smoked herring

blob n soft mass or drop; indistinct or shapeless form

bloc n people or countries combined by a common interest

block n large solid piece of wood, stone, etc.; large building of offices, flats, etc.; group of buildings enclosed by intersecting streets; obstruction or hindrance; (slang) person's head ▸ v obstruct or impede by introducing an obstacle ▸ **blockage** n ▸ **blockhead** n stupid person ▸ **block letter** plain capital letter

blockade *n* sealing off of a place to prevent the passage of goods ▷ *v* impose a blockade on

blockie *n* (*Aust*) owner of a small property, esp. a farm

blog *n* person's online journal (also **weblog**)

blogosphere *n* (*informal*) blogs on the internet as a whole

bloke *n* (*informal*) man

blonde, (*masc*) **blond** *adj*, *n* fair-haired (person)

blood *n* red fluid that flows around the body; race or kinship **in cold blood** done deliberately ▶ **bloodless** *adj* ▶ **blood bath** massacre ▶ **bloodhound** *n* large dog formerly used for tracking ▶ **bloodshed** *n* slaughter or killing ▶ **bloodshot** *adj* (of an eye) inflamed ▶ **blood sport** sport involving the killing of animals ▶ **bloodstream** *n* flow of blood round the body ▶ **bloodsucker** *n* animal that sucks blood; (*informal*) person who extorts money from other people ▶ **bloodthirsty** *adj* taking pleasure in violence

bloody *adj* covered with blood; marked by much killing ▷ *adj*, *adv* (*slang*) extreme or extremely ▷ *v* stain with blood ▶ **bloody-minded** *adj* deliberately unhelpful

bloom *n* blossom on a flowering plant; youthful or healthy glow ▷ *v* bear flowers; be in a healthy, glowing condition

bloomer *n* (*Brit informal*) stupid mistake

bloomers *pl n* woman's baggy knickers

blooper *n* (*chiefly US informal*) stupid mistake

blossom *n* flowers of a plant ▷ *v* (of plants) flower; come to a promising stage

blot *n* spot or stain; something that spoils ▷ *v* **blotting, blotted** cause a blemish in or on; soak up (ink) by using blotting paper ▶ **blotter** *n* ▶ **blot out** darken or hide completely ▶ **blotting paper** soft absorbent paper for soaking up ink

blotch *n* discoloured area or stain ▶ **blotchy** *adj*

blotto *adj* (*Brit, Aust & NZ slang*) extremely drunk

blouse *n* woman's shirtlike garment

blow¹ *v* **blowing, blew, blown** (of air, the wind, etc.) move; move or be carried as if by the wind; expel (air etc.) through the mouth or nose; cause (a musical instrument) to sound by forcing air into it; burn out (a fuse etc.); (*slang*) spend (money) freely ▶ **blower** *n* ▶ **blowy** *adj* windy ▶ **blow-dry** *v* style (the hair) with a hand-held dryer ▶ **blowout** *n* sudden loss of air in a tyre; escape of oil or gas from a well; (*slang*) filling meal; (*Aust informal*) large financial deficit ▶ **blow up** *v* explode; fill with air; (*informal*) lose one's temper; (*informal*) enlarge (a photograph)

blow² *n* hard hit; sudden setback; attacking action

blowie *n* (*Aust informal*) bluebottle

blown *v* past participle of **blow¹**

blowsy *adj* fat, untidy, and red-faced

blubber *n* fat of whales, seals, etc. ▷ *v* sob without restraint

bludge (*informal*) *v* (*Aust & NZ*) evade work; (*Aust & NZ*) scrounge ▷ *n* (*Aust*) easy task ▶ **bludger** ▷ *n* person who scrounges

bludgeon *n* short thick club ▷ *v* hit with a bludgeon; force or bully

blue *n* colour of a clear unclouded sky ▷ *pl* feeling of depression; type of folk music of Black American origin ▷ *adj* **bluer, bluest** of the colour blue; depressed; pornographic **out of the blue** unexpectedly ▶ **bluish** adj ▶ **bluebell** n flower with blue bell-shaped flowers ▶ **bluebottle** n large fly with a dark-blue body ▶ **blue-collar** adj denoting manual industrial workers ▶ **blue heeler** (Aust & NZ informal) dog that controls cattle by biting their heels ▶ **blueprint** n photographic print of a plan; description of how a plan is expected to work ▶ **bluetongue** n Australian lizard with a blue tongue ▶ **Bluetooth®** n technology allowing short-range wireless communication

bluff¹ v pretend to be confident in order to influence (someone) ▷ n act of bluffing

bluff² n steep cliff or bank ▷ adj good-naturedly frank and hearty

blunder n clumsy mistake ▷ v make a blunder; act clumsily

blunderbuss n obsolete gun with a wide flared muzzle

blunt adj not having a sharp edge or point; (of people, speech, etc.) straightforward or uncomplicated ▷ v make less sharp ▶ **bluntly** adv

blur v blurring, blurred make or become vague or less distinct ▷ n something vague, hazy, or indistinct ▶ **blurry** adj

blurb n promotional description, as on the jacket of a book

blurt v (foll by out) utter suddenly and involuntarily

blush v become red in the face, esp. from embarrassment or shame ▷ n reddening of the face

bluster v speak loudly or in a bullying way ▷ n empty threats or protests ▶ **blustery** adj (of weather) rough and windy

BMA British Medical Association

BO (informal) body odour

boa n large nonvenomous snake; long scarf of fur or feathers ▶ **boa constrictor** large snake that kills its prey by crushing

boab [boh-ab] n (Aust informal) short for baobab

boar n uncastrated male pig; wild pig

board n long flat piece of sawn timber; smaller flat piece of rigid material for a specific purpose e.g. ironing board; chess board; group of people who administer a company, trust, etc.; meals provided for money ▷ v go aboard (a train, aeroplane, etc.); cover with boards; receive meals and lodgings in return for money **on board** on or in a ship, aeroplane, etc. ▶ **boarder** n person who pays rent in return for accommodation in someone else's home; (Brit) pupil who lives at school during the school term ▶ **boarding house** private house that provides meals and accommodation for paying guests ▶ **boardroom** n room where the board of a company meets

boast v speak too proudly about one's talents etc.; possess (something to be proud of) ▷ n bragging statement ▶ **boastful** adj

boat n small vehicle for travelling across water ▶ **boater** n flat straw hat ▶ **boating** n

boatswain n same as bosun

bob¹ v bobbing, bobbed move up and down repeatedly ▷ n short abrupt movement

bob² n hairstyle in which the hair is cut short evenly all round the head ▷ v bobbing, bobbed cut (the hair) in a bob

bobbin n reel on which thread is wound

bobble n small ball of material, usu. for decoration

bobby n, pl **-bies** (Brit informal) policeman

bobotie [ba-**boot**-ee] n (SAfr) dish of curried mince

bobsleigh n sledge for racing down an icy track ▷ v ride on a bobsleigh

bode v be an omen of (good or ill)

bodice n upper part of a dress

bodkin n blunt large-eyed needle

body n, pl **bodies** entire physical structure of an animal or human; trunk or torso; corpse; group regarded as a single entity; main part of anything; woman's one-piece undergarment ▶ **bodily** adj relating to the body ▷ adv by taking hold of the body ▶ **body-board** n small polystyrene surfboard ▶ **body-boarder** n ▶ **bodyguard** n person or group of persons employed to protect someone ▶ **body mass index** measure used to gauge whether a person is overweight: a person's weight in kilograms divided by the square of his or her height in metres ▶ **bodywork** n outer shell of a motor vehicle

Boer n descendant of the Dutch settlers in S Africa ▶ **boerewors** n (SAfr) spiced sausage

boffin n (Brit, Aust, NZ & SAfr informal) scientist or expert

bog n wet spongy ground; (slang) toilet ▶ **boggy** adj ▶ **bog down** v **bogging, bogged** impede physically or mentally

bogan n (Aust & NZ slang) youth who dresses and behaves rebelliously

bogey, bogy n something that worries or annoys; (Golf) score of one stroke over par on a hole

boggle v be surprised, confused, or alarmed

bogong, bugong n large nocturnal Australian moth

bogus adj not genuine

bogy n, pl **-gies** same as **bogey**

bohemian n, adj (person) leading an unconventional life

boil¹ v (cause to) change from a liquid to a vapour so quickly that bubbles are formed; cook by the process of boiling ▷ n state or action of boiling ▶ **boiler** n piece of equipment which provides hot water

boil² n red pus-filled swelling on the skin

boisterous adj noisy and lively ▶ **boisterously** adv

bold adj confident and fearless; immodest or impudent ▶ **boldly** adv ▶ **boldness** n

bole n tree trunk

bolero n, pl **-ros** (music for) traditional Spanish dance; short open jacket

bollard n short thick post used to prevent the passage of motor vehicles

boloney n same as **baloney**

Bolshevik n (formerly) Russian Communist ▶ **bolshie, bolshy** adj (informal) difficult or rebellious

bolster v support or strengthen ▷ n long narrow pillow

bolt n sliding metal bar for fastening a door etc.; metal pin which screws into a nut; flash (of lightning) ▷ v run away suddenly; fasten with a bolt; eat hurriedly **bolt upright** stiff and rigid ▶ **bolt hole** place of escape

bomb n container fitted with explosive material; (slang) large amount of money ▷ v attack with bombs; move very quickly **the bomb** nuclear bomb ▶ **bomber** n aircraft that drops bombs; person who throws or puts a bomb in a particular place ▶ **bomb out** v (Aust, NZ & SAfr informal) fail disastrously ▸ **bombshell** n shocking or unwelcome surprise

bombard v attack with heavy gunfire or bombs; attack verbally, esp. with questions ▸ **bombardment** n

bombast n pompous language ▸ **bombastic** adj

bona fide [bone-a fide-ee] adj genuine

bonanza n sudden good luck or wealth

bond n something that binds, fastens or holds together; something that unites people; written or spoken agreement; (Finance) certificate of debt issued to raise funds; (S Afr) conditional pledging of property, esp. a house, as security for the repayment of a loan ▸ pl something that restrains or imprisons ▸ v bind ▸ **bonded** adj

bondage n slavery

bone n any of the hard parts in the body that form the skeleton ▸ v remove the bones from (meat for cooking etc.) ▸ **boneless** adj ▸ **bony** adj having many bones; thin or emaciated ▸ **bone-dry** adj completely dry ▸ **bone-idle** adj extremely lazy

bonfire n large outdoor fire

bongo n, pl **-gos** or **-goes** small drum played with the fingers

bonhomie [bon-om-ee] n cheerful friendliness

bonito [ba-nee-toh] n, pl **-os** small tunny-like marine food fish; related fish, whose flesh is dried and flaked and used in Japanese cookery

bonk v (informal) have sex with; hit

bonnet n metal cover over a vehicle's engine; hat which ties under the chin

bonny adj **-nier**, **-niest** (Scot) beautiful

bonsai n, pl **-sai** ornamental miniature tree or shrub

bonus n something given, paid, or received above what is due or expected

boo interj shout of disapproval ▸ v **booing**, **booed** shout 'boo' to show disapproval

boob (slang) n foolish mistake; female breast; (Aust) prison

boobook [boo-book] n small spotted Australian brown owl

booby n, pl **-bies** foolish person ▸ **booby prize** prize given for the lowest score in a competition ▸ **booby trap** hidden bomb primed to be set off by an unsuspecting victim; trap for an unsuspecting person, intended as a joke

boogie v (informal) dance to fast pop music

book n number of pages bound together between covers; long written work; number of tickets, stamps, etc. fastened together ▸ pl record of transactions of a business or society ▸ v reserve (a place, passage, etc.) in advance; record the name of (a person) who has committed an offence ▸ **booklet** n thin book with paper covers

bookie n (informal) short for **bookmaker**

book-keeping n systematic recording of business transactions

bookmaker n person whose occupation is taking bets

bookmark n person whose occupation is taking bets; (Computing) marker on a website that enables the user to return to it quickly and easily ▸ v (Computing) identify and store (a website) so that one can return to it quickly and easily

bookworm n person devoted to reading

boom¹ v make a loud deep echoing sound; prosper vigorously and rapidly ▷ n loud deep echoing sound; period of high economic growth ▶ **boomer** n (Aust) large male kangaroo

boom² n pole to which the foot of a sail is attached; pole carrying an overhead microphone; barrier across a waterway

boomerang n curved wooden missile which can be made to return to the thrower ▷ v (of a plan) recoil unexpectedly

boon n something helpful or beneficial

boongary [boong-gar-ree] n, pl **-garies** tree kangaroo of NE Queensland, Australia

boor n rude or insensitive person ▶ **boorish** adj

boost n encouragement or help; increase ▷ v improve; increase ▶ **booster** n small additional injection of a vaccine

boot¹ n outer covering for the foot that extends above the ankle; space in a car for luggage; (informal) kick ▷ v (informal) kick; start up (a computer) ▶ **bootee** n baby's soft shoe ▶ **boot camp** centre for young offenders, with strict discipline and hard physical exercise ▶ **boot-cut** adj (of trousers) slightly flared at the bottom of the legs

boot² n to boot in addition

booth n small partly enclosed cubicle; stall at a fair or market

bootleg adj produced, distributed, or sold illicitly ▷ v **-legging**, **-legged** make, carry, or sell (illicit goods) ▶ **bootlegger** n

booty n, pl **-ties** valuable articles obtained as plunder

booze v, n (informal) (consume) alcoholic drink ▶ **boozy** adj ▶ **boozer** n (informal) person who is fond of drinking; (Brit, Aust & NZ)

pub ▶ **booze bus** (Aust informal) mobile police unit for drug and alcohol testing ▶ **booze-up** n (informal) drinking spree

bop v **bopping**, **bopped** (informal) dance to pop music

bora n (Aust) Aboriginal ceremony

borax n white mineral used in making glass

border n dividing line between political or geographical regions; band around or along the edge of something ▷ v provide with a border; be nearly the same as e.g. resentment that borders on hatred

bore¹ v make (a hole) with a drill etc. ▷ n (diameter of) the hollow of a gun barrel or other tube

bore² v make weary by being dull or repetitious ▷ n dull or repetitious person or thing ▶ **bored** adj ▶ **boredom** n

bore³ n high wave in a narrow estuary, caused by the tide

bore⁴ v past tense of **bear¹**

boree [baw-ree] n (Aust) same as **myall**

born v a past participle of **bear¹** ▷ adj possessing certain qualities from birth e.g. a born musician

borne v a past participle of **bear¹**

boron n (Chem) element used in hardening steel

boronia n Australian aromatic flowering shrub

borough n (chiefly Brit) town or district with its own council

borrow v obtain (something) temporarily; adopt (ideas etc.) from another source ▶ **borrower** n

USAGE NOTE
Borrow is followed by from: Borrow a pound from Frank. The use of borrow followed by off or of is nonstandard

borscht, borsch n Russian soup based on beetroot

borstal n (formerly in Britain) prison for young criminals

borzoi n tall dog with a long silky coat

bosh n (Brit, Aust & NZ informal) empty talk, nonsense

Bosnian n, adj (person) from Bosnia

bosom n chest of a person, esp. the female breasts ▷ adj very dear to you ▷ a bosom friend

boss¹ n person in charge of or employing others ▷ v **boss around, about** be domineering towards ▶ **bossy** adj

boss² n raised knob or stud

bosun n officer responsible for the maintenance of a ship

botany n study of plants ▶ **botanical, botanic** adj ▶ **botanist** n

botch v spoil through clumsiness ▷ n (also **botch-up**) badly done piece of work or repair

both adj, pron two considered together

bother v take the time or trouble; give annoyance or trouble to; pester ▷ n trouble, fuss, or difficulty ▶ **bothersome** adj

Botox® n drug used to treat muscle spasms and to reduce wrinkles ▷ v apply Botox to (a person or a body part)

bottle n container for holding liquids; (Brit informal) courage ▷ v put in a bottle ▶ **bottleneck** n narrow stretch of road where traffic is held up ▶ **bottle shop** (Aust & NZ) shop licensed to sell alcohol for drinking elsewhere ▶ **bottle store** (SAfr) shop licensed to sell alcohol for drinking elsewhere ▶ **bottle tree** Australian tree with a bottle-shaped swollen trunk ▶ **bottle up** v restrain (powerful emotion)

bottom n lowest, deepest, or farthest removed part of a thing; buttocks ▷ adj lowest or last ▶ **bottomless** adj

botulism n severe food poisoning

boudoir [boo-dwahr] n woman's bedroom or private sitting room

bougainvillea n climbing plant with red or purple flowers

bough n large branch of a tree

bought v past of **buy**

boulder n large rounded rock

boulevard n wide, usu. tree-lined, street

bounce v (of a ball etc.) rebound from an impact; (slang) (of a cheque) be returned uncashed owing to a lack of funds in the account ▷ n act of rebounding; springiness; (informal) vitality or vigour ▶ **bouncer** n person employed at a disco etc. to remove unwanted people ▶ **bouncing** adj vigorous and robust

bound¹ v past of **bind** ▷ adj destined or certain; compelled or obliged

bound² v move forwards by jumps ▷ n jump upwards or forwards

bound³ v form a boundary of ▷ pl n limit ▶ **boundary** n dividing line that indicates the farthest limit

bound⁴ adj going or intending to go towards e.g. homeward bound

bounty n, pl **-ties** generosity; generous gift or reward ▶ **bountiful, bounteous** adj

bouquet n bunch of flowers; aroma of wine

bourbon [bur-bn] n whiskey made from maize

bourgeois [boor-zhwah] adj, n (offens) middle-class (person)

bout n period of activity or illness; boxing or wrestling match

boutique n small clothes shop

bovine adj relating to cattle; rather slow and stupid

bow¹ [rhymes with *now*] *v* lower (one's head) or bend (one's knee or body) as a sign of respect or shame; comply or accept ▷ *n* movement made when bowing

bow² [rhymes with *go*] *n* knot with two loops and loose ends; weapon for shooting arrows; long stick stretched with horsehair for playing stringed instruments ▶ **bow-legged** *adj* having legs that curve outwards at the knees

bow³ [rhymes with *now*] *n* front end of a ship

bowdlerize *v* remove words regarded as indecent from (a play, novel, etc.)

bowel *n* intestine, esp. the large intestine ▷ *pl* innermost part

bower *n* shady leafy shelter ▶ **bowerbird** *n* songbird of Australia and New Guinea, the males of which build bower-like display grounds to attract females

bowl¹ *n* round container with an open top; hollow part of an object

bowl² *n* large heavy ball ▷ *pl* game played on smooth grass with wooden bowls ▷ *v* (*Cricket*) send (a ball) towards the batsman ▶ **bowling** *n* game in which bowls are rolled at a group of pins

bowler¹ *n* (*Cricket*) player who sends (a ball) towards the batsman; person who plays bowls or bowling

bowler² *n* stiff felt hat with a rounded crown

box¹ *n* container with a firm flat base and sides; separate compartment in a theatre, stable, etc. ▷ *v* put into a box ▶ **the box** (*informal*) television ▶ **box jellyfish** highly venomous jellyfish with a cuboidal body that lives in Australian tropical waters ▶ **box office** place where theatre or cinema tickets are sold

box² *v* fight (an opponent) in a boxing match ▶ **boxer** *n* person who participates in the sport of boxing; medium-sized dog with smooth hair and a short nose ▶ **boxer shorts, boxers** *pl n* men's underpants shaped like shorts but with a front opening ▶ **boxing** *n* sport of fighting with the fists

box³ *n* evergreen tree with shiny leaves; eucalyptus tree with similar timber and foliage, and with rough bark

boy *n* male child ▶ **boyish** *adj* ▶ **boyhood** *n* ▶ **boyfriend** *n* male friend with whom a person is romantically or sexually involved

boycott *v* refuse to deal with (an organization or country) ▷ *n* instance of boycotting

> **SPELLING TIP**
> The word **boycott** has two ts, whether or not it has an ending such as in **boycotting**

boykie *n* (*S Afr informal*) chap or fellow

bra *n* woman's undergarment for supporting the breasts

braaivleis [brye-flayss], **braai** (*S Afr*) *n* grill on which food is cooked over hot charcoal, usu. outdoors; outdoor party at which food like this is served ▶ *v* cook (food) on in this way

brace *n* object fastened to something to straighten or support it; pair, esp. of game birds ▷ *pl* straps worn over the shoulders to hold up trousers ▷ *v* steady or prepare (oneself) for something unpleasant; strengthen or fit with a brace ▶ **bracing** *adj* refreshing and invigorating

bracelet *n* ornamental chain or band for the wrist

bracken *n* large fern

bracket n pair of characters used to enclose a section of writing; group falling within certain defined limits; support fixed to a wall ▷ v **-eting, -eted** put in brackets; class together

brackish adj (of water) slightly salty

bract n leaf at the base of a flower

brag v **bragging, bragged** speak arrogantly and boastfully ▶ **braggart** n

braid v interweave (hair, thread, etc.) ▷ n length of hair etc. that has been braided; narrow ornamental tape of woven silk etc.

Braille n system of writing for the blind, consisting of raised dots interpreted by touch

brain n soft mass of nervous tissue in the head; intellectual ability ▷ v hit (someone) hard on the head ▶ **brainless** adj stupid ▶ **brainy** adj (informal) clever ▶ **brainchild** n idea produced by creative thought ▶ **brainfood** n food thought to promote brain function ▶ **brain up** v (Brit) make (something) more intellectually demanding or sophisticated ▶ **brainwash** v cause (a person) to alter his or her beliefs, esp. by methods based on isolation, sleeplessness, etc. ▶ **brainwave** n sudden idea

braise v cook slowly in a covered pan with a little liquid

brake n device for slowing or stopping a vehicle ▷ v slow down or stop by using a brake

bramble n prickly shrub that produces blackberries

bran n husks of cereal grain

branch n secondary stem of a tree; offshoot or subsidiary part of something larger or more complex ▷ v (of stems, roots, etc.) divide, then develop in different directions ▶ **branch out** v expand one's interests

brand n particular product; particular kind or variety; identifying mark burnt onto the skin of an animal ▷ v mark with a brand; denounce as being e.g. he was branded a fascist ▶ **brand-new** adj absolutely new

brandish v wave (a weapon etc.) in a threatening way

brandy n, pl **-dies** alcoholic spirit distilled from wine

brash adj offensively loud, showy, or self-confident ▶ **brashness** n

brass n alloy of copper and zinc; family of wind instruments made of brass; (N English dialect) money ▶ **brassy** adj brazen or flashy; like brass, esp. in colour

brasserie n restaurant serving drinks and cheap meals

brassiere n bra

brat n unruly child

bravado n showy display of self-confidence

brave adj having or showing courage, resolution, and daring ▷ n Native American warrior ▷ v confront with resolution or courage ▶ **bravery** n

bravo interj well done!

brawl n noisy fight ▷ v fight noisily

brawn n physical strength; pressed meat from the head of a pig or calf ▶ **brawny** adj

bray v (of a donkey) utter its loud harsh sound ▷ n a donkey's loud harsh sound

brazen adj shameless and bold ▶ **brazenly** adv

brazier [bray-zee-er] n portable container for burning charcoal or coal

breach n breaking of a promise, obligation, etc.; gap or break

▷ v break (a promise, law, etc.); make a gap in

bread n food made by baking a mixture of flour and water or milk; (slang) money ▶ **breadwinner** n person whose earnings support a family

breadth n extent of something from side to side

break v **breaking, broke, broken** separate or become separated into two or more pieces; damage or become damaged so as to be inoperative; fail to observe (an agreement etc.); disclose or be disclosed e.g. he broke the news; bring or come to an end e.g. the good weather broke at last; weaken or be weakened, as in spirit; improve on or surpass e.g. break a record; (of the male voice) become permanently deeper at puberty ▷ n act or result of breaking; gap or interruption in continuity; (informal) fortunate opportunity ▶ **break even** make neither a profit nor a loss ▶ **breakable** adj ▶ **breakage** n ▶ **breaker** n large wave ▶ **break down** v cease to function; yield to strong emotion ▶ **breakdown** n act or instance of breaking down; nervous breakdown ▶ **break-in** n illegal entering of a building, esp. by thieves ▶ **breakneck** adj fast and dangerous ▶ **break off** v sever or detach; end (a relationship etc.) ▶ **break out** v begin or arise suddenly ▶ **breakthrough** n important development of discovery ▶ **break up** v (cause to) separate; come to an end; (of a school) close for the holidays; (of a caller) to become inaudible on the telephone ▶ **breakwater** n wall that extends into the sea to protect a harbour or beach from the force of waves

breakfast v, n (eat) the first meal of the day

bream n freshwater fish with silvery scales; food fish of European seas

breast n either of the two soft fleshy milk-secreting glands on a woman's chest; chest ▶ **breastbone** n long flat bone in the front of the body, to which most of the ribs are attached ▶ **breaststroke** n swimming stroke in which the arms are extended in front of the head and swept back on either side

breath n taking in and letting out of air during breathing; air taken in and let out during breathing ▶ **breathless** adj ▶ **breathtaking** adj causing awe or excitement ▶ **breathe** v take in oxygen and give out carbon dioxide; whisper ▶ **breather** n (informal) short rest ▶ **breathing** n

Breathalyser® n device for estimating the amount of alcohol in the breath ▶ **breathalyse** v

bred v past of **breed**

breech n buttocks; back part of gun where bullet or shell is loaded ▶ **breech birth** n birth of a baby with the feet or buttocks appearing first

breeches pl n trousers extending to just below the knee

breed v **breeding, bred** produce new or improved strains of (domestic animals or plants); bear (offspring); produce or be produced e.g. breed trouble ▷ n group of animals etc. within a species that have certain clearly defined characteristics; kind or sort ▶ **breeder** n ▶ **breeding** n result of good upbringing or training

breeze n gentle wind ▷ v move quickly or casually ▶ **breezy** adj windy; casual or carefree

brethren pl n (old-fashioned) (used in religious contexts) brothers

Breton adj of Brittany ▷ n person from Brittany; language of Brittany

brevity n shortness

brew v make (beer etc.) by steeping, boiling, and fermentation; prepare (a drink) by infusing; be about to happen or forming ▷ n beverage produced by brewing

brewer n person or company that brews beer ▶ **brewery** n, pl -eries place where beer etc. is brewed

briar¹, brier n European shrub with a hard woody root; tobacco pipe made from this root

briar² n same as **brier¹**

bribe v offer or give something to someone to gain favour, influence, etc. ▷ n something given or offered as a bribe ▶ **bribery** n

bric-a-brac n miscellaneous small ornamental objects

brick n (rectangular block of) baked clay used in building ▷ v (foll by up or over) build, enclose, or fill with bricks ▶ **bricklayer** n person who builds with bricks

bride n woman who has just been or is about to be married ▶ **bridal** adj ▶ **bridegroom** n man who has just been or is about to be married ▶ **bridesmaid** n girl or woman who attends a bride at her wedding

bridge¹ n structure for crossing a river etc.; platform from which a ship is steered or controlled; upper part of the nose; piece of wood supporting the strings of a violin etc. ▷ v build a bridge over (something) ▶ **bridgehead** n fortified position at the end of a bridge nearest the enemy

bridge² n card game based on whist, played between two pairs

bridle n headgear for controlling a horse ▷ v show anger or indignation ▶ **bridle path** path suitable for riding horses

Brie [bree] n soft creamy white cheese

brief adj short in duration ▷ n condensed statement or written synopsis; (also **briefing**) set of instructions ▷ pl men's or women's underpants ▷ v give information and instructions to (a person) ▶ **briefly** adv ▶ **briefcase** n small flat case for carrying papers, books, etc.

brier¹, briar n wild rose with long thorny stems

brier² n same as **briar¹**

brig n two-masted square-rigged ship

brigade n army unit smaller than a division; group of people organized for a certain task

brigadier n high-ranking army officer

brigalow n (Aust) type of acacia tree

brigand n (lit) bandit

brigantine n two-masted sailing ship

bright adj emitting or reflecting much light; (of colours) intense; clever ▶ **brightly** adv ▶ **brightness** n ▶ **brighten** v

brilliant adj shining with light; splendid; extremely clever ▶ **brilliance, brilliancy** n

brim n upper rim of a cup etc.; projecting edge of a hat ▷ v brimming, brimmed be full to the brim

brimstone n (obs) sulphur

brine n salt water ▷ **briny** adj very salty **the briny** (informal) the sea

bring v bringing, brought carry, convey, or take to a designated place or person; cause to happen; (Law) put forward (charges) officially ▶ **bring about** v cause to happen ▶ **bring off** v succeed in achieving ▶ **bring out** v publish or have (a book) published; reveal or cause to be seen ▶ **bring up** v rear (a child); mention; vomit (food)

brinjal n (S Afr) dark purple tropical fruit, cooked and eaten as a vegetable

brink n edge of a steep place

brisk adj lively and quick ▶ **briskly** adv

brisket n beef from the breast of a cow

bristle n short stiff hair ▷ v (cause to) stand up like bristles; show anger ▶ **bristly** adj

Brit n (informal) British person

British adj of Great Britain or the British Commonwealth ▷ pl n **the British** people of Great Britain

Briton n native or inhabitant of Britain

brittle adj hard but easily broken ▶ **brittleness** n

broach v introduce (a topic) for discussion; open (a bottle or barrel)

broad adj having great breadth or width; not detailed; extensive e.g. broad support; strongly marked e.g. a broad American accent ▶ **broadly** adv ▶ **broaden** v ▶ **broadband** n telecommunication transmission technique using a wide range of frequencies ▶ **broad bean** thick flat edible bean ▶ **broad-minded** adj tolerant ▶ **broadside** n strong verbal or written attack; (Naval) firing of all the guns on one side of a ship at once

broadcast n programme or announcement on radio or television ▷ v transmit (a programme or announcement) on radio or television; make widely known ▶ **broadcaster** n ▶ **broadcasting** n

brocade n rich fabric woven with a raised design

broccoli n type of cabbage with greenish flower heads

> **SPELLING TIP**
> Remember **broccoli** has two cs and one l

brochure n booklet that contains information about a product or service

broekies [brook-eez] pl n (S Afr informal) underpants

brogue[1] n sturdy walking shoe

brogue[2] n strong accent, esp. Irish

broil v (Aust, NZ, US & Canadian) cook by direct heat under a grill

broke v past tense of **break** ▷ adj (informal) having no money

broken v past participle of **break** ▷ adj fractured or smashed; (of the speech of a foreigner) noticeably imperfect e.g. broken English ▶ **brokenhearted** adj overwhelmed by grief

broker n agent who buys or sells goods, securities, etc.

brolga n large grey Australian crane with a trumpeting call (also **native companion**)

brolly n, pl -**lies** (informal) umbrella

bromance n (informal) close, non-sexual friendship between two men

bromide n chemical compound used in medicine and photography

bromine n (Chem) dark red liquid element that gives off a pungent vapour

bronchial [bronk-ee-al] adj of the bronchi

bronchitis [bronk-eye-tiss] n inflammation of the bronchi

bronchus [bronk-uss] n, pl **bronchi** [bronk-eye] either of the two branches of the windpipe

bronco n, pl -**cos** (in the US) wild or partially tamed pony

brontosaurus n very large plant-eating four-footed dinosaur

b

bronze n alloy of copper and tin; statue, medal, etc. made of bronze ▷ adj made of, or coloured like, bronze ▷ v (esp. of the skin) make or become brown ▸ **Bronze Age** era when bronze tools and weapons were used

brooch n ornament with a pin, worn fastened to clothes

brood n number of birds produced at one hatching; all the children of a family ▷ v think long and unhappily ▸ **broody** adj moody and sullen; (informal) (of a woman) wishing to have a baby

brook¹ n small stream

brook² v bear or tolerate

broom n long-handled sweeping brush; yellow-flowered shrub ▸ **broomstick** n handle of a broom

broth n soup, usu. containing vegetables

brothel n house where men pay to have sex with prostitutes

brother n boy or man with the same parents as another person; member of a male religious order ▸ **brotherly** adj ▸ **brotherhood** n fellowship; association, such as a trade union ▸ **brother-in-law** n, pl **brothers-in-law** brother of one's husband or wife; husband of one's sibling

brought v past of **bring**

brow n part of the face from the eyes to the hairline; eyebrow; top of a hill

browbeat v frighten (someone) with threats

brown n colour of earth or wood ▷ adj of the colour brown ▷ v make or become brown ▸ **brownish** adj ▸ **browned-off** adj (informal) bored and depressed

Brownie Guide, Brownie n junior Guide

browse v look through (a book or articles for sale) in a casual manner; nibble on young shoots or leaves ▷ n instance of browsing ▸ **browser** n (Computing) software package that enables a user to read hypertext, esp. on the internet

bruise n discoloured area on the skin caused by an injury ▷ v cause a bruise on ▸ **bruiser** n strong tough person

brumby n, pl -bies (Aust) wild horse; unruly person

brunch n (informal) breakfast and lunch combined

brunette n girl or woman with dark brown hair

brunt n main force or shock of a blow, attack, etc.

brush¹ n device made of bristles, wires, etc. used for cleaning, painting, etc.; brief unpleasant encounter; fox's tail ▷ v clean, scrub, or paint with a brush; touch lightly and briefly ▸ **brush off** v (slang) dismiss or ignore (someone) ▸ **brush up** v refresh one's knowledge of (a subject)

brush² n thick growth of shrubs

brush turkey n bird of New Guinea and Australia resembling the domestic fowl, with black plumage

brusque adj blunt or curt in manner or speech ▸ **brusquely** adv ▸ **brusqueness** n

Brussels sprout n vegetable like a tiny cabbage

brute n brutal person; animal other than man ▷ adj wholly instinctive or physical, like an animal; without reason ▸ **brutish** adj of or like an animal ▸ **brutal** adj cruel and vicious; extremely honest in speech or manner ▸ **brutally** adv ▸ **brutality** n ▸ **brutalize** v

BSc Bachelor of Science

BSE bovine spongiform encephalopathy: fatal virus disease of cattle

BST British Summer Time

bubble n ball of air in a liquid or solid ▷ v form bubbles; move or flow with a gurgling sound ▶ **bubbly** adj excited and lively; full of bubbles ▶ **bubble over** v express an emotion freely

bubonic plague [bew-**bonn**-ik] n acute infectious disease characterized by swellings

buccaneer n (Hist) pirate

buck¹ n male of the goat, hare, kangaroo, rabbit, and reindeer ▷ v (of a horse etc.) jump with legs stiff and back arched ▶ **buck up** v make or become more cheerful

buck² n (US, Canadian, Aust & NZ slang) dollar; (SAfr) rand

buck³ n **pass the buck** (informal) shift blame or responsibility onto someone else

bucket n open-topped round container with a handle ▷ v **-eting, -eted** rain heavily ▶ **bucketful** n ▶ **bucket list** (informal) list of things wanted to be experienced before one dies

buckle n clasp for fastening a belt or strap ▷ v fasten or be fastened with a buckle; (cause to) bend out of shape through pressure or heat ▶ **buckle down** v (informal) apply oneself with determination

buckshee adj (slang) free

buckteeth pl n projecting upper front teeth ▶ **buck-toothed** adj

buckwheat n small black grain used for making flour

bucolic [bew-**koll**-ik] adj of the countryside or country life

bud n swelling on a tree or plant that develops into a leaf or flower ▷ v **budding, budded** produce buds ▶ **budding** adj beginning to develop or grow

Buddhism n eastern religion founded by Buddha ▶ **Buddhist** n, adj

buddleia n shrub with long spikes of purple flowers

buddy n, pl **-dies** (informal) friend

budge v move slightly

budgerigar n small cage bird bred in many different-coloured varieties

budget n financial plan for a period of time; money allocated for a specific purpose ▷ v **-eting, -eted** plan the expenditure of (money or time) ▷ adj cheap ▶ **budgetary** adj

> **SPELLING TIP**
> Many words ending in et, have two ts when you add an ending like -ing, but **budget** is not one of them: **budgeting** and **budgeted** have one t

budgie n (informal) short for **budgerigar**

buff¹ adj dull yellowish-brown ▷ v clean or polish with soft material

buff² n (informal) expert on or devotee of a given subject

buffalo n type of cattle; (US) bison

buffer n something that lessens shock or protects from damaging impact, circumstances, etc.

buffet¹ [**boof**-ay, **buff**-ay] n counter where drinks and snacks are served

buffet² [**buff**-it] v **-feting, -feted** knock against or about

buffoon n clown or fool ▶ **buffoonery** n

bug n small insect; (informal) minor illness; small mistake in a computer program; concealed microphone; (Aust) flattish edible shellfish ▷ v **bugging, bugged** (informal) irritate (someone); conceal a microphone in (a room or phone)

bugbear n thing that causes obsessive anxiety

bugger (slang) n unpleasant or difficult person or thing; person who practises buggery ▷ v tire; practise buggery ▶ **buggery** ▷ n anal intercourse

bugle n instrument like a small trumpet ▶ **bugler** n

build v **building, built** make, construct, or form by joining parts or materials ▶ n shape of the body ▶ **builder** n ▶ **building** n structure with walls and a roof ▶ **building society** organization where money can be borrowed or invested ▶ **build-up** n gradual increase

built v past of **build** ▶ **built-up** adj having many buildings

bulb n same as **light bulb**; onion-shaped root which grows into a flower or plant ▶ **bulbous** adj round and fat

bulge n swelling on a normally flat surface; sudden increase in number ▶ v swell outwards ▶ **bulging** adj

bulimia n disorder characterized by compulsive overeating followed by vomiting ▶ **bulimic** adj, n

bulk n size or volume, esp. when great; main part **in bulk** in large quantities ▶ **bulky** adj

bulkhead n partition in a ship or aeroplane

bull¹ n male of some animals, such as cattle, elephants, and whales ▶ **bullock** n castrated bull ▶ **bulldog** n thickset dog with a broad head and a muscular body ▶ **bulldozer** n powerful tractor for moving earth ▶ **bulldoze** v ▶ **bullfight** n public show in which a matador kills a bull ▶ **bull's-eye** n central disc of a target ▶ **bullswool** n (Aust & NZ slang) nonsense

bull² n (informal) complete nonsense

bull³ n papal decree

bullet n small piece of metal fired from a gun

bulletin n short official report or announcement

bullion n gold or silver in the form of bars

bully n, pl **-lies** person who hurts, persecutes, or intimidates another person, often repeatedly ▶ v **-lying, -lied** hurt, intimidate, or persecute (another person)

bulrush n tall stiff reed

bulwark n wall used as a fortification; person or thing acting as a defence

bum¹ n (slang) buttocks or anus

bum² (informal) n disreputable idler ▶ adj of poor quality

bumble v speak, do, or move in a clumsy way ▶ **bumbling** adj, n

bumblebee n large hairy bee

bumf, bumph n (informal) official documents or forms

bump v knock or strike with a jolt; travel in jerks and jolts ▶ n dull thud from an impact or collision; raised uneven part ▶ **bumpy** adj, n ▶ **bump off** v (informal) murder

bumper¹ n bar on the front and back of a vehicle to protect against damage

bumper² adj unusually large or abundant

bumph n same as **bumf**

bumpkin n awkward simple country person

bumptious adj offensively self-assertive

bun n small sweet bread roll or cake; hair gathered into a bun shape at the back of the head

bunch n number of things growing, fastened, or grouped together ▶ v group or be grouped together in a bunch

bundle n number of things gathered loosely together ▶ v cause to go roughly or unceremoniously ▶ **bundle up** v make into a bundle

bung n stopper for a cask etc. ▶ v (foll by up) (informal) close with a bung; (Brit slang) throw (something) somewhere in a careless manner

bungalow n one-storey house

bungee jumping, bungy jumping n sport of leaping from a high bridge, tower, etc., to which one is connected by a rubber rope

bungle v spoil through incompetence ▶ **bungler** n ▶ **bungling** adj, n

bunion n inflamed swelling on the big toe

bunk¹ n narrow shelflike bed ▶ **bunk bed** n one of a pair of beds constructed one above the other

bunk² n same as **bunkum**

bunk³ (slang) n (Brit) **do a bunk** make a hurried and secret departure ▶ v (Brit, NZ & S Afr) be absent without permission

bunker n sand-filled hollow forming an obstacle on a golf course; underground shelter; large storage container for coal etc.

bunkum n nonsense

bunny n, pl -**nies** child's word for a rabbit

Bunsen burner n gas burner used in laboratories

bunting n decorative flags

bunya n tall dome-shaped Australian coniferous tree (also **bunya-bunya**)

bunyip n (Aust) legendary monster said to live in swamps and lakes

buoy n floating marker anchored in the sea ▶ v prevent from sinking; encourage or hearten ▶ **buoyant** adj able to float; cheerful or resilient ▶ **buoyancy** n

bur n same as **burr¹**

burble v make a bubbling sound; talk quickly and excitedly

burden¹ n heavy load; something difficult to cope with ▶ v put a burden on; oppress ▶ **burdensome** adj

burden² n theme of a speech etc.

bureau n, pl -**reaus** or -**reaux** office that provides a service; writing desk with shelves and drawers ▶ **bureau de change** place where foreign currencies can be exchanged

bureaucracy n, pl -**cies** administrative system based on complex rules and procedures; excessive adherence to complex procedures ▶ **bureaucrat** n ▶ **bureaucratic** adj

burgeon v develop or grow rapidly

burgh n Scottish borough

burglar n person who enters a building to commit a crime, esp. theft ▶ **burglary** n ▶ **burgle** v

Burgundy n type of French wine ▶ **burgundy** adj dark-purplish red

burial n burying of a dead body

burlesque n artistic work which satirizes a subject by caricature

burly adj -**lier**, -**liest** (of a person) broad and strong

burn¹ v **burning**, **burnt** or **burned** be or set on fire; destroy or be destroyed by fire; damage, injure, or mark by heat; feel strong emotion; record data on (a compact disc) ▶ n injury or mark caused by fire or exposure to heat ▶ **burning** adj intense; urgent or crucial

> **USAGE NOTE**
> Either burnt or burned may be used as a past form

burn² n (Scot) small stream

burnish v make smooth and shiny by rubbing

burp v, n (informal) belch

burr¹ n head of a plant with prickles or hooks

burr² n soft trilling sound given to the letter r in some dialects; whirring sound

burrawang n Australian plant with fernlike leaves and an edible nut

burrow n hole dug in the ground by a rabbit etc. ▷ v dig holes in the ground

bursar n treasurer of a school, college, or university ▶ **bursary** n scholarship

burst v **bursting, burst** (cause to) break open or apart noisily and suddenly; come or go suddenly and forcibly; be full to the point of breaking open ▷ n instance of breaking open suddenly; sudden outbreak or occurrence ▶ **burst into** v give vent to (an emotion) suddenly

bury v **burying, buried** place in a grave; place in the earth and cover with soil; conceal or hide

bus n large motor vehicle for carrying passengers ▷ v **bussing, bussed** travel or transport by bus

busby n, pl **-bies** tall fur hat worn by some soldiers

bush n dense woody plant, smaller than a tree; wild uncultivated part of a country ▶ **bushy** adj (of hair) thick and shaggy ▶ **bushbaby** n small African tree-living mammal with large eyes ▶ **bushfire** n uncontrolled scrub or forest fire

bushel n obsolete unit of measure equal to 8 gallons (36.4 litres)

business n purchase and sale of goods and services; commercial establishment; trade or profession; proper concern or responsibility; affair e.g. it's a dreadful business ▶ **businesslike** adj efficient and methodical ▶ **businessman**, **businesswoman** n

busker n street entertainer ▶ **busk** v act as a busker

bust[1] n treasurer's bosom; sculpture of the head and shoulders

bust[2] (informal) v **busting, bust** or **busted** burst or break; (of the police) raid (a place) or arrest

(someone) ▷ adj broken **go bust** become bankrupt

bustard n bird with long strong legs, a heavy body, a long neck, and speckled plumage

bustle[1] v hurry with a show of activity or energy ▷ n energetic and noisy activity ▶ **bustling** adj

bustle[2] n cushion or framework formerly worn under the back of a woman's skirt to hold it out

busy adj **busier, busiest** actively employed; crowded or full of activity ▷ v **busying, busied** keep (someone, esp. oneself) busy ▶ **busily** adv ▶ **busybody** n meddlesome or nosy person

but conj contrary to expectation; in contrast; other than; without it happening ▷ prep except ▷ adv only **but for** were it not for

butane n gas used for fuel

butch adj (slang) markedly or aggressively masculine

butcher n person who slaughters animals or sells their meat; brutal murderer ▷ v kill and prepare (animals) for meat; kill (people) brutally or indiscriminately ▶ **butchery** n ▶ **butcherbird** n Australian magpie that impales its prey on thorns

butler n chief male servant

butt[1] n thicker end of something; unused end of a cigar or cigarette; (slang) buttocks

butt[2] n person or thing that is the target of ridicule

butt[3] v strike with the head or horns ▶ **butt in** v interrupt a conversation

butt[4] n large cask

butter n edible fatty solid made by churning cream ▷ v put butter on ▶ **buttery** adj ▶ **butter up** v flatter ▶ **butter bean** large pale flat edible bean

buttercup n small yellow flower

butterfingers n (informal) person who drops things by mistake

butterfly n insect with brightly coloured wings; swimming stroke in which both arms move together in a forward circular action

buttermilk n sourish milk that remains after the butter has been separated from milk

butterscotch n kind of hard brittle toffee

buttock n either of the two fleshy masses that form the human rump

button n small disc or knob sewn to clothing, which can be passed through a slit in another piece of fabric to fasten them; knob that operates a piece of equipment when pressed ▷ v fasten with buttons ▶ **buttonhole** n slit in a garment through which a button is passed; flower worn on a lapel ▷ v detain (someone) in conversation

buttress n structure to support a wall ▷ v support with, or as if with, a buttress

buxom adj (of a woman) healthily plump and full-bosomed

buy v buying, bought acquire by paying money for; (slang) accept as true ▷ n thing acquired through payment ▶ **buyer** n customer; person employed to buy merchandise

buzz n rapidly vibrating humming sound; (informal) sense of excitement ▷ v make a humming sound; be filled with an air of excitement ▶ **buzzer** n ▶ **buzz around** v move around quickly and busily ▶ **buzz word** jargon word which becomes fashionably popular

buzzard n bird of prey of the hawk family

by prep indicating the doer of an action, nearness, movement past, time before or during which, etc. e.g. bitten by a dog; down by the river; driving by the school; in bed by midnight ▷ adv near; past **by and by** eventually **by and large** in general

bye, bye-bye interj (informal) goodbye

by-election n election held during parliament to fill a vacant seat

bygone adj past or former

bylaw, bye-law n rule made by a local authority

BYO, BYOG n (Aust & NZ) unlicensed restaurant at which diners may bring their own alcoholic drink

bypass n main road built to avoid a city; operation to divert blood flow away from a damaged part of the heart ▷ v go round or avoid

by-product n secondary or incidental product of a process

byre n (Brit) shelter for cows

bystander n person present but not involved

byte n (Computing) group of bits processed as one unit of data

byway n minor road

byword n person or thing regarded as a perfect example of something

C (Chem) carbon; Celsius; centigrade; century

c. circa

cab n taxi; enclosed driver's compartment in a train, truck, etc. ▶ **cabbie, cabby** n, pl **-bies** (informal) taxi driver

cabal [kab-**bal**] n small group of political plotters; secret plot

cabaret [kab-a-**ray**] n dancing and singing show in a nightclub

cabbage n vegetable with a large head of green leaves ▶ **cabbage tree** (NZ) palm-like tree with a bare trunk and spiky leaves

caber n tree trunk tossed in competition in Highland games

cabin n compartment in a ship or aircraft; small hut ▶ **cabin cruiser** motorboat with a cabin

cabinet n piece of furniture with drawers or shelves; (**C-**) committee of senior government ministers ▶ **cabinet-maker** n person who makes fine furniture

cable n strong thick rope; bundle of wires that carries electricity or electronic signals; telegram sent abroad ▷ v send (someone) a message by cable ▶ **cable car** vehicle pulled up a steep slope by a moving cable ▶ **cable television** television service conveyed by cable to subscribers

caboodle n **the whole caboodle** (informal) the whole lot

cabriolet [kab-ree-oh-**lay**] n small horse-drawn carriage with a folding hood

cacao [kak-**kah**-oh] n tropical tree with seed pods from which chocolate and cocoa are made

cache [**kash**] n hidden store of weapons or treasure

cachet [**kash**-shay] n prestige, distinction

cack-handed adj (informal) clumsy

cackle v laugh shrilly; (of a hen) squawk with shrill broken notes ▷ n cackling noise

cacophony [kak-**koff**-on-ee] n harsh discordant sound ▶ **cacophonous** adj

cactus n, pl **-tuses** or **-ti** fleshy desert plant with spines but no leaves

cad n (old-fashioned) dishonourable man ▶ **caddish** adj

cadaver [kad-**dav**-ver] n corpse ▶ **cadaverous** adj pale, thin, and haggard

caddie, caddy n, pl **-dies** person who carries a golfer's clubs ▷ v **-dying, -died** act as a caddie

caddis fly n insect whose larva (**caddis worm**) lives underwater in a protective case of sand and stones

caddy n, pl **-dies** small container for tea

cadence [**kade**-enss] n rise and fall in the pitch of the voice; close of a musical phrase

cadenza n complex solo passage in a piece of music

cadet n young person training for the armed forces or police

cadge v (informal) get (something) by taking advantage of someone's generosity ▶ **cadger** n

cadmium n (Chem) bluish-white metallic element used in alloys

cadre [**kah**-der] n small group of people selected and trained to form the core of a political organization or military unit

caecum [seek-um] *n, pl* **-ca** [-ka] pouch at the beginning of the large intestine

Caesar [seez-ar] *n* title of Roman emperors ▸ **Caesar salad** salad of lettuce, cheese, and croutons, with olive oil, garlic, and lemon juice

Caesarean section [see-zair-ee-an] *n* surgical incision into the womb to deliver a baby

caesium *n* (*Chem*) silvery-white metallic element used in photocells

café *n* small or inexpensive restaurant serving light refreshments; (*S Afr*) corner shop or grocer ▸ **cafeteria** *n* self-service restaurant

caffeine *n* stimulant found in tea and coffee

caftan *n* same as **kaftan**

cage *n* enclosure of bars or wires, for keeping animals or birds; enclosed platform of a lift in a mine ▸ **caged** *adj* kept in a cage

cagey *adj* **cagier, cagiest** (*informal*) reluctant to go into details

cagoule *n* (*Brit*) lightweight hooded waterproof jacket

cahoots *pl n* **in cahoots** (*informal*) conspiring together

cairn *n* mound of stones erected as a memorial or landmark

cajole *v* persuade by flattery ▸ **cajolery** *n*

cake *n* sweet food baked from a mixture of flour, eggs, etc.; flat compact mass of something, such as soap ▹ *v* form into a hardened mass or crust

calamine *n* pink powder consisting chiefly of zinc oxide, used in skin lotions and ointments

calamity *n, pl* **-ties** disaster ▸ **calamitous** *adj*

calcify *v* **-fying, -fied** harden by the depositing of calcium salts ▸ **calcification** *n*

calcium *n* (*Chem*) silvery-white metallic element found in bones, teeth, limestone, and chalk

calculate *v* solve or find out by a mathematical procedure or by reasoning; aim to have a particular effect ▸ **calculable** *adj* ▸ **calculating** *adj* selfishly scheming ▸ **calculation** *n* ▸ **calculator** *n* small electronic device for making calculations

calculus *n* branch of mathematics dealing with infinitesimal changes to a variable number or quantity; (*pl* **-li**) (*Pathology*) hard deposit in kidney or bladder

Caledonian *adj* Scottish

calendar *n* chart showing a year divided up into months, weeks, and days; system for determining the beginning, length, and division of years; schedule of events or appointments

calendula *n* marigold

calf¹ *n, pl* **calves** young cow, bull, elephant, whale, or seal; leather made from calf skin ▸ **calve** *v* give birth to a calf

calf² *n, pl* **calves** back of the leg between the ankle and knee

calibre *n* person's ability or worth; diameter of the bore of a gun or of a shell or bullet ▸ **calibrate** *v* mark the scale or check the accuracy of (a measuring instrument) ▸ **calibration** *n*

calico *n, pl* **-coes** or **-cos** white cotton fabric

caliph *n* (*Hist*) Muslim ruler

call *v* name; shout to attract attention; telephone; summon; (often foll by *on*) visit; arrange (a meeting, strike, etc.) ▹ *n* cry, shout; animal's or bird's cry; telephone communication; short visit; summons, invitation; need, demand ▸ **caller** *n* ▸ **calling** *n* vocation, profession ▸ **call box** kiosk for a public telephone ▸ **call centre** office where staff carry out an organization's telephone transactions ▸ **call for** v require ▸ **call off** v cancel ▸ **call up** v summon to serve in the armed forces; cause one to remember

calligraphy n (art of) beautiful handwriting ▸ **calligrapher** n

calliper n metal splint for supporting the leg; instrument for measuring diameters

callisthenics pl n light keep-fit exercises

callous adj showing no concern for other people's feelings ▸ **calloused** adj (of skin) thickened and hardened ▸ **callously** adv ▸ **callousness** n

callow adj young and inexperienced

callus n, pl -**luses** area of thick hardened skin

calm adj not agitated or excited; not ruffled by the wind; windless ▷ n peaceful state ▷ v (often foll by down) make or become calm ▸ **calmly** adv ▸ **calmness** n

calorie n unit of measurement for the energy value of food; unit of heat ▸ **calorific** adj of calories or heat

calumny n, pl -**nies** false or malicious statement

calypso n, pl -**sos** West Indian song with improvised topical lyrics

calyx n, pl **calyxes** or **calyces** outer leaves that protect a flower bud

cam n device that converts a circular motion to a to-and-fro motion ▸ **camshaft** n part of an engine consisting of a rod to which cams are fixed

camaraderie n comradeship

camber n slight upward curve to the centre of a surface

cambric n fine white linen fabric

camcorder n combined portable video camera and recorder

came v past tense of **come**

camel n humped mammal that can survive long periods without food or water in desert regions

camellia [kam-**meal**-ya] n evergreen ornamental shrub with white, pink, or red flowers

Camembert [kam-mem-**bare**] n soft creamy French cheese

cameo n, pl **cameos** brooch or ring with a profile head carved in relief; small part in a film or play performed by a well-known actor or actress

camera n apparatus used for taking photographs or pictures for television or cinema **in camera** in private session ▸ **cameraman** n man who operates a camera for television or cinema ▸ **camera phone** mobile phone incorporating a camera

camiknickers pl n (Brit) woman's undergarment consisting of knickers attached to a camisole

camisole n woman's bodice-like garment

camomile n aromatic plant, used to make herbal tea

camouflage [kam-moo-**flahzh**] n use of natural surroundings or artificial aids to conceal or disguise something ▷ v conceal by camouflage

camp[1] n (place for) temporary lodgings consisting of tents, huts, or cabins; group supporting a particular doctrine ▷ v stay in a camp ▸ **camper** n

camp[2] adj (informal) effeminate or homosexual; consciously artificial or affected **camp it up** (informal) behave in a camp way

campaign n series of coordinated activities designed to achieve a goal ▷ v take part in a campaign

campanology n art of ringing bells

campanula n plant with blue or white bell-shaped flowers

camphor n aromatic crystalline substance used medicinally and in mothballs

campion n red, pink, or white wild flower

campus n, pl **-puses** grounds of a university or college

can¹ v, past tense **could** be able to; be allowed to

can² n metal container for food or liquids ▷ v **canning, canned** put (something) into a can ▶ **canned** adj preserved in a can; (of music) prerecorded ▶ **cannery** n, pl **-neries** factory where food is canned

Canadian n, adj (person) from Canada

canal n artificial waterway; passage in the body

canapé [kan-nap-pay] n a small piece of bread or toast with a savoury topping

canary n, pl **-ries** small yellow songbird often kept as a pet

canasta n card game like rummy, played with two packs

cancan n lively high-kicking dance performed by a female group

cancel v **-celling, -celled** stop (something that has been arranged) from taking place; mark (a cheque or stamp) with an official stamp to prevent further use ▶ **cancellation** n ▶ **cancel out** v counterbalance, neutralize

cancer n serious disease resulting from a malignant growth or tumour; malignant growth or tumour ▶ **cancerous** adj

candela [kan-dee-la] n unit of luminous intensity

candelabrum n, pl **-bra** large branched candle holder

candid adj honest and straightforward ▶ **candidly** adv

candidate n person seeking a job or position; person taking an examination ▶ **candidacy, candidature** n

candle n stick of wax enclosing a wick, which is burned to produce light ▶ **candlestick** n holder for a candle ▶ **candlewick** n cotton fabric with a tufted surface

candour n honesty and straightforwardness

candy n, pl **-dies** (US) sweet or sweets ▶ **candied** adj coated with sugar ▶ **candyfloss** n light fluffy mass of spun sugar on a stick ▶ **candy-striped** adj having coloured stripes on a white background

cane n stem of the bamboo or similar plant; flexible rod used to beat someone; slender walking stick ▷ v beat with a cane ▶ **cane toad** large toad used to control insects and other pests of sugar-cane plantations

canine adj of or like a dog ▷ n sharp pointed tooth between the incisors and the molars

canister n metal container

canker n ulceration, ulcerous disease; something evil that spreads and corrupts

cannabis n Asian plant with tough fibres; drug obtained from the dried leaves and flowers of this plant, which can be smoked or chewed

cannelloni pl n tubular pieces of pasta filled with meat etc.

cannibal n person who eats human flesh; animal that eats others of its own kind ▶ **cannibalism** n ▶ **cannibalize** v use parts from (one machine) to repair another

cannon n large gun on wheels; billiard stroke in which the cue ball hits two balls successively ▸ **cannonade** n continuous heavy gunfire ▸ **cannonball** n heavy metal ball fired from a cannon ▸ **cannon into** v collide with

cannot can not

canny adj **-nier, -niest** shrewd, cautious ▸ **cannily** adv

canoe n light narrow open boat propelled by a paddle or paddles ▸ **canoeing** n sport of rowing in a canoe ▸ **canoeist** n

canon[1] n priest serving in a cathedral

canon[2] n Church decree regulating morals or religious practices; general rule or standard; list of the works of an author that are accepted as authentic ▸ **canonical** adj ▸ **canonize** v declare (a person) officially to be a saint ▸ **canonization** n

canoodle v (slang) kiss and cuddle

canopy n, pl **-pies** covering above a bed, door, etc.; any large or wide covering ▸ **canopied** adj covered with a canopy

cant[1] n insincere talk; specialized vocabulary of a particular group

cant[2] n tilted position ▸ v tilt, overturn

can't can not

cantaloupe, cantaloup n kind of melon with sweet orange flesh

cantankerous adj quarrelsome, bad-tempered

cantata n musical work consisting of arias, duets, and choruses

canteen n restaurant attached to a workplace or school; box containing a set of cutlery

canter n horse's gait between a trot and a gallop ▸ v move at a canter

canticle n short hymn with words from the Bible

cantilever n beam or girder fixed at one end only

canto n, pl **-tos** main division of a long poem

canton n political division of a country, esp. Switzerland

cantor n man employed to lead services in a synagogue

canvas n heavy coarse cloth used for sails and tents, and for oil painting; oil painting on canvas

canvass v try to get votes or support (from); find out the opinions of (people) by conducting a survey ▸ n canvassing

canyon n deep narrow valley ▸ **canyoning** n sport of swimming, scrambling, and abseiling through a canyon

cap n soft close-fitting covering for the head; small lid; small explosive device used in a toy gun ▸ v **capping, capped** cover or top with something; select (a player) for a national team; impose an upper limit on (a tax); outdo, excel

capable adj (foll by of) having the ability (for); competent and efficient ▸ **capably** adv ▸ **capability** n, pl **-ties**

capacity n, pl **-ties** ability to contain, absorb, or hold; maximum amount that can be contained or produced; physical or mental ability; position, function ▸ **capacious** adj roomy ▸ **capacitance** n (measure of) the ability of a system to store electrical charge ▸ **capacitor** n device for storing electrical charge

caparisoned adj magnificently decorated

cape[1] n short cloak

cape[2] n large piece of land that juts out into the sea

caper n high-spirited prank ▸ v skip about

capercaillie, capercailzie
[kap-per-**kale**-yee] n large black European grouse

capers pl n pickled flower buds of a Mediterranean shrub used in sauces

capillary n, pl **-laries** very fine blood vessel

capital[1] n chief city of a country; accumulated wealth; wealth used to produce more wealth; large letter, as used at the beginning of a name or sentence ▷ adj involving or punishable by death; (old-fashioned) excellent ▶ **capitalize** v write or print (words) in capitals; convert into or provide with capital ▶ **capitalize on** v take advantage of (a situation)

capital[2] n top part of a pillar

capitalism n economic system based on the private ownership of industry ▶ **capitalist** adj of capitalists or capitalism; supporting capitalism ▷ n supporter of capitalism; person who owns a business

capitation n tax of a fixed amount per person

capitulate v surrender on agreed terms ▶ **capitulation** n

capon n castrated cock fowl fattened for eating

cappuccino [kap-poo-**cheen**-oh] n, pl **-nos** coffee with steamed milk, sprinkled with powdered chocolate

caprice [kap-**reess**] n sudden change of attitude ▶ **capricious** adj tending to have sudden changes of attitude ▶ **capriciously** adv

capsicum n kind of pepper used as a vegetable or as a spice

capsize v (of a boat) overturn accidentally

capstan n rotating cylinder round which a ship's rope is wound

capsule n soluble gelatine case containing a dose of medicine; plant's seed case; detachable crew compartment of a spacecraft

captain n commander of a ship or civil aircraft; middle-ranking naval officer; junior officer in the army; leader of a team or group ▷ v be captain of ▶ **captaincy** n

caption n title or explanation accompanying an illustration ▷ v provide with a caption

captious adj tending to make trivial criticisms

captivate v attract and hold the attention of ▶ **captivating** adj

captive n person kept in confinement ▷ adj kept in confinement; (of an audience) unable to leave ▶ **captivity** n

captor n person who captures a person or animal

capture v take by force; succeed in representing (something elusive) artistically ▷ n capturing

car n motor vehicle designed to carry a small number of people; passenger compartment of a cable car, lift, etc.; (US) railway carriage ▶ **car park** area or building reserved for parking cars

carafe [kar-**raff**] n glass bottle for serving water or wine

caramel n chewy sweet made from sugar and milk; burnt sugar, used for colouring and flavouring food ▶ **caramelize** v turn into caramel

carapace n hard upper shell of tortoises and crustaceans

carat n unit of weight of precious stones; measure of the purity of gold in an alloy

caravan n large enclosed vehicle for living in, designed to be towed by a car or horse; group travelling together in Eastern countries

caraway n plant whose seeds are used as a spice

carb n short for **carbohydrate**, **carburettor**

carbide n compound of carbon with a metal

carbine n light automatic rifle

carbohydrate n any of a large group of energy-producing compounds in food, such as sugars and starches

carbolic acid n disinfectant derived from coal tar

carbon n nonmetallic element occurring as charcoal, graphite, and diamond, found in all organic matter ▶ **carbonate** n salt or ester of carbonic acid ▶ **carbonated** adj (of a drink) containing carbon dioxide ▶ **carbonize** v turn into carbon as a result of heating; coat with carbon ▶ **carbon copy** copy made with carbon paper; very similar person or thing ▶ **carbon dioxide** colourless gas exhaled by people and animals ▶ **carbon footprint** measure of the carbon dioxide produced by an individual or company ▶ **carbonic acid** weak acid formed from carbon dioxide and water ▶ **carbon-neutral** adj not affecting the overall volume of carbon dioxide in the atmosphere ▶ **carbon offset** act which compensates for carbon emissions of an individual or company ▶ **carbon paper** paper coated on one side with a dark waxy pigment, used to make a copy of something as it is typed or written

Carborundum® n compound of silicon and carbon, used for grinding and polishing

carbuncle n inflamed boil

carburettor n device which mixes petrol and air in an internal-combustion engine

carcass, carcase n dead body of an animal

carcinogen n substance that produces cancer ▶ **carcinogenic** adj ▶ **carcinoma** n malignant tumour

card n piece of thick stiff paper or cardboard used for identification, reference, or sending greetings or messages; one of a set of cards with a printed pattern, used for playing games; small rectangle of stiff plastic with identifying numbers for use as a credit card, cheque card, or charge card; (old-fashioned) witty or eccentric person ▷ pl any card game, or card games in general ▶ **cardboard** n thin stiff board made from paper pulp ▶ **cardholder** n person who owns a credit or debit card ▶ **cardsharp** n professional card player who cheats

cardamom n spice obtained from the seeds of a tropical plant

cardiac adj of the heart ▶ **cardiogram** n electrocardiogram ▶ **cardiograph** n electrocardiograph ▶ **cardiology** n study of the heart and its diseases ▶ **cardiologist** n ▶ **cardiovascular** adj of the heart and the blood vessels

cardigan n knitted jacket

cardinal n any of the high-ranking clergymen of the RC Church who elect the Pope and act as his counsellors ▷ adj fundamentally important ▶ **cardinal number** number denoting quantity but not order in a group, for example four as distinct from fourth ▶ **cardinal points** the four main points of the compass

care v be concerned; like (to do something); (foll by for) like, be fond of; (foll by for) look after

▷ *n* careful attention, caution; protection, charge; trouble, worry ▶ **careful** *adj* ▶ **carefully** *adv* ▶ **carefulness** *n* ▶ **careless** *adj* ▶ **carelessly** *adv* ▶ **carelessness** *n*

careen *v* tilt over to one side

> **USAGE NOTE**
> Be careful not to confuse *careen*, 'tilt', with *career*, 'rush headlong'.

career *n* series of jobs or occupation in a profession or occupation that a person has through their life; part of a person's life spent in a particular occupation ▷ *v* move in an uncontrolled way ▶ **careerist** *n* person who seeks advancement by any possible means

carefree *adj* without worry or responsibility

caress *n* gentle affectionate touch or embrace ▷ *v* touch gently and affectionately

caret [kar-rett] *n* symbol (∧) indicating a place in written or printed matter where something is to be inserted

caretaker *n* person employed to look after a place

careworn *adj* showing signs of worry

cargo *n*, *pl* **-goes** goods carried by a ship, aircraft, etc. ▶ **cargo pants**, **cargo trousers** loose trousers with a large pocket on the outside of each leg

caribou *n*, *pl* **-bou** or **-bous** large N American reindeer

caricature *n* drawing or description of a person that exaggerates features for comic effect ▷ *v* make a caricature of

caries [care-reez] *n* tooth decay

carillon [kar-**rill**-yon] *n* set of bells played by keyboard or mechanically; tune played on such bells

cark *v* **cark it** (*Aust & NZ slang*) die

carmine *adj* vivid red

carnage *n* extensive slaughter of people

carnal *adj* of a sexual or sensual nature ▶ **carnal knowledge** sexual intercourse

carnation *n* cultivated plant with fragrant white, pink, or red flowers

carnival *n* festive period with processions, music, and dancing in the street

carnivore *n* meat-eating animal ▶ **carnivorous** *adj*

carob *n* pod of a Mediterranean tree, used as a chocolate substitute

carol *n* joyful Christmas hymn ▷ *v* **-olling**, **-olled** sing carols; sing joyfully

carotid *adj*, *n* (of) either of the two arteries supplying blood to the head

carouse *v* have a merry drinking party

carousel [kar-roo-**sell**] *n* revolving conveyor belt for luggage or photographic slides; (*US*) merry-go-round

carp¹ *n* large freshwater fish

carp² *v* complain, find fault

carpel *n* female reproductive organ of a flowering plant

carpenter *n* person who makes or repairs wooden structures ▶ **carpentry** *n*

carpet *n* heavy fabric for covering floors ▷ *v* **carpeting**, **carpeted** cover with a carpet **on the carpet** (*informal*) being reprimanded ▶ **carpet snake**, **carpet python** large nonvenomous Australian snake with a carpet-like pattern on its back

carpus *n*, *pl* **-pi** set of eight bones of the wrist

carriage *n* one of the sections of a train for passengers; way a person holds his or her head and body;

four-wheeled horse-drawn vehicle; moving part of a machine that supports and shifts another part; charge made for conveying goods ▶ **carriageway** n (Brit) part of a road along which traffic passes in one direction

carrier n person or thing that carries something; person or animal that does not suffer from a disease but can transmit it to others ▶ **carrier pigeon** homing pigeon used for carrying messages

carrion n dead and rotting flesh

carrot n long tapering orange root vegetable; something offered as an incentive ▶ **carroty** adj (of hair) reddish-orange

carry v -rying, -ried take from one place to another; have with one habitually, in one's pocket etc.; transmit (a disease); have as a factor or result; hold (one's head or body) in a specified manner; secure the adoption of (a bill or motion); (of sound) travel a certain distance ▶ **carry on** v continue; (informal) cause a fuss ▶ **carry out** v follow, accomplish

cart n open two-wheeled horse-drawn vehicle for carrying goods or passengers ▷ v carry, usu. with some effort ▶ **carthorse** n large heavily built horse ▶ **cartwheel** n sideways somersault supported by the hands with legs outstretched; large spoked wheel of a cart

carte blanche [kaht blahntsh] n (French) complete authority

cartel n association of competing firms formed to fix prices

cartilage n strong flexible tissue forming part of the skeleton ▶ **cartilaginous** adj

cartography n map making ▶ **cartographer** n ▶ **cartographic** adj

carton n container made of cardboard or waxed paper.

cartoon n humorous or satirical drawing; sequence of these telling a story; film made by photographing a series of drawings which give the illusion of movement when projected ▶ **cartoonist** n

cartridge n casing containing an explosive charge and bullet for a gun; sealed container of film, tape, ink, etc. ▶ **cartridge paper** strong thick drawing paper

carve v cut to form an object; form (an object or design) by cutting; slice (cooked meat) ▶ **carving** n

caryatid [kar-ree-at-id] n supporting column in the shape of a female figure

Casanova n promiscuous man

casbah n citadel of a N African city

cascade n waterfall; something flowing or falling like a waterfall ▷ v flow or fall in a cascade

case[1] n instance, example; condition, state of affairs; set of arguments supporting an action or cause; person or problem dealt with by a doctor, social worker, or solicitor; action, lawsuit; (Grammar) form of a noun, pronoun, or adjective showing its relation to other words in the sentence **in case** so as to allow for the possibility that

case[2] n container, protective covering ▷ v (slang) inspect (a building) with the intention of burgling it ▶ **case-hardened** adj having been made callous by experience

casement n window that is hinged on one side

cash n banknotes and coins ▷ v obtain cash for ▶ **cash in on** v (informal) gain profit or advantage from ▸ **cash register** till that displays and adds the prices of the goods sold ▸ **cash-strapped** adj short of money

cashew n edible kidney-shaped nut

cashier¹ n person responsible for handling cash in a bank, shop, etc.

cashier² v dismiss with dishonour from the armed forces

cashmere n fine soft wool obtained from goats

casing n protective case, covering

casino n, pl **-nos** public building where gambling games are played

cask n barrel used to hold alcoholic drink; (Aust) cubic carton containing wine, with a tap for dispensing

casket n small box for valuables; (US) coffin

cassava n starch obtained from the roots of a tropical American plant, used to make tapioca

casserole n covered dish in which food is cooked slowly, usu. in an oven; dish cooked in this way ▷ v cook in a casserole

cassette n plastic case containing a reel of film or magnetic tape

cassock n long tunic, usu. black, worn by priests

cassowary n, pl **-waries** large flightless bird of Australia and New Guinea

cast n actors in a play or film collectively; object shaped by a mould while molten; mould used to shape such an object; rigid plaster-of-Paris casing for immobilizing broken bones while they heal; sort, kind; slight squint in the eye ▷ v select (an actor) to play a part in a play or film; give (a vote); let fall, shed; shape (molten material) in a mould; throw with force; direct (a glance) ▸ **castaway** n shipwrecked person ▸ **casting**

vote deciding vote used by the chairperson of a meeting when the votes on each side are equal ▸ **cast-iron** adj made of a hard but brittle type of iron; definite, unchallengeable ▸ **cast-off** adj, n discarded (person or thing)

castanets pl n musical instrument, used by Spanish dancers, consisting of curved pieces of hollow wood clicked together in the hand

caste n any of the hereditary classes into which Hindu society is divided; social rank

castellated adj having battlements

caster sugar n finely ground white sugar

castigate v reprimand severely ▸ **castigation** n

castle n large fortified building, often built as a ruler's residence; rook in chess

castor n small swivelling wheel fixed to the bottom of a piece of furniture for easy moving

castor oil n oil obtained from an Indian plant, used as a lubricant and purgative

castrate v remove the testicles of; deprive of vigour or masculinity ▸ **castration** n

casual adj careless, nonchalant; (of work or workers) occasional or not permanent; for informal wear; happening by chance ▸ **casually** adv

casualty n, pl **-ties** person killed or injured in an accident or war; person or thing that has suffered as the result of something

casuarina [kass-yew-a-reen-a] n Australian tree with jointed green branches

casuistry n reasoning that is misleading or oversubtle

cat n small domesticated furry mammal; related wild mammal, such as the lion or tiger ▸ **catty** adj (informal) spiteful ▸ **catcall** n derisive whistle or cry ▸ **catfish** n fish with whisker-like barbels round the mouth ▸ **catgut** n strong cord used to string musical instruments and sports rackets ▸ **catkin** n drooping flower spike of certain trees ▸ **catnap** n, v doze ▸ **Catseyes®** pl n glass reflectors set in the road to indicate traffic lanes ▸ **cat's paw** person used by another to do unpleasant things for him or her ▸ **catwalk** n narrow platform where models display clothes in a fashion show

cataclysm [kat-**a**-kliz-zum] n violent upheaval; disaster, such as an earthquake ▸ **cataclysmic** adj

catacombs [kat-a-koomz] pl n underground burial place consisting of tunnels with recesses for tombs

catafalque [kat-a-falk] n raised platform on which a body lies in state before or during a funeral

catalepsy n trancelike state in which the body is rigid ▸ **cataleptic** adj

catalogue n book containing details of items for sale; systematic list of items ▸ v make a systematic list of

catalyst n substance that speeds up a chemical reaction without itself changing ▸ **catalyse** v speed up (a chemical reaction) by a catalyst ▸ **catalysis** n ▸ **catalytic** adj

catamaran n boat with twin parallel hulls

catapult n Y-shaped device with a loop of elastic, used by children for firing stones ▸ v shoot forwards or upwards violently

cataract n eye disease in which the lens becomes opaque; opaque area of an eye; large waterfall

catarrh [kat-**tar**] n excessive mucus in the nose and throat, during or following a cold ▸ **catarrhal** adj

catastrophe [kat-**ass**-trof-fee] n great and sudden disaster ▸ **catastrophic** adj

catch v **catching, caught** seize, capture; surprise in an act e.g. *two boys were caught stealing*; hit unexpectedly; be in time for (a bus, train, etc.); see or hear; be infected with (an illness); entangle; understand, make out ▸ n device for fastening a door, window, etc.; (informal) concealed or unforeseen drawback **catch it** (informal) be punished ▸ **catching** adj infectious ▸ **catchy** adj (of a tune) pleasant and easily remembered ▸ **catchcry** n (Aust) well-known phrase associated with a person, group, etc. ▸ **catchment area** area served by a particular school or hospital; area of land draining into a river, basin, or reservoir ▸ **catch on** v (informal) become popular; understand ▸ **catch out** v (informal) trap (someone) in an error or lie ▸ **catch phrase** well-known phrase associated with a particular entertainer ▸ **catch 22** inescapable dilemma ▸ **catchword** n well-known and frequently used phrase

catechism [kat-ti-kiz-zum] n instruction on the doctrine of a Christian Church in a series of questions and answers

category n, pl -**ries** class, group ▸ **categorical** adj absolutely clear and certain ▸ **categorically** adv ▸ **categorize** v put in a category ▸ **categorization** n

cater v provide what is needed or wanted, esp. food or services ▶ **caterer** n

caterpillar n wormlike larva of a moth or butterfly; **(C-)** ® endless track, driven by cogged wheels, used to propel a heavy vehicle

caterwaul v wail, yowl

catharsis [kath-**thar**-siss] n, pl -**ses** relief of strong suppressed emotions ▶ **cathartic** adj

cathedral n principal church of a diocese

Catherine wheel n rotating firework

catheter [kath-it-er] n tube inserted into a body cavity to drain fluid

cathode n negative electrode, by which electrons leave a circuit ▶ **cathode rays** stream of electrons from a cathode in a vacuum tube

catholic adj (of tastes or interests) covering a wide range ▶ n, adj **(C-)** (member) of the Roman Catholic Church ▶ **Catholicism** n

cation [kat-eye-on] n positively charged ion

cattle pl n domesticated cows and bulls

Caucasian n, adj (member) of the light-skinned racial group of humankind

caucus n, pl -**cuses** local committee or faction of a political party; political meeting to decide future plans

caught v past of **catch**

cauldron n large pot used for boiling

cauliflower n vegetable with a large head of white flower buds surrounded by green leaves

caulk v fill in (cracks) with paste

causal adj of or being a cause ▶ **causally** adv ▶ **causation,**

causality n relationship of cause and effect

cause n something that produces a particular effect; (foll by for) reason, motive; aim or principle supported by a person or group ▶ v be the cause of

cause célèbre [kawz sill-**leb**-ra] n, pl **causes célèbres** [kawz sill-**leb**-ra] controversial legal case or issue

causeway n raised path or road across water or marshland

caustic adj capable of burning by chemical action; bitter and sarcastic ▶ **caustically** adv

cauterize v burn (a wound) with heat or a caustic agent to prevent infection

caution n care, esp. in the face of danger; warning ▶ v warn, advise ▶ **cautionary** adj warning ▶ **cautious** adj showing caution ▶ **cautiously** adv

cavalcade n procession of people on horseback or in cars

cavalier adj showing haughty disregard ▶ n **(C-)** supporter of Charles I in the English Civil War

cavalry n, pl -**ries** part of the army orig. on horseback, but now often using fast armoured vehicles

cave n hollow in the side of a hill or cliff ▶ **caving** n sport of exploring caves ▶ **cave in** v collapse inwards; (informal) yield under pressure ▶ **caveman** n prehistoric cave dweller

caveat [kav-vee-at] n warning

cavern n large cave ▶ **cavernous** adj

caviar, caviare n salted sturgeon roe, regarded as a delicacy

cavil v -**illing**, -**illed** make petty objections ▶ n petty objection

cavity n, pl -**ties** hollow space; decayed area on a tooth

cavort v skip about

caw n cry of a crow, rook, or raven ▷ v make this cry

cayenne pepper, cayenne n hot red spice made from capsicum seeds

cayman n, pl **-mans** S American reptile similar to an alligator

CB Citizens' Band

CBE (in Britain) Commander of the Order of the British Empire

CBI Confederation of British Industry

cc cubic centimetre; carbon copy; (in S Africa) closed corporation

CD compact disc

CD-ROM compact disc read-only memory

CE Common Era

cease v bring or come to an end ▶ **ceaseless** adj ▶ **ceaselessly** adv ▶ **ceasefire** n temporary truce

cedar n evergreen coniferous tree; its wood

cede v surrender (territory or legal rights)

cedilla n character (,) placed under a c in some languages, to show that it is pronounced s, not k

ceilidh [kay-lee] n informal social gathering for singing and dancing, esp. in Scotland

ceiling n inner upper surface of a room; upper limit set on something

celandine n wild plant with yellow flowers

celebrate v hold festivities to mark (a happy event, anniversary, etc.); perform (a religious ceremony) ▶ **celebrated** adj well known ▶ **celebration** n ▶ **celebrant** n person who performs a religious ceremony ▶ **celebrity** n, pl **-rities** famous person; state of being famous

celeriac [sill-ler-ee-ak] n variety of celery with a large turnip-like root

celerity [sill-ler-rit-tee] n swiftness

celery n vegetable with long green crisp edible stalks

celestial adj heavenly, divine; of the sky

celibate adj unmarried or abstaining from sex, esp. because of a religious vow of chastity ▷ n celibate person ▶ **celibacy** n

cell n smallest unit of an organism that is able to function independently; small room for a prisoner, monk, or nun; small compartment of a honeycomb etc.; small group operating as the core of a larger organization; device that produces electrical energy by chemical reaction ▶ **cellular** adj or consisting of cells

cellar n underground room for storage; stock of wine

cello [chell-oh] n, pl **-los** large low-pitched instrument of the violin family ▶ **cellist** n

Cellophane® n thin transparent cellulose sheeting used as wrapping

cellulite n fat deposits under the skin alleged to resist dieting

celluloid n kind of plastic used to make toys and, formerly, photographic film

cellulose n main constituent of plant cell walls, used in making paper, plastics, etc.

Celsius adj of the temperature scale in which water freezes at 0° and boils at 100°

Celt [kelt, selt] n person from Scotland, Ireland, Wales, Cornwall, or Brittany

Celtic [kel-tik, sel-tik] n group of languages including Gaelic and Welsh ▷ adj of the Celts or the Celtic languages

cement n fine grey powder mixed with water and sand to make mortar or concrete; something that unites, binds, or joins;

material used to fill teeth ▷ v join, bind, or cover with cement; make (a relationship) stronger

cemetery n, pl **-teries** place where dead people are buried

cenotaph n monument honouring soldiers who died in a war

censer n container for burning incense

censor n person authorized to examine films, books, etc., to ban or cut anything considered obscene or objectionable ▷ v ban or cut parts of (a film, book, etc.) ▷ **censorship** n ▷ **censorious** adj harshly critical

censure n severe disapproval ▷ v criticize severely

census n, pl **-suses** official count of a population

cent n hundredth part of a monetary unit such as the dollar or euro

centaur n mythical creature with the head, arms, and torso of a man, and the lower body and legs of a horse

centenary [sen-teen-a-ree] n, pl **-naries** (chiefly Brit) 100th anniversary or its celebration ▷ **centenarian** n person at least 100 years old

centennial n 100th anniversary or its celebration

centi- prefix one hundredth

centigrade adj same as **Celsius**

> **USAGE NOTE**
> Scientists now use Celsius in preference to centigrade

centigram, centigramme n one hundredth of a gram

centilitre n one hundredth of a litre

centimetre n one hundredth of a metre

centipede n small wormlike creature with many legs

central adj of, at, or forming the centre; main, principal ▷ **centrally**

adv ▷ **centrality** n ▷ **centralism** n principle of central control of a country or organization ▷ **centralize** v bring under central control ▷ **centralization** n ▷ **central heating** system for heating a building from one central source of heat

centre n middle point or part; place for a specified activity; political party or group favouring moderation; (Sport) player who plays in the middle of the field ▷ v put in the centre of something ▷ **centred** adj emotionally stable and confident ▷ **centrist** n person favouring political moderation ▷ **centre on** v have as a centre or main theme

centrifugal adj moving away from a centre ▷ **centrifuge** n machine that separates substances by centrifugal force

centripetal adj moving towards a centre

centurion n (in ancient Rome) officer commanding 100 men

century n, pl **-ries** period of 100 years; cricket score of 100 runs

CEO chief executive officer

cephalopod [seff-a-loh-pod] n sea mollusc with a head and tentacles, such as the octopus

ceramic n hard brittle material made by heating clay to a very high temperature; object made of this ▷ pl art of producing ceramic objects ▷ adj made of ceramic

cereal n grass plant with edible grain, such as oat or wheat; this grain; breakfast food made from this grain, eaten mixed with milk

cerebral [ser-rib-ral, ser-reeb-ral] adj of the brain; intellectual

cerebrum [serr-rib-rum] n, pl **-brums** or **-bra** main part of the brain

ceremony *n, pl* **-nies** formal
act or ritual; formally polite
behaviour ▶ **ceremonial**
adj, n ▶ **ceremonious** *adj*
excessively polite or formal
▶ **ceremoniously** *adv*

cerise [ser-**reess**] *adj* cherry-red

certain *adj* positive and confident;
definite; some but not much
▶ **certainly** *adv* ▶ **certainty** *n* state
of being sure; *(pl* **-ties)** something
that is inevitable

certificate *n* official document
stating the details of a birth,
academic course, etc.

certify *v* **-fying, -fied** confirm,
attest to; guarantee; declare
legally insane ▶ **certifiable**
adj considered legally insane
▶ **certification** *n*

certitude *n* confidence, certainty

cervix *n, pl* **cervixes** *or* **cervices**
narrow entrance of the womb;
neck ▶ **cervical** *adj*

cessation *n* ceasing

cesspit, cesspool *n* covered tank
or pit for sewage

cetacean [sit-**tay**-shun] *n* fish-
shaped sea mammal such as a
whale or dolphin

cf compare

CFC chlorofluorocarbon

CGI computer-generated image(s)

ch. chapter; church

chafe *v* make sore or worn by
rubbing; be annoyed or impatient

chaff¹ *n* grain husks

chaff² *v (old-fashioned)* tease good-
naturedly

chaffinch *n* small European
songbird

chagrin [**shag**-grin] *n* annoyance
and disappointment ▶ **chagrined**
adj annoyed and disappointed

chain *n* flexible length of connected
metal links; series of connected

facts or events; group of shops,
hotels, etc. owned by one firm ▶ *v*
restrict or fasten with or as if with
a chain ▶ **chain reaction** series
of events, each of which causes
the next ▶ **chain-smoke** *v* smoke
(cigarettes) continuously ▶ **chain
smoker**

chair *n* seat with a back, for
one person; official position of
authority; person holding this;
professorship ▶ *v* preside over (a
meeting) ▶ **chairlift** *n* series of
chairs suspended from a moving
cable for carrying people up a
slope ▶ **chairman, chairwoman**
n person in charge of a company's
board of directors or a meeting
(also **chairperson**)

chaise [shaze] *n (Hist)* light horse-
drawn carriage

chaise longue [long] *n* couch with
a back and a single armrest

chalcedony [kal-**sed**-don-ee] *n, pl*
-nies variety of quartz

chalet [**shal**-ay] *n* kind of Swiss
wooden house with a steeply
sloping roof; similar house, used as
a holiday home

chalice *n* large goblet

chalk *n* soft white rock consisting of
calcium carbonate; piece of chalk,
often coloured, used for drawing
and writing on blackboards
▶ *v* draw or mark with chalk
▶ **chalky** *adj*

challenge *n* demanding or
stimulating situation; call to
take part in a contest or fight;
questioning of a statement of
fact; demand by a sentry for
identification or a password ▶ *v*
issue a challenge to ▶ **challenged**
adj disabled as specified e.g.
physically challenged ▶ **challenger** *n*

chamber *n* hall used for formal
meetings; legislative or judicial

assembly; (*old-fashioned*) bedroom; compartment, cavity ▷ *pl* set of rooms used as offices by a barrister ▶ **chambermaid** *n* woman employed to clean bedrooms in a hotel ▶ **chamber music** classical music to be performed by a small group of musicians ▶ **chamber pot** bowl for urine, formerly used in bedrooms

chamberlain *n* (*Hist*) officer who managed the household of a king or nobleman

chameleon [kam-**meal**-yon] *n* small lizard that changes colour to blend in with its surroundings

chamfer [**cham**-fer] *v* bevel the edge of

chamois [**sham**-wah] *n*, *pl* -**ois** small mountain antelope; [**sham**-ee] soft suede leather; piece of this, used for cleaning or polishing

chamomile [**kam**-mo-mile] *n* same as **camomile**

champ¹ *v* chew noisily **champ at the bit** (*informal*) be impatient to do something

champ² *n* short for **champion**

champagne *n* sparkling white French wine

champion *n* overall winner of a competition; (foll by *of*) someone who defends a person or cause ▷ *v* support ▷ *adj* (*dialect*) excellent ▶ **championship** *n*

chance *n* likelihood, probability; opportunity to do something; risk, gamble; unpredictable element that causes things to happen one way rather than another ▷ *v* risk, hazard ▶ **chancy** *adj* uncertain, risky

chancel *n* part of a church containing the altar and choir

chancellor *n* head of government in some European countries; honorary head of a university ▶ **chancellorship** *n*

Chancery *n* division of the British High Court of Justice

chandelier [shan-dill-**eer**] *n* ornamental light with branches and holders for several candles or bulbs

chandler *n* dealer, esp. in ships' supplies

change *n* becoming different; variety or novelty; different set, esp. of clothes; balance received when the amount paid is more than the cost of a purchase; coins of low value ▷ *v* make or become different; give and receive (something) in return; exchange (money) for its equivalent in a smaller denomination or different currency; put on other clothes; leave one vehicle and board another ▶ **changeable** *adj* changing often ▶ **changeling** *n* a child believed to have been exchanged by fairies for another

channel *n* band of broadcasting frequencies; means of access or communication; broad strait connecting two areas of sea; bed or course of a river, stream, or canal; groove ▷ *v*-**nelling**, -**nelled** direct or convey through a channel

chant *v* utter or sing (a slogan or psalm) ▷ *n* rhythmic or repetitious slogan; psalm that has a short simple melody with several words sung on one note

chanter *n* (on bagpipes) pipe on which the melody is played

chaos *n* complete disorder or confusion ▶ **chaotic** *adj* ▶ **chaotically** *adv*

chap *n* (*informal*) man or boy

chapati, chapatti *n* (in Indian cookery) flat thin unleavened bread

chapel *n* place of worship with its own altar, within a church; similar place of worship in a large house or institution; Nonconformist place of worship

chaperone [shap-per-rone] n older person who accompanies and supervises a young person or young person on a social occasion ▷ v act as a chaperone to

chaplain n clergyman attached to a chapel, military body, or institution ▸ **chaplaincy** n, pl -**cies**

chaplet n garland for the head

chapped adj (of the skin) raw and cracked, through exposure to cold

chapter n division of a book; period in a life or history; branch of a society or club

char[1] v **charring, charred** blacken by partial burning

char[2] (Brit informal) n charwoman ▷ v **charring, charred** clean other people's houses as a job

char[3] n (Brit old-fashioned slang) tea

charabanc [shar-rab-bang] n (old-fashioned) coach for sightseeing

character n combination of qualities distinguishing a person, group, or place; reputation, esp. good reputation; person represented in a play, film, or story; unusual or amusing person; letter, numeral, or symbol used in writing or printing ▸ **characteristic** n distinguishing feature or quality ▷ adj typical of ▸ **characteristically** adv ▸ **characterize** v be a characteristic of; (foll by as) describe ▸ **characterization** n

charade [shar-rahd] n absurd pretence ▷ pl game in which one team acts out a word or phrase, which the other team has to guess

charcoal n black substance formed by partially burning wood

charge v ask as a price; enter a debit against a person's account for (a purchase); accuse formally; make

a rush at or sudden attack upon; fill (a glass); fill (a battery) with electricity; command, assign ▷ n price charged; formal accusation; attack; command; exhortation; custody, guardianship; person or thing entrusted to someone's care; amount of electricity stored in a battery **in charge of** in control of ▸ **chargeable** adj ▸ **charger** n device for charging an accumulator; (in the Middle Ages) warhorse

chargé d'affaires [shar-zhay daf-fair] n, pl **chargés d'affaires** head of a diplomatic mission in the absence of an ambassador or in a small mission

chariot n two-wheeled horse-drawn vehicle used in ancient times in wars and races ▸ **charioteer** n chariot driver

charisma [kar-rizz-ma] n person's power to attract or influence people ▸ **charismatic** [kar-rizz-mat-ik] adj

charity n, pl -**ties** organization that gives help, such as money or food, to those in need; giving of help to those in need; help given to those in need; kindly attitude towards people ▸ **charitable** adj ▸ **charitably** adv

charlady n (Brit informal) same as **charwoman**

charlatan [shar-lat-tan] n person who claims expertise that he or she does not have

charleston n lively dance of the 1920s

charm n attractive quality; trinket worn on a bracelet; magic spell ▷ v attract, delight; influence by personal charm; protect or influence as if by magic ▸ **charmer** n ▸ **charming** adj attractive

charnel house n (Hist) building or vault for the bones of the dead

chart *n* graph, table, or diagram showing information; map of the sea or stars ▷ *v* plot the course of; make a chart of **the charts** (*informal*) weekly lists of the bestselling pop records

charter *n* document granting or demanding certain rights; fundamental principles of an organization; hire of transport for private use ▷ *v* hire by charter; grant a charter to ▶ **chartered** *adj* officially qualified to practise a profession

chartreuse [shar-**trerz**] *n* sweet-smelling green or yellow liqueur

charwoman *n* woman whose job is to clean other people's homes

chary [**chair**-ee] *adj* **-rier, -riest** wary, careful

chase¹ *v* run after quickly in order to catch or drive away; (*informal*) rush, run; (*informal*) try energetically to obtain ▷ *n* chasing, pursuit ▶ **chaser** *n* milder drink drunk after another stronger one

chase² *v* engrave or emboss (metal)

chasm [**kaz**-zum] *n* deep crack in the earth

chassis [**shass**-ee] *n*, *pl* **-sis** frame, wheels, and mechanical parts of a vehicle

chaste *adj* abstaining from sex outside marriage or altogether; (of style) simple ▶ **chastely** *adv* ▶ **chastity** *n*

chasten [**chase**-en] *v* subdue by criticism

chastise *v* scold severely; punish by beating ▶ **chastisement** *n*

chat *n* informal conversation ▷ *v* **chatting, chatted** have an informal conversation ▶ **chatty** *adj* ▶ **chatroom** *n* site on the internet where users have group discussions by e-mail

chateau [**shat**-toe] *n*, *pl* **-teaux** or **-teaus** French castle

chatelaine [**shat**-tell-lane] *n* (formerly) mistress of a large house or castle

chattels *pl n* possessions

chatter *v* speak quickly and continuously about unimportant things; (of the teeth) rattle with cold or fear ▷ *n* idle talk ▶ **chatterbox** *n* person who chatters a lot

chauffeur *n* person employed to drive a car for someone

chauvinism [**show**-vin-iz-zum] *n* irrational belief that one's own country, race, group, or sex is superior ▶ **chauvinist** *n*, *adj* ▶ **chauvinistic** *adj*

chav *n* (*Brit derogatory informal*) young working-class person considered to have vulgar tastes

cheap *adj* costing relatively little; of poor quality; not valued highly; mean, despicable ▶ **cheaply** *adv* ▶ **cheapen** *v* lower the reputation of; reduce the price of ▶ **cheapskate** *n* (*informal*) miserly person

cheat *v* act dishonestly to gain profit or advantage ▷ *n* person who cheats; fraud, deception

check *v* examine, investigate; slow the growth or progress of; correspond, agree ▷ *n* test to ensure accuracy or progress; break in progress; (*US*) cheque; pattern of squares or crossed lines; (*Chess*) position of a king under attack ▶ **check in** *v* register one's arrival ▶ **checkmate** *n* (*Chess*) winning position in which an opponent's king is under attack and unable to escape; utter defeat ▷ *v* (*Chess*) place the king of (one's opponent) in checkmate; thwart, defeat ▶ **check out** *v* pay the bill and leave a hotel; examine, investigate; (*informal*) have a look at ▸ **checkout** *n* counter in a supermarket, where customers pay ▸ **checkup** *n* thorough medical examination

Cheddar *n* firm orange or yellow-white cheese

cheek *n* either side of the face below the eye; (*informal*) impudence, boldness ▸ *v* (*Brit, Aust & NZ informal*) speak impudently to ▸ **cheeky** *adj* impudent, disrespectful ▸ **cheekily** *adv* ▸ **cheekiness** *n*

cheep *n* young bird's high-pitched cry ▸ *v* utter a cheep

cheer *v* applaud or encourage with shouts; make or become happy ▸ *n* shout of applause or encouragement ▸ **cheerful** *adj* ▸ **cheerfully** *adv* ▸ **cheerfulness** *n* ▸ **cheerless** *adj* dreary, gloomy ▸ **cheery** *adj* ▸ **cheerily** *adv*

cheerio (*informal*) *interj* goodbye ▸ *n* (*Aust & NZ*) small red cocktail sausage

cheese *n* food made from coagulated milk curd; block of this ▸ **cheesy** *adj* ▸ **cheeseburger** *n* hamburger topped with melted cheese ▸ **cheesecake** *n* dessert with a biscuit-crumb base covered with a sweet cream-cheese mixture; (*slang*) photographs of naked or near-naked women ▸ **cheesecloth** *n* light cotton cloth ▸ **cheesed off** bored, annoyed

cheetah *n* large fast-running spotted African wild cat

chef *n* cook in a restaurant

chef-d'oeuvre [shay-**durv**] *n, pl* **chefs-d'oeuvre** masterpiece

chemical *n* substance used in or resulting from a reaction involving changes to atoms or molecules ▸ *adj* of chemistry or chemicals ▸ **chemically** *adv*

chemise [shem-**meez**] *n* (*old-fashioned*) woman's loose-fitting slip

chemistry *n* science of the composition, properties, and reactions of substances ▸ **chemist** *n* shop selling medicines and cosmetics; qualified dispenser of prescribed medicines; specialist in chemistry

chemotherapy *n* treatment of disease, often cancer, using chemicals

chenille [shen-**neel**] *n* (fabric of) thick tufty yarn

cheque *n* written order to one's bank to pay money from one's account ▸ **cheque card** (*Brit*) plastic card issued by a bank guaranteeing payment of a customer's cheques

chequer *n* piece used in Chinese chequers ▸ *pl* game of draughts ▸ **chequered** *adj* marked by varied fortunes; having a pattern of squares

cherish *v* cling to (an idea or feeling); care for

cheroot [sher-**root**] *n* cigar with both ends cut flat

cherry *n, pl* **-ries** small red or black fruit with a stone; tree on which it grows ▸ *adj* deep red

cherub *n, pl* **-ubs** *or* **-ubim** angel, often represented as a winged child; sweet child ▸ **cherubic** [cher-**rew**-bik] *adj*

chervil *n* aniseed-flavoured herb

chess *n* game for two players with 16 pieces each, played on a chequered board of 64 squares ▸ **chessman** *n* piece used in chess

chest *n* front of the body, from neck to waist; large strong box ▸ **chest of drawers** piece of furniture consisting of drawers in a frame

chesterfield *n* couch with high padded sides and back

chestnut *n* reddish-brown edible nut; tree on which it grows; reddish-brown horse; (*informal*) old joke ▷ *adj* (of hair or a horse) reddish-brown

chevron [**shev**-ron] *n* V-shaped pattern, esp. on the sleeve of a military uniform to indicate rank

chew *v* grind (food) between the teeth ▶ **chewy** *adj* requiring a lot of chewing ▶ **chewing gum** flavoured gum to be chewed but not swallowed

chianti [kee-**ant**-ee] *n* dry red Italian wine

chiaroscuro [kee-ah-roh-**skew**-roh] *n*, *pl* -**ros** distribution of light and shade in a picture

chic [**sheek**] *adj* stylish, elegant ▷ *n* stylishness, elegance

chicane [shik-**kane**] *n* obstacle in a motor-racing circuit

chicanery *n* trickery, deception

chick *n* baby bird ▶ **chickpea** *n* edible yellow pealike seed ▶ **chickweed** *n* weed with small white flowers

chicken *n* domestic fowl; its flesh, used as food; (*slang*) coward ▷ *adj* (*slang*) cowardly ▶ **chicken feed** (*slang*) trifling amount of money ▶ **chicken out** *v* (*informal*) fail to do something through cowardice ▶ **chickenpox** *n* infectious disease with an itchy rash

chicory *n*, *pl* -**ries** plant whose leaves are used in salads; root of this plant, used as a coffee substitute

chide *v* chiding, chided *or* chid, chid *or* chidden rebuke, scold

chief *n* head of a group of people ▷ *adj* most important ▶ **chiefly** *adv* especially; mainly ▶ **chieftain** *n* leader of a tribe

chiffon [**shif**-fon] *n* fine see-through fabric

chignon [**sheen**-yon] *n* knot of hair pinned up at the back of the head

chihuahua [chee-**wah**-wah] *n* tiny short-haired dog

chilblain *n* inflammation of the fingers or toes, caused by exposure to cold

child *n*, *pl* **children** young human being, boy or girl; son or daughter ▶ **childhood** *n* ▶ **childish** *adj* immature, silly; of or like a child ▶ **childishly** *adv* ▶ **childless** *adj* ▶ **childlike** *adj* innocent, trustful ▶ **childbirth** *n* giving birth to a child ▶ **child's play** very easy task

USAGE NOTE
Note the difference between *childish* and *childlike*.

chill *n* feverish cold; moderate coldness ▷ *v* make (something) cool or cold; cause (someone) to feel cold or frightened ▷ *adj* unpleasantly cold ▶ **chilly** *adj* moderately cold; unfriendly ▶ **chilly-bin** (*NZ informal*) insulated container for carrying food and drink ▶ **chilliness** *n* ▶ **chill (out)** *v* (*informal*) relax ▶ **chill-out** *adj* (*informal*) suitable for relaxation, esp. after energetic activity

chilli, chili *n* small red or green hot-tasting capsicum pod, used in cooking; (also **chilli con carne**) hot-tasting Mexican dish of meat, onions, beans, and chilli powder

chime *n* musical ringing sound of a bell or clock ▷ *v* make a musical ringing sound; indicate (the time) by chiming; (foll by *with*) be consistent with

chimera [kime-**meer**-a] *n* unrealistic hope or idea; fabled monster with a lion's head, goat's body, and serpent's tail

chimney *n* hollow vertical structure for carrying away smoke from a fire ▸ **chimney pot** short pipe on the top of a chimney ▸ **chimney sweep** person who cleans soot from chimneys

chimp n (informal) short for **chimpanzee**

chimpanzee n intelligent black African ape

chin n part of the face below the mouth ▸ **chinwag** n (Brit, Aust & NZ informal) chat

china n fine earthenware or porcelain; dishes or ornaments made of this; (Brit, Aust, NZ & S Afr informal) friend

chinchilla n S American rodent bred for its soft grey fur; its fur

chine n cut of meat including part of the backbone

Chinese adj of China ▷ n (pl -nese) person from China; any of the languages of China

chink¹ n small narrow opening

chink² v, n (make) a light ringing sound

chintz n printed cotton fabric with a glazed finish

chip n strip of potato, fried in deep fat; tiny wafer of semiconductor material forming an integrated circuit; counter used to represent money in gambling games; small piece removed by chopping, breaking, etc.; mark left where a small piece has been broken off something ▷ v **chipping, chipped** break small pieces from ▸ **have a chip on one's shoulder** (informal) bear a grudge ▸ **chip in** v (informal) contribute (money); interrupt with a remark ▸ **chippie** n (Brit, Aust & NZ informal) carpenter ▸ **chip and PIN** n system in which a credit- or debit-card payment is validated by a customer entering a unique identification number instead of a signature

chipboard n thin board made of compressed wood particles

chipmunk n small squirrel-like N American rodent with a striped back

chiropodist [kir-rop-pod-ist] n person who treats minor foot complaints ▸ **chiropody** n

chiropractic [kire-oh-prak-tik] n system of treating bodily disorders by manipulation of the spine ▸ **chiropractor** n

chirp v (of a bird or insect) make a short high-pitched sound ▷ n chirping sound ▸ **chirpy** adj (informal) lively and cheerful

chisel n metal tool with a sharp end for shaping wood or stone ▷ v **-elling, -elled** carve with a chisel

chit¹ n short official note, such as a receipt

chit² n (Brit, Aust & NZ old-fashioned) pert or impudent girl

chitchat n chat, gossip

chitterlings pl n pig's intestines cooked as food

chivalry n courteous behaviour, esp. by men towards women; medieval systems and principles of knighthood ▸ **chivalrous** adj

chives pl n herbs with a mild onion flavour

chivvy v **-vying, -vied** (informal) harass, nag

chlorine n strong-smelling greenish-yellow gaseous element, used to disinfect water ▸ **chlorinate** v disinfect (water) with chlorine ▸ **chlorination** n ▸ **chloride** n compound of chlorine and another substance

chlorofluorocarbon n any of various gaseous compounds of carbon, hydrogen, chlorine, and fluorine, used in refrigerators and aerosol propellants, some of which break down the ozone in the atmosphere

chloroform n strong-smelling liquid formerly used as an anaesthetic

chlorophyll n green colouring matter of plants, which enables them to convert sunlight into energy

chock n block or wedge used to prevent a heavy object from moving ▷ **chock-full, chock-a-block** adj completely full

chocolate n sweet food made from cacao seeds; sweet or drink made from this ▷ adj dark brown

choice n choosing; opportunity or power of choosing; person or thing chosen or that may be chosen; alternative action or possibility ▷ adj of high quality

choir n organized group of singers, esp. in church; part of a church occupied by the choir

choke v hinder or stop the breathing of (a person) by strangling or smothering; have trouble in breathing; block, clog up ▷ n device controlling the amount of air that is mixed with the fuel in a petrol engine ▷ **choker** n tight-fitting necklace ▷ **choke back** v suppress (tears or anger)

cholera [kol-er-a] n serious infectious disease causing severe vomiting and diarrhoea

choleric [kol-ler-ik] adj bad-tempered

cholesterol [kol-lest-er-oll] n fatty substance found in animal tissue, an excess of which can cause heart disease

chomp v chew noisily

chook n (Aust & NZ informal) hen or chicken ▷ **chook raffle** n raffle for which the main prize is a roast chicken **not be able to run a chook raffle** to be totally incompetent

choose v choosing, chose, chosen select from a number of alternatives; decide (to do something) because one wants to ▷ **choosy** adj (informal) fussy

chop¹ v chopping, chopped cut with a blow from an axe or knife; cut into pieces; dispense with; (Boxing, Karate) hit (an opponent) with a short sharp blow ▷ n cutting or sharp blow; slice of lamb or pork, usu. with a rib ▷ **chopper** n (informal) helicopter; small axe ▷ **choppy** adj (of the sea) fairly rough

chop² v chopping, chopped ▷ **chop and change** change one's mind repeatedly

chops pl n (Brit, Aust & NZ informal) jaws, cheeks

chopsticks pl n pair of thin sticks used to eat Chinese food

chop suey n Chinese dish of chopped meat and vegetables in a sauce

choral adj of a choir

chorale [kor-rahl] n slow stately hymn tune

chord¹ n (Maths) straight line joining two points on a curve

chord² n simultaneous sounding of three or more musical notes

chore n routine task

choreography n composition of steps and movements for dancing ▷ **choreographer** n ▷ **choreographic** adj

chorister n singer in a choir

chortle v chuckle in amusement ▷ n amused chuckle

chorus n, pl **-ruses** large choir; part of a song repeated after each verse; something expressed by many people at once; group of singers or dancers who perform together in a show ▷ v **chorusing, chorused** sing or say together **in chorus** in unison

chose v past tense of **choose**
▶ **chosen** v past participle of **choose**

choux pastry [shoo] n very light pastry made with eggs

chow n thick-coated dog with a curled tail, orig. from China

chowder n thick soup containing clams or fish

chow mein n Chinese-American dish of chopped meat or vegetables fried with noodles

Christ n Jesus of Nazareth, regarded by Christians as the Messiah

christen v baptize; give a name to; (informal) use for the first time
▶ **christening** n

Christendom n all Christian people or countries

Christian n person who believes in and follows Christ ▷ adj of Christ or Christianity; kind, good
▶ **Christianity** n religion based on the life and teachings of Christ
▶ **Christian name** personal name given to Christians at baptism: loosely used to mean a person's first name ▶ **Christian Science** religious system which emphasizes spiritual healing

Christmas n annual festival on December 25 commemorating the birth of Christ; period around this time ▶ **Christmassy** adj
▶ **Christmas Day** December 25
▶ **Christmas Eve** December 24
▶ **Christmas tree** evergreen tree or imitation of one, decorated as part of Christmas celebrations

chromatic adj of colour or colours; (Music) (of a scale) proceeding by semitones

chromatography n separation and analysis of the components of a substance by slowly passing it through an adsorbing material

chromium, chrome n (Chem) grey metallic element used in steel alloys and for electroplating

chromosome n microscopic gene-carrying body in the nucleus of a cell

chronic adj (of an illness) lasting a long time; habitual e.g. chronic drinking; (Brit, Aust & NZ informal) of poor quality ▶ **chronically** adv

chronicle n record of events in order of occurrence ▷ v record in or as if in a chronicle ▶ **chronicler** n

chronology n, pl -gies arrangement or list of events in order of occurrence
▶ **chronological** adj
▶ **chronologically** adv

chronometer n timepiece designed to be accurate in all conditions

chrysalis [kriss-a-liss] n insect in the stage between larva and adult, when it is in a cocoon

chrysanthemum n flower with a large head made up of thin petals

chub n European freshwater fish

chubby adj -bier, -biest plump and round

chuck[1] v (informal) throw; (informal) give up, reject; touch (someone) affectionately under the chin; (Aust & NZ informal) vomit

chuck[2] n cut of beef from the neck to the shoulder; device that holds a workpiece in a lathe or a tool in a drill

chuckle v laugh softly ▷ n soft laugh

chuffed adj (informal) very pleased

chug n short dull sound like the noise of an engine ▷ v **chugging, chugged** operate or move with this sound

chukka n period of play in polo

chum (informal) n close friend ▷ v **chumming, chummed** ▶ **chum**

up with form a close friendship with ► **chummy** ▷ *adj*

chump *n* (*informal*) stupid person; thick piece of meat

chunk *n* thick solid piece; considerable amount ► **chunky** *adj* (of a person) broad and heavy; (of an object) large and thick

church *n* building for public Christian worship; particular Christian denomination; (**C-**) Christians collectively; clergy ► **churchgoer** *n* person who attends church regularly ► **churchwarden** *n* member of a congregation who assists the vicar ► **churchyard** *n* grounds round a church, used as a graveyard

churlish *adj* surly and rude

churn *n* machine in which cream is shaken to make butter; large container for milk ▷ *v* stir (cream) vigorously to make butter; move about violently ► **churn out** *v* (*informal*) produce (things) rapidly in large numbers

chute¹ [**shoot**] *n* steep slope down which things may be slid

chute² [**shoot**] *n* (*informal*) short for **parachute**

chutney *n* pickle made from fruit, vinegar, spices, and sugar

CIA (in the US) Central Intelligence Agency

cicada [sik-**kah**-da] *n* large insect that makes a high-pitched drone

cicatrix [sik-a-trix] *n*, *pl* **-trices** scar

CID (in Britain) Criminal Investigation Department

cider *n* alcoholic drink made from fermented apple juice

cigar *n* roll of cured tobacco leaves for smoking

cigarette *n* thin roll of shredded tobacco in thin paper, for smoking

cinch [sinch] *n* (*informal*) easy task

cinder *n* piece of material that will not burn, left after burning coal

cine camera *n* camera for taking moving pictures

cinema *n* place for showing films; films collectively ► **cinematic** *adj* ► **cinematography** *n* technique of making films ► **cinematographer** *n*

cineraria *n* garden plant with daisy-like flowers

cinnamon *n* spice obtained from the bark of an Asian tree

> **SPELLING TIP**
> The correct spelling of **cinnamon** has two *n*s in the middle and only one *m*

cipher [**sife**-er] *n* system of secret writing; unimportant person

circa [**sir**-ka] *prep* (*Latin*) approximately, about

circle *n* perfectly round geometric figure, line, or shape; group of people sharing an interest or activity; (*Theatre*) section of seats above the main level of the auditorium ▷ *v* move in a circle (round); enclose in a circle

circlet *n* circular ornament worn on the head

circuit *n* complete route or course, esp. a circular one; complete path through which an electric current can flow; periodical journey round a district, as made by judges; motor-racing track ► **circuitous** [sir-**kew**-it-uss] *adj* indirect and lengthy ► **circuitry** [**sir**-kit-tree] *n* electrical circuit(s)

circular *adj* in the shape of a circle; moving in a circle ▷ *n* letter for general distribution ► **circularity** *n*

circulate *v* send, go, or pass from place to place or person to person ► **circulation** *n* flow of blood around the body; number of copies of a newspaper or magazine

sold; sending or moving round
▶ **circulatory** adj

circumcise v remove the foreskin of ▶ **circumcision** n

circumference n boundary of a specified area or shape, esp. of a circle; distance round this

circumflex n mark (^) over a vowel to show that it is pronounced in a particular way

circumlocution n indirect way of saying something

circumnavigate v sail right round ▶ **circumnavigation** n

circumscribe v limit, restrict; draw a line round ▶ **circumscription** n

circumspect adj cautious and careful not to take risks ▶ **circumspectly** adv ▶ **circumspection** n

circumstance n (usu. pl) occurrence or condition that accompanies or influences a person or event ▶ **circumstantial** adj (of evidence) strongly suggesting something but not proving it; very detailed

circumvent v avoid or get round (a rule etc.) ▶ **circumvention** n

circus n, pl **-cuses** (performance given by) a travelling company of acrobats, clowns, performing animals, etc.

cirrhosis [sir-roh-siss] n serious liver disease, often caused by drinking too much alcohol

cirrus n, pl **-ri** high wispy cloud

cistern n water tank, esp. one that holds water for flushing a toilet

citadel n fortress in a city

cite v quote, refer to; bring forward as proof ▶ **citation** n

citizen n native or naturalized member of a state or nation; inhabitant of a city or town ▶ **citizenship** n ▶ **Citizens' Band** range of radio frequencies for private communication by the public

citric acid n weak acid found in citrus fruits

citrus fruit n juicy sharp-tasting fruit such as an orange or lemon

city n, pl **-ties** large or important town; **the City** (Brit) area of London as a financial centre

civet [siv-vit] n spotted catlike African mammal; musky fluid from its glands used in perfume

civic adj of a city or citizens ▶ **civics** n study of the rights and responsibilities of citizenship

civil adj relating to the citizens of a state as opposed to the armed forces or the Church; polite, courteous ▶ **civilly** adv ▶ **civility** n polite or courteous behaviour ▶ **civilian** n, adj (person) not belonging to the armed forces ▶ **civil servant** member of the civil service ▶ **civil service** service responsible for the administration of the government ▶ **civil war** war between people of the same country

civilize v refine or educate (a person); make (a place) more pleasant or more acceptable ▶ **civilization** n high level of human cultural and social development; particular society which has reached this level

civvies pl n (Brit, Aust & NZ slang) ordinary clothes that are not part of a uniform

clack n sound made by two hard objects striking each other ▶ v make this sound

clad v a past of **clothe**

cladding n material used to cover the outside of a building

claim v assert as a fact; demand as a right; need, require ▶ n assertion that something is true; assertion of a right; something claimed as a right ▶ **claimant** n

clairvoyance n power of perceiving things beyond the natural range of the senses ▶ **clairvoyant** n, adj

clam n edible shellfish with a hinged shell ▷ v **clamming, clammed** ▶ **clam up** (informal) stop talking, esp. through nervousness

clamber v climb awkwardly

clammy adj **-mier, -miest** unpleasantly moist and sticky

clamour n loud protest; loud persistent noise or outcry ▷ v make a loud noise or outcry ▶ **clamorous** adj ▶ **clamour for** v demand noisily

clamp n tool with movable jaws for holding things together tightly ▷ v fasten with a clamp ▶ **clamp down on** v become stricter about; suppress

clan n group of families with a common ancestor, esp. among Scottish Highlanders; close group ▶ **clannish** adj (of a group) tending to exclude outsiders

clandestine adj secret and concealed

clang v make a loud ringing metallic sound ▷ n ringing metallic sound

clanger n (informal) obvious mistake

clangour n loud continuous clanging sound

clank n harsh metallic sound ▷ v make such a sound

clap[1] v **clapping, clapped** applaud by hitting the palms of one's hands sharply together; put quickly or forcibly ▷ n act or sound of clapping; sudden loud noise e.g. a clap of thunder ▶ **clapped out** (slang) worn out, dilapidated

clap[2] n (slang) gonorrhoea

clapper n piece of metal inside a bell, which causes it to sound

when struck against the side ▶ **clapperboard** n pair of hinged boards clapped together during filming to help in synchronizing sound and picture

claptrap n (informal) foolish or pretentious talk

claret [klar-rit] n dry red wine from Bordeaux

clarify v **-fying, -fied** make (a matter) clear and unambiguous ▶ **clarification** n

clarinet n keyed woodwind instrument with a single reed ▶ **clarinettist** n

clarion n obsolete high-pitched trumpet; its sound ▶ **clarion call** strong encouragement to do something

clarity n clearness

clash v come into conflict; (of events) happen at the same time; (of colours) look unattractive together; (of objects) make a loud harsh sound by being hit together ▷ n fight, argument; fact of two events happening at the same time

clasp n device for fastening things; firm grasp or embrace ▷ v grasp or embrace firmly; fasten with a clasp

class n group of people sharing a similar social position; system of dividing society into such groups; group of people or things sharing a common characteristic; group of pupils or students taught together; standard of quality; (informal) elegance or excellence e.g. a touch of class ▷ v place in a class

classic adj being a typical example of something; of lasting interest because of excellence; attractive because of simplicity of form ▷ n author, artist, or work of art of recognized excellence ▷ pl study of ancient Greek and Roman literature and culture ▶ **classical** adj of or in a restrained conservative style; denoting serious art music; of or influenced by ancient Greek and Roman culture ▶ **classically** adv ▶ **classicism** n artistic style showing emotional restraint and regularity of form ▶ **classicist** n

classify v -**fying, -fied** divide into groups with similar characteristics; declare (information) to be officially secret ▶ **classifiable** adj ▶ **classification** n

classy adj **classier, classiest** (informal) stylish and elegant

clatter v, n (make) a rattling noise

clause n section of a legal document; part of a sentence, containing a verb

claustrophobia n abnormal fear of confined spaces ▶ **claustrophobic** adj

clavichord n early keyboard instrument

clavicle n same as **collarbone**

claw n sharp hooked nail of a bird or beast; similar part, such as a crab's pincer ▷ v tear with claws or nails

clay n fine-grained earth, soft when moist and hardening when baked, used to make bricks and pottery ▶ **clayey** adj ▶ **clay pigeon** baked clay disc hurled into the air as a target for shooting

claymore n large two-edged sword formerly used by Scottish Highlanders

clean adj free from dirt or impurities; not yet used; morally acceptable, inoffensive; (of a reputation or record) free from dishonesty or corruption; complete e.g. a clean break; smooth and regular ▷ v make (something) free from dirt ▷ adv (not standard) completely e.g. I clean forgot

come clean (informal) reveal or admit something ▶ **cleaner** n ▶ **cleanly** adv ▶ **cleanliness** n ▶ **clean technology, clean-tech** n manufacturing processes that minimize damage caused to the environment ▷ adj using clean technology

cleanse v make clean ▶ **cleanser** n

clear adj free from doubt or confusion; easy to see or hear; able to be seen through; free of obstruction; (of weather) free from clouds; (of skin) without blemish ▷ adv out of the way ▷ v make or become clear; pass by or over (something) without contact; prove (someone) innocent of a crime or mistake; make as profit ▶ **clearly** adv ▶ **clearance** n clearing; official permission ▶ **clearing** n treeless area in a wood ▶ **clear off** (Brit, Aust & NZ informal) go away ▶ **clear out** v remove and sort the contents of; (Brit, Aust & NZ informal) go away ▶ **clear-sighted** adj having good judgment ▶ **clearway** n stretch of road on which motorists may stop in an emergency

cleat n wedge; piece of wood, metal, or plastic with two projecting ends round which ropes are fastened

cleave¹ v **cleaving, cleft, cleaved** or **clove, cleft, cleaved** or **cloven** split apart ▶ **cleavage** n space between a woman's breasts, as revealed by a low-cut dress; division, split

cleave² v cling or stick

cleaver n butcher's heavy knife with a square blade

clef n (Music) symbol at the beginning of a stave to show the pitch

cleft n narrow opening or crack ▷ v a past of **cleave**¹; **in a cleft stick** in a very difficult position

clematis n climbing plant with large colourful flowers

clement adj (of weather) mild ▶ **clemency** n kind or lenient treatment

clementine n small orange citrus fruit

clench v close or squeeze (one's teeth or fist) tightly; grasp firmly

clerestory [clear-store-ee] n, pl **-ries** row of windows at the top of a wall above an adjoining roof

clergy n priests and ministers as a group ▶ **clergyman** n

cleric n member of the clergy

clerical adj of clerks or office work; of the clergy

clerk n employee in an office, bank, or court who keeps records, files, and accounts

clever adj intelligent, quick at learning; showing skill ▶ **cleverly** adv ▶ **cleverness** n

clianthus [klee-anth-us] n Australian or NZ plant with slender scarlet flowers

cliché [klee-shay] n expression or idea that is no longer effective because of overuse ▶ **clichéd** adj

click n short sharp sound ▶ v make this sound; (informal) (of two people) get on well together; (informal) become suddenly clear; (Computing) press and release (a button on a mouse); (slang) be a success

client n person who uses the services of a professional person or company; (Computing) program or work station that requests data from a server ▶ **clientele** [klee-on-tell] n clients collectively

cliff n steep rock face, esp. along the sea shore ▶ **cliffhanger** n film, game, etc., that is tense and exciting whose outcome is uncertain

climate n typical weather conditions of an area ▶ **climatic** adj

climax n most intense point of an experience, series of events, or story ▶ **climactic** adj

climb v go up, ascend; rise to a higher point or intensity ▶ n climbing; place to be climbed ▶ **climber** n ▶ **climb down** v retreat from an opinion or position

clime n (poetic) place or its climate

clinch v settle (an argument or agreement) decisively ▶ **clincher** n (informal) something decisive

cling v **clinging, clung** hold tightly or stick closely ▶ **clingfilm** n thin polythene material for wrapping food

clinic n building where outpatients receive medical treatment or advice; private or specialized hospital ▶ **clinical** adj of a clinic; logical and unemotional ▶ **clinically** adv

clink v, n (make) a light sharp metallic sound

clink n (Brit, Aust & NZ slang) prison

clinker n fused coal left over in a fire or furnace

clinker-built adj (of a boat) made of overlapping planks

clip v **clipping, clipped** cut with shears or scissors; (informal) hit sharply ▶ n short extract of a film; (informal) sharp blow ▶ **clippers** pl n tool for clipping ▶ **clipping** n something cut out, esp. an article from a newspaper

clip n device for attaching or holding things together ▶ v **clipping, clipped** attach or hold together with a clip

clipper n fast commercial sailing ship

clique [kleek] n small exclusive group

clitoris [klit-or-iss] *n* small sexually sensitive organ at the front of the vulva ▷ **clitoral** *adj*

cloak *n* loose sleeveless outer garment ▷ *v* cover or conceal ▷ **cloakroom** *n* room where coats may be left temporarily

clobber¹ *v* (*informal*) hit; defeat utterly

clobber² *n* (*Brit, Aust & NZ informal*) belongings, esp. clothes

cloche [klosh] *n* cover to protect young plants; woman's close-fitting hat

clock *n* instrument for showing the time; device with a dial for recording or measuring ▷ **clockwise** *adv, adj* in the direction in which the hands of a clock rotate ▷ **clock in, clock on** *v* register arrival at work on an automatic time recorder ▷ **clock off, clock out** *v* register departure from work ▷ **clock up** *v* reach (a total) ▷ **clockwork** *n* mechanism similar to the kind in a clock, used in wind-up toys

clod *n* lump of earth; (*Brit, Aust & NZ*) stupid person

clog *v* **clogging, clogged** obstruct ▷ *n* wooden or wooden-soled shoe

cloister *n* covered pillared arcade, usu. in a monastery ▷ **cloistered** *adj* sheltered

clone *n* animal or plant produced artificially from the cells of another animal or plant, and identical to the original; (*informal*) person who closely resembles another ▷ *v* produce as a clone

close¹ *v* [rhymes with **nose**] shut; prevent access to; end; terminate; bring or come nearer together ▷ *n* end; conclusion; [rhymes with **dose**] street closed at one end; [rhymes with **dose**] (*Brit*)

courtyard, quadrangle ▷ **closed shop** place of work in which all workers must belong to a particular trade union

close² *adj* [rhymes with **dose**] near; intimate; careful, thorough; compact, dense; oppressive, stifling; secretive ▷ *adv* closely, tightly ▷ **closely** *adv* ▷ **closeness** *n* ▷ **close season** period when it is illegal to kill certain game or fish ▷ **close shave** (*informal*) narrow escape ▷ **close-up** *n* photograph or film taken at close range

closet *n* (*US*) cupboard; small private room ▷ *adj* private, secret ▷ *v* **closeting, closeted** shut (oneself) away in private

closure *n* closing

clot *n* soft thick lump formed from liquid; (*Brit, Aust & NZ informal*) stupid person ▷ *v* **clotting, clotted** form soft thick lumps

cloth *n* (piece of) woven fabric

clothe *v* **clothing, clothed** or **clad** put clothes on; provide with clothes ▷ **clothes** *pl n* articles of dress; bed coverings ▷ **clothing** *n* clothes collectively

cloud *n* mass of condensed water vapour floating in the sky; floating mass of smoke, dust, etc.; (*Computing*) internet server used to store data and services ▷ *v* (foll by *over*) become cloudy; confuse; make gloomy or depressed ▷ *adj* of or relating to cloud computing ▷ **cloudless** *adj* ▷ **cloudy** *adj* having a lot of clouds; (of liquid) not clear ▷ **cloudburst** *n* heavy fall of rain ▷ **cloud computing** *n* model of computer use in which internet services are provided to users on a temporary basis

clout (*informal*) *n* hard blow; power, influence ▷ *v* hit hard

clove[1] n dried flower bud of a tropical tree, used as a spice

clove[2] n segment of a bulb of garlic

clove[3] v a past tense of **cleave**[1] ▶ **clove hitch** knot used to fasten a rope to a spar

cloven v a past participle of **cleave**[1] ▶ **cloven hoof** divided hoof of a cow, goat, etc.

clover n plant with three-lobed leaves **in clover** in luxury

clown n comic entertainer in a circus; amusing person; stupid person ▶ v behave foolishly; perform as a clown ▶ **clownish** adj ▶ **clownfish** n brightly coloured striped fish

club n association of people with common interests; building used by such a group; thick stick used as a weapon; stick with a curved end used to hit the ball in golf; playing card with black three-leaved symbols ▶ v **clubbing**, **clubbed** hit with a club ▶ **club together** combine resources for a common purpose

club foot n deformity of the foot causing inability to put the foot flat on the ground

cluck n low clicking noise made by a hen ▶ v make this noise

clue n something that helps to solve a mystery or puzzle **not have a clue** to be completely baffled ▶ **clueless** adj stupid

clump n small group of things or people; dull heavy tread ▶ v walk heavily; form into clumps

clumsy adj -**sier**, -**siest** lacking skill or physical coordination; badly made or done ▶ **clumsily** adv ▶ **clumsiness** n

clung v past of **cling**

clunk n dull metallic sound ▶ v make such a sound

cluster n small close group ▶ v gather in clusters

clutch[1] v grasp tightly; (foll by at) try to get hold of ▶ n device enabling two revolving shafts to be connected and disconnected, esp. in a motor vehicle; tight grasp

clutch[2] n set of eggs laid at the same time

clutter v scatter objects about (a place) untidily ▶ n untidy mess

cm centimetre

CND Campaign for Nuclear Disarmament

CO Commanding Officer

Co. Company; County

co- prefix together, joint, or jointly e.g. coproduction

c/o care of; (Book-keeping) carried over

coach n long-distance bus; railway carriage; large four-wheeled horse-drawn carriage; trainer, instructor ▶ v train, teach

coagulate [koh-**ag**-yew-late] v change from a liquid to a semisolid mass ▶ **coagulation** n ▶ **coagulant** n substance causing coagulation

coal n black rock consisting mainly of carbon, used as fuel ▶ **coalfield** n area with coal under the ground

coalesce [koh-a-**less**] v come together, merge ▶ **coalescence** n

coalition [koh-a-**lish**-un] n temporary alliance, esp. between political parties

coarse adj rough in texture; unrefined, indecent ▶ **coarsely** adv ▶ **coarseness** n ▶ **coarsen** v ▶ **coarse fish** any freshwater fish not of the salmon family

coast n place where the land meets the sea ▶ v move by momentum, without the use of power ▶ **coastal** adj ▶ **coaster** n small mat placed underneath a glass ▶ **coastguard** n organization that aids ships and swimmers in trouble and prevents smuggling; member of this ▶ **coastline** n outline of a coast

coat n outer garment with long sleeves; animal's fur or hair; covering that is applied e.g. *a coat of paint* ▷ v cover with a layer ▶ **coating** n covering layer ▶ **coat of arms** heraldic emblem of a family or institution

coax v persuade gently; obtain by persistent coaxing

coaxial [koh-ax-ee-al] adj (of a cable) transmitting by means of two concentric conductors separated by an insulator

cob n stalk of an ear of maize; thickset type of horse; round loaf of bread; male swan

cobalt n (Chem) brittle silvery-white metallic element

cobber n (Aust & NZ, old-fashioned informal) friend

cobble n cobblestone ▶ **cobblestone** n rounded stone used for paving ▶ **cobble together** v put together clumsily

cobbler n shoe mender

cobia [koh-bee-a] n large dark-striped game fish of tropical and subtropical seas

cobra n venomous hooded snake of Asia and Africa

cobweb n spider's web

cocaine n addictive drug used as a narcotic and as an anaesthetic

coccyx [kok-six] n, pl coccyges [kok-sije-eez] bone at the base of the spinal column

cochineal n red dye obtained from a Mexican insect, used for food colouring

cock n male bird, esp. of domestic fowl; stopcock ▷ v draw back (the hammer of a gun) to firing position; lift and turn (part of the body) ▶ **cockerel** n young

domestic cock ▶ **cock-a-hoop** adj (Brit, Aust & NZ) in high spirits ▶ **cock-and-bull story** highly improbable story

cockade n feather or rosette worn on a hat as a badge

cockatiel, cockateel n crested Australian parrot with a greyish-brown and yellow plumage

cockatoo n crested parrot of Australia or the East Indies

cocker spaniel n small spaniel

cockeyed adj (informal) crooked, askew; foolish, absurd

cockie, cocky n, pl -kies (Aust & NZ informal) farmer

cockle n edible shellfish

Cockney n native of the East End of London; London dialect

cockpit n pilot's compartment in an aircraft; driver's compartment in a racing car

cockroach n beetle-like insect which is a household pest

cocksure adj overconfident, arrogant

cocktail n mixed alcoholic drink; appetizer of seafood or mixed fruits

cocky adj cockier, cockiest conceited and overconfident ▶ **cockily** adv ▶ **cockiness** n

cocoa n powder made from the seed of the cacao tree; drink made from this powder

coconut n large hard fruit of a type of palm tree; edible flesh of this fruit

cocoon n silky protective covering of a silkworm; protective covering ▷ v wrap up tightly for protection

COD cash on delivery

cod n large food fish of the North Atlantic; any other Australian fish of the same family

coda n final part of a musical composition

coddle v pamper, overprotect

code n system of letters, symbols, or prearranged signals by which messages can be communicated secretly or briefly; set of principles or rules ▷ v put into code; write computer programs ▶ **codify** v -**fying**, -**fied** organize (rules or procedures) systematically ▶ **codification** n

codeine [kode-een] n drug used as a painkiller

codex n, pl **codices** volume of manuscripts of an ancient text

codger n (Brit, Aust & NZ informal) old man

codicil [kode-iss-ill] n addition to a will

coeducation n education of boys and girls together ▶ **coeducational** adj

coefficient n (Maths) number or constant placed before and multiplying a quantity

coelacanth [seel-a-kanth] n primitive marine fish

coeliac disease [seel-ee-ak] n disease which hampers the digestion of food containing gluten

coerce [koh-urss] v compel, force ▶ **coercion** n ▶ **coercive** adj

coeval [koh-eev-al] adj, n contemporary

coexist v exist together, esp. peacefully despite differences ▶ **coexistence** n

C of E Church of England

coffee n drink made from the roasted and ground seeds of a tropical shrub; beanlike seeds of this shrub ▷ adj medium-brown ▶ **coffee bar** café, snack bar ▶ **coffee table** small low table

coffer n chest for valuables ▷ pl store of money

coffin n box in which a corpse is buried or cremated

cog n one of the teeth on the rim of a gearwheel; unimportant person in a big organization

cogent [koh-jent] adj forcefully convincing ▶ **cogency** n

cogitate [koj-it-tate] v think deeply about ▶ **cogitation** n

cognac [kon-yak] n French brandy

cognate adj derived from a common original form

cognition n act or experience of knowing or acquiring knowledge ▶ **cognitive** adj

cognizance n knowledge, understanding ▶ **cognizant** adj

cognoscenti [kon-yo-shen-tee] pl n connoisseurs

cohabit v live together without being married ▶ **cohabitation** n

cohere v hold or stick together; be logically connected or consistent

coherent adj logical and consistent; capable of intelligible speech ▶ **coherence** n ▶ **coherently** adv

cohesion n sticking together

cohesive adj sticking together to form a whole

cohort n band of associates; tenth part of an ancient Roman legion

coiffure n hairstyle ▶ **coiffeur, coiffeuse** n hairdresser

coil v wind in loops; move in a winding course ▷ n something coiled; single loop of this; coil-shaped contraceptive device inserted in the womb

coin n piece of metal money; metal currency collectively ▷ v invent (a word or phrase) **coin it in** (informal) earn money quickly ▶ **coinage** n coins collectively; word or phrase coined; coining

coincide v happen at the same time; agree or correspond exactly ▶ **coincidence** n occurrence of simultaneous or apparently connected events; coinciding ▶ **coincident** adj in agreement ▶ **coincidental** adj resulting from coincidence ▶ **coincidentally** adv

coir n coconut fibre, used for matting

coitus [koh-it-uss], **coition** [koh-ish-un] n sexual intercourse ▶ **coital** adj

coke¹ n solid fuel left after gas has been distilled from coal

coke² n (slang) cocaine

col n high mountain pass

cola n dark brown fizzy soft drink

colander n perforated bowl for straining or rinsing foods

cold adj lacking heat; lacking affection or enthusiasm; (of a colour) giving an impression of coldness; (slang) unconscious e.g. out cold ▷ n lack of heat; mild illness causing a runny nose, sneezing, and coughing ▶ **coldly** adv ▶ **coldness** n ▶ **cold-blooded** adj cruel, unfeeling; having a body temperature that varies according to the surrounding temperature ▶ **cold cream** creamy preparation for softening and cleansing the skin ▶ **cold feet** (slang) nervousness, fear ▶ **cold-shoulder** v treat with indifference ▶ **cold war** political hostility between countries without actual warfare

coleslaw n salad dish of shredded raw cabbage in a dressing

coley n codlike food fish of the N Atlantic

colic n severe pains in the stomach and bowels ▶ **colicky** adj

colitis [koh-lie-tiss] n inflammation of the colon

collaborate v work with another on a project; cooperate with an enemy invader ▶ **collaboration** n ▶ **collaborative** adj ▶ **collaborator** n

collage [kol-lahzh] n art form in which various materials or objects are glued onto a surface; picture made in this way

collapse v fall down suddenly; fail completely; fold compactly ▷ n collapsing; sudden failure or breakdown ▶ **collapsible** adj

collar n part of a garment round the neck; band put round an animal's neck; cut of meat from an animal's neck ▷ v (Brit, Aust & NZ informal) seize, arrest; catch in order to speak to ▶ **collarbone** n bone joining the shoulder blade to the breastbone

collate v gather together, examine, and put in order ▶ **collation** n collating; light meal

collateral n security pledged for the repayment of a loan

colleague n fellow worker, esp. in a profession

collect¹ v gather together; accumulate (stamps etc.) as a hobby; fetch ▶ **collected** adj calm and controlled ▶ **collection** n things collected; collecting; sum of money collected ▶ **collector** n

collect² n short prayer

collective adj or done by a group ▷ n group of people working together on an enterprise and sharing the benefits from it ▶ **collectively** adv

colleen n (Irish) girl

college n place of higher education; group of people of the same profession or with special duties ▶ **collegiate** adj

collide v crash together violently; have an argument ▶ **collision** n

collie n silky-haired sheepdog

collier n coal miner; coal ship ▶ **colliery** n, pl -lieries coal mine

collocate v (of words) occur together regularly ▶ **collocation** n

colloid n suspension of particles in a solution

colloquial adj suitable for informal speech or writing ▶ **colloquialism** n colloquial word or phrase

collusion n secret or illegal cooperation ▶ **collude** v act in collusion

collywobbles pl n (slang) nervousness

cologne n mild perfume

colon¹ n punctuation mark (:)

colon² n part of the large intestine connected to the rectum

colonel n senior commissioned army or air-force officer

colonnade n row of columns

colony n, pl **-nies** group of people who settle in a new country but remain under the rule of their homeland; territory occupied by a colony; group of people or animals of the same kind living together ▶ **colonial** adj, n (inhabitant) of a colony ▶ **colonialism** n policy of acquiring and maintaining colonies ▶ **colonist** n settler in a colony ▶ **colonize** v make into a colony ▶ **colonization** n

Colorado beetle n black-and-yellow beetle that is a serious pest of potatoes

coloration n arrangement of colours

colossal adj very large

colossus n, pl **-si** or **-suses** huge statue; huge or important person or thing

colostomy n, pl **-mies** operation to form an opening from the colon onto the surface of the body, for emptying the bowel

colour n appearance of things as a result of reflecting light; substance that gives colour; complexion ▷ pl flag of a country or regiment; (Sport) badge or symbol denoting

membership of a team ▷ v apply colour to; influence (someone's judgment); blush ▶ **coloured** adj having colour; (C-) (in S Africa) of mixed White and non-White parentage ▶ **colourful** adj with bright or varied colours; vivid, distinctive ▶ **colourfully** adv ▶ **colourless** adj ▶ **colour-blind** adj unable to distinguish between certain colours

colt n young male horse

columbine n garden flower with five petals

column n pillar; vertical division of a newspaper page; regular feature in a newspaper; vertical arrangement of numbers; narrow formation of troops ▶ **columnist** n journalist who writes a regular feature in a newspaper

coma n state of deep unconsciousness ▶ **comatose** adj in a coma; sound asleep

comb n toothed implement for arranging the hair; cock's crest; honeycomb ▷ v use a comb on; search with great care

combat n, v **-bating, -bated** fight, struggle ▶ **combatant** n ▶ **combative** adj ▶ **combat trousers, combats** loose casual trousers with large pockets on the legs

combine v join together ▷ n association of people or firms for a common purpose ▶ **combination** n combining; people or things combined; set of numbers that opens a special lock ▷ pl (Brit) old-fashioned undergarment with long sleeves and long legs ▶ **combine harvester** machine that reaps and threshes grain in one process

combustion n process of burning ▶ **combustible** adj burning easily

come v **coming, came, come** move towards a place, arrive; occur; reach a specified point or condition; be produced; (foll by *from*) be born in; become e.g. *a dream come true* ▶ **come across** v meet or find by accident; (often foll by *as*) give an impression of (being) ▶ **comeback** n (*informal*) return to a former position; retort ▶ **comedown** n decline in status; disappointment ▶ **comeuppance** n (*informal*) deserved punishment

comedy n, pl -**dies** humorous play, film, or programme ▶ **comedian, comedienne** n entertainer who tells jokes; person who performs in comedy

comely adj -**lier, -liest** (*old-fashioned*) nice-looking

comestibles pl n (*formal*) food

comet n heavenly body with a long luminous tail

comfit n (*old-fashioned*) sugar-coated sweet

comfort n physical ease or wellbeing; consolation; means of consolation ▷ v soothe, console ▶ **comfortable** adj giving comfort; free from pain; (*informal*) well-off financially ▶ **comfortably** adv ▶ **comforter** n ▶ **comfort food** simple food that makes the eater feel better emotionally

comfrey n tall plant with bell-shaped flowers

comfy adj -**fier, -fiest** (*informal*) comfortable

comic adj humorous, funny; of comedy ▷ n comedian; magazine containing strip cartoons ▶ **comical** adj amusing ▶ **comically** adv

comma n punctuation mark (,)

command v order; have authority over; deserve and get; look down over ▷ n authoritative

instruction that something must be done; authority to command; knowledge; military or naval unit with a specific function ▶ **commandant** n officer commanding a military group ▶ **commandeer** v seize for military use ▶ **commandment** n command from God

commander n military officer in command of a group or operation; middle-ranking naval officer ▶ **commander-in-chief** n, pl **commanders-in-chief** supreme commander of a nation's armed forces

commando n, pl -**dos** or -**does** (member of) a military unit trained for swift raids in enemy territory

commemorate v honour the memory of ▶ **commemoration** n ▶ **commemorative** adj

> **SPELLING TIP**
> Remember that **commemorate** has only three *m*s; a double *m* shortly followed by a single

commence v begin ▶ **commencement** n

commend v praise; recommend ▶ **commendable** adj ▶ **commendably** adv ▶ **commendation** n

commensurable adj measurable by the same standards

commensurate adj corresponding in degree, size, or value

comment n remark; talk, gossip; explanatory note ▷ v make a comment ▶ **commentary** n, pl -**taries** spoken accompaniment to a broadcast or film; explanatory notes ▶ **commentate** v provide a commentary ▶ **commentator** n

commerce n buying and selling, trade ▶ **commercial** adj of commerce; (of television or radio) paid for by advertisers; having profit as the main aim ▷ *n* television or radio advertisement ▶ **commercialize** *v* make commercial ▶ **commercialization** *n*

commiserate *v* (foll by with) express sympathy (for) ▶ **commiseration** *n*

SPELLING TIP
There are two *m*s in **commiserate**, but only one *s*

commissar *n* (formerly) official responsible for political education in Communist countries

commissariat *n* (Brit, Aust & NZ) military department in charge of food supplies

commission *n* piece of work that an artist is asked to do; duty, task; percentage paid to a salesperson for each sale made; group of people appointed to perform certain duties; committing of a crime; (Mil) rank or authority officially given to an officer ▷ *v* place an order for; (Mil) give a commission to; grant authority to **out of commission** not in working order ▶ **commissioner** *n* appointed official in a government department; member of a commission

commissionaire *n* uniformed doorman at a hotel, theatre, etc.

commit *v* -mitting, -mitted perform (a crime or error); pledge (oneself) to a course of action; send (someone) to prison or hospital ▶ **committal** *n* sending someone to prison or hospital ▶ **commitment** *n* dedication to a cause; responsibility that restricts freedom of action

SPELLING TIP
The correct spelling of **commitment** has three *m*s

altogether, but only two *t*s (which are not next to each other)

committee *n* group of people appointed to perform a specified service or function

SPELLING TIP
Remember **committee** has two *m*s and two *t*s

commode *n* seat with a hinged flap concealing a chamber pot; chest of drawers

commodious *adj* roomy

commodity *n*, *pl* -**ities** something that can be bought or sold

commodore *n* senior commissioned officer in the navy

common *adj* occurring often; belonging to two or more people; public, general; lacking in taste or manners ▷ *n* area of grassy land belonging to a community **House of Commons, the Commons** lower chamber of the British parliament ▶ **commonly** *adv* ▶ **commoner** *n* person who does not belong to the nobility ▶ **common-law** *adj* (of a relationship) regarded as a marriage through being long-standing ▶ **Common Market** former name for **European Union** ▶ **commonplace** *adj* ordinary, everyday ▷ *n* trite remark ▶ **common sense** good practical understanding

commonwealth *n* state or nation viewed politically; (**C-**) association of independent states that used to be ruled by Britain

commotion *n* noisy disturbance

commune[1] *n* group of people who live together and share everything ▶ **communal** *adj* shared ▶ **communally** *adv*

commune[2] *v* (foll by with) feel very close (to) e.g. *communing with* *nature* ▶ **communion** *n* sharing of thoughts or feelings; (**C-**) Christian ritual of sharing consecrated bread and wine; religious group with shared beliefs and practices

communicate *v* make known or share (information, thoughts, or feelings) ▶ **communicable** *adj* (of a disease) able to be passed on ▶ **communicant** *n* person who receives Communion ▶ **communicating** *adj* (of a door) joining two rooms ▶ **communication** *n* communicating; thing communicated ▶ *pl* means of travelling or sending messages ▶ **communicative** *adj* talking freely

communiqué [kom-**mune**-ik-kay] *n* official announcement

communism *n* belief that all property and means of production should be shared by the community; (**C-**) system of state control of the economy and society in some countries ▶ **communist** *n, adj*

community *n, pl* **-ties** all the people living in one district; group with shared origins or interests; the public, society ▶ **community centre** building used by a community for activities

commute *v* travel daily to and from work; reduce (a sentence) to a less severe one ▶ **commutator** *n* device used to change alternating electric current into direct current ▶ **commuter** *n* person who commutes to and from work

compact[1] *adj* closely packed; neatly arranged; concise, brief ▷ *n* small flat case containing a mirror and face powder ▷ *v* pack closely together ▶ **compactly** *adv* ▶ **compactness** *n* ▶ **compact disc** small digital audio disc on which the sound is read by an optical laser system

compact[2] *n* contract, agreement

companion *n* person who associates with or accompanies someone ▶ **companionable** *adj* friendly ▶ **companionship** *n*

companionway *n* ladder linking the decks of a ship

company *n, pl* **-nies** business organization; group of actors; fact of being with someone; guest or guests

compare *v* examine (things) and point out the resemblances or differences; (foll by *to*) declare to be (like); (foll by *with*) be worthy of comparison ▶ **comparable** *adj* ▶ **comparability** *n* ▶ **comparative** *adj* relative; involving comparison; (*Grammar*) denoting the form of an adjective or adverb indicating *more* ▷ *n* (*Grammar*) comparative form of a word ▶ **comparatively** *adv* ▶ **comparison** *n* comparing; similarity or equivalence

compartment *n* section of a railway carriage; separate section

compass *n* instrument for showing direction, with a needle that points north; limits, range ▷ *pl* hinged instrument used for drawing circles

compassion *n* pity, sympathy ▶ **compassionate** *adj*

compatible *adj* able to exist, work, or be used together ▶ **compatibility** *n*

compatriot *n* fellow countryman or countrywoman

compel *v* **-pelling, -pelled** force (to be or do)

compendium *n, pl* **-diums** *or* **-dia** selection of board games in one box ▶ **compendious** *adj* brief but comprehensive

compensate *v* make amends to (someone), esp. for injury or loss;

(foll by *for*) cancel out (a bad effect)
▸ **compensation** *n* payment
to make up for loss or injury
▸ **compensatory** *adj*

compere *n* person who presents a
stage, radio, or television show ▹ *v*
be the compere of

compete *v* try to win or achieve (a
prize, profit, etc.) ▸ **competition** *n*
competing; event in which people
compete; people against whom
one competes ▸ **competitive** *adj*
▸ **competitor** *n*

competent *adj* having the skill
or knowledge to do something
well ▸ **competently** *adv*
▸ **competence** *n*

compile *v* collect and arrange
(information), esp. to make a book
▸ **compilation** *n* ▸ **compiler** *n*

complacent *adj* self-
satisfied ▸ **complacently** *adv*
▸ **complacency** *n*

complain *v* express resentment
or displeasure; (foll by *of*) say that
one is suffering from (an illness)
▸ **complaint** *n* complaining; mild
illness ▸ **complainant** *n* (*Law*)
plaintiff

complaisant [kom-**play**-
zant] *adj* willing to please
▸ **complaisance** *n*

complement *n* thing that
completes something; complete
amount or number; (*Grammar*)
word or words added to a verb to
complete the meaning ▹ *v* make
complete ▸ **complementary** *adj*

USAGE NOTE
Avoid confusing this word with
compliment

complete *adj* thorough, absolute;
finished; having all the necessary
parts ▹ *v* finish; make whole
or perfect ▸ **completely** *adv*
▸ **completeness** *n* ▸ **completion**
n finishing

complex *adj* made up of parts;
complicated ▹ *n* whole made up
of parts; group of unconscious
feelings that influences behaviour
▸ **complexity** *n*

complexion *n* skin of the face;
character, nature

compliance *n* complying;
tendency to do what others want
▸ **compliant** *adj*

complicate *v* make or become
complex or difficult to deal with
▸ **complication** *n*

complicity *n* fact of being an
accomplice in a crime

compliment *n* expression of praise
▹ *pl* formal greetings ▹ *v* praise
▸ **complimentary** *adj* expressing
praise; free of charge

USAGE NOTE
Avoid confusing this word with
complement

compline *n* last service of the day in
the Roman Catholic Church

comply *v* -plying, -plied (foll by
with) act in accordance (with)

component *n*, *adj* (being) part of
a whole

comport *v* (*formal*) behave (oneself)
in a specified way

compose *v* put together; be the
component parts of; create (a
piece of music or writing); calm
(oneself); arrange artistically

composer *n* person who writes
music

composite *n*, *adj* (something)
made up of separate parts

composition *n* way that
something is put together or
arranged; work of art, esp. a
musical one; essay; composing

compositor *n* person who
arranges type for printing

compos mentis *adj* (*Latin*) sane

compost *n* decayed plants used
as a fertilizer

composure n calmness

compote n fruit stewed with sugar

compound¹ n, adj (thing, esp. chemical) made up of two or more combined parts or elements ▷ v combine or make by combining; intensify, make worse

compound² n fenced enclosure containing buildings

comprehend v understand
▶ **comprehensible** adj
▶ **comprehension** n
▶ **comprehensive** adj of broad scope, fully inclusive ▷ n (Brit) comprehensive school
▶ **comprehensive school** (Brit) secondary school for children of all abilities

compress v [kum-**press**] squeeze together; make shorter ▷ n [**kom**-press] pad applied to stop bleeding or cool inflammation
▶ **compression** n ▶ **compressor** n machine that compresses gas or air

comprise v be made up of or make up

> **USAGE NOTE**
> Comprise is not followed by of but directly by its object

compromise [**kom**-prom-mize] n settlement reached by concessions on each side ▷ v settle a dispute by making concessions; put in a dishonourable position

comptroller n (in titles) financial controller

compulsion n irresistible urge; forcing by threats or violence ▶ **compulsive** adj
▶ **compulsively** adv
▶ **compulsory** adj required by rules or laws

compute v calculate, esp. using a computer ▶ **computation** n

computer n electronic machine that stores and processes data

▶ **computerize** v adapt (a system) to be handled by computer; store or process in a computer
▶ **computerization** n

comrade n fellow member of a union or socialist political party; companion ▶ **comradeship** n

con¹ (informal) v short for **confidence trick** ▷ v **conning, conned** deceive, swindle

con² n pros and cons see **pro¹**

concatenation n series of linked events

concave adj curving inwards

conceal v cover and hide; keep secret ▶ **concealment** n

concede v admit to be true; acknowledge defeat in (a contest or argument); grant as a right

conceit n too high an opinion of oneself; far-fetched or clever comparison ▶ **conceited** adj

conceive v imagine, think; form in the mind; become pregnant
▶ **conceivable** adj imaginable, possible ▶ **conceivably** adv

concentrate v fix one's attention or efforts on something; bring or come together in large numbers in one place; make (a liquid) stronger by removing water from it ▷ n concentrated liquid ▶ **concentration** n concentrating; proportion of a substance in a mixture or solution
▶ **concentration camp** prison camp for civilian prisoners, esp. in Nazi Germany

concentric adj having the same centre

concept n abstract or general idea ▶ **conceptual** adj of or based on concepts
▶ **conceptualize** v form a concept of

conception n general idea; becoming pregnant

concern n anxiety, worry; something that is of importance to someone; business, firm ▷ v worry (someone); involve (oneself); be relevant or important to ▶ **concerned** adj interested, involved; anxious, worried ▶ **concerning** prep about, regarding

concert n musical entertainment **in concert** working together; (of musicians) performing live ▶ **concerted** adj done together

concertina n small musical instrument similar to an accordion ▷ v **-naing, -naed** collapse or fold up like a concertina

concerto [kon-**chair**-toe] n, pl **-tos** or **-ti** large-scale composition for a solo instrument and orchestra

concession n grant of rights, land, or property; reduction in price for a specified category of people; conceding; thing conceded ▶ **concessionary** adj

conch n shellfish with a large spiral shell; its shell

concierge [kon-see-**airzh**] n (in France) caretaker in a block of flats

conciliate v try to end a disagreement (with) ▶ **conciliation** n ▶ **conciliator** n

conciliatory adj intended to end a disagreement

concise adj brief and to the point ▶ **concisely** adv ▶ **concision, conciseness** n

conclave n secret meeting; private meeting of cardinals to elect a new pope

conclude v decide by reasoning; end, finish; arrange or settle finally ▶ **conclusion** n decision based on reasoning; ending; final arrangement or settlement ▶ **conclusive** adj ending doubt, convincing ▶ **conclusively** adv

concoct v make up (a story or plan); make by combining ingredients ▶ **concoction** n

concomitant adj existing along with something else

concord n state of peaceful agreement, harmony ▶ **concordance** n similarity or consistency; index of words in a book ▶ **concordant** adj agreeing

concourse n large open public place where people can gather; large crowd

concrete n mixture of cement, sand, stone, and water, used in building ▷ adj made of concrete; particular, specific; real or solid, not abstract

concubine [kon-kew-bine] n (Hist) woman living in a man's house but not married to him and kept for his sexual pleasure

concupiscence [kon-kew-**piss**-enss] n (formal) lust

concur v **-curring, -curred** agree ▶ **concurrence** n ▶ **concurrent** adj happening at the same time or place ▶ **concurrently** adv at the same time

concussion n period of unconsciousness caused by a blow to the head ▶ **concussed** adj having concussion

condemn v express disapproval of; sentence e.g. he was condemned to death; force into an unpleasant situation; declare unfit for use ▶ **condemnation** n ▶ **condemnatory** adj

condense v make shorter; turn from gas into liquid ▶ **condensation** n ▶ **condenser** n (Electricity) capacitor

condescend v behave patronizingly towards someone; agree to do something, but as if doing someone a favour ▶ **condescension** n

condiment n seasoning for food, such as salt or pepper

condition n particular state of being; necessary requirement for something else to happen; restriction, qualification; state of health, physical fitness; medical problem ▷ pl circumstances ▷ v train or influence to behave in a particular way; treat with conditioner; control **on condition that** only if ▶ **conditional** adj depending on circumstances ▶ **conditioner** n thick liquid used when washing to make hair or clothes feel softer

condolence n sympathy ▷ pl expression of sympathy

condom n rubber sheath worn on the penis or in the vagina during sexual intercourse to prevent conception or infection

condominium n (Aust, US & Canadian) (also **condo**) block of flats in which each flat is owned by the occupant

condone v overlook or forgive (wrongdoing)

condor n large vulture of S America

conducive adj (foll by to) likely to lead (to)

conduct n management of an activity; behaviour ▷ v carry out (a task); behave (oneself); direct (musicians) by moving the hands or a baton; lead, guide; transmit (heat or electricity) ▶ **conduction** n transmission of heat or electricity ▶ **conductivity** n ability to transmit heat or electricity ▶ **conductive** adj ▶ **conductor** n person who conducts musicians; (fem **conductress**) official on a bus who collects fares; something that conducts heat or electricity

conduit [kon-dew-it] n channel or tube for fluid or cables

cone n object with a circular base,

tapering to a point; cone-shaped ice-cream wafer; (Brit, Aust & NZ) plastic cone used as a traffic marker on the roads; scaly fruit of a conifer tree

coney n same as **cony**

confab n (informal) conversation (also **confabulation**)

confection n any sweet food; (old-fashioned) elaborate article of clothing ▶ **confectioner** n maker or seller of confectionery ▶ **confectionery** n sweets

confederate n member of a confederacy; accomplice ▷ adj united, allied ▷ v unite in a confederacy ▶ **confederacy** n, pl **-cies** union of states or people for a common purpose ▶ **confederation** n alliance of political units

confer v **-ferring, -ferred** discuss together; grant, give ▶ **conferment** n granting, giving

conference n meeting for discussion

confess v admit (a fault or crime); admit to be true; declare (one's sins) to God or a priest, in hope of forgiveness ▶ **confession** n something confessed; confessing ▶ **confessional** n small stall in which a priest hears confessions ▶ **confessor** n priest who hears confessions

confetti n small pieces of coloured paper thrown at weddings

confidant n person confided in ▶ **confidante** n fem

confide v tell someone (a secret); entrust

confidence n trust; self-assurance; something confided **in confidence** as a secret ▶ **confidence trick** swindle involving gaining a person's trust in order to cheat him or her

confident *adj* sure, esp. of oneself ▶ **confidently** *adv*
▶ **confidential** *adj* private, secret; entrusted with someone's secret affairs ▶ **confidentially** *adv*
▶ **confidentiality** *n*

configuration *n* arrangement of parts

confine *v* keep within bounds; restrict the free movement of ▶ **confines** *pl n* boundaries, limits ▶ **confinement** *n* being confined; period of childbirth

confirm *v* prove to be true; reaffirm, strengthen; administer the rite of confirmation to ▶ **confirmation** *n* confirming; something that confirms; (Christianity) rite that admits a baptized person to full church membership ▶ **confirmed** *adj* firmly established in a habit or condition

confiscate *v* seize (property) by authority ▶ **confiscation** *n*

conflagration *n* large destructive fire

conflate *v* combine or blend into a whole ▶ **conflation** *n*

conflict *n* disagreement; struggle or fight ▷ *v* be incompatible

confluence *n* place where two rivers join

conform *v* comply with accepted standards or customs; (foll by *to* or *with*) be like or in accordance with ▶ **conformist** *n, adj* (person) complying with accepted standards and customs ▶ **conformity** *n* compliance with accepted standards or customs

confound *v* astound, bewilder; confuse ▶ **confounded** *adj* (old-fashioned) damned

confront *v* come face to face with ▶ **confrontation** *n* serious argument

confuse *v* mix up; perplex, disconcert; make unclear ▶ **confusion** *n*

confute *v* prove wrong

conga *n* dance performed by a number of people in single file; large single-headed drum played with the hands

congeal *v* (of a liquid) become thick and sticky

congenial *adj* pleasant, agreeable; having similar interests and attitudes ▶ **congeniality** *n*

congenital *adj* (of a condition) existing from birth ▶ **congenitally** *adv*

conger *n* large sea eel

congested *adj* crowded to excess ▶ **congestion** *n*

conglomerate *n* large corporation made up of many companies; thing made up of several different elements ▷ *v* form into a mass ▷ *adj* made up of several different elements ▶ **conglomeration** *n*

congratulate *v* express one's pleasure to (someone) at his or her good fortune or success ▶ **congratulations** *pl n, interj* ▶ **congratulatory** *adj*

congregate *v* gather together in a crowd ▶ **congregation** *n* people who attend a church ▶ **congregational** *adj* ▶ **Congregationalism** *n* Protestant denomination in which each church is self-governing ▶ **Congregationalist** *adj, n*

congress *n* formal meeting for discussion; (C-) federal parliament of the US ▶ **congressional** *adj* ▶ **Congressman, Congresswoman** *n* member of Congress

congruent *adj* similar, corresponding; (Geom) identical in shape and size ▶ **congruence** *n*

conical adj cone-shaped

conifer n cone-bearing tree, such as the fir or pine ▶ **coniferous** adj

conjecture n, v guess ▶ **conjectural** adj

conjugal adj of marriage

conjugate v give the inflections of (a verb) ▶ **conjugation** n complete set of inflections of a verb

conjunction n combination; simultaneous occurrence of events; part of speech joining words, phrases, or clauses

conjunctivitis n inflammation of the membrane covering the eyeball and inner eyelid ▶ **conjunctiva** n this membrane

conjure v perform tricks that appear to be magic ▶ **conjuror** n ▶ **conjure up** v produce as if by magic

conk n (Brit, Aust & NZ slang) nose

conker n (informal) nut of the horse chestnut

conk out v (informal) (of a machine) break down

connect v join together; associate in the mind ▶ **connection**, **connexion** n relationship, association; link or bond; opportunity to transfer from one public vehicle to another; influential acquaintance ▶ **connective** adj

conning tower n raised observation tower containing the periscope on a submarine

connive v (foll by at) allow (wrongdoing) by ignoring it; conspire ▶ **connivance** n

connoisseur [kon-noss-**sir**] n person with special knowledge of the arts, food, or drink

connotation n associated idea conveyed by a word ▶ **connote** v

connubial adj (formal) of marriage

conquer v defeat; overcome (a difficulty); take (a place) by force

▶ **conqueror** n ▶ **conquest** n conquering; person or thing conquered

conscience n sense of right or wrong as regards thoughts and actions

conscientious adj painstaking ▶ **conscientiously** adv ▶ **conscientious objector** person who refuses to serve in the armed forces on moral or religious grounds

conscious adj alert and awake; aware; deliberate, intentional ▶ **consciously** adv ▶ **consciousness** n

conscript n person enrolled for compulsory military service ▷ v enrol (someone) for compulsory military service ▶ **conscription** n

consecrate v make sacred; dedicate to a specific purpose ▶ **consecration** n

consecutive adj in unbroken succession ▶ **consecutively** adv

consensus n general agreement

> **SPELLING TIP**
>
> Note the spelling of this word; it is often misspelled *concensus*, perhaps because people think it is related to the word *census*

> **USAGE NOTE**
>
> In view of its meaning, it is redundant to say 'a consensus of opinion'

consent n agreement, permission ▷ v (foll by to) permit, agree to

consequence n result, effect; importance ▶ **consequent** adj resulting ▶ **consequently** adv as a result, therefore ▶ **consequential** adj important

conservative adj opposing change; moderate, cautious; conventional in style; (**C-**) of the Conservative Party, the British right-wing political party which believes in private enterprise and capitalism; *n* conservative person; (**C-**) supporter or member of the Conservative Party ▸ **conservatism** *n*

conservatoire [kon-**serv**-a-twahr] *n* school of music

conservatory *n, pl* -**ries** room with glass walls and a glass roof, attached to a house; (*chiefly US*) conservatoire

conserve *v* protect from harm, decay, or loss; preserve (fruit) with sugar ▸ *n* jam containing large pieces of fruit ▸ **conservancy** *n* environmental conservation ▸ **conservation** *n* protection of natural resources and the environment; conserving ▸ **conservationist** *n*

consider *v* regard as; think about; be considerate of; discuss; look at ▸ **considerable** *adj* large in amount or degree ▸ **considerably** *adv* ▸ **considerate** *adj* thoughtful towards others ▸ **considerately** *adv* ▸ **consideration** *n* careful thought; fact that should be considered; thoughtfulness; payment for a service ▸ **considering** *prep* taking a (specified fact) into account

consign *v* put somewhere; send (goods) ▸ **consignment** *n* shipment of goods

consist *v* **consist of** be made up of ▸ **consist in** have as its main or only feature

consistent *adj* unchanging, constant; (*foll by* **with**) in agreement ▸ **consistently** *adv* ▸ **consistency** *n, pl* -**cies** being consistent; degree of thickness or smoothness

console¹ *v* comfort in distress ▸ **consolation** *n* consoling; person or thing that consoles

console² *n* panel of controls for electronic equipment; same as **games console**; cabinet for a television or audio equipment; ornamental wall bracket; part of an organ containing the pedals, stops, and keys

consolidate *v* make or become stronger or more stable; combine into a whole ▸ **consolidation** *n*

consommé [kon-**som**-may] *n* thin clear meat soup

consonant *n* speech sound made by partially or completely blocking the breath stream, such as *b* or *f*; letter representing this ▸ *adj* (*foll by* **with**) agreeing (with) ▸ **consonance** *n* agreement, harmony

consort *v* (*foll by* **with**) keep company (with) ▸ *n* husband or wife of a monarch

consortium *n, pl* -**tia** association of business firms

conspectus *n* (*formal*) survey or summary

conspicuous *adj* clearly visible; noteworthy, striking ▸ **conspicuously** *adv*

conspire *v* plan a crime together in secret; act together as if by design ▸ **conspiracy** *n, pl* -**cies** conspiring; plan made by conspiring ▸ **conspirator** *n* ▸ **conspiratorial** *adj*

constable *n* police officer of the lowest rank ▸ **constabulary** *n, pl* -**laries** police force of an area

constant *adj* continuous; unchanging; faithful ▸ *n* unvarying quantity; something that stays the same ▸ **constantly** *adv* ▸ **constancy** *n*

constellation *n* group of stars

constermation *n* anxiety or dismay

constipation n difficulty in defecating ▸ **constipated** adj having constipation

constituent n member of a constituency; component part ▷ adj forming part of a whole ▸ **constituency** n, pl **-cies** area represented by a Member of Parliament; voters in such an area

constitute v form, make up ▸ **constitution** n principles on which a state is governed; physical condition; structure ▸ **constitutional** adj of a constitution; in accordance with a political constitution ▷ n walk taken for exercise ▸ **constitutionally** adv

constrain v compel, force; limit, restrict ▸ **constraint** n

constrict v make narrower by squeezing ▸ **constriction** n ▸ **constrictive** adj ▸ **constrictor** n large snake that squeezes its prey to death; muscle that compresses an organ

construct v build or put together ▸ **construction** n constructing; thing constructed; interpretation; (Grammar) way in which words are arranged in a sentence, clause, or phrase ▸ **constructive** adj (of advice, criticism, etc.) useful and helpful ▸ **constructively** adv

construe v **-struing, -strued** interpret

consul n official representing a state in a foreign country; one of the two chief magistrates in ancient Rome ▸ **consular** adj ▸ **consulate** n workplace or position of a consul ▸ **consulship** n

consult v go to for advice or information ▸ **consultant** n specialist doctor with a senior position in a hospital; specialist who gives professional advice

▸ **consultancy** n, pl **-cies** work or position of a consultant ▸ **consultation** n (meeting for) consulting ▸ **consultative** adj giving advice

consume v eat or drink; use up; destroy; obsess ▸ **consumption** n amount consumed; consuming; (old-fashioned) tuberculosis ▸ **consumptive** n, adj (old-fashioned) (person) having tuberculosis

consumer n person who buys goods or uses services

consummate v [kon-sum-mate] make (a marriage) legal by sexual intercourse; complete or fulfil ▷ adj [kon-**sum**-mit] supremely skilled; complete, extreme ▸ **consummation** n

cont. continued

contact n communicating; touching; useful acquaintance; connection between two electrical conductors in a circuit ▷ v get in touch with ▸ **contact lens** lens placed on the eyeball to correct defective vision

contagion n passing on of disease by contact; disease spread by contact; spreading of a harmful influence ▸ **contagious** adj spreading by contact, catching

contain v hold or be capable of holding; consist of; control, restrain ▸ **container** n object used to hold or store things in; large standard-sized box for transporting cargo by truck or ship ▸ **containment** n prevention of the spread of something harmful

contaminate v make impure, pollute; make radioactive ▸ **contaminant** n contaminating substance ▸ **contamination** n

contemplate v think deeply; consider as a possibility;

gaze at ▶ **contemplation** n
▶ **contemplative** adj

contemporary adj present-day, modern; living or occurring at the same time ▷ n, pl **-raries** person or thing living or occurring at the same time as another ▶ **contemporaneous** adj happening at the same time

SPELLING TIP

It's easy to miss a syllable out when you say **contemporary**. Syllables get lost from spellings too - for example, *contempory* is a common mistake. But remember that the correct spelling ends in *orary*

contempt n dislike and disregard; open disrespect for the authority of a court ▶ **contemptible** adj deserving contempt ▶ **contemptuous** adj showing contempt ▶ **contemptuously** adv

contend v (foll by *with*) deal with; state, assert; compete ▶ **contender** n competitor, esp. a strong one

content[1] n meaning or substance of a piece of writing; amount of a substance in a mixture ▷ pl what something contains; list of chapters at the front of a book

content[2] adj satisfied with things as they are ▷ v make (someone) content ▷ n happiness and satisfaction ▶ **contented** adj ▶ **contentment** n

contention n disagreement or dispute; point asserted in argument ▶ **contentious** adj causing disagreement; quarrelsome

contest n competition or struggle ▷ v dispute, object to; fight or compete for ▶ **contestant** n

context n circumstances of an event or fact; words before and after a word or sentence that help make its meaning clear ▶ **contextual** adj

contiguous adj very near or touching

continent[1] n one of the earth's large masses of land **the Continent** mainland of Europe ▶ **continental** adj ▶ **continental breakfast** light breakfast of coffee and rolls

continent[2] adj able to control one's bladder and bowels; sexually restrained ▶ **continence** n

contingent n group of people that represents or is part of a larger group ▷ adj (foll by *on*) dependent on (something uncertain) ▶ **contingency** n, pl **-cies** something that may happen

continue v **-tinuing, -tinued** (cause to) remain in a condition or place; carry on (doing something); resume ▶ **continual** adj constant; recurring frequently ▶ **continually** adv ▶ **continuance** n continuing ▶ **continuation** n continuing; part added ▶ **continuity** n, pl **-ties** smooth development or sequence ▶ **continuous** adj continuing uninterrupted ▶ **continuously** adv

continuo n, pl **-tinuos** (*Music*) continuous bass part, usu. played on a keyboard instrument

continuum n, pl **-tinua** or **-tinuums** continuous series

contort v twist out of shape ▶ **contortion** n ▶ **contortionist** n performer who contorts his or her body to entertain

contour n outline; (also **contour line**) line on a map joining places of the same height

contra- prefix against or contrasting e.g. *contraflow*

contraband n, adj smuggled (goods)

contraception n prevention of pregnancy by artificial means ▶ **contraceptive** n device or pill taken to prevent pregnancy ▷ adj preventing pregnancy

contract n (document setting out) a formal agreement ▷ v make a formal agreement (to do something); make or become smaller or shorter; catch (an illness) ▶ **contraction** n ▶ **contractor** n firm that supplies materials or labour ▶ **contractual** adj

contradict v declare the opposite of (a statement) to be true; be at variance with ▶ **contradiction** n ▶ **contradictory** adj

contraflow n flow of traffic going alongside but in an opposite direction to the usual flow

contralto n, pl -**tos** (singer with) the lowest female voice

contraption n strange-looking device

contrapuntal adj (Music) of or in counterpoint

contrary n complete opposite ▷ adj opposed, completely different; perverse, obstinate ▷ adv in opposition ▶ **contrarily** adv ▶ **contrariness** n ▶ **contrariwise** adv

contrast n obvious difference; person or thing very different from another ▷ v compare in order to show differences; (foll by with) be very different (from)

contravene v break (a rule or law) ▶ **contravention** n

contretemps [kon-tra-tahn] n, pl -**temps** embarrassing minor disagreement

contribute v give for a common purpose or fund; (foll by to) be partly responsible (for) ▶ **contribution** n ▶ **contributor** n ▶ **contributory** adj

contrite adj sorry and apologetic ▶ **contritely** adv ▶ **contrition** n

contrive v make happen; devise or construct ▶ **contrivance** n device; plan ▶ **contrived** adj planned or artificial

control n power to direct something; curb or check ▷ pl instruments used to operate a machine ▷ v -**trolling**, -**trolled** have power over; limit, restrain; regulate, operate ▶ **controllable** adj ▶ **controller** n

controversy n, pl -**sies** fierce argument or debate ▶ **controversial** adj causing controversy

contumely [kon-tume-mill-ee] n (lit) scornful or insulting treatment

contusion n (formal) bruise

conundrum n riddle

conurbation n large urban area formed by the growth and merging of towns

convalesce v recover after an illness or operation ▶ **convalescence** n ▶ **convalescent** n, adj

convection n transmission of heat in liquids or gases by the circulation of currents ▶ **convector** n heater that gives out hot air

convene v gather or summon for a formal meeting ▶ **convener, convenor** n person who calls a meeting

convenient adj suitable or opportune; easy to use; nearby ▶ **conveniently** adv ▶ **convenience** n quality of being convenient; useful object; (euphemistic) public toilet

convent n building where nuns live; school run by nuns

convention n widely accepted view of proper behaviour; assembly or meeting; formal agreement ▸ **conventional** adj (unthinkingly) following the accepted customs; customary; (of weapons or warfare) not nuclear ▸ **conventionally** adv ▸ **conventionality** n

converge v meet or join ▸ **convergence** n

conversant adj **conversant with** having knowledge or experience of

conversation n informal talk ▸ **conversational** adj ▸ **conversationalist** n person with a specified ability at conversation

converse¹ v have a conversation

converse² adj, n opposite or contrary ▸ **conversely** adv

convert v change in form, character, or function; cause to change in opinion or belief ▸ n person who has converted to a different belief or religion ▸ **conversion** n (thing resulting from) converting; (Rugby) score made after a try by kicking the ball over the crossbar ▸ **convertible** adj capable of being converted ▸ n car with a folding or removable roof

convex adj curving outwards

convey v communicate (information); carry, transport ▸ **conveyance** n (old-fashioned) vehicle; transfer of the legal title to property ▸ **conveyancing** n branch of law dealing with the transfer of ownership of property ▸ **conveyor belt** continuous moving belt for transporting things, esp. in a factory

convict v declare guilty ▸ n person serving a prison sentence ▸ **conviction** n firm belief; instance of being convicted

convince v persuade by argument or evidence ▸ **convincing** adj ▸ **convincingly** adv

convivial adj sociable, lively ▸ **conviviality** n

convocation n calling together; large formal meeting ▸ **convoke** v call together

convoluted adj coiled, twisted; (of an argument or sentence) complex and hard to understand ▸ **convolution** n

convolvulus n twining plant with funnel-shaped flowers

convoy n group of vehicles or ships travelling together

convulse v (of part of the body) undergo violent spasms; (informal) (be) overcome with laughter ▸ **convulsion** n violent muscular spasm ▸ pl uncontrollable laughter ▸ **convulsive** adj

cony n, pl **conies** (Brit) rabbit; rabbit fur

coo v **cooing, cooed** (of a dove or pigeon) make a soft murmuring sound

cooee interj (Brit, Aust & NZ) call to attract attention

cook v prepare (food) by heating; (of food) be cooked ▸ n person who cooks food **cook the books** falsify accounts ▸ **cooker** n (chiefly Brit) apparatus for cooking heated by gas or electricity; apple suitable for cooking ▸ **cookery** n art of cooking ▸ **cookie** n (US) biscuit; (Computing) item of data allowing user of a website to be identified on future visits ▸ **cook up** v (informal) devise (a story or scheme)

Cooktown orchid n purple Australian orchid

cool adj moderately cold; calm and unemotional; indifferent or unfriendly; (informal) sophisticated or excellent; (informal) (of a large sum of money) without exaggeration e.g. a cool million ▸ v make or become cool ▸ n coolness; (slang) calmness, composure ▸ **coolly** adv ▸ **coolness** n

c

▶ **coolant** n fluid used to cool machinery while it is working
▶ **cool drink** (S Afr) nonalcoholic drink ▶ **cooler** n container for making or keeping things cool

coolibah n Australian eucalypt that grows beside rivers

coolie n (old-fashioned offens) unskilled Oriental labourer

coomb, coombe n (S English) short valley or deep hollow

coon n (SAfr offens) person of mixed race

coop¹ n cage or pen for poultry
▶ **coop up** v confine in a restricted place

coop² [koh-op] n (Brit, US & Aust) (shop run by) a cooperative society

cooper n person who makes or repairs barrels

cooperate v work or act together
▶ **cooperation** n ▶ **cooperative** adj willing to cooperate; (of an enterprise) owned and managed collectively ▷ n cooperative organization

coopt [koh-opt] v add (someone) to a group by the agreement of the existing members

coordinate v bring together and cause to work together efficiently ▷ n (Maths) any of a set of numbers defining the location of a point ▷ pl clothes designed to be worn together ▶ **coordination** n ▶ **coordinator** n

coot n small black water bird

cop (slang) n policeman ▷ v copping, copped take or seize ▶ **cop it** get into trouble or be punished ▶ **cop out** v avoid taking responsibility or committing oneself

cope¹ v (often foll by with) deal successfully (with)

cope² n large ceremonial cloak worn by some Christian priests

coping n sloping top row of a wall

copious [kope-ee-uss] adj abundant, plentiful
▶ **copiously** adv

copper¹ n soft reddish-brown metal; copper or bronze coin
▶ **copper-bottomed** adj financially reliable ▶ **copperplate** n fine handwriting style

copper² n (Brit slang) policeman

coppice, copse n small group of trees growing close together

copra n dried oil-yielding kernel of the coconut

copulate v have sexual intercourse
▶ **copulation** n

copy n, pl **copies** thing made to look exactly like another; single specimen of a book etc.; material for printing ▷ v copying, copied make a copy of; act or try to be like ▶ **copyright** n exclusive legal right to reproduce and control a book, work of art, etc. ▷ v take out a copyright on ▷ adj protected by copyright ▶ **copywriter** n person who writes advertising copy

coquette n woman who flirts
▶ **coquettish** adj

coracle n small round boat of wicker covered with skins

coral n hard substance formed from the skeletons of very small sea animals ▷ adj orange-pink

cor anglais n, pl **cors anglais** woodwind instrument similar to the oboe

cord n thin rope or thick string; cordlike structure in the body; corduroy ▷ pl corduroy trousers

cordial adj warm and friendly ▷ n drink with a fruit base ▶ **cordially** adv ▶ **cordiality** n

cordite n explosive used in guns and bombs

cordon n chain of police, soldiers, etc., guarding an area ▶ **cordon off** v form a cordon round

cordon bleu [bluh] *adj* (of cookery or cooks) of the highest standard

corduroy *n* cotton fabric with a velvety ribbed surface

core *n* central part of certain fruits, containing the seeds; essential part ▷ *v* remove the core from

corella *n* white Australian cockatoo

co-respondent *n* (Brit, Aust & NZ) person with whom someone being sued for divorce is claimed to have committed adultery

corgi *n* short-legged sturdy dog

coriander *n* plant grown for its aromatic seeds and leaves

cork *n* thick light bark of a Mediterranean oak; piece of this used as a stopper ▷ *v* seal with a cork ▶ **corkage** *n* restaurant's charge for serving wine bought elsewhere ▶ **corkscrew** *n* spiral metal tool for pulling corks from bottles

corm *n* bulblike underground stem of certain plants

cormorant *n* large dark-coloured long-necked sea bird

corn¹ *n* cereal plant such as wheat or oats; grain of such plants; (US, Canadian, Aust & NZ) maize; (slang) something unoriginal or oversentimental ▶ **corny** *adj* (slang) unoriginal or oversentimental ▶ **cornflakes** *pl n* breakfast cereal made from toasted maize ▶ **cornflour** *n* (chiefly Brit) fine maize flour; (NZ) fine wheat flour ▶ **cornflower** *n* plant with blue flowers ▶ **cornmeal** *n* powder made from maize ▶ **corn on the cob** maize cooked and eaten still attached to the plant head

corn² *n* painful hard skin on the toe

cornea [korn-ee-a] *n*, *pl* **-neas** or **-neae** transparent membrane covering the eyeball ▶ **corneal** *adj*

corned beef *n* beef preserved in salt

corner *n* area or angle where two converging lines or surfaces meet; place where two streets meet; remote place; (Sport) free kick or shot from the corner of the field ▷ *v* force into a difficult or inescapable position; (of a vehicle) turn a corner; obtain a monopoly of ▶ **cornerstone** *n* indispensable part or basis

cornet *n* brass instrument similar to the trumpet; cone-shaped ice-cream wafer

cornice *n* decorative moulding round the top of a wall

Cornish *pl n*, *adj* (people) of Cornwall ▶ **Cornish pasty** pastry case with a filling of meat and vegetables

cornucopia [korn-yew-kope-ea] *n* great abundance; symbol of plenty, consisting of a horn overflowing with fruit and flowers

corolla *n* petals of a flower collectively

corollary *n*, *pl* **-laries** idea, fact, or proposition which is the natural result of something else

corona *n*, *pl* **-nas** or **-nae** ring of light round the moon or sun

coronary [kor-ron-a-ree] *adj* of the arteries surrounding the heart ▷ *n*, *pl* **-naries** coronary thrombosis ▶ **coronary thrombosis** condition in which the flow of blood to the heart is blocked by a blood clot

coronation *n* ceremony of crowning a monarch

coronavirus *n* type of airborne virus that causes colds

coroner *n* (Brit, Aust & NZ) official responsible for the investigation of violent or sudden deaths

coronet *n* small crown

corpora *n* plural of **corpus**

corporal¹ *n* noncommissioned officer in an army

corporal[1] adj of the body
▶ **corporal punishment** physical punishment, such as caning

corporation n large business or company; city or town council
▶ **corporate** adj of business corporations; shared by a group

corporeal [kore-pore-ee-al] adj physical or tangible

corps [kore] n, pl **corps** military unit with a specific function; organized body of people

corpse n dead body

corpulent adj fat or plump
▶ **corpulence** n

corpus n, pl **corpora** collection of writings, esp. by a single author

corpuscle n red or white blood cell

corral (US) n enclosure for cattle or horses ▶ v **-ralling, -ralled** put in a corral

correct adj free from error, true; in accordance with accepted standards ▶ v put right; indicate the errors in; rebuke or punish
▶ **correctly** adv ▶ **correctness** n ▶ **correction** n correcting; alteration correcting something
▶ **corrective** adj intended to put right something wrong

correlate v place or be placed in a mutual relationship
▶ **correlation** n

correspond v be consistent or compatible (with); be the same or similar; communicate by letter ▶ **corresponding** adj
▶ **correspondingly** adv ▶ **correspondence** n communication by letters; letters so exchanged; relationship or similarity ▶ **correspondent** n person employed by a newspaper etc. to report on a special subject or from a foreign country; letter writer

corridor n passage in a building or train; strip of land or airspace providing access through foreign territory

corrigendum [kor-rij-**end**-um] n, pl **-da** error to be corrected

corroborate v support (a fact or opinion) by giving proof ▶ **corroboration** n
▶ **corroborative** adj

corroboree n (Aust) Aboriginal gathering or dance

corrode v eat or be eaten away by chemical action or rust
▶ **corrosion** n ▶ **corrosive** adj

corrugated adj folded into alternate grooves and ridges

corrupt adj open to or involving bribery; morally depraved; (of a text or data) unreliable through errors or alterations ▶ v make corrupt ▶ **corruptly** adv
▶ **corruption** n ▶ **corruptible** adj

corsage [kor-**sahzh**] n small bouquet worn on the bodice of a dress

corsair n pirate; pirate ship

corset n women's close-fitting undergarment worn to shape the torso

cortege [kor-**tayzh**] n funeral procession

cortex n, pl **-tices** (Anat) outer layer of the brain or other internal organ
▶ **cortical** adj

cortisone n steroid hormone used to treat various diseases

corundum n hard mineral used as an abrasive

coruscate v (Literary) sparkle

corvette n lightly armed escort warship

cos (Maths) cosine

cosh n (Brit) heavy blunt weapon ▶ v hit with a cosh

cosine [**koh**-sine] n (in trigonometry) ratio of the length of the adjacent side to that of the hypotenuse in a right-angled triangle

cosmetic n preparation used to improve the appearance of a person's skin ▷ adj improving the appearance only

cosmic adj of the whole universe ▶ **cosmic rays** electromagnetic radiation from outer space

cosmonaut n Russian name for an astronaut

cosmopolitan adj composed of people or elements from many countries; having lived and travelled in many countries ▷ n cosmopolitan person ▶ **cosmopolitanism** n

cosmos n the universe ▶ **cosmology** n study of the origin and nature of the universe ▶ **cosmological** adj

Cossack n member of a S Russian people famous as horsemen and dancers

cosset v cosseting, cosseted pamper

cost n amount of money, time, labour, etc., required for something ▷ pl expenses of a lawsuit ▷ v **costing, cost** have as its cost; involve the loss or sacrifice of; (past tense **costed**) estimate the cost of ▶ **costly** adj expensive; involving great loss or sacrifice ▶ **costliness** n

costermonger n (Brit) person who sells fruit and vegetables from a street barrow

costume n style of dress of a particular place or time, or for a particular activity; clothes worn by an actor or performer ▶ **costumier** n maker or seller of costumes ▶ **costume jewellery** inexpensive artificial jewellery

cosy adj -sier, -siest warm and snug; intimate, friendly ▷ n cover for keeping things warm e.g. a tea cosy ▶ **cosily** adv ▶ **cosiness** n

cot n baby's bed with high sides; small portable bed ▶ **cot death** unexplained death of a baby while asleep

cote n shelter for birds or animals

coterie [kote-er-ee] n exclusive group, clique

cotoneaster [kot-tone-ee-**ass**-ter] n garden shrub with red berries

cottage n small house in the country ▶ **cottage cheese** soft mild white cheese ▶ **cottage industry** craft industry in which employees work at home ▶ **cottage pie** dish of minced meat topped with mashed potato

cotter n pin or wedge used to secure machine parts

cotton n white downy fibre covering the seeds of a tropical plant; cloth or thread made from this ▶ **cottony** adj ▶ **cotton on (to)** v (informal) understand ▶ **cotton wool** fluffy cotton used for surgical dressings etc.

cotyledon [kot-ill-**ee**-don] n first leaf of a plant embryo

couch n piece of upholstered furniture for seating more than one person ▷ v express in a particular way ▶ **couch potato** (slang) lazy person whose only hobby is watching television

couchette [koo-**shett**] n bed converted from seats on a train or ship

couch grass n quickly spreading grassy weed

cougan n (Aust slang) drunk and rowdy person

cougar n puma

cough v expel air from the lungs abruptly and noisily ▷ n act or sound of coughing; illness which causes coughing

could v past tense of **can**[1]

couldn't could not

coulomb [koo-lom] n SI unit of electric charge

coulter n blade at the front of a ploughshare

council n group meeting for discussion or consultation; local governing body of a town or region ▷ adj of or by a council ▶ **councillor** n member of a council ▶ **council tax** (in Britain) tax based on the value of property, to fund local services

counsel n advice or guidance; barrister or barristers ▷ v -**selling**, -**selled** give guidance to; urge, recommend ▶ **counsellor** n

count¹ v say numbers in order; find the total of; be important; regard as; take into account ▷ n counting; number reached by counting; (Law) one of a number of charges ▶ **countless** adj too many to count ▶ **count on** v rely or depend on

count² n European nobleman

countdown n counting backwards to zero of the seconds before an event

countenance n (expression of) the face ▷ v allow or tolerate

counter¹ n long flat surface in a bank or shop, on which business is transacted; small flat disc used in board games

counter² v oppose, retaliate against ▷ adv in the opposite direction; in direct contrast ▷ n opposing or retaliatory action

counter- prefix opposite, against e.g. counterattack; complementary, corresponding e.g. counterpart

counteract v act against or neutralize ▶ **counteraction** n

counterattack n, v attack in response to an attack

counterbalance n weight or force balancing or neutralizing another ▷ v act as a counterbalance to

counterblast n aggressive response to a verbal attack

counterfeit adj fake, forged ▷ n fake, forgery ▷ v fake, forge

counterfoil n part of a cheque or receipt kept as a record

countermand v cancel (a previous order)

counterpane n bed covering

counterpart n person or thing complementary to or corresponding to another

counterpoint n (Music) technique of combining melodies

counterpoise n, v counterbalance

counterproductive adj having an effect opposite to the one intended

countersign v sign (a document already signed by someone) as confirmation

countersink v drive (a screw) into a shaped hole so that its head is below the surface

countertenor n male alto

counterterrorism n measures to prevent terrorist attacks or eradicate terrorist groups

countess n woman holding the rank of count or earl; wife or widow of a count or earl

country n, pl -**tries** nation; nation's territory; nation's people; part of the land away from cities ▶ **countrified** adj rustic in manner or appearance ▶ **country and western, country music** popular music based on American White folk music ▶ **countryman, countrywoman** n person from one's native land; (Brit, Aust & NZ) person who lives in the country ▶ **countryside** n land away from cities

county n, pl -**ties** (in some countries) division of a country

coup [koo] n successful action; coup d'état

coup de grâce [koo de **grahss**] n final or decisive action

coup d'état [koo day-**tah**] n sudden violent overthrow of a government

coupé [koo-pay] n sports car with two doors and a sloping fixed roof

couple n two people who are married or romantically involved; two partners in a dance or game ▷ v connect, associate **a couple** pair; (*informal*) small number ▶ **couplet** n two consecutive lines of verse, usu. rhyming and of the same metre ▶ **coupling** n device for connecting things, such as railway carriages

coupon n piece of paper entitling the holder to a discount or gift; detachable order form; football pools entry form

courage n ability to face danger or pain without fear ▶ **courageous** adj ▶ **courageously** adv

courgette n type of small vegetable marrow

courier n person employed to look after holiday-makers; person employed to deliver urgent messages

course n series of lessons or medical treatment; route or direction taken; area where golf is played or a race is run; any of the successive parts of a meal; mode of conduct or action; natural development of events ▷ v (of liquid) run swiftly **of course** as expected, naturally

court n body which decides legal cases; place where it meets; marked area for playing a racket game; courtyard; residence, household, or retinue of a sovereign ▷ v (*old-fashioned*) try to gain the love of; try to win the favour of; invite e.g. *to court disaster* ▶ **courtier** n attendant at a royal

court ▶ **courtly** adj ceremoniously polite ▶ **courtliness** n ▶ **courtship** n courting of an intended spouse or mate ▶ **court martial** n, pl **courts martial** court for trying naval or military offences ▶ **court shoe** woman's low-cut shoe without straps or laces ▶ **courtyard** n paved space enclosed by buildings or walls

courtesan [kor-tiz-**zan**] n (*Hist*) mistress or high-class prostitute

courtesy n politeness, good manners; (pl **-sies**) courteous act **(by) courtesy of** by permission of ▶ **courteous** adj polite ▶ **courteously** adv

cousin n child of one's uncle or aunt

couture [koo-**toor**] n high-fashion designing and dressmaking ▶ **couturier** n person who designs women's fashion clothes

cove n small bay or inlet

coven [kuv-ven] n meeting of witches

covenant [kuv-ven-ant] n contract; (*chiefly Brit*) formal agreement to make an annual (charitable) payment ▷ v agree by a covenant

Coventry n **send someone to Coventry** punish someone by refusing to speak to him or her

cover v place something over, to protect or conceal; extend over or lie on the surface of; travel over; insure against loss or risk; include; report (an event) for a newspaper; be enough to pay for ▷ n anything that covers; outside of a book or magazine; insurance; shelter or protection ▶ **coverage** n amount or extent covered ▶ **coverlet** n bed cover

covert adj concealed, secret ▷ n thicket giving shelter to game birds or animals ▶ **covertly** adv

covet v **coveting, coveted** long to possess (what belongs to someone else) ▸ **covetous** adj ▸ **covetousness** n

covey [kuv-vee] n small flock of grouse or partridge

cow[1] n mature female of cattle and of certain other mammals, such as the elephant or seal; (informal offens) disagreeable woman ▸ **cowboy** n (in the US) ranch worker who herds and tends cattle, usu. on horseback; (informal) irresponsible or unscrupulous worker

cow[2] v intimidate, subdue

coward n person who lacks courage ▸ **cowardly** adj ▸ **cowardice** n lack of courage

cower v cringe in fear

cowl n loose hood; monk's hooded robe; cover on a chimney to increase ventilation

cowling n cover on an engine

cowrie n brightly-marked sea shell

cowslip n small yellow wild European flower

cox n coxswain ▸ v act as cox of (a boat)

coxswain [kok-sn] n person who steers a rowing boat

coy adj affectedly shy or modest ▸ **coyly** adv ▸ **coyness** n

coyote [koy-ote-ee] n prairie wolf of N America

coypu n beaver-like aquatic rodent native to S America, bred for its fur

cozen v (lit) cheat, trick

CPU (Computing) central processing unit

crab n edible shellfish with ten legs, the first pair modified into pincers

crab apple n small sour apple

crabbed adj (of handwriting) hard to read; (also **crabby**) bad-tempered

crack v break or split partially; (cause to) make a sharp noise;

break down or yield under strain; hit suddenly; solve (a code or problem); tell (a joke) ▸ n sudden sharp noise; narrow gap; sharp blow; (informal) gibe, joke; (slang) highly addictive form of cocaine ▸ adj (informal) first-rate, excellent e.g. a crack shot ▸ **cracking** adj very good ▸ **crackdown** n severe disciplinary measures ▸ **crack down on** v take severe measures against

cracker n thin dry biscuit; decorated cardboard tube, pulled apart with a bang, containing a paper hat and a joke or toy; small explosive firework; (slang) outstanding thing or person

crackers adj (slang) insane

crackle v make sharp popping noises ▸ n crackling sound ▸ **crackling** n crackle; crisp skin of roast pork

crackpot n, adj (informal) eccentric (person)

cradle n baby's bed on rockers; place where something originates; supporting structure ▸ v hold gently as if in a cradle

craft n occupation requiring skill with the hands; skill or ability; (pl **craft**) boat, aircraft, or spaceship ▸ **crafty** adj skilled in deception ▸ **craftily** adv ▸ **craftiness** n ▸ **craftsman, craftswoman** n skilled worker ▸ **craftsmanship** n

crag n steep rugged rock ▸ **craggy** adj

cram v **cramming, crammed** force into too little space; fill too full; study hard just before an examination

cramp[1] n painful muscular contraction; clamp for holding masonry or timber together

cramp[2] v confine, restrict

crampon n spiked plate strapped to a boot for climbing on ice

cranberry n sour edible red berry

crane n machine for lifting and moving heavy weights; large wading bird with a long neck and legs ▷ v stretch (one's neck) to see something ▶ **crane fly** long-legged insect with slender wings

cranium n, pl -**niums** or -**nia** (Anat) skull ▶ **cranial** adj

crank n arm projecting at right angles from a shaft, for transmitting or converting motion; (informal) eccentric person ▷ v start (an engine) with a crank ▶ **cranky** adj (informal) eccentric; bad-tempered ▶ **crankshaft** n shaft driven by a crank

cranny n, pl -**nies** narrow opening

crape n same as **crepe**

craps n gambling game played with two dice

crash n collision involving a vehicle or vehicles; sudden loud smashing noise; financial collapse ▷ v (cause to) collide violently with a vehicle, a stationary object, or the ground; (cause to) make a loud smashing noise; (cause to) fall with a crash; collapse or fail financially ▶ **crash course** short, very intensive course in a particular subject ▶ **crash helmet** protective helmet worn by a motorcyclist ▶ **crash-land** v (of an aircraft) land in an emergency, causing damage ▶ **crash-landing** n

crass adj stupid and insensitive ▶ **crassly** adv ▶ **crassness** n

crate n large wooden container for packing goods

crater n very large hole in the ground or in the surface of the moon

cravat n man's scarf worn like a tie

crave v desire intensely; beg or plead for ▶ **craving** n

craven adj cowardly

crawfish n same as **crayfish**

crawl v move on one's hands and knees; move very slowly; (foll by to) flatter in order to gain some advantage; feel as if covered with crawling creatures ▷ n crawling motion or pace; overarm swimming stroke ▶ **crawler** n

crayfish n edible shellfish like a lobster; Australian freshwater crustacean

crayon v, n (draw or colour with) a stick or pencil of coloured wax or clay

craze n short-lived fashion or enthusiasm ▶ **crazed** adj wild and uncontrolled; (of porcelain) having fine cracks

crazy adj ridiculous; (foll by about) very fond (of); insane ▶ **craziness** n ▶ **crazy paving** paving made of irregularly shaped slabs of stone

creak v, n (make) a harsh squeaking sound ▶ **creaky** adj

cream n fatty part of milk; food or cosmetic resembling cream in consistency; best part (of something) ▷ adj yellowish-white ▷ v beat to a creamy consistency ▶ **creamy** adj ▶ **cream cheese** rich soft white cheese ▶ **cream off** v take the best part from

crease n line made by folding or pressing; (Cricket) line marking the bowler's and batsman's positions ▷ v crush or line

create v make, cause to exist; appoint to a new rank or position; (slang) make an angry fuss ▶ **creation** n ▶ **creationism** n doctrine that ascribes the origins of all things to God's acts of creation rather than to evolution ▶ **creationist** n ▶ **creative**

adj imaginative or inventive
▶ **creativity** *n* ▶ **creator** *n*

creature *n* animal, person, or other being

crèche *n* place where small children are looked after while their parents are working, shopping, etc.

credence *n* belief in the truth or accuracy of a statement

credentials *pl n* document giving evidence of a person's identity or qualifications

credible *adj* believable; trustworthy ▶ **credibly** *adv* ▶ **credibility** *n*

credit *n* system of allowing customers to receive goods and pay later; reputation for trustworthiness in paying debts; money at one's disposal in a bank account; side of an account book on which such sums are entered; (source or cause of) praise or approval; influence or reputation based on the good opinion of others; belief or trust ▷ *pl* list of people responsible for the production of a film, programme, or record ▷ *v* **crediting, credited** enter as a credit in an account; (foll by *with*) attribute (to); believe ▶ **creditable** *adj* praiseworthy ▶ **creditably** *adv* ▶ **creditor** *n* person to whom money is owed ▶ **credit card** card allowing a person to buy on credit ▶ **credit crunch** period during which there is a sudden reduction in the availability of credit from banks, mortgage lenders, etc.

credo *n, pl* **-dos** creed

credulous *adj* too willing to believe ▶ **credulity** *n*

creed *n* statement or system of (Christian) beliefs or principles

creek *n* narrow inlet or bay; (Aust, NZ, US & Canadian) small stream

creel *n* wicker basket used by anglers

creep *v* **creeping, crept** move quietly and cautiously; crawl with the body near to the ground; (of a plant) grow along the ground or over rocks ▷ *n* (slang) obnoxious or servile person **give one the creeps** (informal) make one feel fear or disgust ▶ **creeper** *n* creeping plant ▶ **creepy** *adj* (informal) causing a feeling of fear or disgust

cremate *v* burn (a corpse) to ash ▶ **cremation** *n* ▶ **crematorium** *n* building where corpses are cremated

crenellated *adj* having battlements

creole *n* language developed from a mixture of languages; (**C-**) native-born W Indian or Latin American of mixed European and African descent

creosote *n* dark oily liquid made from coal tar and used for preserving wood ▷ *v* treat with creosote

crepe [krayp] *n* fabric or rubber with a crinkled texture; very thin pancake ▶ **crepe paper** paper with a crinkled texture

crept *v* past of **creep**

crepuscular *adj* (lit) of or like twilight

crescendo [krish-**end**-oh] *n, pl* **-dos** gradual increase in loudness, esp. in music

crescent *n* (curved shape of) the moon as seen in its first or last quarter; crescent-shaped street

cress *n* plant with strong-tasting leaves, used in salads

crest *n* top of a mountain, hill, or wave; tuft or growth on a bird's or animal's head; heraldic design used on a coat of arms and elsewhere

▶ **crested** adj ▶ **crestfallen** adj disheartened

cretin n (informal offens) stupid person; (obs) person with physical and mental disability caused by a thyroid deficiency ▶ **cretinous** adj

crevasse n deep open crack in a glacier

crevice n narrow crack or gap in rock

crew n people who work on a ship or aircraft; group of people working together; (informal) any group of people ▷ v serve as a crew member (on) ▶ **crew cut** n man's closely cropped haircut

crewel n fine worsted yarn used in embroidery

crib n piece of writing stolen from elsewhere; translation or list of answers used by students, often illicitly; baby's cradle; rack for fodder; short for **cribbage** ▷ v **cribbing, cribbed** copy (someone's work) dishonestly ▶ **crib-wall** n (NZ) retaining wall built against an earth bank

cribbage n card game for two to four players

crick n muscle spasm or cramp in the back or neck ▷ v cause a crick in

cricket[1] n outdoor game played with bats, a ball, and wickets by two teams of eleven ▶ **cricketer** n

cricket[2] n chirping insect like a grasshopper

crime n unlawful act; unlawful acts collectively ▶ **criminal** n person guilty of a crime ▷ adj of crime; (informal) deplorable ▶ **criminally** adv ▶ **criminality** n ▶ **criminology** n study of crime ▶ **criminologist** n

crimp v fold or press into ridges

crimson adj deep purplish-red

cringe v flinch in fear; behave in a submissive or timid way

crinkle v, n wrinkle, crease, or fold

crinoline n hooped petticoat

cripple n (offens) person who is lame or disabled ▷ v make lame or disabled; damage (something)

crisis n, pl **-ses** crucial stage, turning point; time of extreme trouble

crisp adj fresh and firm; dry and brittle; clean and neat; (of weather) cold but invigorating; lively or brisk ▷ n (Brit) very thin slice of potato fried till crunchy ▶ **crisply** adv ▶ **crispness** n ▶ **crispy** adj hard and crunchy ▶ **crispbread** n thin dry biscuit

crisscross v move in or mark with a crosswise pattern ▷ adj (of lines) crossing in different directions

criterion n, pl **-ria** standard of judgment

> **USAGE NOTE**
> It is incorrect to use *criteria* as a singular, though this use is often found

critic n professional judge of any of the arts; person who finds fault ▶ **critical** adj very important or dangerous; fault-finding; able to examine and judge carefully; of a critic or criticism ▶ **critically** adv ▶ **criticism** n fault-finding; analysis of a book, work of art, etc. ▶ **criticize** v find fault with ▶ **critique** n critical essay

croak v (of a frog or crow) give a low hoarse cry; utter or speak with a croak ▷ n low hoarse sound ▶ **croaky** adj hoarse

Croatian [kroh-ay-shun], **Croat** adj of Croatia ▷ n person from Croatia; dialect of Serbo-Croat spoken in Croatia

crochet [kroh-shay] v **-cheting, -cheted** make by looping and intertwining with a hooked needle ▷ n work made in this way

crock¹ n earthenware pot or jar ▸ **crockery** n dishes

crock² n (Brit, Aust & NZ informal) old or decrepit person or thing

crocodile n large amphibious tropical reptile; (Brit, Aust & NZ) line of people, esp. schoolchildren, walking two by two ▸ **crocodile tears** insincere show of grief

crocus n, pl -**cuses** small plant with yellow, white, or purple flowers in spring

croft n small farm worked by one family in Scotland ▸ **crofter** n

croissant [krwah-son] n rich flaky crescent-shaped roll

cromlech n (Brit) circle of prehistoric standing stones

crone n witchlike old woman

crony n, pl -**nies** close friend

crook n (informal) criminal; bent or curved part; hooked pole ▸ adj (Aust & NZ slang) unwell, injured **go crook** (Aust & NZ slang) become angry ▸ **crooked** adj bent or twisted; set at an angle; (informal) dishonest

croon v sing, hum, or speak in a soft low tone

crooner n male singer of sentimental ballads

crop n cultivated plant; season's total yield of produce; group of things appearing at one time; (handle of) a whip; pouch in a bird's gullet; very short haircut ▸ v **cropping, cropped** cut very short; produce or harvest as a crop; (of animals) feed on (grass) ▸ **cropper** n **come a cropper** (informal) have a disastrous failure or heavy fall ▸ **crop-top** n short T-shirt or vest that reveals the wearer's midriff ▸ **crop up** v (informal) happen unexpectedly

croquet [kroh-kay] n game played on a lawn in which balls are hit through hoops

croquette [kroh-kett] n fried cake of potato, meat, or fish

crosier n same as **crozier**

cross v move or go across (something); meet and pass; (foll by out) place with a cross or lines; place (one's arms or legs) crosswise ▸ n structure, symbol, or mark of two intersecting lines; such a structure of wood as a means of execution; representation of the Cross as an emblem of Christianity; mixture of two things ▸ adj angry, annoyed **the Cross** (Christianity) the cross on which Christ was crucified ▸ **crossing** n place where a street may be crossed safely; place where one thing crosses another; journey across water ▸ **crossly** adv ▸ **crossbar** n horizontal bar across goalposts or on a bicycle ▸ **crossbow** n weapon consisting of a bow fixed across a wooden stock ▸ **crossbred** adj bred from two different types of animal or plant ▸ **crossbreed** n crossbred animal or plant ▸ **cross-check** v check using a different method ▸ **cross-country** adj, adv by way of open country or fields ▸ **cross-examine** v (Law) question (a witness for the opposing side) to check his or her testimony ▸ **cross-examination** n ▸ **cross-eyed** adj with eyes looking towards each other ▸ **cross-fertilize** v fertilize (an animal or plant) from one of a different kind ▸ **cross-fertilization** n ▸ **crossfire** n gunfire crossing another line of fire ▸ **cross-purposes** pl n **at cross-purposes** misunderstanding each other ▸ **cross-reference** n reference within a text to another part ▸ **crossroads** n place where roads intersect ▸ **cross section** (diagram of) a surface made by cutting across something; representative sample ▶ **crosswise** adj, adv across; in the shape of a cross ▶ **crossword puzzle, crossword** n puzzle in which words suggested by clues are written into a grid of squares

crotch n part of the body between the tops of the legs

crotchet n musical note half the length of a minim

crotchety adj (informal) bad-tempered

crouch v bend low with the legs and body close ▷ n this position

croup¹ [kroop] n throat disease of children, with a cough

croup² [kroop] n hind quarters of a horse

croupier [kroop-ee-ay] n person who collects bets and pays out winnings at a gambling table in a casino

crouton [kroot-on] n small piece of fried or toasted bread served in soup

crow¹ n large black bird with a harsh call **as the crow flies** in a straight line **stone the crows!** (Brit & Aust slang) expression of surprise, dismay, etc. ▶ **crow's feet** wrinkles at the corners of the eyes ▶ **crow's nest** lookout platform at the top of a ship's mast

crow² v (of a cock) make a shrill squawking sound; boast or gloat

crowbar n iron bar used as a lever

crowd n large group of people or things; particular group of people ▷ v gather together in large numbers; press together in a confined space; fill or occupy fully

crown n monarch's headdress of gold and jewels; wreath for the head, given as an honour; top of the head or of a hill; artificial cover for a broken or decayed tooth; former British coin worth 25 pence ▷ v put a crown on the head of (someone) to proclaim him or her monarch; put on or form the top of; put the finishing touch to (a series of events); (informal) hit on the head **the Crown** power of the monarchy ▶ **crown court** local criminal court in England and Wales ▶ **crown-of-thorns** n starfish with a spiny outer covering that feeds on living coral ▶ **crown prince, crown princess** heir to a throne

crozier n bishop's hooked staff

crucial adj very important ▶ **crucially** adv

cruciate ligament [kroo-shee-it] n either of a pair of ligaments in the knee, connecting the tibia and the femur

crucible n pot in which metals are melted

crucify v -fying, -fied put to death by fastening to a cross ▶ **crucifix** n model of Christ on the Cross ▶ **crucifixion** n crucifying **the Crucifixion** (Christianity) crucifying of Christ ▶ **cruciform** adj cross-shaped

crude adj rough and simple; tasteless, vulgar; in a natural or unrefined state ▶ **crudely** adv ▶ **crudity** n

cruel adj delighting in others' pain; causing pain or suffering ▶ **cruelly** adv ▶ **cruelty** n

cruet n small container for salt, pepper, etc., at table

cruise n sail for pleasure ▷ v sail from place to place for pleasure; (of a vehicle) travel at a moderate and economical speed ▶ **cruiser** n fast warship; motorboat with a cabin ▶ **cruise missile** low-flying guided missile

crumb n small fragment of bread or other dry food; small amount

crumble v break into fragments; fall apart or decay ▷ n pudding of stewed fruit with a crumbly topping ▶ **crumbly** adj

crummy adj -mier, -miest (slang) of poor quality

crumpet n round soft yeast cake, eaten buttered; (Brit, Aust & NZ slang) sexually attractive women collectively

crumple v crush, crease; collapse, esp. from shock ▶ **crumpled** adj

crunch v bite or chew with a noisy crushing sound; make a crisp or brittle sound ▷ n crunching sound; (informal) critical moment ▶ **crunchy** adj

crupper n strap that passes from the back of a saddle under a horse's tail

crusade n medieval Christian war to recover the Holy Land from the Muslims; vigorous campaign in favour of a cause ▷ v take part in a crusade

crusader n person who took part in the medieval Christian war to recover the Holy Land from the Muslims; person who campaigns vigorously in favour of a cause

crush v compress so as to injure, break, or crumple; break into small pieces; defeat or humiliate utterly ▷ n dense crowd; (informal) infatuation; drink made by crushing fruit

crust n hard outer part of something, esp. bread ▷ v cover with or form a crust ▶ **crusty** adj having a crust; irritable

crustacean n hard-shelled, usu. aquatic animal with several pairs of legs, such as the crab or lobster

crutch n long sticklike support with a rest for the armpit, used by a lame person; person or thing that gives support; crotch

crux n, pl **cruxes** crucial or decisive point

cry v **crying**, **cried** shed tears; call or utter loudly ▷ n, pl **cries** fit of weeping; loud utterance; urgent appeal e.g. a cry for help ▶ **crybaby** n person, esp. a child, who cries too readily ▶ **cry off** v (informal) withdraw from an arrangement ▶ **cry out for** v need urgently

cryogenics n branch of physics concerned with very low temperatures ▶ **cryogenic** adj

crypt n vault under a church, esp. one used as a burial place

cryptic adj obscure in meaning, secret ▶ **cryptically** adv ▶ **cryptography** n art of writing in and deciphering codes

crystal n (single grain of) a symmetrically shaped solid formed naturally by some substances; very clear and brilliant glass, usu. with the surface cut in many planes; tumblers, vases, etc., made of crystal ▷ adj bright and clear ▶ **crystalline** adj of or like crystal or crystals; clear ▶ **crystallize** v make or become definite; form into crystals ▶ **crystallization** n

cu. cubic

cub n young wild animal such as a bear or fox; (C-) Cub Scout ▷ v **cubbing**, **cubbed** give birth to cubs ▶ **Cub Scout** member of a junior branch of the Scout Association

cubbyhole n small enclosed space or room

cube n object with six equal square sides; number resulting from multiplying a number by itself twice ▷ v cut into cubes; find the cube of (a number) ▶ **cubic** adj having three dimensions; cube-shaped ▶ **cubism** n style of art in

which objects are represented by geometrical shapes ▶ **cubist** adj, n
▶ **cube root** number whose cube is a given number

cubicle n enclosed part of a large room, screened for privacy

cuckold n man whose spouse has been unfaithful ▷ v be unfaithful to (one's spouse)

cuckoo n migratory bird with a characteristic two-note call, which lays its eggs in the nests of other birds ▷ adj (informal) insane or foolish

cucumber n long green-skinned fleshy fruit used in salads

cud n partially digested food which a ruminant brings back into its mouth to chew again **chew the cud** think deeply

cuddle v, n hug ▶ **cuddly** adj

cudgel n short thick stick used as a weapon

cue¹ n signal to an actor or musician to begin speaking or playing; signal or reminder ▷ v **cueing, cued** give a cue to

cue² n long tapering stick used in billiards, snooker, or pool ▷ v **cueing, cued** hit (a ball) with a cue

cuff¹ n end of a sleeve **off the cuff** (informal) without preparation ▶ **cuff link** one of a pair of decorative fastenings for shirt cuffs

cuff² (Brit, Aust & NZ) v hit with an open hand ▷ n blow with an open hand

cuisine [quiz-zeen] n style of cooking

cul-de-sac n road with one end blocked off

culinary adj of kitchens or cookery

cull v choose, gather; remove or kill (inferior or surplus animals) from a herd ▷ n culling

culminate v reach the highest point or climax ▶ **culmination** n

culottes pl n women's knee-length trousers cut to look like a skirt

culpable adj deserving blame ▶ **culpability** n

culprit n person guilty of an offence or misdeed

cult n specific system of worship; devotion to a person, idea, or activity; popular fashion

cultivate v prepare (land) to grow crops; grow (plants); develop or improve (something); try to develop a friendship with (someone) ▶ **cultivated** adj well-educated ▶ **cultivation** n

culture n ideas, customs, and art of a particular society; particular society; developed understanding of the arts; cultivation of plants or rearing of animals; growth of bacteria for study ▶ **cultural** adj ▶ **cultured** adj showing good taste or manners ▶ **cultured pearl** pearl artificially grown in an oyster shell

culvert n drain under a road or railway

cumbersome adj awkward because of size or shape

cumin, cummin n sweet-smelling seeds of a Mediterranean plant, used in cooking

cummerbund n wide sash worn round the waist

cumulative adj increasing steadily

cumulus [kew-myew-luss] n, pl **-li** thick white or dark grey cloud

cuneiform [kew-nif-form] n, adj (written in) an ancient system of writing using wedge-shaped characters

cunjevoi n (Aust) plant of tropical Asia and Australia with small flowers, cultivated for its edible rhizome; sea squirt

cunning adj clever at deceiving; ingenious ▷ n cleverness at deceiving; ingenuity ▷ **cunningly** adv

cup n small bowl-shaped drinking container with a handle; contents of a cup; (competition with) a cup-shaped trophy given as a prize; hollow rounded shape ▷ v **cupping, cupped** form (one's hands) into the shape of a cup; hold in cupped hands ▷ **cupful** n

cupboard n piece of furniture or alcove with a door, for storage

cupidity [kew-**pid**-it-ee] n greed for money or possessions

cupola [kew-pol-la] n domed roof or ceiling

cur n (lit) mongrel dog; contemptible person

curaçao [kew-rah-so] n orange-flavoured liqueur

curare [kew-rah-ree] n poisonous resin of a S American tree, used as a muscle relaxant in medicine

curate n clergyman who assists a parish priest ▷ **curacy** [kew-rah-see] n, pl **-cies** work or position of a curate

curative n, adj (something) able to cure

curator n person in charge of a museum or art gallery ▷ **curatorship** n

curb n something that restrains ▷ v control, restrain

curd n coagulated milk, used to make cheese ▷ **curdle** v turn into curd, coagulate

cure v get rid of (an illness or problem); make (someone) well again; preserve by salting, smoking, or drying ▷ n (treatment causing) curing of an illness or person; remedy or solution ▷ **curable** adj

curette n surgical instrument for scraping tissue from body

cavities ▷ v scrape with a curette ▷ **curettage** n

curfew n law ordering people to stay inside their homes after a specific time at night; time set as a deadline by such a law

curie n standard unit of radioactivity

curio n, pl **-rios** rare or unusual object valued as a collector's item

curious adj eager to learn or know; eager to find out private details; unusual or peculiar ▷ **curiously** adv ▷ **curiosity** n eagerness to know or find out; (pl **-ties**) rare or unusual object

curl n curved piece of hair; curved spiral shape ▷ v make (hair) into curls or (of hair) grow in curls; make into a curved spiral shape ▷ **curly** adj ▷ **curling** n game like bowls, played with heavy stones on ice

curlew n long-billed wading bird

curmudgeon n bad-tempered person

currajong n same as **kurrajong**

currant n small dried grape; small round berry, such as a redcurrant

currawong n Australian songbird

current adj of the immediate present; most recent, up-to-date; commonly accepted ▷ n flow of water or air in one direction; flow of electricity; general trend ▷ **currently** adv ▷ **currency** n, pl **-cies** money in use in a particular country; general acceptance or use

curriculum n, pl **-la** or **-lums** all the courses of study offered by a school or college ▷ **curriculum vitae** [vee-tie] outline of someone's educational and professional history, prepared for job applications

SPELLING TIP

The only letter that is doubled in **curriculum** is the r in the middle

curry¹ *n, pl* **-ries** Indian dish of meat or vegetables in a hot spicy sauce ▷ *v* **-rying, -ried** prepare (food) with curry powder ▶ **curry powder** mixture of spices for making curry

curry² *v* **-rying, -ried** groom (a horse) **curry favour** ingratiate oneself with an important person ▶ **curry comb** ridged comb for grooming a horse

curse *v* swear (at); ask a supernatural power to cause harm to ▷ *n* swear word; (result of) a call to a supernatural power to cause harm to someone; something causing trouble or harm ▶ **cursed** *adj*

cursive *n, adj* (handwriting) done with joined letters

cursor *n* movable point of light that shows a specific position on a visual display unit

cursory *adj* quick and superficial ▶ **cursorily** *adv*

curt *adj* brief and rather rude ▶ **curtly** *adv* ▶ **curtness** *n*

curtail *v* cut short; restrict ▶ **curtailment** *n*

curtain *n* piece of cloth hung at a window or opening as a screen; hanging cloth separating the audience and the stage in a theatre; fall or closing of the curtain at the end, or the rise or opening of the curtain at the start of a theatrical performance; something forming a barrier or screen ▷ *v* provide with curtains; (foll by *off*) separate with a curtain

curtsy, curtsey *n, pl* **-sies** or **-seys** woman's gesture of respect made by bending the knees and bowing the head ▷ *v* **-sying, -sied** or **-seying, -seyed** make a curtsy

curve *n* continuously bending line with no straight parts ▷ *v* form

or move in a curve ▶ **curvy** *adj* ▶ **curvaceous** *adj* (*informal*) (of a woman) having a shapely body ▶ **curvature** *n* curved shape ▶ **curvilinear** *adj* consisting of or bounded by a curve

cuscus *n, pl* **-cuses** large Australian nocturnal possum

cushion *n* bag filled with soft material, to make a seat more comfortable; something that provides comfort or absorbs shock ▷ *v* lessen the effects of; protect from injury or shock

cushy *adj* **cushier, cushiest** (*informal*) easy

cusp *n* pointed end, esp. on a tooth; (*Astrol*) division between houses or signs of the zodiac

cuss (*informal*) *n* curse, oath; annoying person ▷ *v* swear (at) ▶ **cussed** [kuss-id] ▷ *adj* (*informal*) obstinate

custard *n* sweet yellow sauce made from milk and eggs

custody *n* protective care; imprisonment prior to being tried ▶ **custodial** *adj* ▶ **custodian** *n* person in charge of a public building

custom *n* long-established activity or action; usual habit; regular use of a shop or business ▷ *pl* duty charged on imports or exports; government department which collects these; area at a port, airport, or border where baggage and freight are examined for dutiable goods ▶ **customary** *adj* usual; established by custom ▶ **customarily** *adv* ▶ **custom-built, custom-made** *adj* made to the specifications of an individual customer ▶ **customize** *v* modify (something) according to a customer's individual requirements ▶ **customization** *n*

customer *n* person who buys goods or services

cut *v* **cutting, cut** open up, penetrate, wound, or divide with a sharp instrument; divide; trim or shape by cutting; abridge, shorten; reduce, restrict; (*informal*) hurt the feelings of; pretend not to recognize ▷ *n* stroke or incision made by cutting; piece cut off; reduction; deletion in a text, film, or play; (*informal*) share, esp. of profits; style in which hair or a garment is cut ▸ **cut in** *v* interrupt; obstruct another vehicle in overtaking it

cutaneous [kew-**tane**-ee-uss] *adj* of the skin

cute *adj* appealing or attractive; (*informal*) clever or shrewd ▸ **cutely** *adv* ▸ **cuteness** *n*

cuticle *n* skin at the base of a fingernail or toenail

cutlass *n* curved one-edged sword formerly used by sailors

cutlery *n* knives, forks, and spoons ▸ **cutler** *n* maker of cutlery

cutlet *n* small piece of meat like a chop; flat croquette of chopped meat or fish

cutter *n* person or tool that cuts; any of various small fast boats

cut-throat *adj* fierce or relentless ▷ *n* murderer

cutting *n* article cut from a newspaper or magazine; piece cut from a plant from which to grow a new plant; passage cut through high ground for a road or railway ▷ *adj* (of a remark) hurtful

cuttlefish *n* squidlike sea mollusc

cuz *n* (*Aust & NZ informal*) term of address for a male friend or family member

CV curriculum vitae

cwt hundredweight

cyanide *n* extremely poisonous chemical compound

cyber- *combining form* computers e.g. *cyberspace*

cybernetics *n* branch of science in which electronic and mechanical systems are studied and compared to biological systems

cyberspace *n* place said to contain all the data stored in computers

cybersquatting *n* registering an internet domain name belonging to another person in the hope of selling it to them for a profit ▸ **cybersquatter** *n*

cyclamen [**sik**-la-men] *n* plant with red, pink, or white flowers

cycle *v* ride a bicycle ▷ *n* (*Brit, Aust & NZ*) bicycle; (*US*) motorcycle; complete series of recurring events; time taken for one such series ▸ **cyclical, cyclic** *adj* occurring in cycles ▸ **cyclist** *n* person who rides a bicycle

cyclone *n* violent wind moving round a central area

cyclotron *n* apparatus that accelerates charged particles by means of a strong vertical magnetic field

cygnet *n* young swan

cylinder *n* solid or hollow body with straight sides and circular ends; chamber within which the piston moves in an internal-combustion engine ▸ **cylindrical** *adj*

cymbal *n* percussion instrument consisting of a brass plate which is struck against another or hit with a stick

cynic [**sin**-ik] *n* person who believes that people always act selfishly ▸ **cynical** *adj* ▸ **cynically** *adv* ▸ **cynicism** *n*

cynosure [**sin**-oh-zyure] *n* centre of attention

cypher *n* same as **cipher**

cypress n evergreen tree with dark green leaves

Cypriot n, adj (person) from Cyprus

cyst [sist] n (abnormal) sac in the body containing fluid or soft matter ▶ **cystic** adj ▶ **cystitis** [siss-**tite**-iss] n inflammation of the bladder

cytology [site-ol-a-jee] n study of plant and animal cells ▶ **cytological** adj ▶ **cytologist** n

czar [zahr] n same as **tsar**

Czech n, adj (person) from the Czech Republic ▷ n language of the Czech Republic

Dd

D (Chem) deuterium

d (Physics) density

d. (Brit)(before decimalization) penny; died

dab[1] v **dabbing, dabbed** pat lightly; apply with short tapping strokes ▷ n small amount of something soft or moist; light stroke or tap ▶ **dab hand** (informal) person who is particularly good at something

dab[2] n small European flatfish with rough scales

dabble v be involved in something superficially; splash about ▶ **dabbler** n

dace n small European freshwater fish

dachshund n dog with a long body and short legs

dad n (informal) father

daddy n, pl -**dies** (informal) father

daddy-longlegs n (Brit) crane fly; (US & Canadian) small web-spinning spider with long legs

dado [**day**-doe] n, pl -**does** or -**dos** lower part of an interior wall, below a rail, decorated differently from the upper part

daffodil n yellow trumpet-shaped flower that blooms in spring

daft adj (informal) foolish or crazy

dag (NZ) n dried dung on a sheep's rear; (informal) amusing person ▷ pl n **rattle one's dags** (informal) hurry up ▷ v remove the dags from a sheep ▶ **daggy** ▷ adj (informal) amusing

dagga n (S Afr informal) cannabis

dagger n short knifelike weapon with a pointed blade

daguerreotype [dag-**gair**-oh-type] n type of early photograph produced on chemically treated silver

dahlia [day-lya] n brightly coloured garden flower

daily adj occurring every day or every weekday ▷ adv every day ▷ n, pl -**lies** daily newspaper; (Brit informal) person who cleans other people's houses

dainty adj -**tier**, -**tiest** delicate or elegant ▷ **daintily** adv

daiquiri [dak-eer-ee] n iced drink containing rum, lime juice, and sugar

dairy n, pl **dairies** place for the processing or sale of milk and its products; (NZ) small shop selling groceries and milk often outside normal trading hours; food containing milk or its products e.g. She can't eat dairy ▷ adj of milk or its products

dais [day-iss, dayss] n raised platform in a hall, used by a speaker

daisy n, pl -**sies** small wild flower with a yellow centre and white petals ▷ **daisy wheel** flat disc in a word processor with radiating spokes for printing letters

Dalai Lama n chief lama and (until 1959) ruler of Tibet

dale n (esp. in N England) valley

dally v -**lying**, -**lied** waste time; (foll by with) deal frivolously (with) ▷ **dalliance** n flirtation

Dalmatian n large dog with a white coat and black spots

dam[1] n barrier built across a river to create a lake; lake created by this ▷ v **damming**, **dammed** build a dam across (a river)

dam[2] n mother of an animal such as a sheep or horse

damage v harm, spoil ▷ n harm to a person or thing; (informal) cost e.g. what's the damage? ▷ pl money awarded as compensation for injury or loss

damask n fabric with a pattern woven into it, used for tablecloths etc.

dame n (chiefly US & Canadian slang) woman; (**D-**) title of a woman who has been awarded the OBE or another order of chivalry

damn interj (slang) exclamation of annoyance ▷ adv, adj (also **damned**) (slang) extreme(ly) ▷ v condemn as bad or worthless; (of God) condemn to hell ▷ **damnable** adj annoying ▷ **damnably** adv ▷ **damnation** interj, n ▷ **damning** adj proving or suggesting guilt e.g. a damning report

damp adj slightly wet ▷ n slight wetness, moisture ▷ v (also **dampen**) make damp; (foll by down) reduce the intensity of (feelings or actions) ▷ **damply** adv ▷ **dampness** n ▷ **damper** n movable plate to regulate the draught in a fire; pad in a piano that deadens the vibration of each string ▷ **put a damper on** have a depressing or inhibiting effect on

damsel n (old-fashioned) young woman

damson n small blue-black plumlike fruit

dance v move the feet and body rhythmically in time to music; perform (a particular dance); skip or leap; move rhythmically ▷ n series of steps and movements in time to music; social meeting arranged for dancing ▷ **dancer** n

D and C n (Med) dilatation and curettage: a minor operation in which the neck of the womb is stretched and the lining of the womb scraped, to clear the womb or remove tissue for diagnosis

dandelion n yellow-flowered wild plant

dander n **get one's dander up** (slang) become angry

dandle v move (a child) up and down on one's knee

dandruff n loose scales of dry dead skin shed from the scalp

dandy n, pl **-dies** man who is overconcerned with the elegance of his appearance ▷ adj **-dier, -diest** (informal) very good ▶ **dandified** adj

Dane n person from Denmark ▶ **Danish** n, adj (language) of Denmark ▶ **Danish blue** strong white cheese with blue veins ▶ **Danish pastry** iced puff pastry filled with fruit, almond paste, etc.

danger n possibility of being injured or killed; person or thing that may cause injury or harm; likelihood that something unpleasant will happen ▶ **dangerous** adj ▶ **dangerously** adv

dangle v hang loosely; display as an enticement

dank adj unpleasantly damp and chilly

dapper adj (of a man) neat in appearance

dappled adj marked with spots of a different colour ▶ **dapple-grey** n horse with a grey coat and darker coloured spots

dare v be courageous enough to try (to do something); challenge to do something risky ▷ n challenge to do something risky ▶ **daring** adj willing to take risks ▷ n courage to do dangerous things ▶ **daringly** adv ▶ **daredevil** adj, n recklessly bold (person)

USAGE NOTE
When *dare* is used in a question

or as a negative, it does not take an -s: *He dare not come*

dark adj having little or no light; (of a colour) reflecting little light; (of hair or skin) brown or black; gloomy, sad; sinister, evil; secret e.g. *keep it dark* ▷ n absence of light; night ▶ **darkly** adv ▶ **darkness** n ▶ **darken** v ▶ **dark horse** person about whom little is known ▶ **darkroom** n darkened room for processing photographic film

darling n much-loved person; favourite ▷ adj much-loved

darn[1] v mend (a garment) with a series of interwoven stitches ▷ n patch of darned work

darn[2] interj, adv, adj, v (euphemistic) damn

dart n small narrow pointed missile that is thrown or shot, esp. in the game of darts; sudden quick movement; tapered tuck made in dressmaking ▷ pl game in which darts are thrown at a circular numbered board ▷ v move or direct quickly and suddenly

Darwinism n theory of the origin of animal and plant species by evolution ▶ **Darwinian, Darwinist** adj, n

dash v move quickly; hurl or crash; frustrate (someone's hopes) ▷ n sudden quick movement; small amount; mixture of style and courage; punctuation mark (–) indicating a change of subject; longer symbol used in Morse code ▶ **dashing** adj stylish and attractive ▶ **dashboard** n instrument panel in a vehicle

dassie n (S Afr) type of hoofed rodent-like animal (also **hyrax**)

dastardly adj wicked and cowardly

dasyure [dass-ee-your] n small marsupial of Australia, New Guinea, and adjacent islands

data n information consisting of observations, measurements, or facts; numbers, digits, etc., stored by a computer ▶ **data base** store of information that can be easily handled by a computer ▶ **data capture** process for converting information into a form that can be handled by a computer ▶ **data processing** series of operations performed on data, esp. by a computer, to extract or interpret information

date¹ n specified day of the month; particular day or year when an event happened; (informal) appointment, esp. with a person to whom one is sexually attracted; (informal) person with whom one has a date ▷ v mark with the date; (informal) go on a date (with); assign a date of occurrence to; become old-fashioned; (foll by from) originate from ▶ **dated** adj old-fashioned

date² n dark-brown sweet-tasting fruit of the date palm ▶ **date palm** tall palm grown in tropical regions for its fruit

dative n (in certain languages) the form of the noun that expresses the indirect object

datum n, pl **data** single piece of information in the form of a fact or statistic

daub v smear or spread quickly or clumsily

daughter n female offspring; woman who comes from a certain place or is connected with a certain thing ▶ **daughterly** adj ▶ **daughter-in-law** n, pl **daughters-in-law** wife of one's child

daunting adj intimidating or worrying ▶ **dauntless** adj fearless

dauphin [doe-fan] n (formerly) eldest son of the king of France

davenport n (chiefly Brit) small writing table with drawers; (Aust, US & Canadian) large couch

davit [dav-vit] n crane, usu. one of a pair, at a ship's side, for lowering and hoisting a lifeboat

Davy lamp n miner's lamp designed to prevent it from igniting gas

dawdle v walk slowly, lag behind

dawn n daybreak; beginning (of something) ▷ v begin to grow light; begin to develop or appear; (foll by on) become apparent (to)

day n period of 24 hours; period of light between sunrise and sunset; part of a day occupied with regular activity, esp. work; period or point in time; time of success ▶ **daybreak** n time in the morning when light first appears ▶ **daydream** n pleasant fantasy indulged in while awake ▷ v indulge in idle fantasy ▶ **daydreamer** n ▶ **daylight** n light from the sun ▶ **day release** (Brit) system in which workers go to college one day a week ▶ **day-to-day** adj routine

daze v stun, by a blow or shock ▷ n state of confusion or shock

dazzle v impress greatly; blind temporarily by sudden excessive light ▷ n bright light that dazzles ▶ **dazzling** adj ▶ **dazzlingly** adv

dB, db decibel(s)

DC direct current

DD Doctor of Divinity

D-day n day selected for the start of some operation, orig. the Allied invasion of Europe in 1944

DDT n kind of insecticide

de- prefix indicating removal e.g. dethrone; indicating reversal e.g. declassify; indicating departure e.g. decamp

deacon n (Christianity) ordained minister ranking immediately below a priest; (in some Protestant churches) lay official who assists the minister

dead adj no longer alive; no longer in use; numb e.g. my leg has gone dead; complete, absolute e.g. dead silence; (informal) very tired; (of a place) lacking activity ▷ n period during which coldness or darkness is most intense e.g. in the dead of night ▷ adv extremely; suddenly e.g. I stopped dead **the dead** dead people ▶ **dead set** firmly decided ▶ **deadbeat** n (informal) lazy useless person ▶ **dead beat** (informal) exhausted ▶ **dead end** road with one end blocked off; situation in which further progress is impossible ▶ **deadline** n time limit ▶ **deadlock** n point in a dispute at which no agreement can be reached ▶ **deadlocked** adj ▶ **deadpan** adj, adv showing no emotion or expression ▶ **dead reckoning** method of establishing one's position using the distance and direction travelled ▶ **dead zone** area where a mobile phone does not receive a signal; area where something does not exist or prosper ▶ **dead weight** heavy weight

deaden v make less intense

deadly adj **-lier, -liest** likely to cause death; (informal) extremely boring ▷ adv extremely ▶ **deadly nightshade** plant with poisonous black berries

deaf adj unable to hear **deaf to** refusing to listen to or take notice of ▶ **deafen** v make deaf, esp. temporarily ▶ **deafness** n

deal¹ n agreement or transaction; kind of treatment e.g. a fair deal; large amount ▷ v **dealing,**

dealt [delt] inflict (a blow) on; (Cards) give out (cards) to the players ▶ **dealer** n ▶ **dealings** pl n transactions or business relations ▶ **deal in** v buy or sell (goods) ▶ **deal out** v distribute ▶ **deal with** v take action on; be concerned with

deal² n plank of fir or pine wood

dean n chief administrative official of a college or university faculty; chief administrator of a cathedral ▶ **deanery** n, pl **-eries** office or residence of a dean; parishes of a dean

dear n someone regarded with affection ▷ adj much-loved; costly ▶ **dearly** adv ▶ **dearness** n

dearth [dirth] n inadequate amount, scarcity

death n permanent end of life in a person or animal; instance of this; ending, destruction ▶ **deathly** adj, adv like death e.g. a deathly silence; deathly pale ▶ **death duty** (in Britain) former name for **inheritance tax** ▶ **death's-head** n human skull or a representation of one ▶ **death trap** place or vehicle considered very unsafe ▶ **deathwatch beetle** beetle that bores into wood and makes a tapping sound

deb n (informal) debutante

debacle [day-bah-kl] n disastrous failure

debar v prevent, bar

debase v lower in value, quality, or character ▶ **debasement** n

debate n discussion ▷ v discuss formally; consider (a course of action) ▶ **debatable** adj not absolutely certain

debauch [dib-bawch] v make (someone) bad or corrupt, esp. sexually ▶ **debauched** adj immoral, sexually corrupt ▶ **debauchery** n

debenture n long-term bond bearing fixed interest, issued by a company or a government agency

debilitate v weaken, make feeble ▶ **debilitation** n ▶ **debility** n weakness, infirmity

debit n acknowledgment of a sum owing by entry on the left side of an account ▷ v **debiting, debited** charge (an account) with a debt

debonair adj (of a man) charming and refined

debouch v move out from a narrow place to a wider one

debrief v receive a report from (a soldier, diplomat, etc.) after an event ▶ **debriefing** n

debris [deb-ree] n fragments of something destroyed

debt n something owed, esp. money **in debt** owing money ▶ **debtor** n

debug v (informal) find and remove defects in (a computer program); remove concealed microphones from (a room or telephone)

debunk v (informal) expose the falseness of

debut [day-byoo] n first public appearance of a performer ▶ **debutante** [day-byoo-tont] n young upper-class woman being formally presented to society

Dec December

decade n period of ten years

decadence n deterioration in morality or culture ▶ **decadent** adj

decaffeinated [dee-kaf-fin-ate-id] adj (of coffee, tea, or cola) with caffeine removed

decagon n geometric figure with ten sides

decahedron [deck-a-heed-ron] n solid figure with ten faces

Decalogue n the Ten Commandments

decamp v depart secretly or suddenly

decant v pour (a liquid) from one container to another; (chiefly Brit) rehouse (people) while their homes are being renovated

decanter n stoppered bottle for wine or spirits

decapitate v behead ▶ **decapitation** n

decathlon n athletic contest with ten events

decay v become weaker or more corrupt; rot ▷ n process of decaying; state brought about by this process

decease n (formal) death

deceased adj (formal) dead **the deceased** dead person

deceive v mislead by lying; be unfaithful to (one's sexual partner) ▶ **deceiver** n ▶ **deceit** n behaviour intended to deceive ▶ **deceitful** adj

decelerate v slow down ▶ **deceleration** n

December n twelfth month of the year

decent adj (of a person) polite and morally acceptable; fitting or proper; conforming to conventions of sexual behaviour; (informal) kind ▶ **decently** adv ▶ **decency** n

decentralize v reorganize into smaller local units ▶ **decentralization** n

deception n deceiving; something that deceives; trick ▶ **deceptive** adj likely or designed to deceive ▶ **deceptively** adv ▶ **deceptiveness** n

deci- combining form one tenth

decibel n unit for measuring the intensity of sound

decide v (cause to) reach a decision; settle (a contest or question) ▶ **decided** adj unmistakable; determined ▶ **decidedly** adv

deciduous adj (of a tree) shedding its leaves annually

decimal n fraction written in the form of a dot followed by one or more numbers ▷ adj relating to or using powers of ten; expressed as a decimal ▶ **decimalization** n ▶ **decimal currency** system of currency in which the units are parts or powers of ten ▶ **decimal point** dot between the unit and the fraction of a number in the decimal system ▶ **decimal system** number system with a base of ten, in which numbers are expressed by combinations of the digits 0 to 9

decimate v destroy or kill a large proportion of ▶ **decimation** n

decipher v work out the meaning of (something illegible or in code) ▶ **decipherable** adj

decision n judgment, conclusion, or resolution; act of making up one's mind; firmness of purpose ▶ **decisive** adj having a definite influence; having the ability to make quick decisions ▶ **decisively** adv ▶ **decisiveness** n

deck n area of a ship that forms a floor; similar area in a bus; platform that supports the turntable and pick-up of a record player ▶ **deck chair** folding chair made of canvas over a wooden frame ▶ **decking** n wooden platform in a garden ▶ **deck out** v decorate

declaim v speak loudly and dramatically; protest loudly ▶ **declamation** n ▶ **declamatory** adj

declare v state firmly and forcefully; announce officially; acknowledge for tax purposes ▶ **declaration** n ▶ **declaratory** adj

declension n (Grammar) changes in the form of nouns, pronouns, or adjectives to show case, number, and gender

decline v become smaller, weaker, or less important; refuse politely to accept or do; (Grammar) list the inflections of (a noun, pronoun, or adjective) ▷ n gradual weakening or loss

declivity n, pl **-ties** downward slope

declutch v disengage the clutch of a motor vehicle

decoct v extract the essence from (a substance) by boiling ▶ **decoction** n

decode v convert from code into ordinary language ▶ **decoder** n

décolleté [day-**kol**-tay] adj (of a woman's garment) low-cut

decommission v dismantle (a nuclear reactor, weapon, etc.) which is no longer needed

decompose v be broken down through chemical or bacterial action ▶ **decomposition** n

decompress v free from pressure; return (a diver) to normal atmospheric pressure ▶ **decompression** n ▶ **decompression sickness** severe pain and difficulty in breathing, caused by a sudden change in atmospheric pressure

decongestant n medicine that relieves nasal congestion

decontaminate v make safe by removing poisons, radioactivity, etc. ▶ **decontamination** n

decor [day-**core**] n style in which a room or house is decorated

decorate v make more attractive by adding something ornamental; paint or wallpaper; award a (military) medal to ▶ **decoration** n ▶ **decorative** adj ▶ **decorator** n

decorous [dek-a-russ] adj polite, calm, and sensible in behaviour ▶ **decorously** adv

decorum [dik-**core**-um] n polite and socially correct behaviour

decoy n person or thing used to lure someone into danger; dummy bird or animal, used to lure game within shooting range ▷ v lure away by means of a trick

decrease v make or become less ▷ n lessening, reduction; amount by which something has decreased

decree n law made by someone in authority; court judgment ▷ v order by decree

decrepit adj weakened or worn out by age or long use ▶ **decrepitude** n

decry v -**crying**, -**cried** express disapproval of

decrypt v decode (a message) with or without previous knowledge of its key; make intelligible (a television or other signal) that has been deliberately distorted for transmission

dedicate v commit (oneself or one's time) wholly to a special purpose or cause; inscribe or address (a book etc.) to someone as a tribute ▶ **dedicated** adj devoted to a particular purpose or cause ▶ **dedication** n

deduce v reach (a conclusion) by reasoning from evidence ▶ **deducible** adj

deduct v subtract

deduction n deducting; something that is deducted; deducing; conclusion reached by deducing ▶ **deductive** adj

deduplicate v remove (duplicated material) from a system

deed n something that is done; legal document

deem v consider, judge

deep adj extending or situated far down, inwards, backwards, or sideways; of a specified dimension downwards, inwards, or backwards; difficult to understand; of great intensity; (foll by in) absorbed in (an activity); (of a colour) strong or dark; low in pitch **the deep** (poetic) the sea ▶ **deeply** adv profoundly or intensely (also **deep down**) ▶ **deepen** v ▶ **deep-freeze** n same as **freezer**

deer n, pl **deer** large wild animal, the male of which has antlers ▶ **deerstalker** n cloth hat with peaks at the back and front and earflaps

deface v deliberately spoil the appearance of ▶ **defacement** n

de facto adv in fact ▷ adj existing in fact, whether legally recognized or not

defame v attack the good reputation of ▶ **defamation** n ▶ **defamatory** [dif-**fam**-a-tree] adj

default n failure to do something; (Computing) instruction to a computer to select a particular option unless the user specifies otherwise ▷ v fail to fulfil an obligation **in default of** in the absence of ▶ **defaulter** n

defeat v win a victory over; thwart, frustrate ▷ n defeating ▶ **defeatism** n ready acceptance or expectation of defeat ▶ **defeatist** adj, n

defecate v discharge waste from the body through the anus ▶ **defecation** n

defect n imperfection, blemish ▷ v desert one's cause or country to join the opposing forces ▶ **defective** adj imperfect, faulty ▶ **defection** n ▶ **defector** n

defence n resistance against attack; argument in support of something; country's military resources; defendant's case in a court of law ▶ **defenceless** adj

defend v protect from harm or danger; support in the face of criticism; represent (a defendant) in court ▸ **defendant** n person accused of a crime ▸ **defensible** adj capable of being defended because believed to be right ▸ **defensibility** n ▸ **defensive** adj intended for defence; overanxious to protect oneself against (threatened) criticism ▸ **defensively** adv

defender n person who supports someone or something in the face of criticism; player whose chief task is to stop the opposition scoring

defer¹ v -ferring, -ferred delay (something) until a future time ▸ **deferment**, **deferral** n

defer² v -ferring, -ferred (foll by to) comply with the wishes (of) ▸ **deference** n polite and respectful behaviour ▸ **deferential** adj ▸ **deferentially** adv

defiance n see **defy**

defibrillator n apparatus for stopping fibrillation of the heart by application of an electric current to the chest wall or directly to the heart

deficient adj lacking some essential thing or quality; inadequate in quality or quantity ▸ **deficiency** n state of being deficient; lack, shortage

deficit n amount by which a sum of money is too small

defile¹ v treat (something sacred or important) without respect ▸ **defilement** n

defile² n narrow valley or pass

define v state precisely the meaning of; show clearly the outline of ▸ **definable** adj ▸ **definite** adj firm, clear, and precise; having precise limits; known for certain ▸ **definitely** adv ▸ **definition** n statement of the meaning of a word or phrase;

quality of being clear and distinct ▸ **definitive** adj providing an unquestionable conclusion; being the best example of something

deflate v (cause to) collapse through the release of air; take away the self-esteem or conceit from; (Economics) cause deflation of (an economy) ▸ **deflation** n (Economics) reduction in economic activity resulting in lower output and investment; feeling of sadness following excitement ▸ **deflationary** adj

deflect v (cause to) turn aside from a course ▸ **deflection** n ▸ **deflector** n

deflower v (lit) deprive (a woman) of her virginity

defoliate v deprive (a plant) of its leaves ▸ **defoliant** n ▸ **defoliation** n

deforestation n destruction of all the trees in an area

deform v put out of shape or spoil the appearance of ▸ **deformation** n ▸ **deformity** n

defraud v cheat out of money, property, etc.

defray v provide money for (costs or expenses)

defriend v remove (a person) from the list of one's friends on a social networking website

defrock v deprive (a priest) of priestly status

defrost v make or become free of ice; thaw (frozen food) by removing it from a freezer

deft adj quick and skilful in movement ▸ **deftly** adv ▸ **deftness** n

defunct adj no longer existing or operative

defuse v remove the fuse of (an explosive device); remove the tension from (a situation)

defy v **-fying, -fied** resist openly and boldly; make impossible e.g. *the condition of the refugees defied description* ▶ **defiance** n ▶ **defiant** adj

degenerate adj having deteriorated to a lower mental, moral, or physical level ▷ n degenerate person ▷ v become degenerate ▶ **degeneracy** n degenerate behaviour ▶ **degeneration** n

degrade v reduce to dishonour or disgrace; reduce in status or quality; (Chem) decompose into smaller molecules ▶ **degradation** n

degree n stage in a scale of relative amount or intensity; academic award given by a university or college on successful completion of a course; unit of measurement for temperature, angles, or latitude and longitude

dehumanize v deprive of human qualities; make (an activity) mechanical or routine ▶ **dehumanization** n

dehydrate v remove water from (food) to preserve it **be dehydrated** become weak through losing too much water from the body ▶ **dehydration** n

de-ice v free of ice ▶ **de-icer** n

deify [day-if-fie] v **-fying, -fied** treat or worship as a god ▶ **deification** n

deign [dane] v agree (to do something), but as if doing someone a favour

deity [dee-it-ee, day-it-ee] n, pl **-ties** god or goddess; state of being divine

déjà vu [day-zhah voo] n feeling of having experienced before something that is actually happening now

dejected adj unhappy ▶ **dejectedly** adv ▶ **dejection** n

de jure adv, adj according to law

dekko n (Brit, Aust & NZ slang) **have a dekko** have a look

delay v put off to a later time; slow up or cause to be late ▷ n act of delaying; interval of time between events

delectable adj delightful, very attractive ▶ **delectation** n (formal) great pleasure

delegate n person chosen to represent others, esp. at a meeting ▷ v entrust (duties or powers) to someone; appoint as a delegate ▶ **delegation** n group chosen to represent others; delegating

delete v remove (something written or printed) ▶ **deletion** n

deleterious [del-lit-eer-ee-uss] adj harmful, injurious

deli n (informal) delicatessen

deliberate adj planned in advance, intentional; careful and unhurried ▷ v think something over ▶ **deliberately** adv ▶ **deliberation** n ▶ **deliberative** adj

delicate adj fine or subtle in quality or workmanship; having a fragile beauty; (of a taste etc.) pleasantly subtle; easily damaged; requiring tact ▶ **delicately** adv ▶ **delicacy** n being delicate; (pl **-cies**) something particularly good to eat

delicatessen n shop selling imported or unusual foods, often already cooked or prepared

delicious adj very appealing to taste or smell ▶ **deliciously** adv

delight n (source of) great pleasure ▷ v please greatly; (foll by in) take great pleasure (in) ▶ **delightful** adj ▶ **delightfully** adv

delimit v mark or lay down the limits of ▶ **delimitation** n

delineate [dill-**lin**-ee-ate] v show by drawing; describe in words ▶ **delineation** n

delinquent n someone, esp. a young person, who repeatedly breaks the law ▷ adj repeatedly breaking the law ▶ **delinquency** n

delirium n state of excitement and mental confusion, often with hallucinations; great excitement ▶ **delirious** adj ▶ **deliriously** adv

deliver v carry (goods etc.) to a destination; hand over; aid in the birth of; present (a lecture or speech); release or rescue; strike (a blow) ▶ **deliverance** n rescue from captivity or evil ▶ **delivery** n, pl **-eries** delivering; something that is delivered; act of giving birth to a baby; style in public speaking

dell n (chiefly Brit) small wooded hollow

Delphic adj ambiguous, like the ancient Greek oracle at Delphi

delphinium n large garden plant with blue flowers

delta n fourth letter in the Greek alphabet; flat area at the mouth of some rivers where the main stream splits up into several branches

delude v deceive

deluge [**del**-lyooj] n great flood; torrential rain; overwhelming number ▷ v flood; overwhelm

delusion n mistaken idea or belief; state of being deluded ▶ **delusive** adj

de luxe adj rich or sumptuous, superior in quality

delve v research deeply (for information)

demagogue n political agitator who appeals to the prejudice and passions of the mob ▶ **demagogic** adj ▶ **demagogy** n

demand v request forcefully; require as just, urgent, etc.; claim

as a right ▷ n forceful request; (Economics) willingness and ability to purchase goods and services; something that requires special effort or sacrifice ▶ **demanding** adj requiring a lot of time or effort

demarcation n (formal) establishing limits or boundaries, esp. between the work performed by different trade unions

demean v **demean oneself** do something unworthy of one's status or character

demeanour n way a person behaves

demented adj mad ▶ **dementedly** adv ▶ **dementia** [dim-**men**-sha] n state of serious mental deterioration

demerara sugar n brown crystallized cane sugar

demerit n fault, disadvantage

demesne [dim-**mane**] n land surrounding a house; (Law) possession of one's own property or land

demi- combining form half

demijohn n large bottle with a short neck, often encased in wicker

demilitarize v remove the military forces from ▶ **demilitarization** n

demimonde n (esp. in the 19th century) class of women considered to be outside respectable society because of promiscuity; group considered not wholly respectable

demise n eventual failure of something successful); (formal) death

demo n, pl **demos** (informal) demonstration, organized expression of public opinion

demob v (Brit, Aust & NZ informal) demobilize

demobilize v release from the armed forces ▶ **demobilization** n

democracy n, pl **-cies** government by the people or their elected representatives; state governed in this way ▶ **democrat** n advocate of democracy; (**D-**) member or supporter of the Democratic Party in the US ▶ **democratic** adj of democracy; upholding democracy; (**D-**) of the Democratic Party, the more liberal of the two main political parties in the US ▶ **democratically** adv

demography n study of population statistics, such as births and deaths ▶ **demographer** n ▶ **demographic** adj

demolish v knock down or destroy (a building); disprove (an argument) ▶ **demolition** n

demon n evil spirit; person who does something with great energy or skill ▶ **demonic** adj evil ▶ **demoniac, demoniacal** adj appearing to be possessed by a devil; frenzied ▶ **demonology** n study of demons

demonstrate v show or prove by reasoning or evidence; display and explain the workings of; reveal the existence of; show support or opposition by public parades or rallies ▶ **demonstrable** adj able to be proved ▶ **demonstrably** adv ▶ **demonstration** n organized expression of public opinion; explanation or display of how something works; proof ▶ **demonstrative** adj tending to show one's feelings unreservedly ▶ **demonstratively** adv ▶ **demonstrator** n person who demonstrates how a device or machine works; person who takes part in a public demonstration

demoralize v undermine the morale of ▶ **demoralization** n

demote v reduce in status or rank ▶ **demotion** n

demur v -murring, -murred show reluctance ▷ n **without demur** without objecting

demure adj quiet, reserved, and rather shy ▶ **demurely** adv

den n home of a wild animal; small secluded room in a home; place where people indulge in criminal or immoral activities

denationalize v transfer (an industry) from public to private ownership ▶ **denationalization** n

denature v change the nature of; make (alcohol) unfit to drink

denier [den-yer] n unit of weight used to measure the fineness of nylon or silk

denigrate v criticize unfairly ▶ **denigration** n

denim n hard-wearing cotton fabric, usu. blue ▷ pl jeans made of denim

denizen n inhabitant

denominate v give a specific name to

denomination n group having a distinctive interpretation of a religious faith; unit in a system of weights, values, or measures ▶ **denominational** adj

denominator n number below the line in a fraction

denote v be a sign of; have as a literal meaning ▶ **denotation** n

denouement [day-noo-mon] n final outcome or solution in a play or book

denounce v speak vehemently against; give information against ▶ **denunciation** n open condemnation

dense adj closely packed; difficult to see through; stupid ▶ **densely** adv ▶ **density** n, pl **-ties** degree to which something is filled or occupied; measure of the compactness of a substance, expressed as its mass per unit volume

dent n hollow in the surface of something, made by hitting it ▷ v make a dent in

dental adj of teeth or dentistry ▶ **dental floss** waxed thread used to remove food particles from between the teeth ▶ **dentine** [den-teen] n hard dense tissue forming the bulk of a tooth ▶ **denture** n false tooth

dentist n person qualified to practise dentistry ▶ **dentistry** n branch of medicine concerned with the teeth and gums

denude v remove the covering or protection from

deny v **-nying, -nied** declare to be untrue; refuse to give or allow; refuse to acknowledge ▶ **deniable** adj ▶ **denial** n statement that something is not true; rejection of a request

deodorant n substance applied to the body to mask the smell of perspiration

deodorize v remove or disguise the smell of

depart v leave; differ, deviate ▶ **departed** adj (euphemistic) dead **the departed** (euphemistic) dead person ▶ **departure** n

department n specialized division of a large organization; major subdivision of the administration of a government ▶ **departmental** adj ▶ **department store** large shop selling many kinds of goods

depend v (foll by on) put trust (in); be influenced or determined (by); rely (on) for income or support ▶ **dependable** adj ▶ **dependably** adv ▶ **dependability** n ▶ **dependant** n person who depends on another for financial

support ▶ **dependence** n state of being dependent ▶ **dependency** n, pl **-cies** country controlled by another country; overreliance on another person or on a drug ▶ **dependent** adj depending on someone or something

SPELLING TIP
The words **dependant** and **dependent** are easy to confuse. The first, ending in -ant, is a noun meaning a person who is dependent (adjective ending in -ent) on someone else

depict v produce a picture of; describe in words ▶ **depiction** n

depilatory [dip-pill-a-tree] n, pl **-tories** ▷ adj (substance) designed to remove unwanted hair

deplete v use up; reduce in number ▶ **depletion** n

deplore v condemn strongly ▶ **deplorable** adj very bad or unpleasant

deploy v organize (troops or resources) into a position ready for immediate action ▶ **deployment** n

depopulate v reduce the population of a ▶ **depopulation** n

deport v remove forcibly from a country **deport oneself** behave in a specified way ▶ **deportation** n ▶ **deportee** n

deportment n way in which a person moves or stands

depose v remove from an office or position of power; (Law) testify on oath

deposit v put down; entrust for safekeeping, esp. to a bank; lay down naturally ▷ n sum of money paid into a bank account; money given in part payment for goods or services; accumulation of sediments, minerals, etc. ▶ **depositary** n person to whom something is entrusted for safety

▶ **depositor** n ▶ **depository** n store for furniture etc.

deposition n (Law) sworn statement of a witness used in court in his or her absence; depositing; depositing; something deposited

depot [dep-oh] n building where goods or vehicles are kept when not in use; (NZ & US) bus or railway station

depraved adj morally bad ▶ **depravity** n

deprecate v express disapproval of ▶ **deprecation** n ▶ **deprecatory** adj

depreciate v decline in value or price; criticize ▶ **depreciation** n

depredation n plundering

depress v make sad; lower (prices or wages); push down ▶ **depressing** adj ▶ **depressingly** adv ▶ **depressant** n, adj (drug) able to reduce nervous activity ▶ **depression** n mental state in which a person has feelings of gloom and inadequacy; economic condition in which there is high unemployment and low output and investment; area of low air pressure; sunken place ▶ **depressive** adj tending to cause depression ▷ n person who suffers from depression

deprive v (foll by of) prevent from (having or enjoying) ▶ **deprivation** n ▶ **deprived** adj lacking adequate living conditions, education, etc.

dept department

depth n distance downwards, backwards, or inwards; intensity of emotion; profundity of character or thought ▶ **depth charge** bomb used to attack submarines by exploding at a preset depth of water

depute v appoint (someone) to act on one's behalf ▶ **deputation** n body of people appointed to represent others

deputy n, pl -ties person appointed to act on behalf of another ▶ **deputize** v act as deputy

derail v cause (a train) to go off the rails ▶ **derailment** n

deranged adj insane or uncontrolled; in a state of disorder ▶ **derangement** n

derby [dah-bee] n, pl -bies sporting event between teams from the same area ▷ n any of various horse races

deregulate v remove regulations or controls from ▶ **deregulation** n

derelict adj unused and falling into ruins ▷ n social outcast, vagrant ▶ **dereliction** n state of being abandoned **dereliction of duty** failure to do one's duty

deride v treat with contempt or ridicule ▶ **derision** n ▶ **derisive** adj mocking, scornful ▶ **derisory** adj too small or inadequate to be considered seriously

de rigueur [de rig-gur] adj required by fashion

derive v (foll by from) take or develop (from) ▶ **derivation** n ▶ **derivative** adj based on other sources, not original ▷ n word, idea, etc., derived from another; (Maths) rate of change of one quantity in relation to another

dermatitis n inflammation of the skin

dermatology n branch of medicine concerned with the skin ▶ **dermatologist** n

derogatory [dir-rog-a-tree] adj intentionally offensive

derrick n simple crane; framework erected over an oil well

derv n (Brit) diesel oil, when used for road transport

dervish n member of a Muslim religious order noted for a frenzied whirling dance

descant n (Music) tune played or sung above a basic melody

descend v move down (a slope etc.); move to a lower level, pitch, etc.; (foll by to) stoop to (unworthy behaviour); (foll by on) visit unexpectedly **be descended from** connected by a blood relationship to ▸ **descendant** n person or animal descended from an individual, race, or species ▸ **descendent** adj descending ▸ **descent** n descending; downward slope; derivation from an ancestor

describe v give an account of (something or someone) in words; trace the outline of (a circle etc.) ▸ **description** n statement that describes something or someone; sort e.g. *flowers of every description* ▸ **descriptive** adj ▸ **descriptively** adv

descry v -scrying, -scried catch sight of; discover by looking carefully

desecrate v damage or insult (something sacred) ▸ **desecration** n

desegregate v end racial segregation in ▸ **desegregation** n

deselect v (Brit Politics) refuse to select (an MP) for re-election ▸ **deselection** n

desert[1] n region with little or no vegetation because of low rainfall

desert[2] v abandon (a person or place) without intending to return; (Mil) leave (a post or duty) with no intention of returning ▸ **deserter** n ▸ **desertion** n

deserts pl n **get one's just deserts** get the punishment one deserves

deserve v be entitled to or worthy of ▸ **deserved** adj rightfully earned

▸ **deservedly** adv ▸ **deserving** adj worthy of help, praise, or reward

deshabille [day-zab-beel] n state of being partly dressed

desiccate v remove most of the water from ▸ **desiccation** n

> **SPELLING TIP**
> The word **desiccate** has two cs because it comes from the Latin word *siccus*, which means 'dry'

design v work out the structure or form of (something), by making a sketch or plans; plan and make artistically; intend for a specific purpose ▸ n preliminary drawing; arrangement or features of an artistic or decorative work; art of designing; intention e.g. *by design* ▸ **designedly** [dee-zine-id-lee] adv intentionally ▸ **designer** n person who draws up original sketches or plans from which things are made ▸ adj designed by a well-known designer ▸ **designing** adj cunning and scheming

designate [dez-zig-nate] v give a name to; select (someone) for an office or duty ▸ adj appointed but not yet in office ▸ **designation** n name

desire v want very much ▸ n wish, longing; sexual appetite; person or thing desired ▸ **desirable** adj worth having; arousing sexual desire ▸ **desirability** n **desirous of** having a desire for

desist v (foll by from) stop (doing something)

desk n piece of furniture with a writing surface and drawers; service counter in a public building; section of a newspaper covering a specific subject e.g. *the sports desk* ▸ **desktop** adj (of a computer) small enough to use at a desk ▸ n main screen display on computer

desolate adj uninhabited and bleak; very sad ▸ v deprive of inhabitants; make (someone) very sad ▸ **desolation** n

despair n total loss of hope ▹ v lose hope

despatch v, n same as **dispatch**

desperado n, pl **-does** or **-dos** reckless person ready to commit any violent illegal act

desperate adj in despair and reckless; (of an action) undertaken as a last resort; having a strong need or desire ▸ **desperately** adv ▸ **desperation** n

> **SPELLING TIP**
>
> Spelling **desperate** as *desparate* (with an *a* in its second syllable) is a common mistake

despise v regard with contempt ▸ **despicable** adj deserving contempt ▸ **despicably** adv

despite prep in spite of

despoil v (formal) plunder ▸ **despoliation** n

despondent adj unhappy ▸ **despondently** adv ▸ **despondency** n

despot n person in power who acts unfairly or cruelly ▸ **despotic** adj ▸ **despotism** n unfair or cruel government or behaviour

dessert n sweet course served at the end of a meal ▸ **dessertspoon** n spoon between a tablespoon and a teaspoon in size

destination n place to which someone or something is going

destined [dess-tind] adj certain to be or to do something

destiny n, pl **-nies** future marked out for a person or thing; the power that predetermines the course of events

destitute adj having no money or possessions ▸ **destitution** n

destroy v ruin, demolish; put an end to; kill (an animal) ▸ **destroyer** n small heavily armed warship; person or thing that destroys

destruction n destroying; cause of ruin ▸ **destructive** adj (capable of) causing destruction ▸ **destructively** adv

desuetude [diss-syoo-it-tude] n condition of not being in use

desultory [dez-zl-tree] adj jumping from one thing to another, disconnected; random ▸ **desultorily** adv

detach v disengage and separate ▸ **detachable** adj ▸ **detached** adj (Brit, Aust & S Afr) (of a house) not joined to another house; showing no emotional involvement ▸ **detachment** n lack of emotional involvement; small group of soldiers

detail n individual piece of information; unimportant item; small individual features of something, considered collectively; (chiefly Mil) (personnel assigned) a specific duty ▹ v list fully

detain v delay (someone); hold (someone) in custody ▸ **detainee** n

detect v notice; discover, find ▸ **detectable** adj ▸ **detection** n ▸ **detective** n policeman or private agent who investigates crime ▸ **detector** n instrument used to find something

detente [day-tont] n easing of tension between nations

detention n imprisonment; form of punishment in which a pupil is detained after school

deter v **-terring**, **-terred** discourage (someone) from doing something by instilling fear or doubt ▸ **deterrent** n something that deters; weapon, esp. nuclear, intended to deter attack ▹ adj tending to deter

detergent n chemical substance for washing clothes or dishes

deteriorate v become worse
▸ **deterioration** n

determine v settle (an argument or a question) conclusively; find out the facts about; make a firm decision (to do something) ▸ **determinant** n factor that determines ▸ **determinate** adj definitely limited or fixed ▸ **determination** n being determined or resolute ▸ **determined** adj firmly decided, unable to be dissuaded ▸ **determinedly** adv ▸ **determiner** n (Grammar) word that determines the object to which a noun phrase refers e.g. all ▸ **determinism** n theory that human choice is not free, but decided by past events ▸ **determinist** n, adj

detest v dislike intensely ▸ **detestable** adj ▸ **detestation** n

dethrone v remove from a throne or position of power

detonate v explode ▸ **detonation** n ▸ **detonator** n small amount of explosive, or a device, used to set off an explosion

detour n route that is not the most direct one

detox v, n (informal) (undergo) treatment to rid the body of poisonous substances, esp. alcohol or drugs

detract v (foll by from) make (something) seem less good ▸ **detractor** n

detriment n disadvantage or damage ▸ **detrimental** adj ▸ **detrimentally** adv

detritus [dit-**trite**-uss] n loose mass of stones and silt worn away from rocks; debris

de trop [de **troh**] adj (French) unwanted, unwelcome

deuce [dyooss] n (Tennis) score of forty all; playing card with two symbols or dice with two spots

deuterium n isotope of hydrogen twice as heavy as the normal atom

Deutschmark [doytch-mark], **Deutsche Mark** [doytch-a] former monetary unit of Germany

devalue v -valuing, -valued reduce the exchange value of (a currency); reduce the value of (something or someone) ▸ **devaluation** n

devastate v destroy ▸ **devastated** adj shocked and extremely upset ▸ **devastation** n

develop v grow or bring to a later, more elaborate, or more advanced stage; come or bring into existence; build houses or factories on (an area of land); produce (photographs) by making negatives or prints from a film ▸ **developer** n person who develops property; chemical used to develop photographs or films ▸ **development** n ▸ **developing country** poor or nonindustrial country that is trying to develop its resources by industrialization

deviate v differ from others in belief or thought; depart from one's previous behaviour ▸ **deviation** n ▸ **deviant** n, adj (person) deviating from what is considered acceptable behaviour ▸ **deviance** n

device n machine or tool used for a specific task; scheme or plan

devil n evil spirit; evil person; person e.g. poor devil; daring person e.g. be a devil!; (informal) something difficult or annoying e.g. a devil of a long time to v -illing, -illed prepare (food) with a highly flavoured spiced mixture **the Devil** (Theology) chief spirit of evil and enemy of God ▸ **devilish** adj cruel or unpleasant ▷ *adv* (also **devilishly**) (*informal*) extremely ▷ **devilment** *n* mischievous conduct ▷ **devilry** *n* mischievousness ▷ **devil-may-care** *adj* carefree and cheerful ▷ **devil's advocate** person who takes an opposing or unpopular point of view for the sake of argument

devious *adj* insincere and dishonest; indirect ▷ **deviously** *adv* ▷ **deviousness** *n*

devise *v* work out (something) in one's mind

devoid *adj* (foll by *of*) completely lacking (in)

devolve *v* (foll by *on* or *to*) pass (power or duties) or (of power or duties) be passed to a successor or substitute ▷ **devolution** *n* transfer of authority from a central government to regional governments

devote *v* apply or dedicate to a particular purpose ▷ **devoted** *adj* showing loyalty or devotion ▷ **devotedly** *adv* ▷ **devotee** *n* person who is very enthusiastic about something; zealous follower of a religion ▷ **devotion** *n* strong affection for or loyalty to someone or something; religious zeal ▷ *pl* prayers ▷ **devotional** *adj*

devour *v* eat greedily; (of an emotion) engulf and destroy; read eagerly

devout *adj* deeply religious ▷ **devoutly** *adv*

dew *n* drops of water that form on the ground at night from vapour in the air ▷ **dewy** *adj*

dewlap *n* loose fold of skin hanging under the throat in dogs, cattle, etc.

dexterity *n* skill in using one's hands; mental quickness ▷ **dexterous** *adj* ▷ **dexterously** *adv*

dextrose *n* glucose occurring in fruit, honey, and the blood of animals

DH (in Britain) Department of Health

DI donor insemination: method of making a woman pregnant by transferring sperm from a man other than her regular partner using artificial means: method of making a woman pregnant by transferring sperm from a man other than her regular partner using artificial means

diabetes [die-a-beet-eez] *n* disorder in which an abnormal amount of urine containing an excess of sugar is excreted ▷ **diabetic** *n* ▷ *adj*

diabolic *adj* of the Devil ▷ **diabolism** *n* witchcraft, devil worship

diabolical *adj* (*informal*) extremely bad ▷ **diabolically** *adv*

diaconate *n* position or period of office of a deacon

diacritic *n* sign above or below a character to indicate phonetic value or stress

diadem *n* (*old-fashioned*) crown

diaeresis *n*, *pl* **-ses** mark (¨) placed over a vowel to show that it is pronounced separately from the preceding one, for example in *Noël*

diagnosis [die-ag-no-siss] *n*, *pl* **-ses** [-seez] discovery and identification of diseases from the examination of symptoms ▷ **diagnose** *v* ▷ **diagnostic** *adj*

diagonal *adj* from corner to corner; slanting ▷ *n* diagonal line ▷ **diagonally** *adv*

diagram *n* sketch showing the form or workings of something ▷ **diagrammatic** *adj*

dial *n* face of a clock or watch; graduated disc on a measuring instrument; control on a radio or television set used to change the station; numbered disc on the front of some telephones ▷ v **dialling, dialled** operate the dial or buttons on a telephone in order to contact (a number)

dialect n form of a language spoken in a particular area ▶ **dialectal** adj

dialectic n logical debate by question and answer to resolve differences between two views ▶ **dialectical** adj

dialogue n conversation between two people, esp. in a book, film, or play; discussion between representatives of two nations or groups ▶ **dialogue box** small window that may open on a computer screen to prompt the user to enter information or select an option

dialysis [die-**al**-iss-iss] n (Med) filtering of blood through a membrane to remove waste products

diamanté [die-a-man-tee] adj decorated with artificial jewels or sequins

diameter n (length of) a straight line through the centre of a circle or sphere ▶ **diametric, diametrical** adj of a diameter; completely opposed e.g. the diametric opposite ▶ **diametrically** adv

diamond n exceptionally hard, usu. colourless, precious stone; (Geom) figure with four sides of equal length forming two acute and two obtuse angles; playing card marked with red diamond-shaped symbols ▶ **diamond wedding** sixtieth anniversary of a wedding

diaper n (US) nappy

diaphanous [die-**af**-an-ous] adj fine and almost transparent

diaphragm [die-a-fram] n muscular partition that separates the abdominal cavity and chest cavity; contraceptive device placed over the neck of the womb

diarrhoea [die-a-ree-a] n frequent discharge of abnormally liquid faeces

> **SPELLING TIP**
> The word **diarrhoea** has two rs and one h.

diary n, pl **-ries** (book for) a record of daily events, appointments, or observations ▶ **diarist** n

diatribe n bitter critical attack

dibble n small hand tool used to make holes in the ground for seeds or plants

dice n, pl **dice** small cube each of whose sides has a different number of spots (1 to 6), used in games of chance ▷ v **dice** (food) into small cubes **dice with death** take a risk ▶ **dicey** adj (informal) dangerous or risky

dichotomy [die-kot-a-mee] n, pl **-mies** division into two opposed groups or parts

dicky[1] n, pl **dickies** false shirt front ▶ **dicky-bird** n child's word for a bird

dicky[2] adj **dickier, dickiest** (informal) shaky or weak

Dictaphone® n tape recorder for recording dictation for subsequent typing

dictate v say aloud for someone else to write down; (foll by to) seek to impose one's will on (other people) ▷ n authoritative command; guiding principle ▶ **dictation** n ▶ **dictator** n ruler who has complete power; person in power who acts unfairly or cruelly ▶ **dictatorship** n ▶ **dictatorial** adj like a dictator

diction n manner of pronouncing words and sounds

dictionary n, pl **-aries** book consisting of an alphabetical list of words with their meanings; alphabetically ordered reference book of terms relating to a particular subject

dictum n, pl **-tums** or **-ta** formal statement; popular saying

did v past tense of **do**[1]

didactic adj intended to instruct ▸ **didactically** adv

diddle v (informal) swindle

didgeridoo n Australian musical instrument made from a long hollow piece of wood

didn't did not

die[1] v **dying, died** (of a person, animal, or plant) cease all biological activity permanently; (of something inanimate) cease to exist or function **be dying for something, be dying to do something** (informal) be eager for or to do something ▸ **die-hard** n person who resists change

die[2] n shaped block used to cut or form metal

dieresis [die-air-iss-iss] n, pl **-ses** [-seez] same as **diaeresis**

diesel n diesel engine; vehicle driven by a diesel engine; diesel oil ▸ **diesel engine** internal-combustion engine in which oil is ignited by compression ▸ **diesel oil** fuel obtained from petroleum distillation

diet[1] n food that a person or animal regularly eats; specific range of foods, to control weight or for health reasons ▸ v follow a special diet so as to lose weight ▸ adj (of food) suitable for a weight-reduction diet ▸ **dietary** adj ▸ **dietary fibre** fibrous substances in fruit and vegetables that aid digestion ▸ **dieter** n ▸ **dietetic** adj prepared for special dietary requirements ▸ **dietetics** n study of diet and nutrition ▸ **dietician** n person who specializes in dietetics

diet[2] n parliament of some countries

differ v be unlike; disagree

different adj unlike; unusual ▸ **difference** n state of being unlike; disagreement; remainder left after subtraction ▸ **differently** adv

> **USAGE NOTE**
>
> The accepted idiom is different from but different to is also used. Different than is used in America

differential adj of or using a difference; (Maths) involving differentials ▸ n factor that differentiates between two comparable things; (Maths) tiny difference between values in a scale; difference between rates of pay for different types of work ▸ **differential calculus** branch of calculus concerned with derivatives and differentials ▸ **differentiate** v perceive or show the difference (between); make (one thing) distinct from other such things ▸ **differentiation** n

difficult adj requiring effort or skill to do or understand; not easily pleased ▸ **difficulty** n

diffident adj lacking self-confidence ▸ **diffidence** n ▸ **diffidently** adv

diffraction n (Physics) deviation in the direction of a wave at the edge of an obstacle in its path; formation of light and dark fringes by the passage of light through a small aperture

diffuse v spread over a wide area ▸ adj widely spread; lacking concision ▸ **diffusion** n

dig v **digging, dug** cut into, break up, and turn over or remove (earth), esp. with a spade; (foll by *out* or *up*) find by effort or searching; (foll by *in* or *into*) thrust or jab ▷ n digging; archaeological excavation; thrust or poke; spiteful remark ▷ pl (Brit, Aust & SAfr informal) lodgings ▸ **digger** n machine used for digging

digest v subject to a process of digestion; absorb mentally ▷ n shortened version of a book, report, or article ▸ **digestible** adj ▸ **digestion** n (body's system for) breaking down food into easily absorbed substances ▸ **digestive** adj ▸ **digestive biscuit** biscuit made from wholemeal flour

digicam n digital camera

digit [dij-it] n finger or toe; numeral from 0 to 9 ▸ **digital** adj displaying information as numbers rather than with hands and a dial e.g. *a digital clock* ▸ **digital recording** sound-recording process that converts audio or analogue signals into a series of pulses ▸ **digital television** television in which the picture is transmitted in digital form and then decoded ▸ **digitally** adv

digitalis n drug made from foxglove leaves, used as a heart stimulant

dignity n, pl **-ties** serious, calm, and controlled behaviour or manner; quality of being worthy of respect; sense of self-importance ▸ **dignify** v add distinction to ▸ **dignitary** n person of high official position

digress v depart from the main subject in speech or writing ▸ **digression** n

dike n same as **dyke¹, dyke²**

dilapidated adj (of a building) having fallen into ruin ▸ **dilapidation** n

dilate v make or become wider or larger ▸ **dilation, dilatation** n

dilatory [dill-a-tree] adj tending or intended to waste time

dildo n, pl **-dos** object used as a substitute for an erect penis

dilemma n situation offering a choice between two equally undesirable alternatives

> **USAGE NOTE**
> If a difficult choice involves more than two courses of action, it is preferable to speak of a *problem* or *difficulty*

dilettante [dill-it-tan-tee] n, pl **-tantes** or **-tanti** person whose interest in a subject is superficial rather than serious ▸ **dilettantism** n

diligent adj careful and persevering in carrying out duties; carried out with care and perseverance ▸ **diligently** adv ▸ **diligence** n

dill n sweet-smelling herb

dilly-dally v **-lying, -lied** (Brit, Aust & NZ informal) dawdle, waste time

dilute v make (a liquid) less concentrated, esp. by adding water; make (a quality etc.) weaker in force ▸ **dilution** n

diluvial, diluvian adj of a flood, esp. the great Flood described in the Old Testament

dim adj **dimmer, dimmest** badly lit; not clearly seen; unintelligent ▷ v **dimming, dimmed** make or become dim **take a dim view** of disapprove of ▸ **dimly** adv ▸ **dimness** n ▸ **dimmer** n device for dimming an electric light

dime n coin of the US and Canada, worth ten cents

dimension n measurement of the size of something in a particular direction; aspect, factor

diminish v make or become smaller, fewer, or less ▶ **diminution** n ▶ **diminutive** adj very small ▶ n a word or affix which implies smallness or unimportance

diminuendo n (Music) gradual decrease in loudness

dimple n small natural dent, esp. in the cheeks or chin ▷ v produce dimples by smiling

din n loud unpleasant confused noise ▷ v **dinning, dinned** (foll by into) instil (something) into someone by constant repetition

dinar [dee-nahr] n monetary unit of various Balkan, Middle Eastern, and North African countries

dine v eat dinner ▶ **diner** n person eating a meal; (chiefly US) small cheap restaurant ▶ **dining car** railway coach where meals are served ▶ **dining room** room where meals are eaten

ding n (Aust & NZ) Australian wild dent in a vehicle

ding-dong n sound of a bell; (informal) lively quarrel or fight

dinghy [ding-ee] n, pl **-ghies** small boat, powered by sails, oars, or a motor

dingo n, pl **-goes** Australian wild dog

dingy [din-jee] adj **-gier, -giest** (Brit, Aust & NZ) dull and drab ▶ **dinginess** n

dinkum adj (Aust & NZ informal) genuine or right

dinky adj **-kier, -kiest** (Brit, Aust & NZ informal) small and neat

dinky-di adj (Aust informal) typical

dinner n main meal of the day, eaten either in the evening or at midday ▶ **dinner jacket** man's semiformal black evening jacket

dinosaur n type of extinct prehistoric reptile, many of which were of gigantic size

dint n **by dint of** by means of

diocese [die-a-siss] n district over which a bishop has control ▶ **diocesan** adj

diode n semiconductor device for converting alternating current to direct current

dioptre [die-op-ter] n unit for measuring the refractive power of a lens

dioxide n oxide containing two oxygen atoms per molecule

dip v **dipping, dipped** plunge quickly or briefly into a liquid; slope downwards; switch (car headlights) from the main to the lower beam; lower briefly ▷ n dipping; brief swim; liquid chemical in which farm animals are dipped to rid them of insects; depression in a landscape; creamy mixture into which pieces of food are dipped before being eaten ▶ **dip into** v read passages at random from (a book or journal)

diphtheria [dif-theer-ya] n contagious disease producing fever and difficulty in breathing and swallowing

diphthong n union of two vowel sounds in a single compound sound

diploma n qualification awarded by a college on successful completion of a course

diplomacy n conduct of the relations between nations by peaceful means; tact or skill in dealing with people ▶ **diplomat** n official engaged in diplomacy ▶ **diplomatic** adj of diplomacy; tactful in dealing with people ▶ **diplomatically** adv

dipper n ladle used for dipping; (also **ousel, ouzel**) European songbird that lives by a river

diprotodont [die-pro-toe-dont] n marsupial with fewer than three upper incisor teeth on each side of the jaw

dipsomania n compulsive craving for alcohol ▷ **dipsomaniac** n, adj

diptych [dip-tik] n painting on two hinged panels

dire adj disastrous, urgent, or terrible

direct adj (of a route) shortest, straight; without anyone or anything intervening; likely to have an immediate effect; honest, frank ▷ adv in a direct manner ▷ v lead and organize; tell (someone) to do something; tell (someone) the way to a place; address (a letter, package, remark, etc.); provide guidance to (actors, cameramen, etc.) in (a play or film) ▶ **directly** adv in a direct manner; at once ▷ conj as soon as ▶ **directness** n ▶ **direct current** electric current that flows in one direction only

direction n course or line along which a person or thing moves, points, or lies; management or guidance ▷ pl instructions for doing something or for reaching a place ▶ **directional** adj

directive n instruction, order

director n person or thing that directs or controls; member of the governing board of a business etc.; person responsible for the artistic and technical aspects of the making of a film etc.
▶ **directorial** adj ▶ **directorship** n ▶ **directorate** n board of directors; position of director

directory n, pl -**tories** book listing names, addresses, and telephone numbers; (Computing) area of a disk containing the names and locations of the files it currently holds

dirge n slow sad song of mourning

dirigible [dir-rij-jib-bl] adj able to be steered ▷ n airship

dirk n dagger, formerly worn by Scottish Highlanders

dirndl n full gathered skirt originating from Tyrolean peasant wear

dirt n unclean substance, filth; earth, soil; obscene speech or writing; (informal) harmful gossip ▶ **dirt track** racetrack made of packed earth or cinders

dirty adj **dirtier, dirtiest** covered or marked with dirt; unfair or dishonest; obscene; displaying dislike or anger e.g. a dirty look ▷ v **dirtying, dirtied** make dirty ▶ **dirtiness** n

dis- prefix indicating reversal e.g. disconnect; indicating negation or lack e.g. dissimilar; disgrace; indicating removal or release e.g. disembowel

disable v make ineffective, unfit, or incapable ▶ **disabled** adj lacking a physical power, such as the ability to walk ▶ **disablement** n ▶ **disability** n, pl -**ties** condition of being disabled; something that disables someone

disabuse v (foll by of) rid (someone) of a mistaken idea

disadvantage n unfavourable or harmful circumstance ▶ **disadvantageous** adj ▶ **disadvantaged** adj socially or economically deprived

disaffected adj having lost loyalty to or affection for someone or something ▶ **disaffection** n

disagree v -**greeing, -greed** argue or have different opinions; be different, conflict; (foll by with) cause physical discomfort (to) e.g. curry disagrees with me ▶ **disagreement** n ▶ **disagreeable** adj unpleasant; (of a person) unfriendly or unhelpful ▶ **disagreeably** adv

disallow v reject as untrue or invalid

disappear v cease to be visible; cease to exist ▶ **disappearance** n

disappoint v fail to meet the expectations or hopes of ▶ **disappointment** n feeling of being disappointed; person or thing that disappoints

> **SPELLING TIP**
> Remember there is only one s in **disappointment**, but two ps

disapprobation n disapproval

disapprove v (foll by of) consider wrong or bad ▶ **disapproval** n

disarm v deprive of weapons; win the confidence or affection of; (of a country) decrease the size of one's armed forces ▶ **disarmament** n ▶ **disarming** adj removing hostility or suspicion ▶ **disarmingly** adv

disarrange v throw into disorder

disarray n confusion and lack of discipline; extreme untidiness

disaster n occurrence that causes great distress or destruction; project etc. that fails ▶ **disastrous** adj ▶ **disastrously** adv

disavow v deny connection with or responsibility for ▶ **disavowal** n

disband v (cause to) cease to function as a group

disbelieve v reject as false; (foll by in) have no faith (in) ▶ **disbelief** n

disburse v pay out ▶ **disbursement** n

disc n flat circular object; gramophone record; (Anat) circular flat structure in the body, esp. between the vertebrae; (Computing) same as **disk** ▶ **disc jockey** person who introduces and plays pop records on a radio programme or at a disco

discard v get rid of (something or someone) as useless or undesirable

discern v see or be aware of (something) clearly ▶ **discernible** adj ▶ **discerning** adj having good judgment ▶ **discernment** n

discharge v release, allow to go; dismiss (someone) from duty or employment; fire (a gun); pour forth, send out; meet the demands of (a duty or responsibility); relieve oneself of (a debt) ▷ n substance that comes out from a place; discharging

disciple [diss-sipe-pl] n follower of the doctrines of a teacher, esp. Jesus Christ

discipline n practice of imposing strict rules of behaviour; area of academic study ▷ v attempt to improve the behaviour of (oneself or another) by training or rules; punish ▶ **disciplined** adj able to behave and work in a controlled way ▶ **disciplinarian** n person who practises strict discipline ▶ **disciplinary** adj

disclaimer n statement denying responsibility ▶ **disclaim** v

disclose v make known; allow to be seen ▶ **disclosure** n

disco n, pl **-cos** nightclub where people dance to amplified pop records; occasion at which people dance to amplified pop records; mobile equipment for providing music for a disco

discolour v change in colour, fade ▶ **discoloration** n

discomfit v make uneasy or confused ▶ **discomfiture** n

discomfort n inconvenience, distress, or mild pain

discommode v cause inconvenience to

disconcert v embarrass or upset

disconnect v undo or break the connection between (two things); stop the supply of electricity or gas of ▸ **disconnected** adj (of speech or ideas) not logically connected ▸ **disconnection** n

disconsolate adj sad beyond comfort ▸ **disconsolately** adv

discontent n lack of contentment ▸ **discontented** adj

discontinue v come or bring to an end ▸ **discontinuous** adj characterized by interruptions ▸ **discontinuity** n

discord n lack of agreement or harmony between people; harsh confused sounds ▸ **discordant** adj ▸ **discordance** n

discotheque n same as **disco**

discount v take no account of (something) because it is considered to be unreliable, prejudiced, or irrelevant; deduct (an amount) from the price of something ▷ n deduction from the full price of something

discourage v deprive of the will to persist in something; oppose by expressing disapproval ▸ **discouragement** n

discourse n conversation; formal treatment of a subject in speech or writing ▷ v (foll by on) speak or write (about) at length

discourteous adj showing bad manners ▸ **discourtesy** n

discover v be the first to find or to find out about; learn about for the first time; find after study or search ▸ **discoverer** n ▸ **discovery** n, pl **-eries** discovering; person, place, or thing that has been discovered

discredit v damage the reputation of; cause (an idea) to be disbelieved or distrusted ▷ n damage to someone's reputation ▸ **discreditable** adj bringing shame

discreet adj careful to avoid embarrassment, esp. by keeping confidences secret; unobtrusive ▸ **discreetly** adv

discrepancy n, pl **-cies** conflict or variation between facts, figures, or claims

discrete adj separate, distinct

discretion n quality of behaving in a discreet way; freedom or authority to make judgments and decide what to do ▸ **discretionary** adj

discriminate v (foll by against or in favour of) single out (a particular person or group) for worse or better treatment than others; (foll by between) recognize or understand the difference (between) ▸ **discriminating** adj showing good taste and judgment ▸ **discrimination** n ▸ **discriminatory** adj based on prejudice

discursive adj passing from one topic to another

discus n heavy disc-shaped object thrown in sports competitions

discuss v consider (something) by talking it over; treat (a subject) in speech or writing ▸ **discussion** n

disdain n feeling of superiority and dislike ▷ v refuse with disdain ▸ **disdainful** adj ▸ **disdainfully** adv

disease n illness, sickness ▸ **diseased** adj

disembark v get off a ship, aircraft, or bus ▸ **disembarkation** n

disembodied adj lacking a body; seeming not to be attached to or coming from anyone

disembowel v -elling, -elled remove the entrails of

disenchanted adj disappointed and disillusioned ▸ **disenchantment** n

disenfranchise v deprive (someone) of the right to vote or of other rights of citizenship

disengage v release from a connection ▸ **disengagement** n

disentangle v release from entanglement or confusion

disfavour n disapproval or dislike

disfigure v spoil the appearance of ▸ **disfigurement** n

disfranchise v same as **disenfranchise**

disgorge v empty out, discharge

disgrace n condition of shame, loss of reputation, or dishonour; shameful person or thing ▸ v bring shame upon (oneself or others) ▸ **disgraceful** adj ▸ **disgracefully** adv

disgruntled adj sulky or discontented ▸ **disgruntlement** n

disguise v change the appearance or manner in order to conceal the identity of (someone or something); misrepresent (something) in order to obscure its actual nature or meaning ▸ n mask, costume, or manner that disguises; state of being disguised

disgust n great loathing or distaste ▸ v sicken, fill with loathing

dish n shallow container used for holding or serving food; particular kind of food; short for **dish aerial**; (informal) attractive person ▸ **dish aerial** aerial consisting of a concave disc-shaped reflector, used esp. for satellite television ▸ **dishcloth** n cloth for washing dishes ▸ **dish out** v (informal) distribute ▸ **dish up** v (informal) serve (food)

dishabille [diss-a-beel] n same as **deshabille**

dishearten v weaken or destroy the hope, courage, or enthusiasm of

dishevelled adj (of a person's hair, clothes, or general appearance) disordered and untidy

dishonest adj not honest or fair ▸ **dishonestly** adv ▸ **dishonesty** n

dishonour v treat with disrespect ▸ n lack of respect; state of shame or disgrace; something that causes a loss of honour ▸ **dishonourable** adj ▸ **dishonourably** adv

disillusion v destroy the illusions or false ideas of ▸ n (also **disillusionment**) state of being disillusioned

disincentive n something that acts as a deterrent

disinclined adj unwilling, reluctant ▸ **disinclination** n

disinfect v rid of harmful germs, chemically ▸ **disinfectant** n substance that destroys harmful germs ▸ **disinfection** n

disinformation n false information intended to mislead

disingenuous adj not sincere ▸ **disingenuously** adv

disinherit v (Law) deprive (an heir) of inheritance ▸ **disinheritance** n

disintegrate v break up ▸ **disintegration** n

disinter v -terring, -terred dig up; reveal, make known

disinterested adj free from bias or involvement ▸ **disinterest** n

USAGE NOTE
People sometimes use *disinterested* where they mean *uninterested*. If you want to say that someone shows a lack of interest, use *uninterested*. *Disinterested* would be used in a sentence such as *We asked him to decide because he was a disinterested observer*

disjointed adj having no coherence, disconnected

disk n (Computing) storage device, consisting of a stack of plates coated with a magnetic layer, which rotates rapidly as a single unit

dislike v consider unpleasant or disagreeable ▷ n feeling of not liking something or someone

dislocate v displace (a bone or joint) from its normal position; disrupt or shift out of place ▶ **dislocation** n

dislodge v remove (something) from a previously fixed position

disloyal adj not loyal, deserting one's allegiance ▶ **disloyalty** n

dismal adj gloomy and depressing; (informal) of poor quality ▶ **dismally** adv

dismantle v take apart piece by piece

dismay v fill with alarm or depression ▷ n alarm mixed with sadness

dismember v remove the limbs of; cut to pieces ▶ **dismemberment** n

dismiss v remove (an employee) from a job; allow (someone) to leave; put out of one's mind; (of a judge) state that (a case) will not be brought to trial ▶ **dismissal** n ▶ **dismissive** adj scornful, contemptuous

dismount v get off a horse or bicycle

disobey v neglect or refuse to obey ▶ **disobedient** adj ▶ **disobedience** n

disobliging adj unwilling to help

disorder n state of untidiness and disorganization; public violence or rioting; illness ▶ **disordered** adj untidy ▶ **disorderly** adj untidy and disorganized; uncontrolled, unruly

disorganize v disrupt the arrangement or system of ▶ **disorganization** n

disorientate, disorient v cause (someone) to lose his or her bearings ▶ **disorientation** n

disown v deny any connection with (someone)

disparage v speak contemptuously of ▶ **disparagement** n

disparate adj completely different ▶ **disparity** n

dispassionate adj not influenced by emotion ▶ **dispassionately** adv

dispatch v send off to a destination or to perform a task; carry out (a duty or a task) with speed; (old-fashioned) kill ▷ n official communication or report, sent in haste; report sent to a newspaper by a correspondent ▶ **dispatch rider** (Brit, Aust & NZ) motorcyclist who carries dispatches

dispel v -pelling, -pelled destroy or remove

dispense v distribute in portions; prepare and distribute (medicine); administer (the law etc.) ▶ **dispensable** adj not essential ▶ **dispensation** n dispensing; exemption from an obligation ▶ **dispenser** n ▶ **dispensary** n, pl -saries place where medicine is dispensed ▶ **dispense with** v do away with, manage without

disperse v scatter over a wide area; (cause to) leave a gathering ▶ **dispersal, dispersion** n

dispirit v make downhearted

displace v move from the usual location; remove from office ▶ **displacement** n ▶ **displaced person** person forced from his or her home or country, esp. by war

display v make visible or noticeable ▷ n displaying; something displayed; exhibition

displease v annoy or upset ▶ **displeasure** n

disport v disport oneself indulge oneself in pleasure

dispose v place in a certain order ▶ **disposed** adj willing or eager; having an attitude as specified e.g. *he felt well disposed towards her* ▸ **disposable** *adj* designed to be thrown away after use; available for use e.g. *disposable income* ▸ **disposal** *n* getting rid of something **at one's disposal** available for use ▸ **disposition** *n* person's usual temperament; desire or tendency to do something; arrangement ▸ **dispose of** *v* throw away, get rid of; deal with (a problem etc.); kill

dispossess *v* (foll by *of*) deprive (someone) of (a possession) ▸ **dispossession** *n*

disproportion *n* lack of proportion or equality

disproportionate *adj* out of proportion ▸ **disproportionately** *adv*

disprove *v* show (an assertion or claim) to be incorrect

dispute *n* disagreement, argument ▸ *v* argue about (something); doubt the validity of; fight over possession of

disqualify *v* stop (someone) officially from taking part in something for wrongdoing ▸ **disqualification** *n*

disquiet *n* feeling of anxiety ▸ *v* make (someone) anxious ▸ **disquietude** *n*

disregard *v* give little or no attention to ▸ *n* lack of attention or respect

disrepair *n* condition of being worn out or in poor working order

disrepute *n* loss or lack of good reputation ▸ **disreputable** *adj* having or causing a bad reputation

disrespect *n* lack of respect ▸ **disrespectful** *adj* ▸ **disrespectfully** *adv*

disrobe *v* undress

disrupt *v* interrupt the progress of ▸ **disruption** *n* ▸ **disruptive** *adj*

dissatisfied *adj* not pleased or contented ▸ **dissatisfaction** *n*

dissect *v* cut open (a corpse) to examine it; examine critically and minutely ▸ **dissection** *n*

dissemble *v* conceal one's real motives or emotions by pretence

disseminate *v* spread (information) ▸ **dissemination** *n*

dissent *v* disagree; (*Christianity*) reject the doctrines of an established church ▸ *n* disagreement; (*Christianity*) separation from an established church ▸ **dissension** *n* ▸ **dissenter** *n*

dissertation *n* written thesis, usu. required for a higher university degree; long formal speech

disservice *n* harmful action

dissident *n* person who disagrees with and criticizes the government ▸ *adj* disagreeing with the government ▸ **dissidence** *n*

dissimilar *adj* not alike, different ▸ **dissimilarity** *n*

dissimulate *v* conceal one's real feelings by pretence ▸ **dissimulation** *n*

dissipate *v* waste or squander; scatter, disappear ▸ **dissipated** *adj* showing signs of overindulgence in alcohol and other physical pleasures ▸ **dissipation** *n*

dissociate *v* regard or treat as separate **dissociate oneself from** deny or break an association with ▸ **dissociation** *n*

dissolute *adj* leading an immoral life

dissolution *n* official breaking up of an organization or institution, such as Parliament; official ending of a formal agreement, such as a marriage

dissolve v (cause to) become liquid; break up or end officially; break down emotionally e.g. *she dissolved into tears*

dissonance n lack of agreement or harmony ▶ **dissonant** adj

dissuade v deter (someone) by persuasion from doing something ▶ **dissuasion** n

distaff n rod on which wool etc. is wound for spinning ▶ **distaff side** female side of a family

distance n space between two points; state of being apart; remoteness in manner **the distance** most distant part of the visible scene **distance oneself from** separate oneself mentally from ▶ **distant** adj far apart; separated by a specified distance; remote in manner ▶ **distantly** adv

distaste n dislike, disgust ▶ **distasteful** adj unpleasant, offensive

distemper¹ n highly contagious viral disease of dogs

distemper² n paint mixed with water, glue, etc., used for painting walls

distend v (of part of the body) swell ▶ **distension** n

distil v -tilling, -tilled subject to or obtain by distillation; give off (a substance) in drops; extract the essence of ▶ **distillation** n process of evaporating a liquid and condensing its vapour; (also **distillate**) concentrated essence

distiller n person or company that makes strong alcoholic drink, esp. whisky ▶ **distillery** n, pl -leries place where a strong alcoholic drink, esp. whisky, is made

distinct adj not the same; easily sensed or understood; clear and definite ▶ **distinctly**

adv ▶ **distinction** n act of distinguishing; distinguishing feature; state of being different; special honour, recognition, or fame ▶ **distinctive** adj easily recognizable ▶ **distinctively** adv ▶ **distinctiveness** n

distinguish v (usu. foll by between) make, show, or recognize a difference (between); be a distinctive feature of; make out by hearing, seeing, etc. ▶ **distinguishable** adj ▶ **distinguished** adj dignified in appearance; highly respected

distort v misrepresent (the truth or facts); twist out of shape ▶ **distortion** n

distract v draw the attention of (a person) away from something; entertain ▶ **distracted** adj unable to concentrate, preoccupied ▶ **distraction** n

distrait [diss-tray] adj absent-minded or preoccupied

distraught [diss-trawt] adj extremely anxious or agitated

distress n extreme unhappiness; great physical pain; poverty ▷ v upset badly ▶ **distressed** adj extremely upset; in financial difficulties ▶ **distressing** adj ▶ **distressingly** adv

distribute v hand out or deliver; share out ▶ **distribution** n distributing; arrangement or spread ▶ **distributor** n wholesaler who distributes goods to retailers in a specific area; device in a petrol engine that sends the electric current to the spark plugs ▶ **distributive** adj

district n area of land regarded as an administrative or geographical unit **district court judge** (Aust & NZ) judge presiding over a lower court

distrust v regard as untrustworthy ▷ n feeling of suspicion or doubt ▶ **distrustful** adj

disturb v intrude on; worry, make anxious; change the position or shape of ▶ **disturbance** n ▶ **disturbing** adj ▶ **disturbingly** adv ▶ **disturbed** adj (Psychiatry) emotionally upset or maladjusted

disunite v cause disagreement among ▶ **disunity** n

disuse n state of being no longer used ▶ **disused** adj

ditch n narrow channel dug in the earth for drainage or irrigation ▷ v (slang) abandon

dither v be uncertain or indecisive ▷ n state of indecision or agitation ▶ **ditherer** n ▶ **dithery** adj

ditto n, pl -tos the same ▷ adv in the same way

ditty n, pl -ties short simple poem or song

diuretic [die-yoor-et-ik] n drug that increases the flow of urine

diurnal [die-urn-al] adj happening during the day or daily

diva n distinguished female singer

divan n low backless bed; backless sofa or couch

dive v **diving, dived** plunge headfirst into water; (of a submarine or diver) submerge under water; fly in a steep nose-down descending path; move quickly in a specified direction; (foll by in or into) start doing (something) enthusiastically ▷ n diving; steep nose-down descent; (slang) disreputable bar or club ▶ **diver** n person who works or explores underwater; person who dives for sport ▶ **dive bomber** military aircraft designed to release bombs during a dive

diverge v separate and go in different directions; deviate (from a prescribed course) ▶ **divergence** n ▶ **divergent** adj

divers adj (old-fashioned) various

diverse adj having variety, assorted; different in kind ▶ **diversity** n, pl -ties quality of being different or varied; range of difference ▶ **diversify** v -fying, -fied ▶ **diversification** n

divert v change the direction of; entertain, distract the attention of ▶ **diversion** n official detour used by traffic when a main route is closed; something that distracts someone's attention; diverting; amusing pastime ▶ **diversionary** adj

divest v strip (of clothes); deprive (of a role or function)

divide v separate into parts; share or be shared out in parts; (cause to) disagree; keep apart; be a boundary between; calculate how many times (one number) can be contained in (another) ▷ n division, split ▶ **dividend** n sum of money representing part of the profit made, paid by a company to its shareholders; extra benefit ▶ **divider** n screen used to divide a room into separate areas ▷ pl compasses with two pointed arms, used for measuring or dividing lines

divine adj of God or a god; godlike; (informal) splendid ▷ v discover (something) by intuition or guessing ▶ **divinely** adv ▶ **divination** n art of discovering future events, as though by supernatural powers ▶ **divinity** n study of religion; (pl -ties) god; state of being divine ▶ **divining rod** forked twig said to move when held over ground in which water or metal is to be found

division n dividing, sharing out; one of the parts into

which something is divided; mathematical operation of dividing; difference of opinion
▶ **divisional** adj of a division in an organization ▶ **divisibility** n ▶ **divisible** adj tending to cause disagreement ▶ **divisor** n number to be divided into another number

divorce n legal ending of a marriage; any separation, esp. a permanent one ▷ v legally end one's marriage (to); separate, consider separately ▶ **divorcée**, (masc) **divorcé** n person who is divorced

divulge v make known, disclose ▶ **divulgence** n

Dixie n southern states of the US (also **Dixieland**)

DIY (Brit, Aust & NZ) do-it-yourself

dizzy adj -zier, -ziest having or causing a whirling sensation; mentally confused ▷ v -zying, -zied make dizzy ▶ **dizzily** adv ▶ **dizziness** n

DJ disc jockey; (Brit) dinner jacket

DNA n deoxyribonucleic acid, the main constituent of the chromosomes of all living things

do v **does, doing, did, done** perform or complete (a deed or action); be adequate e.g. that one will do; suit or improve e.g. that style does nothing for you; find the answer to (a problem or puzzle); cause, produce e.g. it does no harm to think ahead; give, grant e.g. do me a favour; work at, as a course of study or a job; used to form questions e.g. how do you know?; used to intensify positive statements and commands e.g. I do like port; do go on; used to form negative statements and commands e.g. I do not know her well; do not get up; used to replace an earlier verb e.g. he gets paid more

than I do ▷ n, pl **dos** or **do's** (informal) party, celebration ▶ **do away with** v get rid of ▶ **do-it-yourself** n constructing and repairing things oneself ▶ **do up** v fasten; decorate and repair ▶ **do with** v find useful or benefit from e.g. I could do with a rest ▶ **do without** v manage without

Doberman pinscher, Doberman n large dog with a black-and-tan coat

dob in v **dobbing, dobbed** (Aust & NZ informal) inform against; contribute to a fund

DOC (in New Zealand) Department of Conservation

docile adj (of a person or animal) easily controlled ▶ **docilely** adv ▶ **docility** n

dock¹ n enclosed area of water where ships are loaded, unloaded, or repaired ▷ v bring or be brought into dock; link (two spacecraft) or (of two spacecraft) be linked together in space ▶ **docker** n (Brit) person employed to load and unload ships ▶ **dockyard** n place where ships are built or repaired

dock² v deduct money from (a person's wages); remove part of (an animal's tail) by cutting through the bone

dock³ n enclosed space in a court of law where the accused person sits or stands

dock⁴ n weed with broad leaves

docket n label on a package or other delivery, stating contents, delivery instructions, etc.

doctor n person licensed to practise medicine; person who has been awarded a doctorate ▷ v alter in order to deceive; poison or drug (food or drink); (informal) castrate (an animal) ▶ **doctoral** adj ▶ **doctorate** n highest academic degree in any field of knowledge

doctrine n body of teachings of a religious, political, or philosophical group; principle or body of principles that is taught or advocated ▸ **doctrinal** adj of doctrines ▸ **doctrinaire** adj stubbornly insistent on the application of a theory without regard to practicality

document n piece of paper providing an official record of something ▷ v record or report (something) in detail; support (a claim) with evidence ▸ **documentation** n

documentary n, pl **-ries** film or television programme presenting the facts about a particular subject ▷ adj (of evidence) based on documents

docu-soap n television documentary series presenting the lives of the people filmed as entertainment

dodder v move unsteadily ▸ **doddery** adj

dodecagon [doe-**deck**-a-gon] n geometric figure with twelve sides

dodge v avoid (a blow, being seen, etc.) by moving suddenly; evade by cleverness or trickery ▷ n cunning or deceitful trick ▸ **dodgy** adj **dodgier, dodgiest** (informal) dangerous, risky; untrustworthy

Dodgem® n small electric car driven and bumped against similar cars in a rink at a funfair

dodger n person who evades by a responsibility or duty

dodo n, pl **dodos** or **dodoes** large flightless extinct bird

doe n female deer, hare, or rabbit

does v third person singular of the present tense of **do¹**

doesn't does not

doff v take off or lift (one's hat) in polite greeting

dog n domesticated four-legged mammal of many different breeds; related wild mammal, such as the dingo or coyote; male animal of the dog family; (informal) person e.g. you lucky dog! ▷ v **dogging, dogged** follow (someone) closely; trouble, plague **go to the dogs** (informal) go to ruin physically or morally **let sleeping dogs lie** leave things undisturbed ▸ **dogging** n (Brit slang) exhibitionist sex in parked cars, often with strangers ▸ **doggy, doggie** n, pl **-gies** child's word for a dog ▸ **dogcart** n light horse-drawn two-wheeled cart ▸ **dog collar** collar for a dog; (informal) white collar fastened at the back, worn by members of the clergy ▸ **dog-eared** adj (of a book) having pages folded down at the corner; shabby, worn ▸ **dogfight** n close-quarters combat between fighter aircraft ▸ **dogfish** n small shark ▸ **doghouse** n (US) kennel **in the doghouse** (informal) in disgrace ▸ **dogleg** n sharp bend ▸ **dog-roll** n (NZ) sausage-shaped roll of meat processed as dog food ▸ **dog rose** wild rose with pink or white flowers ▸ **dog-tired** adj (informal) exhausted

doge [doje] n (formerly) chief magistrate of Venice or Genoa

dogged [dog-gid] adj obstinately determined ▸ **doggedly** adv ▸ **doggedness** n

doggerel n poorly written poetry, usu. comic

doggo adv lie doggo (informal) hide and keep quiet

dogma n doctrine or system of doctrines proclaimed by authority as true ▸ **dogmatic** adj habitually stating one's opinions forcefully

or arrogantly ▸ **dogmatically** *adv*
▸ **dogmatism** *n*

dogsbody *n*, *pl* -**bodies** (*informal*)
person who carries out boring
tasks for others

doily *n*, *pl* -**lies** decorative lacy
paper mat, laid on a plate

doldrums *pl n* depressed state of
mind; state of inactivity

dole *n* (*Brit, Aust & NZ informal*)
money received from the state
while unemployed ▸ *v* (foll by *out*)
distribute in small quantities

doleful *adj* dreary, unhappy
▸ **dolefully** *adv*

doll *n* small model of a human
being, used as a toy; (*slang*) pretty
girl or young woman

dollar *n* standard monetary unit of
many countries

dollop *n* (*informal*) lump (of food)

dolly *n*, *pl* -**lies** child's word for a
doll; wheeled support on which a
camera may be moved

dolman sleeve *n* sleeve that is
very wide at the armhole, tapering
to a tight wrist

dolmen *n* prehistoric monument
consisting of a horizontal stone
supported by vertical stones

dolomite *n* mineral consisting of
calcium magnesium carbonate

dolorous *adj* sad, mournful

dolphin *n* sea mammal of the
whale family, with a beaklike snout
▸ **dolphinarium** *n* aquarium for
dolphins

dolt *n* stupid person ▸ **doltish** *adj*

domain *n* field of knowledge or
activity; land under one ruler
or government; (*Computing*)
group of computers with the
same name on the internet; (*NZ*)
public park

dome *n* rounded roof built on a
circular base; something shaped
like this ▸ **domed** *adj*

domestic *adj* of one's own
country or a specific country;
of the home or family; enjoying
running a home; (of an animal)
kept as a pet or to produce food
▸ *n* person whose job is to do
housework in someone else's
house ▸ **domestically** *adv*
▸ **domesticity** *n* ▸ **domesticate**
v bring or keep (a wild animal or
plant) under control or cultivation;
accustom (someone) to home life
▸ **domestication** *n* ▸ **domestic
science** study of household skills

domicile [**dom**-miss-ile] *n* place
where one lives

dominant *adj* having authority
or influence; main, chief
▸ **dominance** *n*

dominate *v* control or govern;
tower above (surroundings); be
very significant in ▸ **domination** *n*

domineering *adj* forceful and
arrogant

Dominican *n*, *adj* (friar or nun) of
an order founded by Saint Dominic

dominion *n* control or authority;
land governed by one ruler or
government; (formerly) self-
governing division of the British
Empire

domino *n*, *pl* -**noes** small
rectangular block marked with
dots, used in dominoes ▸ *pl* game
in which dominoes with matching
halves are laid together

don[1] *v* **donning, donned** put on
(clothing)

don[2] *n* (*Brit*) member of the
teaching staff at a university or
college; Spanish gentleman or
nobleman ▸ **donnish** *adj* serious
and academic

donate *v* give, esp. to a charity
or organization ▸ **donation** *n*
donating; thing donated ▸ **donor**
n (*Med*) person who gives blood or

organs for use in the treatment of another person; person who makes a donation

done v past participle of **do¹**

doner kebab n see kebab

donga [dong-ga] n (S Afr, Aust & NZ) steep-sided gully created by soil erosion

dongle n (Computing) device that allows a computer user to access the internet via mobile broadband

donkey n long-eared member of the horse family ▶ **donkey jacket** (Brit, Aust & NZ) man's long thick jacket with a waterproof panel across the shoulders ▶ **donkey's years** (informal) long time ▶ **donkey-work** n tedious hard work

don't do not

doodle v scribble or draw aimlessly ▷ n shape or picture drawn aimlessly

doom n death or a terrible fate ▷ v destine or condemn to death or a terrible fate ▶ **doomsday** n (Christianity) day on which the Last Judgment will occur; any dreaded day

door n hinged or sliding panel for closing the entrance to a building, room, etc.; entrance ▶ **doormat** n mat for wiping dirt from shoes before going indoors; (informal) person who offers little resistance to ill-treatment ▶ **doorway** n opening into a building or room

dope n (slang) illegal drug, usu. cannabis; medicine, drug; (informal) stupid person ▷ v give a drug to, esp. in order to improve performance in a race ▶ **dopey, dopy** adj half-asleep, drowsy; (slang) silly

dorba n (Aust slang) stupid, inept, or clumsy person (also **dorb**)

dork n (slang) stupid person

dormant adj temporarily quiet, inactive, or not being used ▶ **dormancy** n

dormer, dormer window n window that sticks out from a sloping roof

dormitory n, pl -ries large room, esp. at a school, containing several beds

dormouse n, pl -mice small mouselike rodent with a furry tail

dorp n (S Afr) small town

dorsal adj of or on the back

dory, John Dory n, pl -ries spiny-finned edible sea fish

dose n specific quantity of a medicine taken at one time; (informal) something unpleasant to experience ▷ v give a dose to ▶ **dosage** n size of a dose

doss v **doss down** (slang) sleep in an uncomfortable place ▶ **dosshouse** n (Brit & S Afr slang) cheap lodging house for homeless people

dossier [doss-ee-ay] n collection of documents about a subject or person

dot n small round mark; shorter symbol used in Morse code ▷ v **dotting, dotted** mark with a dot; scatter, spread around **on the dot** at exactly the arranged time ▶ **dotty** adj (slang) rather eccentric ▶ **dotcom, dot.com** n company that does most of its business on the internet

dote v **dote on** love to an excessive degree ▶ **dotage** n weakness as a result of old age

double adj as much again in number, amount, size, etc.; composed of two equal or similar parts; designed for two users e.g. double room; folded in two ▷ adv twice over ▷ n twice the number, amount, size, etc.; person who looks almost exactly like another ▷ pl game between two pairs of players ▷ v make or become twice as much or as many; bend or fold (material etc.); play two parts; turn sharply **at the double, on the double** quickly or immediately ▸ **doubly** adv employed by two parties ▸ **double agent** spy employed by two enemy countries at the same time ▸ **double bass** stringed instrument, largest and lowest member of the violin family ▸ **double chin** fold of fat under the chin ▸ **double cream** (Brit) thick cream with a high fat content ▸ **double-cross** v cheat or betray ▷ n double-crossing ▸ **double-dealing** n treacherous or deceitful behaviour ▸ **double-decker** n bus with two passenger decks one on top of the other ▸ **double Dutch** (informal) incomprehensible talk, gibberish ▸ **double glazing** two panes of glass in a window, fitted to reduce heat loss ▸ **double talk** deceptive or ambiguous talk ▸ **double whammy** (informal) devastating setback made up of two elements

double entendre [doob-bl ont-tond-ra] n word or phrase that can be interpreted in two ways, one of which is rude

doublet [dub-lit] n (Hist) man's close-fitting jacket, with or without sleeves

doubloon n former Spanish gold coin

doubt n uncertainty about the truth, facts, or existence of something; unresolved difficulty or point ▷ v question the truth of; distrust or be suspicious of (someone) ▸ **doubter** n ▸ **doubtful** adj unlikely; feeling doubt ▸ **doubtfully** adv ▸ **doubtless** adv probably or certainly

USAGE NOTE
If doubt is followed by a clause it is connected by whether: I doubt

whether she means it. If it is used with a negative it is followed by that: I don't doubt that he is sincere

douche [doosh] n (instrument for applying) a stream of water directed onto or into the body for cleansing or medical purposes ▷ v cleanse or treat by means of a douche

dough n thick mixture of flour and water or milk, used for making bread etc.; (slang) money ▸ **doughnut** n small cake of sweetened dough fried in deep fat

doughty [dowt-ee] adj -tier, -tiest (old-fashioned) brave and determined

dour [door-er] adj sullen and unfriendly ▸ **dourness** n

douse [rhymes with mouse] v drench with water or other liquid; put out (a light)

dove n bird with a heavy body, small head, and short legs; (Politics) person opposed to war ▸ **dovecote, dovecot** n structure for housing pigeons ▸ **dovetail** n joint containing wedge-shaped tenons ▷ v fit together neatly

dowager n widow possessing property or a title obtained from her husband

dowdy adj -dier, -diest dull and old-fashioned ▸ **dowdiness** n

dowel n wooden or metal peg that fits into two corresponding holes to join two adjacent parts

dower n life interest in a part of her husband's estate allotted to a widow by law

down¹ prep, adv indicating movement to or position in a lower place ▷ adv indicating completion of an action, lessening of intensity, etc. e.g. calm down ▷ adj depressed, unhappy ▷ v (informal) drink quickly **have a down on** (informal) feel

d

hostile towards ▸ **down under** (*informal*) (in or to) Australia or New Zealand ▸ **downward** *adj*, *adv* (descending) from a higher to a lower level, condition, or position ▸ **downwards** *adv* from a higher to a lower level, condition, or position ▸ **down-and-out** *n* person who is homeless and destitute ▸ *adj* without any means of support ▸ **down-to-earth** *adj* sensible or practical

down² *n* soft fine feathers ▸ **downy** *adj*

downbeat *adj* (*informal*) gloomy; (*Brit*, *Aust & NZ*) relaxed

downcast *adj* sad, dejected; (of the eyes) directed downwards

downfall *n* (cause of) a sudden loss of position or reputation

downgrade *v* reduce in importance or value

downhearted *adj* sad and discouraged

downhill *adj* going or sloping down ▸ *adv* towards the bottom of a hill

download *v* transfer (data) from the memory of one computer to that of another, especially over the internet ▸ *n* file transferred in such a way

downpour *n* heavy fall of rain

downright *adj*, *adv* extreme(ly)

downs *pl n* low grassy hills, esp. in S England

Down's syndrome *n* genetic disorder characterized by a flat face, slanting eyes, and learning difficulties

downstairs *adv* to or on a lower floor ▸ *n* lower or ground floor

downtrodden *adj* oppressed and lacking the will to resist

dowry *n*, *pl* **-ries** property brought by a woman to her husband at marriage

dowse [rhymes with **cows**] *v* search for underground water or minerals using a divining rod

doxology *n*, *pl* **-gies** short hymn of praise to God

doyen [doy-en] *n* senior member of a group, profession, or society ▸ **doyenne** [doy-en] *n fem*

doze *v* sleep lightly or briefly ▸ *n* short sleep ▸ **dozy** *adj* **dozier, doziest** feeling sleepy; (*informal*) stupid ▸ **doze off** *v* fall into a light sleep

dozen *adj*, *n* twelve ▸ **dozenth** *adj*

DPB (in New Zealand) Domestic Purposes Benefit

DPP (in Britain) Director of Public Prosecutions

Dr Doctor; Drive

drab *adj* **drabber, drabbest** dull and dreary ▸ **drabness** *n*

drachm [dram] *n* (*Brit*) one eighth of a fluid ounce

drachma *n*, *pl* **-mas** or **-mae** former monetary unit of Greece

draconian *adj* severe, harsh

draft *n* plan, sketch, or drawing of something; preliminary outline of a book, speech, etc.; written order for payment of money by a bank; (*US & Aust*) selection for compulsory military service ▸ *v* draw up an outline or plan of; send (people) from one place to another to do a specific job; (*US & Aust*) select for compulsory military service

drag *v* **dragging, dragged** pull with force, esp. along the ground; trail on the ground; persuade or force (oneself or someone else) to go somewhere; (foll by *on* or *out*) last or be prolonged tediously; search (a river) with a dragnet or hook; (*Computing*) move (an image) on the screen by use of the mouse ▸ *n* person or thing that slows up progress; (*informal*) tedious thing or person; (*slang*) women's clothes worn by a man ▸ **dragnet** n net used to scour the bottom of a pond or river to search for something

▸ **drag race** race in which specially built cars or motorcycles are timed over a measured course

dragon n mythical fire-breathing monster like a huge lizard; (*informal*) fierce woman

▸ **dragonfly** n brightly coloured insect with a long slender body and two pairs of wings

dragoon n heavily armed cavalryman ▸ v coerce, force

drain n pipe or channel that carries off water or sewage; cause of a continuous reduction in energy or resources ▸ v draw off or remove liquid from; flow away or filter off; drink the entire contents of (a glass or cup); make constant demands on (energy or resources), exhaust

▸ **drainage** n system of drains; process or method of draining

drake n male duck

dram n small amount of a strong alcoholic drink, e.g. whisky; one sixteenth of an ounce

drama n serious play for theatre, television, or radio; writing, producing, or acting in plays; situation that is exciting or highly emotional ▸ **dramatic** adj of or like drama; behaving flamboyantly ▸ **dramatically** adv ▸ **dramatist** n person who writes plays ▸ **dramatize** v rewrite (a book) in the form of a play; express (something) in a dramatic or exaggerated way ▸ **dramatization** n

drank v past tense of **drink**

drape v cover with material, usu. in folds; place casually ▸ n (*Aust, US & Canadian*) piece of cloth hung at a window or opening as a screen

▸ **drapery** n, pl **-peries** fabric or clothing arranged and draped; fabrics and cloth collectively

draper n (*Brit*) person who sells fabrics and sewing materials

drastic adj strong and severe

draught n current of cold air, esp. in an enclosed space; portion of liquid to be drunk, esp. medicine; gulp or swallow; one of the flat discs used in the game of draughts ▸ pl game for two players using a chessboard and twelve draughts each ▸ adj (of an animal) used for pulling heavy loads ▸ **draughty** adj exposed to draughts of air ▸ **draughtsman** n person employed to prepare detailed scale drawings of machinery, buildings, etc.

▸ **draughtsmanship** n ▸ **draught beer** beer stored in a cask

draw v **drawing**, **drew**, **drawn** sketch (a figure, picture, etc.) with a pencil or pen; pull (a person or thing) closer to or further away from a place; move in a specified direction e.g. the car drew near; take from a source e.g. draw money from bank accounts; attract, interest; formulate or decide e.g. to draw conclusions; (of two teams or contestants) finish a game with an equal number of points ▸ n raffle or lottery; contest or game ending in a tie; event, act, etc., that attracts a large audience

▸ **drawing** n picture or plan made by means of lines on a surface; art of making drawings ▸ **drawing pin** short tack with a broad smooth head ▸ **drawing room** (*old-fashioned*) room where visitors are received and entertained

▸ **drawback** n disadvantage

▸ **drawbridge** n bridge that may be raised to prevent access to or to enable vessels to pass ▸ **draw out** v encourage (someone) to talk freely; make longer ▸ **drawstring** n cord run through a hem around an opening, so that when it is pulled tighter, the opening closes ▸ **draw up** v prepare and write out (a contract); (of a vehicle) come to a stop

drawer n sliding box-shaped part of a piece of furniture, used for storage ▷ pl (old-fashioned) undergarment worn on the lower part of the body

drawl v speak slowly, with long vowel sounds ▷ n drawling manner of speech

drawn v past participle of **draw** ▷ adj haggard, tired, or tense in appearance

dray n low cart used for carrying heavy loads

dread v anticipate with apprehension or fear ▷ n great fear ▸ **dreadful** adj very disagreeable or shocking; extreme ▸ **dreadfully** adv

dreadlocks pl n hair worn in the Rastafarian style of tightly twisted strands

dream n imagined series of events experienced in the mind while asleep; cherished hope; (informal) wonderful person or thing ▷ v **dreaming, dreamed** or **dreamt** see imaginary pictures in the mind while asleep; (often foll by of or about) have an image (of) or fantasy (about); (foll by of) consider the possibility (of) ▷ adj ideal e.g. a dream house ▸ **dreamer** n ▸ **dreamy** adj vague or impractical; (informal) wonderful ▸ **dreamily** adv

dreary adj **drearier, dreariest** dull, boring ▸ **drearily** adv ▸ **dreariness** n

dredge[1] v clear or search (a river bed or harbour) by removing silt or mud ▸ **dredger** n boat fitted with machinery for dredging

dredge[2] v sprinkle (food) with flour etc.

dregs pl n solid particles that settle at the bottom of some liquids; most despised elements

drench v make completely wet

dress n one-piece garment for a woman or girl, consisting of a skirt and bodice and sometimes sleeves; complete style of clothing ▷ v put clothes on; put on formal clothes; apply a protective covering to (a wound); arrange or prepare ▸ **dressing** n sauce for salad; covering for a wound ▸ **dressing-down** n (informal) severe scolding ▸ **dressing gown** coat-shaped garment worn over pyjamas or nightdress ▸ **dressing room** room used for changing clothes, esp. backstage in a theatre ▸ **dressy** adj (of clothes) elegant ▸ **dress circle** first gallery in a theatre ▸ **dressmaker** n person who makes women's clothes ▸ **dressmaking** n ▸ **dress rehearsal** last rehearsal of a play or show, using costumes, lighting, etc.

dressage [dress-ahzh] n training of a horse to perform manoeuvres in response to the rider's body signals

dresser[1] n piece of furniture with shelves and cupboards, for storing or displaying dishes

dresser[2] n (Theatre) person employed to assist actors with their costumes

drew v past tense of **draw**

drey n squirrel's nest

dribble v (allow to) flow in drops; allow saliva to trickle from the mouth; (Sport) propel (a ball) by repeatedly tapping with the foot, hand, or a stick ▷ n small quantity of liquid falling in drops ▶ **dribbler** n

dried v past of **dry**

drier¹ adj a comparative of **dry**

drier² n same as **dryer**

driest adj a superlative of **dry**

drift v be carried along by currents of air or water; move aimlessly from one place or activity to another ▷ n something piled up by the wind or current, such as a snowdrift; general movement or development; point, meaning e.g. *catch my drift?* ▶ **drifter** n person who moves aimlessly from place to place or job to job ▶ **driftwood** n wood floating on or washed ashore by the sea

drill¹ n tool or machine for boring holes; strict and often repetitive training; (informal) correct procedure ▷ v bore a hole in (something) with or as if with a drill; teach by rigorous exercises or training

drill² n machine for sowing seed in rows; small furrow for seed

drill³ n hard-wearing cotton cloth

drily adv see **dry**

drink v **drinking, drank, drunk** swallow (a liquid); consume alcohol, esp. to excess ▷ n (portion of) a liquid suitable for drinking; alcohol, or its habitual or excessive consumption ▶ **drinkable** adj ▶ **drinker** n ▶ **drink in** v pay close attention to ▶ **drink to** v drink a toast to

drip v **dripping, dripped** (let) fall in drops ▷ n falling of drops of liquid; sound made by falling drops; (informal) weak dull person; (Med) device by which a solution is passed in small drops through

a tube into a vein ▶ **drip-dry** adj denoting clothing that will dry free of creases if hung up when wet

dripping n fat that comes from meat while it is being roasted or fried

drive v **driving, drove, driven** guide the movement of (a vehicle); transport in a vehicle; goad into a specified state; push or propel; (Sport) hit (a ball) very hard and straight ▷ n journey by car, van, etc.; (also **driveway**) path for vehicles connecting a building to a public road; united effort towards a common goal; energy and ambition; (Psychol) motive or interest e.g. *sex drive*; means by which power is transmitted in a mechanism ▶ **drive at** v (informal) intend or mean e.g. *what was he driving at?* ▶ **drive-in** adj, n (denoting) a cinema, restaurant, etc., used by people in their cars

drivel n foolish talk ▷ v **-elling, -elled** speak foolishly

driver n person who drives a vehicle

drizzle n very light rain ▷ v rain lightly ▶ **drizzly** adj

droll adj quaintly amusing ▶ **drolly** adv ▶ **drollery** n

dromedary [drom-mid-er-ee] n, pl **-daries** camel with a single hump

drone¹ n male bee

drone² v, n (make) a monotonous low dull sound ▶ **drone on** v talk for a long time in a monotonous tone

drongo n, pl **-gos** tropical songbird with a glossy black plumage, a forked tail, and a stout bill

drool v (foll by over) show excessive enthusiasm (for); allow saliva to flow from the mouth

droop v hang downwards loosely ▶ **droopy** adj

drop v **dropping, dropped** (allow to) fall vertically; decrease in amount, strength, or value; mention (a hint or name) casually; discontinue ▷ n small quantity of liquid forming a round shape; any small quantity of liquid; decrease in amount, strength, or value; vertical distance that something may fall ▷ pl liquid medication applied in small drops ▶ **droplet** n ▶ **droppings** pl n faeces of certain animals, such as rabbits or birds ▶ **drop in, drop by** v pay someone a casual visit ▶ **drop off** v (informal) fall asleep; grow smaller or less ▶ **dropout** n person who rejects conventional society; person who does not complete a course of study ▶ **drop out (of)** v abandon or withdraw from (a school, job, etc.)

dropsy n illness in which watery fluid collects in the body

dross n scum formed on the surfaces of molten metals; anything worthless

drought n prolonged shortage of rainfall

drove[1] v past tense of **drive**

drove[2] n very large group, esp. of people ▶ **drover** n person who drives sheep or cattle

drown v die or kill by immersion in liquid; forget (one's sorrows) temporarily by drinking alcohol; drench thoroughly; make (a sound) inaudible by being louder

drowse v be sleepy, dull, or sluggish ▶ **drowsy** adj ▶ **drowsily** adv ▶ **drowsiness** n

drubbing n utter defeat in a contest etc.

drudge n person who works hard at uninteresting tasks ▶ **drudgery** n

drug n substance used in the treatment or prevention of disease; chemical substance, esp. a narcotic, taken for the effects it produces ▷ v **drugging,**

drugged give a drug to (a person or animal) to cause sleepiness or unconsciousness; mix a drug with (food or drink) ▶ **drugstore** n (US) pharmacy where a wide range of goods are available

Druid n member of an ancient order of Celtic priests ▶ **Druidic, Druidical** adj

drum n percussion instrument sounded by striking a membrane stretched across the opening of a hollow cylinder; cylindrical object or container ▷ v **drumming, drummed** play (music) on a drum; tap rhythmically or regularly ▶ **drum into** v instil into (someone) by constant repetition ▶ **drumstick** n stick used for playing a drum; lower joint of the leg of a cooked chicken etc. ▶ **drum up** v obtain (support or business) by making requests or canvassing

drummer n person who plays a drum or drums

drunk v past participle of **drink** ▷ adj intoxicated with alcohol to the extent of losing control over normal functions; overwhelmed by a strong influence or emotion ▷ n person who is drunk or who frequently gets drunk ▶ **drunkard** n person who frequently gets drunk ▶ **drunken** adj drunk or frequently drunk; caused by or relating to alcoholic intoxication ▶ **drunkenly** adv ▶ **drunkenness** n

dry adj **drier, driest** or **dryer, dryest** lacking moisture; having little or no rainfall; (informal) thirsty; (of wine) not sweet; uninteresting; (of humour) subtle and sarcastic; prohibiting the sale of alcohol e.g. a dry town ▷ v **drying, dried** make or become dry; preserve (food) by removing the moisture ▶ **drily,**

dryly adv ▸ **dryness** n ▸ **dryer** n apparatus for removing moisture ▸ **dry-clean** v clean (clothes etc.) with chemicals rather than water ▸ **dry-cleaner** n ▸ **dry-cleaning** n ▸ **dry out** v make or become dry; (cause to) undergo treatment for alcoholism ▸ **dry rot** crumbling and drying of timber, caused by certain fungi ▸ **dry run** (informal) rehearsal ▸ **dry stock** (NZ) cattle raised for meat

dryad n wood nymph

DSS (in Britain) Department of Social Security

dual adj having two parts, functions, or aspects ▸ **duality** n ▸ **dual carriageway** (Brit, Aust & NZ) road on which traffic travelling in opposite directions is separated by a central strip of grass or concrete

dub[1] v **dubbing, dubbed** give (a person or place) a name or nickname

dub[2] v **dubbing, dubbed** provide (a film) with a new soundtrack, esp. in a different language; provide (a film or tape) with a soundtrack

dubbin (Brit) thick grease applied to leather to soften and waterproof it

dubious [dew-bee-uss] adj feeling or causing doubt ▸ **dubiously** adv ▸ **dubiety** [dew-by-it-ee] n

ducal [duke-al] adj of a duke

ducat [duck-it] n former European gold or silver coin

duchess n woman who holds the rank of duke; wife or widow of a duke

duchesse n (NZ) dressing table with a mirror

duchy n, pl **duchies** territory of a duke or duchess

duck[1] n water bird with short legs, webbed feet, and a broad blunt bill;

its flesh, used as food; female of this bird; (Cricket) score of nothing ▸ **duckling** n baby duck

duck[2] v move (the head or body) quickly downwards, to avoid being seen or to dodge a blow; plunge suddenly under water; (informal) dodge (a duty or responsibility)

duct n tube, pipe, or channel through which liquid or gas is conveyed; bodily passage conveying secretions or excretions

ductile adj (of a metal) able to be shaped into sheets or wires

dud (informal) n ineffectual person or thing ▸ adj bad or useless

dude n (US informal) man; (old-fashioned) dandy; any member

dudgeon n **in high dudgeon** angry, resentful

due adj expected or scheduled to be present or arrive; owed as a debt; fitting, proper ▸ n something that is owed or required ▸ pl charges for membership of a club or organization ▸ adv directly or exactly e.g. due south **due to** attributable to or caused by

USAGE NOTE
The use of due to as a compound preposition as in the performance has been cancelled due to bad weather was formerly considered incorrect, but is now acceptable

duel n formal fight with deadly weapons between two people, to settle a quarrel ▸ v **duelling, duelled** fight in a duel ▸ **duellist** n

duet n piece of music for two performers

duff adj (chiefly Brit) broken or useless ▸ **duff up** v (Brit informal) beat (someone) severely

duffel, duffle n short for **duffel coat** ▸ **duffel bag** cylindrical canvas bag fastened with a drawstring ▸ **duffel coat** wool coat with toggle fastenings, usu. with a hood

duffer n (informal) dull or incompetent person

dug¹ v past of **dig**

dug² n teat or udder

dugite [doo-gyte] n medium-sized Australian venomous snake

dugong n whalelike mammal of tropical waters

dugout n (Brit) (at a sports ground) covered bench where managers and substitutes sit; canoe made by hollowing out a log; (Mil) covered excavation to provide shelter

duke n nobleman of the highest rank; prince or ruler of a small principality or duchy ▷ **dukedom** n

dulcet [dull-sit] adj (of a sound) soothing or pleasant

dulcimer n tuned percussion instrument consisting of a set of strings stretched over a sounding board and struck with hammers

dull adj not interesting; (of an ache) not acute; (of weather) not bright or clear; lacking in spirit; not very intelligent; (of a blade) not sharp ▷ v make or become dull ▷ **dullness** n ▷ **dully** adv ▷ **dullard** n dull or stupid person

duly adv in a proper manner; at the proper time

dumb adj (offens) lacking the power to speak; silent; (informal) stupid ▷ **dumbly** adv ▷ **dumbness** n ▷ **dumbbell** n short bar with a heavy ball or disc at each end, used for physical exercise ▷ **dumbfounded** adj speechless with astonishment ▷ **dumb down** make less intellectually demanding or sophisticated ▷ **dumb show** meaningful gestures without speech

dumdum n soft-nosed bullet that expands on impact and causes serious wounds

dummy n, pl **-mies** figure representing the human form, used for displaying clothes etc.; copy of an object, often lacking some essential feature of the original; rubber teat for a baby to suck; (slang) stupid person ▷ adj imitation, substitute ▷ **dummy run** rehearsal

dump v drop or let fall in a careless manner; (informal) get rid of (someone or something no longer wanted) ▷ n place where waste materials are left; (informal) dirty unattractive place; (informal) place where weapons or supplies are stored **down in the dumps** (informal) depressed and miserable

dumpling n small ball of dough cooked and served with stew; round pastry case filled with fruit

dumpy adj **dumpier, dumpiest** short and plump

dun adj brownish-grey

dunce n person who is stupid or slow to learn

dunderhead n slow-witted person

dune n mound or ridge of drifted sand

dung n faeces from animals such as cattle

dungarees pl n trousers with a bib attached

dungeon n underground prison cell

dunk v dip (a biscuit or bread) in a drink or soup before eating it; put (something) in liquid

dunny n, pl **-nies** (Aust & NZ informal) toilet

duo n, pl **duos** pair of performers; (informal) pair of closely connected people

duodenum [dew-oh-**deen**-um] n, pl **-na** or **-nums** first part of the small intestine, just below the stomach ▶ **duodenal** adj

dupe v deceive or cheat ▷ n person who is easily deceived

duple adj (Music) having two beats in a bar

duplex n (chiefly US) apartment on two floors

duplicate adj copied exactly from an original ▷ n exact copy ▷ v make an exact copy of; do again (something that has already been done) ▶ **duplication** n ▶ **duplicator** n

duplicity n deceitful behaviour

durable adj long-lasting ▶ **durability** n ▶ **durable goods**, **durables** pl n goods that require infrequent replacement

duration n length of time that something lasts

duress n compulsion by use of force or threats

during prep throughout or within the limit of (a period of time)

dusk n time just before nightfall, when it is almost dark ▶ **dusky** adj dark in colour; shadowy

dust n small tiny particles of earth, sand, or dirt ▷ v remove dust from (furniture) by wiping; sprinkle (something) with a powdery substance ▶ **duster** n cloth used for dusting ▶ **dusty** adj covered with dust ▶ **dustbin** n large container for household rubbish ▶ **dust bowl** dry area in which the surface soil is exposed to wind erosion ▶ **dust jacket** removable paper cover used to protect a book ▶ **dustman** n (Brit) man whose job is to collect household rubbish ▶ **dustpan** n short-handled shovel into which dust is swept from floors

Dutch adj of the Netherlands **go Dutch** (informal) share the

expenses on an outing ▶ **Dutch courage** false courage gained from drinking alcohol

duty n, pl **-ties** work or a task performed as part of one's job; task that a person feels morally bound to do; government tax on imports **on duty** at work ▶ **dutiable** adj (of goods) requiring payment of duty ▶ **dutiful** adj doing what is expected ▶ **dutifully** adv

duvet [doo-vay] n kind of quilt used in bed instead of a top sheet and blankets

DVD Digital Versatile (or Video) Disk

DVT deep-vein thrombosis

dwang n (NZ & SAfr) short piece of wood inserted in a timber-framed wall

dwarf n, pl **dwarfs** or **dwarves** person who is smaller than average; (in folklore) small ugly manlike creature, often possessing magical powers ▷ adj (of an animal or plant) much smaller than the usual size for the species ▷ v cause (someone or something) to seem small by being much larger

dwell v **dwelling**, **dwelt** or **dwelled** live, reside ▶ **dwelling** n place of residence ▶ **dwell on**, **dwell upon** v think, speak, or write at length about

dweller n person who lives in a specified place e.g. city dweller

dwindle v grow less in size, strength, or number

dye n colouring substance; colour produced by dyeing ▷ v **dyeing**, **dyed** colour (hair or fabric) by applying a dye ▶ **dyer** n ▶ **dyed-in-the-wool** adj uncompromising or unchanging in opinion

dying v present participle of **die**¹

dyke¹ n wall built to prevent flooding

dyke² n (offens slang) lesbian

dynamic *adj* full of energy, ambition, and new ideas; (*Physics*) of energy or forces that produce motion ▶ **dynamically** *adv*
▶ **dynamism** *n* great energy and enthusiasm

dynamics *n* branch of mechanics concerned with the forces that change or produce the motions of bodies ▷ *pl n* forces that produce change in a system

dynamite *n* explosive made of nitroglycerine; (*informal*) dangerous or exciting person or thing ▷ *v* blow (something) up with dynamite

dynamo *n*, *pl* **-mos** device for converting mechanical energy into electrical energy

dynasty *n*, *pl* **-ties** sequence of hereditary rulers ▶ **dynastic** *adj*

dysentery *n* infection of the intestine causing severe diarrhoea

dysfunction *n* (*Med*) disturbance or abnormality in the function of an organ or part ▶ **dysfunctional** *adj*

dyslexia *n* disorder causing impaired ability to read
▶ **dyslexic** *adj*

dysmenorrhoea *n* painful menstruation

dyspepsia *n* indigestion
▶ **dyspeptic** *adj*

dystrophy [diss-trof-fee] *n* see **muscular dystrophy**

Ee

E East(ern) *n*, *pl* **Es** or **E's** (*slang*) ecstasy (the drug)

e- *prefix* electronic e.g. *e-mail*

each *adj*, *pron* every (one) taken separately

eager *adj* showing or feeling great desire, keen ▶ **eagerly** *adv*
▶ **eagerness** *n*

eagle *n* large bird of prey with keen eyesight; (*Golf*) score of two strokes under par for a hole ▶ **eaglet** *n* young eagle

ear¹ *n* organ of hearing, esp. the external part of it; sensitivity to musical or other sounds
▶ **earache** *n* pain in the ear
▶ **earbash** *v* (*Aust & NZ informal*) talk incessantly ▶ **earbashing** *n* ▶ **eardrum** *n* thin piece of skin inside the ear which enables one to hear sounds ▶ **earmark** *v* set (something) aside for a specific purpose ▶ **earphone** *n* receiver for a radio etc., held to or put in the ear ▶ **earring** *n* ornament for the lobe of the ear ▶ **earshot** *n* hearing range

ear² *n* head of corn

earl *n* British nobleman ranking next below a marquess
▶ **earldom** *n*

early *adj*, *adv* **-lier, -liest** before the expected or usual time; in the first part of a period; in a period far back in time

early adopter *n* one of the first people or organizations to make use of a new technology

earn v obtain by work or merit; (of investments etc.) gain (interest) ▶ **earnings** pl n money earned

earnest¹ adj serious and sincere **in earnest** seriously ▶ **earnestly** adv

earnest² n part payment given in advance, esp. to confirm a contract

earth n planet that we live on; land, the ground; soil; fox's hole; wire connecting an electrical apparatus with the earth ▶ v connect (a circuit) to earth ▶ **earthen** adj made of baked clay or earth ▶ **earthenware** n pottery made of baked clay ▶ **earthly** adj conceivable or possible ▶ **earthy** adj coarse or crude; of or like earth ▶ **earthquake** n violent vibration of the earth's surface ▶ **earthwork** n fortification made of earth ▶ **earthworm** n worm which burrows in the soil

earwig n small insect with a pincer-like tail

ease n freedom from difficulty, discomfort, or worry; rest or leisure ▶ v give bodily or mental ease to; lessen (severity, tension, pain, etc.); move carefully or gradually

easel n frame to support an artist's canvas or a blackboard

east n (direction towards) the part of the horizon where the sun rises; region lying in this direction ▶ adj to or in the east; (of a wind) from the east ▶ adv in, to, or towards the east ▶ **easterly** adj ▶ **eastern** adj ▶ **eastward** adj, adv ▶ **eastwards** adv

Easter n Christian spring festival commemorating the Resurrection of Jesus Christ ▶ **Easter egg** chocolate egg given at Easter

easy adj **easier, easiest** not needing much work or effort; free from pain, care, or anxiety; easy-going ▶ **easily** adv ▶ **easiness** n ▶ **easy chair** comfortable armchair ▶ **easy-going** adj relaxed in attitude, tolerant

eat v **eating, ate, eaten** take (food) into the mouth and swallow it; have a meal; (foll by away or up) destroy ▶ **eatable** adj fit or suitable for eating

eau de Cologne [oh de kol-**lone**] n (French) light perfume

eaves pl n overhanging edges of a roof

eavesdrop v **-dropping, -dropped** listen secretly to a private conversation ▶ **eavesdropper** n ▶ **eavesdropping** n

ebb v (of tide water) flow back; fall away or decline ▶ n flowing back of the tide **at a low ebb** in a state of weakness

Ebola virus n virus that causes severe infectious disease

ebony n, pl **-onies** hard black wood ▶ adj deep black

e-book n electronic book ▶ v book (airline tickets, appointments, etc.) on the internet

ebullient adj full of enthusiasm or excitement ▶ **ebullience** n

EC European Commission; European Community: a former name for the European Union

e-card, eCard n greetings card created and sent electronically

eccentric adj odd or unconventional; (of circles) not having the same centre ▶ n eccentric person ▶ **eccentrically** adv ▶ **eccentricity** n

ecclesiastic n member of the clergy ▶ adj (also **ecclesiastical**) of the Christian Church or clergy

ECG electrocardiogram

echelon [esh-a-**lon**] n level of power or responsibility; (Mil) formation in which units follow one another but are spaced out sideways to allow each a line of fire ahead

echidna [ik-kid-na] n, pl -nas or -nae Australian spiny egg-laying mammal (also **spiny anteater**)

echo n, pl -oes repetition of sound by reflection of sound waves off a surface; close imitation ▷ v -oing, -oed repeat or be repeated as an echo; imitate (what someone else has said) ▶ **echo sounder** sonar

e-cigarette n electronic vaporizer that simulates the effect of smoking

éclair n finger-shaped pastry filled with cream and covered with chocolate

éclat [ake-lah] n brilliant success; splendour

eclectic adj selecting from various styles, ideas, or sources ▶ **eclecticism** n

eclipse n temporary obscuring of one star or planet by another ▷ v surpass or outclass ▶ **ecliptic** n apparent path of the sun

ecological adj of ecology; intended to protect the environment ▶ **ecologically** adv ▶ **ecology** n study of the relationships between living things and their environment ▶ **ecologist** n

e-commerce, ecommerce n business transactions done on the internet

economy n, pl -mies system of interrelationship of money, industry, and employment in a country; careful use of money or resources to avoid waste ▶ **economic** adj of economics; profitable; (informal) inexpensive or cheap ▶ **economics** n social science concerned with the production and consumption of goods and services ▷ pl n financial aspects ▶ **economical** adj not wasteful, thrifty ▶ **economically** adv ▶ **economist** n specialist in economics ▶ **economize** v reduce

expense or waste ▶ **economic migrant** person emigrating to improve his or her standard of living

ecosystem n system involving interactions between a community and its environment

ecru adj pale creamy-brown

ecstasy n state of intense delight; (slang) powerful drug that can produce hallucinations ▶ **ecstatic** adj ▶ **ecstatically** adv

> **SPELLING TIP**
> Spelling **ecstasy** with an extra c, as ecstacy, is a common mistake

ectoplasm n (Spiritualism) substance that supposedly is emitted from the body of a medium during a trance

ecumenical adj of the Christian Church throughout the world, esp. with regard to its unity

eczema [ek-sim-a, ig-zeem-a] n skin disease causing intense itching

Edam n round Dutch cheese with a red waxy cover

eddy n, pl **eddies** circular movement of air, water, etc. ▷ v **eddying, eddied** move with a circular motion

edelweiss [ade-el-vice] n alpine plant with white flowers

Eden n (Bible) garden in which Adam and Eve were placed at the Creation

edge n border or line where something ends or begins; cutting side of a blade; sharpness of tone ▷ v provide an edge or border for; push (one's way) gradually **have the edge on** have an advantage over **on edge** nervous or irritable ▶ **edgeways** adv with the edge forwards or uppermost ▶ **edging** n anything placed along an edge to finish it ▶ **edgy** adj nervous or irritable

edible adj fit to be eaten ▶ **edibility** n

edict [ee-dikt] n order issued by an authority

edifice [ed-if-iss] n large building

edify [ed-if-fie] v **-fying, -fied** improve morally by instruction ▶ **edification** n

edit v prepare (a book, film, etc.) for publication or broadcast ▶ **edition** n number of copies of a new publication printed at one time ▶ **editor** n person who edits; person in charge of one section of a newspaper or magazine ▶ **editorial** n newspaper article stating the opinion of the editor ▷ adj of editing or editors

educate v teach; provide schooling for ▶ **education** n ▶ **educational** adj ▶ **educationally** adv ▶ **educationalist** n expert in the theory of education ▶ **educative** adj educating

Edwardian adj of the reign of King Edward VII of Great Britain and Ireland (1901–10)

EEG electroencephalogram

eel n snakelike fish

eerie adj **eerier, eeriest** uncannily frightening or disturbing ▶ **eerily** adv

efface v remove by rubbing; make (oneself) inconspicuous ▶ **effacement** n

effect n change or result caused by someone or something; condition of being operative e.g. *the law comes into effect next month;* overall impression ▷ pl personal belongings; lighting, sounds, etc. to accompany a film or a broadcast ▷ v cause to happen, accomplish ▶ **effective** adj producing a desired result; operative; impressive ▶ **effectively** adv ▶ **effectual** adj producing the intended result ▶ **effectually** adv

USAGE NOTE
Note the difference between *effect* meaning 'accomplish' and *affect* meaning 'influence'

effeminate adj (of a man) displaying characteristics thought to be typical of a woman ▶ **effeminacy** n

effervescent adj (of a liquid) giving off bubbles of gas; (of a person) lively and enthusiastic ▶ **effervescence** n

effete [if-feet] adj powerless, feeble

efficacious adj producing the intended result ▶ **efficacy** n

efficient adj functioning effectively with little waste of effort ▶ **efficiently** adv ▶ **efficiency** n

effigy [ef-fij-ee] n, pl **-gies** image or likeness of a person

efflorescence n flowering

effluent n liquid discharged as waste

effluvium n, pl **-via** unpleasant smell, as of decaying matter or gaseous waste

effort n physical or mental exertion; attempt ▶ **effortless** adj

effrontery n brazen impudence

effusion n unrestrained outburst ▶ **effusive** adj openly emotional, demonstrative ▶ **effusively** adv

EFTA European Free Trade Association

e.g. for example

egalitarian adj upholding the equality of all people ▷ n person who holds egalitarian beliefs ▶ **egalitarianism** n

egg[1] n oval or round object laid by the females of birds and other creatures, containing a developing embryo; hen's egg used as food; (also **egg cell**) ovum ▶ **egghead** n (informal) intellectual person ▶ **eggplant** n (US, Canadian, Aust & NZ) dark purple tropical fruit, cooked and eaten as a vegetable

egg² v **egg on** encourage or incite, esp. to do wrong

ego n, pl **egos** the conscious mind of an individual; self-esteem ▶ **egoism, egotism** n excessive concern for one's own interests; excessively high opinion of oneself ▶ **egotist, egoist** n ▶ **egotistic, egoistic** adj ▶ **egocentric** adj self-centred

egregious [ig-**greej**-uss] adj outstandingly bad

egress [**ee**-gress] n departure; way out

egret [**ee**-grit] n lesser white heron

Egyptian adj relating to Egypt ▷ n person from Egypt

Egyptology n study of the culture of ancient Egypt

eider n Arctic duck ▶ **eiderdown** n quilt (orig. stuffed with eider feathers)

eight adj, n one more than seven ▷ n eight-oared boat; its crew ▶ **eighth** adj, n (of) number eight in a series ▶ **eighteen** adj, n eight and ten ▶ **eighteenth** adj, n ▶ **eighty** adj, n eight times ten ▶ **eightieth** adj, n

eisteddfod [ice-**sted**-fod] n Welsh festival with competitions in music and other performing arts

either adj, pron one or the other (of two); each of two ▷ conj used preceding two or more possibilities joined by or ▷ adv likewise e.g. I don't eat meat and he doesn't either

ejaculate v eject (semen); utter abruptly ▶ **ejaculation** n

eject v force out, expel ▶ **ejection** n ▶ **ejector** n

eke out v make (a supply) last by frugal use; make (a living) with difficulty

elaborate adj with a lot of fine detail ▷ v expand upon ▶ **elaboration** n

élan [ale-**an**] n style and vigour

eland [**eel**-and] n large antelope of southern Africa

elapse v (of time) pass by

elastic adj resuming normal shape after distortion; adapting easily to change ▷ n tape or fabric containing interwoven strands of flexible rubber ▶ **elasticity** n

elated v extremely happy and excited ▶ **elation** n

elbow n joint between the upper arm and the forearm ▷ v shove or strike with the elbow ▶ **elbow grease** vigorous physical labour ▶ **elbow room** sufficient room to move freely

elder¹ adj older ▷ n older person; (in certain Protestant Churches) lay officer ▶ **elderly** adj (fairly) old ▶ **eldest** adj oldest

> **USAGE NOTE**
> Elder (eldest) is used for age comparison in families, older for other age comparisons

elder² n small tree with white flowers and black berries

El Dorado [el dor-**rah**-doe] n fictitious country rich in gold

eldritch adj (Scot) weird, uncanny

elect v choose by voting; decide (to do something) ▷ adj appointed but not yet in office e.g. president elect ▶ **election** n choosing of representatives by voting; act of choosing ▶ **electioneering** n active participation in a political campaign ▶ **elective** adj chosen by election; optional ▶ **elector** n someone who has the right to vote in an election ▶ **electoral** adj ▶ **electorate** n people who have the right to vote

electricity n form of energy associated with stationary or moving electrons or other charged particles; electric current or charge ▶ **electric** adj produced by, transmitting, or powered by electricity; exciting or tense ▶ **electrical** adj using or concerning electricity ▶ **electrician** n person trained to install and repair electrical equipment ▶ **electrics** pl n (Brit) electric appliances ▶ **electric chair** (US) chair in which criminals who have been sentenced to death are electrocuted

electrify v -fying, -fied adapt for operation by electric power; charge with electricity; startle or excite intensely ▶ **electrification** n

electro- combining form operated by or caused by electricity

electrocardiograph n instrument for recording the electrical activity of the heart ▶ **electrocardiogram** n tracing produced by this

electrocute v kill or injure by electricity ▶ **electrocution** n

electrode n conductor through which an electric current enters or leaves a battery, vacuum tube, etc.

electrodynamics n branch of physics concerned with the interactions between electrical and mechanical forces

electroencephalograph [ill-lek-tro-en-**sef**-a-loh-graf] n instrument for recording the electrical activity of the brain ▶ **electroencephalogram** n tracing produced by this

electrolysis [ill-lek-**troll**-iss-iss] n conduction of electricity by an electrolyte, esp. to induce chemical change; destruction of living tissue such as hair roots by an electric current

electrolyte n solution or molten substance that conducts electricity ▶ **electrolytic** adj

electromagnet n magnet containing a coil of wire through which an electric current is passed

electromagnetic adj of or operated by an electromagnet ▶ **electromagnetism** n

electron n elementary particle in all atoms that has a negative electrical charge ▶ **electron microscope** microscope that uses electrons, rather than light, to produce a magnified image ▶ **electronvolt** n unit of energy used in nuclear physics

electronic adj (of a device) dependent on the action of electrons; (of a process) using electronic devices ▶ **electronic mail** see **e-mail** ▶ **electronics** n technology concerned with the development of electronic devices and circuits

electroplate v coat with silver etc. by electrolysis

elegant adj pleasing or graceful in dress, style, or design ▶ **elegance** n

elegy [el-lij-ee] n, pl -**gies** mournful poem, esp. a lament for the dead ▶ **elegiac** adj mournful or plaintive

element n component part; substance which cannot be separated into other substances by ordinary chemical techniques; section of people within a larger group e.g. the rowdy element; heating wire in an electric kettle, stove, etc. ▶ pl basic principles of something; weather conditions, esp. wind, rain, and cold **in one's element** in a situation where one is happiest ▶ **elemental** adj of primitive natural forces or passions ▶ **elementary** adj simple and straightforward

elephant n huge four-footed thick-skinned animal with ivory tusks and a long trunk ▶ **elephantine** adj unwieldy, clumsy ▶ **elephantiasis** [el-lee-fan-**tie**-a-siss] n disease with hardening of the skin and enlargement of the legs etc.

elevate v raise in rank or status; lift up ▶ **elevation** n raising; height above sea level; scale drawing of one side of a building ▶ **elevator** n (Aust, US & Canadian) lift for carrying people

eleven adj, n one more than ten ▷ n (Sport) team of eleven people ▶ **eleventh** adj, n (of) number eleven in a series ▶ **elevenses** n (Brit & S Afr informal) mid-morning snack

elf n, pl **elves** (in folklore) small mischievous fairy ▶ **elfin** adj small and delicate

elicit v bring about (a response or reaction); find out (information) by careful questioning

elide v omit (a vowel or syllable) from a spoken word ▶ **elision** n

eligible adj meeting the requirements or qualifications needed; desirable as a spouse ▶ **eligibility** n

eliminate v get rid of ▶ **elimination** n

elite [ill-**eet**] n most powerful, rich, or gifted members of a group ▶ **elitism** n belief that society should be governed by a small group of superior people ▶ **elitist** n, adj

elixir [ill-**ix**-er] n imaginary liquid that can prolong life or turn base metals into gold

Elizabethan adj of the reign of Elizabeth I of England (1558–1603)

elk n large deer of N Europe and Asia

ellipse n oval shape ▶ **elliptical** adj oval-shaped; (of speech or writing) obscure or ambiguous

ellipsis n, pl **-ses** omission of letters or words in a sentence

elm n tree with serrated leaves

elocution n art of speaking clearly in public

elongate [eel-**long**-gate] v make or become longer ▶ **elongation** n

elope v (of two people) run away secretly to get married ▶ **elopement** n

eloquence n fluent powerful use of language ▶ **eloquent** adj ▶ **eloquently** adv

else adv in addition or more e.g. what else can I do?; other or different e.g. it was unlike anything else that had happened ▶ **elsewhere** adv in or to another place

elucidate v make (something difficult) clear ▶ **elucidation** n

elude v escape from by cleverness or quickness; baffle ▶ **elusive** adj difficult to catch or remember

elver n young eel

elves n plural of **elf**

emaciated [im-**mace**-ee-ate-id] adj abnormally thin ▶ **emaciation** n

e-mail, email n (also **electronic mail**) sending of messages between computer terminals ▷ v communicate in this way

emanate [**em**-a-nate] v issue, proceed from a source ▶ **emanation** n

emancipate v free from social, political, or legal restraints ▶ **emancipation** n

emasculate v deprive of power ▶ **emasculation** n

embalm v preserve (a corpse) from decay by the use of chemicals etc.

embankment n man-made ridge that carries a road or railway or holds back water

embargo n, pl **-goes** order by a government prohibiting trade with a country ▷ v **-going, -goed** put an embargo on

embark v board a ship or aircraft; (foll by *on*) begin (a new project)
▶ **embarkation** n

embarrass v cause to feel self-conscious or ashamed
▶ **embarrassed** adj
▶ **embarrassing** adj
▶ **embarrassment** n

SPELLING TIP
Remember **embarrass** and all the other words in this family should each have two rs and two ss

embassy n, pl **-sies** offices or official residence of an ambassador; ambassador and his or her staff

embattled adj having a lot of difficulties

embed v **-bedding, -bedded** fix firmly in something solid
▶ **embedded** adj (of a journalist) assigned to accompany an active military unit

embellish v decorate; embroider (a story) ▶ **embellishment** n

ember n glowing piece of wood or coal in a dying fire

embezzle v steal money that has been entrusted to one
▶ **embezzlement** n ▶ **embezzler** n

embittered adj feeling anger as a result of misfortune

emblazon v decorate with bright colours; proclaim or publicize

emblem n object or design that symbolizes a quality, type, or group
▶ **emblematic** adj

embody v **-bodying, -bodied** be an example or expression of; comprise, include
▶ **embodiment** n

embolden v encourage (someone)

embolism n blocking of a blood vessel by a blood clot or air bubble

embossed adj (of a design or pattern) standing out from a surface

embrace v clasp in the arms, hug; accept (an idea) eagerly; comprise
▷ n act of embracing

embrasure n door or window having splayed sides so that the opening is larger on the inside; opening like this in a fortified wall, for shooting through

embrocation n lotion for rubbing into the skin to relieve pain

embroider v decorate with needlework; make (a story) more interesting with fictitious detail
▶ **embroidery** n

embroil v involve (a person) in problems

embryo [em-bree-oh] n, pl **-bryos** unborn creature in the early stages of development; something at an undeveloped stage
▶ **embryonic** adj at an early stage
▶ **embryology** n

emend v remove errors from
▶ **emendation** n

emerald n bright green precious stone ▷ adj bright green

emerge v come into view; (foll by *from*) come out of; become known
▶ **emergence** n ▶ **emergent** adj

emergency n, pl **-cies** sudden unforeseen occurrence needing immediate action

emeritus [im-mer-rit-uss] adj retired, but retaining an honorary title e.g. *emeritus professor*

emery n hard mineral used for smoothing and polishing
▶ **emery board** cardboard strip coated with crushed emery, for filing the nails

emetic [im-met-ik] n substance that causes vomiting ▷ adj causing vomiting

emigrate v go and settle in another country ▶ **emigrant** n
▶ **emigration** n

émigré [em-mig-gray] *n* man who has left his native country for political reasons ▸ **émigrée** *n fem*

eminent *adj* distinguished, well-known ▸ **eminently** *adv* ▸ **eminence** *n* position of superiority or fame; (**E-**) title of a cardinal

emir [em-**meer**] *n* Muslim ruler ▸ **emirate** *n* his country

emissary *n, pl* **-saries** agent sent on a mission by a government

emit *v* **emitting, emitted** give out (heat, light, or a smell); utter ▸ **emission** *n*

emollient *adj* softening, soothing ▸ *n* substance which softens or soothes the skin

emolument *n* (formal) fees or wages from employment

emoji [im-**moh**-jee] *n* image used in electronic messages

emoticon [i-**mote**-i-kon] *n* (Computing) same as **smiley**

emotion *n* strong feeling ▸ **emotional** *adj* readily affected by or appealing to the emotions ▸ **emotionally** *adv* ▸ **emotive** *adj* tending to arouse emotion

empathy *n* ability to understand someone else's feelings as if they were one's own

emperor *n* ruler of an empire ▸ **empress** *n fem*

emphasis *n, pl* **-ses** special importance or significance; stress on a word or phrase in speech ▸ **emphasize** *v* ▸ **emphatic** *adj* showing emphasis ▸ **emphatically** *adv*

emphysema [em-fiss-**see**-ma] *n* condition in which the air sacs of the lungs are grossly enlarged, causing breathlessness

empire *n* group of territories under the rule of one state or person; large organization that is directed by one person or group

empirical *adj* relying on experiment or experience, not on theory ▸ **empirically** *adv* ▸ **empiricism** *n* doctrine that all knowledge derives from experience ▸ **empiricist** *n*

emplacement *n* prepared position for a gun

employ *v* hire (a person); provide work or occupation for; use ▸ *n* **in the employ of** doing regular paid work for ▸ **employee** *n* ▸ **employment** *n* state of being employed; work done by a person to earn money

employer *n* person or organization that employs someone

emporium *n, pl* **-riums** *or* **-ria** (old-fashioned) large general shop

empower *v* enable, authorize

empress *n see* **emperor**

empty *adj* **-tier, -tiest** containing nothing; unoccupied; without purpose or value; (of words) insincere ▸ *v* **-tying, -tied** make or become empty ▸ **empties** *pl n* empty boxes, bottles, etc. ▸ **emptiness** *n*

emu *n* large Australian flightless bird with long legs ▸ **emu oil** *n* derived from emu fat, used as a liniment by native Australians

emulate *v* attempt to equal or surpass by imitating ▸ **emulation** *n*

emulsion *n* light-sensitive coating on photographic film; type of water-based paint ▸ *v* paint with emulsion paint ▸ **emulsify** *v* (of two liquids) join together or join (two liquids) together ▸ **emulsifier** *n*

enable *v* provide (a person) with the means, opportunity, or authority (to do something)

enact *v* establish by law; perform (a story or play) by acting ▸ **enactment** *n*

enamel *n* glasslike coating applied to metal etc. to preserve the surface; hard white coating on a tooth ▷ v -elling, -elled cover with enamel

enamoured adj inspired with love

en bloc adv (French) as a whole, all together

encamp v set up in a camp
▶ **encampment** n

encapsulate v summarize; enclose as in a capsule

encephalitis [en-sef-a-lite-iss] n inflammation of the brain

encephalogram n short for **electroencephalogram**

enchant v delight and fascinate
▶ **enchantment** n ▶ **enchanter** n
▶ **enchantress** n fem

encircle v form a circle around
▶ **encirclement** n

enclave n part of a country entirely surrounded by foreign territory

enclose v surround completely; include along with something else
▶ **enclosure** n

encomium n, pl **-miums** or **-mia** formal expression of praise

encompass v surround; include comprehensively

encore interj again, once more
▷ n extra performance due to enthusiastic demand

encounter v meet unexpectedly; be faced with ▷ n unexpected meeting; game or battle

encourage v inspire with confidence; spur on
▶ **encouragement** n

encroach v intrude gradually on a person's rights or land
▶ **encroachment** n

encrust v cover with a layer of something

encrypt v put (a message) into code; distort (a television or other signal) so that it cannot be viewed or understood without decryption equipment

encumber v hinder or impede
▶ **encumbrance** n something that impedes or is burdensome

encyclical [en-sik-lik-kl] n letter sent by the Pope to all bishops

encyclopedia, encyclopaedia n book or set of books containing facts about many subjects, usu. in alphabetical order ▶ **encyclopedic, encyclopaedic** adj

end n furthest point or part; limit; last part of something; fragment; death or destruction; purpose; (Sport) either of the two defended areas of a playing field ▷ v bring or come to a finish **make ends meet** have just enough money for one's needs ▶ **ending** n ▶ **endless** adj
▶ **endways** adv having the end forwards or upwards

endanger v put in danger

endear v cause to be liked
▶ **endearing** adj ▶ **endearment** n affectionate word or phrase

endeavour v try ▷ n effort

endemic adj present within a localized area or peculiar to a particular group of people

endive n curly-leaved plant used in salads

endocrine adj relating to the glands which secrete hormones directly into the bloodstream

endogenous [en-dodge-in-uss] adj originating from within

endorse v give approval to; sign the back of (a cheque); record a conviction on (a driving licence)
▶ **endorsement** n

endow v provide permanent income for **endowed with** provided with ▶ **endowment** n

endure v bear (hardship) patiently; last for a long time ▶ **endurable** adj ▶ **endurance** n act or power of enduring

enema [en-im-a] n medicine injected into the rectum to empty the bowels

enemy n, pl **-mies** hostile person or nation, opponent

energy n, pl **-gies** capacity for intense activity; capacity to do work and overcome resistance; source of power, such as electricity ▸ **energetic** adj ▸ **energetically** adv ▸ **energize** v give vigour to ▸ **energy drink** soft drink supposed to boost the drinker's energy levels

enervate v deprive of strength or vitality ▸ **enervation** n

enfant terrible [on-fon ter-reeb-la] n, pl **enfants terribles** (French) clever but unconventional or indiscreet person

enfeeble v weaken

enfold v cover by wrapping something around; embrace

enforce v impose obedience (to a law etc.); impose (a condition) ▸ **enforceable** adj ▸ **enforcement** n

enfranchise v grant (a person) the right to vote ▸ **enfranchisement** n

engage v take part, participate; involve (a person or his or her attention) intensely; employ (a person); begin a battle with; bring (a mechanism) into operation ▸ **engaged** adj pledged to be married; in use ▸ **engagement** n ▸ **engaging** adj charming

engender v produce, cause to occur

engine n any machine which converts energy into mechanical work; railway locomotive

engineer n person trained in any branch of engineering ▸ v plan in a clever manner; design or construct as an engineer

engineering n profession of applying scientific principles to the

design and construction of engines, cars, buildings, or machines

English n official language of Britain, Ireland, Australia, New Zealand, South Africa, Canada, the US, and several other countries ▸ adj relating to England **the English** the people of England

engrave v carve (a design) onto a hard surface; fix deeply in the mind ▸ **engraver** n ▸ **engraving** n print made from an engraved plate

engross [en-groce] v occupy the attention of (a person) completely

engulf v cover or surround completely

enhance v increase in quality, value, or attractiveness ▸ **enhancement** n

enigma n puzzling thing or person ▸ **enigmatic** adj ▸ **enigmatically** adv

enjoin v order (someone) to do something

enjoy v take joy in; have the benefit of; experience ▸ **enjoyable** adj ▸ **enjoyment** n

enlarge v make or grow larger; (foll by on) speak or write about in greater detail ▸ **enlargement** n

enlighten v give information to ▸ **enlightenment** n

enlist v enter the armed forces; obtain the support of ▸ **enlistment** n

enliven v make lively or cheerful

en masse [on mass] adv (French) in a group, all together

enmeshed adj deeply involved

enmity n, pl **-ties** ill will, hatred

ennoble v make noble, elevate

ennui [on-nwee] n boredom, dissatisfaction

enormous adj very big, vast ▸ **enormity** n, pl **-ties** great wickedness; gross offence; (informal) great size

enough adj as much as or as many as necessary ▷ n sufficient quantity ▷ adv sufficiently; fairly or quite e.g. *that's a common enough experience*

en passant [on pass-on] adv (French) in passing, by the way

enquire v same as **inquire** ▶ **enquiry** n

enraptured adj filled with delight and fascination

enrich v improve in quality; make wealthy or wealthier

enrol v -**rolling**, -**rolled** (cause to) become a member ▶ **enrolment** n

en route adv (French) on the way

ensconce v settle firmly or comfortably

ensemble [on-som-bl] n all the parts of something taken together; complete outfit of clothes; company of actors or musicians; (Music) group of musicians playing together

enshrine v cherish or treasure

ensign n naval flag; banner; (US) naval officer

enslave v make a slave of (someone) ▶ **enslavement** n

ensnare v catch in or as if in a snare

ensue v come next, result

en suite adv (French) connected to a bedroom and entered directly from it

ensure v make certain or sure; make safe or protect

entail v bring about or impose inevitably

entangle v catch or involve in or as if in a tangle ▶ **entanglement** n

entente [on-tont] n friendly understanding between nations

enter v come or go in; join; become involved in, take part in; record (an item) in a journal etc.; begin ▶ **entrance** n way into a place; act of entering; right of entering ▶ **entrant** n person who enters a university, contest, etc. ▶ **entry** n,

pl -**tries** entrance; entering; item entered in a journal etc.

enteric [en-ter-ik] adj intestinal ▶ **enteritis** [en-ter-rite-iss] n inflammation of the intestine, causing diarrhoea

enterprise n company or firm; bold or difficult undertaking; boldness and energy ▶ **enterprising** adj full of boldness and initiative

entertain v amuse; receive as a guest; consider (an idea) ▶ **entertainer** n ▶ **entertainment** n

enthral [en-thrawl] v -**thralling**, -**thralled** hold the attention of ▶ **enthralling** adj

enthusiasm n ardent interest, eagerness ▶ **enthuse** v (cause to) show enthusiasm ▶ **enthusiast** n ardent supporter of something ▶ **enthusiastic** adj ▶ **enthusiastically** adv

entice v attract by exciting hope or desire, tempt ▶ **enticement** n

entire adj including every detail, part, or aspect of something ▶ **entirely** adv ▶ **entirety** n

entitle v give a right to; give a title to ▶ **entitlement** n

entity n, pl -**ties** separate distinct thing

entomology n study of insects ▶ **entomological** adj ▶ **entomologist** n

entourage [on-toor-ahzh] n group of people who assist an important person

entrails pl n intestines; innermost parts of something

entrance¹ n see **enter**

entrance² v delight; put into a trance

entreat v ask earnestly ▶ **entreaty** n, pl -**ties** earnest request

entrée [on-tray] n dish served before a main course; main course; right of admission

entrench v establish firmly; establish in a fortified position with trenches ▶ **entrenchment** n

entrepreneur n business person who attempts to make a profit by risk and initiative

entropy [en-trop-ee] n lack of organization

entrust v put into the care or protection of

entwine v twist together or around

E number n any of a series of numbers with the prefix E indicating a specific food additive recognized by the EU

enumerate v name one by one ▶ **enumeration** n

enunciate v pronounce clearly; state precisely or formally ▶ **enunciation** n

envelop v enveloping, enveloped wrap up, enclose ▶ **envelopment** n

envelope n folded gummed paper cover for a letter

environment [en-vire-on-ment] n external conditions and surroundings in which people, animals, or plants live ▶ **environmental** adj ▶ **environmentalist** n person concerned with the protection of the natural environment

> **SPELLING TIP**
> Note the n before the m in the spelling of environment

environs pl n surrounding area, esp. of a town

envisage v conceive of as a possibility

envoy n messenger; diplomat ranking below an ambassador

envy n feeling of discontent aroused by another's good fortune ▷ v -vying, -vied grudge (another's good fortune, success, or qualities) ▶ **enviable** adj arousing envy, fortunate ▶ **envious** adj full of envy

enzyme n any of a group of complex proteins that act as catalysts in specific biochemical reactions

Eolithic adj of the early part of the Stone Age

epaulette n shoulder ornament on a uniform

ephemeral adj short-lived

epic n long poem, book, or film about heroic events or actions ▷ adj very impressive or ambitious

epicentre n point on the earth's surface immediately above the origin of an earthquake

epicure n person who enjoys good food and drink ▶ **epicurean** adj devoted to sensual pleasures, esp. food and drink ▷ n epicure

epidemic n widespread occurrence of a disease; rapid spread of something

epidermis n outer layer of the skin

epidural [ep-pid-dure-al] adj, n (of) spinal anaesthetic injected to relieve pain during childbirth

epiglottis n thin flap that covers the opening of the larynx during swallowing

epigram n short witty remark or poem ▶ **epigrammatic** adj

epigraph n quotation at the start of a book; inscription

epilepsy n disorder of the nervous system causing loss of consciousness and sometimes convulsions ▶ **epileptic** adj of or having epilepsy ▷ n (offens) person who has epilepsy

epilogue n short speech or poem at the end of a literary work, esp. a play

Epiphany n Christian festival held on January 6 commemorating the manifestation of Christ to the Magi

episcopal [ip-piss-kop-al] adj of or governed by bishops ▶ **episcopalian** adj advocating

Church government by bishops
▷ *n* advocate of such Church
government

episode *n* incident in a series of
incidents; section of a serialized
book, television programme,
etc. ▶ **episodic** *adj* occurring at
irregular intervals

epistemology [ip-iss-stem-ol-a-
jee] *n* study of the source, nature,
and limitations of knowledge
▶ **epistemological** *adj*

epistle *n* letter, esp. of an apostle
▶ **epistolary** *adj*

epitaph *n* commemorative
inscription on a tomb;
commemorative speech or
passage

epithet *n* descriptive word or name

epitome [ip-pit-a-mee] *n* typical
example ▶ **epitomize** *v* be the
epitome of

> **USAGE NOTE**
> Avoid the use of *epitome* to mean
> 'the peak' of something

epoch [ee-pok] *n* period of notable
events ▶ **epoch-making** *adj*
extremely important

eponymous [ip-pon-im-uss] *adj*
after whom a book, play, etc. is
named

equable [ek-wab-bl] *adj* even-
tempered ▶ **equably** *adv*

equal *adj* identical in size, quantity,
degree, etc.; having identical
rights or status; evenly balanced;
(foll by *to*) having the necessary
ability (for) ▷ *n* person or thing
equal to another ▶ *v* **equalling,
equalled** be equal to ▶ **equally** *adv*
▶ **equality** *n* state of being equal
▶ **equalize** *v* make or become
equal; reach the same score as
one's opponent ▶ **equalization** *n*
▶ **equal opportunity**
nondiscrimination as to sex, race,
etc. in employment

equanimity *n* calmness of mind

equate *v* make or regard as
equivalent ▶ **equation** *n*
mathematical statement that
two expressions are equal; act of
equating

equator *n* imaginary circle round
the earth, equidistant from the
poles ▶ **equatorial** *adj*

equerry [ek-kwer-ee] *n, pl*
-ries (*Brit*) officer who acts as an
attendant to a member of a royal
family

equestrian *adj* of horses and riding

equidistant *adj* equally distant

equilateral *adj* having equal sides

equilibrium *n, pl* **-ria** steadiness
or stability

equine *adj* of or like a horse

equinox *n* time of year when day
and night are of equal length
▶ **equinoctial** *adj*

equip *v* **equipping, equipped**
provide with supplies,
components, etc. ▶ **equipment**
n set of tools or devices used
for a particular purpose; act of
equipping

equipoise *n* perfect balance

equity *n, pl* **-ties** fairness;
legal system, founded on the
principles of natural justice, that
supplements common law ▷ *pl*
interest of ordinary shareholders in
a company ▶ **equitable** *adj* fair and
reasonable ▶ **equitably** *adv*

equivalent *adj* equal in value;
having the same meaning or result
▷ *n* something that is equivalent
▶ **equivalence** *n*

equivocal *adj* ambiguous;
deliberately misleading; of
doubtful character or sincerity
▶ **equivocally** *adv* ▶ **equivocate**
v use vague or ambiguous
language to mislead people
▶ **equivocation** *n*

ER Queen Elizabeth

era n period of time considered as distinctive

eradicate v destroy completely ▶ **eradication** n

erase v rub out; remove sound or information from (a magnetic tape or disk) ▶ **eraser** n object for erasing something written ▶ **erasure** n erasing; place or mark where something has been erased

ere prep, conj (poetic) before

e-reader, eReader n portable device that allows users to download and read texts in electronic form

erect v build; found or form ▷ adj upright; (of the penis, clitoris, or nipples) rigid as a result of sexual excitement ▶ **erectile** adj capable of becoming erect from sexual excitement ▶ **erection** n

erg n unit of work or energy

ergonomics n study of the relationship between workers and their environment ▶ **ergonomic** adj

ergot n fungal disease of cereal; dried fungus used in medicine

ermine n stoat in northern regions; its white winter fur

erode v wear away ▶ **erosion** n

erogenous [ir-roj-in-uss] adj sensitive to sexual stimulation

erotic adj relating to sexual pleasure or desire ▶ **eroticism** n ▶ **erotica** n sexual literature or art

err v make a mistake ▶ **erratum** n, pl -**ta** error in writing or printing ▶ **erroneous** adj incorrect, mistaken

errand n short trip to do something for someone

errant adj behaving in a manner considered to be unacceptable

erratic adj irregular or unpredictable ▶ **erratically** adv

error n mistake, inaccuracy, or misjudgment

ersatz [air-zats] adj made in imitation of something written e.g. ersatz coffee

erstwhile adj former

erudite adj having great academic knowledge ▶ **erudition** n

erupt v eject (steam, water, or volcanic material) violently; burst forth suddenly and violently; (of a blemish) appear on the skin ▶ **eruption** n

erysipelas [err-riss-sip-pel-ass] n acute skin infection causing purplish patches

escalate v increase in extent or intensity ▶ **escalation** n

escalator n moving staircase

escalope [ess-kal-lop] n thin slice of meat, esp. veal

escapade n mischievous adventure

escape v get free (of); avoid e.g. escape attention; (of a gas, liquid, etc.) leak gradually ▷ n act of escaping; means of relaxation ▶ **escapee** n person who has escaped ▶ **escapism** n taking refuge in fantasy to avoid unpleasant reality ▶ **escapologist** n entertainer who specializes in freeing himself or herself from confinement ▶ **escapology** n

escarpment n steep face of a ridge or mountain

eschew [iss-chew] v abstain from, avoid

escort n people or vehicles accompanying another person for protection or as an honour; person who accompanies a person of the opposite sex to a social event ▷ v act as an escort to

escudo [ess-kyoo-doe] n, pl -**dos** former monetary unit of Portugal

escutcheon n shield with a coat of arms **blot on one's escutcheon** stain on one's honour

Eskimo n member of the aboriginal race inhabiting N Canada, Greenland, Alaska, and E Siberia; their language

> **USAGE NOTE**
> Eskimo is considered by many people to be offensive, and in North America the term *Inuit* is often used for the peoples native to the area from W Greenland to NW Canada

esoteric [ee-so-**ter**-rik] adj understood by only a small number of people with special knowledge

ESP extrasensory perception

esp. especially

espadrille [**ess**-pad-drill] n light canvas shoe with a braided cord sole

espalier [ess-**pal**-yer] n shrub or fruit tree trained to grow flat; trellis for this

esparto n, pl -tos grass of S Europe and N Africa used for making rope etc.

especial adj (formal) special

especially adv particularly

Esperanto n universal artificial language

espionage [ess-pyon-ahzh] n spying

esplanade n wide open road used as a seaside promenade

espouse v adopt or give support to (a cause etc.) ▶ **espousal** n

espresso n, pl -sos strong coffee made by forcing steam or boiling water through ground coffee beans

> **SPELLING TIP**
> Note that the second letter is s not x

esprit [ess-**pree**] n spirit, liveliness, or wit ▶ **esprit de corps** [de **core**] pride in and loyalty to a group

espy v spying, espied catch sight of

Esq. esquire

esquire n courtesy title placed after a man's name

essay n short literary composition; short piece of writing on a subject done as an exercise by a student ▷ v attempt ▶ **essayist** n

essence n most important feature of a thing which determines its identity; concentrated liquid used to flavour food ▶ **essential** adj vitally important; basic or fundamental ▷ n something fundamental or indispensable ▶ **essentially** adv

establish v set up on a permanent basis; make secure or permanent in a certain place, job, etc.; prove; cause to be accepted ▶ **establishment** n act of establishing; commercial or other institution **the Establishment** group of people having authority within a society

estate n landed property; large area of property development, esp. of new houses or factories; property of a deceased person ▶ **estate agent** agent concerned with the valuation, lease, and sale of property ▶ **estate car** car with a rear door and luggage space behind the rear seats

esteem n high regard ▷ v think highly of; judge or consider

ester n (Chem) compound produced by the reaction between an acid and an alcohol

estimate v calculate roughly; form an opinion about ▷ n approximate calculation; statement from a workman etc. of the likely charge for a job; opinion ▶ **estimable** adj worthy of respect ▶ **estimation** n considered opinion

estranged adj no longer living with one's spouse ▶ **estrangement** n

estuary n, pl -aries mouth of a river

ETA estimated time of arrival

et al. and others

etc. et cetera

et cetera [et **set**-ra] (*Latin*) and the rest, and others; or the like ▶ **etceteras** ▷ *pl n* miscellaneous extra things or people

etch *v* wear away or cut the surface of (metal, glass, etc.) with acid; imprint vividly (on someone's mind) ▶ **etching** *n*

eternal *adj* without beginning or end; unchanging ▶ **eternally** *adv* ▶ **eternity** *n* infinite time; timeless existence after death ▶ **eternity ring** ring given as a token of lasting affection

ether *n* colourless sweet-smelling liquid used as an anaesthetic; region above the clouds ▶ **ethereal** [eth-**eer**-ee-al] *adj* extremely delicate

ethic *n* moral principle ▶ **ethical** *adj* ▶ **ethically** *adv* ▶ **ethics** *n* code of behaviour; study of morals

ethnic *adj* relating to a people or group that shares a culture, religion, or language; belonging or relating to such a group, esp. one that is a minority group in a particular place ▶ **ethnic cleansing** practice, by the dominant ethnic group in an area, of removing other ethnic groups by expulsion or extermination ▶ **ethnology** *n* study of human races ▶ **ethnological** *adj* ▶ **ethnologist** *n*

ethos [**eeth**-oss] *n* distinctive spirit and attitudes of a people, culture, etc.

ethyl [**eeth**-ile] *adj* of, consisting of, or containing the hydrocarbon group C_2H_5 ▶ **ethylene** *n* poisonous gas used as an anaesthetic and as fuel

etiolate [ee-tee-oh-late] *v* become pale and weak; (*Botany*) whiten through lack of sunlight

etiology *n* study of the causes of diseases

etiquette *n* conventional code of conduct

étude [ay-**tewd**] *n* short musical composition for a solo instrument, esp. intended as a technical exercise

etymology *n, pl* -**gies** study of the sources and development of words ▶ **etymological** *adj*

EU European Union

eucalyptus, eucalypt *n* tree, mainly grown in Australia, that provides timber, gum, and medicinal oil from the leaves

Eucharist [**yew**-kar-ist] *n* Christian sacrament commemorating Christ's Last Supper; consecrated elements of bread and wine ▶ **Eucharistic** *adj*

eugenics [yew-**jen**-iks] *n* study of methods of improving the human race

eulogy *n, pl* -**gies** speech or writing in praise of a person ▶ **eulogize** *v* praise (a person or thing) highly in speech or writing ▶ **eulogistic** *adj*

eunuch *n* castrated man, esp. (formerly) a guard in a harem

euphemism *n* inoffensive word or phrase substituted for one considered offensive or upsetting ▶ **euphemistic** *adj* ▶ **euphemistically** *adv*

euphony *n, pl* -**nies** pleasing sound ▶ **euphonious** *adj* pleasing to the ear ▶ **euphonium** *n* brass musical instrument, tenor tuba

euphoria *n* sense of elation ▶ **euphoric** *adj*

Eurasian *adj* of Europe and Asia; of mixed European and Asian parentage ▷ *n* person of Eurasian parentage

eureka [yew-**reek**-a] *interj* exclamation of triumph at finding something

euro n, pl **euros** unit of the single currency of the European Union

European n, adj (person) from Europe ▶ **European Union** economic and political association of a number of European nations

Eustachian tube n passage leading from the ear to the throat

euthanasia n act of killing someone painlessly, esp. to relieve his or her suffering

evacuate v send (someone) away from a place of danger; empty ▶ **evacuation** n ▶ **evacuee** n

evade v get away from or avoid; elude ▶ **evasion** n ▶ **evasive** adj not straightforward ▶ **evasively** adv

evaluate v find or judge the value of ▶ **evaluation** n

evanescent adj quickly fading away ▶ **evanescence** n

evangelical adj of or according to gospel teaching; of certain Protestant sects which maintain the doctrine of salvation by faith ▷ n member of an evangelical sect ▶ **evangelicalism** n

evangelist n writer of one of the four gospels; travelling preacher ▶ **evangelism** n teaching and spreading of the Christian gospel ▶ **evangelize** v preach the gospel ▶ **evangelization** n

evaporate v change from a liquid or solid to a vapour; disappear ▶ **evaporation** n ▶ **evaporated milk** thick unsweetened tinned milk

eve n evening or day before some special event; period immediately before an event ▶ **evensong** n evening prayer

even adj flat or smooth; (foll by with) on the same level (as); constant; calm; equally balanced; divisible by two ▷ adv equally; simply; nevertheless ▷ v make even

evening n end of the day or early part of the night ▷ adj of or in the evening

event n anything that takes place; planned and organized occasion; contest in a sporting programme ▶ **eventful** adj full of exciting incidents

eventing n (Brit, Aust & NZ) riding competitions, usu. involving cross-country, jumping, and dressage

eventual adj ultimate ▶ **eventuality** n possible event

eventually adv at the end of a situation or process

ever adv at any time; always ▶ **evergreen** n, adj (tree or shrub) having leaves throughout the year ▶ **everlasting** adj ▶ **evermore** adv for all time to come

every adj each without exception; all possible ▶ **everybody** pron every person ▶ **everyday** adj usual or ordinary ▶ **everyone** pron every person ▶ **everything** pron ▶ **everywhere** adv in all places

evict v legally expel (someone) from his or her home ▶ **eviction** n

evidence n ground for belief; matter produced before a law court to prove or disprove a point; sign, indication ▷ v demonstrate, prove **in evidence** conspicuous ▶ **evident** adj easily seen or understood ▶ **evidently** adv ▶ **evidential** adj of, serving as, or based on evidence

evil n wickedness; wicked deed ▷ adj harmful; morally bad; very unpleasant ▶ **evilly** adv ▶ **evildoer** n wicked person

evince v make evident

eviscerate v disembowel ▶ **evisceration** n

evoke v call or summon up (a memory, feeling, etc.) ▶ **evocation** n ▶ **evocative** adj

evolve v develop gradually; (of an animal or plant species) undergo evolution ▸ **evolution** n gradual change in the characteristics of living things over successive generations, esp. to a more complex form ▸ **evolutionary** adj

ewe n female sheep

ewer n large jug with a wide mouth

ex n (informal) former wife or husband

ex- prefix out of, outside, from e.g. exodus; former e.g. ex-wife

exacerbate [ig-zass-er-bate] v make (pain, emotion, or a situation) worse ▸ **exacerbation** n

exact adj correct and complete in every detail; precise, as opposed to approximate ▸ v demand (payment or obedience) ▸ **exactly** adv precisely, in every respect ▸ **exactness, exactitude** n ▸ **exacting** adj making rigorous or excessive demands

exaggerate v regard or represent as greater than is true; make greater or more noticeable ▸ **exaggeratedly** adv ▸ **exaggeration** n

> **SPELLING TIP**
> Note that **exaggerate** has a double g but only one r

exalt v praise highly; raise to a higher rank ▸ **exalted** adj ▸ **exaltation** n

exam n short for **examination**

examine v look at closely; test the knowledge of; ask questions of ▸ **examination** n examining; test of a candidate's knowledge or skill ▸ **examinee** n ▸ **examiner** n

example n specimen typical of its group; person or thing worthy of imitation; punishment regarded as a warning to others

exasperate v cause great irritation to ▸ **exasperation** n

excavate v unearth buried objects from (a piece of land) methodically to learn about the past; make (a hole) in solid matter by digging ▸ **excavation** n ▸ **excavator** n large machine used for digging

exceed v be greater than; go beyond (a limit) ▸ **exceedingly** adv very

excel v -celling, -celled be superior to; be outstandingly good at something

Excellency n title used to address a high-ranking official, such as an ambassador

excellent adj exceptionally good ▸ **excellence** n

except prep (sometimes foll by for) other than, not including ▸ v not include **except that** but for the fact that ▸ **excepting** prep except ▸ **exception** n excepting; thing that is excluded from or does not conform to the general rule ▸ **exceptional** adj not ordinary; much above the average

excerpt n passage taken from a book, speech, etc.

excess n state or act of exceeding the permitted limits; immoderate amount; amount by which a thing exceeds the permitted limits ▸ **excessive** adj ▸ **excessively** adv

exchange v give or receive (something) in return for something else ▸ n act of exchanging; thing given or received in place of another; centre in which telephone lines are interconnected; (Finance) place where securities or commodities are traded; transfer of sums of money of equal value between different currencies ▸ **exchangeable** adj

Exchequer n (Brit) government department in charge of state money

excise[1] *n* tax on goods produced for the home market

excise[2] *v* cut out or away ▶ **excision** *n*

excite *v* arouse to strong emotion; arouse or evoke (an emotion); arouse sexually ▶ **excitement** *n* ▶ **excitable** *adj* easily excited ▶ **excitability** *n*

exclaim *v* speak suddenly, cry out ▶ **exclamation** *n* ▶ **exclamation mark** punctuation mark (!) used after exclamations ▶ **exclamatory** *adj*

exclude *v* keep out, leave out; leave out of consideration ▶ **exclusion** *n* ▶ **exclusive** *adj* excluding everything else; not shared; catering for a privileged minority ▶ *n* story reported in only one newspaper ▶ **exclusively** *adv* ▶ **exclusive, exclusiveness** *n*

excommunicate *v* exclude from membership and the sacraments of the Church ▶ **excommunication** *n*

excoriate *v* censure severely; strip skin from ▶ **excoriation** *n*

excrement *n* waste matter discharged from the body

excrescence *n* lump or growth on the surface of an animal or plant

excrete *v* discharge (waste matter) from the body ▶ **excretion** *n* ▶ **excreta** [ik-skree-ta] *n* excrement ▶ **excretory** *adj*

excruciating *adj* agonizing; hard to bear ▶ **excruciatingly** *adv*

exculpate *v* free from blame or guilt

excursion *n* short journey, esp. for pleasure

excuse *n* explanation offered to justify (a fault etc.) ▶ *v* put forward a reason or justification for (a fault etc.); forgive (a person) or overlook (a fault etc.); make allowances

for; exempt; allow to leave ▶ **excusable** *adj*

ex-directory *adj* not listed in a telephone directory by request

execrable [eks-sik-rab-bl] *adj* of very poor quality

execute *v* put (a condemned person) to death; carry out or accomplish; produce (a work of art); render (a legal document) effective, as by signing ▶ **execution** *n* ▶ **executioner** *n*

executive *n* person or group in an administrative position; branch of government responsible for carrying out laws etc. ▷ *adj* having the function of carrying out plans, orders, laws, etc.

executor, (*fem*) **executrix** *n* person appointed to perform the instructions of a will

exegesis [eks-sij-jee-siss] *n*, *pl* **-ses** [-seez] explanation of a text, esp. of the Bible

exemplar *n* person or thing to be copied, model; example ▶ **exemplary** *adj* being a good example; serving as a warning

exemplify *v* **-fying, -fied** show an example of; be an example of ▶ **exemplification** *n*

exempt *adj* not subject to an obligation etc. ▷ *v* release from an obligation etc. ▶ **exemption** *n*

exequies [eks-wik-wiz] *pl n* funeral rites

exercise *n* activity to train the body or mind; set of movements or tasks designed to improve or test a person's ability; performance of a function ▷ *v* make use of e.g. *to exercise one's rights*; take exercise or perform exercises

exert *v* use (influence, authority, etc.) forcefully or effectively **exert oneself** make a special effort ▶ **exertion** *n*

exeunt [eks-see-unt] (Latin) they go out: used as a stage direction

exfoliate v scrub away dead skin cells, esp by washing with abrasive lotion

ex gratia [eks gray-sha] adj given as a favour where no legal obligation exists

exhale v breathe out
▶ **exhalation** n

exhaust v tire out; use up; discuss (a subject) thoroughly ▷ n gases ejected from an engine as waste products; pipe through which an engine's exhaust fumes pass ▶ **exhaustion** n extreme tiredness; exhausting
▶ **exhaustive** adj comprehensive
▶ **exhaustively** adv

exhibit v display to the public; show (a quality or feeling) ▷ n object exhibited to the public; (Law) document or object produced in court as evidence ▶ **exhibitor** n
▶ **exhibition** n public display of art, skills, etc.; exhibiting
▶ **exhibitionism** n compulsive desire to draw attention to oneself; compulsive desire to display one's genitals in public ▶ **exhibitionist** n

exhilarate v make lively and cheerful ▶ **exhilaration** n

> **SPELLING TIP**
> A common error is to spell the third syllable of **exhilarate**, **exhilaration**, etc. with an e rather than the correct a

exhort v urge earnestly
▶ **exhortation** n

exhume [ig-zyume] v dig up (something buried, esp. a corpse)
▶ **exhumation** n

exigency n, pl -cies urgent demand or need ▶ **exigent** adj

exiguous adj scanty or meagre

exile n prolonged, usu. enforced, absence from one's country; person banished or living away from his or her country ▷ v expel from one's country

exist v have reality or being; eke out a living; live ▶ **existence** n
▶ **existent** adj

> **SPELLING TIP**
> There is much confusion about whether certain words end with -ance or with -ence: **existence** is one that ends with the latter

existential adj of or relating to existence, esp. human existence
▶ **existentialism** n philosophical movement stressing the personal experience and responsibility of the individual, who is seen as a free agent ▶ **existentialist** adj, n

exit n way out; going out; actor's going off stage ▷ v go out; go offstage: used as a stage direction

exocrine adj relating to a gland, such as the sweat gland, that secretes externally through a duct

exodus [eks-so-duss] n departure of a large number of people

ex officio [eks off-fish-ee-oh] adv, adj (Latin) by right of position or office

exonerate v free from blame or a criminal charge ▶ **exoneration** n

exorbitant adj (of prices, demands, etc.) excessive, immoderate
▶ **exorbitantly** adv

exorcize v expel (evil spirits) by prayers and religious rites
▶ **exorcism** n ▶ **exorcist** n

exotic adj having a strange allure or beauty; originating in a foreign country ▷ n non-native plant
▶ **exotically** adv ▶ **exotica** pl n (collection of) exotic objects

expand v make or become larger; spread out; (foll by on) enlarge (on); become more relaxed, friendly, and talkative ▶ **expansion** n
▶ **expanse** n uninterrupted wide

area ▶ **expansive** adj wide or extensive; friendly and talkative

expat adj, n short for **expatriate**

expatiate [iks-**pay**-shee-ate] v (foll by on) speak or write at great length (on)

expatriate [eks-**pat**-ree-it] adj living outside one's native country ▷ n person living outside his or her native country ▶ **expatriation** n

expect v regard as probable; look forward to, await; require as an obligation ▶ **expectancy** n something expected on the basis of an average e.g. life expectancy; feeling of anticipation ▶ **expectant** adj expecting or hopeful; pregnant ▶ **expectantly** adv ▶ **expectation** n act or state of expecting; something looked forward to; attitude of anticipation or hope

expectorant n medicine that helps to bring up phlegm from the respiratory passages

expectorate v spit out (phlegm etc.) ▶ **expectoration** n

expedient n something that achieves a particular purpose ▷ adj suitable to the circumstances, appropriate ▶ **expediency** n

expedite v hasten the progress of ▶ **expedition** n organized journey, esp. for exploration; people and equipment comprising an expedition; pleasure trip or excursion ▶ **expeditionary** adj relating to an expedition, esp. a military one ▶ **expeditious** adj done quickly and efficiently

expel v -**pelling**, -**pelled** drive out with force; dismiss from a school etc. permanently ▶ **expulsion** n

expend v spend, use up ▶ **expendable** adj able to be sacrificed to achieve an objective ▶ **expenditure** n something

expended, esp. money; amount expended

expense n cost; (cause of) spending ▷ pl charges, outlay incurred

expensive adj high-priced

experience n direct personal participation; particular incident, feeling, etc. that a person has undergone; accumulated knowledge ▷ v participate in; be affected by (an emotion) ▶ **experienced** adj skilful from extensive participation

experiment n test to provide evidence to prove or disprove a theory; attempt at something new ▷ v carry out an experiment ▶ **experimental** adj ▶ **experimentally** adv ▶ **experimentation** n

expert n person with extensive skill or knowledge in a particular field ▷ adj skilful or knowledgeable ▶ **expertise** [eks-per-**teez**] n special skill or knowledge

expiate v make amends for ▶ **expiation** n

expire v finish or run out; breathe out; (lit) die ▶ **expiration** n ▶ **expiry** n end, esp. of a contract period

explain v make clear and intelligible; account for ▶ **explanation** n ▶ **explanatory** adj

expletive n swearword

explicable adj able to be explained ▶ **explicate** v (formal) explain ▶ **explication** n

explicit adj precisely and clearly expressed; shown in realistic detail ▶ **explicitly** adv

explode v burst with great violence, blow up; react suddenly with emotion; increase rapidly; show (a theory etc.) to be baseless

▸ **explosion** n ▸ **explosive** adj
tending to explode ▸ n substance
that causes explosions

exploit v take advantage of for
one's own purposes; make the
best use of ▸ n notable feat or deed
▸ **exploitation** n ▸ **exploiter** n

explore v investigate; travel into
(unfamiliar regions), esp. for
scientific purposes ▸ **exploration**
n ▸ **exploratory** adj ▸ **explorer** n

expo n, pl **expos** (informal)
exposition, large public exhibition

exponent n person who advocates
an idea, cause, etc.; skilful
performer, esp. a musician

exponential adj (informal) very
rapid ▸ **exponentially** adv

export n selling or shipping of
goods to a foreign country;
product shipped or sold to a
foreign country ▸ v sell or ship
(goods) to a foreign country
▸ **exporter** n

expose v uncover or reveal; make
vulnerable, leave unprotected;
subject (a photographic film) to
light **expose oneself** display one's
sexual organs in public ▸ **exposure**
n exposing; lack of shelter from the
weather, esp. the cold; appearance
before the public, as on television

exposé [iks-pose-ay] n bringing of a
crime, scandal, etc. to public notice

exposition n see **expound**

expostulate v (foll by with) reason
(with), esp. to dissuade

expound v explain in detail
▸ **exposition** n explanation; large
public exhibition

express v put into words; show
(an emotion); indicate by a symbol
or formula; squeeze out (juice
etc.) ▸ adj explicitly stated; (of a
purpose) particular; of or for rapid
transportation of people, mail,
etc. ▸ n fast train or bus stopping
at only a few stations ▸ adv by
express delivery ▸ **expression**
n expressing; word or phrase;
showing or communication of
emotion; look on the face that
indicates mood; (Maths) variable,
function, or some combination
of these ▸ **expressionless** adj
▸ **expressive** adj

expressionism n early 20th-
century artistic movement which
sought to express emotions rather
than represent the physical world
▸ **expressionist** n, adj

expropriate v deprive an owner of
(property) ▸ **expropriation** n

expunge v delete, erase, blot out

expurgate v remove objectionable
parts from (a book etc.)

exquisite adj of extreme beauty
or delicacy; intense in feeling
▸ **exquisitely** adv

extant adj still existing

extemporize v speak, perform, or
compose without preparation

extend v draw out or be drawn
out, stretch; last for a certain
time; (foll by to) include; increase
in size or scope; offer e.g. extend
one's sympathy ▸ **extendable** adj
▸ **extension** n room or rooms
added to an existing building;
additional telephone connected
to the same line as another;
extending ▸ **extensive** adj
having a large extent, widespread
▸ **extensor** n muscle that extends
a part of the body ▸ **extent** n range
over which something extends,
area

> **SPELLING TIP**
>
> Lots of nouns in English end with
> -tion, but **extension** is not one
> of them

extenuate v make (an offence
or fault) less blameworthy
▸ **extenuation** n

exterior n part or surface on the outside; outward appearance ▷ adj of, on, or coming from the outside

exterminate v destroy (animals or people) completely ▶ **extermination** n ▶ **exterminator** n

external adj of, situated on, or coming from the outside ▶ **externally** adv

extinct adj having died out; (of a volcano) no longer liable to erupt ▶ **extinction** n

extinguish v put out (a fire or light); remove or destroy entirely

extinguisher n device for extinguishing a fire or light

extirpate v destroy utterly

extol v -**tolling, -tolled** praise highly

extort v get (something) by force or threats ▶ **extortion** n ▶ **extortionate** adj (of prices) excessive

extra adj more than is usual, expected or needed ▷ n additional person or thing; something for which an additional charge is made; (Films) actor hired for crowd scenes ▷ adv unusually or exceptionally

extra- prefix outside or beyond an area or scope e.g. extrasensory; extraterrestrial

extract v pull out by force; remove; derive; copy out (an article, passage, etc.) from a publication ▷ n something extracted, such as a passage from a book etc.; preparation containing the concentrated essence of a substance e.g. beef extract ▶ **extraction** n ▶ **extractor** n

extradite v send (an accused person) back to his or her own country for trial ▶ **extradition** n

extramural adj connected with but outside the normal courses of a university or college

extraneous [iks-train-ee-uss] adj irrelevant

extraordinary adj very unusual; (of a meeting) specially arranged to deal with a particular subject ▶ **extraordinarily** adv

extraordinary rendition n process by which a country seizes a suspected terrorist and transports them for interrogation to a country where due process of law is unlikely to be respected

extrapolate v infer (something not known) from the known facts; (Maths) estimate (a value of a function or measurement) beyond the known values by the extension of a curve ▶ **extrapolation** n

extrasensory adj ▶ **extrasensory perception** supposed ability to obtain information other than through the normal senses

extravagant adj spending money excessively; going beyond reasonable limits ▶ **extravagance** n ▶ **extravaganza** n elaborate and lavish entertainment, display, etc.

> **SPELLING TIP**
> Make sure that **extravagant** ends in -ant, even though -ent sounds like a possibility

extreme adj of a high or the highest degree or intensity; severe; immoderate; farthest or outermost ▷ n either of the two limits of a scale or range ▶ **extremely** adv ▶ **extreme sport** sport with a high risk of injury or death ▶ **extremist** n person who favours immoderate methods ▷ adj holding extreme opinions ▶ **extremity** n, pl -**ties** farthest point; extreme condition, as of misfortune ▷ pl hands and feet

e

extricate v free from complication or difficulty ▸ **extrication** n

extrovert adj lively and outgoing; concerned more with external reality than inner feelings ▸ n extrovert person

extrude v squeeze or force out ▸ **extrusion** n

exuberant adj high-spirited; growing luxuriantly ▸ **exuberance** n

exude v (of a liquid or smell) seep or flow out slowly and steadily; make apparent by mood or behaviour e.g. *exude confidence*

exult v be joyful or jubilant ▸ **exultation** n ▸ **exultant** adj

eye n organ of sight; ability to judge or appreciate e.g. *a good eye for detail*; one end of a sewing needle; dark spot on a potato from which a stem grows ▸ v **eyeing** or **eying**, **eyed** look at carefully or warily ▸ **eyeless** adj ▸ **eyelet** n small hole for a lace or cord to be passed through; ring that strengthens this ▸ **eyeball** n ball-shaped part of the eye ▸ **eyebrow** n line of hair on the bony ridge above the eye ▸ **eyeglass** n lens for aiding defective vision ▸ **eyelash** n short hair that grows out from the eyelid ▸ **eyelid** n fold of skin that covers the eye when it is closed ▸ **eyeliner** n cosmetic used to outline the eyes ▸ **eye-opener** n (informal) something startling or revealing ▸ **eye shadow** coloured cosmetic worn on the upper eyelids ▸ **eyesight** n ability to see ▸ **eyesore** n ugly object ▸ **eye tooth** canine tooth ▸ **eyewitness** n person who was present at an event and can describe what happened

eyrie n nest of an eagle; high isolated place

Ff

F Fahrenheit; farad

f (Music) forte

FA Football Association (of England)

fable n story with a moral; false or fictitious account; legend ▸ **fabled** adj made famous in legend

fabric n knitted or woven cloth; framework or structure

fabricate v make up (a story or lie); make or build ▸ **fabrication** n

fabulous adj (informal) excellent; astounding; told of in fables ▸ **fabulously** adv

facade [fas-**sahd**] n front of a building; (false) outward appearance

face n front of the head; facial expression; distorted expression; outward appearance; front or main side; dial of a clock; dignity, self-respect ▸ v look or turn towards; be opposite; be confronted by; provide with a surface ▸ **faceless** adj impersonal, anonymous ▸ **face-lift** n operation to tighten facial skin, to remove wrinkles ▸ **face-saving** adj maintaining dignity or self-respect ▸ **face up to** v accept (an unpleasant fact or reality) ▸ **face value** apparent worth or meaning

facet n aspect; surface of a cut gem

facetious [fas-**see-shuss**] adj funny or trying to be funny, esp. at inappropriate times

facia n, pl -ciae same as **fascia**

facial adj of the face ▸ n beauty treatment for the face

facile [fas-sile] *adj* (of a remark, argument, etc.) superficial and showing lack of real thought

facilitate *v* make easy ▸ **facilitation** *n*

facility *n, pl* **-ties** skill; easiness ▷ *pl* means or equipment for an activity

facing *n* lining or covering for decoration or reinforcement ▷ *pl* contrasting collar and cuffs on a jacket

facsimile [fak-sim-ill-ee] *n* exact copy

fact *n* event or thing known to have happened or existed; provable truth **facts of life** details of sex and reproduction ▸ **factual** *adj*

faction *n* (dissenting) minority group within a larger body; dissension ▸ **factious** *adj* of or producing factions

factitious *adj* artificial

factor *n* element contributing to a result; (*Maths*) one of the integers multiplied together to give a given number; (*Scot*) property manager ▸ **factorial** *n* product of all the integers from one to a given number ▸ **factorize** *v* calculate the factors of (a number)

factory *n, pl* **-ries** building where goods are manufactured

factotum *n* person employed to do all sorts of work

faculty *n, pl* **-ties** physical or mental ability; department in a university or college

fad *n* short-lived fashion; whim ▸ **faddy, faddish** *adj*

fade *v* (cause to) lose brightness, colour, or strength; vanish slowly

faeces [fee-seez] *pl n* waste matter discharged from the anus ▸ **faecal** [fee-kl] *adj*

fag¹ *n* (*informal*) boring task; (*Brit*) young public schoolboy who does menial chores for a senior boy ▷ *v*

(*Brit*) do menial chores in a public school

fag² *n* (*Brit slang*) cigarette ▸ **fag end** last and worst part; (*slang*) cigarette stub

faggot¹ *n* (*Brit, Aust & NZ*) ball of chopped liver, herbs, and bread; bundle of sticks for fuel

faggot² *n* (*offens*) male homosexual

Fahrenheit [far-ren-hite] *adj* of a temperature scale with the freezing point of water at 32° and the boiling point at 212°

faïence [fie-ence] *n* tin-glazed earthenware

fail *v* be unsuccessful; stop operating; be or judge to be below the required standard in a test; disappoint or be useless to (someone); neglect or be unable to do (something); go bankrupt ▷ *n* instance of not passing an exam or test **without fail** regularly; definitely ▸ **failing** *n* weak point ▷ *prep* in the absence of ▸ **failure** *n* act or instance of failing; unsuccessful person or thing

fain *adv* (*obs*) gladly

faint *adj* lacking clarity, brightness, or volume; feeling dizzy or weak; lacking conviction or force ▷ *v* lose consciousness temporarily ▷ *n* temporary loss of consciousness

fair¹ *adj* unbiased and reasonable; light in colour; beautiful; quite good e.g. *a fair attempt*; quite large e.g. *a fair amount of money*; (of weather) fine ▷ *adv* fairly ▸ **fairly** *adv* moderately; to a great degree or extent; as deserved, reasonably ▸ **fairness** *n* ▸ **fair trade** practice of directly benefiting producers in the developing world by buying straight from them at a guaranteed price ▸ **fairway** *n* (*Golf*) smooth area between the tee and the green

fair² n travelling entertainment with sideshows, rides, and amusements; exhibition of commercial or industrial products ▸**fairground** n open space used for a fair

Fair Isle n intricate multicoloured knitted pattern

fairy n, pl **fairies** imaginary small creature with magic powers; (offens) male homosexual ▸**fairy godmother** person who helps in time of trouble ▸**fairyland** n ▸**fairy lights** small coloured electric bulbs used as decoration ▸**fairy penguin** small penguin with a bluish head and back, found on the Australian coast ▸**fairy tale, fairy story** story about fairies or magic; unbelievable story or explanation

fait accompli [fate ak-kom-plee] n (French) something already done that cannot be altered

faith n strong belief, esp. without proof; religion; complete confidence or trust; allegiance to a person or cause ▸**faithful** adj loyal; consistently reliable; accurate in detail ▸**faithfully** adv ▸**faithless** adj disloyal or dishonest ▸**faith school** (Brit) school that provides a general education within a framework of a specific religious belief

fake v cause something not genuine to appear real or more valuable by fraud; pretend to have (an illness, emotion, etc.) ▸ n person, thing, or act that is not genuine ▸ adj not genuine

fakir [fay-keer] n Muslim who spurns worldly possessions; Hindu holy man

falcon n small bird of prey ▸**falconry** n art of training falcons; sport of hunting with falcons ▸**falconer** n

fall v **falling, fell, fallen** drop from a higher to a lower place through the force of gravity; collapse to the ground; decrease in number or quality; pass into a specified condition; occur ▸ n falling; thing or amount that falls; decrease in value or number; decline in power or influence; (US) autumn ▸ pl waterfall ▸**fall for** v (informal) fall in love with; be deceived by (a lie or trick) ▸**fall guy** (informal) victim of a confidence trick; scapegoat ▸**fallout** n radioactive particles spread as a result of a nuclear explosion

fallacy n, pl **-cies** false belief; unsound reasoning ▸**fallacious** adj

fallible adj (of a person) liable to make mistakes ▸**fallibility** n

Fallopian tube n either of a pair of tubes through which egg cells pass from the ovary to the womb

fallow adj (of land) ploughed but left unseeded to regain fertility

false adj not true or correct; artificial; fake; deceptive e.g. false promises ▸**falsely** adv ▸**falseness** n ▸**falsity** n ▸**falsehood** n quality of being untrue; lie

falsetto n, pl **-tos** voice pitched higher than one's natural range

falsify v **-fying, -fied** alter fraudulently ▸**falsification** n

falter v be hesitant, weak, or unsure; lose power momentarily; utter hesitantly; move unsteadily

fame n state of being widely known or recognized ▸**famed** adj famous

familiar adj well-known; intimate, friendly; too friendly ▸ n demon supposed to attend a witch; friend ▸**familiarly** adv ▸**familiarity** n ▸**familiarize** v acquaint

fully with a particular subject
▶ **familiarization** n

family n, pl **-lies** group of parents
and their children; one's spouse
and children; group descended
from a common ancestor; group
of related objects or beings ▷ adj
suitable for parents and children
together ▶ **familial** adj ▶ **family
planning** control of the number
of children in a family by the use of
contraception

famine n severe shortage of food

famished adj very hungry

famous adj very well-known
▶ **famously** adv (informal)
excellently

fan[1] n hand-held or mechanical
object used to create a current of
air for ventilation or cooling ▷ v
fanning, fanned blow or cool
with a fan; spread out like a fan
▶ **fanbase** n body of admirers of
a particular pop singer, sports
team, etc. ▶ **fan belt** belt that
drives a cooling fan in a car engine
▶ **fantail** n small New Zealand bird
with a tail like a fan

fan[2] n (informal) devotee of a pop
star, sport, or hobby

fanatic n person who is excessively
enthusiastic about something
▶ **fanatical** adj ▶ **fanatically** adv
▶ **fanaticism** n

fancy adj **-cier, -ciest** elaborate,
not plain; (of prices) higher than
usual ▷ n, pl **-cies** sudden irrational
liking or desire; uncontrolled
imagination ▷ v **-cying, -cied**
(informal) be sexually attracted to;
(informal) have a wish for; picture
in the imagination; suppose
fancy oneself (informal) have a
high opinion of oneself ▶ **fanciful**
adj not based on fact; excessively
elaborate ▶ **fancifully** adv ▶ **fancy
dress** party costume representing

a historical figure, animal, etc.
▶ **fancy-free** adj not in love

fandango n, pl **-gos** lively Spanish
dance

fanfare n short loud tune played on
brass instruments

fang n snake's tooth which injects
poison; long pointed tooth

fantasia n musical composition of
an improvised nature

fantastic adj (informal) very good;
unrealistic or absurd; strange or
difficult to believe ▶ **fantastically**
adv

fantasy n, pl **-sies** far-fetched
notion; imagination unrestricted
by reality; daydream; fiction with a
large fantasy content ▶ **fantasize**
v indulge in daydreams

FAQ (Computing) frequently asked
question or questions

far adv **farther** or **further, farthest**
or **furthest** at, to, or from a great
distance; at or to a remote time;
very much ▷ adj remote in space or
time **the Far East** East Asia ▶ **far-
fetched** adj hard to believe

> **USAGE NOTE**
> Farther is often used, rather than
> further, when distance, not time
> or effort, is involved

farad n unit of electrical
capacitance

farce n boisterous comedy;
ludicrous situation ▶ **farcical** adj
ludicrous ▶ **farcically** adv

fare n charge for a passenger's
journey; passenger; food provided
▷ v get on (as specified) e.g. we
fared badly

farewell interj goodbye ▷ n act of
saying goodbye and leaving ▷ v
(NZ) say goodbye

farinaceous adj containing starch
or having a starchy texture

farm n area of land for growing
crops or rearing livestock; area

of land or water for growing or rearing a specified animal or plant e.g. *fish farm* ▷ v cultivate (land); rear (stock) ▶ **farmhouse** n ▶ **farm out** v send (work) to be done by others ▶ **farmstead** n farm and its buildings ▶ **farmyard** n

farmer n person who owns or runs a farm ▶ **farmers' market** market at which farm produce is sold directly to the public by the producer

farrago [far-rah-go] n, pl -gos or -goes jumbled mixture of things

farrier n person who shoes horses

farrow n litter of piglets ▷ v (of a sow) give birth

fart (vulgar slang) n emission of gas from the anus ▷ v emit gas from the anus

farther, farthest adv, adj see far

farthing n former British coin equivalent to a quarter of a penny

fascia [fay-shya] n, pl -ciae or -cias outer surface of a dashboard; flat surface above a shop window; mobile phone casing with spaces for the buttons

fascinate v attract and interest strongly; make motionless from fear or awe ▶ **fascinating** adj ▶ **fascination** n

> **SPELLING TIP**
> Remember that there is a silent c after the s in **fascinate**, **fascinated**, and **fascinating**

fascism [fash-iz-zum] n right-wing totalitarian political system characterized by state control and extreme nationalism ▶ **fascist** adj, n

fashion n style in clothes, hairstyle, etc.; popular at a particular time; way something happens or is done ▷ v form or make into a particular shape ▶ **fashionable** adj currently popular ▶ **fashionably** adv

fast¹ adj (capable of) acting or moving quickly; done in or lasting a short time; adapted to or allowing rapid movement; (of a clock or watch) showing a time later than the correct one; dissipated; firmly fixed, fastened, or shut ▷ adv quickly; soundly, deeply e.g. *fast asleep*; tightly and firmly ▶ **fast food** food, such as hamburgers, prepared and served very quickly ▶ **fast-track** adj taking the quickest but most competitive route to success e.g. *fast-track executives* ▷ v speed up the progress of (a project or person)

fast² v go without food, esp. for religious reasons ▷ n period of fasting

fasten v make or become firmly fixed or joined; close by fixing in place or locking; (foll by on) direct (one's attention) towards ▶ **fastener, fastening** n device that fastens

fastidious adj very fussy about details; excessively concerned with cleanliness ▶ **fastidiously** adv ▶ **fastidiousness** n

fastness n fortress, safe place

fat adj fatter, fattest having excess flesh on the body; (of meat) containing a lot of fat; thick; profitable ▷ n extra flesh on the body; oily substance obtained from animals or plants ▶ **fatness** n ▶ **fatten** v (cause to) become fat ▶ **fatty** adj containing fat ▶ **fathead** n (informal) stupid person ▶ **fat-headed** adj

fatal adj causing death or ruin ▶ **fatally** adv ▶ **fatality** n, pl -ties death caused by an accident or disaster

fatalism n belief that all events are predetermined and people are powerless to change their destinies ▶ **fatalist** n ▶ **fatalistic** adj

fate n power supposed to predetermine events; inevitable fortune that befalls a person or thing ▶**fated** adj destined; doomed to death or destruction ▶**fateful** adj having important, usu. disastrous, consequences

father n male parent; person who founds a line or family; man who starts, creates, or invents something; (F-) God; (F-) title of some priests ▶ v be the father of (offspring) ▶**fatherhood** n ▶**fatherless** adj ▶**fatherly** adj ▶**father-in-law**, pl **fathers-in-law** father of one's husband or wife ▶**fatherland** n one's native country

fathom n unit of length, used in navigation, equal to six feet (1.83 metres) ▶ v understand ▶**fathomable** adj ▶**fathomless** adj too deep or difficult to fathom

fatigue [fat-**eeg**] n extreme physical or mental tiredness; weakening of a material due to stress; soldier's nonmilitary duty ▶ v tire out

fatuous adj foolish ▶**fatuously** adv ▶**fatuity** n

faucet [**faw**-set] n (US) tap

fault n responsibility for something wrong; defect or flaw; mistake or error; (Geology) break in layers of rock; (Tennis, squash etc.) invalid serve ▶ v criticize or blame **at fault** guilty of error **find fault with** seek out minor imperfections in **to a fault** excessively ▶**faulty** adj ▶**faultless** adj ▶**faultlessly** adv

faun n (in Roman legend) creature with a human face and torso and a goat's horns and legs

fauna n, pl **-nas** or **-nae** animals of a given place or time

faux pas [foe **pah**] n, pl **faux pas** social blunder

favicon n (Computing) small icon associated with a particular website, usu displayed before the URL in a web browser

favour n approving attitude; act of goodwill or generosity; partiality ▶ v prefer; regard or treat with especial kindness; support or advocate

favourable adj encouraging or advantageous; giving consent; useful or beneficial ▶**favourably** adv

favourite adj most liked ▶ n preferred person or thing; (Sport) competitor expected to win ▶**favouritism** n practice of giving special treatment to a person or group

fawn¹ n young deer ▶ adj light yellowish-brown

fawn² v (foll by on) seek attention from (someone) by insincere flattery; (of a dog) try to please by a show of extreme affection

fax n electronic system for sending facsimiles of documents by telephone; document sent by this system ▶ v send (a document) by this system

FBI (US) Federal Bureau of Investigation

FC (in Britain) Football Club

Fe (Chem) iron

fealty n (in feudal society) subordinate's loyalty to his ruler or lord

fear n distress or alarm caused by impending danger or pain; something that causes distress ▶ v be afraid of (something or someone) **fear for** feel anxiety about something ▶**fearful** adj feeling fear; causing fear; (informal) very unpleasant ▶**fearfully** adv ▶**fearless** adj ▶**fearlessly** adv ▶**fearsome** adj terrifying

feasible adj able to be done, possible ▸ **feasibly** adv ▸ **feasibility** n

feast n lavish meal; something extremely pleasing; annual religious celebration ▷ v eat a feast; give a feast to; (foll by on) eat a large amount of

feat n remarkable, skilful, or daring action

feather n one of the barbed shafts forming the plumage of birds ▷ v fit or cover with feathers; turn (an oar) edgeways **feather in one's cap** achievement one can be pleased with **feather one's nest** make one's life comfortable ▸ **feathered** adj ▸ **feathery** adj ▸ **featherweight** n boxer weighing up to 126lb (professional) or 57kg (amateur); insignificant person or thing

feature n part of the face, such as the eyes; prominent or distinctive part; special article in a newspaper or magazine; main film in a cinema programme ▷ v have as a feature or be a feature in; give prominence to ▸ **featureless** adj

Feb. February

febrile [fee-brile] adj feverish

February n second month of the year

feckless adj ineffectual or irresponsible

fecund adj fertile ▸ **fecundity** n

fed v past of **feed** ▸ **fed up** (informal) bored, dissatisfied

federal adj of a system in which power is divided between one central government and several regional governments; of the central government of a federation ▸ **federalism** n ▸ **federalist** n ▸ **federate** v unite in a federation ▸ **federation** n union of several states, provinces, etc.; association

fedora [fid-**or**-a] n man's soft hat with a brim

fee n charge paid to be allowed to do something; payment for professional services

feeble adj lacking physical or mental power; unconvincing ▸ **feebleness** n ▸ **feebly** adv ▸ **feeble-minded** adj unable to think or understand effectively

feed v **feeding**, **fed** give food to; give (something) as food; eat; supply or prepare food for; supply (what is needed) ▷ n act of feeding; food, esp. for babies or animals; (informal) meal ▸ **feeder** n road or railway line linking outlying areas to the main traffic network ▸ **feedback** n information received in response to something done; return of part of the output of an electrical circuit or loudspeaker to its source

feel v **feeling**, **felt** have a physical or emotional sensation of; become aware of or examine by touch; believe ▷ n act of feeling; impression; way something feels; sense of touch; instinctive aptitude **feel like** wish for, want ▸ **feeler** n organ of touch in some animals; remark made to test others' opinion ▸ **feeling** n emotional reaction; intuitive understanding; opinion; sympathy, understanding; ability to experience physical sensations; sensation experienced ▷ pl emotional sensitivities

feet n plural of **foot**

feign [fane] v pretend

feint[1] [faint] n sham attack or blow meant to distract an opponent ▷ v make a feint

feint[2] [faint] n narrow lines on ruled paper

feldspar n hard mineral that is the main constituent of igneous rocks

felicity n happiness; (pl **-ties**) appropriate expression or style
▶ **felicitations** pl n congratulations
▶ **felicitous** adj

feline adj of cats; catlike ▷ n member of the cat family

fell¹ v past tense of **fall**

fell² v cut down (a tree); knock down

fell³ adj **in one fell swoop** in a single action or occurrence

fell⁴ n (Scot & N English) mountain, hill, or moor

felloe n (segment of) the rim of a wheel

fellow n man or boy; comrade or associate; person in the same group or condition; member of a learned society or the governing body of a college ▷ adj in the same group or condition ▶ **fellowship** n sharing of aims or interests; group with shared aims or interests; feeling of friendliness; paid research post in a college or university

felon n (Criminal law) (formerly) person guilty of a felony
▶ **felony** n, pl **-nies** serious crime
▶ **felonious** adj

felspar n same as **feldspar**

felt¹ v past of **feel**

felt² n matted fabric made by bonding fibres by pressure ▶ **felt-tip pen** n pen with a writing point made from pressed fibres

fem. feminine

female adj of the sex which bears offspring; (of plants) producing fruits ▷ n female person or animal

feminine adj having qualities traditionally regarded as suitable for, or typical of, women; of women; belonging to a particular class of grammatical inflection in some languages ▶ **femininity** n
▶ **feminism** n advocacy of equal rights for women ▶ **feminist** n, adj

femme fatale [fam fat-**tahl**] n, pl **femmes fatales** alluring woman who leads men into dangerous situations by her charm

femur [fee-mer] n thighbone
▶ **femoral** adj of the thigh

fen n (Brit) low-lying flat marshy land

fence n barrier of posts linked by wire or wood, enclosing an area; (slang) dealer in stolen property ▷ v enclose with or as if with a fence; fight with swords as a sport; avoid a question ▶ **fencing** n sport of fighting with swords; material for making fences ▶ **fencer** n

fend v **fend for oneself** provide for oneself ▶ **fend off** v defend oneself against (verbal or physical attack)

fender n low metal frame in front of a fireplace; soft but solid object hung over a ship's side to prevent damage when docking; (chiefly US) wing of a car

feng shui [fung **shway**] n Chinese art of deciding the best design of a building, etc., in order to bring good luck

fennel n fragrant plant whose seeds, leaves, and root are used in cookery

fenugreek n Mediterranean plant grown for its heavily scented seeds

feral adj wild

ferment n commotion, unrest ▷ v undergo or cause to undergo fermentation ▶ **fermentation** n reaction in which an organic molecule splits into simpler substances, esp. the conversion of sugar to alcohol

fern n flowerless plant with fine fronds

ferocious adj savagely fierce or cruel ▶ **ferocity** n

ferret n tamed polecat used to catch rabbits or rats ▷ v **ferreting,**

ferreted hunt with ferrets; search around ▶ **ferret out** v find by searching

ferric, ferrous adj of or containing iron

Ferris wheel n large vertical fairground wheel with hanging seats for riding in

ferry n, pl **-ries** boat for transporting people and vehicles ▷ v **-rying, -ried** carry by ferry; convey (goods or people) ▶ **ferryman** n

fertile adj capable of producing young, crops, or vegetation; highly productive e.g. *a fertile mind* ▶ **fertility** n ▶ **fertilize** v provide (an animal or plant) with sperm or pollen to bring about fertilization; supply (soil) with nutrients ▶ **fertilization** n

fertilizer n substance added to the soil to increase its productivity

fervent, fervid adj intensely passionate and sincere ▶ **fervently** adv ▶ **fervour** n intensity of feeling

fescue n pasture and lawn grass with stiff narrow leaves

fester v grow worse and increasingly hostile; (of a wound) form pus; rot and decay

festival n organized series of special events or performances; day or period of celebration ▶ **festive** adj of or like a celebration ▶ **festivity** n, pl **-ties** happy celebration ▶ pl celebrations

festoon v hang decorations in loops

feta n white salty Greek cheese

fetch v go after and bring back; be sold for; (informal) deal (a blow) ▶ **fetching** adj attractive ▶ **fetch up** v (informal) arrive or end up

fete [fate] n gala, bazaar, etc., usu. held outdoors ▷ v honour or entertain regally

fetid adj stinking

fetish n form of behaviour in which sexual pleasure is derived from looking at or handling an inanimate object; thing with which one is excessively concerned; object believed to have magical powers ▶ **fetishism** n ▶ **fetishist** n

fetlock n projection behind and above a horse's hoof

fetter n chain or shackle for the foot ▷ pl restrictions ▷ v restrict; bind in fetters

fettle n state of health or spirits

fetus [fee-tuss] n, pl **-tuses** embryo of a mammal in the later stages of development ▶ **fetal** adj

feu n (in Scotland) right of use of land in return for a fixed annual payment

feud n long bitter hostility between two people or groups ▷ v carry on a feud

feudalism n medieval system in which people held land from a lord, and in return worked and fought for him ▶ **feudal** adj of or like feudalism

fever n (illness causing) high body temperature; nervous excitement ▶ **fevered** adj ▶ **feverish** adj suffering from fever; in a state of nervous excitement ▶ **feverishly** adv

few adj not many **a few** small number **quite a few, a good few** several

USAGE NOTE
Few(er) is used of things that can be counted: *Fewer than five visits.* Compare *less*, which is used for quantity: *It uses less sugar*

fey adj whimsically strange; having the ability to look into the future

fez n, pl **fezzes** brimless tasselled cap, orig. from Turkey

ff (Music) fortissimo

fiancé [fee-on-say] *n* man engaged to be married ► **fiancée** *n fem*

fiasco *n, pl* **-cos** *or* **-coes** ridiculous or humiliating failure

fiat [fee-at] *n* arbitrary order; official permission

fib *n* trivial lie ▷ *v* **fibbing, fibbed** tell a lie ► **fibber** *n*

fibre *n* thread that can be spun into yarn; threadlike animal or plant tissue; fibrous material in food; strength of character; essential substance or nature ► **fibrous** *adj* ► **fibreglass** *n* material made of fine glass fibres ► **fibre optics** transmission of information by light along very thin flexible fibres of glass

fibrillation *n* uncontrollable twitching of muscles, esp. those in the heart

fibro *n* (*Aust*) mixture of cement and asbestos fibre, used in sheets for building (also **fibrocement**)

fibroid [fibe-royd] *n* benign tumour composed of fibrous connective tissue ► **fibrositis** [fibe-roh-site-iss] *n* inflammation of the tissues of muscle sheaths

fibula *n, pl* **-lae** *or* **-las** slender outer bone of the lower leg

fiche [feesh] *n* sheet of film for storing publications in miniaturized form

fickle *adj* changeable, inconstant ► **fickleness** *n*

fiction *n* literary works of the imagination, such as novels; invented story ► **fictional** *adj* ► **fictionalize** *v* turn into fiction ► **fictitious** *adj* not genuine; or of in fiction

fiddle *n* violin; (*informal*) dishonest action or scheme ▷ *v* play the violin; falsify (accounts); move or touch something restlessly ► **fiddling** *adj* trivial ► **fiddly** *adj* awkward to do or use

► **fiddlesticks** *interj* expression of annoyance or disagreement

fidelity *n* faithfulness; accuracy in detail; quality of sound reproduction

fidget *v* move about restlessly ▷ *n* person who fidgets ▷ *pl* restlessness ► **fidgety** *adj*

fiduciary [fid-yew-she-er-ee] (*Law*) *n, pl* **-aries** person bound to act for someone else's benefit, as a trustee ▷ *adj* of a trust or trustee

fief [feef] *n* (*Hist*) land granted by a lord in return for war service

field *n* enclosed piece of agricultural land; marked off area for sports; area rich in a specified natural resource; sphere of knowledge or activity; place away from the laboratory or classroom where practical work is done ▷ *v* (*Sport*) catch and return (a ball); deal with (a question) successfully ► **fielder** *n* (*Sport*) player whose task is to field the ball ► **field day** day or time of exciting activity ► **field events** throwing and jumping events in athletics ► **fieldfare** *n* type of large Old World thrush ► **field glasses** binoculars ► **field marshal** army officer of the highest rank ► **field sports** hunting, shooting, and fishing ► **fieldwork** *n* investigation made in the field as opposed to the classroom or the laboratory

fiend [feend] *n* evil spirit; cruel or wicked person; (*informal*) person devoted to something e.g. fitness fiend ► **fiendish** *adj* ► **fiendishly** *adv*

fierce *adj* wild or aggressive; intense or strong ► **fiercely** *adv* ► **fierceness** *n*

fiery *adj* **fierier, fieriest** consisting of or like fire; easily angered; (of food) very spicy

fiesta n religious festival, carnival

FIFA [fee-fa] Fédération Internationale de Football Association (International Association Football Federation)

fife n small high-pitched flute

fifteen adj, n five and ten ▶ **fifteenth** adj, n

fifth adj, n (of) number five in a series ▶ **fifth column** group secretly helping the enemy

fifty adj, n, pl **-ties** five times ten ▶ **fiftieth** adj, n

fig n soft pear-shaped fruit; tree bearing it

fight v **fighting, fought** struggle (against) in battle or physical combat; struggle to overcome someone or obtain something; carry on (a battle or contest); make (one's way) somewhere with difficulty ▷ n aggressive conflict between two (groups of) people; quarrel or contest; resistance; boxing match ▶ **fighter** n boxer; determined person; aircraft designed to destroy other aircraft ▶ **fight off** v drive away (an attacker); struggle to avoid

figment n **figment of one's imagination** imaginary thing

figure n numerical symbol; amount expressed in numbers; bodily shape; well-known person; representation in painting or sculpture of a human form; (Maths) any combination of lines, planes, points, or curves ▷ v consider, conclude; (usu. foll by in) be included (in) ▶ **figure of speech** expression in which words do not have their literal meaning ▶ **figurative** adj (of language) abstract, imaginative, or symbolic ▶ **figuratively** adv ▶ **figurine** n statuette ▶ **figurehead** n nominal leader; carved bust at the bow of a ship ▶ **figure out** v solve or understand

filament n fine wire in a light bulb that gives out light; fine thread

filbert n hazelnut

filch v steal (small amounts)

file¹ n box or folder used to keep documents in order; documents in a file; information about a person or subject; line of people one behind the other; (Computing) organized collection of related material ▷ v place (a document) in a file; place (a legal document) on official record; bring a lawsuit, esp. for divorce; walk or march in a line ▶ **file sharing** sharing computer data on a network, esp. the internet

file² n tool with a roughened blade for smoothing or shaping ▷ v shape or smooth with a file ▶ **filings** pl n shavings removed by a file

filial adj of or befitting a son or daughter

filibuster n obstruction of legislation by making long speeches; person who filibusters ▷ v obstruct (legislation) with such delaying tactics

filigree n delicate ornamental work of gold or silver wire ▷ adj made of filigree

Filipino [fill-lip-**pee**-no] adj of the Philippines ▷ n (fem **Filipina**) person from the Philippines

fill v make or become full; occupy completely; plug (a gap); satisfy (a need); hold and perform the duties of (a position); appoint to (a job or position) **one's fill** sufficient for one's needs or wants ▶ **filler** n substance that fills a gap or increases bulk ▶ **filling** n substance that fills a gap or cavity, esp. in a tooth ▷ adj (of food) substantial and satisfying ▶ **filling station**

(chiefly Brit) garage selling petrol, oil, etc.

fillet n boneless piece of meat or fish ▷ v **filleting, filleted** remove the bones from

fillip n something that adds stimulation or enjoyment

filly n, pl **-lies** young female horse

film n sequence of images projected on a screen, creating the illusion of movement; story told in such a sequence of images; thin strip of light-sensitive cellulose used to make photographic negatives and transparencies; thin sheet or layer ▷ v photograph with a movie or video camera; make a film of (a scene, story, etc.); cover or become covered with a thin layer ▷ adj connected with films or the cinema ▶ **filmy** adj very thin, delicate ▶ **film strip** set of pictures on a strip of film, projected separately as slides

filter n material or device permitting fluid to pass but retaining solid particles; device that blocks certain frequencies of sound or light; (Brit) traffic signal that allows vehicles to turn either left or right while the main signals are at red ▷ v remove impurities from (a substance) with a filter; pass slowly or faintly

filth n disgusting dirt; offensive material or language ▶ **filthy** adj ▶ **filthiness** n

filtrate n filtered gas or liquid ▷ v remove impurities with a filter ▶ **filtration** n

fin n projection from a fish's body enabling it to balance and swim; vertical tailplane of an aircraft

finagle [fin-**nay**-gl] v get or achieve by craftiness or trickery

final adj at the end; having no possibility of further change, action, or discussion ▷ n deciding contest between winners of previous rounds in a competition ▷ pl (Brit & S Afr) last examinations in an educational course ▶ **finally** adv ▶ **finality** n ▶ **finalist** n competitor in a final ▶ **finalize** v put into final form ▶ **finale** [fin-**nah**-lee] n concluding part of a dramatic performance or musical work

finance v provide or obtain funds for ▷ n management of money, loans, or credits; (provision of) funds ▷ pl money resources ▶ **financial** adj ▶ **financially** adv ▶ **financier** n person involved in large-scale financial business ▶ **financial year** twelve-month period used for financial calculations

finch n, pl **finches** small songbird with a short strong beak

find v **finding, found** discover by chance; discover by search or effort; become aware of; consider to have a particular quality; experience (a particular feeling); (Law) pronounce (the defendant) guilty or not guilty; provide, esp. with difficulty ▷ n person or thing found, esp. when valuable ▶ **finder** n ▶ **finding** n conclusion from an investigation ▶ **find out** v gain knowledge of; detect (a crime, deception, etc.)

fine[1] adj very good; (of weather) clear and dry; in good health; satisfactory; of delicate workmanship; thin or slender; subtle or abstruse e.g. *a fine distinction* ▶ **finely** adv ▶ **fineness** n ▶ **finery** n showy clothing ▶ **fine art** art produced to appeal to the sense of beauty ▶ **fine-tune** v make small adjustments to (something) so that it works really well

fine² n payment imposed as a penalty ▷ v impose a fine on

finesse [fin-**ness**] n delicate skill; subtlety and tact

finger n one of the four long jointed parts of the hand; part of a glove that covers a finger; quantity of liquid in a glass as deep as a finger is wide ▷ v touch or handle with the fingers ▶ **fingering** n technique of using the fingers in playing a musical instrument ▶ **fingerboard** n part of a stringed instrument against which the strings are pressed ▶ **fingerprint** n impression of the ridges on the tip of the finger ▷ v take the fingerprints of (someone) ▶ **finger stall** cover to protect an injured finger

finicky adj excessively particular, fussy; overelaborate

finish v bring to an end, stop; use up; bring to a desired or completed condition; put a surface texture on (wood, cloth, or metal); defeat or destroy ▷ n end, last part; death or defeat; surface texture

finite adj having limits in space, time, or size

Finn n native of Finland ▶ **Finnish** adj of Finland ▷ n official language of Finland

fiord n same as **fjord**

fir n pyramid-shaped tree with needle-like leaves and erect cones

fire n state of combustion producing heat, flames, and smoke; (Brit) burning coal or wood, or a gas or electric device, used to heat a room; uncontrolled destructive burning; shooting of guns; intense passion, ardour ▷ v operate (a weapon) so that a bullet or missile is released; (informal) dismiss from employment; bake (ceramics etc.) in a kiln; excite ▶ **firearm** n rifle, pistol, or shotgun ▶ **firebrand** n person who causes

unrest ▶ **firebreak** n strip of cleared land to stop the advance of a fire ▶ **fire brigade** organized body of people whose job is to put out fires ▶ **firedamp** n explosive gas, composed mainly of methane, formed in mines ▶ **fire drill** rehearsal of procedures for escape from a fire ▶ **fire engine** vehicle with apparatus for extinguishing fires ▶ **fire escape** metal staircase or ladder down the outside of a building for escape in the event of fire ▶ **firefighter** n member of a fire brigade ▶ **firefly** n, pl -**flies** beetle that glows in the dark ▶ **fireguard** n protective grating in front of a fire ▶ **fire irons** tongs, poker, and shovel for tending a domestic fire ▶ **fireplace** n recess in a room for a fire ▶ **fire power** (Mil) amount a weapon or unit can fire ▶ **fire station** building where firefighters are stationed ▶ **firewall** n (Computing) computer that prevents unauthorized access to a computer network from the internet ▶ **firework** n device containing chemicals that is ignited to produce spectacular explosions and coloured sparks ▷ pl show of fireworks; (informal) outburst of temper ▶ **firing squad** group of soldiers ordered to execute an offender by shooting

firie n (Aust informal) firefighter

firm¹ adj not soft or yielding; securely in position; definite; having determination or strength ▷ adv in an unyielding manner e.g. hold firm ▷ v make or become firm ▶ **firmly** adv ▶ **firmness** n

firm² n business company

firmament n (lit) sky or the heavens

first adj earliest in time or order; graded or ranked above all others ▷ n person or thing coming before all others; outset or beginning; first-class honours degree at university; lowest forward gear in a motor vehicle ▷ *adv* before anything else; for the first time ▶ **firstly** *adv* ▶ **first aid** immediate medical assistance given in an emergency ▶ **first-class** *adj* of the highest class or grade; excellent ▶ **first-hand** *adj, adv* (obtained) directly from the original source ▶ **first mate** officer of a merchant ship second in command to the captain ▶ **first person** (*Grammar*) category of verbs and pronouns used by a speaker to refer to himself or herself ▶ **first-rate** *adj* excellent ▶ **first-strike** *adj* (of a nuclear missile) for use in an opening attack to destroy enemy weapons

firth *n* narrow inlet of the sea, esp. in Scotland

fiscal *adj* of government finances, esp. taxes

fish *n, pl* **fish** or **fishes** cold-blooded vertebrate with gills, that lives in water; its flesh as food ▷ *v* try to catch fish; fish in (a particular area of water); (foll by *for*) grope for and find with difficulty; (foll by *for*) seek indirectly ▶ **fisherman** *n* person who catches fish for a living or for pleasure ▶ **fishery** *n, pl* **-eries** area of the sea used for fishing ▶ **fishy** *adj* of or like fish; (*informal*) suspicious or questionable ▶ **fishfinger** *n* oblong piece of fish covered in breadcrumbs ▶ **fishmeal** *n* dried ground fish used as animal feed or fertilizer ▶ **fishmonger** *n* seller of fish ▶ **fishnet** *n* open mesh fabric resembling netting ▶ **fishwife** *n, pl* **-wives** coarse scolding woman

fishplate *n* metal plate holding rails together

fission *n* splitting; (*Biol*) asexual reproduction involving a division into two or more equal parts; splitting of an atomic nucleus with the release of a large amount of energy ▶ **fissionable** *adj* ▶ **fissile** *adj* capable of undergoing nuclear fission; tending to split

fissure [fish-er] *n* long narrow cleft or crack

fist *n* clenched hand ▶ **fisticuffs** *pl n* fighting with the fists

fit¹ *v* **fitting, fitted** be appropriate or suitable for; be of the correct size or shape (for); adjust so as to make appropriate; try (clothes) on and note any adjustments needed; make competent or ready; correspond with the facts or circumstances ▷ *adj* appropriate; in good health; worthy or deserving ▷ *n* way in which something fits ▶ **fitness** *n* ▶ **fitter** *n* person skilled in the installation and adjustment of machinery; person who fits garments ▶ **fitting** *adj* appropriate, suitable ▷ *n* accessory or part; trying on of clothes for size ▷ *pl* furnishings and accessories in a building ▶ **fitment** *n* detachable part of the furnishings of a room ▶ **fit in** *v* give a place or time to; belong or conform ▶ **fit out** *v* provide with the necessary equipment

fit² *n* sudden attack or convulsion, such as an epileptic seizure; sudden short burst or spell

fitful *adj* occurring in irregular spells ▶ **fitfully** *adv*

five *adj, n* one more than four ▶ **fiver** *n* (*informal*) five-pound note ▶ **fives** *n* ball game resembling squash but played with bats or the hands

fix *v* make or become firm, stable, or secure; repair; place permanently; settle definitely; direct (the eyes etc.) steadily; (*informal*) unfairly influence the outcome of ▷ *n* (*informal*) difficult situation; ascertaining the position of a ship by radar etc.; (*slang*) injection of a narcotic drug ▶ **fixed** *adj* ▶ **fixedly** *adv* steadily ▶ **fixer** *n* solution used to make a photographic image permanent; (*slang*) person who arranges things ▶ **fix up** *v* arrange; provide (with)

fixation *n* obsessive interest in something ▶ **fixated** *adj* obsessed

fixative *n* liquid used to preserve or hold things in place

fixture *n* permanently fitted piece of household equipment; person whose presence seems permanent; sports match or the date fixed for it

fizz *v* make a hissing or bubbling noise; give off small bubbles ▷ *n* hissing or bubbling noise; releasing of small bubbles of gas by a liquid; effervescent drink ▶ **fizzy** *adj*

fizzle *v* make a weak hissing or bubbling sound ▶ **fizzle out** *v* (*informal*) come to nothing, fail

fjord [fee-ord] *n* long narrow inlet of the sea between cliffs, esp. in Norway

flab *n* (*informal*) unsightly body fat

flabbergasted *adj* completely astonished

flabby *adj* **-bier, -biest** having flabby flesh; loose or limp

flaccid [flas-sid] *adj* soft and limp

flag[1] *n* piece of cloth attached to a pole as an emblem or signal ▷ *v* **flagging, flagged** mark with a flag or sticker; (often foll by *down*) signal (a vehicle) to stop by waving the arm ▶ **flag day** (*Brit*) day on which small stickers are sold in the streets for charity ▶ **flagpole, flagstaff** *n* pole for a flag ▶ **flagship** *n*

admiral's ship; most important product of an organization ▶ **flag up** *v* bring (something) to someone's attention

flag[2] *v* **flagging, flagged** lose enthusiasm or vigour

flag[3], **flagstone** *n* flat paving-stone ▶ **flagged** *adj* paved with flagstones

flagellate [flaj-a-late] *v* whip, esp. in religious penance or for sexual pleasure ▶ **flagellation** *n* ▶ **flagellant** *n* person who whips himself or herself

flageolet [flaj-a-**let**] *n* small instrument like a recorder

flagon *n* wide bottle for wine or cider; narrow-necked jug for liquid

flagrant [flayg-rant] *adj* openly outrageous ▶ **flagrantly** *adv*

flail *v* wave about wildly; beat or thrash ▷ *n* tool formerly used for threshing grain by hand

flair *n* natural ability; stylishness

flak *n* anti-aircraft fire; (*informal*) severe criticism

flake[1] *n* small thin piece, esp. chipped off something; (*Aust & NZ informal*) unreliable person ▷ *v* peel off in flakes ▶ **flaky** *adj* ▶ **flake out** *v* (*informal*) collapse or fall asleep from exhaustion

flake[2] *n* (in Australia) the commercial name for the meat of the gummy shark

flambé [flahm-bay] *v* **flambéing, flambéed** cook or serve (food) in flaming brandy

flamboyant *adj* behaving in a very noticeable, extravagant way; very bright and showy ▶ **flamboyance** *n*

flame *n* luminous burning gas coming from burning material; (*informal*) abusive e-mail message ▷ *v* burn brightly; become bright red; (*informal*) send an abusive e-mail message **old flame** (*informal*) former sweetheart

flamenco *n*, *pl* **-cos** rhythmical Spanish dance accompanied by a guitar and vocalist; music for this dance

flamingo *n*, *pl* **-gos** *or* **-goes** large pink wading bird with a long neck and legs

flammable *adj* easily set on fire ▶ **flammability** *n*

> **USAGE NOTE**
> This now replaces *inflammable* in labelling and packaging because *inflammable* was often mistaken to mean 'not flammable'

flan *n* open sweet or savoury tart

flange *n* projecting rim or collar

flank *n* part of the side between the hips and ribs; side of a body of troops ▷ *vb* be at or move along the side of

flannel *n* (*Brit*) small piece of cloth for washing the face; soft woollen fabric for clothing; (*informal*) evasive talk ▷ *pl* trousers made of flannel ▷ *v* **-nelling, -nelled** (*informal*) talk evasively ▶ **flannelette** *n* cotton imitation of flannel ▶ **flanny, flannie** *n* (*Aust*) shirt made of flannel or flannelette

flap *v* **flapping, flapped** move back and forwards or up and down ▷ *n* action or sound of flapping; piece of something attached by one edge only; (*informal*) state of panic

flapjack *n* chewy biscuit made with oats

flare *v* blaze with a sudden unsteady flame; (*informal*) (of temper, violence, or trouble) break out suddenly; (of a skirt or trousers) become wider towards the hem ▷ *n* sudden unsteady flame; signal light ▷ *pl* flared trousers ▶ **flared** *adj* (of a skirt or trousers) becoming wider towards the hem

flash *n* sudden burst of light or flame; sudden occurrence (of intuition or emotion); very short time; brief unscheduled news announcement; (*Photog*) small bulb that produces an intense flash of light ▷ *adj* (*also* **flashy**) vulgarly showy ▷ *v* (cause to) burst into flame; (cause to) emit light suddenly or intermittently; move very fast; come rapidly (to mind or view); (*informal*) display ostentatiously; (*slang*) expose oneself indecently ▶ **flasher** *n* (*slang*) man who exposes himself indecently ▶ **flashback** *n* scene in a book, play, or film, that shows earlier events ▶ **flash drive** portable computer hard drive and data storage device ▶ **flash flood** sudden short-lived flood ▶ **flashlight** *n* (US) torch ▶ **flash point** critical point beyond which a situation will inevitably erupt into violence; lowest temperature at which vapour given off by a liquid can ignite

flashing *n* watertight material used to cover joins in a roof

flask *n* same as **vacuum flask**; flat bottle for carrying alcoholic drink in the pocket; narrow-necked bottle

flat¹ *adj* **flatter, flattest** level and horizontal; even and smooth; (of a tyre) deflated; outright; fixed; without variation or emotion; (of a drink) no longer fizzy; (of a battery) with no electrical charge; (*Music*) below the true pitch ▷ *adv* in or into a flat position; completely or absolutely; exactly; (*Music*) too low in pitch ▷ *n* (*Music*) symbol lowering the pitch of a note by a semitone; mud bank exposed at low tide **flat out** with maximum speed or effort ▶ **flatly** *adv*

▶ **flatness** n ▶ **flatten** v ▶ **flatfish** n sea fish, such as the sole, which has a flat body ▶ **flat-pack** adj (of furniture, etc.) supplied in pieces in a flat box for assembly by the buyer ▶ **flatscreen** n slim lightweight TV set or computer with a flat screen ▶ **flat racing** horse racing over level ground with no jumps

flat² n set of rooms for living in which are part of a larger building ▷ v **flatting, flatted** (Aust & NZ) live in a flat ▶ **flatlet** n (Brit, Aust & S Afr) small flat ▶ **flatmate** n person with whom one shares a flat

flatter v praise insincerely; show to advantage; make (a person) appear more attractive in a picture than in reality ▶ **flatterer** n ▶ **flattery** n

flattie n (NZ & S Afr informal) flat tyre

flatulent adj suffering from or caused by too much gas in the intestines ▶ **flatulence** n

flaunt v display (oneself or one's possessions) arrogantly

> **USAGE NOTE**
> Be careful not to confuse this with *flout* meaning 'disobey'

flautist n flute player

flavour n distinctive taste; distinctive characteristic or quality ▷ v give flavour to ▶ **flavouring** n substance used to flavour food ▶ **flavourless** adj

flaw n imperfection or blemish; mistake that makes a plan or argument invalid ▶ **flawed** adj ▶ **flawless** adj

flax n plant grown for its stem fibres and seeds; its fibres, spun into linen thread ▶ **flaxen** adj (of hair) pale yellow

flay v strip the skin off; criticize severely

flea n small wingless jumping bloodsucking insect ▶ **flea market**

market for cheap goods ▶ **fleapit** n (informal) shabby cinema or theatre

fleck n small mark, streak, or speck ▷ v speckle

fled v past of **flee**

fledged adj (of young birds) able to fly; (of people) fully trained ▶ **fledgling, fledgeling** n young bird ▷ adj new or inexperienced

flee v **fleeing, fled** run away (from)

fleece n sheep's coat of wool; sheepskin used as a lining for coats etc.; warm polyester fabric; (Brit) jacket or top made of this fabric ▷ v defraud or overcharge ▶ **fleecy** adj made of or like fleece

fleet¹ n number of warships organized as a unit; number of vehicles under the same ownership

fleet² adj swift in movement ▶ **fleeting** adj rapid and soon passing ▶ **fleetingly** adv

Flemish n one of two official languages of Belgium ▷ adj of Flanders, in Belgium

flesh n soft part of a human or animal body; (informal) excess fat; meat of animals as opposed to fish or fowl; thick soft part of a fruit or vegetable; human body as opposed to the soul ▷ the flesh in person, actually present **one's own flesh and blood** one's family ▶ **flesh-coloured** adj yellowish-pink ▶ **fleshly** adj carnal; worldly ▶ **fleshy** adj plump; like flesh ▶ **flesh wound** wound affecting only superficial tissue

fleur-de-lys, fleur-de-lis [flur-de-lee] n, pl **fleurs-de-lys** or **fleurs-de-lis** heraldic lily with three petals

flew v past tense of **fly¹**

flex n flexible insulated electric cable ▷ v bend ▶ **flexible** adj easily bent; adaptable ▶ **flexibly** adv ▶ **flexibility** n ▶ **flexitime,**

flextime n system permitting variation in starting and finishing times of work

flick v touch or move with the finger or hand in a quick movement; move with a short sudden movement, often repeatedly ▷ n tap or quick stroke ▷ pl (slang) the cinema ▶ **flick knife** knife with a spring-loaded blade which shoots out when a button is pressed ▶ **flick through** v look at (a book or magazine) quickly or idly

flicker v shine unsteadily or intermittently; move quickly to and fro ▷ n unsteady brief light; brief faint indication

flier n see **fly¹**

flight¹ n journey by air; act or manner of flying through the air; group of birds or aircraft flying together; aircraft flying on a scheduled journey; set of stairs between two landings; stabilizing feathers or plastic fins on an arrow or dart ▶ **flightless** adj (of certain birds or insects) unable to fly ▶ **flight attendant** person who looks after passengers on an aircraft ▶ **flight deck** crew compartment in an airliner; runway deck on an aircraft carrier ▶ **flight recorder** electronic device in an aircraft storing information about its flight

flight² n act of running away

flighty adj **flightier, flightiest** frivolous and fickle

flimsy adj **-sier, -siest** not strong or substantial; thin; not very convincing ▶ **flimsily** adv ▶ **flimsiness** n

flinch v draw back or wince, as from pain ▶ **flinch from** v shrink from or avoid

fling v **flinging, flung** throw, send, or move forcefully or hurriedly ▷ n

spell of self-indulgent enjoyment; brief romantic or sexual relationship ▶ **fling oneself into** (start to) do with great vigour

flint n hard grey stone; piece of this; small piece of an iron alloy, used in cigarette lighters ▶ **flinty** adj cruel; of or like flint

flip v **flipping, flipped** throw (something small or light) carelessly; turn (something) over; (also **flip one's lid**) (slang) fly into an emotional state ▷ n snap or tap ▷ adj (informal) flippant ▶ **flipper** n limb of a sea animal adapted for swimming; one of a pair of paddle-like rubber devices worn on the feet to help in swimming ▶ **flip-flop** n (Brit & SAfr) rubber-soled sandal held on by a thong between the big toe and the next toe ▶ **flip through** v look at (a book or magazine) quickly or idly

flippant adj treating serious things lightly ▶ **flippancy** n

flirt v behave as if sexually attracted to someone; consider lightly, toy (with) ▷ n person who flirts ▶ **flirtation** n ▶ **flirtatious** adj

flit v **flitting, flitted** move lightly and rapidly; (Scot) move house; (informal) depart hurriedly and secretly ▷ n act of flitting

float v rest on the surface of a liquid; move lightly and freely; move about aimlessly; launch (a company); offer for sale on the stock market; allow (a currency) to fluctuate against other currencies ▷ n light object used to help someone or something float; indicator on a fishing line that moves when a fish bites; decorated truck in a procession; (Brit) small delivery vehicle; sum of money used for minor expenses or to provide change ▶ **floating**

adj moving about, changing e.g. *floating population*; (of a voter) not committed to one party

flock[1] *n* number of animals of one kind together; large group of people; (Christianity) congregation ▷ *v* gather in a crowd

flock[2] *n* wool or cotton waste used as stuffing ▷ *adj* (of wallpaper) with a velvety raised pattern

floe *n* sheet of floating ice

flog *v* **flogging, flogged** beat with a whip or stick; (sometimes foll by *off*) (Brit, NZ & S Afr *informal*) sell; (NZ *informal*) steal ▶ **flogging** *n*

flood *n* overflow of water onto a normally dry area; large amount of water; rising of the tide ▷ *v* cover or become covered with water; fill to overflowing; come in large numbers or quantities ▶ **floodgate** *n* gate used to control the flow of water ▶ **floodlight** *n* lamp that casts a broad intense beam of light ▷ *v* **-lighting, -lit** illuminate by floodlight

floor *n* lower surface of a room; level of a building; flat bottom surface; (right to speak in) a legislative hall ▷ *v* knock down; (*informal*) disconcert or defeat ▶ **floored** *adj* covered with a floor ▶ **flooring** *n* material for floors ▶ **floor show** entertainment in a nightclub

floozy *n, pl* **-zies** (old-fashioned *slang*) disreputable woman

flop *v* **flopping, flopped** bend, fall, or collapse loosely or carelessly; (*informal*) fail ▷ *n* (*informal*) failure; flopping movement ▶ **floppy** *adj* hanging downwards, loose ▶ **floppy disk** (Computing) flexible magnetic disk that stores information

flora *n* plants of a given place or time

floral *adj* consisting of or decorated with flowers

floret *n* small flower forming part of a composite flower head

floribunda *n* type of rose whose flowers grow in large clusters

florid *adj* with a red or flushed complexion; ornate

florin *n* former British and Australian coin

florist *n* seller of flowers

floss *n* fine silky fibres

flotation *n* launching or financing of a business enterprise

flotilla *n* small fleet or fleet of small ships

flotsam *n* floating wreckage

flotsam and jetsam odds and ends; (Brit) homeless or vagrant people

flounce[1] *v* go with emphatic movements ▷ *n* flouncing movement

flounce[2] *n* ornamental frill on a garment

flounder[1] *v* move with difficulty, as in mud; behave or speak in a bungling or hesitating manner

flounder[2] *n* edible flatfish

flour *n* powder made by grinding grain, esp. wheat ▷ *v* sprinkle with flour ▶ **floury** *adj*

flourish *v* be active, successful, or widespread; be at the peak of development; wave (something) dramatically ▷ *n* dramatic waving motion; ornamental curly line in writing ▶ **flourishing** *adj*

flout *v* deliberately disobey (a rule, law, etc.)

USAGE NOTE
Be careful not to confuse this with *flaunt* meaning 'display'

flow *v* (of liquid) move in a stream; (of blood or electricity) circulate; proceed smoothly; hang loosely; be abundant ▷ *n* act, rate, or manner of flowing; continuous stream or discharge ▶ **flow chart** diagram showing a sequence of operations in a process

flower n part of a plant that produces seeds; plant grown for its colourful flowers; best or finest part ▷ v produce flowers, bloom; reach full growth or maturity **in flower** with flowers open ▶ **flowered** adj decorated with a floral design ▶ **flowery** adj decorated with a floral design (of language or style) elaborate ▶ **flowerbed** n piece of ground for growing flowers

flown v past participle of **fly**[1]

fl. oz. fluid ounce(s)

flu n short for **influenza**

fluctuate v change frequently and erratically ▶ **fluctuation** n

flue n passage or pipe for smoke or hot air

fluent adj able to speak or write with ease; spoken or written with ease ▶ **fluently** adv ▶ **fluency** n

fluff n soft fibres; (Brit, Aust & NZ informal) mistake ▷ v make or become soft and puffy; (informal) make a mistake ▶ **fluffy** adj

fluid n substance able to flow and change its shape; a liquid or a gas ▷ adj able to flow or change shape easily ▶ **fluidity** n ▶ **fluid ounce** (Brit) one twentieth of a pint (28.4 ml)

fluke[1] n accidental stroke of luck

fluke[2] n flat triangular point of an anchor; lobe of a whale's tail

fluke[3] n parasitic worm

flume n narrow sloping channel for water; enclosed water slide at a swimming pool

flummox v puzzle or confuse

flung v past of **fling**

flunk v (US, Aust, NZ & S Afr informal) fail

flunky, flunkey n, pl **flunkies** or **flunkeys** servile person; manservant who wears a livery

fluorescence n emission of light from a substance bombarded by particles, such as electrons, or by radiation ▶ **fluoresce** v exhibit fluorescence

fluorescent adj of or resembling fluorescence

fluoride n compound containing fluorine ▶ **fluoridate** v add fluoride to (water) as protection against tooth decay ▶ **fluoridation** n

fluorine n (Chem) toxic yellow gas, most reactive of all the elements

flurry n, pl **-ries** sudden commotion; gust of rain or wind or fall of snow ▷ v **-rying, -ried** confuse

flush[1] v blush or cause to blush; send water through (a toilet or pipe) so as to clean it; elate ▷ n blush; rush of water; excitement or elation

flush[2] adj level with the surrounding surface; (informal) having plenty of money

flush[3] v drive out of a hiding place

flush[4] n (in card games) hand all of one suit

fluster v make nervous or upset ▷ n nervous or upset state

flute n wind instrument consisting of a tube with sound holes and a mouth hole in the side; tall narrow wineglass ▶ **fluted** adj having decorative grooves

flutter v wave rapidly; flap the wings; move quickly and irregularly; (of the heart) beat abnormally quickly ▷ n flapping movement; nervous agitation; (informal) small bet; abnormally fast heartbeat

fluvial adj of rivers

flux n constant change or instability; flow or discharge; substance mixed with metal to assist in fusion

fly¹ v **flying, flew, flown** move through the air on wings or in an aircraft; control the flight of; float, flutter, display, or be displayed in the air; transport or be transported by air; move quickly or suddenly; (of time) pass rapidly; flee ▷ n, pl **flies** (often pl) (Brit) fastening at the front of trousers; flap forming the entrance to a tent ▷ pl space above a stage, used for storage ▶ **flyer, flier** n small advertising leaflet; aviator ▶ **flyleaf** n blank leaf at the beginning or end of a book ▶ **flyover** n road passing over another by a bridge ▶ **flywheel** n heavy wheel regulating the speed of a machine

fly² n, pl **flies** two-winged insect ▶ **flycatcher** n small insect-eating songbird ▶ **fly-fishing** n fishing with an artificial fly as a lure ▶ **flypaper** n paper with a sticky poisonous coating, used to kill flies ▶ **flyweight** n boxer weighing up to 112lb (professional) or 51kg (amateur)

fly³ adj (slang) sharp and cunning

flying adj hurried and brief ▶ **flying boat** aircraft fitted with floats instead of landing wheels ▶ **flying colours** conspicuous success ▶ **flying fish** fish with winglike fins used for gliding above the water ▶ **flying fox** large fruit-eating bat; (Aust & NZ) platform suspended from an overhead cable, used for transporting people or materials ▶ **flying phalanger** phalanger with black-striped greyish fur, which moves with gliding leaps ▶ **flying saucer** unidentified disc-shaped flying object, supposedly from outer space ▶ **flying squad** small group of police, soldiers, etc., ready to act quickly ▶ **flying start** very good start

FM (in Scotland) First Minister; frequency modulation

foal n young of a horse or related animal ▷ v give birth to a foal

foam n mass of small bubbles on a liquid; frothy saliva; light spongelike solid used for insulation, packing, etc. ▷ v produce foam ▶ **foamy** adj

fob n short watch chain; small pocket in a waistcoat

fob off v **fobbing, fobbed** pretend to satisfy (a person) with lies or excuses; sell or pass off (inferior goods) as valuable

fo'c's'le n same as **forecastle**

focus n, pl **-cuses** or **-ci** point at which light or sound waves converge; state of an optical image when it is clearly defined; state of an instrument producing such an image; centre of interest or activity ▷ v **-cusing, -cused** or **-cussing, -cussed** bring or come into focus; concentrate (on) ▶ **focal** adj of or at a focus ▶ **focus group** group of people gathered by a market-research company to discuss and assess a product or service

fodder n feed for livestock

foe n enemy, opponent

foefie slide n (S Afr) rope, fixed at an incline, along which a person suspended on a pulley may cross a space, esp a river

foetid adj same as **fetid**

foetus n, pl **-tuses** same as **fetus**

fog n mass of condensed water vapour in the lower air, often greatly reducing visibility ▷ v **fogging, fogged** cover with steam ▶ **foggy** adj ▶ **foghorn** n large horn sounded to warn ships in fog

fogey, fogy n, pl **-geys** or **-gies** old-fashioned person

foible n minor weakness or slight peculiarity

foil¹ v ruin (someone's plan)

foil² n metal in a thin sheet, esp. for wrapping food; anything which sets off another thing to advantage

foil³ n light slender flexible sword tipped with a button

foist v (foll by on or upon) force or impose on

fold¹ v bend so that one part covers another; interlace (the arms); clasp (in the arms); (Cooking) mix gently; (informal) fail or go bankrupt ▷ n folded piece or part; mark, crease, or hollow made by folding ▶ **folder** n piece of folded cardboard for holding loose papers

fold² n (Brit, Aust & S Afr) enclosure for sheep; church or its members

foliage n leaves ▶ **foliation** n process of producing leaves

folio n, pl **-lios** sheet of paper folded in half to make two leaves of a book; book made up of such sheets; page number

folk n people in general; race of people ▷ pl relatives ▶ **folksy** adj simple and unpretentious ▶ **folk dance** traditional country dance ▶ **folklore** n traditional beliefs and stories of a people ▶ **folk song** song handed down among the common people; modern song like this ▶ **folk singer**

follicle n small cavity in the body, esp. one from which a hair grows

follow v go or come after; be a logical or natural consequence of; keep to the course or track of; act in accordance with; accept the ideas or beliefs of; understand; have a keen interest in; (Computing) choose to receive messages posted by (a blogger or microblogger) ▶ **follower** n disciple or supporter ▶ **following** adj about to be mentioned; next in time ▷ n group of supporters ▷ prep as a result of ▶ **follow up** v investigate; do a

second, often similar, thing after (a first) ▶ **follow-up** n something done to reinforce an initial action

folly n, pl **-lies** foolishness; foolish action or idea; useless extravagant building

foment [foam-**ent**] v encourage or stir up (trouble)

fond adj tender, loving; unlikely to be realized e.g. a fond hope ▶ **fond of** having a liking for ▶ **fondly** adv ▶ **fondness** n

fondant n (sweet made from) flavoured paste of sugar and water

fondle v caress

fondue n Swiss dish of a hot melted cheese sauce into which pieces of bread are dipped

font¹ n bowl in a church for baptismal water

font² n set of printing type of one style and size

fontanelle n soft membranous gap between the bones of a baby's skull

food n what one eats, solid nourishment ▶ **foodie** n (informal) gourmet ▶ **food group** category of food based on its nutritional content ▶ **foodstuff** n substance used as food

fool¹ n person lacking sense or judgment; person made to appear ridiculous; (Hist) jester, clown ▷ v deceive (someone) ▶ **foolish** adj unwise, silly, or absurd ▶ **foolishly** adv ▶ **foolishness** n ▶ **foolery** n foolish behaviour ▶ **fool around** v act or play irresponsibly or aimlessly ▶ **foolproof** adj unable to fail

fool² n dessert of puréed fruit mixed with cream

foolhardy adj recklessly adventurous ▶ **foolhardiness** n

foolscap n size of paper, 34.3 × 43.2 centimetres

foosball n (US & Canadian) same as **table football**

foot n, pl **feet** part of the leg below the ankle; unit of length of twelve inches (0.3048 metre); lowest part of anything; unit of poetic rhythm **foot it** (informal) walk **foot the bill** pay the entire cost ▸ **footage** n amount of film used ▸ **foot-and-mouth disease** n infectious viral disease of sheep, cattle, etc. ▸ **footbridge** n bridge for pedestrians ▸ **footfall** n sound of a footstep ▸ **foothills** pl n hills at the foot of a mountain ▸ **foothold** n secure position from which progress may be made; small place giving a secure grip for the foot ▸ **footlights** pl n lights across the front of a stage ▸ **footloose** adj free from ties ▸ **footman** n male servant in uniform ▸ **footnote** n note printed at the foot of a page ▸ **footpath** n narrow path for walkers only; (Aust) raised space alongside a road, for pedestrians ▸ **footplate** n platform in the cab of a locomotive for the driver ▸ **footprint** n mark left by a foot ▸ **footstep** n step in walking; sound made by walking ▸ **footstool** n low stool used to rest the feet on while sitting ▸ **footwear** n anything worn to cover the feet ▸ **footwork** n skilful use of the feet, as in sport or dancing

football n game played by two teams of eleven players kicking a ball in an attempt to score goals; any of various similar games, such as rugby; ball used for this ▸ **footballer** n ▸ **football pools** form of gambling on the results of soccer matches

footing n basis or foundation; relationship between people; secure grip by or for the feet

footling adj (chiefly Brit informal) trivial

footsie n (informal) flirtation involving the touching together of feet

fop n man excessively concerned with fashion ▸ **foppery** n ▸ **foppish** adj

for prep indicating a person intended to benefit from or receive something, span of time or distance, person or thing represented by someone, etc. e.g. a gift for you; for five miles; playing for his country ▷ conj because **for it** (informal) liable for punishment or blame

forage v search about (for) ▷ n food for cattle or horses

foray n brief raid or attack; first attempt or new undertaking

forbear v cease or refrain (from doing something) ▸ **forbearance** n tolerance, patience

forbid v prohibit, refuse to allow ▸ **forbidden** adj ▸ **forbidding** adj severe, threatening

force n strength or power; compulsion; (Physics) influence tending to produce a change in a physical system; mental or moral strength; person or thing with strength or influence; vehemence or intensity; group of people organized for a particular task or duty ▷ v compel, make (someone) do something; acquire or produce through effort, strength, etc.; propel or drive; break open; impose or inflict; cause to grow at an increased rate **in force** having legal validity; in great numbers ▸ **forced** adj compulsory; false or unnatural; due to an emergency ▸ **forceful** adj emphatic and confident; effective ▸ **forcefully** adv ▸ **forcible** adj involving physical force or violence; strong and emphatic ▸ **forcibly** adv

forceps pl n surgical pincers

ford n shallow place where a river may be crossed ▸ v cross (a river) at a ford

fore adj in, at, or towards the front ▸ n front part **to the fore** in a conspicuous position

fore- prefix before in time or rank e.g. *forefather*; at the front e.g. *forecourt*

fore-and-aft adj located at both ends of a ship

forearm¹ n arm from the wrist to the elbow

forearm² v prepare beforehand

forebear n ancestor

foreboding n feeling that something is about to happen

forecast v -casting, -cast or -casted predict (weather, events, etc.) ▸ n prediction

forecastle [foke-sl] n raised front part of a ship

foreclose v take possession of (property bought with borrowed money which has not been repaid) ▸ **foreclosure** n

forecourt n courtyard or open space in front of a building

forefather n ancestor

forefinger n finger next to the thumb

forefront n most active or prominent position; very front

foregather v meet together or assemble

forego v same as **forgo**

foregoing adj going before, preceding ▸ **foregone conclusion** inevitable result

foreground n part of a view, esp. in a picture, nearest the observer

forehand n (Tennis etc.) stroke played with the palm of the hand facing forward

forehead n part of the face above the eyebrows

foreign adj not of, or in, one's own country; relating to or connected with other countries; unfamiliar, strange; in an abnormal place or position e.g. *foreign matter* ▸ **foreigner** n

foreleg n either of the front legs of an animal

forelock n lock of hair over the forehead

foreman n person in charge of a group of workers; leader of a jury

foremast n mast nearest the bow of a ship

foremost adj, adv first in time, place, or importance

forename n first name

forenoon n (chiefly US & Canadian) morning

forensic adj used in or connected with courts of law ▸ **forensic medicine** use of medical knowledge for the purposes of the law

foreplay n sexual stimulation before intercourse

forerunner n person or thing that goes before, precursor

foresail n main sail on the foremast of a ship

foresee v see or know beforehand ▸ **foreseeable** adj

> **SPELLING TIP**
> According to the Collins Corpus, the word **unforeseen** is misspelled as *unforseen* once for every eight occurrences

foreshadow v show or indicate beforehand

foreshore n part of the shore between high- and low-tide marks

foreshorten v represent (an object) in a picture as shorter than it really is, in accordance with perspective

foresight n ability to anticipate and provide for future needs

foreskin n fold of skin covering the tip of the penis

forest n large area with a thick growth of trees ▶ **forested** adj ▶ **forestry** n science of planting and caring for trees; management of forests ▶ **forester** n person skilled in forestry

forestall v prevent or guard against in advance

foretaste n early limited experience of something to come

foretell v tell or indicate beforehand

forethought n thoughtful planning for future events

forever, for ever adv without end; at all times; (informal) for a long time

forewarn v warn beforehand

foreword n introduction to a book

forfeit [for-fit] n thing lost or given up as a penalty for a fault or mistake ▷ v lose as a forfeit ▷ adj lost as a forfeit ▶ **forfeiture** n

forge¹ n place where metal is worked, smithy; furnace for melting metal ▷ v make a fraudulent imitation of (something); shape (metal) by heating and hammering it; create (an alliance etc.)

forge² v advance steadily **forge ahead** increase speed or take the lead

forger n person who makes an illegal copy of something

forgery n, pl -ries illegal copy of something; crime of making an illegal copy

forget v -getting, -got, -gotten fail to remember; neglect; leave behind by mistake ▶ **forgetful** adj tending to forget ▶ **forgetfulness** n ▶ **forget-me-not** n plant with clusters of small blue flowers

forgive v -giving, -gave, -given cease to blame or hold resentment against, pardon ▶ **forgiveness** n

forgo v do without or give up

forgot v past tense of **forget** ▶ **forgotten** v past participle of **forget**

fork n tool for eating food, with prongs and a handle; large similarly-shaped tool for digging or lifting; point where a road, river, etc. divides into branches; one of the branches ▷ v pick up, dig, etc. with a fork; branch; take one or other branch at a fork in the road ▶ **forked** adj ▶ **fork-lift truck** vehicle with a forklike device at the front which can be raised or lowered to move loads ▶ **fork out** v (informal) pay

forlorn adj lonely and unhappy ▶ **forlorn hope** hopeless enterprise ▶ **forlornly** adv

form n shape or appearance; mode in which something appears; type or kind; printed document with spaces for details; physical or mental condition; previous record of an athlete, racehorse, etc.; class in school ▷ v give a (particular) shape to or take a (particular) shape; come or bring into existence; make or be made; train; acquire or develop ▶ **formless** adj

formal adj of or characterized by established conventions of ceremony and behaviour; of or for formal occasions; stiff in manner; organized; symmetrical ▶ **formally** adv ▶ **formality** n, pl -ties requirement of custom or etiquette; necessary procedure without real importance ▶ **formalize** v make official or formal

formaldehyde [for-mal-de-hide] n colourless pungent gas used to make formalin ▶ **formalin** n solution of formaldehyde in water, used as a disinfectant or a preservative for biological specimens

format n style in which something is arranged ▷ v **-matting, -matted** arrange in a format

formation n forming; thing formed; structure or shape; arrangement of people or things acting as a unit

formative adj of or relating to development; shaping

former adj of an earlier time, previous **the former** first mentioned of two ▶ **formerly** adv

Formica ® n kind of laminated sheet used to make heat-resistant surfaces

formic acid n acid derived from ants

formidable adj frightening because difficult to overcome or manage; extremely impressive ▶ **formidably** adv

formula n, pl **-las** or **-lae** group of numbers, letters, or symbols expressing a scientific or mathematical rule; method or rule for doing or producing something; set form of words used in religion, law, etc.; specific category of car in motor racing ▶ **formulaic** adj ▶ **formulate** v plan or describe precisely and clearly ▶ **formulation** n

fornicate v have sexual intercourse without being married ▶ **fornication** n ▶ **fornicator** n

forsake v **-saking, -sook, -saken** withdraw support or friendship from; give up, renounce

forsooth adv (obs) indeed

forswear v **-swearing, -swore, -sworn** renounce or reject

forsythia [for-syth-ee-a] n shrub with yellow flowers in spring

fort n fortified building or place **hold the fort** (informal) keep things going during someone's absence

forte[1] [for-tay] n thing at which a person excels

forte[2] [for-tay] adv (Music) loudly

forth adv forwards, out, or away

forthcoming adj about to appear or happen; available; (of a person) communicative

forthright adj direct and outspoken

forthwith adv at once

fortieth adj, n see forty

fortify v **-fying, -fied** make (a place) defensible, as by building walls; strengthen; add vitamins etc. to (food); add alcohol to (wine) to make sherry or port ▶ **fortification** n

fortissimo adv (Music) very loudly

fortitude n courage in adversity or pain

fortnight n two weeks ▶ **fortnightly** adv, adj

FORTRAN n (Computing) programming language for mathematical and scientific purposes

fortress n large fort or fortified town

fortuitous [for-tyew-it-uss] adj happening by (lucky) chance ▶ **fortuitously** adv

fortunate adj having good luck; occurring by good luck ▶ **fortunately** adv

fortune n luck, esp. when favourable; power regarded as influencing human destiny; wealth, large sum of money ▷ pl person's destiny ▶ **fortune-teller** n person who claims to predict the future of others

forty adj, n, pl **-ties** four times ten ▶ **fortieth** adj, n

forum n meeting or medium for open discussion or debate

forward adj directed or moving ahead; in, at, or near the front; presumptuous; well developed or advanced; relating to the future ▷ n attacking player in various team games, such as soccer or hockey ▷ adv forwards ▷ v send (a letter etc.) on to an ultimate destination; advance or promote ▶ **forwards** adv towards or at a place further ahead in space or time; towards the front

fossick v (Aust & NZ) search, esp. for gold or precious stones

fossil n hardened remains of a prehistoric animal or plant preserved in rock ▶ **fossilize** v turn into a fossil; become out-of-date or inflexible

foster v promote the growth or development of; bring up (a child not one's own) ▷ adj of or involved in fostering a child e.g. foster parents

fought v past of **fight**

foul adj loathsome or offensive; stinking or dirty; (of language) obscene or vulgar; unfair ▷ n (Sport) violation of the rules ▷ v make dirty or polluted; make or become entangled or clogged; (Sport) commit a foul against (an opponent) **foul of** come into conflict with ▶ **foul-mouthed** adj habitually using foul language ▶ **foul play** unfair conduct, esp. involving violence

found¹ v past of **find**

found² v establish or bring into being; lay the foundation of; (foll by on or upon) have a basis (in) ▶ **founder** n

found³ v cast (metal or glass) by melting and setting in a mould; make (articles) by this method

foundation n basis or base; part of a building or wall below the ground; act of founding; institution supported by an endowment; cosmetic used as a base for make-up

founder v break down or fail; (of a ship) sink; stumble or fall

foundling n (chiefly Brit) abandoned baby

foundry n, pl -ries place where metal is melted and cast

fount¹ n (lit) fountain; source

fount² n set of printing type of one style and size

fountain n jet of water; structure from which such a jet spurts; source ▶ **fountainhead** n original source ▶ **fountain pen** pen supplied with ink from a container inside it

four adj, n one more than three ▷ n (crew of) four-oared rowing boat **on all fours** on hands and knees ▶ **four-letter word** short obscene word referring to sex or excrement ▶ **four-poster** n bed with four posts supporting a canopy ▶ **foursome** n group of four people

fourteen adj, n four and ten ▶ **fourteenth** adj, n

fourth adj, n (of) number four in a series ▷ n quarter ▶ **fourth dimension** time ▶ **fourth estate** the press

fowl n domestic cock or hen; any bird used for food or hunted as game

fox n reddish-brown bushy-tailed animal of the dog family; its fur; cunning person ▷ v (informal) perplex or deceive ▶ **foxy** adj of or like a fox, esp. in craftiness ▶ **foxglove** n tall plant with purple or white flowers ▶ **foxhole** n (Mil) small pit dug for protection

▶**foxhound** n dog bred for hunting foxes ▶**fox terrier** small short-haired terrier ▶**foxtrot** n ballroom dance with slow and quick steps; music for this

foyer [foy-ay] n entrance hall in a theatre, cinema, or hotel

fracas [frak-ah] n, pl **-cas** noisy quarrel

fracking n extraction of oil or gas by forcing liquid into rock at high pressure

fraction n numerical quantity that is not a whole number; fragment; piece; (Chem) substance separated by distillation ▶**fractional** adj ▶**fractionally** adv

fractious adj easily upset and angered

fracture n breaking, esp. of a bone ▷v break

fragile adj easily broken or damaged; in a weakened physical state ▶**fragility** n

fragment n piece broken off; incomplete piece ▷v break into pieces ▶**fragmentary** adj ▶**fragmentation** n

fragrant adj sweet-smelling ▶**fragrance** n pleasant smell; perfume; scent

frail adj physically weak; easily damaged ▶**frailty** n, pl **-ties** physical or moral weakness

frame n structure giving shape or support; enclosing case or border, as round a picture; person's build; individual exposure on a strip of film; individual game of snooker in a match ▷v put together, construct; put into words; put into a frame; (slang) incriminate (a person) on a false charge **frame of mind** mood or attitude ▶**frame-up** n (slang) false incrimination ▶**framework** n supporting structure

franc n monetary unit of Switzerland, various African countries, and formerly of France and Belgium

franchise n right to vote; authorization to sell a company's goods

Franciscan n, adj (friar or nun) of the order founded by St. Francis of Assisi

francium n (Chem) radioactive metallic element

Franco- combining form of France or the French

frangipani [fran-jee-**pah**-nee] n Australian evergreen tree with large yellow fragrant flowers; tropical shrub with fragrant white or pink flowers

frank adj honest and straightforward in speech or attitude ▷n official mark on a letter permitting delivery ▷v put a mark on (a letter) ▶**frankly** adv ▶**frankness** n

frankfurter n smoked sausage

frankincense n aromatic gum resin burned as incense

frantic adj distracted with rage, grief, joy, etc.; hurried and disorganized ▶**frantically** adv

fraternal adj of a brother, brotherly ▶**fraternally** adv ▶**fraternity** n group of people with shared interests, aims, etc.; brotherhood; (US) male social club at college ▶**fraternize** v associate on friendly terms ▶**fraternization** n ▶**fratricide** n crime of killing one's brother; person who does this

Frau [rhymes with **how**] n, pl **Fraus** or **Frauen** German title, equivalent to Mrs ▶**Fräulein** [**froy**-line] n, pl **-leins** or **-lein** German title, equivalent to Miss

fraud n (criminal) deception, swindle; person who acts in a deceitful way ▶**fraudulent** adj ▶**fraudulence** n

fraught [frawt] adj tense or anxious **fraught with** involving, filled with

fray¹ n (Brit, Aust & NZ) noisy quarrel or conflict

fray² v make or become ragged at the edge; become strained

frazzle n (informal) exhausted state

freak n abnormal person or thing; person who is excessively enthusiastic about something ▷ adj abnormal ▸ **freakish** adj ▸ **freak out** v (informal) (cause to) be in a heightened emotional state

freckle n small brown spot on the skin ▸ **freckled** adj marked with freckles

free adj **freer, freest** able to act at will, not compelled or restrained; not subject (to); provided without charge; not in use; (of a person) not busy; not fixed or joined ▷ v freeing, freed release, liberate; remove (obstacles, pain, etc.) from; make available or usable **a free hand** unlimited freedom to act ▸ **freely** adv ▸ **Freecycle** n® informal network of citizens who promote recycling by offering one another unwanted items free of charge ▷ v (**f-**) recycle (an unwanted item) by offering it to someone free of charge ▸ **free fall** part of a parachute descent before the parachute opens ▸ **free-for-all** n (informal) brawl ▸ **freehand** adj drawn without guiding instruments ▸ **freehold** n tenure of land for life without restrictions ▸ **freeholder** n ▸ **free house** (Brit) public house not bound to sell only one brewer's products ▸ **freelance** adj, n (of) a self-employed person doing specific pieces of work for various employers ▸ **freeloader** n (slang) habitual scrounger ▸ **free-range** adj kept or produced in natural conditions ▸ **Freeview®** n (in Britain) free service providing digital terrestrial television

▸ **freeway** n (US & Aust) motorway ▸ **freewheel** v travel downhill on a bicycle without pedalling

-free combining form without e.g. a trouble-free journey

freedom n being free; right or privilege of unlimited access e.g. the freedom of the city

Freemason n member of a secret fraternity pledged to help each other

freesia n plant with fragrant tubular flowers

freeze v **freezing, froze, frozen** change from a liquid to a solid by the reduction of temperature, as water to ice; preserve (food etc.) by extreme cold; (cause to) be very cold; become motionless with fear, shock, etc.; fix (prices or wages) at a particular level; ban the exchange or collection of (loans, assets, etc.) ▷ n period of very cold weather; freezing of prices or wages ▸ **freezer** n insulated cabinet for cold-storage of perishable foods ▸ **freeze-dry** v preserve (food) by rapid freezing and drying in a vacuum ▸ **freezing** adj (informal) very cold

freight [frate] n commercial transport of goods; cargo transported; cost of this ▷ v send by freight ▸ **freighter** n ship or aircraft for transporting goods

French n language of France, also spoken in parts of Belgium, Canada, and Switzerland ▷ adj of France, its people, or their language ▸ **French bread** white bread in a long thin crusty loaf ▸ **French dressing** salad dressing of oil and vinegar ▸ **French fries** potato chips ▸ **French horn** brass wind instrument with a coiled tube ▸ **French letter** (slang) condom ▸ **French polish** shellac varnish for wood ▸ **French**

window window extending to floor level, used as a door

frenetic [frin-net-ik] *adj* uncontrolled, excited ▸ **frenetically** *adv*

frenzy *n, pl* **-zies** violent mental derangement; wild excitement ▸ **frenzied** *adj* ▸ **frenziedly** *adv*

frequent *adj* happening often; habitual ▹ *v* visit habitually ▸ **frequently** *adv* ▸ **frequency** *n, pl* **-cies** rate of occurrence; (*Physics*) number of times a wave repeats itself in a given time

fresco *n, pl* **-coes** or **-cos** watercolour painting done on wet plaster on a wall

fresh *adj* newly made, acquired, etc.; novel, original; further, additional; (of food) not preserved; (of water) not salty; (of weather) brisk or invigorating; not tired ▸ **freshly** *adv* ▸ **freshness** *n* ▸ **freshen** *v* make or become fresh or fresher ▸ **fresher**, (US) **freshman** *n* (Brit & US) first-year student

fret[1] *v* **fretting, fretted** be worried ▸ **fretful** *adj* irritable

fret[2] *n* small bar on the fingerboard of a guitar etc.

fretwork *n* decorative carving in wood ▸ **fretsaw** *n* fine saw with a narrow blade, used for fretwork

Freudian [froy-dee-an] *adj* of or relating to the psychoanalyst Sigmund Freud or his theories

friable *adj* easily crumbled

friar *n* member of a male Roman Catholic religious order ▸ **friary** *n, pl* **-ries** house of friars

fricassee *n* stewed meat served in a thick white sauce

friction *n* resistance met with by a body moving over another; rubbing; clash of wills or personalities ▸ **frictional** *adj*

Friday *n* sixth day of the week **Good Friday** Friday before Easter

fridge *n* apparatus in which food and drinks are kept cool

fried *v* past of **fry**[1]

friend *n* person whom one knows well and likes; supporter or ally; (**F-**) Quaker ▸ **friendly** *adj* showing or expressing liking; not hostile; on the same side ▹ *n, pl* **-lies** (*Sport*) match played for its own sake and not as part of a competition ▸ **-friendly** *combining form* good or easy for the person or thing specified e.g. user-friendly ▸ **friendly society** (in Britain) association of people who pay regular dues in return for pensions, sickness benefits, etc. ▸ **friendliness** *n* ▸ **friendless** *adj* ▸ **friendship** *n*

Friesian [free-zhan] *n* breed of black-and-white dairy cattle

frieze [freeze] *n* ornamental band on a wall

frigate [frig-it] *n* medium-sized fast warship

fright *n* sudden fear or alarm; sudden alarming shock ▸ **frightful** *adj* horrifying; (*informal*) very great ▸ **frightfully** *adv*

frighten *v* scare or terrify; force (someone) to do something from fear ▸ **frightening** *adj*

frigid [frij-id] *adj* (of a woman) sexually unresponsive; very cold; excessively formal ▸ **frigidity** *n*

frill *n* gathered strip of fabric attached at one edge ▹ *pl* superfluous decorations or details ▸ **frilled** *adj* ▸ **frilled lizard** large tree-living Australian lizard with an erectile fold of skin round the neck ▸ **frilly** *adj*

fringe *n* hair cut short and hanging over the forehead; ornamental edge of hanging threads, tassels, etc.; outer edge; less important parts of an activity or group ▷ v decorate with a fringe ▷ adj (of theatre) unofficial or unconventional ▶ **fringed** adj ▶ **fringe benefit** benefit given in addition to a regular salary

frippery n, pl -**peries** useless ornamentation; trivia

frisk v move or leap playfully; (informal) search (a person) for concealed weapons etc. ▶ **frisky** adj lively or high-spirited

frisson [frees-**sonn**] n shiver of fear or excitement

fritter n piece of food fried in batter

fritter away v waste

frivolous adj not serious or sensible; enjoyable but trivial ▶ **frivolity** n

frizz v form (hair) into stiff wiry curls ▶ **frizzy** adj

frizzle v cook or heat until crisp and shrivelled

frock n dress ▶ **frock coat** man's skirted coat as worn in the 19th century

frog n smooth-skinned tailless amphibian with long back legs used for jumping **frog in one's throat** phlegm on the vocal cords, hindering speech ▶ **frogman** n swimmer with a rubber suit and breathing equipment for working underwater ▶ **frogspawn** n jelly-like substance containing frog's eggs

frolic v -**icking**, -**icked** run and play in a lively way ▷ n lively and merry behaviour ▶ **frolicsome** adj playful

from prep indicating the point of departure, source, distance, cause, change of state, etc.

USAGE NOTE
The use of off to mean from is very informal: They bought milk from (rather than off) a farmer

frond n long leaf or leaflike part of a fern, palm, or seaweed

front n fore part; position directly before or ahead; battle line or area; (Meteorol) dividing line between two different air masses; outward appearance; (informal) cover for another, usu. criminal, activity; particular field of activity e.g. on the economic front ▷ adj of or at the front ▷ v (foll. by onto); be the presenter of (a television show) ▶ **frontal** adj ▶ **frontage** n facade of a building ▶ **front bench** (in Britain) parliamentary leaders of the government or opposition ▶ **front-bencher** n ▶ **frontrunner** n (informal) person regarded as most likely to win a race, election, etc.

frontier n area of a country bordering on another

frontispiece n illustration facing the title page of a book

frost n white frozen dew or mist; atmospheric temperature below freezing point ▷ v become covered with frost ▶ **frosted** adj (of glass) having a rough surface to make it opaque ▶ **frosting** n (chiefly US) sugar icing ▶ **frosty** adj characterized by or covered by frost; unfriendly ▶ **frostily** adv ▶ **frostiness** n ▶ **frostbite** n destruction of tissue, esp. of the fingers or ears, by cold ▶ **frostbitten** adj

froth n mass of small bubbles ▷ v foam ▶ **frothy** adj

frown v wrinkle one's brows in worry, anger, or thought; look disapprovingly (on) ▷ n frowning expression

frowsty adj (Brit) stale or musty

frowzy, frowsy adj -**zier**, -**ziest** or -**sier**, -**siest** dirty or unkempt

froze v past tense of **freeze** ▶ **frozen** v past participle of **freeze**

frugal *adj* thrifty, sparing; meagre and inexpensive ▸ **frugally** *adv* ▸ **frugality** *n*

fruit *n* part of a plant containing seeds, esp. if edible; any plant product useful to humans; (often pl) result of an action or effort ▸ *v* bear fruit ▸ **fruiterer** *n* person who sells fruit ▸ **fruit fly** small fly that feeds on and lays its eggs in plant tissues; similar fly that feeds on plant sap, decaying fruit, etc., and is widely used in genetics experiments ▸ **fruitful** *adj* useful or productive ▸ **fruitfully** *adv* ▸ **fruitless** *adj* useless or unproductive ▸ **fruitlessly** *adv* ▸ **fruity** *adj* of or like fruit; (of a voice) mellow; (*Brit informal*) mildly bawdy ▸ **fruit machine** coin-operated gambling machine

fruition [froo-**ish**-on] *n* fulfilment of something worked for or desired

frump *n* dowdy woman ▸ **frumpy** *adj*

frustrate *v* upset or anger; hinder or prevent ▸ **frustrated** *adj* ▸ **frustrating** *adj* ▸ **frustration** *n*

fry[1] *v* **frying, fried** cook or be cooked in fat or oil ▸ *n*, *pl* **fries** (also **fry-up**) dish of fried food ▸ *pl* potato chips

fry[2] *pl n* young fishes **small fry** young or insignificant people

ft. foot; feet

ftp file transfer protocol: standard protocol for transferring files across a network, esp. the internet

fuchsia [**fyew**-sha] *n* ornamental shrub with hanging flowers

fuddle *v* cause to be intoxicated or confused ▸ **fuddled** *adj*

fuddy-duddy *adj*, *n*, *pl* **-dies** (*informal*) old-fashioned (person)

fudge[1] *n* soft caramel-like sweet

fudge[2] *v* avoid making a firm statement or decision

fuel *n* substance burned or treated to produce heat or power; something that intensifies (a feeling etc.) ▸ *v* **fuelling, fuelled** provide with fuel

fug *n* hot stale atmosphere ▸ **fuggy** *adj*

fugitive [**fyew**-jit-iv] *n* person who flees, esp. from arrest or pursuit ▸ *adj* fleeing; transient

fugue [**fyewg**] *n* musical composition in which a theme is repeated in different parts

fulcrum *n*, *pl* **-crums** or **-cra** pivot about which a lever turns

fulfil *v* **-filling, -filled** bring about the achievement of (a desire or promise); carry out (a request or order); do what is required ▸ **fulfilment** *n* ▸ **fulfil oneself** *v* achieve one's potential

full *adj* containing as much or as many as possible; abundant in supply; having had enough to eat; plump; complete, whole; (of a garment) of ample cut; (of a sound or flavour) rich and strong ▸ *adv* completely; directly; very ▸ **fully** *adv* ▸ **fullness** *n* in full without shortening ▸ **full-blooded** *adj* vigorous or enthusiastic ▸ **full-blown** *adj* fully developed ▸ **full moon** phase of the moon when it is visible as a full illuminated disc ▸ **full-scale** *adj* (of a plan) of actual size; using all resources ▸ **full stop** punctuation mark (.) at the end of a sentence and after abbreviations

fulmar *n* Arctic sea bird

fulminate *v* (foll by *against*) criticize or denounce angrily

fulsome *adj* distastefully excessive or insincere

fumble *v* handle awkwardly; say awkwardly ▸ *n* act of fumbling

fume v be very angry; give out smoke or vapour ▷ pl n pungent smoke or vapour

fumigate [fyew-mig-gate] v disinfect with fumes ▶ **fumigation** n

fun n enjoyment or amusement **make fun of** mock or tease ▶ **funny** adj comical, humorous; odd ▶ **funny bone** part of the elbow where the nerve is near the surface ▶ **funnily** adv

function n purpose something exists for; way something works; large or formal social event; (Maths) quantity whose value depends on the varying value of another; sequence of operations performed by a computer at a key stroke ▷ v operate or work; (foll by as) fill the role of ▶ **functional** adj of or as a function; practical rather than decorative; in working order ▶ **functionally** adv ▶ **functionary** n, pl **-aries** official

fund n stock of money for a special purpose; supply or store ▷ pl money resources ▷ v provide money to ▶ **funding** n

fundamental adj essential or primary; basic ▷ n basic rule or fact ▶ **fundamentally** adv ▶ **fundamentalism** n literal or strict interpretation of a religion ▶ **fundamentalist** n, adj

fundi n (S Afr) expert or boffin

funeral n ceremony of burying or cremating a dead person

funerary adj of or for a funeral

funereal [fyew-neer-ee-al] adj gloomy or sombre

funfair n entertainment with machines to ride on and stalls

fungus n, pl **-gi** or **-guses** plant without leaves, flowers, or roots, such as a mushroom or mould ▶ **fungal, fungous** adj

▶ **fungicide** n substance that destroys fungi

funicular n cable railway on a mountainside or cliff

funk¹ n style of dance music with a strong beat ▶ **funky** adj (of music) having a strong beat

funk² (informal) n nervous or fearful state ▷ v avoid (doing something) through fear

funnel n cone-shaped tube for pouring liquids into a narrow opening; chimney of a ship or locomotive ▷ v **-nelling, -nelled** (cause to) move through or as if through a funnel ▶ **funnel-web** n large poisonous black spider that builds funnel-shaped webs

fur n soft hair of a mammal; animal skin with the fur left on; garment made of this; whitish coating on the tongue or inside a kettle ▷ v cover or become covered with fur ▶ **furry** adj ▶ **furrier** n dealer in furs

furbish v smarten up

furious adj very angry; violent or unrestrained ▶ **furiously** adv

furl v roll up and fasten (a sail, umbrella, or flag)

furlong n unit of length equal to 220 yards (201.168 metres)

furlough [fur-loh] n leave of absence

furnace n enclosed chamber containing a very hot fire

furnish v provide (a house or room) with furniture; supply, provide ▶ **furnishings** pl n furniture, carpets, and fittings ▶ **furniture** n large movable articles such as chairs and wardrobes

furore [fyew-ror-ee] n very excited or angry reaction

furrow n trench made by a plough; groove, esp. a wrinkle on the forehead ▷ v make or become wrinkled; make furrows in

further adv in addition; to a greater distance or extent ▷ adj additional; more distant ▷ v assist the progress of ▸ **further education** (Brit) education beyond school other than at a university ▸ **furthest** adv to the greatest distance or extent ▷ adj most distant ▸ **furtherance** n ▸ **furthermore** adv besides ▸ **furthermost** adj most distant

furtive adj sly and secretive ▸ **furtively** adv

fury n, pl -**ries** wild anger; uncontrolled violence

furze n gorse

fuse¹ n cord containing an explosive for detonating a bomb

fuse² n safety device for electric circuits, containing a wire that melts and breaks the connection when the circuit is overloaded ▷ v (cause to) fail as a result of a blown fuse; join or combine; unite by melting; melt with heat

fuselage [fyew-zill-lahzh] n body of an aircraft

fusilier [fyew-zill-leer] n soldier of certain regiments

fusillade [fyew-zill-lade] n continuous discharge of firearms; outburst of criticism, questions, etc.

fusion n melting; product of fusing; combination of the nucleus of two atoms with the release of energy; something new created by a mixture of qualities, ideas, or things; popular music blending styles, esp. jazz and funk ▷ adj of a style of cooking that combines traditional Western techniques and ingredients with those used in Eastern cuisine

fuss n needless activity or worry; complaint or objection; great display of attention ▷ v make a fuss ▸ **fussy** adj inclined to fuss;

overparticular; overelaborate ▸ **fussily** adv ▸ **fussiness** n

fusty adj -**tier**, -**tiest** stale-smelling; behind the times ▸ **fustiness** n

futile adj unsuccessful or useless ▸ **futility** n

futon [foo-tonn] n Japanese-style bed

futsal [foot-sal] n form of soccer played indoors with five players on each side

future n time to come; what will happen; prospects ▷ adj yet to come or be; of or relating to time to come; (of a verb tense) indicating that the action specified has not yet taken place ▸ **futuristic** adj of a design appearing to belong to some future time

fuzz¹ n mass of fine or curly hairs or fibres ▸ **fuzzy** adj of, like, or covered with fuzz; blurred or indistinct; (of hair) tightly curled ▸ **fuzzily** adv ▸ **fuzziness** n

fuzz² n (slang) police

Gg

g gram(s); (acceleration due to) gravity

gab n, v **gabbing, gabbed** (*informal*) talk or chatter **gift of the gab** eloquence ▸ **gabby** *adj* **-bier, -biest** (*informal*) talkative

gabardine, gaberdine n strong twill cloth used esp. for raincoats

gabble v speak rapidly and indistinctly ▸ n rapid indistinct speech

gable n triangular upper part of a wall between sloping roofs ▸ **gabled** *adj*

gad v **gadding, gadded** ▸ **gad about, around** go around in search of pleasure ▸ **gadabout** n pleasure-seeker

gadfly n fly that bites cattle; constantly annoying person

gadget n small mechanical device or appliance ▸ **gadgetry** n gadgets

Gael [gayl] n speaker of Gaelic ▸ **Gaelic** [gal-lik, gay-lik] n any of the Celtic languages of Ireland and the Scottish Highlands ▸ *adj* of the Gaels or their language

gaff n stick with an iron hook for landing large fish

gaff[2] v **blow the gaff** (*slang*) divulge a secret

gaffe n social blunder

gaffer n (*Brit informal*) foreman or boss; (*informal*) old man; senior electrician on a TV or film set

gag[1] v **gagging, gagged** choke or retch; stop up the mouth of (a person) with cloth etc.; deprive of free speech ▸ n cloth etc. put into or tied across the mouth

gag[2] n (*informal*) joke

gaga [gah-gah] *adj* (*slang*) senile

gaggle n (*informal*) disorderly crowd; flock of geese

gaiety n cheerfulness; merrymaking ▸ **gaily** *adv* merrily; colourfully

gain v acquire or obtain; increase or improve; reach; (of a watch or clock) be or become too fast ▸ n profit or advantage; increase or improvement ▸ **gainful** *adj* useful or profitable ▸ **gainfully** *adv* ▸ **gain on, gain upon** v get nearer to or catch up with

gainsay v **-saying, -said** deny or contradict

gait n manner of walking

gaiter n cloth or leather covering for the lower leg

gala [gah-la] n festival; competitive sporting event

galaxy n, pl **-axies** system of stars; gathering of famous people ▸ **galactic** *adj*

gale n strong wind; (*informal*) loud outburst

gall[1] [gawl] n (*informal*) impudence; bitter feeling ▸ **gall bladder** sac attached to the liver, storing bile ▸ **gallstone** n hard mass formed in the gall bladder or its ducts

gall[2] [gawl] v annoy; make sore by rubbing

gall[3] [gawl] n abnormal outgrowth on a tree or plant

gallant *adj* brave and noble; (of a man) attentive to women ▸ **gallantly** *adv* ▸ **gallantry** n showy, attentive treatment of women; bravery

galleon n large three-masted sailing ship of the 15th–17th centuries

gallery n, pl **-ries** room or building for displaying works of art; balcony in a church, theatre, etc.; passage in a mine; long narrow room for a specific purpose e.g. *shooting gallery*

galley *n* kitchen of a ship or aircraft; (*Hist*) ship propelled by oars, usu. rowed by slaves ▸ **galley slave** (*Hist*) slave forced to row in a galley; (*informal*) drudge

Gallic *adj* French; of ancient Gaul

gallium *n* (*Chem*) soft grey metallic element used in semiconductors

gallivant *v* go about in search of pleasure

gallon *n* liquid measure of eight pints, equal to 4.55 litres

gallop *n* horse's fastest pace; galloping ▸ *v* **galloping, galloped** go or ride at a gallop; move or progress rapidly

> **SPELLING TIP**
> Although **gallop** has two *l*s, remember that **galloping** and **galloped** have only one *p*

gallows *n* wooden structure used for hanging criminals

Gallup poll *n* public opinion poll carried out by questioning a cross section of the population

galore *adv* in abundance

galoshes *pl n* (*Brit, Aust & NZ*) waterproof overshoes

galumph *v* (*Brit, Aust & NZ informal*) leap or move about clumsily

galvanic *adj* of or producing an electric current generated by chemical means; (*informal*) stimulating or startling ▸ **galvanize** *v* stimulate into action; coat (metal) with zinc

gambit *n* opening line or move intended to secure an advantage; (*Chess*) opening move involving the sacrifice of a pawn

gamble *v* play games of chance to win money; act on the expectation of something ▸ *n* risky undertaking; bet or wager ▸ **gambler** *n* ▸ **gambling** *n*

gamboge [gam-**boje**] *n* gum resin used as a yellow pigment and purgative

gambol *v* **-bolling, -bolled** jump about playfully, frolic ▸ *n* frolic

> **SPELLING TIP**
> Although the pronunciation is the same as 'gamble', both the verb and the noun **gambol** must always contain an *o*

game¹ *n* amusement or pastime; contest for amusement; single period of play in a contest; animals or birds hunted for sport or food; their flesh; scheme or trick ▸ *v* gamble ▸ *adj* brave; willing ▸ **gamely** *adv* ▸ **gaming** *n* gambling ▸ **gamekeeper** *n* (*Brit, Aust & S Afr*) person employed to breed game and prevent poaching ▸ **gamer** *n* person who plays computer games ▸ **games console** *n* electronic device enabling computer games to be played on a TV screen ▸ **gamesmanship** *n* art of winning by cunning practices without actually cheating

game² *adj* (*Brit, Aust & NZ*) lame, crippled

gamete *n* (*Biol*) reproductive cell

gamine [gam-**een**] *n* slim boyish young woman

gamma *n* third letter of the Greek alphabet ▸ **gamma ray** electromagnetic ray of shorter wavelength and higher energy than an x-ray

gammon *n* cured or smoked ham

gammy *adj* **-mier, -miest** same as **game²**

gamut *n* whole range or scale (of music, emotions, etc.)

gander *n* male goose; (*informal*) quick look

gang *n* (criminal) group; organized group of workmen ▸ **gangland** *n* criminal underworld ▶ **gang up** v form an alliance (against)

gangling adj lanky and awkward

ganglion n group of nerve cells; small harmless tumour

gangplank n portable bridge for boarding or leaving a ship

gangrene n decay of body tissue as a result of disease or injury ▶ **gangrenous** adj

gangsta rap n style of rap music originating from US Black street culture

gangster n member of a criminal gang

gangway n passage between rows of seats; gangplank

gannet n large sea bird; (Brit slang) greedy person

gantry n, pl -**tries** structure supporting something such as a crane or rocket

gaol [jayl] n same as **jail**

gap n break or opening; interruption or interval; divergence or difference ▶ **gappy** adj

gape v stare in wonder; open the mouth wide; be or become wide open ▶ **gaping** adj

garage n building used to house cars; place for the refuelling, sale, and repair of cars ▷ v put or keep a car in a garage

garb n clothes ▷ v clothe

garbage n rubbish

garbled adj (of a story etc.) jumbled and confused

garden n piece of land for growing flowers, fruit, or vegetables ▷ pl ornamental park ▷ v cultivate a garden ▶ **gardener** n ▶ **gardening** n ▶ **garden centre** place selling plants and gardening equipment

gardenia [gar-deen-ya] n large fragrant white waxy flower; shrub bearing this

garfish n freshwater fish with a long body and very long toothed jaws; sea fish with similar characteristics

gargantuan adj huge

gargle v wash the throat with (a liquid) by breathing out slowly through the liquid ▷ n liquid used for gargling

gargoyle n waterspout carved in the form of a grotesque face, esp. on a church

garish adj crudely bright or colourful ▶ **garishly** adv ▶ **garishness** n

garland n wreath of flowers worn or hung as a decoration ▷ v decorate with garlands

garlic n pungent bulb of a plant of the onion family, used in cooking

garment n article of clothing ▷ pl clothes

garner v collect or store

garnet n red semiprecious stone

garnish v decorate (food) ▷ n decoration for food

garret n attic in a house

garrison n troops stationed in a town or fort; fortified place ▷ v station troops in

garrotte, garotte n Spanish method of execution by strangling; cord or wire used for this ▷ v kill by this method

garrulous adj talkative

garter n band worn round the leg to hold up a sock or stocking

gas n, pl **gases** or **gasses** airlike substance that is not liquid or solid; fossil fuel in the form of a gas, used for heating; gaseous anaesthetic; (chiefly US) petrol ▷ v **gassing, gassed** poison or render unconscious with gas; (informal) talk idly or boastfully ▶ **gassy** adj filled with gas ▶ **gaseous** adj of or like gas ▶ **gasbag** n (informal) person who talks too much ▶ **gas chamber** airtight room which is filled with poison gas to kill people or animals ▶ **gasholder, gasometer** [gas-som-it-ar] *n* large tank for storing gas ▶ **gas mask** mask with a chemical filter to protect the wearer against poison gas

gash *v* make a long deep cut in ▷ *n* long deep cut

gasket *n* piece of rubber etc. placed between the faces of a metal joint to act as a seal

gasoline *n* (US) petrol

gasp *v* draw in breath sharply or with difficulty; utter breathlessly ▷ *n* convulsive intake of breath

gastric *adj* of the stomach ▶ **gastritis** *n* inflammation of the stomach lining

gastroenteritis *n* inflammation of the stomach and intestines

gastronomy *n* art of good eating ▶ **gastronomic** *adj*

gastropod *n* mollusc, such as a snail, with a single flattened muscular foot

gate *n* movable barrier, usu. hinged, in a wall or fence; opening with a gate; any entrance or way in; (entrance money paid by) those attending a sporting event ▶ **gatecrash** *v* enter (a party) uninvited ▶ **gatehouse** *n* building at or above a gateway ▶ **gateway** *n* entrance with a gate; means of access to: *London's gateway to Scotland*

gâteau [gat-toe] *n, pl* **-teaux** [-toes] rich elaborate cake

gather *v* assemble; collect gradually; increase gradually; learn from information given; pick or harvest; draw (material) into small tucks or folds ▶ **gathers** *pl n* gathered folds in material ▶ **gathering** *n* assembly

gatvol [hhut-fol] *adj* (S Afr slang) disenchanted; fed up

gauche [gohsh] *adj* socially awkward ▶ **gaucheness** *n*

gaucho [gow-choh] *n, pl* **-chos** S American cowboy

gaudy *adj* **gaudier, gaudiest** vulgarly bright or colourful ▶ **gaudily** *adv* ▶ **gaudiness** *n*

gauge [gayj] *v* estimate or judge; measure the amount or condition of ▷ *n* measuring instrument; scale or standard of measurement; distance between the rails of a railway track

> **SPELLING TIP**
>
> The vowels in **gauge** are often confused so that the misspelling *guage* is common in the Collins Corpus

gaunt *adj* lean and haggard ▶ **gauntness** *n*

gauntlet[1] *n* heavy glove with a long cuff **throw down the gauntlet** offer a challenge

gauntlet[2] *n* **run the gauntlet** be exposed to criticism or unpleasant treatment

gauze *n* transparent loosely-woven fabric, often used for surgical dressings ▶ **gauzy** *adj*

gave *v* past tense of **give**

gavel [gav-el] *n* small hammer banged on a table by a judge, auctioneer, or chairman to call for attention

gavotte *n* old formal dance; music for this

gawk *v* stare stupidly ▶ **gawky** *adj* clumsy or awkward ▶ **gawkiness** *n*

gawp *v* (slang) stare stupidly

gay *adj* homosexual; carefree and merry; colourful ▷ *n* homosexual ▶ **gayness** *n* homosexuality

gaze *v* look fixedly ▷ *n* fixed look

gazebo [gaz-zee-boh] *n, pl* **-bos** or **-boes** summerhouse with a good view

gazelle *n* small graceful antelope

gazette *n* official publication containing announcements
▸ **gazetteer** *n* (part of) a book that lists and describes places

gazillion *n* (*informal*) extremely large, unspecified amount
▸ **gazillionaire** *n* (*informal*) enormously rich person

gazump *v* (*Brit & Aust*) raise the price of a property after verbally agreeing with it (with a prospective buyer)

GB Great Britain; (also **Gb**) gigabyte

GBH (in Britain) grievous bodily harm

GCE (in Britain) General Certificate of Education

GCSE (in Britain) General Certificate of Secondary Education

GDP Gross Domestic Product

gear *n* set of toothed wheels connecting with another or with a rack to change the direction or speed of transmitted motion; mechanism for transmitting motion by gears; setting of a gear to suit engine speed e.g. *first gear*; clothing or belongings; equipment ▸ *v* prepare or organize for something **in gear, out of gear** with the gear mechanism engaged or disengaged ▸ **gearbox** *n* case enclosing a set of gears in a motor vehicle ▸ **gear up** *v* prepare for an activity

gecko *n*, *pl* **geckos** *or* **geckoes** small tropical lizard

geebung [gee-bung] *n* Australian tree or shrub with an edible but tasteless fruit; fruit of this tree

geek *n* (*informal*) person who is knowledgeable and enthusiastic about a subject; boring, unattractive person ▸ **geeky** *adj*

geelbek *n* (*SAfr*) edible marine fish

geese *n* plural of **goose**

geezer *n* (*Brit, Aust & NZ informal*) man

Geiger counter [guy-ger] *n* instrument for detecting and measuring radiation

geisha [gay-sha] *n*, *pl* **-sha** *or* **-shas** (in Japan) professional female companion for men

gel [jell] *n* jelly-like substance, esp. one used to secure a hairstyle ▸ *v* **gelling, gelled** form a gel; (*informal*) take on a definite form

gelatine [jel-at-teen], **gelatin** *n* substance made by boiling animal bones; edible jelly made of this ▸ **gelatinous** [jel-**at**-in-uss] *adj* of or like jelly

geld *v* castrate

gelding *n* castrated horse

gelignite *n* type of dynamite used for blasting

gem *n* precious stone or jewel; highly valued person or thing ▸ **gemfish** *n* Australian food fish with a delicate flavour

gen *n* (*informal*) information ▸ **gen up on** *v* **genning, genned** (*Brit informal*) make or become fully informed about

gendarme [zhohn-darm] *n* member of the French police force

gender *n* state of being male or female; (*Grammar*) classification of nouns in certain languages as masculine, feminine, or neuter

gene [jean] *n* part of a cell which determines inherited characteristics ▸ **gene therapy** (*Med*) replacement or alteration of defective genes to prevent inherited diseases

genealogy [jean-ee-**al**-a-gee] *n*, *pl* **-gies** (study of) the history and descent of a family or families ▸ **genealogical** *adj* ▸ **genealogist** *n*

genera [**jen**-er-a] *n* plural of **genus**

general *adj* common or widespread; of or affecting all or most; not specific; including or dealing with various or miscellaneous items; highest in authority or rank e.g. *general manager* ▸ n very senior army officer **in general** mostly or usually ▸ **generally** adv ▸ **generality** n, pl **-ties** general principle; state of being general ▸ **generalize** v draw general conclusions; speak in generalities; make widely known or used ▸ **generalization** n ▸ **general election** election in which representatives are chosen for every constituency ▸ **general practitioner** nonspecialist doctor serving a local area

generate v produce or bring into being ▸ **generative** adj capable of producing ▸ **generator** n machine for converting mechanical energy into electrical energy

generation n all the people born about the same time; average time between two generations (about 30 years); production

generic [jin-ner-ik] adj of a class, group, or genus ▸ **generically** adv

generous adj free in giving; free from pettiness; plentiful ▸ **generously** adv ▸ **generosity** n

genesis [jen-iss-iss] n, pl **-eses** [-iss-eez] beginning or origin

genetic [jin-net-ik] adj of genes or genetics ▸ **genetics** n study of heredity and variation in organisms ▸ **geneticist** n ▸ **genetic engineering** alteration of the genetic structure of an organism for a particular purpose ▸ **genetic fingerprinting** use of a person's unique DNA pattern as a means of identification

genial [jean-ee-al] adj cheerful and friendly ▸ **genially** adv ▸ **geniality** n

genie [jean-ee] n (in fairy tales) servant who appears by magic and grants wishes

genital adj of the sexual organs or reproduction ▸ **genitals, genitalia** [jen-it-**ail**-ya] pl n external sexual organs

genitive n grammatical case indicating possession or association

genius [jean-yuss] n (person with) exceptional ability in a particular field

genocide [jen-no-side] n murder of a race of people

genome n (Biol) full complement of genetic material within an organism

genre [zhohn-ra] n style of literary, musical, or artistic work

gent n (Brit, Aust & NZ informal) gentleman ▸ **gents** n men's public toilet

genteel adj affectedly proper and polite ▸ **genteelly** adv

gentian [jen-shun] n mountain plant with deep blue flowers

gentile adj, n non-Jewish (person)

gentle adj mild or kindly; not rough or severe; gradual; easily controlled, tame ▸ **gentleness** n ▸ **gently** adv ▸ **gentleman** n polite well-bred man; man of high social position; polite name for a man ▸ **gentlemanly** adj ▸ **gentlewoman** n fem

gentry n people just below the nobility in social rank ▸ **gentrification** n taking-over of a traditionally working-class area by middle-class incomers ▸ **gentrify** v

genuflect v bend the knee as a sign of reverence or deference ▸ **genuflection, genuflexion** n

genuine adj not fake, authentic; sincere ▸ **genuinely** adv ▸ **genuineness** n

genus [jean-uss] n, pl **genera** group into which a family of animals or plants is divided; kind, type

geocentric *adj* having the earth as a centre; measured as from the earth's centre

geography *n* study of the earth's physical features, climate, population, etc. ▸ **geographer** *n* ▸ **geographical**, **geographic** *adj* ▸ **geographically** *adv*

geology *n* study of the earth's origin, structure, and composition ▸ **geological** *adj* ▸ **geologically** *adv* ▸ **geologist** *n*

geometry *n* branch of mathematics dealing with points, lines, curves, and surfaces ▸ **geometric**, **geometrical** *adj* ▸ **geometrically** *adv*

Geordie *n* person from, or dialect of, Tyneside, an area of NE England

Georgian *adj* of the time of any of the four kings of Britain called George who ruled from 1714 to 1830

geostationary *adj* (of a satellite) orbiting so as to remain over the same point of the earth's surface

geothermal *adj* of or using the heat in the earth's interior

geranium *n* cultivated plant with red, pink, or white flowers

gerbil [jer-bill] *n* burrowing desert rodent of Asia and Africa

geriatrics *n* branch of medicine dealing with old age and its diseases ▸ **geriatric** *adj*, *n* old (person)

germ *n* microbe, esp. one causing disease; beginning from which something may develop; simple structure that can develop into a complete organism

German *n* language of Germany, Austria, and part of Switzerland; person from Germany ▷ *adj* of Germany or its language ▸ **Germanic** *adj* ▸ **German measles** contagious disease accompanied by a cough, sore throat, and red spots ▸ **German shepherd dog** Alsatian

germane *adj* germane to relevant to

germanium *n* (Chem) brittle grey element that is a semiconductor

germinate *v* (cause to) sprout or begin to grow ▸ **germination** *n* ▸ **germinal** *adj* of or in the earliest stage of development

gerrymandering *n* alteration of voting constituencies in order to give an unfair advantage to one party

gerund [jer-rund] *n* noun formed from a verb e.g. *living*

Gestapo *n* secret state police of Nazi Germany

gestation *n* (period of) carrying of young in the womb between conception and birth; developing of a plan or idea in the mind

gesticulate *v* make expressive movements with the hands and arms ▸ **gesticulation** *n*

gesture *n* movement to convey meaning; thing said or done to show one's feelings ▷ *v* gesticulate

get *v* **getting**, **got** obtain or receive; bring or fetch; contract (an illness); (cause to) become as specified e.g. *get wet*; understand; (often foll by *to*) come (to) or arrive (at); go on board (a plane, bus, etc.); persuade; (*informal*) annoy ▸ **get across** *v* (cause to) be understood ▸ **get at** *v* gain access to; imply or mean; criticize ▸ **getaway** *adj*, *n* (used in) escape ▸ **get by** *v* manage in spite of difficulties ▸ **get off** *v* (cause to) avoid the consequences of, or punishment for, an action ▸ **get off with** *v* (*informal*) start a romantic or sexual relationship with ▸ **get over** *v* recover from ▸ **get through** *v* (cause to) succeed; use up (money or supplies) ▸ **get through to** v make (a person) understand; contact by telephone ▸ **get-up** n (informal) costume ▸ **get up to** v be involved in

geyser [geez-er] n spring that discharges steam and hot water; (Brit & S Afr) domestic gas water heater

ghastly adj **-lier, -liest** (informal) unpleasant; deathly pale; (informal) unwell; (informal) horrible ▸ **ghastliness** n

ghat n (in India) steps leading down to a river; mountain pass

ghee [gee] n (in Indian cookery) clarified butter

gherkin n small pickled cucumber

ghetto n, pl **-tos** or **-toes** slum area inhabited by a deprived minority ▸ **ghetto-blaster** n (informal) large portable cassette recorder or CD player

ghillie n same as **gillie**

ghost n disembodied spirit of a dead person; faint trace ▸ v ghostwrite ▸ **ghost gum** (Aust) eucalyptus with white trunk and branches ▸ **ghostly** adj ▸ **ghost town** deserted town ▸ **ghostwriter** n writer of a book or article on behalf of another person who is credited as the author

ghoul [gool] n person with morbid interests; demon that eats corpses ▸ **ghoulish** adj

GI n (informal) US soldier; glycaemic index: index showing the effects of various foods on blood sugar

giant n mythical being of superhuman size; very large person or thing ▸ adj huge

gibber[1] [jib-ber] v speak rapidly and unintelligibly ▸ **gibberish** n rapid unintelligible talk

gibber[2] [gib-ber] n (Aust) boulder; barren land covered with stones

gibbet [jib-bit] n gallows for displaying executed criminals

gibbon [gib-bon] n agile tree-dwelling ape of S Asia

gibbous adj (of the moon) more than half but less than fully illuminated

gibe [jibe] v, n same as **jibe**[1]

giblets [jib-lets] pl n gizzard, liver, heart, and neck of a fowl

gidday, g'day interj (Aust & NZ) expression of greeting

giddy adj **-dier, -diest** having or causing a feeling of dizziness ▸ **giddily** adv ▸ **giddiness** n

gift n present; natural talent ▸ v make a present of ▸ **gifted** adj talented

gig[1] n single performance by pop or jazz musicians ▸ v **gigging, gigged** play a gig or gigs

gig[2] n light two-wheeled horse-drawn carriage

gig[3] n (informal) short for **gigabyte**

gigabyte n (Computing) 1024 megabytes

gigantic adj enormous

giggle v laugh nervously ▸ n such a laugh ▸ **giggly** adj

gigolo [jig-a-lo] n, pl **-los** man paid by an older woman to be her escort or lover

gigot n (chiefly Brit) leg of lamb or mutton

gild v **gilding, gilded** or **gilt** put a thin layer of gold on; make falsely attractive

gill[1] [jill] n liquid measure of quarter of a pint, equal to 0.142 litres

gillie n (in Scotland) attendant for hunting or fishing

gills [gilz] pl n breathing organs in fish and other water creatures

gilt adj covered with a thin layer of gold ▸ n thin layer of gold used as decoration ▸ **gilt-edged** adj denoting government stocks on which interest payments and final repayments are guaranteed

gimbals pl n set of pivoted rings which allow nautical instruments to remain horizontal at sea

gimcrack [**jim**-krak] adj showy but cheap; shoddy

gimlet [**gim**-let] n small tool with a screwlike tip for boring holes in wood ▶ **gimlet-eyed** adj having a piercing glance

gimmick n something designed to attract attention or publicity ▶ **gimmickry** n ▶ **gimmicky** adj

gin[1] n spirit flavoured with juniper berries

gin[2] n wire noose used to trap small animals; machine for separating seeds from raw cotton

gin[3] n (Aust offens) Aboriginal woman

ginger n root of a tropical plant, used as a spice; light orange-brown colour ▶ **gingery** adj ▶ **ginger ale**, **ginger beer** fizzy ginger-flavoured soft drink ▶ **gingerbread** n moist cake flavoured with ginger ▶ **ginger group** (Brit, Aust & NZ) group within a larger group that agitates for a more active policy ▶ **ginger nut**, **ginger snap** crisp ginger-flavoured biscuit

gingerly adv cautiously

gingham n cotton cloth, usu. checked or striped

gingivitis [jin-jiv-**vite**-iss] n inflammation of the gums

ginkgo [**gink**-go] n, pl **-goes** ornamental Chinese tree

ginseng [**jin**-seng] n (root of a) plant believed to have tonic and energy-giving properties

Gipsy n, pl **-sies** same as **Gypsy**

giraffe n African ruminant mammal with a spotted yellow skin and long neck and legs

gird v **girding**, **girded** or **girt** put a belt round; secure with or as if with a belt; surround **gird (up) one's loins** prepare for action

girder n large metal beam

girdle[1] n woman's elastic corset; belt; (Anat) encircling structure or part ▶ v surround or encircle

girdle[2] n (Scot) griddle

girl n female child; young woman; girlfriend; (informal) any woman ▶ **girlhood** n ▶ **girlish** adj ▶ **girlie**, **girly** adj (informal) featuring photographs of naked or scantily clad women ▶ **girlfriend** n girl or woman with whom a person is romantically or sexually involved; female friend

giro [**jire**-oh] n, pl **-ros** (in some countries) system of transferring money directly from one account to another; (Brit informal) social security payment by giro cheque

girt v a past of **gird**

girth n measurement round something; band round a horse to hold the saddle in position

gist [jist] n substance or main point of a matter

give v **giving**, **gave**, **given** present (something) to another person; impart; administer; utter or emit; sacrifice or devote; organize or host; yield or break under pressure ▶ n resilience or elasticity ▶ **give away** v donate as a gift; reveal ▶ **giveaway** n something that reveals hidden feelings or intentions ▶ adj very cheap or free ▶ **give in** v admit defeat ▶ **give off** v emit ▶ **give out** v distribute; emit; come to an end or fail ▶ **give over** v set aside for a specific purpose; (informal) cease ▶ **give up** v abandon; acknowledge defeat

gizzard n part of a bird's stomach

GL glycaemic load: index showing the amount of carbohydrate contained in a specified serving of a particular food

glacé [glass-say] *adj* preserved in a thick sugary syrup

glacier *n* slow-moving mass of ice formed by accumulated snow ▸ **glacial** *adj* of ice or glaciers; very cold; unfriendly ▸ **glaciated** *adj* covered with or affected by glaciers ▸ **glaciation** *n*

glad *adj* **gladder, gladdest** pleased and happy; causing happiness **glad to** very willing to (do something) **the glad eye** (chiefly Brit informal) seductive glance ▸ **gladly** *adv* ▸ **gladness** *n* ▸ **gladden** *v* make glad ▸ **glad rags** (informal) best clothes

glade *n* open space in a forest

gladiator *n* (in ancient Rome) man trained to fight in arenas to provide entertainment

gladiolus *n*, *pl* **-lus** or **-li, -luses** garden plant with sword-shaped leaves

gladwrap® (Aust, NZ & S Afr) *n* thin polythene material for wrapping food ▸ *v* wrap in gladwrap

glamour *n* alluring charm or fascination ▸ **glamorous** *adj* alluring ▸ **glamorize** *v*

> **SPELLING TIP**
> People often forget to drop the *u* in **glamour** when they add *ous*, writing **glamourous** instead of the correct **glamorous**

glance *v* look rapidly or briefly; glint or gleam ▸ *n* brief look ▸ **glancing** *adj* hitting at an oblique angle ▸ **glance off** *v* strike and be deflected off (an object) at an oblique angle

gland *n* organ that produces and secretes substances in the body ▸ **glandular** *adj*

glare *v* stare angrily; be unpleasantly bright ▸ *n* angry

stare; unpleasant brightness ▸ **glaring** *adj* conspicuous; unpleasantly bright ▸ **glaringly** *adv*

glass *n* hard brittle, usu. transparent substance consisting of metal silicates or similar compounds; tumbler; its contents; objects made of glass; mirror; barometer ▸ *pl* spectacles ▸ **glassy** *adj* like glass; expressionless ▸ **glasshouse** *n* greenhouse; (Brit informal) army prison

glaucoma *n* eye disease

glaze *v* fit or cover with glass; cover with a protective shiny coating ▸ *n* transparent coating; substance used for this ▸ **glazier** *n* person who fits windows with glass

gleam *n* small beam or glow of light; brief or faint indication ▸ *v* emit a gleam ▸ **gleaming** *adj*

glean *v* gather (facts etc.) bit by bit; gather (the useful remnants of a crop) after harvesting ▸ **gleaner** *n*

glee *n* triumph and delight ▸ **gleeful** *adj* ▸ **gleefully** *adv*

glen *n* deep narrow valley, esp. in Scotland

glib *adj* **glibber, glibbest** fluent but insincere or superficial ▸ **glibly** *adv* ▸ **glibness** *n*

glide *v* move easily and smoothly; (of an aircraft) move without the use of engines ▸ *n* smooth easy movement ▸ **glider** *n* plane without an engine; (Aust) flying phalanger ▸ **gliding** *n* sport of flying gliders

glimmer *v* shine faintly, flicker ▸ *n* faint gleam; faint indication

glimpse *n* brief or incomplete view ▸ *v* catch a glimpse of

glint *v* gleam brightly ▸ *n* bright gleam

glissando *n* (Music) slide between two notes in which all intermediate notes are played

glisten v gleam by reflecting light

glitch n small problem that stops something from working properly

glitter v shine with bright flashes; be showy ▷ n sparkle or brilliance; tiny pieces of shiny decorative material

glitzy adj glitzier, glitziest (slang) showily attractive

gloaming n (Scot poetic) twilight

gloat v (often foll by over) regard one's own good fortune or the misfortune of others with smug or malicious pleasure

glob n rounded mass of thick fluid

globe n sphere with a map of the earth on it; spherical object; (S Afr) light bulb **the globe** the earth ▶ **global** adj worldwide; total or comprehensive ▶ **globalization** n process by which a company, etc., expands to operate internationally ▶ **global warming** increase in the overall temperature worldwide believed to be caused by the greenhouse effect ▶ **globally** adv ▶ **globetrotter** n habitual worldwide traveller ▶ **globetrotting** n, adj

globule n small round drop ▶ **globular** adj

glockenspiel n percussion instrument consisting of small metal bars played with hammers

gloom n melancholy or depression; darkness ▶ **gloomy** adj ▶ **gloomily** adv

glory n, pl-ries praise or honour; splendour; praiseworthy thing ▷ v -rying, -ried (foll by in) triumph or exalt ▶ **glorify** v make (something) seem more worthy than it is; praise ▶ **glorification** n ▶ **glorious** adj brilliantly beautiful; delightful; full of or conferring glory ▶ **gloriously** adv ▶ **glory hole** (informal) untidy cupboard or storeroom

gloss¹ n surface shine or lustre; paint or cosmetic giving a shiny finish ▶ **glossy** adj -sier, -siest smooth and shiny; (of a magazine) printed on shiny paper ▶ **glossily** adv ▶ **glossiness** n ▶ **gloss over** v (try to) conceal (a fault or error)

gloss² n explanatory comment added to the text of a book ▷ v add glosses to

glossary n, pl-ries list of special or technical words with definitions

glottal adj of the glottis

glottis n, pl-tises or-tides vocal cords and the space between them

glove n covering for the hand with individual sheaths for each finger and the thumb ▶ **gloved** adj covered by a glove or gloves ▶ **glove compartment, glove box** small storage area in the dashboard of a car

glow v emit light and heat without flames; shine; have a feeling of wellbeing or satisfaction; (of a colour) look warm; be hot ▷ n glowing light; warmth of colour; feeling of wellbeing ▶ **glow-worm** n insect giving out a green light

glower [rhymes with **power**] v, n scowl

gloxinia n tropical plant with large bell-shaped flowers

glucose n kind of sugar found in fruit

glue n natural or synthetic sticky substance used as an adhesive ▷ v gluing or glueing, glued fasten with glue; (foll by to) pay full attention to e.g. her eyes were glued to the TV ▶ **gluey** adj ▶ **glue-sniffing** n inhaling of glue fumes for intoxicating or hallucinatory effects

glum adj glummer, glummest sullen or gloomy ▶ **glumly** adv

glut n excessive supply ▷ v glutting, glutted oversupply

gluten [gloo-ten] n protein found in cereal grain

glutinous [gloo-tin-uss] adj sticky or gluey

glutton n greedy person; person with a great capacity for something ▶ **gluttonous** adj ▶ **gluttony** n

glycerine, glycerin n colourless sweet liquid used widely in chemistry and industry

glycerol [gliss-ser-ol] n technical name for **glycerine**

GM genetically modified

gm gram

GMO genetically modified organism

GMT Greenwich Mean Time

gnarled adj rough, twisted, and knobbly

gnash v grind (the teeth) together in anger or pain

gnat n small biting two-winged fly

gnaw v **gnawing, gnawed, gnawed** or **gnawn** bite or chew steadily; (foll by at) cause constant distress (to)

gneiss n coarse-grained metamorphic rock

gnome n imaginary creature like a little old man

gnomic [no-mik] adj of pithy sayings

Gnosticism n religious movement believing in intuitive spiritual knowledge ▶ **Gnostic** n, adj

gnu [noo] n oxlike S African antelope

go v **going, went, gone** move to or from a place; be in regular attendance; depart; be, do, or become as specified; be allotted to a specific purpose or recipient; blend or harmonize; fail or break down; elapse; be got rid of; attend; be acceptable ▷ n attempt; verbal attack; turn **make a go of** be successful at ▶ **go back on** v break (a promise etc.) ▶ **go-between** n intermediary ▶ **go for** v (informal)

choose; attack; apply to equally ▶ **go-getter** n energetically ambitious person ▶ **go-go dancer** scantily dressed erotic dancer ▶ **go off** v explode; ring or sound; (informal) become stale or rotten; (informal) stop liking ▶ **go out** v go to entertainments or social functions; be romantically involved (with); be extinguished ▶ **go over** v examine or check ▶ **go-slow** n deliberate slowing of work-rate as an industrial protest ▶ **go through** v suffer or undergo; examine or search

goad v provoke (someone) to take some kind of action, usu. in anger ▷ n spur or provocation; spiked stick for driving cattle

goal n (Sport) posts through which the ball or puck has to be propelled to score; score made in this way; aim or purpose ▶ **goalie** n (informal) goalkeeper ▶ **goalkeeper** n player whose task is to stop shots entering the goal ▶ **goalpost** n one of the two posts marking the limit of a goal **move the goalposts** change the aims of an activity to ensure the desired result

goanna n large Australian lizard

goat n sure-footed ruminant animal with horns **get someone's goat** (slang) annoy someone ▶ **goatee** n pointed tuftlike beard

gob n lump of a soft substance; (Brit, Aust & NZ slang) mouth

gobbet n lump, esp. of food

gobble¹ v eat hastily and greedily

gobble² n rapid gurgling cry of the male turkey ▷ v make this noise

gobbledegook, gobbledygook n unintelligible (official) language or jargon

goblet n drinking cup without handles

goblin n (in folklore) small malevolent creature

goby n, -by or -bies small spiny-finned fish

god n spirit or being worshipped as having supernatural power; object of worship, idol; (**G-**) (in monotheistic religions) the Supreme Being, creator and ruler of the universe **the gods** top balcony in a theatre ▸ **goddess** n fem ▸ **godlike** adj ▸ **godly** adj devout or pious ▸ **godliness** n ▸ **godfearing** adj pious and devout ▸ **godforsaken** adj desolate or dismal ▸ **godsend** n something unexpected but welcome

godetia n plant with showy flowers

godparent n person who promises at a child's baptism to bring the child up as a Christian ▸ **godchild** n child for whom a person stands as godparent ▸ **goddaughter** n ▸ **godfather** n male godparent; head of a criminal, esp. Mafia, organization ▸ **godmother** n ▸ **godson** n

gogga n (S Afr informal) any small insect

goggle v (of the eyes) bulge; stare ▸ **goggles** pl n protective spectacles

going n condition of the ground for walking or riding over; speed or progress; departure ▷ adj thriving; current or accepted ▸ **going-over** n, pl **goings-over** (informal) investigation or examination; scolding or thrashing ▸ **goings-on** pl n mysterious or unacceptable events

goitre [goy-ter] n swelling of the thyroid gland in the neck

go-kart n small low-powered racing car

gold n yellow precious metal; coins or articles made of this; colour of gold ▷ adj made of gold; gold-coloured ▸ **goldcrest** n small bird with a yellow crown ▸ **gold-digger** n (informal) woman who uses her sexual attractions to get money from a man ▸ **goldfinch** n kind of finch, the male of which has yellow-and-black wings ▸ **goldfish** n orange fish kept in ponds or aquariums ▸ **gold leaf** thin gold sheet used for gilding ▸ **gold medal** medal given to the winner of a competition or race

golden adj made of gold; gold-coloured; very successful or promising ▸ **golden eagle** large mountain eagle of the N hemisphere ▸ **golden handshake** (informal) payment to a departing employee ▸ **golden rule** important principle ▸ **golden wattle** Australian plant with yellow flowers that yields a useful gum and bark ▸ **golden wedding** fiftieth wedding anniversary

golf n outdoor game in which a ball is struck with clubs into a series of holes ▷ v play golf ▸ **golfer** n

golliwog n soft black-faced doll

gonad n organ producing reproductive cells, such as a testicle or ovary

gondola n long narrow boat used in Venice; suspended cabin of a cable car, airship, etc. ▸ **gondolier** n person who propels a gondola

gone v past participle of **go** ▸ **goner** n (informal) person or thing beyond help or recovery

gong n rimmed metal disc that produces a note when struck; (slang) medal

gonorrhoea [gon-or-**ree**-a] n venereal disease with a discharge from the genitals

good adj better, best giving pleasure; morally excellent;

beneficial; kindly; talented; well-behaved; valid; reliable; complete or full ▷ n benefit; positive moral qualities ▷ pl merchandise; property; **as good as** virtually for **good** permanently ▶ **goodness** n ▶ **goodly** adj considerable ▶ **goody** n (informal) hero in a book or film; enjoyable thing ▷ adj ▶ **goody-goody** adj, n smugly virtuous (person) ▶ **good-for-nothing** adj, n irresponsible or worthless (person) ▶ **Good Samaritan** person who helps another in distress ▶ **goodwill** n kindly feeling; value of a business in reputation etc. over and above its tangible assets

> **USAGE NOTE**
> Note that good is an adjective. To modify a verb, use well: She did well

goodbye interj, n expression used on parting

gooey adj **gooier, gooiest** (informal) sticky and soft

goof (informal) n mistake ▷ v make a mistake

Google® n popular internet search engine ▷ v (**g-**) search for (something on the internet) using a search engine

googly n, pl **-lies** (Cricket) ball that spins unexpectedly from off to leg on the bounce

goon n (informal) stupid person; (chiefly US) hired thug ▶ **goon bag** n (Aust informal) plastic bladder inside a box of (usu cheap) wine

goose n, pl **geese** web-footed bird like a large duck; female of this bird ▶ **goose flesh, goose pimples** bumpy condition of the skin and bristling of the hair due to cold or fright ▶ **goose step** march step in which the leg is raised rigidly

gooseberry n edible yellowy-green berry; (Brit informal) unwanted

third person accompanying a couple

gopher [go-fer] n American burrowing rodent

gore¹ n blood from a wound

gore² v pierce with horns

gorge n deep narrow valley ▷ v eat greedily **make one's gorge rise** cause feelings of disgust or nausea

gorgeous adj strikingly beautiful or attractive; (informal) very pleasant ▶ **gorgeously** adv

gorgon n terrifying or repulsive woman

Gorgonzola n sharp-flavoured blue-veined Italian cheese

gorilla n largest of the apes, found in Africa

gormless adj (informal) stupid

gorse n prickly yellow-flowered shrub

gory adj **gorier, goriest** horrific or bloodthirsty; involving bloodshed

goshawk n large hawk

gosling n young goose

gospel n (**G-**) any of the first four books of the New Testament; unquestionable truth; Black religious music originating in the churches of the Southern US

gossamer n very fine fabric; filmy cobweb

gossip n idle talk, esp. about other people; person who engages in gossip ▷ v **gossiping, gossiped** engage in gossip ▶ **gossipy** adj

got v past of **get**; **have got** possess **have got to** need or be required to

Goth n member of an East Germanic people who invaded the Roman Empire

Gothic adj (of architecture) of or in the style common in Europe from the 12th–16th centuries, with pointed arches; of or in an 18th-century literary style characterized by gloom and the supernatural;

(of print) using a heavy ornate typeface

gouache n (painting using) watercolours mixed with glue

Gouda n mild-flavoured Dutch cheese

gouge [gowj] v scoop or force out; cut (a hole or groove) in (something) ▷ n hole or groove; chisel with a curved cutting edge

goulash [goo-lash] n rich stew seasoned with paprika

gourd [goord] n fleshy fruit of a climbing plant; its dried shell, used as a container

gourmand [goor-mand] n person who is very keen on food and drink

gourmet [goor-may] n connoisseur of food and drink

gout [gowt] n disease causing inflammation of the joints

govern v rule, direct, or control; exercise restraint over (temper etc.) ▶ **governable** adj ▶ **governance** n governing ▶ **governess** n woman teacher in a private household ▶ **governor** n official governing a province or state; senior administrator of a society, institution, or prison ▶ **governor general** n representative of the Crown in a Commonwealth country

government n executive policy-making body of a state; exercise of political authority over a country or state; system by which a country or state is ruled ▶ **governmental** adj

> **SPELLING TIP**
> Note the n between the r and the m in **government**

gown n woman's long formal dress; surgeon's overall; official robe worn by judges, clergymen, etc.

goy n, pl **goyim** or **goys** (slang) Jewish word for a non-Jew

GP general practitioner

GPS Global Positioning System: a satellite-based navigation system

grab v **grabbing, grabbed** grasp suddenly, snatch ▷ n sudden snatch; mechanical device for gripping

grace n beauty and elegance; polite, kind behaviour; goodwill or favour; delay granted; short prayer of thanks for a meal; (**G-**) title of a duke, duchess, or archbishop ▷ v add grace to ▶ **graceful** adj ▶ **gracefully** adv ▶ **graceless** adj ▶ **gracious** adj kind and courteous; condescendingly polite; elegant ▶ **graciously** adv ▶ **grace note** (Music) note ornamenting a melody

grade n place on a scale of quality, rank, or size; mark or rating; (US, Aust & S Afr) class in school ▷ v arrange in grades; assign a grade to **make the grade** succeed ▶ **gradation** n (stage in) a series of degrees or steps; arrangement in stages

gradient n (degree of) slope

gradual adj occurring, developing, or moving in small stages ▶ **gradually** adv

graduate v receive a degree or diploma; group by type or quality; mark (a container etc.) with units of measurement ▷ n holder of a degree ▶ **graduation** n

graffiti [graf-fee-tee] pl n words or drawings scribbled or sprayed on walls etc.

> **SPELLING TIP**
> Remember that the correct spelling of **graffiti** has two fs and only one t

graft n surgical transplant of skin or tissue; shoot of a plant set in the stalk of another ▷ v transplant (living tissue) surgically;

insert (a plant shoot) in another stalk

graft² (Brit informal) n hard work; obtaining of money by misusing one's position ▸ v work hard ▸ **grafter** n

grail n same as **Holy Grail**

grain n seedlike fruit of a cereal plant; cereal plants in general; small hard particle; very small amount; arrangement of fibres, as in wood; texture or pattern resulting from this **go against the grain** be contrary to one's natural inclination ▸ **grainy** adj

gram, gramme n metric unit of mass equal to one thousandth of a kilogram

grammar n branch of linguistics dealing with the form, function, and order of words; use of words; book on the rules of grammar ▸ **grammarian** n ▸ **grammatical** adj according to the rules of grammar ▸ **grammatically** adv ▸ **grammar school** (Brit) esp. formerly, a secondary school providing an education with a strong academic bias

gramophone n old-fashioned type of record player

grampus n, pl **-puses** dolphin-like mammal

gran n (Brit, Aust & NZ informal) grandmother

granary n, pl **-ries** storehouse for grain

grand adj large or impressive; imposing; dignified or haughty; (informal) excellent; (of a total) final ▸ n (slang) thousand pounds or dollars; grand piano ▸ **grandchild** n child of one's child ▸ **granddaughter** n female grandchild ▸ **grandfather** n male grandparent ▸ **grandfather clock** tall standing clock with a pendulum and wooden case ▸ **grandmother** n female grandparent ▸ **grandparent** n parent of one's parent ▸ **grand piano** large harp-shaped piano with the strings set horizontally ▸ **grand slam** winning of all the games or major tournaments in a sport in one season ▸ **grandson** n male grandchild ▸ **grandstand** n terraced block of seats giving the best view at a sports ground

grandee n person of high station

grandeur n magnificence; nobility or dignity

grandiloquent adj using pompous language ▸ **grandiloquence** n

grandiose adj imposing; pretentiously grand ▸ **grandiosity** n

grange n (Brit) country house with farm buildings

granite [gran-nit] n very hard igneous rock often used in building

granny, grannie n, pl **-nies** (informal) grandmother ▸ **granny flat** flat in or added to a house, suitable for an elderly parent

grant v consent to fulfil (a request); give formally; admit ▸ n sum of money provided by a government for a specific purpose, such as education **take for granted** accept as true without proof; take advantage of without due appreciation

granule n small grain ▸ **granular** adj of or like grains ▸ **granulated** adj (of sugar) in the form of coarse grains

grape n small juicy green or purple berry, eaten raw or used to produce wine, raisins, currants, or sultanas ▸ **grapevine** n grape-bearing vine; (informal) unofficial way of spreading news

grapefruit n large round yellow citrus fruit

graph n drawing showing the relation of different numbers or quantities plotted against a set of axes

graphic adj vividly descriptive; of or using drawing, painting, etc. ▸ **graphics** pl n diagrams, graphs, etc., esp. as used on a television programme or computer screen ▸ **graphically** adv

graphite n soft black form of carbon, used in pencil leads

graphology n study of handwriting ▸ **graphologist** n

grapnel n device with several hooks, used to grasp or secure things

grapple v try to cope with (something difficult); come to grips with (a person) ▸ **grappling iron** grapnel

grasp v grip something firmly; understand; try to seize ▸ n grip or clasp; understanding; total rule or possession ▸ **grasping** adj greedy or avaricious

grass n common type of plant with jointed stems and long narrow leaves, including cereals and bamboo; lawn; pasture land; (slang) marijuana; (Brit slang) person who informs, esp. on criminals ▸ v cover with grass; (often foll by **on**) (Brit slang) inform on ▸ **grassy** adj **-sier, -siest** ▸ **grasshopper** n jumping insect with long hind legs ▸ **grass roots** ordinary members of a group, as distinct from its leaders; essentials ▸ **grassroots** adj ▸ **grass tree** Australian plant with stiff grasslike leaves and small white flowers

grate¹ v rub into small bits on a rough surface; scrape with a harsh rasping noise; annoy ▸ **grater** n

▸ **grating** adj harsh or rasping; annoying

grate² n framework of metal bars for holding fuel in a fireplace ▸ **grating** n framework of metal bars covering an opening

grateful adj feeling or showing gratitude ▸ **gratefully** adv

gratify v **-fying, -fied** satisfy or please; indulge (a desire or whim) ▸ **gratification** n

gratis adv, adj free, for nothing

gratitude n feeling of being thankful for a favour or gift

gratuitous adj unjustified e.g. gratuitous violence; given free ▸ **gratuitously** adv

gratuity n, pl **-ties** money given for services rendered; tip

grave¹ n hole for burying a corpse ▸ **gravestone** n stone marking a grave ▸ **graveyard** n cemetery

grave² adj causing concern; serious and solemn ▸ **gravely** adv

grave³ [rhymes with **halve**] n accent (ˋ) over a vowel to indicate a special pronunciation

gravel n mixture of small stones and coarse sand ▸ **gravelled** adj covered with gravel ▸ **gravelly** adj covered with gravel; rough-sounding

graven adj carved or engraved

gravid [grav-id] adj (Med) pregnant

gravitate v be influenced or drawn towards; (Physics) move by gravity ▸ **gravitation** n ▸ **gravitational** adj

gravity n, pl **-ties** force of attraction of one object for another, esp. of objects to the earth; seriousness or importance; solemnity

gravy n, pl **-vies** juices from meat in cooking; sauce made from these

gray adj (chiefly US) grey

grayling n fish of the salmon family

graze¹ v feed on grass

graze² v scratch or scrape the skin; touch lightly in passing ▷ n slight scratch or scrape

grease n soft melted animal fat; any thick oily substance ▷ v apply grease to ▶ **greasy** adj **greasier, greasiest** covered with or containing grease ▶ **greasiness** n ▶ **greasepaint** n theatrical make-up

great adj large in size or number; important; pre-eminent; (informal) excellent ▶ **great-** prefix one generation older or younger than e.g. great-grandfather ▶ **greatly** adv ▶ **greatness** n ▶ **greatcoat** n heavy overcoat ▶ **Great Dane** very large dog with short smooth coat

greave n piece of armour for the shin

grebe n diving water bird

Grecian [gree-shan] adj of ancient Greece

greed n excessive desire for food, wealth, etc. ▶ **greedy** adj ▶ **greedily** adv ▶ **greediness** n

Greek n language of Greece; person from Greece ▷ adj of Greece, the Greeks, or the Greek language

green adj of a colour between blue and yellow; characterized by green plants or foliage; **(G-)** of or concerned with environmental issues; unripe; envious or jealous; immature or gullible ▷ n colour between blue and yellow; area of grass kept for a special purpose; **(G-)** person concerned with environmental issues ▷ pl green vegetables ▷ v make or become green ▶ **greenness** n ▶ **greenish, greeny** adj ▶ **greenery** n vegetation ▶ **green belt** protected area of open country around a town ▶ **greenfinch** n European finch with dull green plumage in the male ▶ **green fingers** skill in gardening ▶ **greenfly** n green aphid, a common garden pest ▶ **greengage** n sweet green plum ▶ **greengrocer** n (Brit) shopkeeper selling vegetables and fruit ▶ **greenhorn** n (chiefly US) novice ▶ **greenhouse** n glass building for rearing plants ▶ **greenhouse effect** rise in the temperature of the earth caused by heat absorbed from the sun being unable to leave the atmosphere ▶ **greenhouse gas** any gas that contributes to the greenhouse effect ▶ **green light** signal to go; permission to proceed with something ▶ **greenshank** n large European sandpiper ▶ **greenstone** n (NZ) type of green jade used for Māori ornaments ▶ **green tax** tax imposed with the aim of regulating activity in a way that benefits the environment

greet v meet with expressions of welcome; receive in a specified manner; be immediately noticeable to ▶ **greeting** n

gregarious adj fond of company; (of animals) living in flocks or herds

gremlin n imaginary being blamed for mechanical malfunctions

grenade n small bomb thrown by hand or fired from a rifle ▶ **grenadier** n soldier of a regiment formerly trained to throw grenades

grenadine [gren-a-deen] n syrup made from pomegranates

grevillea n any of various Australian evergreen trees and shrubs

grew v past tense of **grow**

grey adj of a colour between black and white; (of hair) partly turned white; dismal or dark; dull or boring ▷ n grey colour; grey or white horse ▶ **greyed out** adj (of an item on a computer screen) unavailable ▸ **greying** adj (of hair) turning grey ▸ **greyish** adj ▸ **greyness** n ▸ **grey matter** (informal) brains

greyhound n swift slender dog used in racing

grid n network of horizontal and vertical lines, bars, etc.; national network of electricity supply cables

griddle n flat iron plate for cooking

gridiron n frame of metal bars for grilling food; American football pitch

gridlock n situation where traffic is not moving; point in a dispute at which no agreement can be reached ▸ **gridlocked** adj

grief n deep sadness ▸ **grieve** v (cause to) feel grief ▸ **grievance** n real or imaginary cause for complaint ▸ **grievous** adj very severe or painful; very serious

griffin n mythical monster with an eagle's head and wings and a lion's body

grill n device on a cooker that radiates heat downwards; grilled food; gridiron ▸ v cook under a grill; question relentlessly ▸ **grilling** n relentless questioning

grille, grill n grating over an opening

grilse [grilss] n salmon on its first return from the sea to fresh water

grim adj grimmer, grimmest stern; harsh or forbidding; very unpleasant ▸ **grimly** adv ▸ **grimness** n

grimace n ugly or distorted facial expression of pain, disgust, etc. ▸ v make a grimace

grime n ingrained dirt ▸ v make very dirty ▸ **grimy** adj

grin v grinning, grinned smile broadly, showing the teeth ▸ n broad smile

grind v grinding, ground crush or rub to a powder; smooth or sharpen by friction; scrape together with a harsh noise; oppress ▸ n (informal) hard work; act or sound of grinding ▸ **grind out** v produce in a routine or uninspired manner ▸ **grindstone** n stone used for grinding

grip n firm hold or grasp; way in which something is grasped; mastery or understanding; (US) travelling bag; handle ▸ v gripping, gripped grasp or hold tightly; hold the interest or attention of ▸ **gripping** adj

gripe v (informal) complain persistently ▸ n (informal) complaint; sudden intense bowel pain

grisly adj -lier, -liest horrifying or ghastly

grist n grain for grinding **grist to one's mill** something which can be turned to advantage

gristle n tough stringy animal tissue found in meat ▸ **gristly** adj

grit n rough particles of sand; courage ▸ pl coarsely ground grain ▸ v gritting, gritted spread grit on (an icy road etc.); clench or grind (the teeth) ▸ **gritty** adj -tier, -tiest ▸ **grittiness** n

grizzle v (Brit, Aust & NZ informal) whine or complain

grizzled adj grey-haired

grizzly n, pl -zlies large American bear (also **grizzly bear**)

groan n deep sound of grief or pain; (informal) complaint ▸ v utter a groan; (informal) complain

groat n (Hist) fourpenny piece

grocer n shopkeeper selling foodstuffs ▸ **grocery** n, pl -ceries business or premises of a grocer ▸ pl goods sold by a grocer

grog n (Brit, Aust & NZ) spirit, usu. rum, and water

groggy adj **-gier, -giest** (informal) faint, shaky, or dizzy

groin n place where the legs join the abdomen

grommet n ring or eyelet; (Med) tube inserted in the ear to drain fluid from the middle ear

groom n person who looks after horses; bridegroom; officer in a royal household ▷ v make or keep one's clothes and appearance neat and tidy; brush or clean a horse; train (someone) for a future role

groove n long narrow channel in a surface

grope v feel about or search uncertainly; (slang) fondle (someone) in a rough sexual way ▶ **groping** n

gross adj flagrant; vulgar; (slang) disgusting or repulsive; repulsively fat; total, without deductions ▷ n twelve dozen ▷ v make as total revenue before deductions ▶ **grossly** adv ▶ **grossness** n

grotesque [grow-**tesk**] adj strangely distorted; absurd ▷ n grotesque person or thing; artistic style mixing distorted human, animal, and plant forms ▶ **grotesquely** adv

grotto n, pl **-toes** or **-tos** small picturesque cave

grotty adj **-tier, -tiest** (informal) nasty or in bad condition

grouch (informal) v grumble or complain ▷ n person who is always complaining; persistent complaint ▶ **grouchy** adj

ground[1] n surface of the earth; soil; area used for a specific purpose e.g. rugby ground; position in an argument or controversy ▷ pl enclosed land round a house; reason or motive; coffee dregs

▷ v base or establish; instruct in the basics; ban an aircraft or pilot from flying; run (a ship) aground ▶ **groundless** adj without reason ▶ **grounding** n basic knowledge of a subject ▶ **ground-breaking** adj innovative ▶ **ground floor** floor of a building level with the ground ▶ **groundnut** n (Brit) peanut ▶ **groundsheet** n waterproof sheet put on the ground under a tent ▶ **groundsman** n person employed to maintain a sports ground or park ▶ **groundswell** n rapidly developing general feeling or opinion ▶ **groundwork** n preliminary work

ground[2] v past of **grind**

group n number of people or things regarded as a unit; small band of musicians or singers ▷ v place or form into a group

grouse[1] n stocky game bird; its flesh

grouse[2] v grumble or complain ▷ n complaint

grout n thin mortar ▷ v fill up with grout

grove n small group of trees

grovel [grov-el] v **-elling, -elled** behave humbly in order to win a superior's favour; crawl on the floor

grow v **growing, grew, grown** develop physically; (of a plant) exist; cultivate (plants); increase in size or degree; originate; become gradually e.g. it was growing dark ▶ **growth** n growing; increase; something grown or growing; tumour ▶ **grown-up** adj, n adult ▶ **grow up** v mature

growl v make a low rumbling sound; utter with a growl ▷ n growling sound

groyne n wall built out from the shore to control erosion

grub n legless insect larva; (slang) food ▷ v **grubbing, grubbed**

search carefully for something by digging or by moving things about; dig up the surface of (soil)

grubby adj **-bier, -biest** dirty
▶ **grubbiness** n

grudge v be unwilling to give or allow ▶ n resentment

gruel n thin porridge

gruelling adj exhausting or severe

gruesome adj causing horror and disgust

gruff adj rough or surly in manner or voice ▶ **gruffly** adv
▶ **gruffness** n

grumble v complain; rumble ▶ n complaint; rumble ▶ **grumbler** n
▶ **grumbling** adj, n

grumpy adj **grumpier, grumpiest** bad-tempered ▶ **grumpily** adv
▶ **grumpiness** n

grunge n style of rock music with a fuzzy guitar sound; deliberately untidy and uncoordinated fashion style

grunt v make a low short gruff sound, like a pig ▶ n pig's sound; gruff noise

Gruyère [grew-yair] n hard yellow Swiss cheese with holes

gryphon n same as **griffin**

GST (in Australia, New Zealand, and Canada) Goods and Services Tax

G-string n small strip of cloth covering the genitals and attached to a waistband

GT (of a sports car) gran turismo

guano [gwah-no] n dried sea-bird manure, used as fertilizer

guarantee n formal assurance, esp. in writing, that a product will meet certain standards; something that makes a specified condition or outcome certain ▶ v **-teeing, -teed** give a guarantee; secure against risk etc.; ensure ▶ **guarantor** n person who gives or is bound by a guarantee

guard v watch over to protect or to prevent escape ▶ n person or group that guards; official in charge of a train; protection; screen for enclosing anything dangerous; posture of defence in sports such as boxing or fencing ▶ pl **(G-)** regiment with ceremonial duties ▶ **guarded** adj cautious or noncommittal
▶ **guardedly** adv ▶ **guard against** v take precautions against
▶ **guardsman** n member of the Guards

guardian n keeper or protector; person legally responsible for a child or vulnerable adult
▶ **guardianship** n

guava [gwah-va] n yellow-skinned tropical American fruit

gudgeon n small freshwater fish

Guernsey [gurn-zee] n breed of dairy cattle

guerrilla, guerilla n member of an unofficial armed force fighting regular forces

guess v estimate or draw a conclusion without proper knowledge; estimate correctly by guessing; suppose ▶ n estimate or conclusion reached by guessing
▶ **guesswork** n process or results of guessing

guest n person entertained at another's house or at another's expense; invited performer or speaker; customer at a hotel or restaurant ▶ v appear as a visiting player or performer ▶ **guesthouse** n boarding house

guff n (Brit, Aust & NZ slang) nonsense

guffaw n crude noisy laugh ▶ v laugh in this way

guide n person who conducts tour expeditions; person who shows the way; book of instruction or information; model for behaviour;

something used to gauge something or to help in planning one's actions; **(G-)** member of an organization for girls equivalent to the Scouts ▷ v act as a guide for; control, supervise, or influence ▶ **guidance** n leadership, instruction, or advice ▶ **guided missile** missile whose flight is controlled electronically ▶ **guide dog** dog trained to lead a blind person ▶ **guideline** n set principle for doing something

guild n organization or club; (Hist) society of men in the same trade or craft

guilder n former monetary unit of the Netherlands

guile [gile] n cunning or deceit ▶ **guileful** adj ▶ **guileless** adj

guillemot [gil-lee-mot] n black-and-white diving sea bird of N hemisphere

guillotine n machine for beheading people; device for cutting paper or sheet metal; method of preventing lengthy debate in parliament by fixing a time for taking the vote ▷ v behead by guillotine; limit debate by the guillotine

guilt n fact or state of having done wrong; remorse for wrongdoing ▶ **guiltless** adj innocent ▶ **guilty** adj responsible for an offence or misdeed; feeling or showing guilt ▶ **guiltily** adv

guinea n former British monetary unit worth 21 shillings (1.05 pounds); former gold coin of this value ▶ **guinea fowl** wild bird related to the pheasant ▶ **guinea pig** tailless S American rodent, commonly kept as a pet; (informal) person used for experimentation

guise [rhymes with size] n false appearance; external appearance

guitar n stringed instrument with a flat back and a long neck, played by plucking or strumming ▶ **guitarist** n

gulch n (US) deep narrow valley

gulf n large deep bay; chasm; large difference in opinion or understanding

gull n long-winged sea bird

gullet n muscular tube through which food passes from the mouth to the stomach

gullible adj easily tricked ▶ **gullibility** n

gully n, pl -lies channel cut by running water

gulp v swallow hastily; gasp ▷ n gulping; thing gulped

gum[1] n firm flesh in which the teeth are set ▶ **gummy** adj -mier, -miest toothless

gum[2] n sticky substance obtained from certain trees; adhesive; chewing gum; gumdrop; gum tree ▷ v **gumming, gummed** stick with gum ▶ **gummy** adj -mier, -miest ▶ **gumboots** pl n (chiefly Brit) Wellington boots ▶ **gumdrop** n hard jelly-like sweet ▶ **gum tree** eucalyptus tree

gumption n (informal) resourcefulness; courage

gun n weapon with a metal tube from which missiles are fired by explosion; device from which a substance is ejected under pressure ▷ v **gunning, gunned** cause (an engine) to run at high speed **jump the gun** act prematurely ▶ **gunner** n artillery soldier ▶ **gunnery** n use or science of large guns ▶ **gunboat** n small warship ▶ **gun dog** dog used to retrieve game ▶ **gun for** v seek or pursue vigorously ▶ **gunman** n armed criminal ▶ **gunmetal** n alloy of copper, tin, and zinc

g

▷ *adj* dark grey ▶ **gunpowder** *n* explosive mixture of potassium nitrate, sulphur, and charcoal ▶ **gunrunning** *n* smuggling of guns and ammunition ▶ **gunrunner** *n* ▶ **gunshot** *n* shot or range of a gun

gunge *n* (*informal*) sticky unpleasant substance ▶ **gungy** *adj* **-gier, -giest**

gunny *n* strong coarse fabric used for sacks

gunwale, gunnel [gun-nel] *n* top of a ship's side

gunyah *n* (*Aust*) hut or shelter in the bush

guppy *n*, *pl* **-pies** small colourful aquarium fish

gurgle *v*, *n* (make) a bubbling noise

Gurkha *n* person, esp. a soldier, belonging to a Hindu people of Nepal

guru *n* Hindu or Sikh religious teacher or leader; leader, adviser, or expert

gush *v* flow out suddenly and profusely; express admiration effusively ▷ *n* sudden copious flow; sudden surge of strong feeling ▶ **gusher** *n* spurting oil well

gusset *n* piece of material sewn into a garment to strengthen it

gust *n* sudden blast of wind ▷ *v* blow in gusts ▶ **gusty** *adj*

gusto *n* enjoyment or zest

gut *n* intestine; (*informal*) fat stomach; short for **catgut** ▷ *pl* internal organs; (*informal*) courage ▷ *v* **gutting, gutted** remove the guts from; (of a fire) destroy the inside of (a building) ▷ *adj* basic or instinctive e.g. *a gut reaction* ▶ **gutsy** *adj* **-sier, -siest** (*informal*) courageous, vigorous or robust e.g. *a gutsy performance* ▶ **gutted** *adj* (*Brit, Aust & NZ informal*) disappointed and upset

gutta-percha *n* whitish rubbery substance obtained from an Asian tree

gutter *n* shallow channel for carrying away water from a roof or roadside ▷ *v* (of a candle) burn unsteadily, with wax running down the sides **the gutter** degraded or criminal environment ▶ **guttering** *n* material for gutters ▶ **gutter press** newspapers that rely on sensationalism ▶ **guttersnipe** *n* (*Brit*) neglected slum child

guttural *adj* (of a sound) produced at the back of the throat; (of a voice) harsh-sounding

guy¹ *n* (*informal*) man or boy; effigy of Guy Fawkes burnt on November 5 (**Guy Fawkes Day**)

guy² *n* rope or chain to steady or secure something ▶ **guy rope**

guzzle *v* eat or drink greedily

gybe [**jibe**] *v* (of a fore-and-aft sail) swing suddenly from one side to the other; (of a boat) change course by letting the sail gybe

gym *n* gymnasium; gymnastics

gymkhana [jim-kah-na] *n* horse-riding competition

gymnasium *n* large room with equipment for physical training ▶ **gymnast** *n* expert in gymnastics ▶ **gymnastic** *adj* ▶ **gymnastics** *pl n* exercises to develop strength and agility

gynaecology [guy-nee-kol-la-jee] *n* branch of medicine dealing with diseases and conditions specific to women ▶ **gynaecological** *adj* ▶ **gynaecologist** *n*

gypsophila *n* garden plant with small white flowers

gypsum *n* chalklike mineral used to make plaster of Paris

Gypsy *n*, *pl* **-sies** member of a travelling people found throughout Europe

gyrate [jire-**rate**] v rotate or spiral about a point or axis ▶ **gyration** n ▶ **gyratory** adj gyrating
gyrocompass n compass using a gyroscope
gyroscope [jire-oh-**skohp**] n disc rotating on an axis that can turn in any direction, so the disc maintains the same position regardless of the movement of the surrounding structure ▶ **gyroscopic** adj

Hh

H (Chem) hydrogen
habeas corpus [hay-bee-ass **kor**-puss] n writ ordering a prisoner to be brought before a court
haberdasher n (Brit, Aust & NZ) dealer in small articles used for sewing ▶ **haberdashery** n
habit n established way of behaving; addiction to a drug; costume of a monk or nun
habitable adj fit to be lived in ▶ **habitation** n (occupation of) a dwelling place
habitat n natural home of an animal or plant
habitual adj done regularly and repeatedly ▶ **habitually** adv
habituate v accustom ▶ **habituation** n ▶ **habitué** [hab-**it**-yew-ay] n frequent visitor to a place
hacienda [hass-ee-**end**-a] n ranch or large estate in Latin America
hack¹ v cut or chop violently; (Brit & NZ informal) tolerate
hack² n (inferior) writer or journalist; horse kept for riding
hacker n computer user who breaks into the computer system of a company or government
hackles pl n **make one's hackles rise** make one feel angry or hostile
hackney n (Brit) taxi
hackneyed adj (of a word or phrase) unoriginal and overused
hacksaw n small saw for cutting metal
had v past of **have**

haddock n edible sea fish of N Atlantic

Hades [hay-deez] n (Greek myth) underworld home of the dead

hadj n same as **hajj**

haematology n study of blood and its diseases

haemoglobin [hee-moh-globe-in] n protein found in red blood cells which carries oxygen

haemophilia [hee-moh-fill-lee-a] n hereditary illness in which the blood does not clot ▶ **haemophiliac** n

haemorrhage [hem-or-ij] n heavy bleeding ▷ v bleed heavily

> **SPELLING TIP**
> The Collins Corpus shows that the most usual mistake in spelling **haemorrhage** is to miss out the second h, which is silent

haemorrhoids [hem-or-oydz] pl n swollen veins in the anus (also **piles**)

hafnium n (Chem) metallic element found in zirconium ores

haft n handle of an axe, knife, or dagger

hag n ugly old woman ▶ **hag-ridden** adj distressed or worried

haggard adj looking tired and ill

haggis n Scottish dish made from sheep's offal, oatmeal, suet, and seasonings, boiled in a bag made from the sheep's stomach

haggle v bargain or wrangle over a price

hagiography n, pl -phies writing about the lives of the saints

hail¹ n (shower of) small pellets of ice; large number of insults, missiles, blows, etc. ▷ v fall as or like hail ▶ **hailstone** n

hail² v call out to; greet; stop (a taxi) by waving; acknowledge publicly ▶ **hail from** v come originally from

hair n threadlike growth on the skin; such growths collectively, esp. on the head ▶ **hairy** adj covered with hair; (slang) dangerous or exciting ▶ **hairiness** n ▶ **hairclip** n small bent metal hairpin ▶ **hairdo** n (informal) hairstyle ▶ **hairdresser** n person who cuts and styles hair ▶ **hairgrip** n (Brit) same as **hairclip** ▶ **hairline** n edge of hair at the top of the forehead ▷ adj very fine or narrow ▶ **hairpin** n U-shaped wire used to hold the hair in place ▶ **hairpin bend** very sharp bend in a road ▶ **hair-raising** adj frightening or exciting ▶ **hair-splitting** n, adj making petty distinctions ▶ **hairstyle** n cut and arrangement of a person's hair

hajj n pilgrimage a Muslim makes to Mecca

haka n (NZ) ceremonial Māori dance with chanting; similar dance performed by a sports team before a match

hake n edible sea fish of N hemisphere; (Aust) same as **barracouta**

hakea [hah-kee-a] n Australian tree or shrub with hard woody fruit

halal n meat from animals slaughtered according to Muslim law

halberd n (Hist) spear with an axe blade

halcyon [hal-see-on] adj peaceful and happy ▶ **halcyon days** time of peace and happiness

hale adj healthy, robust

half n, pl **halves** either of two equal parts; (informal) half-pint of beer etc.; half-price ticket ▷ adj denoting one of two equal parts ▷ adv to the extent of half; partially ▶ **half-baked** adj (informal) not properly thought out ▶ **half-brother,**

half-sister n brother or sister related through one parent only
▶ **half-caste** adj (offens) person with parents of different races
▶ **half-cocked** adj **go off half-cocked, go off at half-cock** fail because of inadequate preparation
▶ **half-hearted** adj unenthusiastic
▶ **half-life** n time taken for half the atoms in radioactive material to decay ▶ **half-nelson** n wrestling hold in which one wrestler's arm is pinned behind his back by his opponent ▶ **half-pie** adj (NZ informal) incomplete ▶ **half-pipe** n large U-shaped ramp used for skateboarding, snowboarding, etc.
▶ **half-timbered** adj (of a house) having an exposed wooden frame filled in with plaster ▶ **half-time** n (Sport) short rest period between two halves of a game ▶ **halftone** n illustration showing lights and shadows by means of very small dots ▶ **halfway** adv, adj at or to half the distance ▶ **halfwit** n foolish or stupid person

halfpenny [hayp-nee] n former British coin worth half an old penny

halibut n large edible flatfish of N Atlantic

halitosis n unpleasant-smelling breath

hall n (also **hallway**) entrance passage; large room or building for public meetings, dances, etc.; (Brit) large country house

hallelujah [hal-ee-loo-ya] interj exclamation of praise to God

hallmark n typical feature; mark indicating the standard of tested gold and silver ▷ v stamp with a hallmark

hallo interj same as **hello**

hallowed adj regarded as holy

Halloween, Hallowe'en n October 31, celebrated by children by dressing up as ghosts, witches, etc.

hallucinate v seem to see something that is not really there ▶ **hallucination** n
▶ **hallucinatory** adj
▶ **hallucinogen** n drug that causes hallucinations ▶ **hallucinogenic** adj

halo [hay-loh] n, pl **-loes** or **-los** ring of light round the head of a sacred figure; circle of refracted light round the sun or moon

halogen [hal-oh-jen] n (Chem) any of a group of nonmetallic elements including chlorine and iodine

halt v come or bring to a stop ▷ n temporary stop; minor railway station without a building
▶ **halting** adj hesitant, uncertain

halter n strap round a horse's head with a rope to lead it with
▶ **halterneck** n woman's top or dress with a strap fastened at the back of the neck

halve v divide in half; reduce by half

halves n plural of **half**

halyard n rope for raising a ship's sail or flag

ham¹ n smoked or salted meat from a pig's thigh ▶ **ham-fisted** adj clumsy

ham² (informal) n amateur radio operator; actor who overacts ▷ v **hamming, hammed** ▶ **ham it up** overact

hamburger n minced beef shaped into a flat disc, cooked and usu. served in a bread roll

hamlet n small village

hammer n tool with a heavy metal head and a wooden handle, used to drive in nails etc.; part of a gun which causes the bullet to be fired; heavy metal ball on a wire, thrown as a sport; auctioneer's mallet; striking mechanism in a piano ▷ v hit (as if) with a hammer; (informal)

punish or defeat utterly **go at it**
hammer and tongs do something,
esp. argue, very vigorously
▶ **hammerhead** n shark with a
wide flattened head ▶ **hammer
toe** condition in which a toe is
permanently bent at the joint
hammock n hanging bed made of
canvas or net
hamper[1] v make it difficult for
(someone or something) to move
or progress
hamper[2] n large basket with a lid;
selection of food and drink packed
as a gift
hamster n small rodent with a
short tail and cheek pouches

> **SPELLING TIP**
> Spelling **hamster** as **hampster**
> is a common mistake. There is
> no p

hamstring n tendon at the back
of the knee ▷ v make it difficult for
(someone) to take any action
hand n part of the body at the end
of the arm, consisting of a palm,
four fingers, and a thumb; style of
handwriting; round of applause;
manual worker; pointer on a
dial, esp. on a clock; cards dealt
to a player in a card game; unit
of length of four inches (10.16
centimetres) used to measure
horses ▷ v pass, give **have a hand
in** be involved in **lend a hand**
help **out of hand** beyond control;
definitely and finally; **to hand, at
hand, on hand** nearby **win hands
down** win easily ▶ **handbag** n
woman's small bag for carrying
personal articles in ▶ **handbill** n
small printed notice ▶ **handbook**
n small reference or instruction
book ▶ **handcuff** n one of a linked
pair of metal rings designed to be
locked round a prisoner's wrists
by the police ▷ v put handcuffs on

▶ **hand-held** adj (of a film camera)
held rather than mounted, as
in close-up action shots; (of a
computer) able to be held in the
hand ▷ n computer that can be
held in the hand ▶ **hand-out** n
clothing, food, or money given to a
needy person; written information
given out at a talk etc. ▶ **hands-
free** adj, n (of) a device allowing
the user to make and receive
phonecalls without holding the
handset ▶ **hands-on** adj involving
practical experience of equipment
▶ **handstand** n act of supporting
the body on the hands in an upside-
down position ▶ **handwriting** n
(style of) writing by hand
handful n amount that can be
held in the hand; small number;
(informal) person or animal that is
difficult to control
handicap n (old-fashioned, offensive)
physical or mental disability;
something that makes progress
difficult; contest in which the
competitors are given advantages
or disadvantages in an attempt to
equalize their chances; advantage
or disadvantage given ▷ v make
it difficult for (someone) to do
something
handicraft n objects made by hand
handiwork n result of someone's
work or activity
handkerchief n small square of
fabric used to wipe the nose

> **SPELLING TIP**
> People often forget to write a
> d in **handkerchief**, probably
> because they don't say it or hear
> it either

handle n part of an object that
is held so that it can be used ▷ v
hold, feel, or move with the hands;
control or deal with ▶ **handler** n
person who controls an animal

▶ **handlebars** pl n curved metal bar used to steer a cycle

handsome adj (esp. of a man) good-looking; large or generous e.g. *a handsome profit*

handy adj **handier, handiest** convenient, useful; good at manual work ▶ **handily** adv ▶ **handyman** n man who is good at making or repairing things

hang v **hanging, hung** attach or be attached at the top with the lower part free; (*past tense* **hanged**) suspend or be suspended by the neck until dead; fasten to a wall **get the hang of** (informal) begin to understand ▶ **hanger** n curved piece of wood, wire, or plastic, with a hook, for hanging up clothes (also **coat hanger**) ▶ **hang back** v hesitate, be reluctant ▶ **hangman** n man who executes people by hanging ▶ **hangover** n headache and nausea as a result of drinking too much alcohol ▶ **hang-up** n (informal) emotional or psychological problem

hangar n large shed for storing aircraft

hangdog adj guilty, ashamed e.g. *a hangdog look*

hang-glider n glider with a light framework from which the pilot hangs in a harness ▶ **hang-gliding** n

hangi n, **-gi** or **-gis** (NZ) Māori oven consisting of a hole in the ground filled with hot stones

hank n coil, esp. of yarn

hanker v (foll by *after* or *for*) desire intensely

hanky, hankie n, pl **hankies** (informal) handkerchief

hanky-panky n (informal) illicit sexual relations

hansom cab n (formerly) two-wheeled horse-drawn carriage for hire

haphazard adj not organized or planned ▶ **haphazardly** adv

hapless adj unlucky

happen v take place; occur; chance (to be or do something) ▶ **happening** n event, occurrence

happy adj **-pier, -piest** feeling or causing joy; lucky, fortunate ▶ **happily** adv ▶ **happiness** n ▶ **happy-go-lucky** adj carefree and cheerful

hara-kiri n (formerly, in Japan) ritual suicide by disembowelling

harangue v address angrily or forcefully ▷ n angry or forceful speech

harass v annoy or trouble constantly ▶ **harassed** adj ▶ **harassment** n

> **SPELLING TIP**
> Note there is only one r but two ss in **harass** and its derivatives

harbinger [har-binj-a] n someone or something that announces the approach of something

harbour n sheltered port ▷ v maintain secretly in the mind; give shelter or protection to

hard adj firm, solid, or rigid; difficult; requiring a lot of effort; unkind, unfeeling; causing pain, sorrow, or hardship; (of water) containing calcium salts which stop soap lathering freely; (of a drug) strong and addictive ▷ adv with great energy or effort; with great intensity **hard of hearing** unable to hear properly **hard up** (informal) short of money ▶ **harden** v ▶ **hardness** n ▶ **hardship** n suffering; difficult circumstances ▶ **hard-bitten** adj tough and determined ▶ **hard-boiled** adj (of an egg) boiled until solid; (informal) tough, unemotional ▶ **hard copy** computer output printed on paper ▶ **hardfill** n (NZ & S Afr) stone waste

h

material used for landscaping ▶ **hard-headed** adj shrewd, practical ▶ **hardhearted** adj unsympathetic, uncaring ▶ **hard sell** aggressive sales technique ▶ **hard shoulder** surfaced verge at the edge of a motorway for emergency stops

hardback n book with a stiff cover

hardboard n thin stiff board made of compressed sawdust and wood chips

hard drive n (Computing) mechanism that handles the reading, writing, and storage of data on the hard disk of a computer

hardly adv scarcely or not at all; with difficulty

hardware n metal tools or implements; machinery used in a computer system; heavy military equipment, such as tanks and missiles

hardwood n wood of a broadleaved tree such as oak or ash

hardy adj **hardier, hardiest** able to stand difficult conditions ▶ **hardiness** n

hare n animal like a large rabbit, with longer ears and legs ▷ v (usu. foll by off) run (away) quickly ▶ **harebell** n blue bell-shaped flower ▶ **harebrained** adj foolish or impractical ▶ **harelip** n slight split in the upper lip

harem n (apartments of) a Muslim man's wives and concubines

haricot bean [har-rik-oh] n small pale edible bean, usu. sold dried

harissa n (in North African cuisine) hot sauce made from chilli peppers, spices, and olive oil

hark v (old-fashioned) listen ▶ **hark back** v return (to an earlier subject)

harlequin n stock comic character with a diamond-patterned costume and mask ▷ adj in many colours

harlot n (lit) prostitute

harm v injure physically, mentally, or morally ▷ n physical, mental, or moral injury ▶ **harmful** adj ▶ **harmless** adj

harmonica n small wind instrument played by sucking and blowing

harmonium n keyboard instrument like a small organ

harmony n, pl **-nies** peaceful agreement and cooperation; pleasant combination of notes sounded at the same time ▶ **harmonious** adj ▶ **harmoniously** adv ▶ **harmonic** adj of harmony ▶ **harmonics** n science of musical sounds ▶ **harmonize** v blend well together ▶ **harmonization** n

harness n arrangement of straps for attaching a horse to a cart or plough; set of straps fastened round someone's body to attach something e.g. a safety harness ▷ v put a harness on; control (something) in order to make use of it

harp n large triangular stringed instrument played with the fingers ▶ **harpist** n ▶ **harp on about** v talk about continuously

harpoon n barbed spear attached to a rope used for hunting whales ▷ v spear with a harpoon

harpsichord n stringed keyboard instrument

harpy n, pl **-pies** nasty or bad-tempered woman

harridan n nagging or vicious woman

harrier n cross-country runner

harrow n implement used to break up lumps of soil ▷ v draw a harrow over

harrowing adj very distressing

harry v **-rying, -ried** keep asking (someone) to do something, pester

harsh adj severe and difficult to cope with; unkind, unsympathetic; extremely hard, bright, or rough ▸ **harshly** adv ▸ **harshness** n

hart n adult male deer

harum-scarum adj reckless

harvest n (season for) the gathering of crops; crops gathered ▷ v gather (a ripened crop) ▸ **harvester** n

has v third person singular of the present tense of **have** ▸ **has-been** n (informal) person who is no longer popular or successful

hash¹ n dish of diced cooked meat and vegetables reheated **make a hash of** (informal) spoil, do badly

hash² n (informal) hashish

hashish [hash-eesh] n drug made from the cannabis plant, smoked for its intoxicating effects

hashtag n (Computing) (on the Twitter website) word or phrase preceded by a hash mark, used to denote the topic of a post

hasp n clasp that fits over a staple and is secured by a bolt or padlock, used as a fastening

hassle (informal) n trouble, bother ▷ v bother or annoy

hassock n cushion for kneeling on in church

haste n (excessive) quickness **make haste** hurry, rush ▸ **hasten** v (cause to) hurry ▸ **hasty** adj (too) quick ▸ **hastily** adv

hat n covering for the head, often with a brim, usu. worn to give protection from the weather **keep something under one's hat** keep something secret ▸ **hat trick** any three successive achievements, esp. in sport

hatch¹ v (cause to) emerge from an egg; devise (a plot)

hatch² n hinged door covering an opening in a floor or wall; opening

in the wall between a kitchen and a dining area; door in an aircraft or spacecraft ▸ **hatchback** n car with a lifting door at the back ▸ **hatchway** n opening in the deck of a ship

hatchet n small axe **bury the hatchet** become reconciled ▸ **hatchet job** malicious verbal or written attack ▸ **hatchet man** (informal) person carrying out unpleasant tasks for an employer

hate v dislike intensely; be unwilling (to do something) ▷ n intense dislike; person or thing hated ▸ **hateful** adj causing or deserving hate ▸ **hater** n ▸ **hatred** n intense dislike

haughty adj -tier, -tiest proud, arrogant ▸ **haughtily** adv ▸ **haughtiness** n

haul v pull or drag with effort ▷ n amount gained by effort or theft **long haul** something that takes a lot of time and effort ▸ **haulage** n (charge for) transporting goods ▸ **haulier** n firm or person that transports goods by road

haunch n human hip or fleshy hindquarter of an animal

haunt v visit in the form of a ghost; remain in the memory or thoughts of ▷ n place visited frequently ▸ **haunted** adj frequented by ghosts; worried ▸ **haunting** adj memorably beautiful or sad

haute couture [oat koo-**ture**] n (French) high fashion

hauteur [oat-**ur**] n haughtiness

have v has, having, had possess, hold; receive, take, or obtain; experience or be affected by; (foll by to) be obliged, e.g. I had to go; cause to be done; give birth to; used to form past tenses (with a past participle) e.g. we have looked; she had done enough **have**

it out (*informal*) settle a matter by argument ▸ **have on** *v* wear; (*informal*) tease or trick ▸ **have up** *v* bring to trial

haven *n* place of safety

haversack *n* canvas bag carried on the back or shoulder

havoc *n* disorder and confusion

haw *n* hawthorn berry

hawk[1] *n* bird of prey with a short hooked bill and very good eyesight; (*Politics*) supporter or advocate of warlike policies ▸ **hawkish, hawklike** *adj* ▸ **hawk-eyed** *adj* having very good eyesight

hawk[2] *v* offer (goods) for sale in the street or door-to-door ▸ **hawker** *n*

hawk[3] *v* cough noisily

hawser *n* large rope used on a ship

hawthorn *n* thorny shrub or tree

hay *n* grass cut and dried as fodder ▸ **hay fever** allergy to pollen, causing sneezing and watery eyes ▸ **haystack** *n* large pile of stored hay ▸ **haywire** *adj* **go haywire** (*informal*) not function properly

hazard *n* something that could be dangerous ▸ *v* put in danger; make (a guess) ▸ **hazardous** *adj*

haze *n* mist, often caused by heat ▸ **hazy** *adj* not clear, misty; confused or vague

hazel *n* small tree producing edible nuts ▸ *adj* (of eyes) greenish-brown ▸ **hazelnut** *n*

H-bomb *n* hydrogen bomb

he *pron* referring to male person or animal ▸ *n* male person or animal e.g. *a he-goat*

head *n* upper or front part of the body, containing the sense organs and the brain; mind and mental abilities; upper or most forward part of anything; person in charge of a group, organization, or school; pus-filled tip of a spot or boil; white froth on beer; (*pl* **head**) person or animal considered as a unit ▸ *adj* chief, principal ▸ *v* be at the top or front of; be in charge of; move (in a particular direction); hit (a ball) with the head; provide with a heading **go to one's head** make one drunk or conceited **head over heels (in love)** very much in love **not make head nor tail of** not understand **off one's head** (*slang*) foolish or insane ▸ **heads** *adv* (*informal*) with the side of a coin which has a portrait of a head on it uppermost ▸ **header** *n* striking a ball with the head; headlong fall ▸ **heading** *n* title written or printed at the top of a page ▸ **heady** *adj* intoxicating or exciting ▸ **headache** *n* continuous pain in the head; cause of worry or annoyance ▸ **headboard** *n* vertical board at the top end of a bed ▸ **headdress** *n* decorative head covering ▸ **head-hunt** *v* (of a company) approach and offer a job to (a person working for a rival company) ▸ **head-hunter** *n* ▸ **headland** *n* area of land jutting out into the sea ▸ **headlight** *n* powerful light on the front of a vehicle ▸ **headline** *n* title at the top of a newspaper article, esp. on the front page ▸ *pl* main points of a news broadcast ▸ **headlong** *adv*, *adj* with the head first; hastily ▸ **headphones** *pl n* two small loudspeakers held against the ears ▸ **headquarters** *pl n* centre from which operations are directed ▸ **head start** advantage in a competition ▸ **headstone** *n* memorial stone on a grave ▸ **headstrong** *adj* self-willed, obstinate ▸ **headway** *n* progress ▸ **headwind** *n* wind blowing against the course of an aircraft or ship

heal v make or become well ▶ **healer** n

health n normal (good) condition of someone's body ▶ **health food** natural food, organically grown and free from additives ▶ **healthy** adj having good health; of or producing good health; functioning well, sound ▶ **healthily** adv

heap n pile of things one on top of another; (also **heaps**) large number or quantity ▷ v gather into a pile; (foll by **on**) give liberally (to)

hear v **hearing, heard** perceive (a sound) by ear; listen to; learn or be informed; (Law) try (a case) **hear! hear!** exclamation of approval or agreement ▶ **hearer** n ▶ **hearing** ability to hear; trial of a case **within hearing** close enough to be heard

hearsay n gossip, rumour

hearse n funeral car used to carry a coffin

heart n organ that pumps blood round the body; centre of emotions, esp. love; courage, spirit; central or most important part; figure representing a heart; playing card with red heart-shaped symbols **break someone's heart** cause someone great grief **by heart** from memory **set one's heart on something** greatly desire something **take something to heart** be upset about something ▶ **hearten** v encourage, make cheerful ▶ **heartless** adj cruel, unkind ▶ **hearty** adj substantial, nourishing; friendly, enthusiastic ▶ **heartily** adv ▶ **heart attack** sudden severe malfunction of the heart ▶ **heart failure** sudden stopping of the heartbeat ▶ **heart-rending** adj causing great sorrow ▶ **heart-throb** n (slang)

very attractive man, esp. a film or pop star

heartache n intense anguish

heartbeat n one complete pulsation of the heart

heartbreak n intense grief

heartburn n burning sensation in the chest caused by indigestion

heartfelt adj felt sincerely or strongly

hearth n floor of a fireplace

heat v make or become hot ▷ n state of being hot; energy transferred as a result of a difference in temperature; hot weather; intensity of feeling; preliminary eliminating contest in a competition **on heat, in heat** (of some female animals) ready for mating ▶ **heated** adj angry and excited ▶ **heatedly** adv ▶ **heater** n

heath n (Brit) area of open uncultivated land

heathen adj, n (of) a person who does not believe in an established religion

heather n low-growing plant with small purple, pinkish, or white flowers, growing on heaths and mountains

heave v lift with effort; throw (something heavy); utter (a sigh); rise and fall; vomit ▷ n heaving

heaven n place believed to be the home of God, where good people go when they die; place or state of bliss **the heavens** sky ▶ **heavenly** adj of or like heaven; of or occurring in space; wonderful or beautiful

heavy adj **heavier, heaviest** of great weight; having a high density; great in degree or amount; (informal) (of a situation) serious ▶ **heavily** adv ▶ **heaviness** n ▶ **heavy industry** large-scale production of raw material or machinery ▶ **heavy metal** very loud rock music featuring guitar riffs ▶ **heavyweight** n boxer weighing over 175lb (professional) or 81kg (amateur)

Hebrew n member of an ancient Semitic people; ancient language of the Hebrews; its modern form, used in Israel ▷ adj of the Hebrews

heckle v interrupt (a public speaker) with comments, questions, or taunts ▶ **heckler** n

hectare n one hundred ares or 10 000 square metres (2.471 acres)

hectic adj rushed or busy

hector n bully

hedge n row of bushes forming a barrier or boundary ▷ v be evasive or noncommittal; (foll by against) protect oneself (from) ▶ **hedgerow** n bushes forming a hedge

hedgehog n small mammal with a protective covering of spines

hedonism n doctrine that pleasure is the most important thing in life ▶ **hedonist** n ▶ **hedonistic** adj

heed n careful attention ▷ v pay careful attention to ▶ **heedless** adj **heedless of** taking no notice of

heel[1] n back part of the foot; part of a shoe supporting the heel; (old-fashioned) contemptible person ▷ v repair the heel of (a shoe) ▶ **heeler** n (Aust & NZ) dog that herds cattle by biting at their heels

heel[2] v (foll by over) lean to one side

hefty adj **heftier, heftiest** large, heavy, or strong

hegemony [hig-em-one-ee] n political domination

Hegira n Mohammed's flight from Mecca to Medina in 622 AD

heifer [hef-fer] n young cow

height n distance from base to top; distance above sea level; highest degree or topmost point ▶ **heighten** v make or become higher or more intense

heinous adj evil and shocking

heir n person entitled to inherit property or rank ▶ **heiress** n fem ▶ **heirloom** n object that has belonged to a family for generations

held v past of **hold**[1]

helical adj spiral

helicopter n aircraft lifted and propelled by rotating overhead blades ▶ **heliport** n airport for helicopters

heliotrope n plant with purple flowers ▷ adj light purple

helium [heel-ee-um] n (Chem) very light colourless odourless gas

helix [heel-iks] n, pl **helices** or **helixes** continuous curve winding as if around the surface of a cone or cylinder, as on a corkscrew

hell n place believed to be where wicked people go when they die; place or state of wickedness, suffering, or punishment **hell for leather** at great speed ▶ **hellish** adj ▶ **hellbent** adj (foll by on) intent

Hellenic adj of the (ancient) Greeks or their language

hello interj expression of greeting or surprise

helm n tiller or wheel for steering a ship

helmet n hard hat worn for protection

help v make something easier, better, or quicker for (someone); improve (a situation); refrain from e.g. I can't help smiling ▷ n assistance or support **help oneself** take something, esp. food or drink, without being served; (informal) steal something ▶ **helper** n ▶ **helpful** adj ▶ **helping** n single portion of food ▶ **helpless** adj weak or incapable ▶ **helplessly** adv ▶ **helpline** n telephone line set aside for callers to contact an organization for help with a problem ▶ **helpmate** n companion and helper, esp. a husband or wife

helter-skelter adj haphazard and careless ▷ adv in a haphazard and careless manner ▷ n high spiral slide at a fairground

hem n bottom edge of a garment, folded under and stitched down ▷ v **hemming, hemmed** provide with a hem ▶ **hem in** v surround and prevent from moving ▶ **hemline** n level to which the hem of a skirt hangs

hemisphere n half of a sphere, esp. the earth ▶ **hemispherical** adj

hemlock n poison made from a plant with spotted stems and small white flowers

hemp n (also **cannabis**) Asian plant with tough fibres; its fibre, used to make canvas and rope; narcotic drug obtained from hemp

hen n female domestic fowl; female of any bird ▶ **hen night, hen party** party for women only ▶ **henpecked** adj (of a man) dominated by his wife

hence conj for this reason ▷ adv from this time ▶ **henceforth** adv from now on

henchman n person employed by someone powerful to carry out orders

henna n reddish dye made from a shrub or tree ▷ v dye (the hair) with henna

henry n, pl **-ry** or **-ries** or **-rys** unit of electrical inductance

hepatitis n inflammation of the liver

heptagon n geometric figure with seven sides

heptathlon n athletic contest for women, involving seven events

her pron refers to a female person or animal or anything personified

as feminine when the object of a sentence or clause ▷ adj belonging to her

herald n person who announces important news; forerunner ▷ v signal the approach of ▶ **heraldry** n study of coats of arms and family trees ▶ **heraldic** adj

herb n plant used for flavouring in cookery, and in medicine ▶ **herbal** adj ▶ **herbalist** n person who grows or specializes in the use of medicinal herbs ▶ **herbaceous** adj (of a plant) soft-stemmed ▶ **herbicide** n chemical used to destroy plants, esp. weeds ▶ **herbivore** n animal that eats only plants ▶ **herbivorous** [her-biv-or-uss] adj

herculean [her-kew-lee-an] adj requiring great strength or effort

herd n group of animals feeding and living together; large crowd of people ▷ v collect into a herd ▶ **herdsman** n man who looks after a herd of animals

Here interj (S Afr) exclamation of surprise or dismay

here adv in, at, or to this place or point ▶ **hereabouts** adv near here. ▶ **hereafter** adv after this point or time the hereafter life after death ▶ **hereby** adv by means of or as a result of this ▶ **herein** adv in this place, matter, or document ▶ **herewith** adv with this

heredity [hir-red-it-ee] n passing on of characteristics from one generation to another ▶ **hereditary** adj passed on genetically from one generation to another; passed on by inheritance

> **SPELLING TIP**
> Remember there are three
> letters where the t in **hereditary**

heresy [herr-iss-ee] n, pl **-sies** opinion contrary to accepted opinion or belief ▶ **heretic** [herr-it-ik] *n* person who holds unorthodox opinions ▶ **heretical** [hir-**ret**-ik-al] *adj*

heritage *n* something inherited; anything from the past, considered as the inheritance of present-day society

hermaphrodite [her-**maf**-roe-dite] *n* animal, plant, or person with both male and female reproductive organs

hermetic *adj* sealed so as to be airtight ▶ **hermetically** *adv*

hermit *n* person living in solitude, esp. for religious reasons ▶ **hermitage** *n* home of a hermit

hernia *n* protrusion of an organ or part through the lining of the surrounding body cavity

hero *n, pl* **heroes** principal character in a film, book, etc.; man greatly admired for his exceptional qualities or achievements ▶ **heroine** *n fem* ▶ **heroic** *adj* courageous; of or like a hero ▶ **heroics** *pl n* extravagant behaviour ▶ **heroically** *adv* ▶ **heroism** [**herr**-oh-izz-um] *n*

heroin *n* highly addictive drug derived from morphine

heron *n* long-legged wading bird

herpes [her-peez] *n* any of several inflammatory skin diseases, including shingles and cold sores

Herr [hair] *n, pl* **Herren** German term of address equivalent to *Mr*

herring *n* important food fish of northern seas ▶ **herringbone** *n* pattern of zigzag lines

herself *pron* emphatic or reflexive form of **she, her**

hertz *n, pl* **hertz** (*Physics*) unit of frequency

hesitate *v* be slow or uncertain in doing something; be reluctant (to do something) ▶ **hesitation** *n*

▶ **hesitant** *adj* undecided or wavering ▶ **hesitantly** *adv* ▶ **hesitancy** *n*

hessian *n* coarse jute fabric

heterodox *adj* differing from accepted doctrines or beliefs ▶ **heterodoxy** *n*

heterogeneous [het-er-oh-**jean**-ee-uss] *adj* composed of diverse elements ▶ **heterogeneity** *n*

heterosexual *n, adj* (person) sexually attracted to members of the opposite sex ▶ **heterosexuality** *n*

heuristic [hew-**rist**-ik] *adj* involving learning by investigation

hew *v* **hewing, hewed** or **hewn** cut with an axe; carve from a substance

hexagon *n* geometrical figure with six sides ▶ **hexagonal** *adj*

hey *interj* expression of surprise or for catching attention

heyday *n* time of greatest success, prime

hiatus [hie-**ay**-tuss] *n, pl* **-tuses** or **-tus** pause or interruption in continuity

hibernate *v* (of an animal) pass the winter as if in a deep sleep ▶ **hibernation** *n*

Hibernian *adj* (*poetic*) Irish

hibiscus *n, pl* **-cuses** tropical plant with large brightly coloured flowers

hiccup, hiccough *n* spasm of the breathing organs with a sharp coughlike sound; (*informal*) small problem, hitch ▶ *v* make a hiccup

hick *n* (*US, Aust & NZ informal*) unsophisticated country person

hickory *n, pl* **-ries** N American nut-bearing tree; its wood

hide¹ *v* **hiding, hid, hidden** put (oneself or an object) somewhere difficult to see or find; keep secret ▶ *n* place of concealment, esp. for a bird-watcher ▶ **hiding** n state of concealment e.g. in hiding ▶ **hide-out** n place to hide in

hide² n skin of an animal ▶ **hiding** n (slang) severe beating ▶ **hidebound** adj unwilling to accept new ideas

hideous [hide-ee-uss] adj ugly, revolting ▶ **hideously** adv

hierarchy [hire-ark-ee] n, pl **-chies** system of people or things arranged in a graded order ▶ **hierarchical** adj

hieroglyphic [hire-oh-gliff-ik] adj of a form of writing using picture symbols, as used in ancient Egypt ▷ n symbol that is difficult to decipher; (also **hieroglyph**) symbol representing an object, idea, or sound

hi-fi n set of high-quality sound-reproducing equipment ▷ adj high-fidelity

higgledy-piggledy adv, adj in a muddle

high adj of a great height; far above ground or sea level; being at its peak; greater than usual in intensity or amount; (of a sound) acute in pitch; of great importance, quality, or rank; (informal) under the influence of alcohol or drugs ▷ adv at or to a high level ▶ **highly** adv ▶ **highly strung** nervous and easily upset ▶ **Highness** n title used to address or refer to a royal person ▶ **High-Church** adj belonging to a section within the Church of England stressing the importance of ceremony and ritual ▶ **higher education** education at colleges and universities ▶ **high-fidelity** adj able to reproduce sound with little or no distortion ▶ **high-flown** adj (of language) extravagant or pretentious ▶ **high-handed** adj excessively forceful ▶ **high-maintenance** adj (of a piece of equipment, vehicle etc.)

requiring regular maintenance; (informal) (of a person) requiring a high level of care and attention ▶ **high-rise** adj (of a building) having many storeys ▶ **high tea** early evening meal consisting of a cooked dish, bread, cakes, and tea ▶ **high time** latest possible time

highbrow adj, n intellectual and serious (person)

Highland adj of the Highlands, a mountainous region of N Scotland ▶ **Highlander** n

highlands pl n area of high ground

highlight n outstanding part or feature; light-toned area in a painting or photograph; lightened streak in the hair ▷ v give emphasis to

highway n (US, Aust & NZ) main road ▶ **Highway Code** regulations and recommendations applying to all road users ▶ **highwayman** n (formerly) robber, usu. on horseback, who robbed travellers at gunpoint

hijab n head covering worn by Muslim women

hijack v seize control of (an aircraft or other vehicle) while travelling ▶ **hijacker** n

hike n long walk in the country, esp. for pleasure ▷ v go for a long walk; (foll by up) pull (up) or raise ▶ **hiker** n

hilarious adj very funny ▶ **hilariously** adv ▶ **hilarity** n

hill n raised part of the earth's surface, less high than a mountain ▶ **hilly** adj ▶ **hillock** n small hill ▶ **hillbilly** n (US) unsophisticated country person

Hills Hoist® n Australian brand of rotary clothesline

hilt n handle of a sword or knife

him pron refers to a male person or animal when the object of a

sentence or clause ▸ **himself**
pron emphatic or reflexive form
of **he, him**

hind[1] *adj* **hinder, hindmost**
situated at the back

hind[2] *n* female deer

hinder *v* get in the way of
▸ **hindrance** *n*

Hindu *n* person who practises
Hinduism ▷ *adj* of Hinduism
▸ **Hindi** *n* language of N central
India ▸ **Hinduism** *n* dominant
religion of India, which involves the
worship of many gods and a belief
in reincarnation

hinge *n* device for holding together
two parts so that one can swing
freely ▷ *v* (foll by **on**) depend (on);
fit a hinge to

hint *n* indirect suggestion; piece of
advice; small amount ▷ *v* suggest
indirectly

hinterland *n* land lying behind a
coast or near a city, esp. a port

hip[1] *n* either side of the pelvis and the thigh

hip[2] *n* rosehip

hip-hop *n* pop-culture movement
originating in the 1980s, comprising
rap music, graffiti, and break dancing

hippie *adj, n* same as **hippy**

hippo *n, pl* **-pos** (*informal*)
hippopotamus

hippodrome *n* music hall, variety
theatre, or circus

hippopotamus *n, pl* **-muses** or **-mi**
large African mammal with thick
wrinkled skin, living near rivers

hippy *adj, n, pl* **-pies** (esp. in
the 1960s) (of) a person whose
behaviour and dress imply a
rejection of conventional values

hipster *n* (*informal*) person pursuing
non-mainstream cultural trends

hire *v* pay to have temporary use of;
employ for wages ▷ *n* hiring **for
hire** available to be hired

▸ **hireling** *n* person who works only
for wages ▸ **hire-purchase** *n*
system of purchase by which the
buyer pays for goods by instalments

hirsute [her-suit] *adj* hairy

his *pron, adj* (something) belonging
to him

Hispanic *adj* Spanish or Latin-
American

hiss *n* sound like that of a long *s* (as
an expression of contempt) ▷ *v*
utter a hiss; show derision or anger
towards

histamine [hiss-ta-meen] *n*
substance released by the body
tissues in allergic reactions

histogram *n* statistical graph in
which the frequency of values is
represented by vertical bars of
varying heights and widths

histology *n* study of the tissues of
an animal or plant

history *n, pl* **-ries** (record or
account of) past events and
developments; study of these;
record of someone's past
▸ **historian** *n* writer of history
▸ **historic** *adj* famous or significant
in history ▸ **historical** *adj*
occurring in the past; based on
history ▸ **historically** *adv*

histrionic *adj* excessively dramatic
▸ **histrionics** *pl n* excessively
dramatic behaviour

hit *v* **hitting, hit** strike, touch
forcefully; come into violent
contact with; affect badly; reach
(a point or place) ▷ *n* hitting;
successful record, film, etc.;
(*Computing*) single visit to a website
hit it off (*informal*) get on well
together **hit the road** (*informal*)
start a journey ▸ **hit-and-miss**
adj sometimes successful and
sometimes not ▸ **hit man**
hired assassin ▸ **hit on** *v* think
of (an idea)

hitch n minor problem ▷ v (informal) obtain (a lift) by hitchhiking; fasten with a knot or tie; fasten (full by up) pull up with a jerk ▶ **hitchhike** v travel by obtaining free lifts ▶ **hitchhiker** n

hi-tech adj using sophisticated technology

hither adv (old-fashioned) to or towards this place

hitherto adv until this time

HIV human immunodeficiency virus: the cause of AIDS

hive n same as **beehive**; **hive of activity** place where people are very busy ▶ **hive off** v separate from a larger group

hives n allergic reaction in which itchy red or whitish patches appear on the skin

HM (in Britain) Her (or His) Majesty

HMS (in Britain) Her (or His) Majesty's Ship

HNC (in Britain) Higher National Certificate

HND (in Britain) Higher National Diploma

hoard n store hidden away for future use ▷ v save or store ▶ **hoarder** n

USAGE NOTE
Do not confuse hoard with horde, which means 'a large group'

hoarding n large board for displaying advertisements

hoarfrost n white ground frost

hoarse adj (of a voice) rough and unclear; having a rough and unclear voice ▶ **hoarsely** adv ▶ **hoarseness** n

hoary adj **hoarier, hoariest** grey or white(-haired); very old

hoax n deception or trick ▷ v deceive or play a trick upon ▶ **hoaxer** n

hob n (Brit) flat top part of a cooker, or a separate flat surface, containing gas or electric rings for cooking on

hobble v walk lamely; tie the legs of (a horse) together

hobby n, pl -**bies** activity pursued in one's spare time ▶ **hobbyhorse** n favourite topic; toy horse

hobgoblin n mischievous goblin

hobnail boots pl n heavy boots with short nails in the soles

hobnob v -**nobbing, -nobbed** (foll by with) be on friendly terms (with)

hobo n, pl -**bos** (US, Aust & NZ) tramp or vagrant

hock¹ n joint in the back leg of an animal such as a horse that corresponds to the human ankle

hock² n white German wine

hock³ v (informal) pawn **in hock** (informal) in debt

hockey n team game played on a field with a ball and curved sticks; (US) ice hockey

hocus-pocus n trickery

hod n open wooden box attached to a pole, for carrying bricks or mortar

hoe n long-handled tool used for loosening soil or weeding ▷ v scrape or weed with a hoe

hog n castrated male pig; (informal) greedy person ▷ v **hogging, hogged** (informal) take more than one's share ▶ **hogshead** n large cask ▶ **hogwash** n (informal) nonsense

Hogmanay n (in Scotland) New Year's Eve

hoick v raise abruptly and sharply

hoi polloi n the ordinary people

hoisin n sweet spicy sauce of soya beans, sugar, garlic, and vinegar, used in Chinese cookery

hoist v raise or lift up ▷ n device for lifting things

hoity-toity adj (informal) arrogant or haughty

hokey-pokey n (NZ) brittle toffee sold in lumps

h

hold v **holding, held** keep or support in or with the hands or arms; arrange for (a meeting, party, etc.) to take place; consider to be as specified e.g. *who are you holding responsible?*; maintain in a specified position or state; have the capacity for; (*informal*) wait, esp. on the telephone; restrain or keep back; own, possess ▷ n act or way of holding; controlling influence ▶ **holder** n ▶ **holding** n property, such as land or stocks and shares ▶ **holdall** n large strong travelling bag ▶ **hold-up** n armed robbery; delay

hold² n cargo compartment in a ship or aircraft

hole n area hollowed out in a solid; opening or hollow; animal's burrow; (*informal*) unattractive place; (*informal*) difficult situation ▷ v make holes in; hit (a golf ball) into the target hole

holiday n time spent away from home for rest or recreation; day or other period of rest from work or studies

holiness n state of being holy; (**H-**) title used to address or refer to the Pope

holistic adj considering the complete person, physically and mentally, in the treatment of an illness ▶ **holism** n

hollow adj having a hole or space inside; (of a sound) as if echoing in a hollow place; without any real value or worth ▷ n cavity or space; dip in the land ▷ v form a hollow in

holly n evergreen tree with prickly leaves and red berries

hollyhock n tall garden plant with spikes of colourful flowers

holocaust n destruction or loss of life on a massive scale

hologram n three-dimensional photographic image

holograph n document handwritten by the author

holster n leather case for a pistol, hung from a belt

holy adj **-lier, -liest** of God or a god; devout or virtuous ▶ **holier-than-thou** adj self-righteous ▶ **Holy Communion** (*Christianity*) service in which people take bread and wine in remembrance of the death and resurrection of Jesus Christ ▶ **Holy Grail** (in medieval legend) the bowl used by Jesus Christ at the Last Supper ▶ **Holy Spirit, Holy Ghost** (*Christianity*) one of the three aspects of God ▶ **Holy Week** (*Christianity*) week before Easter

homage n show of respect or honour towards someone or something

home n place where one lives; institution for the care of the elderly, orphans, etc. ▷ adj of one's home, birthplace, or native country; (*Sport*) played on one's own ground ▷ adv to or at home ▷ v (foll by *in* or *in on*) direct towards (a point or target) **at home** at ease **bring home to** make clear to **home and dry** (*informal*) safe or successful ▶ **homeless** adj having nowhere to live ▷ pl n people who have nowhere to live ▶ **homelessness** n ▶ **homely** adj simple, ordinary, and comfortable; (*US*) unattractive ▶ **homeward** adj, adv ▶ **homewards** adv ▶ **home-brew** n beer made at home ▶ **home-made** adj made at home or on the premises ▶ **home page** (*Computing*) introductory information about a website with links to the information or services provided ▶ **home truths** unpleasant facts told to a person about himself or herself

homeland n country from which a person's ancestors came

homeopathy [home-ee-op-ath-ee] n treatment of disease by small doses of a drug that produces symptoms of the disease in healthy people ▶ **homeopath** n person who practises homeopathy ▶ **homeopathic** adj

homesick adj sad because missing one's home and family ▶ **homesickness** n

homestead n farmhouse plus the adjoining land

homework n school work done at home

homicide n killing of a human being; person who kills someone ▶ **homicidal** adj

homily n, pl -**lies** speech telling people how they should behave

hominid n man or any extinct forerunner of man

homo- combining form same, like e.g. homosexual

homogeneous [home-oh-**jean**-ee-uss] adj formed of similar parts ▶ **homogeneity** n ▶ **homogenize** v break up fat globules in (milk or cream) to distribute them evenly; make homogeneous

homograph n word spelt the same as another, but with a different meaning

homologous [hom-**ol**-log-uss] adj having a related or similar position or structure

homonym n word spelt or pronounced the same as another, but with a different meaning

homophobia n hatred or fear of homosexuals ▶ **homophobic** adj

homophone n word pronounced the same as another, but with a different meaning or spelling

Homo sapiens [hoe-moh **sap**-ee-enz] n human beings as a species

homosexual n, adj (person) sexually attracted to members of the same sex ▶ **homosexuality** n

Hon. Honourable

hone v sharpen

honest adj truthful and moral; open and sincere ▶ **honestly** adv ▶ **honesty** n quality of being honest; plant with silvery seed pods

honey n sweet edible sticky substance made by bees from nectar; term of endearment ▶ **honeycomb** n waxy structure of six-sided cells in which honey is stored by bees in a beehive ▶ **honeydew melon** melon with a yellow skin and sweet pale flesh ▶ **honeymoon** n holiday taken by a newly married couple ▶ **honeysuckle** n climbing shrub with sweet-smelling flowers; Australian tree or shrub with nectar-rich flowers

hongi [hong-jee] n (NZ) Māori greeting in which people touch noses

honk n sound made by a car horn; sound made by a goose ▶ v (cause to) make this sound

honour n sense of honesty and fairness; (award given out of) respect; pleasure or privilege ▶ pl university degree of a higher standard than an ordinary degree ▶ v give praise and attention to; give an award to (someone) out of respect; accept or pay (a cheque or bill); keep (a promise) **do the honours** act as host or hostess by pouring drinks or giving out food ▶ **honourable** adj worthy of respect or esteem ▶ **honourably** adv ▶ **honorary** adj held or given only as an honour; unpaid ▶ **honorific** adj showing respect

hood¹ n head covering, often attached to a coat or jacket; folding roof of a convertible car or a pram; (US & Aust) car bonnet ▸ **hooded** adj (of a garment) having a hood; (of eyes) having heavy eyelids that appear to be half-closed ▸ **hoodie** n (informal) hooded sweatshirt; person who wears a hooded sweatshirt

hood² n (chiefly US slang) hoodlum

hoodlum n (slang) violent criminal, gangster

hoodoo n, pl **-doos** (cause of) bad luck

hoodwink v trick, deceive

hoof n, pl **hooves** or **hoofs** horny covering of the foot of a horse, deer, etc. **hoof it** (slang) walk

hoo-ha n fuss or commotion

hook n curved piece of metal, plastic, etc., used to hang, hold, or pull something; short swinging punch ▸ v fasten or catch (as if) with a hook ▸ **hooked** adj bent like a hook; (foll by **on**) (slang) addicted (to) or obsessed (with) ▸ **hooker** n (chiefly US slang) prostitute; (Rugby) player who uses his or her feet to get the ball in a scrum ▸ **hook-up** n linking of radio or television stations ▸ **hookworm** n blood-sucking worm with hooked mouthparts

hookah n oriental pipe through which smoke is drawn through water and a long tube

hooligan n rowdy young person ▸ **hooliganism** n

hoon n (Aust & NZ slang) loutish youth who drives irresponsibly

hoop n rigid circular band, used esp. as a child's toy or for animals to jump through in the circus **jump through the hoops**, **be put through the hoops** go through an ordeal or test ▸ **hoop pine** Australian tree or shrub with flowers in dense spikes

hoopla n fairground game in which hoops are thrown over objects in an attempt to win them

hooray interj same as **hurrah**

hoot n sound of a car horn; cry of an owl; cry of derision; (informal) amusing person or thing ▸ v sound (a car horn); jeer or yell contemptuously (at someone) ▸ **hooter** n device that hoots; (chiefly Brit slang) nose

Hoover® n vacuum cleaner ▸ v (**h-**) clean with a vacuum cleaner

hooves n a plural of **hoof**

hop¹ v **hopping**, **hopped** jump on one foot; move in short jumps; (informal) move quickly ▸ n instance of hopping; (informal) dance; short journey, esp. by air **catch someone on the hop** (informal) catch someone unprepared

hop² n (often pl) climbing plant, the dried flowers of which are used to make beer

hope v want (something) to happen or be true ▸ n expectation of something desired; thing that gives cause for hope or is desired ▸ **hopeful** adj having, expressing, or inspiring hope ▸ n person considered to be on the brink of success ▸ **hopefully** adv in a hopeful manner; it is hoped ▸ **hopeless** adj

hopper n container for storing substances such as grain or sand

hopscotch n children's game of hopping in a pattern drawn on the ground

horde n large crowd

> **USAGE NOTE**
> Do not confuse **horde** with **hoard**, which means 'to store hidden away for future use'

horizon n apparent line that divides the earth and the sky

▷ *pl* limits of scope, interest, or knowledge

horizontal *adj* parallel to the horizon, level, flat ▶ **horizontally** *adv*

hormone *n* substance secreted by certain glands which stimulates certain organs of the body; synthetic substance with the same effect ▶ **hormonal** *adj*

horn *n* one of a pair of bony growths sticking out of the heads of cattle, sheep, etc.; substance of which horns are made; musical instrument with a tube or pipe of brass fitted with a mouthpiece; device on a vehicle sounded as a warning ▶ **horned** *adj* ▶ **horny** *adj* of or like horn; (*slang*) (easily) sexually aroused ▶ **hornbeam** *n* tree with smooth grey bark ▶ **hornbill** *n* bird with a bony growth on its large beak ▶ **hornpipe** *n* (music for) a solo dance, traditionally performed by sailors

hornblende *n* mineral containing aluminium, calcium, sodium, magnesium, and iron

hornet *n* large wasp with a severe sting

horoscope *n* prediction of a person's future based on the positions of the planets, sun, and moon at his or her birth

horrendous *adj* very unpleasant and shocking

horrible *adj* disagreeable, unpleasant; causing horror ▶ **horribly** *adv*

horrid *adj* disagreeable, unpleasant; (*informal*) nasty

horrify *v* **-fying, -fied** cause to feel horror or shock ▶ **horrific** *adj* causing horror

horror *n* (thing or person causing) terror or hatred

hors d'oeuvre [or **durv**] *n* appetizer served before a main meal

horse *n* large animal with hooves, a mane, and a tail, used for riding and pulling carts etc.; piece of gymnastic equipment used for vaulting over **(straight) from the horse's mouth** from the original source ▶ **horsey, horsy** *adj* very keen on horses; of or like a horse ▶ **horse around** *v* (*informal*) play roughly or boisterously ▶ **horse chestnut** tree with broad leaves and inedible large brown shiny nuts in spiky cases ▶ **horsefly** *n* large bloodsucking fly ▶ **horsehair** *n* hair from the tail or mane of a horse ▶ **horse laugh** loud coarse laugh ▶ **horseman, horsewoman** *n* person riding a horse ▶ **horseplay** *n* rough or rowdy play ▶ **horsepower** *n* unit of power (equivalent to 745.7 watts), used to measure the power of an engine ▶ **horseradish** *n* strong-tasting root of a plant, usu. made into a sauce ▶ **horseshoe** *n* protective U-shaped piece of iron nailed to a horse's hoof, regarded as a symbol of good luck

horticulture *n* art or science of cultivating gardens ▶ **horticultural** *adj* ▶ **horticulturalist, horticulturist** *n*

hosanna *interj* exclamation of praise to God

hose¹ *n* flexible pipe for conveying liquid ▷ *v* water with a hose

hose² *n* stockings, socks, and tights ▶ **hosiery** *n* stockings, socks, and tights collectively

hospice [**hoss-piss**] *n* nursing home for the terminally ill

hospital *n* place where people who are ill are looked after and treated

▶ **hospitalize** v send or admit to hospital ▶ **hospitalization** n

hospitality n kindness in welcoming strangers or guests ▶ **hospitable** adj welcoming to strangers or guests

host¹ n (fem **hostess**) person who entertains guests, esp. in his own home; place or country providing the facilities for an event; compere of a show; animal or plant on which a parasite lives ▷ v be the host of

host² n large number

Host n (Christianity) bread used in Holy Communion

hostage n person who is illegally held prisoner until certain demands are met by other people

hostel n building providing accommodation at a low cost for a specific group of people such as students, travellers, homeless people, etc.

hostelry n, pl -ries (old-fashioned or facetious) inn, pub

hostile adj unfriendly; (foll by to) opposed (to); of an enemy ▶ **hostility** n, pl -ties unfriendly and aggressive feelings or behaviour ▷ pl acts of warfare

hot adj **hotter, hottest** having a high temperature; strong, spicy; (of news) very recent; (of a contest) fiercely fought; (of a temper) quick to rouse; liked very much e.g. a hot favourite; (slang) stolen **in hot water** (informal) in trouble ▶ **hotly** adv ▶ **hot air** (informal) empty talk ▶ **hot-blooded** adj passionate or excitable ▶ **hot dog** long roll split lengthways with a hot frankfurter inside ▶ **hot-headed** adj rash, having a hot temper ▶ **hotline** n direct telephone link for emergency use ▶ **hot pool** (NZ) geothermally heated pool ▶ **hot**

spot place where there is a lot of exciting activity; area where there is a concentration of violence or political unrest; (Computing) place (esp. a public building or commercial premises) offering a Wi-Fi connection

hotbed n any place encouraging a particular activity e.g. hotbeds of unrest

hotchpotch n jumbled mixture

hotel n commercial establishment providing lodging and meals ▶ **hotelier** n owner or manager of a hotel

hotfoot adv (informal) quickly and eagerly **hotfoot it** (informal) go quickly and eagerly

hothouse n greenhouse

hotplate n a heated metal surface on an electric cooker; portable device for keeping food warm

hound n hunting dog ▷ v pursue relentlessly

hour n twenty-fourth part of a day, sixty minutes; time of day ▷ pl period regularly appointed for work or business ▶ **hourly** adj, adv (happening) every hour; frequent(ly) ▶ **hourglass** n device with two glass compartments, containing a quantity of sand that takes an hour to trickle from the top section to the bottom one

houri n (Islam) any of the nymphs of paradise

house n building used as a home; building used for some specific . purpose e.g. the opera house; business firm; law-making body or the hall where it meets; family or dynasty; theatre or cinema audience **get on like a house on fire** (informal) get on very well together **on the house** (informal) provided free by the management ▷ v give accommodation to;

contain or cover ▸ **housing** n (providing of) houses; protective case or covering of a machine
▸ **house arrest** confinement to one's home rather than in prison
▸ **houseboat** n stationary boat used as a home ▸ **housebreaker** n burglar ▸ **housecoat** n woman's long loose coat-shaped garment for wearing at home ▸ **household** n all the people living in a house
▸ **householder** n person who owns or rents a house ▸ **housekeeper** n person employed to run someone else's household ▸ **housekeeping** n (money for) running a household
▸ **housemaid** n female servant employed to do housework
▸ **house-train** v train (a pet) to urinate and defecate outside
▸ **house-warming** n party to celebrate moving into a new home
▸ **housewife** n woman who runs her own household and does not have a job ▸ **housework** n work of running a home, such as cleaning, cooking, and shopping

House music, House n electronic funk-based disco music with samples of other recordings edited in

hovea n Australian plant with purple flowers

hovel n small dirty house or hut

hover v (of a bird etc.) remain suspended in one place in the air; loiter; be in a state of indecision
▸ **hovercraft** n vehicle which can travel over both land and sea on a cushion of air

how adv in what way, by what means; to what degree e.g. I know how hard it is ▸ **however** adv nevertheless; by whatever means; no matter how e.g. however much it hurt, he could do it

howdah n canopied seat on an elephant's back

howitzer n large gun firing shells at a steep angle

howl n loud wailing cry; loud burst of laughter ▸ v utter a howl ▸ **howler** n (informal) stupid mistake

hoyden n (old-fashioned) wild or boisterous girl

HP, h.p. hire-purchase; horsepower

HQ headquarters

HR human resources

HRH Her (or His) Royal Highness

HRT hormone replacement therapy

HSDPA high-speed download packet access

HTML hypertext markup language: text description language used on the internet

hub n centre of a wheel, through which the axle passes; central point of activity

hubbub n confused noise of many voices

hubby n, pl -**bies** (informal) husband

hubris [hew-briss] n (formal) pride, arrogance

huckster n person using aggressive methods of selling

huddle v hunch (oneself) through cold or fear; crowd closely together ▸ n small group; (informal) impromptu conference

hue n colour, shade

hue and cry n public outcry

huff n passing mood of anger or resentment ▸ v blow or puff heavily
▸ **huffy** adj ▸ **huffily** adv

hug v **hugging, hugged** clasp tightly in the arms, usu. with affection; keep close to (the ground, kerb, etc.) ▸ n tight or fond embrace

huge adj very big ▸ **hugely** adv

huh interj exclamation of derision, bewilderment, or inquiry

hui [hoo-ee] n (NZ) meeting of Māori people; meeting to discuss Māori matters

hula n swaying Hawaiian dance ▶ **Hula Hoop®** plastic hoop twirled round the body by gyrating the hips

hulk n body of an abandoned ship; (offens) large heavy person or thing ▶ **hulking** adj bulky, unwieldy

hull n main body of a boat; leaves round the stem of a strawberry, raspberry, etc. ▷ v remove the hulls from

hullabaloo n, pl **-loos** loud confused noise or clamour

hum v humming, hummed make a low continuous vibrating sound; sing with the lips closed; (slang) (of a place) to be very busy ▷ n humming sound ▶ **hummingbird** n very small American bird whose powerful wings make a humming noise as they vibrate

human adj of or typical of people ▷ n human being ▶ **humanly** adv by human powers or means ▶ **human being** man, woman, or child

humane adj kind or merciful ▶ **humanely** adv

humanism n belief in human effort rather than religion ▶ **humanist** n

humanitarian n, adj (person) having the interests of humankind at heart

humanity n, pl **-ties** human race; the quality of being human; kindness or mercy ▷ pl study of literature, philosophy, and the arts

humanize v make human or humane

humankind n human race

humble adj conscious of one's failings; modest, unpretentious, unimportant ▷ v cause to feel humble, humiliate ▶ **humbly** adv

humbug n (Brit) hard striped peppermint sweet; nonsense; dishonest person

humdinger n (slang) excellent person or thing

humdrum adj ordinary, dull

humerus [hew-mer-uss] n, pl **-meri** [-mer-rye] bone from the shoulder to the elbow

humid adj damp and hot ▶ **humidity** n ▶ **humidify** v -fying, -fied make humid ▶ **humidifier** n device for increasing the amount of water vapour in the air in a room

humiliate v lower the dignity or hurt the pride of ▶ **humiliating** adj ▶ **humiliation** n

humility n quality of being humble

hummock n very small hill

humour n ability to say or perceive things that are amusing; amusing quality in a situation, film, etc.; state of mind, mood; (old-fashioned) fluid in the body ▷ v be kind and indulgent to ▶ **humorous** adj ▶ **humorously** adv ▶ **humorist** n writer or entertainer who uses humour in his or her work

SPELLING TIP
Don't forget to drop the second u in **humour** when forming the derivatives **humorous** and **humorist**

hump n raised piece of ground; large lump on the back of an animal or person ▷ v (slang) carry or heave **get the hump, take the hump** (informal) be annoyed, sulk ▶ **hump-back bridge, humpbacked bridge** road bridge with a sharp slope on each side

humus [hew-muss] n decomposing vegetable and animal mould in the soil

hunch n feeling or suspicion not based on facts ▷ v draw (one's shoulders) up or together

▶ **hunchback** n (offens) person with an abnormal curvature of the spine

hundred adj, n ten times ten ▷ n (often pl) large but unspecified number ▶ **hundredth** adj, n
▶ **hundredweight** n (Brit) unit of weight of 112 pounds (50.8 kilograms)

hung v past of **hang** ▷ adj (of a parliament or jury) with no side having a clear majority **hung over** (informal) suffering the effects of a hangover

Hungarian adj of Hungary ▷ n person from Hungary; language of Hungary

hunger n discomfort or weakness from lack of food; desire or craving ▷ v (foll by for) want very much
▶ **hunger strike** refusal of all food, as a means of protest

hungry adj hungrier, hungriest desiring food; (foll by for) having a desire or craving (for)
▶ **hungrily** adv

hunk n large piece; (slang) sexually attractive man

hunt v seek out and kill (wild animals) for food or sport; (foll by for) search (for) ▷ n hunting; (party organized for) hunting wild animals for sport ▶ **huntaway** n (NZ) sheepdog trained to drive sheep by barking ▶ **huntsman** n man who hunts wild animals, esp. foxes

hunter n person or animal that hunts wild animals for food or sport

hurdle n (Sport) light barrier for jumping over in some races; problem or difficulty ▷ pl race involving hurdles ▷ v jump over (something) ▶ **hurdler** n

hurdy-gurdy n, pl **-dies** mechanical musical instrument, such as a barrel organ

hurl v throw or utter forcefully

hurling, hurley n Irish game like hockey

hurly-burly n loud confusion

hurrah, hurray interj exclamation of joy or applause

hurricane n very strong, often destructive, wind or storm
▶ **hurricane lamp** paraffin lamp with a glass covering

hurry v **-rying, -ried** (cause to) move or act very quickly ▷ n doing something quickly or the need to do something quickly
▶ **hurriedly** adv

hurt v hurting, hurt cause physical or mental pain to; be painful; (informal) feel pain ▷ n physical or mental pain ▶ **hurtful** adj unkind

hurtle v move quickly or violently

husband n male partner in marriage ▷ v use economically
▶ **husbandry** n farming; management of resources

hush v make or be silent ▷ n stillness or silence ▶ **hush-hush** adj (informal) secret ▶ **hush up** v suppress information about

husk n outer covering of certain seeds and fruits ▷ v remove the husk from

husky[1] adj huskier, huskiest slightly hoarse; (informal) big and strong ▶ **huskily** adv

husky[2] n, pl huskies Arctic sledge dog with thick hair and a curled tail

hussar [hoo-**zar**] n (Hist) lightly armed cavalry soldier

hussy n, pl **-sies** immodest or promiscuous woman

hustings pl n political campaigns and speeches before an election

hustle v push about, jostle ▷ n lively activity or bustle

hut n small house, shelter, or shed

hutch n cage for pet rabbits etc.

hyacinth n sweet-smelling spring flower that grows from a bulb

hyaena n same as **hyena**

hybrid n offspring of two plants or animals of different species; anything of mixed origin ▷ adj of mixed origin

hydra n mythical many-headed water serpent

hydrangea n ornamental shrub with clusters of pink, blue, or white flowers

hydrant n outlet from a water main with a nozzle for a hose

hydrate n chemical compound of water with another substance

hydraulic adj operated by pressure forced through a pipe by a liquid such as water or oil ▷ **hydraulics** n study of the mechanical properties of fluids as they apply to practical engineering ▷ **hydraulically** adv

hydro¹ n, pl **hydros** hotel offering facilities for hydropathy

hydro² adj short for **hydroelectric**

hydro- combining form water e.g. hydroelectric; hydrogen e.g. hydrochloric acid

hydrocarbon n compound of hydrogen and carbon

hydrochloric acid n strong colourless acid used in many industrial and laboratory processes

hydroelectric adj of the generation of electricity by water pressure

hydrofoil n fast light boat with its hull raised out of the water on one or more pairs of fins

hydrogen n (Chem) light flammable colourless gas that combines with oxygen to form water ▷ **hydrogen bomb** extremely powerful bomb in which energy is released by fusion of hydrogen nuclei to give helium nuclei ▷ **hydrogen peroxide** colourless liquid used as a hair bleach and as an antiseptic

hydrolysis [hie-**drol**-iss-iss] n decomposition of a chemical compound reacting with water

hydrometer [hie-**drom**-it-er] n instrument for measuring the density of a liquid

hydropathy n method of treating disease by the use of large quantities of water both internally and externally

hydrophobia n rabies; fear of water

hydroplane n light motorboat that skims the water

hydroponics n method of growing plants in water rather than soil

hydrotherapy n (Med) treatment of certain diseases by exercise in water

hyena n scavenging doglike mammal of Africa and S Asia

hygiene n principles and practice of health and cleanliness ▷ **hygienic** adj ▷ **hygienically** adv

hymen n membrane partly covering the opening of a girl's vagina, which breaks before puberty or at the first occurrence of sexual intercourse

hymn n Christian song of praise sung to God or a saint ▷ **hymnal** n book of hymns (also **hymn book**)

hype n intensive or exaggerated publicity or sales promotion ▷ v promote (a product) using intensive or exaggerated publicity

hyper adj (informal) overactive or overexcited

hyper- prefix over, above, excessively e.g. hyperactive

hyperbola [hie-**per**-bol-a] n (Geom) curve produced when a cone is cut by a plane at a steeper angle to its base than its side

hyperbole [hie-**per**-bol-ee] n deliberate exaggeration for effect ▷ **hyperbolic** adj

hyperlink (Computing) n link from a hypertext file that gives users instant access to related material in another file ▷ v link (files) in this way

hypermarket n very large self-service store

hypersensitive adj extremely sensitive to certain drugs, extremes of temperature, etc.; very easily upset

hypersonic adj having a speed of at least five times the speed of sound

hypertension n very high blood pressure

hypertext n computer software and hardware that allows users to store and view text and move between related items easily

hyphen n punctuation mark (-) indicating that two words or syllables are connected ▶ **hyphenated** adj (of two words or syllables) having a hyphen between them ▶ **hyphenation** n

hypnosis n artificially induced state of relaxation in which the mind is more than usually receptive to suggestion ▶ **hypnotic** adj of or (as if) producing hypnosis ▶ **hypnotism** n inducing hypnosis in someone ▶ **hypnotist** n ▶ **hypnotize** v

hypo- prefix beneath, less than e.g. hypothermia

hypoallergenic adj (of cosmetics) not likely to cause an allergic reaction

hypochondria n undue preoccupation with one's health ▶ **hypochondriac** n

hypocrisy [hip-ok-rass-ee] n, pl -sies (instance of) pretence of having standards or beliefs that are contrary to one's real character or actual behaviour

▶ **hypocrite** [hip-oh-krit] n person who pretends to be what he or she is not ▶ **hypocritical** adj ▶ **hypocritically** adv

hypodermic adj, n (denoting) a syringe or needle used to inject a drug beneath the skin

hypotension n very low blood pressure

hypotenuse [hie-pot-a-news] n side of a right-angled triangle opposite the right angle

hypothalamus n area at the base of the brain, which controls hunger, thirst, etc.

hypothermia n condition in which a person's body temperature is dangerously low as a result of prolonged exposure to severe cold

hypothesis [hie-poth-iss-iss] n, pl -ses [-seez] suggested but unproved explanation of something ▶ **hypothetical** adj based on assumption rather than fact or reality ▶ **hypothetically** adv

hyrax n, pl -raxes or -races type of hoofed rodent-like animal of Africa and Asia

hysterectomy n, pl -mies surgical removal of the womb

hysteria n state of uncontrolled excitement, anger, or panic ▶ **hysterical** adj ▶ **hysterically** adv ▶ **hysterics** pl n attack of hysteria; (informal) uncontrollable laughter

Hz hertz

Ii

I *pron* used by a speaker or writer to refer to himself or herself as the subject of a verb

Iberian *adj* of Iberia, the peninsula comprising Spain and Portugal

ibex [ibe-eks] *n* wild goat of N Africa with large backward-curving horns

ibid. (referring to a book, page, or passage already mentioned) in the same place

ibis [ibe-iss] *n* large wading bird with long legs

ice *n* frozen water; (*chiefly Brit*) portion of ice cream **the Ice** (*NZ informal*) Antarctica ▷ *v* (foll by *up* or *over*) become covered with ice; cover with icing **break the ice** create a relaxed atmosphere, esp. between people meeting for the first time ▶ **iced** *adj* covered with icing; (of a drink) containing ice ▶ **icy** *adj* **icier, iciest** very cold; covered with ice; aloof and unfriendly ▶ **icily** *adv* ▶ **iciness** *n* ▶ **Ice Age** period when much of the earth's surface was covered in glaciers ▶ **iceberg** *n* large floating mass of ice ▶ **icebox** *n* (*US*) refrigerator ▶ **icecap** *n* mass of ice permanently covering an area ▶ **ice cream** sweet creamy frozen food ▶ **ice cube** small square block of ice added to a drink to cool it ▶ **ice floe** sheet of ice floating in the sea ▶ **ice hockey** team game like hockey played on ice with a puck ▶ **ice lolly** flavoured ice on a stick ▶ **ice pick** pointed tool for breaking ice ▶ **ice skate** boot with a steel blade fixed to the sole, to enable the wearer to glide over ice ▶ **ice-skate** *v* ▶ **ice-skater** *n*

Icelandic *adj* of Iceland ▷ *n* language of Iceland ▶ **Icelander** *n*

ichthyology [ik-thi-ol-a-jee] *n* scientific study of fish

icicle *n* tapering spike of ice hanging where water has dripped

icing *n* mixture of sugar and water etc., used to cover and decorate cakes ▶ **icing sugar** finely ground sugar for making icing

icon *n* picture of Christ or another religious figure, regarded as holy in the Orthodox Church; picture on a computer screen representing a function that can be activated by moving the cursor over it

iconoclast *n* person who attacks established ideas or principles ▶ **iconoclastic** *adj*

ICT Information and Communications Technology

ID identification

id *n* (*Psychoanalysis*) the mind's instinctive unconscious energies

idea *n* plan or thought formed in the mind; thought of something; belief or opinion

ideal *adj* most suitable; perfect ▷ *n* conception of something that is perfect; perfect person or thing ▶ **ideally** *adv* ▶ **idealism** *n* tendency to seek perfection in everything ▶ **idealist** *n* ▶ **idealistic** *adj* ▶ **idealize** *v* regard or portray as perfect or nearly perfect ▶ **idealization** *n*

idem *pron, adj* (*Latin*) the same: used to refer to an article, chapter, or book already quoted

identical *adj* exactly the same ▶ **identically** *adv*

identify *v* **-fying, -fied** prove or recognize as being a certain person or thing; (foll by with) understand and sympathize with (a person or group that one regards as being similar or similarly situated); treat as being the same ▶ **identifiable** adj ▶ **identification** n

Identikit® n composite picture, assembled from descriptions given, of a person wanted by the police

identity n, pl -**ties** state of being a specified person or thing; individuality or personality; state of being the same ▶ **identity theft** crime of using financial accounts fraudulently in another person's name without their knowledge

ideology n, pl -**gies** body of ideas and beliefs of a group, nation, etc. ▶ **ideological** adj ▶ **ideologist** n

idiocy n utter stupidity

idiom n group of words which when used together have a different meaning from the words individually e.g. raining cats and dogs; way of expression natural or peculiar to a language or group ▶ **idiomatic** adj ▶ **idiomatically** adv

idiosyncrasy n, pl -**sies** personal peculiarity of mind, habit, or behaviour

idiot n foolish or stupid person; (offens) person with learning difficulties ▶ **idiotic** adj ▶ **idiotically** adv

idle adj not doing anything; not willing to work, lazy; not being used; useless or meaningless e.g. an idle threat ▶ v (usu. foll by away) spend (time) doing very little; (of an engine) run slowly with the gears disengaged ▶ **idleness** n ▶ **idler** n ▶ **idly** adv

idol n object of excessive devotion; image of a god as an object of worship ▶ **idolatry** n worship of idols ▶ **idolatrous** adj ▶ **idolize** v love or admire excessively

idyll [id-**ill**] n scene or time of great peace and happiness ▶ **idyllic** adj ▶ **idyllically** adv

i.e. that is to say

IED improvised explosive device

if conj on the condition or supposition that; whether; even though ▶ n uncertainty or doubt e.g. no ifs, ands, or buts ▶ **iffy** adj (informal) doubtful, uncertain

igloo n, pl -**loos** dome-shaped Inuit house made of snow and ice

igneous [ig-nee-uss] adj (of rock) formed as molten rock cools and hardens

ignite v catch fire or set fire to

ignition n system that ignites the fuel-and-air mixture to start an engine; igniting

ignoble adj dishonourable

ignominy [ig-nom-in-ee] n humiliating disgrace ▶ **ignominious** adj ▶ **ignominiously** adv

ignoramus n, pl -**muses** ignorant person

ignorant adj lacking knowledge; rude through lack of knowledge of good manners ▶ **ignorance** n

ignore v refuse to notice, disregard deliberately

iguana n large tropical American lizard

ileum n lowest part of the small intestine

ilk n type e.g. others of his ilk

ill adj not in good health; harmful or unpleasant e.g. ill effects ▶ n evil, harm ▶ adv badly; hardly, with difficulty e.g. I can ill afford to lose him **ill at ease** uncomfortable, unable to relax ▶ **illness** n ▶ **ill-advised** adj badly thought out; unwise ▶ **ill-disposed** adj (often foll by towards) unfriendly, unsympathetic ▶ **ill-fated** adj doomed to end unhappily

▶ **ill-gotten** adj obtained dishonestly ▶ **ill-health** n condition of being unwell ▶ **ill-mannered** adj having bad manners ▶ **ill-treat** v treat cruelly ▶ **ill will** unkind feeling, hostility

illegal adj against the law ▶ **illegally** adv ▶ **illegality** n, pl **-ties**

illegible adj unable to be read or deciphered

illegitimate adj born of parents not married to each other; not lawful ▶ **illegitimacy** n

illicit adj illegal; forbidden or disapproved of by society

illiterate adj, n (person) unable to read or write ▶ **illiteracy** n

illogical adj unreasonable; not logical ▶ **illogicality** n

illuminate v light up; make clear, explain; decorate with lights; (Hist) decorate (a manuscript) with designs of gold and bright colours ▶ **illumination** n ▶ **illuminating** adj

illusion n deceptive appearance or belief ▶ **illusionist** n conjuror ▶ **illusory** adj seeming to be true, but actually false

illustrate v explain by use of examples; provide (a book or text) with pictures; be an example of ▶ **illustration** n picture or diagram; example ▶ **illustrative** adj ▶ **illustrator** n

illustrious adj famous and distinguished

IM (Computing) instant messaging

image n mental picture of someone or something; impression people have of a person, organization, etc.; representation of a person or thing in a work of art; optical reproduction of someone or something, for example in a mirror; person or thing that looks almost exactly like another; figure of speech, esp. a metaphor or simile ▶ **imagery** n images collectively, esp. in the arts

imagine v form a mental image of; think, believe, or guess ▶ **imaginable** adj ▶ **imaginary** adj existing only in the imagination ▶ **imagination** n ability to make mental images of things that may not exist in real life; creative mental ability ▶ **imaginative** adj having or showing a lot of creative mental ability ▶ **imaginatively** adv

> **SPELLING TIP**
> Remembering that an e changes to an a to form **imagination** is a good way of getting **imaginary** right, because it has an a instead of an e too

imago [im-**may**-go] n, pl **imagoes** or **imagines** sexually mature adult insect

imam n leader of prayers in a mosque; title of some Islamic leaders

IMAX® [**eye**-max] n film projection process which produces an image ten times larger than standard

imbalance n lack of balance or proportion

imbecile [**imb**-ess-eel] n stupid person ▷ adj (also **imbecilic**) stupid or senseless ▶ **imbecility** n

imbibe v drink (alcoholic drinks); (lit) absorb (ideas etc.)

imbroglio [imb-**role**-ee-oh] n, pl **-ios** confusing and complicated situation

imbue v -buing, -bued (usu. foll by with) fill or inspire with (ideals or principles)

IMF International Monetary Fund

imitate v take as a model; copy the voice and mannerisms of, esp. for entertainment ▶ **imitation** n

n copy of an original; imitating
▶ **imitative** *adj* ▶ **imitator** *n*

immaculate *adj* completely
clean or tidy; completely flawless
▶ **immaculately** *adv*

immanent *adj* present within
and throughout something
▶ **immanence** *n*

immaterial *adj* not important,
not relevant

immature *adj* not fully developed;
lacking wisdom or stability
because of youth ▶ **immaturity** *n*

immediate *adj* occurring at once;
next or nearest in time, space, or
relationship ▶ **immediately** *adv*
▶ **immediacy** *n*

immemorial *adj* since, from time
immemorial longer than anyone
can remember

immense *adj* extremely large
▶ **immensely** *adv* to a very great
degree ▶ **immensity** *n*

immerse *v* involve deeply, engross;
plunge (something or someone)
into liquid ▶ **immersion** *n*
▶ **immersion heater** electrical
device in a domestic hot-water
tank for heating water

immigration *n* coming to a foreign
country in order to settle there
▶ **immigrant** *n*

imminent *adj* about to happen
▶ **imminently** *adv* ▶ **imminence** *n*

immobile *adj* not moving;
unable to move ▶ **immobility** *n*
▶ **immobilize** *v* make unable to
move or work

immoderate *adj* excessive or
unreasonable

immolate *v* kill as a sacrifice
▶ **immolation** *n*

immoral *adj* morally wrong,
corrupt; sexually depraved or
promiscuous ▶ **immorality** *n*

USAGE NOTE
Do not confuse *immoral* with

amoral, which means 'having no
moral standards'

immortal *adj* living forever;
famous for all time ▷ *n* person
whose fame will last for all time;
immortal being ▶ **immortality** *n*
▶ **immortalize** *v*

immune *adj* protected against a
specific disease; (foll by *to*) secure
(against); (foll by *from*) exempt
(from) ▶ **immunity** *n*, *pl* **-ties**
ability to resist disease; freedom
from prosecution, tax, etc.
▶ **immunize** *v* make immune to a
disease ▶ **immunization** *n*

immunodeficiency *n* deficiency
in or breakdown of a person's
ability to fight diseases

immunology *n* branch of medicine
concerned with the study of
immunity ▶ **immunological** *adj*
▶ **immunologist** *n*

immutable [im-**mute**-a-bl] *adj*
unchangeable ▶ **immutability** *n*

imp *n* (in folklore) mischievous small
creature with magical powers;
mischievous child

impact *n* strong effect; (force of)
a collision ▷ *v* press firmly into
something

impair *v* weaken or damage
▶ **impairment** *n*

impala [imp-**ah**-la] *n* southern
African antelope

impale *v* pierce with a sharp object

impalpable *adj* difficult to define or
understand

impart *v* communicate
(information); give

impartial *adj* not favouring one
side or the other ▶ **impartially** *adv*
▶ **impartiality** *n*

impassable *adj* (of a road etc.)
impossible to travel through
or over

impasse [am-**pass**] *n* situation in
which progress is impossible

impassioned adj full of emotion

impassive adj showing no emotion, calm

impatient adj irritable at any delay or difficulty; restless (to have or do something) ▶ **impatiently** adv ▶ **impatience** n

impeach v charge with a serious crime against the state ▶ **impeachment** n

impeccable adj without fault, excellent ▶ **impeccably** adv

impecunious adj penniless, poor

impedance [imp-eed-anss] n (Electricity) measure of the opposition to the flow of an alternating current

impede v hinder in action or progress ▶ **impediment** n something that makes action, speech, or progress difficult ▶ **impedimenta** pl n objects impeding progress, esp. baggage or equipment

impel v -pelling, -pelled push or force (someone) to do something

impending adj (esp. of something bad) about to happen

impenetrable adj impossible to get through; impossible to understand

imperative adj extremely urgent, vital; (Grammar) denoting a mood of verbs used in commands ▷ n (Grammar) imperative mood

imperceptible adj too slight or gradual to be noticed ▶ **imperceptibly** adv

imperfect adj having faults or mistakes; not complete; (Grammar) denoting a tense of verbs describing continuous, incomplete, or repeated past actions ▷ n (Grammar) imperfect tense ▶ **imperfection** n

imperial adj of or like an empire or emperor; denoting a system of

weights and measures formerly used in Britain ▶ **imperialism** n rule by one country over many others ▶ **imperialist** adj, n

imperil v -illing, -illed put in danger

imperious adj proud and domineering

impersonal adj not relating to any particular person, objective; lacking human warmth or sympathy; (Grammar) (of a verb) without a personal subject e.g. it is snowing ▶ **impersonality** n

impersonate v pretend to be (another person); copy the voice and mannerisms of, esp. for entertainment ▶ **impersonation** n ▶ **impersonator** n

impertinent adj disrespectful or rude ▶ **impertinently** adv ▶ **impertinence** n

imperturbable adj calm, not excitable

impervious adj (foll by to) not letting (water etc.) through; not influenced by (a feeling, argument, etc.)

impetigo [imp-it-tie-go] n contagious skin disease

impetuous adj done or acting without thought, rash ▶ **impetuously** adv ▶ **impetuosity** n

impetus [imp-it-uss] n, pl -tuses incentive, impulse; force that starts a body moving

impinge v (foll by on) affect or restrict

impious [imp-ee-uss] adj showing a lack of respect or reverence

impish adj mischievous

implacable adj not prepared to be appeased, unyielding ▶ **implacably** adv ▶ **implacability** n

implant n (Med) something put into someone's body, usu. by surgical operation ▷ v put (something) into someone's body, usu. by surgical operation; fix firmly in someone's mind ▶ **implantation** n

implement v carry out (instructions etc.) ▷ n tool, instrument ▶ **implementation** n

implicate v show to be involved, esp. in a crime ▶ **implication** n something implied

implicit adj expressed indirectly; absolute and unquestioning e.g. implicit support ▶ **implicitly** adv

implode v collapse inwards

implore v beg earnestly

imply v -plying, -plied indicate by hinting, suggest; involve as a necessary consequence

impolitic adj unwise or inadvisable

imponderable n, adj (something) impossible to assess

import v bring in (goods) from another country ▷ n something imported; importance; meaning ▶ **importation** n ▶ **importer** n

important adj of great significance or value; having influence or power ▶ **importance** n

importunate adj persistent or demanding ▶ **importune** v harass with persistent requests ▶ **importunity** n, pl -ties

impose v force the acceptance of; (foll by on) take unfair advantage (of) ▶ **imposing** adj grand, impressive ▶ **imposition** n unreasonable demand

impossible adj not able to be done or to happen; absurd or unreasonable ▶ **impossibly** adv ▶ **impossibility** n, pl -ties

imposter, impostor n person who cheats or swindles by pretending to be someone else

impotent [imp-a-tent] adj powerless; (of a man) incapable of sexual intercourse ▶ **impotence** n ▶ **impotently** adv

impound v take legal possession of, confiscate

impoverish v make poor or weak ▶ **impoverishment** n

impracticable adj incapable of being put into practice

impractical adj not sensible

imprecation n curse

impregnable adj impossible to break into ▶ **impregnability** n

impregnate v saturate, spread all through; make pregnant ▶ **impregnation** n

impresario n, pl -ios person who runs theatre performances, concerts, etc.

SPELLING TIP
Note that impresario has only one s

impress v affect strongly, usu. favourably; stress, emphasize; imprint, stamp ▶ **impression** n effect, esp. a strong or favourable one; vague idea; impersonation for entertainment; mark made by pressing ▶ **impressionable** adj easily impressed or influenced

Impressionism n art style that gives a general effect or mood rather than form or structure ▶ **Impressionist** n ▶ **Impressionistic** adj

impressive adj making a strong impression, esp. through size, importance, or quality

imprimatur [imp-rim-**ah**-ter] n official approval to print a book

imprint n mark made by printing or stamping; publisher's name and address on a book ▷ v produce (a mark) by printing or stamping

imprison v put in prison ▶ **imprisonment** n

improbable adj not likely to be true or to happen ▶ **improbability** n, pl **-ties**

impromptu adj without planning or preparation

improper adj indecent; incorrect or irregular ▶ **improper fraction** fraction in which the numerator is larger than the denominator, as in 5/3

impropriety [imp-roe-**pry**-a-tee] n, pl **-ties** unsuitable or slightly improper behaviour

improve v make or become better ▶ **improvement** n

improvident adj not planning for future needs ▶ **improvidence** n

improvise v make use of whatever materials are available; make up (a piece of music, speech, etc.) as one goes along ▶ **improvisation** n

impudent adj cheeky, disrespectful ▶ **impudently** adv ▶ **impudence** n

impugn [imp-**yoon**] v challenge the truth or validity of

impulse n sudden urge to do something; short electrical signal passing along a wire or nerve or through the ear **on impulse** suddenly and without planning ▶ **impulsive** adj acting or done without careful consideration ▶ **impulsively** adv

impunity [imp-**yoon**-it-ee] n **with impunity** without punishment

impure adj having dirty or unwanted substances mixed in; immoral, obscene ▶ **impurity** n

impute v attribute responsibility to ▶ **imputation** n

in prep indicating position inside, state or situation, etc. e.g. in the net; in tears ▶ adv indicating position inside, entry into, etc. e.g. she stayed in; come in ▶ adj fashionable ▶ **inward** adj directed towards the middle; situated

within; spiritual or mental ▶ adv (also **inwards**) towards the inside or middle ▶ **inwardly** adv

inability n lack of means or skill to do something

inaccessible adj impossible or very difficult to reach ▶ **inaccessibility** n

inaccurate adj not correct ▶ **inaccuracy** n, pl **-cies**

inadequate adj not enough; not good enough ▶ **inadequacy** n

inadvertent adj unintentional ▶ **inadvertently** adv

inalienable adj not able to be taken away e.g. an inalienable right

inane adj senseless, silly ▶ **inanity** n

inanimate adj not living

inappropriate adj not suitable

inarticulate adj unable to express oneself clearly or well

inasmuch as conj because or in so far as

inaugurate v open or begin the use of, esp. with ceremony; formally establish (a new leader) in office ▶ **inaugural** adj ▶ **inauguration** n

inauspicious adj unlucky, likely to have an unfavourable outcome

inauthentic adj not authentic, false

inboard adj (of a boat's engine) inside the hull

inborn adj existing from birth, natural

inbox n (on a computer) folder in a mailbox in which incoming messages are stored and displayed

inbred adj produced as a result of inbreeding; inborn or ingrained

inbreeding n breeding of animals or people that are closely related

inbuilt adj present from the start

Inc. (US & Aust) (of a company) incorporated

incalculable adj too great to be estimated

in camera adv see **camera**

incandescent adj glowing with heat ▸ **incandescence** n

incantation n ritual chanting of magic words or sounds

incapable adj (foll by of) unable (to do something); incompetent

incapacitate v deprive of strength or ability ▸ **incapacity** n

incarcerate v imprison ▸ **incarceration** n

incarnate adj in human form ▸ **incarnation** n ▸ **Incarnation** n (Christianity) God's coming to earth in human form as Jesus Christ

incendiary [in-send-ya-ree] adj (of a bomb, attack, etc.) designed to cause fires ▸ n, pl **-aries** bomb designed to cause fires

incense[1] v make very angry

incense[2] n substance that gives off a sweet perfume when burned

incentive n something that encourages effort or action

inception n beginning

incessant adj never stopping ▸ **incessantly** adv

incest n sexual intercourse between two people too closely related to marry ▸ **incestuous** adj

inch n unit of length equal to one twelfth of a foot or 2.54 centimetres ▸ v move slowly and gradually

inchoate [in-koe-ate] adj just begun and not yet properly developed

incidence n extent or frequency of occurrence

incident n something that happens; event involving violence

incidental adj occurring in connection with or resulting from something more important ▸ **incidentally** adv ▸ **incidental music** background music for a film or play

incinerate v burn to ashes ▸ **incineration** n ▸ **incinerator** n furnace for burning rubbish

incipient adj just starting to appear or happen

incise v cut into with a sharp tool ▸ **incision** n ▸ **incisor** n front tooth, used for biting into food

incisive adj direct and forceful

incite v stir up, provoke ▸ **incitement** n

incivility n, pl **-ties** rudeness or a rude remark

inclement adj (of weather) stormy or severe

incline v lean, slope; (cause to) have a certain disposition or tendency ▸ n slope ▸ **inclination** n liking, tendency, or preference; slope

include v have as part of the whole; put in as part of a set or group ▸ **inclusion** n ▸ **inclusive** adj including everything (specified) ▸ **inclusively** adv

incognito [in-kog-nee-toe] adj, adv having adopted a false identity ▸ n, pl **-tos** false identity

incoherent adj unclear and impossible to understand ▸ **incoherence** n ▸ **incoherently** adv

income n amount of money earned from work, investments, etc. ▸ **income support** (in New Zealand) allowance paid by the government to people with a very low income ▸ **income tax** personal tax levied on annual income

incoming adj coming in; about to come into office

incommode v cause inconvenience to

incommunicado adj, adv deprived of communication with other people

incomparable adj beyond comparison, unequalled ▸ **incomparably** adv

incompatible adj inconsistent or conflicting ▸ **incompatibility** n

incompetent adj not having the necessary ability or skill to do something ▸ **incompetence** n

inconceivable adj extremely unlikely, unimaginable

inconclusive adj not giving a final decision or result

incongruous adj inappropriate or out of place ▸ **incongruously** adv ▸ **incongruity** n, pl -**ties**

inconsequential adj unimportant, insignificant

inconsiderable adj **not inconsiderable** fairly large

inconstant adj liable to change one's loyalties or opinions

incontinent adj unable to control one's bladder or bowels ▸ **incontinence** n

incontrovertible adj impossible to deny or disprove

inconvenience n trouble or difficulty ▷ v cause trouble or difficulty to ▸ **inconvenient** adj

incorporate v include or be included as part of a larger unit

incorporeal adj without material form

incorrigible adj beyond correction or reform

incorruptible adj too honest to be bribed or corrupted; not subject to decay

increase v make or become greater in size, number, etc. ▷ n rise in number, size, etc.; amount by which something increases ▸ **increasingly** adv

incredible adj hard to believe or imagine; (informal) marvellous, amazing ▸ **incredibly** adv

incredulous adj not willing to believe something ▸ **incredulity** n

increment n increase in money or value, esp. a regular salary increase ▸ **incremental** adj

incriminate v make (someone) seem guilty of a crime ▸ **incriminating** adj

incubate [in-cube-ate] v (of a bird) hatch (eggs) by sitting on them; grow (bacteria); (of bacteria) remain inactive in an animal or person before causing disease ▸ **incubation** n ▸ **incubator** n heated enclosed apparatus for rearing premature babies; apparatus for artificially hatching birds' eggs

incubus [in-cube-uss] n, pl -**bi** or -**buses** (in folklore) demon believed to have sex with sleeping women; nightmarish burden or worry

inculcate v fix in someone's mind by constant repetition ▸ **inculcation** n

incumbent n person holding a particular office or position ▷ adj **it is incumbent on** it is the duty of ▸ **incumbency** n, pl -**cies**

incur v -**curring**, -**curred** cause (something unpleasant) to happen

incurable adj not able to be cured ▸ **incurably** adv

incurious adj showing no curiosity or interest

incursion n sudden brief invasion

indebted adj owing gratitude for help or favours; owing money ▸ **indebtedness** n

indecent adj morally or sexually offensive; unsuitable or unseemly e.g. indecent haste ▸ **indecently** adv ▸ **indecency** n ▸ **indecent assault** sexual attack which does not include rape ▸ **indecent exposure** showing of one's genitals in public

indecipherable adj impossible to read

indeed adv really, certainly ▷ interj expression of indignation or surprise

indefatigable adj never getting tired ▸ **indefatigably** adv

indefensible adj unable to be justified; impossible to defend

indefinite adj without exact limits e.g. *for an indefinite period*; vague, unclear ▸ **indefinite article** (Grammar) the word *a* or *an* ▸ **indefinitely** adv

indelible adj impossible to erase or remove; making indelible marks ▸ **indelibly** adv

indelicate adj offensive or embarrassing

indemnify v **-ifying, -ified** secure against loss, damage, or liability; compensate for loss or damage

indemnity n, pl **-ties** insurance against loss or damage; compensation for loss or damage

indent v start (a line of writing) further from the margin than the other lines; order (goods) using a special order form ▸ **indentation** n dent in a surface or edge

indenture n contract, esp. one binding an apprentice to his or her employer

independent adj free from the control or influence of others; separate; financially self-reliant; capable of acting for oneself or on one's own ▷ n politician who does not represent any political party ▸ **independently** adv ▸ **independence** n

> **SPELLING TIP**
> **Independent** should be spelt with an *e* at the end in the same way as the noun it is related to: **independence**

in-depth adj detailed, thorough

indescribable adj too intense or extreme for words ▸ **indescribably** adv

indeterminate adj uncertain in extent, amount, or nature ▸ **indeterminacy** n

index n, pl **indices** [in-diss-eez] alphabetical list of names or subjects dealt with in a book; file or catalogue used to find things ▷ v provide (a book) with an index; enter in an index; make index-linked ▸ **index finger** finger next to the thumb ▸ **index-linked** adj (of pensions, wages, etc.) rising or falling in line with the cost of living

Indian n, adj (person) from India; Native American ▸ **Indian summer** period of warm sunny weather in autumn

indicate v be a sign or symptom of; point out; state briefly; (of a measuring instrument) show a reading of ▸ **indication** n ▸ **indicative** adj (foll by *of*) suggesting; (Grammar) denoting a mood of verbs used to make a statement ▷ n (Grammar) indicative mood ▸ **indicator** n something acting as a sign or indication; flashing light on a vehicle showing the driver's intention to turn; dial or gauge

indict [in-**dite**] v formally charge with a crime ▸ **indictable** adj ▸ **indictment** n

indie adj (informal) (of rock music) released by an independent record company

indifferent adj showing no interest or concern; of poor quality ▸ **indifference** n ▸ **indifferently** adv

indigenous [in-**dij**-in-uss] adj born in or natural to a country

indigent adj extremely poor ▸ **indigence** n

i

indigestion n (discomfort or pain caused by) difficulty in digesting food ▸ **indigestible** adj

indignation n anger at something unfair or wrong ▸ **indignant** adj feeling or showing indignation ▸ **indignantly** adv

indignity n, pl -**ties** embarrassing or humiliating treatment

indigo adj deep violet-blue ▷ n dye of this colour

indirect adj done or caused by someone or something else; not by a straight route ▸ **indirect object** (Grammar) person or thing indirectly affected by the action of a verb, e.g. Amy in I bought Amy a bag ▸ **indirect tax** tax added to the price of something

indiscreet adj incautious or tactless in revealing secrets ▸ **indiscreetly** adv ▸ **indiscretion** n

indiscriminate adj showing lack of careful thought

indispensable adj absolutely essential

> **SPELLING TIP**
>
> For every twenty examples of the word **indispensable** in the Collins Corpus, there is one example of the misspelling indispensible

indisposed adj unwell, ill ▸ **indisposition** n

indisputable adj beyond doubt ▸ **indisputably** adv

indissoluble adj permanent

indium n (Chem) soft silvery-white metallic element

individual adj characteristic of or meant for a single person or thing; separate, distinct; distinctive, unusual ▷ n single person or thing ▸ **individually** adv ▸ **individuality** n ▸ **individualism** n principle of living one's life in

one's own way ▸ **individualist** n ▸ **individualistic** adj

indoctrinate v teach (someone) to accept a doctrine or belief uncritically ▸ **indoctrination** n

Indo-European adj, n (of) a family of languages spoken in most of Europe and much of Asia, including English, Russian, and Hindi

indolent adj lazy ▸ **indolence** n

indomitable adj too strong to be defeated or discouraged ▸ **indomitably** adv

indoor adj inside a building ▸ **indoors** adv

indubitable adj beyond doubt, certain ▸ **indubitably** adv

induce v persuade or influence; cause; (Med) cause (a woman) to go into labour or bring on (labour) by the use of drugs etc. ▸ **inducement** n something used to persuade someone to do something

induct v formally install (someone, esp. a clergyman) in office

inductance n property of an electric circuit creating voltage by a change of current

induction n reasoning process by which general conclusions are drawn from particular instances; process by which electrical or magnetic properties are produced by the proximity of an electrified or magnetic object; formal introduction into an office or position ▸ **inductive** adj ▸ **induction coil** transformer for producing a high voltage from a low voltage ▸ **induction course** training course to help familiarize someone with a new job

indulge v allow oneself pleasure; allow (someone) to have or do everything he or she wants ▸ **indulgence** n something allowed because it gives pleasure; act of indulging oneself or someone else; liberal or tolerant treatment ▸ **indulgent** adj ▸ **indulgently** adv

industrial adj of, used in, or employed in industry ▸ **industrialize** v develop large-scale industry in (a country or region) ▸ **industrialization** n ▸ **industrial action** ways in which workers can protest about their conditions, e.g. by striking or working to rule ▸ **industrial estate** area of land set aside for factories and warehouses ▸ **industrial relations** relations between management and workers

industry n, pl **-tries** manufacture of goods; branch of this e.g. *the music industry*; quality of working hard ▸ **industrious** adj hard-working

inebriate n, adj (person who is) habitually drunk ▸ **inebriated** adj drunk ▸ **inebriation** n

inedible adj not fit to be eaten

ineffable adj too great for words ▸ **ineffably** adv

ineffectual adj having very little effect

ineligible adj not qualified for or entitled to something

ineluctable adj impossible to avoid

inept adj clumsy, lacking skill ▸ **ineptitude** n

inequitable adj unfair

ineradicable adj impossible to remove

inert adj without the power of motion or resistance; chemically unreactive ▸ **inertness** n

inertia n feeling of unwillingness to do anything; (*Physics*) tendency of a body to remain still or continue moving unless a force is applied to it

inescapable adj unavoidable

inestimable adj too great to be estimated ▸ **inestimably** adv

inevitable adj unavoidable, sure to happen **the inevitable** something that cannot be prevented ▸ **inevitably** adv ▸ **inevitability** n

inexorable adj unable to be prevented from continuing or progressing ▸ **inexorably** adv

inexpert adj lacking skill

inexplicable adj impossible to explain ▸ **inexplicably** adv

in extremis adj (*Latin*) in great difficulty; on the point of death

inextricable adj impossible to escape from; impossible to disentangle or separate

infallible adj never wrong ▸ **infallibly** adv ▸ **infallibility** n

infamous [in-fam-uss] adj well-known for something bad ▸ **infamously** adv ▸ **infamy** n

infant n very young child ▸ **infancy** n early childhood; early stage of development ▸ **infantile** adj childish

infanticide n murder of an infant; person guilty of this

infantry n soldiers who fight on foot

infatuated adj feeling intense unreasoning passion ▸ **infatuation** n intense unreasoning passion

infect v affect with a disease; affect with a feeling ▸ **infection** n ▸ **infectious** adj (of a disease) spreading without actual contact; spreading from person to person e.g. *infectious enthusiasm*

infer v **-ferring, -ferred** work out from evidence ▸ **inference** n

> **USAGE NOTE**
> Someone *infers* something by 'reading between the lines' of a remark. Do not confuse with *imply*, which means 'to hint'

inferior adj lower in quality, position, or status ▷ n person of lower position or status ▶ **inferiority** n

infernal adj of hell; (informal) irritating ▶ **infernally** adv

inferno n, pl **-nos** intense raging fire

infertile adj unable to produce offspring; (of soil) barren, not productive ▶ **infertility** n

infest v inhabit or overrun in unpleasantly large numbers ▶ **infestation** n

infidel n person with no religion; person who rejects a particular religion, esp. Christianity or Islam

infidelity n, pl **-ties** (act of) sexual unfaithfulness to one's husband, wife, or lover

infighting n quarrelling within a group

infiltrate v enter gradually and secretly ▶ **infiltration** n ▶ **infiltrator** n

infinite [in-fin-it] adj without any limit or end ▶ **infinitely** adv

infinitesimal adj extremely small

infinitive n (Grammar) form of a verb not showing tense, person, or number e.g. to sleep

infinity n endless space, time, or number

infirm adj physically or mentally weak ▶ **infirmity** n, pl **-ties**

infirmary n, pl **-ries** hospital

inflame v make angry or excited ▶ **inflamed** adj (of part of the body) red, swollen, and painful because of infection ▶ **inflammation** n

inflammable adj easily set on fire

> **USAGE NOTE**
> Inflammable means the same as flammable but is falling out of general use as it was often mistaken to mean 'not flammable'

inflammatory adj likely to provoke anger

inflate v expand by filling with air or gas; cause economic inflation in ▶ **inflatable** adj able to be inflated ▷ n plastic or rubber object which can be inflated

inflation n inflating; increase in prices and fall in the value of money ▶ **inflationary** adj

inflection, inflexion n change in the pitch of the voice; (Grammar) change in the form of a word to show grammatical use

inflexible adj unwilling to be persuaded, obstinate; (of a policy etc.) firmly fixed, unalterable ▶ **inflexibly** adv ▶ **inflexibility** n

inflict v impose (something unpleasant) on ▶ **infliction** n

inflorescence n (Botany) arrangement of flowers on a stem

influence n effect of one person or thing on another; (person with) the power to have such an effect ▷ v have an effect on ▶ **influential** adj

influenza n contagious viral disease causing headaches, muscle pains, and fever

influx n arrival or entry of many people or things

info n (informal) information

inform v tell; give incriminating information to the police ▶ **informant** n person who gives information ▶ **information** n knowledge or facts ▶ **informative** adj giving useful information ▶ **information superhighway** worldwide network of computers transferring information at high speed ▶ **information technology** use of computers and electronic technology to store and communicate information ▶ **informer** n person who informs to the police

informal adj relaxed and friendly; appropriate for everyday life or use ▶ **informally** adv ▶ **informality** n

infra dig (informal) beneath one's dignity

infrared adj of or using rays below the red end of the visible spectrum

infrastructure n basic facilities, services, and equipment needed for a country or organization to function properly

infringe v break (a law or agreement) ▶ **infringement** n

infuriate v make very angry

infuse v fill (with an emotion or quality); soak to extract flavour ▶ **infusion** n infusing; liquid obtained by infusing

ingenious [in-jean-ee-uss] adj showing cleverness and originality ▶ **ingeniously** adv ▶ **ingenuity** [in-jen-**new**-it-ee] n

ingénue [an-jay-new] n naive young woman, esp. as a role played by an actress

ingenuous [in-jen-new-uss] adj unsophisticated and trusting ▶ **ingenuously** adv

ingest v take (food or liquid) into the body ▶ **ingestion** n

inglorious adj dishonourable, shameful

ingot n oblong block of cast metal

ingrained adj firmly fixed

ingratiate v try to make (oneself) popular with someone ▶ **ingratiating** adj ▶ **ingratiatingly** adv

ingredient n component of a mixture or compound

ingress n act or right of entering

ingrowing adj (of a toenail) growing abnormally into the flesh

inhabit v -habiting, -habited live in ▶ **inhabitable** adj ▶ **inhabitant** n

inhale v breathe in (air, smoke, etc.) ▶ **inhalation** n ▶ **inhalant** n medical preparation inhaled to help breathing problems ▶ **inhaler** n container for an inhalant

inherent adj existing as an inseparable part ▶ **inherently** adv

inherit v -heriting, -herited receive (money etc.) from someone who has died; receive (a characteristic) from an earlier generation; receive from a predecessor ▶ **inheritance** n ▶ **inheritance tax** tax paid on property left at death ▶ **inheritor** n

inhibit v -hibiting, -hibited restrain (an impulse or desire); hinder or prevent (action) ▶ **inhibited** adj ▶ **inhibition** n feeling of fear or embarrassment that stops one from behaving naturally

inhospitable adj not welcoming, unfriendly; difficult to live in, harsh

inhuman adj cruel or brutal; not human

inhumane adj cruel or brutal ▶ **inhumanity** n

inimical adj unfavourable or hostile

inimitable adj impossible to imitate, unique

iniquity n, pl -ties injustice or wickedness; wicked act ▶ **iniquitous** adj

initial adj first, at the beginning ▷ n first letter, esp. of a person's name ▷ v -tialling, -tialled sign with one's initials ▶ **initially** adv

initiate v begin or set going; admit (someone) into a closed group; instruct in the basics of something ▷ n recently initiated person ▶ **initiation** n ▶ **initiator** n

initiative n first step, commencing move; ability to act independently

inject v put (a fluid) into the body with a syringe; introduce (a new element) e.g. to inject a bit of humour ▶ **injection** n

injudicious adj showing poor judgment, unwise

injunction n court order not to do something

injure v hurt physically or mentally ▸ **injury** n, pl **-ries** ▸ **injury time** (Sport) playing time added at the end of a match to compensate for time spent treating injured players ▸ **injurious** adj

injustice n unfairness; unfair action

ink n coloured liquid used for writing or printing ▸ v (foll by in) mark in ink (something already marked in pencil) ▸ **inky** adj dark or black; covered in ink

inkhosi [in-koh-see] n, pl **amakhosi** (SAfr) tribal chief

inkling n slight idea or suspicion

inlaid adj set in another material so that the surface is smooth; made like this e.g. an inlaid table

inland adj, adv in or towards the interior of a country, away from the sea ▸ **Inland Revenue** (in Britain) government department that collects taxes

in-laws pl n one's husband's or wife's family

inlay n inlaid substance or pattern

inlet n narrow strip of water extending from the sea into the land; valve etc. through which liquid or gas enters

in loco parentis (Latin) in place of a parent

inmate n person living in an institution such as a prison

inmost adj innermost

inn n pub or small hotel, esp. in the country ▸ **innkeeper** n

innards pl n (informal) internal organs; working parts of a machine

innate adj being part of someone's nature, inborn

inner adj happening or located inside; relating to private feelings

e.g. the inner self ▸ **innermost** adj furthest inside ▸ **inner city** parts of a city near the centre, esp. having severe social and economic problems

innings n (Sport) player's or side's turn of batting; period of opportunity

innocent adj not guilty of a crime; without experience of evil; without malicious intent ▸ n innocent person, esp. a child ▸ **innocently** adv ▸ **innocence** n

innocuous adj not harmful ▸ **innocuously** adv

> **SPELLING TIP**
> Note there are two ns in **innocuous**, and only one c

innovation n new idea or method; introduction of new ideas or methods ▸ **innovate** v ▸ **innovative** adj ▸ **innovator** n

innuendo n, pl **-does** (remark making) an indirect reference to something rude or unpleasant

innumerable adj too many to be counted

innumerate adj having no understanding of mathematics or science ▸ **innumeracy** n

inoculate v protect against disease by injecting with a vaccine ▸ **inoculation** n

> **SPELLING TIP**
> It is tempting to spell **inoculate** with two ns but remember there is only one n, and only one c

inoperable adj (of a tumour or cancer) unable to be surgically removed

inopportune adj badly timed, unsuitable

inordinate adj excessive

inorganic adj not having the characteristics of living organisms; of chemical substances that do not contain carbon

inpatient n patient who stays in a hospital for treatment

input n resources put into a project etc.; data fed into a computer. ▷ v **-putting, -put** enter (data) in a computer

inquest n official inquiry into a sudden death

inquire v seek information or ask (about) ▶ **inquirer** n

inquiry n, pl **-ries** question; investigation

inquisition n thorough investigation; (**I-**) (Hist) organization within the Catholic Church for suppressing heresy ▶ **inquisitor** n ▶ **inquisitorial** adj

inquisitive adj excessively curious about other people's affairs ▶ **inquisitively** adv

inquorate adj without enough people present to make a quorum

inroads pl n **make inroads into** start affecting or reducing

insalubrious adj unpleasant, unhealthy, or sordid

insane adj mentally deranged; stupidly irresponsible ▶ **insanely** adv ▶ **insanity** n

insanitary adj dirty or unhealthy

insatiable [in-saysh-a-bl] adj unable to be satisfied

inscribe v write or carve words on ▶ **inscription** n words inscribed

inscrutable adj mysterious, enigmatic ▶ **inscrutably** adv

insect n small animal with six legs and usu. wings, such as an ant or fly ▶ **insecticide** n substance for killing insects ▶ **insectivorous** adj insect-eating

insecure adj anxious, not confident; not safe or well-protected

insemination n putting semen into a woman's or female animal's body to try to make her pregnant ▶ **inseminate** v

insensate adj without sensation, unconscious; unfeeling

insensible adj unconscious, without feeling; (foll by to or of) not aware (of) or affected (by)

insensitive adj unaware of or ignoring other people's feelings ▶ **insensitivity** n

inseparable adj (of two people) spending most of the time together; (of two things) impossible to separate

> **SPELLING TIP**
>
> Note that the third syllable of the word **inseparable** is spelt with an a rather than an e

insert v put inside or include ▷ n something inserted ▶ **insertion** n

inset n small picture inserted within a larger one

inshore adj close to the shore ▷ adj, adv towards the shore

inside prep in or to the interior of ▷ adj in or of the inside; by or from someone within an organization e.g. inside information ▷ adv on, in, or to the inside, indoors; (Brit, Aust & NZ slang) in(to) prison ▷ n inner side, surface, or part ▷ pl (informal) stomach and bowels **inside out** with the inside facing outwards **know inside out** know thoroughly ▶ **insider** n member of a group who has privileged knowledge about it

> **USAGE NOTE**
>
> Avoid using the expression inside of, as the second preposition of is superfluous

insidious adj subtle or unseen but dangerous ▶ **insidiously** adv

insight n deep understanding

insignia n, pl **-nias** or **-nia** badge or emblem of honour or office

insignificant adj not important ▶ **insignificance** n

insincere adj showing false feelings, not genuine ▶ **insincerely** adv ▶ **insincerity** n, pl **-ties**

insinuate v suggest indirectly; work (oneself) into a position by gradual manoeuvres ▶ **insinuation** n

insipid adj lacking interest, spirit, or flavour

insist v demand or state firmly ▶ **insistent** adj making persistent demands; demanding attention ▶ **insistently** adv ▶ **insistence** n

in situ adv, adj (Latin) in its original position

in so far as, insofar as prep to the extent that

insole n inner sole of a shoe or boot

insolent adj rude and disrespectful ▶ **insolence** n ▶ **insolently** adv

insoluble adj incapable of being solved; incapable of being dissolved

insolvent adj unable to pay one's debts ▶ **insolvency** n

insomnia n inability to sleep ▶ **insomniac** n

insouciant adj carefree and unconcerned ▶ **insouciance** n

inspect v check closely or officially ▶ **inspection** n ▶ **inspector** n person who inspects; high-ranking police officer

inspire v fill with enthusiasm, stimulate; arouse (an emotion) ▶ **inspiration** n creative influence or stimulus; brilliant idea ▶ **inspirational** adj

instability n lack of steadiness or reliability

install v put in and prepare (equipment) for use; place (a person) formally in a position or rank ▶ **installation** n installing; equipment installed; place containing equipment for a particular purpose e.g. oil installations

instalment n any of the portions of a thing presented or a debt paid in successive parts

instance n particular example ▷ v mention as an example **for instance** as an example

instant n very brief time; particular moment ▷ adj happening at once; (of foods) requiring little preparation ▶ **instantly** adv

instantaneous adj happening at once ▶ **instantaneously** adv

instead adv as a replacement or substitute

instep n part of the foot forming the arch between the ankle and toes; part of a shoe or boot covering this

instigate v cause to happen ▶ **instigation** n ▶ **instigator** n

instil v -stilling, -stilled introduce (an idea etc.) gradually into someone's mind

instinct n inborn tendency to behave in a certain way ▶ **instinctive** adj ▶ **instinctively** adv

institute n organization set up for a specific purpose, esp. research or teaching ▷ v start or establish

institution n large important organization such as a university or bank; hospital etc. for people with special needs; long-established custom ▶ **institutional** adj ▶ **institutionalize** v

instruct v order to do something; teach (someone) how to do something ▶ **instruction** n order to do something; teaching ▷ pl information on how to do or use something ▶ **instructive** adj informative or helpful ▶ **instructor** n

instrument n tool used for particular work; object played to produce a musical sound; measuring device to show height, speed, etc.; (informal) someone or something used to achieve an aim ▶ **instrumental** adj (foll by in) having an important function (in); played by or composed for musical instruments ▶ **instrumentalist** n player of a musical instrument ▶ **instrumentation** n set of instruments in a car etc.; arrangement of music for instruments

insubordinate adj not submissive to authority ▶ **insubordination** n

insufferable adj unbearable

insular adj not open to new ideas, narrow-minded ▶ **insularity** n

insulate v prevent or reduce the transfer of electricity, heat, or sound by surrounding or lining with a nonconducting material; isolate or set apart ▶ **insulation** n ▶ **insulator** n

insulin n hormone produced in the pancreas that controls the amount of sugar in the blood

insult v behave rudely to, offend ▷ n insulting remark or action ▶ **insulting** adj

insuperable adj impossible to overcome

insupportable adj impossible to tolerate; impossible to justify

insurance n agreement by which one makes regular payments to a company who pay an agreed sum if damage, loss, or death occurs; money paid to or by an insurance company; means of protection ▶ **insure** v protect by insurance ▶ **insurance policy** contract of insurance

insurgent n, adj (person) in revolt against an established authority ▶ **insurgency** n

insurmountable adj impossible to overcome e.g. an insurmountable problem

insurrection n rebellion

intact adj not changed in any way

intaglio [in-tah-lee-oh] n, pl -**lios** (gem carved with) an engraved design

intake n amount or number taken in

intangible adj not clear or definite enough to be seen or felt easily e.g. an intangible quality

integer n positive or negative whole number or zero

integral adj being an essential part of a whole ▷ n (Maths) sum of a large number of very small quantities

integrate v combine into a whole; amalgamate (a religious or racial group) into a community ▶ **integration** n ▶ **integrated circuit** tiny electronic circuit on a chip of semiconducting material

integrity n quality of having high moral principles; quality of being united

intellect n power of thinking and reasoning

intellectual adj of or appealing to the intellect; clever ▷ n intellectual person ▶ **intellectually** adv

intelligent adj able to understand, learn, and think things out quickly; (of a computerized device) able to initiate or modify action in the light of ongoing events ▶ **intelligently** adv ▶ **intelligent design** theory that a sentient being designed and created the universe and all life ▶ **intelligence** n quality of being intelligent; secret government or military information; people or department collecting such information

intelligentsia n intellectual or cultured people in a society

intelligible adj able to be understood ▶ **intelligibility** n

intemperate adj unrestrained, uncontrolled; drinking alcohol to excess ▶ **intemperance** n

intend v propose or plan (to do something); have as one's purpose

intense adj of great strength or degree; deeply emotional ▶ **intensity** n ▶ **intensify** v **-fying, -fied** make or become more intense ▶ **intensification** n

intensive adj using or needing concentrated effort or resources ▶ **intensively** adv

intent n intention ▷ adj paying close attention **intent on doing something** determined to do something ▶ **intently** adv ▶ **intentness** n

intention n something intended ▶ **intentional** adj deliberate, planned in advance ▶ **intentionally** adv

inter [in-ter] v **-terring, -terred** bury (a corpse) ▶ **interment** n

inter- prefix between or among e.g. international

interact v act on or in close relation with each other ▶ **interaction** n ▶ **interactive** adv

interbreed v breed within a related group

intercede v try to end a dispute between two people or groups ▶ **intercession** n

intercept v seize or stop in transit ▶ **interception** n

interchange v (cause to) exchange places ▷ n motorway junction ▶ **interchangeable** adj

Intercity® adj (in Britain) denoting a fast train (service) travelling between cities

intercom n internal communication system with loudspeakers

intercontinental adj travelling between or linking continents

intercourse n sexual intercourse; communication or dealings between people or groups

interdiction, interdict n formal order forbidding something

interdisciplinary adj involving more than one branch of learning

interest n desire to know or hear more about something; something in which one is interested; (often pl) advantage, benefit; sum paid for the use of borrowed money; (often pl) right or share ▷ v arouse the interest of ▶ **interested** adj feeling or showing interest; involved in or affected by something ▶ **interesting** adj ▶ **interestingly** adv

interface n area where two things interact or link; circuit linking a computer and another device

interfaith adj relating to or involving different religions

interfere v try to influence other people's affairs where one is not involved or wanted; (foll by with) clash (with); (foll by with) (Brit, Aust & NZ euphemistic) abuse (a child) sexually ▶ **interfering** adj ▶ **interference** n interfering; (Radio) interruption of reception by atmospherics or unwanted signals

interferon n protein that stops the development of an invading virus

interim adj temporary or provisional

interior n inside; inland region ▷ adj inside, inner; mental or spiritual

interject v make (a remark) suddenly or as an interruption ▶ **interjection** n

interlace v join together as if by weaving

interlink v connect together

interlock v join firmly together

interlocutor [in-ter-lok-yew-ter] n person who takes part in a conversation

interloper [in-ter-lope-er] *n* person in a place or situation where he or she has no right to be

interlude *n* short rest or break in an activity or event

intermarry *v* (of families, races, or religions) become linked by marriage ▸ **intermarriage** *n*

intermediary *n, pl* **-ries** person trying to create agreement between others

intermediate *adj* coming between two points or extremes

intermezzo [in-ter-met-so] *n, pl* **-zos** short piece of music, esp. one performed between the acts of an opera

interminable *adj* seemingly endless because boring ▸ **interminably** *adv*

intermingle *v* mix together

intermission *n* interval between parts of a play, film, etc.

intermittent *adj* occurring at intervals ▸ **intermittently** *adv*

intern *v* imprison, esp. during a war ▸ *n* trainee doctor in a hospital ▸ **internment** *n* ▸ **internee** *n* person who is interned

internal *adj* of or on the inside; within a country or organization; spiritual or mental ▸ **internally** *adv* ▸ **internal-combustion engine** engine powered by the explosion of a fuel-and-air mixture within the cylinders

international *adj* of or involving two or more countries ▸ *n* game or match between teams of different countries; player in such a match ▸ **internationally** *adv*

internecine *adj* mutually destructive

internet, Internet *n* large international computer network

interplanetary *adj* of or linking planets

interplay *n* action and reaction of two things upon each other

interpolate [in-ter-pole-ate] *v* insert (a comment or passage) into (a conversation or text) ▸ **interpolation** *n*

interpose *v* insert between or among things; say as an interruption

interpret *v* explain the meaning of; translate orally from one language into another; convey the meaning of (a poem, song, etc.) in performance ▸ **interpretation** *n*

interpreter *n* person who translates orally from one language into another

interregnum *n, pl* **-nums** or **-na** interval between reigns

interrogate *v* question closely ▸ **interrogation** *n* ▸ **interrogative** *adj* questioning ▸ *n* word used in asking a question, such as *how* or *why* ▸ **interrogator** *n*

interrupt *v* break into (a conversation etc.); stop (a process or activity) temporarily ▸ **interruption** *n*

intersect *v* (of roads) meet and cross; divide by passing across or through ▸ **intersection** *n*

intersex *n* condition of having physiological characteristics between those of a male and a female

interspersed *adj* scattered (among, between, or on)

interstellar *adj* between or among stars

interstice [in-ter-stiss] *n* small crack or gap between things

intertwine *v* twist together

interval *n* time between two particular moments or events; break between parts of a play, concert, etc.; difference in pitch between musical notes **at**

intervals repeatedly; with spaces left between

intervene v involve oneself in a situation, esp. to prevent conflict; happen so as to stop something ▸ **intervention** n

interview n formal discussion, esp. between a job-seeker and an employer; questioning of a well-known person about his or her career, views, etc., by a reporter ▸ v conduct an interview with ▸ **interviewee** n ▸ **interviewer** n

interweave v weave together

intestate adj not having made a will ▸ **intestacy** n

intestine n (often pl) lower part of the alimentary canal between the stomach and the anus ▸ **intestinal** adj ▸ **intestinally** adv

intimate¹ adj having a close personal relationship; personal or private; (of knowledge) extensive and detailed; (foll by with) (euphemistic) having a sexual relationship (with); having a friendly quiet atmosphere ▸ n close friend ▸ **intimately** adv ▸ **intimacy** n

intimate² v hint at or suggest; announce ▸ **intimation** n

intimidate v subdue or influence by fear ▸ **intimidating** adj ▸ **intimidation** n

into prep indicating motion towards the centre, a change, a division, etc. e.g. *into the valley; turned into a madman; cut into pieces*; (informal) interested in

intolerable adj more than can be endured ▸ **intolerably** adv

intolerant adj refusing to accept practices and beliefs different from one's own ▸ **intolerance** n

intonation n sound pattern produced by variations in the voice

intone v speak or recite in an unvarying tone of voice

intoxicate v make drunk; excite to excess ▸ **intoxicant** n intoxicating drink

intoxication n state of being drunk; overexcited state

intractable adj (of a person) difficult to control; (of a problem or issue) difficult to deal with

intranet n (Computing) internal network that makes use of internet technology

intransigent adj refusing to change one's attitude ▸ **intransigence** n

intransitive adj (of a verb) not taking a direct object

intrauterine [in-tra-yoo-ter-rine] adj within the womb

intravenous [in-tra-vee-nuss] adj into a vein ▸ **intravenously** adv

intrepid adj fearless, bold ▸ **intrepidity** n

intricate adj involved or complicated; full of fine detail ▸ **intricately** adv ▸ **intricacy** n, pl -cies

intrigue v make interested or curious; plot secretly ▸ n secret plotting; secret love affair ▸ **intriguing** adj

intrinsic adj essential to the basic nature of something ▸ **intrinsically** adv

introduce v present (someone) by name (to another person); present (a radio or television programme); bring forward for discussion; bring into use; insert ▸ **introduction** n presentation of one person to another; preliminary part or treatment ▸ **introductory** adj

introspection n examination of one's own thoughts and feelings ▸ **introspective** adj

introvert n person concerned more with his or her thoughts and feelings than with the outside world ▸ **introverted** adj ▸ **introversion** n

intrude v come in or join in without being invited ▸ **intrusion** n ▸ **intrusive** adj

intruder n person who enters a place without permission

intuition n instinctive knowledge or insight without conscious reasoning ▸ **intuitive** adj ▸ **intuitively** adv

Inuit n indigenous inhabitant of North America or Greenland

inundate v flood; overwhelm ▸ **inundation** n

inured adj accustomed, esp. to hardship or danger

invade v enter (a country) by military force; enter in large numbers; disturb (someone's privacy) ▸ **invader** n

invalid[1] adj, n disabled or chronically ill (person) ▷ v (often foll by out) dismiss from active service because of illness or injury ▸ **invalidity** n

invalid[2] adj having no legal force; (of an argument etc.) not valid because based on a mistake ▸ **invalidate** v make or show to be invalid

invaluable adj of very great value or worth

invasion n invading; intrusion e.g. an invasion of privacy

invective n abusive speech or writing

inveigh [in-**vay**] v (foll by against) criticize strongly

inveigle v coax by cunning or trickery

invent v think up or create (something new); make up (a story, excuse, etc.) ▸ **invention** n something invented; ability to invent ▸ **inventive** adj creative and resourceful ▸ **inventiveness** n ▸ **inventor** n

inventory n, pl **-tories** detailed list of goods or furnishings

inverse adj reversed in effect, sequence, direction, etc.; (Maths) linking two variables in such a way that one increases as the other decreases ▸ **inversely** adv

invert v turn upside down or inside out ▸ **inversion** n ▸ **inverted commas** quotation marks

invertebrate n animal with no backbone

invest v spend (money, time, etc.) on something with the expectation of profit; (foll by with) give (power or rights) to ▸ **investment** n money invested; something invested in ▸ **investor** n ▸ **invest in** v buy

investigate v inquire into, examine ▸ **investigation** n ▸ **investigative** adj ▸ **investigator** n

investiture n formal installation of a person in an office or rank

inveterate adj firmly established in a habit or condition

invidious adj likely to cause resentment

invigilate v supervise people sitting an examination ▸ **invigilator** n

invigorate v give energy to, refresh

invincible adj impossible to defeat ▸ **invincibility** n

inviolable adj unable to be broken or violated

inviolate adj unharmed, unaffected

invisible adj not able to be seen ▸ **invisibly** adv ▸ **invisibility** n

invite v request the company of; ask politely for; encourage or provoke e.g. the two works inevitably invite comparison ▷ n (informal) invitation ▸ **inviting** adj tempting, attractive ▸ **invitation** n

in-vitro *adj* happening outside the body in an artificial environment

invoice *v, n* (present with) a bill for goods or services supplied

invoke *v* put (a law or penalty) into operation; prompt or cause (a certain feeling); call on (a god or spirit) for help, inspiration, etc. ▶ **invocation** *n*

involuntary *adj* not done consciously, unintentional ▶ **involuntarily** *adv*

involve *v* include as a necessary part; affect, concern; implicate (a person) ▶ **involved** *adj* complicated; concerned, taking part ▶ **involvement** *n*

invulnerable *adj* not able to be wounded or harmed

inward *adj, adv* see **in**

iodine *n* (Chem) bluish-black element used in medicine and photography ▶ **iodize** *v* treat with iodine

ion *n* electrically charged atom ▶ **ionic** *adj* ▶ **ionize** *v* change into ions ▶ **ionization** *n* ▶ **ionosphere** *n* region of ionized air in the upper atmosphere that reflects radio waves

iota *n* very small amount

IOU *n* signed paper acknowledging debt

IPA International Phonetic Alphabet

iPad® *n* type of small portable computer activated by touching the screen

IP address (Computing) internet protocol address: unique code that identifies each computer connected to the internet

iPhone® *n* type of mobile phone which includes a music player and internet browser

i-Player® *n* (in Britain) service provided by the BBC, allowing its recently broadcast television programmes to be viewed over the internet

iPod® *n* small portable digital audio player able to store thousands of tracks downloaded from the internet or transferred from CD

ipso facto *adv* (Latin) by that very fact

IQ intelligence quotient

IRA Irish Republican Army

Iranian *n, adj* (person) from Iran

Iraqi *n, adj* (person) from Iraq

irascible *adj* easily angered ▶ **irascibility** *n*

irate *adj* very angry

ire *n* (lit) anger

iridescent *adj* having shimmering changing colours like a rainbow ▶ **iridescence** *n*

iridium *n* (Chem) very hard corrosion-resistant metal

iris *n* coloured circular membrane of the eye containing the pupil; tall plant with purple, yellow, or white flowers

Irish *adj* of Ireland

irk *v* irritate, annoy ▶ **irksome** *adj* irritating, annoying

iron *n* strong silvery-white metallic element, widely used for structural and engineering purposes; appliance used, when heated, to press clothes; metal-headed golf club ▷ *pl* chains, restraints ▷ *adj* made of iron; strong, inflexible e.g. *iron will* ▷ *v* smooth (clothes or fabric) with an iron ▶ **ironbark** *n* Australian eucalyptus with hard rough bark ▶ **ironing** *n* clothes to be ironed ▶ **ironing board** long cloth-covered board with folding legs, for ironing clothes on ▶ **Iron Age** era when iron tools were used ▶ **iron out** *v* settle (a problem) through discussion

ironic, ironical *adj* using irony; odd or amusing because the opposite of what one would expect ▶ **ironically** adv

ironmonger n shopkeeper or shop dealing in hardware ▶ **ironmongery** n

ironstone n rock consisting mainly of iron ore

irony n, pl **-nies** mildly sarcastic use of words to imply the opposite of what is said; aspect of a situation that is odd or amusing because the opposite of what one would expect

irradiate v subject to or treat with radiation ▶ **irradiation** n

irrational adj not based on or not using logical reasoning

irredeemable adj not able to be reformed or corrected

irreducible adj impossible to put in a simpler form

irrefutable adj impossible to deny or disprove

irregular adj not regular or even; not conforming to accepted practice; (of a word) not following the typical pattern of formation in a language ▶ **irregularly** adv ▶ **irregularity** n, pl **-ties**

irrelevant adj not connected with the matter in hand ▶ **irrelevantly** adv ▶ **irrelevance** n

irreparable adj not able to be repaired or put right ▶ **irreparably** adv

irreplaceable adj impossible to replace

irrepressible adj lively and jolly ▶ **irrepressibly** adv

irreproachable adj blameless, faultless

irresistible adj too attractive or strong to resist ▶ **irresistibly** adv

irrespective of prep without taking account of

irresponsible adj not showing or not done with due care for the consequences of one's actions or attitudes; not capable of accepting responsibility ▶ **irresponsibility** n

irreverent adj not showing due respect ▶ **irreverence** n

irreversible adj not able to be reversed or put right again e.g. irreversible change ▶ **irreversibly** adv

irrevocable adj not possible to change or undo ▶ **irrevocably** adv

irrigate v supply (land) with water by artificial channels or pipes ▶ **irrigation** n

irritate v annoy, anger; cause (a body part) to itch or become inflamed ▶ **irritable** adj easily annoyed ▶ **irritably** adv ▶ **irritant** n, adj (person or thing) causing irritation ▶ **irritation** n

is v third person singular present tense of **be**

ISA (in Britain) Individual Savings Account

isinglass [ize-ing-glass'] n kind of gelatine obtained from some freshwater fish

Islam n Muslim religion teaching that there is one God and that Mohammed is his prophet; Muslim countries and civilization ▶ **Islamic** adj

island n piece of land surrounded by water ▶ **islander** n person who lives on an island; (**I-**) (NZ) Pacific Islander

isle n (poetic) island ▶ **islet** n small island

isobar [ice-oh-bar'] n line on a map connecting places of equal atmospheric pressure

isolate v place apart or alone; (Chem) obtain (a substance) in uncombined form ▶ **isolation** n ▶ **isolationism** n policy of not participating in international affairs ▶ **isolationist** n, adj

isomer [ice-oh-mer] n substance whose molecules contain the same atoms as another but in a different arrangement

isometric adj relating to muscular contraction without shortening of the muscle ▶ **isometrics** pl n isometric exercises

isosceles triangle [ice-soss-ill-eez] n triangle with two sides of equal length

isotherm [ice-oh-therm] n line on a map connecting points of equal temperature

isotope [ice-oh-tope] n one of two or more atoms with the same number of protons in the nucleus but a different number of neutrons

ISP internet service provider

Israeli n, pl **-lis** or **-li** ▷ adj (person) from Israel

issue n topic of interest or discussion; reason for quarrelling; particular edition of a magazine or newspaper; outcome or result; (Law) child or children ▷ v make (a statement etc.) publicly; supply officially (with); produce and make available **take issue with** disagree with

isthmus [iss-muss] n, pl **-muses** narrow strip of land connecting two areas of land

IT information technology

it pron refers to a nonhuman, animal, plant, or inanimate object; refers to a thing mentioned or being discussed; used as the subject of impersonal verbs e.g. it's windy; (informal) crucial or ultimate point ▶ **its** adj, pron belonging to it ▶ **it's** it is; it has ▶ **itself** pron emphatic form of **it**

USAGE NOTE
Beware of mistaking the possessive *its* (no apostrophe) as in *The cat has hurt its paw*, for the

abbreviation *it's* meaning *it is* or *it has*: *It's been a long time*

Italian n language of Italy and one of the languages of Switzerland; person from Italy ▷ adj of Italy

italic adj (of printing type) sloping to the right ▶ **italics** pl n this type, used for emphasis ▶ **italicize** v put in italics

itch n skin irritation causing a desire to scratch; restless desire ▷ v have an itch ▶ **itchy** adj

item n single thing in a list or collection; piece of information ▶ **itemize** v make a list of

iterate v repeat ▶ **iteration** n

itinerant adj travelling from place to place

itinerary n, pl **-aries** detailed plan of a journey

ITV (in Britain) Independent Television

IUD intrauterine device: a coil-shaped contraceptive fitted into the womb

IVF in-vitro fertilization

ivory n hard white bony substance forming the tusks of elephants ▷ adj yellowish-white ▶ **ivory tower** remoteness from the realities of everyday life

ivy n, pl **ivies** evergreen climbing plant

iwi [ee-wee] n (NZ) Māori tribe

Jj

jab v **jabbing, jabbed** poke sharply
▷ n quick punch or poke; (*informal*)
injection

jabber v talk rapidly or
incoherently

jabiru n large white-and-black
Australian stork

jacaranda n tropical tree with
sweet-smelling wood

jack n device for raising a motor
vehicle or other heavy object;
playing card with a picture of a
pageboy; (*Bowls*) small white bowl
aimed at by the players; socket in
electrical equipment into which a
plug fits; flag flown at the bow of a
ship, showing nationality ▶ **jack-
up** n (*NZ informal*) something
achieved dishonestly ▶ **jack up**
v raise with a jack; (*NZ informal*)
organize by dishonest means

jackal n dog-like wild animal of
Africa and Asia

jackaroo, jackeroo n, pl **-roos**
(*Aust*) trainee on a sheep station

jackass n fool; male of the ass
laughing jackass same as
kookaburra

jackboot n high military boot

jackdaw n black-and-grey Eurasian
bird of the crow family

jacket n short coat; skin of a baked
potato; outer paper cover on a
hardback book

jackknife v (of an articulated truck)
go out of control so that the trailer
swings round at a sharp angle to
the cab ▷ n large clasp knife

jackpot n largest prize that may be
won in a game **hit the jackpot** be
very successful through luck

Jacobean [jak-a-bee-an] adj of the
reign of James I of England

Jacobite n supporter of James II of
England and his descendants

Jacquard [jak-ard] n fabric in which
the design is incorporated into
the weave

Jacuzzi® [jak-oo-zee] n circular
bath with a device that swirls
the water

jade n ornamental semiprecious
stone, usu. dark green ▷ adj
bluish-green

jaded adj tired and unenthusiastic

jagged [jag-gid] adj having an
uneven edge with sharp points

jaguar n large S American spotted
cat

jail n prison ▷ v send to prison
▶ **jailer** n ▶ **jailbird** n (*informal*)
person who has often been in
prison

jalopy [jal-lop-ee] n, pl **-lopies**
(*informal*) old car

jam¹ v **jamming, jammed** pack
tightly into a place; crowd or
congest; make or become stuck;
(*Radio*) block (another station)
with impulses of equal wavelength
▷ n hold-up of traffic; (*informal*)
awkward situation **jam on the
brakes** apply brakes fiercely
▶ **jam-packed** adj filled to capacity
▶ **jam session** informal rock or jazz
performance

jam² n food made from fruit boiled
with sugar

jamb n side post of a door or
window frame

jamboree n large gathering or
celebration

Jan. January

jandal n (*NZ*) sandal with a strap
between the toes

jangle v (cause to) make a harsh ringing noise; (of nerves) be upset or irritated

janitor n caretaker of a school or other building

January n first month of the year

japan n very hard varnish, usu. black ▷ v **-panning, -panned** cover with this varnish

Japanese n, adj (native or language) of Japan

jape n (old-fashioned) joke or prank

japonica n shrub with red flowers

jar[1] n wide-mouthed container, usu. round and made of glass

jar[2] v **jarring, jarred** have a disturbing or unpleasant effect; jolt or bump ▷ n jolt or shock

jargon n specialized technical language of a particular subject

jarrah n Australian eucalyptus yielding valuable timber

jasmine n shrub with sweet-smelling yellow or white flowers

jasper n red, yellow, dark green, or brown variety of quartz

jaundice n disease marked by yellowness of the skin ▶ **jaundiced** adj (of an attitude or opinion) bitter or cynical; having jaundice

jaunt n short journey for pleasure

jaunty adj **-tier, -tiest** sprightly and cheerful; smart ▶ **jauntily** adv

javelin n light spear thrown in sports competitions

jaw n one of the bones in which the teeth are set ▷ pl mouth; gripping part of a tool; narrow opening of a gorge or valley ▷ v (slang) talk lengthily

jay n bird with a pinkish body and blue-and-black wings

jaywalker n person who crosses the road in a careless or dangerous manner ▶ **jaywalking** n

jazz n kind of music with an exciting rhythm, usu. involving improvisation ▶ **jazzy** adj flashy or showy ▶ **jazz up** v make more lively

JCB® n (Brit) construction machine with a shovel at the front and an excavator at the rear

jealous adj fearful of losing a partner or possession to a rival; envious; suspiciously watchful ▶ **jealously** adv ▶ **jealousy** n, pl **-sies**

jeans pl n casual denim trousers

Jeep® n four-wheel-drive vehicle

jeer v scoff or deride ▷ n cry of derision

Jehovah n God

jejune adj simple or naive; dull or boring

jell v form into a jelly-like substance; take on a definite form

jelly n, pl **-lies** soft food made of liquid set with gelatine; jam made from fruit juice and sugar ▶ **jellied** adj prepared in a jelly

jellyfish n small jelly-like sea animal

jemmy n, pl **-mies** short steel crowbar used by burglars

jenny n, pl **-nies** female ass or wren

jeopardy n danger ▶ **jeopardize** v place in danger

jerboa n small mouse-like rodent with long hind legs

jerk v move or throw abruptly ▷ n sharp or abruptly stopped movement; (slang) contemptible person ▶ **jerky** adj sudden or abrupt ▶ **jerkily** adv ▶ **jerkiness** n

jerkin n sleeveless jacket

jerry-built adj built badly using flimsy materials

jerry can n flat-sided can for carrying petrol etc.

jersey n knitted jumper; machine-knitted fabric; (J-) breed of cow

Jerusalem artichoke n small yellowish-white root vegetable

jest n, v joke ▶ **jester** n (Hist) professional clown at court

Jesuit [jezz-yoo-it] n member of the Society of Jesus, a Roman Catholic order

jet¹ n aircraft driven by jet propulsion; stream of liquid or gas, esp. one forced from a small hole; nozzle from which gas or liquid is forced ▷ v **jetting, jetted** fly by jet aircraft ▶ **jetboat** n motorboat propelled by a jet of water ▶ **jet lag** fatigue caused by crossing time zones in an aircraft ▶ **jet propulsion** propulsion by thrust provided by a jet of gas or liquid ▶ **jet-propelled** adj ▶ **jet set** rich and fashionable people who travel the world for pleasure

jet² n hard black mineral ▶ **jet-black** adj glossy black

jetsam n goods thrown overboard to lighten a ship

jettison v **-soning, -soned** abandon; throw overboard

jetty n, pl **-ties** small pier

Jew n person whose religion is Judaism; descendant of the ancient Hebrews ▶ **Jewish** adj ▶ **Jewry** n Jews collectively ▶ **jew's-harp** n musical instrument held between the teeth and played by plucking a metal strip with one's finger

jewel n precious stone; special person or thing ▶ **jeweller** n dealer in jewels ▶ **jewellery** n objects decorated with precious stones

jewfish n (Aust) freshwater catfish

jib¹ n triangular sail set in front of a mast

jib² v **jibbing, jibbed** (of a horse, person, etc.) stop and refuse to go on ▶ **jib at** v object to (a proposal etc.)

jib³ n projecting arm of a crane or derrick

jibe¹ n, v taunt or jeer

jibe² v same as **gybe**

jiffy n, pl **-fies** (informal) very short period of time

jig n type of lively dance; music for it; device that holds a component in place for cutting etc. ▷ v **jigging, jigged** make jerky up-and-down movements

jiggery-pokery n (informal) trickery or mischief

jiggle v move up and down with short jerky movements

jigsaw n (also **jigsaw puzzle**) picture cut into interlocking pieces, which the user tries to fit together again; mechanical saw for cutting along curved lines

jihad n Islamic holy war against unbelievers

jilt v leave or reject (one's lover)

jingle n catchy verse or song used in a radio or television advert; gentle ringing sound ▷ v (cause to) make a gentle ringing sound

jingoism n aggressive nationalism ▶ **jingoistic** adj

jinks pl n **high jinks** boisterous merrymaking

jinni n, pl **jinn** spirit in Muslim mythology

jinx n person or thing bringing bad luck ▷ v be or put a jinx on

jitters pl n worried nervousness ▶ **jittery** adj nervous

jive n lively dance of the 1940s and 1950s ▷ v dance the jive

job n occupation or paid employment; task to be done; (informal) difficult task; (Brit, Aust & NZ informal) crime, esp. robbery ▶ **jobbing** adj doing individual jobs for payment ▶ **jobless** adj, pl n unemployed (people) ▶ **job lot** assortment sold together ▶ **job sharing** splitting of one post between two people working part-time

jockey n (professional) rider of racehorses ▷ v **jockey for position** manoeuvre to obtain an advantage

jockstrap n belt with a pouch to support the genitals, worn by male athletes

jocose [joke-kohss] adj playful or humorous

jocular adj fond of joking; meant as a joke ▶ **jocularity** n ▶ **jocularly** adv

jocund [jok-kund] adj (lit) merry or cheerful

jodhpurs pl n riding trousers, loose-fitting above the knee but tight below

joey n (Aust) young kangaroo

jog v **jogging, jogged** run at a gentle pace, esp. for exercise; nudge slightly ▷ n slow run ▶ **jogger** n ▶ **jogging** n

joggle v shake or move jerkily

joie de vivre [jwah de veev-ra] n (French) enjoyment of life

join v become a member (of); come into someone's company; take part (in); come or bring together ▷ n place where two things are joined ▶ **join up** v enlist in the armed services ▶ **joined-up** adj integrated by an overall strategy e.g. joined-up government

joiner n maker of finished woodwork ▶ **joinery** n joiner's work

joint adj shared by two or more ▷ n place where bones meet but can move; junction of two or more parts or objects; piece of meat for roasting; (slang) house or place, esp. a disreputable bar or nightclub; (slang) marijuana cigarette ▷ v divide meat into joints **out of joint** disorganized; dislocated ▶ **jointed** adj ▶ **jointly** adv

joist n horizontal beam that helps support a floor or ceiling

jojoba [hoe-hoe-ba] n shrub of SW North America whose seeds yield oil used in cosmetics

joke n thing said or done to cause laughter; amusing or ridiculous person or thing ▷ v make jokes ▶ **jokey** adj ▶ **jokingly** adv ▶ **joker** n person who jokes; (slang) fellow; extra card in a pack, counted as any other in some games

jolly adj **-lier, -liest** (of a person) happy and cheerful; (of an occasion) merry and festive ▷ v **-lying, -lied** ▶ **jolly along** try to keep (someone) cheerful by flattery or coaxing ▶ **jollity** n ▶ **jollification** n merrymaking

jolt n unpleasant surprise or shock; sudden jerk or bump ▷ v surprise or shock; move or shake with a jerk

jonquil n fragrant narcissus

josh v (chiefly US slang) tease

joss stick n stick of incense giving off a sweet smell when burnt

jostle v knock or push against

jot v **jotting, jotted** write briefly ▷ n very small amount ▶ **jotter** n notebook ▶ **jottings** pl n notes jotted down

joule [jool] n (Physics) unit of work or energy

journal n daily newspaper or magazine; daily record of events ▶ **journalese** n superficial style of writing, found in some newspapers ▶ **journalism** n writing or editing of newspapers and magazines ▶ **journalist** n ▶ **journalistic** adj

journey n act or process of travelling from one place to another ▷ v travel

journeyman n qualified craftsman employed by another

joust (*Hist*) *n* combat with lances between two mounted knights ▷ *v* fight on horseback using lances

jovial *adj* happy and cheerful ▶ **jovially** *adv* ▶ **joviality** *n*

jowl¹ *n* lower jaw ▷ *pl* cheeks

jowl² *n* fatty flesh hanging from the lower jaw

joy *n* feeling of great delight or pleasure; cause of this feeling ▶ **joyful** *adj* ▶ **joyless** *adj* ▶ **joyous** *adj* extremely happy and enthusiastic ▶ **joyriding** *n* driving for pleasure, esp. in a stolen car ▶ **joyride** *n* ▶ **joyrider** *n* ▶ **joystick** *n* control device for an aircraft or computer

JP (in Britain) Justice of the Peace

JPEG (jay-peg) (*Computing*) *n* standard compressed file format used for pictures; picture held in this file format

Jr Junior

JSA jobseeker's allowance: (in Britain) a payment made to unemployed people

jubilant *adj* feeling or expressing great joy ▶ **jubilantly** *adv* ▶ **jubilation** *n*

jubilee *n* special anniversary, esp. 25th (**silver jubilee**) or 50th (**golden jubilee**)

Judaism *n* religion of the Jews, based on the Old Testament and the Talmud ▶ **Judaic** *adj*

judder *v* vibrate violently ▷ *n* violent vibration ▶ **judder bar** (*NZ*) raised strip across a road designed to slow down vehicles

judge *n* public official who tries cases and passes sentence in a court of law; person who decides the outcome of a contest ▷ *v* act as a judge; appraise critically; consider something to be the case ▶ **judgment, judgement** *n* opinion reached after careful thought; verdict of a judge; ability to appraise critically ▶ **judgmental, judgemental** *adj*

USAGE NOTE
The alternative spellings with or without an 'e' between 'g' and 'm' are equally acceptable

judicial *adj* of or by a court or judge; showing or using judgment ▶ **judicially** *adv*

judiciary *n* system of courts and judges

judicious *adj* well-judged and sensible ▶ **judiciously** *adv*

judo *n* sport in which two opponents try to throw each other to the ground

jug *n* container for liquids, with a handle and small spout ▶ **jugged hare** stewed in an earthenware pot

juggernaut *n* (*Brit*) large heavy truck; any irresistible destructive force

juggle *v* throw and catch (several objects) so that most are in the air at the same time; manipulate (figures, situations, etc.) to suit one's purposes ▶ **juggler** *n*

jugular, jugular vein *n* one of three large veins of the neck that return blood from the head to the heart

juice *n* liquid part of vegetables, fruit, or meat; (*Brit, Aust & NZ informal*) petrol ▷ *pl* fluids secreted by an organ of the body ▶ **juicy** *adj* full of juice; interesting

jujitsu *n* Japanese art of wrestling and self-defence

juju *n* W African magic charm or fetish

jukebox *n* coin-operated machine on which records, CDs, or videos can be played

Jul. July

julep *n* sweet alcoholic drink

July n seventh month of the year

jumble n confused heap or state; articles for a jumble sale ▷ v mix in a disordered way ▶ **jumble sale** sale of miscellaneous second-hand items

jumbo adj (informal) very large ▷ n (also **jumbo jet**) large jet airliner

jumbuck n (Aust, old-fashioned slang) sheep

jump v leap or spring into the air using the leg muscles; move quickly and suddenly; jerk with surprise; increase suddenly; change the subject abruptly; (informal) attack without warning; pass over or miss out (intervening material) ▷ n act of jumping; sudden rise; break in continuity **jump the gun** act prematurely **jump the queue** not wait one's turn ▶ **jumpy** adj nervous ▶ **jump at** v accept (a chance etc.) gladly ▶ **jumped-up** adj arrogant because of recent promotion ▶ **jump jet** fixed-wing jet that can take off and land vertically ▶ **jump leads** electric cables to connect a flat car battery to an external battery to aid starting an engine ▶ **jump on** v attack suddenly and forcefully ▶ **jump suit** one-piece garment of trousers and top

jumper n sweater or pullover

Jun. June; Junior

junction n place where routes, railway lines, or roads meet

juncture n point in time, esp. a critical one

June n sixth month of the year

jungle n tropical forest of dense tangled vegetation; confusion or mess; place of intense struggle for survival

junior adj of lower standing; younger ▷ n junior person

juniper n evergreen shrub with purple berries

junk¹ n discarded or useless objects; (informal) rubbish; (slang) narcotic drug, esp. heroin ▶ **junkie, junky** n, pl **junkies** (slang) drug addict ▶ **junk food** snack food of low nutritional value ▶ **junk mail** unwanted mail advertising goods or services

junk² n flat-bottomed Chinese sailing boat

junket n excursion by public officials paid for from public funds; sweetened milk set with rennet

junta n group of military officers holding power in a country, esp. after a coup

Jupiter n king of the Roman gods; largest of the planets

juridical adj of law or the administration of justice

jurisdiction n right or power to administer justice and apply laws; extent of this right or power

jurisprudence n science or philosophy of law

jurist n expert in law

jury n, pl **-ries** group of people sworn to deliver a verdict in a court of law ▶ **juror** n member of a jury

just adv very recently; at this instant; merely; only; exactly; barely; really ▷ adj fair or impartial in action or judgment; proper or right ▶ **justly** adv ▶ **justness** n

justice n quality of being just; judicial proceedings; judge or magistrate ▶ **Justice of the Peace** (in Britain) person who is authorized to act as a judge in a local court of law

justify v **-fying, -fied** prove right or reasonable; explain the reasons for an action; align (text) so the margins are straight

▶**justifiable** adj ▶**justifiably** adv
▶**justification** n
jut v **jutting, jutted** project or stick out
jute n plant fibre, used for rope, canvas, etc.
juvenile adj young; of or suitable for young people; immature and rather silly ▷ n young person or child ▶**juvenilia** pl n works produced in an author's youth ▶**juvenile delinquent** young person guilty of a crime
juxtapose v put side by side ▶**juxtaposition** n

Kk

K (informal) thousand(s)
Kaffir [kaf-fer] n (S Afr offens) Black African
kaftan n long loose Eastern garment; woman's dress resembling this
kai n (NZ informal) food
kaiser [kize-er] n (Hist) German or Austro-Hungarian emperor
kak n (S Afr slang) faeces; rubbish
Kalashnikov n Russian-made automatic rifle
kale n cabbage with crinkled leaves
kaleidoscope n tube-shaped toy containing loose coloured pieces reflected by mirrors so that intricate patterns form when the tube is twisted ▶**kaleidoscopic** adj
kamikaze [kam-mee-kah-zee] n (in World War II) Japanese pilot who performed a suicide mission ▷ adj (of an action) undertaken in the knowledge that it will kill or injure the person performing it
kangaroo n, pl **-roos** Australian marsupial which moves by jumping with its powerful hind legs ▶**kangaroo court** unofficial court set up by a group to discipline its members ▶**kangaroo paw** Australian plant with green-and-red flowers
kaolin n fine white clay used to make porcelain and also used in some medicines
kapok n fluffy fibre from a tropical tree, used to stuff cushions etc.

kaput [kap-**poot**] *adj* (*informal*) ruined or broken

karaoke *n* form of entertainment in which people sing over a prerecorded backing tape

karate *n* Japanese system of unarmed combat using blows with the feet, hands, elbows, and legs

karma *n* (*Buddhism, Hinduism*) person's actions affecting his or her fate in the next reincarnation

karoo *n* (*SAfr*) high arid plateau

karri *n*, *pl* **-ris** Australian eucalyptus; its wood, used for building

katipo *n* small poisonous New Zealand spider

kauri *n* large NZ conifer that yields valuable timber and resin

kayak *n* Inuit canoe made of sealskins stretched over a frame; fibreglass or canvas-covered canoe of this design

kbps (*Computing*) kilobits per second: measure of the rate of data transfer

kbyte (*Computing*) kilobyte

kebab *n* dish of small pieces of meat grilled on skewers; (also **doner kebab**) grilled minced lamb served in a split slice of unleavened bread

kedgeree *n* dish of fish with rice and eggs

keel *n* main lengthways timber or steel support along the base of a ship ▸ **keel over** *v* turn upside down; (*informal*) collapse suddenly

keen¹ *adj* eager or enthusiastic; intense or strong; intellectually acute; (of the senses) capable of recognizing small distinctions; sharp; cold and penetrating; competitive ▸ **keenly** *adv* ▸ **keenness** *n*

keen² *v* wail over the dead

keep *v* **keeping, kept** have or retain possession of; store; stay or cause to stay (in, on, or at a place

or position); continue or persist; detain (someone); look after or maintain ▸ *n* cost of food and everyday expenses ▸ **keeper** *n* person who looks after animals in a zoo; person in charge of a museum or collection; short for **goalkeeper** ▸ **keeping** *n* care or charge **in keeping with**, **out of keeping with** appropriate or inappropriate for ▸ **keep fit** exercises designed to promote physical fitness ▸ **keepsake** *n* gift treasured for the sake of the giver ▸ **keep up** *v* maintain at the current level ▸ **keep up with** *v* maintain a pace set by (someone)

keg *n* small metal beer barrel

kelp *n* large brown seaweed

kelpie *n* Australian sheepdog with a smooth coat and upright ears

kelvin *n* SI unit of temperature ▸ **Kelvin scale** temperature scale starting at absolute zero (-273.15° Celsius)

ken *v* **kenning, kenned** *or* **kent** (*Scot*) know **beyond one's ken** beyond one's range of knowledge

kendo *n* Japanese sport of fencing using wooden staves

kennel *n* hutlike shelter for a dog ▷ *pl* place for breeding, boarding, or training dogs

kept *v* past of **keep**

keratin *n* fibrous protein found in the hair and nails

kerb *n* edging to a footpath ▸ **kerb crawling** (*Brit*) act of driving slowly beside a pavement to pick up a prostitute

kerchief *n* piece of cloth worn over the head or round the neck

kerfuffle *n* (*informal*) commotion or disorder

kernel *n* seed of a nut, cereal, or fruit stone; central and essential part of something

kerosene n (US, Canadian, Aust & NZ) liquid mixture distilled from petroleum and used as a fuel or solvent

kestrel n type of small falcon

ketch n two-masted sailing ship

ketchup n thick cold sauce, usu. made of tomatoes

kettle n container with a spout and handle used for boiling water ▷ v (Brit informal) (of a police force) to contain (people involved in a public demonstration) in an enclosed space ▶ **kettledrum** n large bowl-shaped metal drum

key n device for operating a lock by moving a bolt; device turned to wind a clock, operate a machine, etc.; any of a set of levers or buttons pressed to operate a typewriter, computer, or musical keyboard instrument; (Music) set of related notes; something crucial in providing an explanation or interpretation; means of achieving a desired end; list of explanations of codes, symbols, etc. ▷ adj of great importance ▷ v (also **key in**) enter (text) using a keyboard **keyed up** very excited or nervous ▶ **key worker** (in Britain) worker in a public sector profession considered to be essential to society

keyboard n set of keys on a piano, computer, etc.; musical instrument played using a keyboard ▷ v enter (text) using a keyboard

keyhole n opening for inserting a key into a lock

keynote n dominant idea of a speech etc.; basic note of a musical key

keypad n panel with a set of buttons for operating an electronic device

keystone n most important part of a process, organization, etc.;

central stone of an arch which locks the others in position

kg kilogram(s)

KGB n (formerly) Soviet secret police

khaki adj dull yellowish-brown ▷ n hard-wearing fabric of this colour used for military uniforms

kHz kilohertz

kia ora [kee-a **aw**-ra] interj (NZ) Māori greeting

kibbutz n, pl **kibbutzim** communal farm or factory in Israel

kibosh n **put the kibosh on** (slang) put a stop to

kick v drive, push, or strike with the foot; (of a gun) recoil when fired; (informal) object or resist; (informal) free oneself of (an addiction); (Rugby) score with a kick ▷ n thrust or blow with the foot; recoil of a gun; (informal) excitement or thrill ▶ **kickback** n money paid illegally for favours done ▶ **kick off** v start a game of soccer; (informal) begin ▶ **kick out** v dismiss or expel forcibly ▶ **kick-start** v start (a motorcycle) by kicking a pedal ▶ **kick up** v (informal) create a fuss)

kid n (informal) child; young goat; leather made from the skin of a young goat

kid v **kidding, kidded** (informal) tease or deceive (someone)

kidnap v **-napping, -napped** seize and hold (a person) to ransom ▶ **kidnapper** n

kidney n either of the pair of organs that filter waste products from the blood to produce urine; animal kidney used as food ▶ **kidney bean** reddish-brown kidney-shaped bean, edible when cooked

kill v cause the death of; (informal) cause (someone) pain or discomfort; put an end to; pass (time) ▷ n act of killing; animals

or birds killed in a hunt ▶ **killer** n
▶ **killing** (informal) adj very tiring;
very funny ▶ n sudden financial
success ▶ **killjoy** n person who
spoils others' pleasure

killer whale n predatory black-
and-white toothed whale with a
large erect dorsal fin

kiln n oven for baking, drying, or
processing pottery, bricks, etc.

kilo n short for **kilogram**

kilo- combining form one thousand
e.g. **kilometre**

kilobyte n (Computing) 1024 units of
information

kilogram, kilogramme n one
thousand grams

kilohertz n one thousand hertz

kilometre n one thousand metres

kilowatt n (Electricity) one
thousand watts

kilt n knee-length pleated tartan
skirt-like garment worn orig. by
Scottish Highlanders ▶ **kilted** adj

kimono n, pl **-nos** loose wide-
sleeved Japanese robe, fastened
with a sash; European dressing
gown resembling this

kin, kinsfolk n person's relatives
collectively ▶ **kinship** n

kind[1] adj considerate, friendly, and
helpful ▶ **kindness** n ▶ **kindly** adj
having a warm-hearted nature;
pleasant or agreeable ▶ adv in a
considerate way; please e.g. will
you kindly be quiet! ▶ **kindliness** n
▶ **kind-hearted** adj

kind[2] n class or group with common
characteristics; essential nature
or character **in kind** (of payment)
in goods rather than money; with
something similar **kind of** to a
certain extent

USAGE NOTE
Note the singular/plural usage:
this (or that) kind of dog; these (or
those) kinds of dog. In the second,

plural example, you can also say
these kinds of dogs

kindergarten n class or school for
children under six years old

kindle v set (a fire) alight; (of a fire)
start to burn; arouse or be aroused
▷ n (K-)® portable electronic device
for downloading and reading
books ▶ **kindling** n dry wood or
straw for starting fires

kindred adj having similar qualities;
related by blood or marriage ▷ n
same as **kin**

kindy, kindie n, pl **-dies** (Aust & NZ
informal) kindergarten

kinetic [kin-net-ik] adj relating to
or caused by motion

king n male ruler of a monarchy;
ruler or chief; best or most
important of its kind; piece in chess
that must be defended; playing
card with a picture of a king on it
▶ **kingdom** n state ruled by a king
or queen; division of the natural
world ▶ **king prawn** large prawn,
fished commercially in Australian
waters ▶ **kingship** n ▶ **king-size,
king-sized** adj larger than standard
size

kingfisher n small bird, often with
a bright-coloured plumage, that
dives for fish

kingpin n most important person
in an organization

kink n twist or bend in rope, wire,
hair, etc.; (informal) quirk in
someone's personality ▶ **kinky**
adj (slang) given to unusual sexual
practices; full of kinks

kiosk n small booth selling drinks,
cigarettes, newspapers, etc.;
public telephone box

kip n, v **kipping, kipped** (informal)
sleep

kipper n cleaned, salted, and
smoked herring

kirk n (Scot) church

Kirsch n brandy made from cherries

kismet n fate or destiny

kiss v touch with the lips in affection or greeting; join lips with a person in love or desire ▷ n touch with the lips **kiss of life** mouth-to-mouth resuscitation ▶ **kisser** n (slang) mouth or face ▶ **kissagram** n greetings service in which a messenger kisses the person celebrating ▶ **kissing crust** (NZ & S Afr) soft end of a loaf of bread where two loaves have been separated

kist n (S Afr) large wooden chest

kit n outfit or equipment for a specific purpose; set of pieces of equipment ready to be assembled; (NZ) flax basket ▶ **kitbag** n bag for a soldier's or traveller's belongings ▶ **kit out** v **kitting, kitted** provide with clothes or equipment needed for a particular activity ▶ **kitset** n (NZ) unassembled pieces for constructing a piece of furniture

kitchen n room used for cooking ▶ **kitchenette** n small kitchen ▶ **kitchen garden** garden for growing vegetables, herbs, etc.

kite n light frame covered with a thin material flown on a string in the wind; large hawk with a forked tail ▶ **Kite mark** (Brit) official mark on articles approved by the British Standards Institution

kith n **kith and kin** friends and relatives

kitsch n art or literature with popular sentimental appeal

kitten n young cat ▶ **kittenish** adj lively and flirtatious

kittiwake n type of seagull

kitty n, pl **-ties** communal fund; total amount wagered in certain gambling games

kiwi n New Zealand flightless bird with a long beak and no tail; (informal) New Zealander ▶ **kiwi fruit** edible fruit with a fuzzy brownish skin and green flesh

klaxon n loud horn used on emergency vehicles as a warning signal

kleptomania n compulsive tendency to steal ▶ **kleptomaniac** n

kloof n (S Afr) mountain pass or gorge

km kilometre(s)

knack n skilful way of doing something; innate ability

knacker n (Brit) buyer of old horses for killing

knackered adj (slang) extremely tired; no longer functioning

knapsack n soldier's or traveller's bag worn strapped on the back

knave n jack at cards; (obs) dishonest man

knead v work (dough) into a smooth mixture with the hands; squeeze or press with the hands

knee n joint between thigh and lower leg; lap; part of a garment covering the knee ▷ v **kneeing, kneed** strike or push with the knee ▶ **kneecap** n bone in front of the knee ▷ v shoot in the kneecap ▶ **kneejerk** adj (of a reply or reaction) automatic and predictable ▶ **knees-up** n (Brit informal) party

kneel v **kneeling, kneeled** or **knelt** fall or rest on one's knees

knell n sound of a bell, esp. at a funeral or death; portent of doom

knew v past tense of **know**

knickerbockers pl n loose-fitting short trousers gathered in at the knee

knickers pl n woman's or girl's undergarment covering the lower trunk and having legs or legholes

k

knick-knack n trifle or trinket

knife n, pl **knives** cutting tool or weapon consisting of a sharp-edged blade with a handle ▷ v cut or stab with a knife

knight n man who has been given a knighthood; (Hist) man who served his lord as a mounted armoured soldier; chess piece shaped like a horse's head ▷ v award a knighthood to ▶ **knighthood** n honorary title given to a man by the British sovereign ▶ **knightly** adj

knit v **knitting, knitted** or **knit** make (a garment) by interlocking a series of loops in wool or other yarn; join closely together; draw (one's eyebrows) together ▶ **knitting** n ▶ **knitwear** n knitted clothes, such as sweaters

knob n rounded projection, such as a switch on a radio; rounded handle on a door or drawer; small amount of butter ▶ **knobbly** adj covered with small bumps

knobkerrie n (SAfr) club with a rounded end

knock v give a blow or push to; rap audibly with the knuckles; make or drive by striking; (informal) criticize adversely; (of an engine) make a regular banging noise as a result of a fault ▷ n blow or rap; knocking sound ▶ **knocker** n metal fitting for knocking on a door ▶ **knock about, knock around** v wander or spend time aimlessly; hit or kick brutally ▶ **knockabout** adj (of comedy) boisterous ▶ **knock back** v (informal) drink quickly; cost; reject or refuse ▶ **knock down** v demolish; reduce the price of ▶ **knockdown** adj (of a price) very low ▶ **knock-knees** pl n legs that curve in at the knees ▶ **knock off** v (informal) cease work; (informal) make or do

(something) hurriedly or easily; take (a specified amount) off a price; (Brit, Aust & NZ informal) steal ▶ **knock out** v render (someone) unconscious; (informal) overwhelm or amaze; defeat in a knockout competition ▶ **knockout** n blow that renders an opponent unconscious; competition from which competitors are progressively eliminated; (informal) overwhelmingly attractive person or thing ▶ **knock up** v (informal) assemble (something) quickly; (informal) waken ▶ **knock-up** n practice session at tennis, squash, or badminton

knoll n small rounded hill

knot n fastening made by looping and pulling tight strands of string, cord, or rope; tangle (of hair); small cluster or huddled group; round lump or spot in timber; feeling of tightness, caused by tension or nervousness; unit of speed used by ships, equal to one nautical mile (1.85 kilometres) per hour ▷ v **knotting, knotted** tie with or into a knot ▶ **knotty** adj full of knots; puzzling or difficult

know v **knowing, knew, known** be or feel certain of the truth of (information etc.); be acquainted with; have a grasp of or understand (a skill or language); be aware of **in the know** (informal) informed or aware ▶ **knowable** adj ▶ **knowing** adj suggesting secret knowledge ▶ **knowingly** adv deliberately; in a way that suggests secret knowledge ▶ **know-all** n (derogatory) person who acts as if knowing more than other people ▶ **know-how** n (informal) ingenuity, aptitude, or skill

knowledge n facts or experiences known by a person; state of knowing; specific information on a subject ▸ **knowledgeable, knowledgable** *adj* intelligent or well-informed

knuckle *n* bone at the finger joint; knee joint of a calf or pig **near the knuckle** (*informal*) rather rude or offensive ▸ **knuckle-duster** *n* metal appliance worn on the knuckles to add force to a blow ▸ **knuckle under** *v* yield or submit

KO knockout

koala *n* tree-dwelling Australian marsupial with dense grey fur

kohanga reo, kohanga *n* (*NZ*) infant class where children are taught in Māori

kohl *n* cosmetic powder used to darken the edges of the eyelids

kookaburra *n* large Australian kingfisher with a cackling cry

koori *n*, *pl* **-ris** Australian Aborigine

kopje, koppie *n* (*SAfr*) small hill

Koran *n* sacred book of Islam

kosher [koh-sher] *adj* conforming to Jewish religious law, esp. (of food) to Jewish dietary law; (*informal*) legitimate or authentic ▹ *n* kosher food

kowhai *n* New Zealand tree with clusters of yellow flowers

kowtow *v* be servile (towards)

kph kilometres per hour

kraal *n* S African village surrounded by a strong fence

Kremlin *n* central government of Russia and, formerly, the Soviet Union

krill *n*, *pl* **krill** small shrimplike sea creature

krypton *n* (*Chem*) colourless gas present in the atmosphere and used in fluorescent lights

kryptonite *n* imaginary substance said to render a person or thing powerless

kudos *n* fame or credit

kugel [koog-el] *n* (*SAfr*) rich, fashion-conscious, materialistic young woman

kumara *n* (*NZ*) tropical root vegetable with yellow flesh

kumquat [kumm-kwott] *n* citrus fruit resembling a tiny orange

kung fu *n* Chinese martial art combining hand, foot, and weapon techniques

kura kaupapa Māori *n* (*NZ*) primary school where the teaching is done in Māori

kurrajong *n* Australian tree or shrub with tough fibrous bark

kW kilowatt

kwela [kway-luh] *n* type of South African popular music using simple instruments

kWh kilowatt-hour

k

Ll

L large; learner (driver)

l litre

laat lammetjie *n* (S Afr informal) a child born many years after its siblings

lab *n* (informal) short for **laboratory**

label *n* piece of card or other material fixed to an object to show its ownership, destination, etc. ▷ *v* **-elling, -elled** give a label to

labia *pl n*, *sing* **labium** four liplike folds of skin forming part of the female genitals ▶ **labial** [lay-bee-al] *adj* of the lips

labor *n* (US & Aust) same as **labour** ▶ **Labor Day** (in the US and Canada) public holiday in honour of labour, held on the first Monday in September; (in Australia) public holiday observed on different days in different states

laboratory *n*, *pl* **-ries** building or room designed for scientific research or for the teaching of practical science

laborious *adj* involving great prolonged effort ▶ **laboriously** *adv*

Labor Party *n* main left-wing political party in Australia

labour, (US & Aust) **labor** *n* physical work or exertion; workers in industry; final stage of pregnancy, leading to childbirth ▷ *v* work hard; stress to excess or too persistently; be at a disadvantage because of a mistake or false belief ▶ **laboured** *adj* uttered or done with difficulty ▶ **labourer** *n* person who labours,

esp. someone doing manual work for wages ▶ **Labour Day** (in Britain) a public holiday in honour of work, held on May 1; (in New Zealand) a public holiday commemorating the introduction of the eight-hour day, held on the fourth Monday in October ▶ **Labour Party** main left-wing political party in a number of countries including Britain and New Zealand

labrador *n* large retriever dog with a usu. gold or black coat

laburnum *n* ornamental tree with yellow hanging flowers

labyrinth [lab-er-inth] *n* complicated network of passages; interconnecting cavities in the internal ear ▶ **labyrinthine** *adj*

lace *n* delicate decorative fabric made from threads woven into an open weblike pattern; cord drawn through eyelets and tied ▷ *v* fasten with laces; thread a cord or string through holes in something; add a small amount of alcohol, a drug, etc. to (food or drink) ▶ **lacy** *adj* fine, like lace ▶ **lace-ups** *pl n* shoes which fasten with laces

lacerate [lass-er-rate] *v* tear (flesh) ▶ **laceration** *n*

lachrymose *adj* tearful; sad

lack *n* shortage or absence of something needed or wanted ▷ *v* need or be short of (something)

lackadaisical *adj* lazy and careless in a dreamy way

lackey *n* servile follower; uniformed male servant

lacklustre *adj* lacking brilliance or vitality

laconic *adj* using only a few words, terse ▶ **laconically** *adv*

lacquer *n* hard varnish for wood or metal; clear sticky substance sprayed onto the hair to hold it in place

lacrimal adj of tears or the glands which produce them

lacrosse n sport in which teams catch and throw a ball using long sticks with a pouched net at the end, in an attempt to score goals

lactation n secretion of milk by female mammals to feed young ▸ **lactic** adj of or derived from milk ▸ **lactose** n white crystalline sugar found in milk

lacuna [lak-**kew**-na] n, pl -**nae** gap or missing part, esp. in a document or series

lad n boy or young man

ladder n frame of two poles connected by horizontal steps used for climbing; line of stitches that have come undone in tights or stockings ▷ v have or cause to have such a line of undone stitches

laden adj loaded; burdened

la-di-da, lah-di-dah adj (informal) affected or pretentious

ladle n spoon with a long handle and a large bowl, used for serving soup etc. ▷ v serve out

lady n, pl -**dies** woman regarded as having characteristics of good breeding or high rank; polite term of address for a woman; (**L-**) title of some female members of the British nobility **Our Lady** the Virgin Mary ▸ **lady-in-waiting** n, pl **ladies-in-waiting** female servant of a queen or princess ▸ **ladykiller** n (informal) man who is or thinks he is irresistible to women ▸ **ladylike** adj polite and dignified

ladybird n small red beetle with black spots

lag¹ v **lagging, lagged** go too slowly, fall behind ▷ n delay between events ▸ **laggard** n person who lags behind

lag² v **lagging, lagged** wrap (a boiler, pipes, etc.) with insulating material ▸ **lagging** n insulating material

lag³ n old lag (Brit, Aust & NZ slang) convict

lager n light-bodied beer

lagoon n body of water cut off from the open sea by coral reefs or sand bars

laid v past of **lay¹** ▸ **laid-back** adj (informal) relaxed

lain v past participle of **lie²**

lair n resting place of an animal

laird n Scottish landowner

laissez-faire [less-ay-**fair**] n principle of nonintervention, esp. by a government in commercial affairs

laity [**lay**-it-ee] n people who are not members of the clergy

lake¹ n expanse of water entirely surrounded by land ▸ **lakeside** n

lake² n red pigment

lama n Buddhist priest in Tibet or Mongolia

lamb n young sheep; its meat ▷ v (of sheep) give birth to a lamb or lambs ▸ **lamb's fry** (Aust & NZ) lamb's liver for cooking ▸ **lambskin** n ▸ **lambswool** n

lambast, lambaste v beat or thrash; reprimand severely

lambent adj (lit) (of a flame) flickering softly

lame adj having an injured or disabled leg or foot; (of an excuse) unconvincing ▷ v make lame ▸ **lamely** adv ▸ **lameness** n ▸ **lame duck** person or thing unable to cope without help

lamé [**lah**-may] n, adj (fabric) interwoven with gold or silver thread

lament v feel or express sorrow (for) ▷ n passionate expression of grief ▸ **lamentable** adj very disappointing ▸ **lamentation** n ▸ **lamented** adj grieved for

laminate v make (a sheet of material) by sticking together thin sheets; cover with a thin sheet of material ▷ n laminated sheet ▶ **laminated** adj

lamington n (Aust & NZ) sponge cake coated with a sweet coating

Lammas n August 1, formerly a harvest festival

lamp n device which produces light from electricity, oil, or gas ▶ **lamppost** n post supporting a lamp in the street ▶ **lampshade** n

lampoon n humorous satire ridiculing someone ▷ v satirize or ridicule

lamprey n eel-like fish with a round sucking mouth

LAN (Computing) local area network

lance n long spear used by a mounted soldier ▷ v pierce (a boil or abscess) with a lancet ▶ **lancer** n formerly, cavalry soldier armed with a lance ▶ **lance corporal** noncommissioned army officer of the lowest rank

lancet n pointed two-edged surgical knife; narrow window in the shape of a pointed arch

land n solid part of the earth's surface; ground, esp. with reference to its type or use; rural or agricultural area; property consisting of land; country or region ▷ v come or bring to earth after a flight, jump, or fall; go or take from a ship at the end of a voyage; come to or touch shore; come or bring to some point or condition; (informal) obtain; take (a hooked fish) from the water; (informal) deliver (a punch) ▶ **landed** adj possessing or consisting of lands ▶ **landless** adj ▶ **landward** adj nearest to or facing the land ▷ adv (also **landwards**) towards land ▶ **landfall** n ship's first landing after a voyage

▶ **landlocked** adj completely surrounded by land ▶ **land up** v arrive at a final point or condition

landau [lan-daw] n four-wheeled carriage with two folding hoods

landing n floor area at the top of a flight of stairs; bringing or coming to land; (also **landing stage**) place where people or goods go onto or come off a boat

landline n fixed telephone communications cable

landlord, landlady n person who rents out land, houses, etc.; owner or manager of a pub or boarding house

landlubber n person who is not experienced at sea

landmark n prominent object in or feature of a landscape; event, decision, etc. considered as an important development

landscape n extensive piece of inland scenery seen from one place; picture of it ▷ v improve natural features of (a piece of land)

landslide n (also **landslip**) falling of soil, rock, etc. down the side of a mountain; overwhelming electoral victory

lane n narrow road; area of road for one stream of traffic; specified route followed by ships or aircraft; strip of a running track or swimming pool for use by one competitor

language n system of sounds, symbols, etc. for communicating thought; particular system used by a nation or people; system of words and symbols for computer programming

languid adj lacking energy or enthusiasm ▶ **languidly** adv

languish v suffer neglect or hardship; lose or diminish in strength or vigour; pine (for)

languor [lang-ger] n state of dreamy relaxation; laziness or weariness ▶ **languorous** adj

lank adj (of hair) straight and limp; thin or gaunt ▶ **lanky** adj ungracefully tall and thin

lanolin n grease from sheep's wool used in ointments etc.

lantana [lan-tay-na] n shrub with orange or yellow flowers, considered a weed in Australia

lantern n light in a transparent protective case ▶ **lantern jaw** long thin jaw ▶ **lantern-jawed** adj

lanthanum n (Chem) silvery-white metallic element ▶ **lanthanide series** class of 15 elements chemically related to lanthanum

lanyard n cord worn round the neck to hold a knife or whistle; (Naut) short rope

lap¹ n part between the waist and knees of a person when sitting ▶ **laptop** adj (of a computer) small enough to fit on a user's lap ▶ n computer small enough to fit on a user's lap

lap² n single circuit of a racecourse or track; stage of a journey ▷ v **lapping**, **lapped** overtake an opponent so as to be one or more circuits ahead

lap³ v **lapping**, **lapped** (of waves) beat softly against ▶ **lap up** v drink by scooping up with the tongue; accept (information or attention) eagerly

lapel [lap-pel] n part of the front of a coat or jacket folded back towards the shoulders

lapidary adj of or relating to stones

lapis lazuli [lap-iss lazz-yoo-lie] n bright blue gemstone

lapse n temporary drop in a standard, esp. through forgetfulness or carelessness; instance of bad behaviour by someone usually well-behaved; break in occurrence or usage ▷ v drop in standard; end or become invalid, esp. through disuse; abandon religious faith; (of time) slip away ▶ **lapsed** adj

lapwing n plover with a tuft of feathers on the head

larboard adj, n (old-fashioned) port (side of a ship)

larceny n, pl **-nies** (Law) theft

larch n deciduous coniferous tree

lard n soft white fat obtained from a pig ▷ v insert strips of bacon in (meat) before cooking; decorate (speech or writing) with strange words unnecessarily

larder n storeroom for food

large adj great in size, number, or extent ▶ **at large** in general; free, not confined; fully ▶ **largely** adv ▶ **largish** adj ▶ **large-scale** adj wide-ranging or extensive

largesse, largess [lar-jess] n generous giving, esp. of money

largo n, pl **-gos** ▷ adv (Music) (piece to be played) in a slow and dignified manner

lariat n lasso

lark¹ n small brown songbird, skylark

lark² n (informal) harmless piece of mischief or fun; unnecessary activity or job ▶ **lark about** v play pranks

larkspur n plant with spikes of blue, pink, or white flowers with spurs

larrikin n (Aust & NZ, old-fashioned slang) mischievous or unruly person

larva n, pl **-vae** insect in an immature stage, often resembling a worm ▶ **larval** adj

larynx n, pl **larynges** part of the throat containing the vocal cords ▶ **laryngeal** adj ▶ **laryngitis** n inflammation of the larynx

lasagne, lasagna [laz-**zan**-ya] *n* pasta in wide flat sheets; dish made from layers of lasagne, meat, and cheese

lascivious [lass-**iv**-ee-uss] *adj* showing or producing sexual desire ▸ **lasciviously** *adv*

laser [**lay**-zer] *n* device that produces a very narrow intense beam of light, used for cutting very hard materials and in surgery etc. ▹ *v* remove by treating with a laser

lash[1] *n* eyelash; sharp blow with a whip ▹ *v* hit with a whip; (of rain or waves) beat forcefully against; attack verbally, scold; flick or wave sharply to and fro ▸ **lash out** *v* make a sudden physical or verbal attack; (*informal*) spend (money) extravagantly

lash[2] *v* fasten or bind tightly with cord etc.

lashings *pl n* (*old-fashioned*) large amounts

lass, lassie *n* (*Scot & N English*) girl

lassitude *n* physical or mental weariness

lasso [lass-**oo**] *n*, *pl* **-sos** or **-soes** rope with a noose for catching cattle and horses ▹ *v* **-soing, -soed** catch with a lasso

last[1] *adj*, *adv* coming at the end or after all others; most recent(ly) ▹ *adj* only remaining ▹ *n* last person or thing ▸ **lastly** *adv* ▸ **last-ditch** *adj* done as a final resort ▸ **last post** army bugle-call played at sunset or funerals ▸ **last straw** small irritation or setback that, coming after others, is too much to bear ▸ **last word** final comment in an argument; most recent or best example of something

last[2] *v* continue; be sufficient for (a specified amount of time); remain fresh, uninjured, or unaltered ▸ **lasting** *adj*

last[3] *n* model of a foot on which shoes and boots are made or repaired

latch *n* fastening for a door with a bar and lever; lock which can only be opened from the outside with a key ▹ *v* fasten with a latch ▸ **latch onto** *v* become attached to (a person or idea)

late *adj* after the normal or expected time; towards the end of a period; being at an advanced time; recently dead; recent; former ▹ *adv* after the normal or expected time; at a relatively advanced age; recently ▸ **lately** *adv* in recent times ▸ **lateness** *n*

latent *adj* hidden and not yet developed ▸ **latency** *n*

lateral [lat-ter-al] *adj* of or relating to the side or sides ▸ **laterally** *adv*

latex *n* milky fluid found in some plants, esp. the rubber tree, used in making rubber

lath *n* thin strip of wood used to support plaster, tiles, etc.

lathe *n* machine for turning wood or metal while it is being shaped

lather *n* froth of soap and water; frothy sweat; (*informal*) state of agitation ▹ *v* make frothy; rub with soap until lather appears

Latin *n* language of the ancient Romans ▹ *adj* of or in Latin; of a people whose language derives from Latin ▸ **Latin America** parts of South and Central America whose official language is Spanish or Portuguese ▸ **Latin American** *n*, *adj*

latitude *n* angular distance measured in degrees N or S of the equator; scope for freedom of action or thought ▹ *pl* regions considered in relation to their distance from the equator

latrine *n* toilet in a barracks or camp

latter adj second of two; later; recent ▶ **latterly** adv ▶ **latter-day** adj modern

> **USAGE NOTE**
> *Latter* is used for the last-mentioned of two items. When there are more, use *last-named*.

lattice [lat-iss] n framework of intersecting strips of wood, metal, etc.; gate, screen, etc. formed of such a framework ▶ **latticed** adj

laud v praise or glorify ▶ **laudable** adj praiseworthy ▶ **laudably** adv ▶ **laudatory** adj praising or glorifying

laudanum [lawd-a-num] n opium-based sedative

laugh v make inarticulate sounds with the voice expressing amusement, merriment, or scorn; utter or express with laughter ▷ n act or instance of laughing; (*informal*) person or thing causing amusement ▶ **laughable** adj ridiculously inadequate ▶ **laughter** n sound or action of laughing ▶ **laughing gas** nitrous oxide as an anaesthetic ▶ **laughing stock** object of general derision ▶ **laugh off** v treat (something serious or difficult) lightly

launch¹ v put (a ship or boat) into the water, esp. for the first time; begin (a campaign, project, etc.); put a new product on the market; send (a missile or spacecraft) into space or the air ▷ n launching ▶ **launcher** n ▶ **launch into** v start doing something enthusiastically ▶ **launch out** v start doing something new

launch² n open motorboat

launder v wash and iron (clothes and linen); make (illegally obtained money) seem legal by passing it through foreign banks or legitimate businesses ▶ **laundry** n, pl **-dries** clothes etc. for washing or which have recently been washed; place for washing clothes and linen ▶ **Launderette®** n shop with coin-operated washing and drying machines

laureate [lor-ee-at] adj see **poet laureate**

laurel n glossy-leaved shrub, bay tree ▷ pl wreath of laurel, an emblem of victory or merit

lava n molten rock thrown out by volcanoes, which hardens as it cools

lavatory n, pl **-ries** toilet

lavender n shrub with fragrant flowers ▷ adj bluish-purple ▶ **lavender water** light perfume made from lavender

lavish adj great in quantity or richness; giving or spending generously; extravagant ▷ v give or spend generously ▶ **lavishly** adv

law n rule binding on a community; system of such rules; (*informal*) police; invariable sequence of events in nature; general principle deduced from facts ▶ **lawful** adj allowed by law ▶ **lawfully** adv ▶ **lawless** adj breaking the law, esp. in a violent way ▶ **lawlessness** n ▶ **law-abiding** adj obeying the laws ▶ **law-breaker** n ▶ **lawsuit** n court case brought by one person or group against another

lawn¹ n area of tended and mown grass ▶ **lawn mower** machine for cutting grass ▶ **lawn tennis** tennis, esp. when played on a grass court

lawn² n fine linen or cotton fabric

lawyer n professionally qualified legal expert

lax adj not strict ▶ **laxity** n

laxative n, adj (medicine) inducing the emptying of the bowels

lay[1] v **laying, laid** cause to lie; devise or prepare; set in a particular place or position; attribute (blame); put forward (a plan, argument, etc.); (of a bird or reptile) produce eggs; arrange (a table) for a meal **lay waste** devastate ▸ **lay-by** n stopping place for traffic beside a road ▸ **lay-off** n ▸ **lay on** v provide or supply ▸ **lay out** v arrange or spread out; prepare (a corpse) for burial; (*informal*) spend money, esp. lavishly; (*informal*) knock unconscious ▸ **layout** n arrangement, esp. of matter for printing or of a building

USAGE NOTE

Lay and *lie* are often confused. The action of the verb *lay* is always done to a person or thing, for example *He laid down his weapon*, while the action of the verb *lie* is not: *I'm going to lie down*

lay[2] v past tense of **lie**[2] ▸ **layabout** n lazy person

lay[3] adj of or involving people who are not clergymen; nonspecialist ▸ **layman** n person who is not a member of the clergy; person without specialist knowledge

lay[4] n short narrative poem designed to be sung

layer n single thickness of some substance, as a cover or coating on a surface; laying hen; shoot of a plant pegged down or partly covered with earth to encourage root growth ⊳ v form a layer; propagate plants by layers ▸ **layered** adj

layette n clothes for a newborn baby

laze v be idle or lazy ⊳ n time spent lazing

lazy adj **lazier, laziest** not inclined to work or exert oneself; done in a relaxed manner without much effort; (of movement) slow and gentle ▸ **lazily** adv ▸ **laziness** n

lb pound (weight)

lbw (Cricket) leg before wicket

lea n (*poetic*) meadow

leach v remove or be removed from a substance by a liquid passing through it

lead[1] v **leading, led** guide or conduct; cause to feel, think, or behave in a certain way; be, go, or play first; (of a road, path, etc.) go towards; control or direct; (foll by to) result in; pass or spend (one's life) ⊳ n first or most prominent place; amount by which a person or group is ahead of another; clue; length of leather or chain attached to a dog's collar to control it; principal role or actor in a film, play, etc.; cable bringing current to an electrical device ⊳ adj acting as a leader or lead ▸ **leading** adj principal; in the first position ▸ **leading question** question worded to prompt the answer desired ▸ **lead-in** n introduction to a subject

lead[2] n soft heavy grey metal; (in a pencil) graphite; lead weight on a line, used for sounding depths of water ▸ **leaded** adj (of windows) made from many small panes of glass held together by lead strips ▸ **leaden** adj heavy or sluggish; dull grey; made from lead

leader n person who leads; article in a newspaper expressing editorial views ▸ **leadership** n

leaf n, pl **leaves** flat usu. green blade attached to the stem of a plant; single sheet of paper in a book; very thin sheet of metal; extending flap on a table ▸ **leafy** adj ▸ **leafless** adj ▸ **leaf mould** rich soil composed of decayed leaves ▶ **leaf through** v turn pages without reading them

leaflet n sheet of printed matter for distribution; small leaf

league¹ n association promoting the interests of its members; association of sports clubs organizing competitions between its members; (informal) class or level

league² n (obs) measure of distance, about three miles

leak n hole or defect that allows the escape or entrance of liquid, gas, radiation, etc.; liquid etc. that escapes or enters; disclosure of secrets ▷ v let liquid etc. in or out; (of liquid etc.) find its way through a leak; disclose secret information ▶ **leakage** n act or instance of leaking ▶ **leaky** adj

lean¹ v leaning, leaned or leant rest against; bend or slope from an upright position; tend (towards) ▶ **leaning** n tendency ▶ **lean on** v (informal) threaten or intimidate; depend on for help or advice ▶ **lean-to** n shed built against an existing wall

lean² adj thin but healthy-looking; (of meat) lacking fat; unproductive ▷ n lean part of meat ▶ **leanness** n

leap v leaping, leapt or leaped make a sudden powerful jump ▷ n sudden powerful jump; abrupt increase, as in costs or prices ▶ **leapfrog** n game in which a player vaults over another bending down ▶ **leap year** year with February 29 as an extra day

learn v learning, learned or learnt gain skill or knowledge by study, practice, or teaching; memorize (something); find out or discover (something) ▶ **learned** adj erudite, deeply read; showing much learning ▶ **learner** n ▶ **learning** n knowledge got by study

lease n contract by which land or property is rented for a stated time by the owner to a tenant ▷ v let or rent by lease ▶ **leasehold** n, adj (land or property) held on lease ▶ **leaseholder** n

leash n lead for a dog

least adj superlative of **little**; smallest ▷ n smallest one ▷ adv in the smallest degree

leather n material made from specially treated animal skins ▷ adj made of leather ▷ v beat or thrash ▶ **leathery** adj like leather, tough

leave¹ v leaving, left go away from; allow to remain, accidentally or deliberately; cause to be or remain in a specified state; discontinue membership of; permit; entrust; bequeath ▶ **leave out** v exclude or omit

leave² n permission to be absent from work or duty; period of such absence; permission to do something; formal parting

leaven [lev-ven] n substance that causes dough to rise; influence that produces a gradual change ▷ v raise with leaven; spread through and influence (something)

lecher n man who has or shows excessive sexual desire ▶ **lechery** n

lecherous [letch-er-uss] adj (of a man) having or showing excessive sexual desire

lectern n sloping reading desk, esp. in a church

lecture n informative talk to an audience on a subject; lengthy rebuke or scolding ▷ v give a talk; scold

lecturer n person who lectures, esp. in a university or college ▶ **lectureship** n appointment as a lecturer

LED light-emitting diode

ledge n narrow shelf sticking out from a wall; shelflike projection from a cliff etc.

ledger n book of debit and credit accounts of a firm

lee n sheltered part or side ▶ **leeward** adj, n (on) the lee side ▷ adv towards this side ▶ **leeway** n room for free movement within limits

leech n species of bloodsucking worm; person who lives off others

leek n vegetable of the onion family with a long bulb and thick stem

leer v look or grin at in a sneering or suggestive manner ▶ n sneering or suggestive look or grin

leery adj (informal) suspicious or wary (of)

lees pl n sediment of wine

left¹ adj of the side that faces west when the front faces north ▷ adv on or towards the left ▷ n left hand or part; (Politics) people supporting socialism rather than capitalism ▶ **leftist** n, adj (person) of the political left ▶ **left-handed** adj more adept with the left hand than with the right ▶ **left-wing** adj socialist; belonging to the more radical part of a political party

left² v past of **leave¹**

leftover n unused portion of food or material

leg n one of the limbs on which a person or animal walks, runs, or stands; part of a garment covering the leg; structure that supports, such as one of the legs of a table; stage of a journey; (Sport) (part of) one game or race in a series **pull someone's leg** tease someone ▶ **leggy** adj having long legs ▶ **legless** adj without legs; (slang) very drunk ▶ **leggings** pl n covering of leather or other material for

the legs; close-fitting trousers for women or children

legacy n, pl -cies thing left in a will; thing handed down to a successor

legal adj established or permitted by law; relating to law or lawyers ▶ **legally** adv ▶ **legality** n ▶ **legalize** v make legal ▶ **legalization** n

legate n messenger or representative, esp. from the Pope ▶ **legation** n diplomatic minister and his or her staff; official residence of a diplomatic minister

legatee n recipient of a legacy

legato [leg-ah-toe] n, pl -tos ▷ adv (Music) (piece to be played) smoothly

legend n traditional story or myth; traditional literature; famous person or event; stories about such a person or event; inscription ▶ **legendary** adj famous; of or in legend

legerdemain [lej-er-de-main] n sleight of hand; cunning deception

legible adj easily read ▶ **legibility** n ▶ **legibly** adv

legion n large military force; large number; association of veterans; infantry unit in the Roman army ▶ **legionary** adj, n ▶ **legionnaire** n member of a legion ▶ **legionnaire's disease** serious bacterial disease similar to pneumonia

legislate v make laws ▶ **legislative** adj ▶ **legislator** n maker of laws ▶ **legislature** n body of people that makes, amends, or repeals laws

legislation n legislating; laws made

legitimate adj authorized by or in accordance with law; fairly deduced; born to parents married to each other ▷ v make legitimate

▶**legitimacy** n ▶ **legitimately** adv
▶ **legitimize** v make legitimate,
legalize ▶ **legitimization** n

Lego® n construction toy of plastic
bricks fitted together by studs

leguaan [leg-oo-ahn] n large S
African lizard

legume n pod of a plant of the pea
or bean family ▷ pl peas or beans
▶ **leguminous** adj (of plants)
pod-bearing

lei n (in Hawaii) garland of flowers

leisure n time for relaxation or
hobbies **at one's leisure** when one
has time ▶ **leisurely** adj deliberate,
unhurried ▷ adv slowly ▶ **leisured**
adj with plenty of spare time
▶ **leisure centre** building with
facilities such as a swimming pool,
gymnasium, and café

leitmotif [lite-mote-eef] n (Music)
recurring theme associated with a
person, situation, or thought

lekker (S Afr slang) attractive or
nice; tasty

lemming n rodent of arctic regions,
reputed to run into the sea and
drown during mass migrations

lemon n yellow oval fruit that
grows on trees; (slang) useless
or defective person or thing
▷ adj pale-yellow ▶ **lemonade** n
lemon-flavoured soft drink, often
fizzy ▶ **lemon curd** creamy spread
made of lemons, butter, etc.
▶ **lemon sole** edible flatfish

lemur n nocturnal animal
like a small monkey, found in
Madagascar

lend v lending, lent give the
temporary use of; provide (money)
temporarily, often for interest; add
(a quality or effect) e.g. her presence
lent beauty to the scene **lend itself
to** be suitable for ▶ **lender** n

length n extent or measurement
from end to end; period of time

for which something happens;
quality of being long; piece of
something narrow and long
at length at last; in full detail
▶ **lengthy** adj very long or tiresome
▶ **lengthily** adv ▶ **lengthen** v make
or become longer ▶ **lengthways,
lengthwise** adj, adv

lenient [lee-nee-ent] adj tolerant,
not strict or severe ▶ **leniency** n
▶ **leniently** adv

lens [lenz] n, pl **lenses** piece of glass
or similar material with one or
both sides curved, used to bring
together or spread light rays in
cameras, spectacles, telescopes,
etc.; transparent structure in the
eye that focuses light

Lent n period from Ash Wednesday
to Easter Saturday ▶ **Lenten** adj of,
in, or suitable to Lent

lent v past of **lend**

lentil n edible seed of a leguminous
Asian plant

lento n, pl **-tos** ▷ adv (Music) (piece
to be played) slowly

leonine adj like a lion

leopard n large spotted
carnivorous animal of the cat
family

leotard n tight-fitting garment
covering the upper body, worn for
dancing or exercise

leper n (offens) person suffering from
leprosy; ignored or despised person

lepidoptera pl n order of insects
with four wings covered with
fine gossamer scales, as moths
and butterflies ▶ **lepidopterist**
n person who studies or collects
butterflies or moths

leprechaun n mischievous elf of
Irish folklore

leprosy n disease attacking the
nerves and skin, resulting in loss
of feeling in the affected parts
▶ **leprous** adj

lesbian n homosexual woman
▷ adj of homosexual women
▶ **lesbianism** n

lese-majesty [lezz-**maj**-est-ee] n treason; taking of liberties against people in authority

lesion n structural change in an organ of the body caused by illness or injury; injury or wound

less adj smaller in extent, degree, or duration; not so much; comparative of **little** ▷ pron smaller part or quantity ▷ adv to a smaller extent or degree ▷ prep after deducting, minus ▶ **lessen** v make or become smaller or not as much ▶ **lesser** adj not as great in quantity, size, or worth

> **USAGE NOTE**
> Avoid confusion with few(er).
> Less is used with amounts that cannot be counted: less time; less fuss. Few(er) is used of things that can be counted

lessee n person to whom a lease is granted

lesson n single period of instruction in a subject; content of this; experience that teaches; portion of Scripture read in church

lest conj so as to prevent any possibility that; for fear that

let¹ v letting, let allow, enable, or cause; used as an auxiliary to express a proposal, command, threat, or assumption; grant use of for rent; allow to escape **let alone** not to mention ▶ **let down** v disappoint; lower; deflate ▶ **letdown** n disappointment ▶ **let off** v excuse from (a duty or punishment); fire or explode (a weapon); emit (gas, steam, etc.) ▶ **let on** v (informal) reveal (a secret) ▶ **let out** v emit; release ▶ **let up** v diminish or stop ▶ **let-up** n lessening

let² n (Tennis) minor infringement or obstruction of the ball requiring a replay of the point; hindrance

lethal adj deadly

lethargy n sluggishness or dullness; abnormal lack of energy ▶ **lethargic** adj ▶ **lethargically** adv

letter n written message, usu. sent by post; alphabetical symbol; strict meaning (of a law etc.) ▷ pl literary knowledge or ability ▶ **lettered** adj learned ▶ **lettering** n ▶ **letter bomb** explosive device in a parcel or letter that explodes on opening ▶ **letter box** slot in a door through which letters are delivered; box in a street or post office where letters are posted ▶ **letterhead** n printed heading on stationery giving the sender's name and address

lettuce n plant with large green leaves used in salads

leucocyte [loo-koh-site] n white blood cell

leukaemia [loo-kee-mee-a] n disease caused by uncontrolled overproduction of white blood cells

levee n (US) natural or artificial river embankment

level adj horizontal; having an even surface; of the same height as something else; equal to or even with (someone or something else); not going above the top edge of (a spoon etc.) ▷ v -**elling**, -**elled** make even or horizontal; make equal in position or status; direct (a gun, accusation, etc.) at; raze to the ground ▷ n horizontal line or surface; device for showing or testing if something is horizontal; position on a scale; standard or grade; flat area of land **on the level** (informal) honest or trustworthy ▶ **level crossing** point where a railway line and road cross ▶ **level-headed** *adj* not apt to be carried away by emotion

lever *n* handle used to operate machinery; bar used to move a heavy object or to open something; rigid bar pivoted about a fulcrum to transfer a force to a load; means of exerting pressure to achieve an aim ▷ *v* prise or move with a lever ▶ **leverage** *n* action or power of a lever; influence or strategic advantage

leveret [lev-ver-it] *n* young hare

leviathan [lev-vie-ath-an] *n* sea monster; anything huge or formidable

Levis ® *pl n* denim jeans

levitat ... *n* raising of a solid bod ... e air supernaturally ▶ **lev** ... ise or cause to rise into ...

levity ... **ies** inclination to make a jok ... ous matters

levy [lev-vee] *v* **levying, levied** impose and collect (a tax); raise (troops) ▷ *n*, *pl* **levies** imposition or collection of taxes; money levied

lewd *adj* lustful or indecent ▶ **lewdly** *adv* ▶ **lewdness** *n*

lexicon *n* dictionary; vocabulary of a language ▶ **lexical** *adj* relating to the vocabulary of a language ▶ **lexicographer** *n* writer of dictionaries ▶ **lexicography** *n*

LGBT lesbian, gay, bisexual, and transgender

liable *adj* legally obliged or responsible; given to or at risk from a condition ▶ **liability** *n* hindrance or disadvantage; state of being liable; financial obligation

USAGE NOTE
The use of *liable* to mean 'likely' is informal. It generally means 'responsible for': *He was liable for any damage*

liaise *v* establish and maintain communication (with) ▶ **liaison** *n* communication and contact between groups; secret or adulterous relationship

SPELLING TIP
Remember to include a second *i* in **liaise** and **liaison**

liana *n* climbing plant in tropical forests

liar *n* person who tells lies

lib *n* (*informal*) short for **liberation**

libation [lie-**bay**-shun] *n* drink poured as an offering to the gods

libel *n* published statement falsely damaging a person's reputation ▷ *v* **-belling, -belled** falsely damage the reputation of (someone) ▶ **libellous** *adj*

liberal *adj* having social and political views that favour progress and reform; generous in behaviour or temperament; tolerant; abundant; (of education) designed to develop general cultural interests ▷ *n* person who has liberal ideas or opinions ▶ **liberally** *adv* ▶ **liberalism** *n* belief in democratic reforms and individual freedom ▶ **liberality** *n* generosity ▶ **liberalize** *v* make (laws, a country, etc.) less restrictive ▶ **liberalization** *n* ▶ **Liberal Democrat, Lib Dem** member of the Liberal Democrats, a British political party favouring a mixed economy and individual freedom ▶ **Liberal Party** main right-wing political party in Australia

liberate *v* set free ▶ **liberation** *n* ▶ **liberator** *n*

libertarian *n* believer in freedom of thought and action ▷ *adj* having such a belief

libertine [lib-er-teen] *n* morally dissolute person

liberty *n*, *pl* **-ties** freedom; act or comment regarded as forward or socially unacceptable **at liberty** free; having the right; **take liberties** be presumptuous

libido [lib-ee-doe] n, pl -**dos** psychic energy; emotional drive, esp. of sexual origin ▶ **libidinous** adj lustful

library n, pl -**braries** room or building where books are kept; collection of books, records, etc. for consultation or borrowing ▶ **librarian** n keeper of or worker in a library ▶ **librarianship** n

libretto n, pl -**tos** or -**ti** words of an opera ▶ **librettist** n

lice n a plural of **louse**

licence n document giving official permission to do something; formal permission; disregard of conventions for effect e.g. poetic licence; excessive liberty ▶ **license** v grant a licence to ▶ **licensed** adj ▶ **licensee** n holder of a licence, esp. to sell alcohol

> **SPELLING TIP**
> Note the -ce ending for the noun licence, with -se for the verb license

licentiate n person licensed as competent to practise a profession

licentious adj sexually unrestrained or promiscuous

lichen n small flowerless plant forming a crust on rocks, trees, etc.

licit adj lawful, permitted

lick v pass the tongue over; touch lightly or flicker round; (slang) defeat ▶ n licking; small amount (of paint etc.); (informal) fast pace

licorice n same as **liquorice**

lid n movable cover; short for **eyelid**

lido [lee-doe] n, pl -**dos** open-air centre for swimming and water sports

lie¹ v lying, lied make a deliberately false statement ▶ n deliberate falsehood **white lie** see **white**

lie² v lying, lay, lain place oneself or be in a horizontal position; be situated; be or remain in a certain state or position; exist or be found ▷ n way something lies ▶ **lie-down** n rest ▶ **lie in** v remain in bed late into the morning ▶ **lie-in** n long stay in bed in the morning

> **USAGE NOTE**
> Note that the past of lie is lay: She lay on the beach all day. Do not confuse with the main verb lay meaning 'put'

lied [leed] n, pl **lieder** (Music) setting for voice and piano of a romantic poem

liege [leej] adj bound to give or receive feudal service ▶ n lord

lien n (Law) right to hold another's property until a debt is paid

lieu [lyew] n **in lieu of** instead of

lieutenant [lef-ten-ant] n junior officer in the army or navy; main assistant

life n, pl **lives** state of living beings, characterized by growth, reproduction, and response to stimuli; period between birth and death or between birth and the present time; way of living; amount of time something is active or functions; biography; liveliness or high spirits; living beings collectively ▶ **lifeless** adj dead; not lively or exciting; unconscious ▶ **lifelike** adj ▶ **lifelong** adj lasting all of a person's life ▶ **life belt, life jacket** buoyant device to keep afloat a person in danger of drowning ▶ **lifeboat** n boat used for rescuing people at sea ▶ **life cycle** series of changes undergone by each generation of an animal or plant ▶ **lifeline** n means of contact or support; rope used in rescuing a person in danger ▶ **life science** any science concerned with living organisms, such as biology, botany, or zoology ▸ **lifestyle** n particular attitudes, habits, etc. ▸ **life-support** adj (of equipment or treatment) necessary to keep a person alive ▸ **lifetime** n length of time a person is alive

lift v move upwards in position, status, volume, etc.; revoke or cancel; take (plants) out of the ground for harvesting; (of fog, etc.) disappear; make or become more cheerful ▷ n cage raised and lowered in a vertical shaft to transport people or goods; ride in a car etc. as a passenger; (informal) feeling of cheerfulness; lifting ▸ **liftoff** n moment a rocket leaves the ground

ligament n band of tissue joining bones

ligature n link, bond, or tie

light¹ n electromagnetic radiation by which things are visible; source of this, lamp; anything that lets in light, such as a window; aspect or view; mental vision; means of setting fire to ▷ pl traffic lights ▷ adj bright; (of a colour) pale ▷ v ignite; illuminate or cause to illuminate ▸ **lighten** v make less dark ▸ **lighting** n apparatus for and use of artificial light in theatres, films, etc. ▸ **light bulb** glass part of an electric lamp ▸ **lighthouse** n tower with a light to guide ships ▸ **light year** (Astronomy) distance light travels in one year, about six million million miles

light² adj not heavy, weighing relatively little; relatively low in strength, amount, density, etc.; not clumsy; not serious or profound; easily digested ▷ adv with little equipment or luggage ▷ v **lighting, lighted, lit** (esp. of birds) settle after flight; come

(upon) by chance ▸ **lightly** adv ▸ **lightness** n ▸ **lighten** v make less heavy or burdensome; make more cheerful or lively ▸ **light-fingered** adj skilful at stealing ▸ **light-headed** adj feeling faint, dizzy ▸ **light-hearted** adj carefree ▸ **lightweight** n, adj (person) of little importance ▷ n boxer weighing up to 135lb (professional) or 60kg (amateur)

lighter¹ n device for lighting cigarettes etc.

lighter² n flat-bottomed boat for unloading ships

lightning n visible discharge of electricity in the atmosphere ▷ adj fast and sudden

> **SPELLING TIP**
> Do not confuse the noun *lightning*, which doesn't have an *e* in the middle, with the verb *lighten*, which has the form *lightening*

lights pl n lungs of animals as animal food

ligneous adj of or like wood

lignite [lig-nite] n woody textured rock used as fuel

like¹ prep, conj, adj, pron indicating similarity, comparison, etc. ▸ **liken** v compare ▸ **likeness** n resemblance; portrait ▸ **likewise** adv similarly

like² v find enjoyable; be fond of; prefer, choose, or wish ▸ **likeable, likable** adj ▸ **liking** n fondness; preference

likely adj tending or inclined; probable; hopeful, promising ▷ adv probably **not likely** (informal) definitely not ▸ **likelihood** n probability

lilac n shrub with pale mauve or white flowers ▷ adj light-purple

Lilliputian [lil-lip-pew-shun] adj tiny

Lilo® n, pl **-los** inflatable rubber mattress

lilt n pleasing musical quality in speaking; jaunty rhythm; graceful rhythmic motion ▶ **lilting** adj

lily n, pl **lilies** plant which grows from a bulb and has large, often white, flowers

limb n arm, leg, or wing; main branch of a tree

limber v (foll by up) loosen stiff muscles by exercising ▶ adj pliant or supple

limbo¹ n in limbo not knowing the result or next stage of something and powerless to influence it

limbo² n, pl **-bos** West Indian dance in which dancers lean backwards to pass under a bar

lime¹ n calcium compound used as a fertilizer or in making cement ▶ **limelight** n glare of publicity ▶ **limestone** n sedimentary rock used in building

lime² n small green citrus fruit ▶ **lime-green** adj greenish-yellow

lime³ n deciduous tree with heart-shaped leaves and fragrant flowers

limerick [lim-mer-ik] n humorous verse of five lines

limey n (US slang) British person

limit n ultimate extent, degree, or amount of something; boundary or edge ▶ v **-iting, -ited** restrict or confine ▶ **limitation** n ▶ **limitless** adj ▶ **limited company** company whose shareholders' liability for debts is restricted

limousine n large luxurious car

limp¹ v walk with an uneven step ▶ n limping walk

limp² adj without firmness or stiffness ▶ **limply** adv

limpet n shellfish which sticks tightly to rocks

limpid adj clear or transparent; easy to understand ▶ **limpidity** n

linchpin, lynchpin n pin to hold a wheel on its axle; essential person or thing

linctus n, pl **-tuses** syrupy cough medicine

linden n same as **lime**³

line¹ n long narrow mark; indented mark or wrinkle; boundary or limit; edge or contour of a shape; string or wire for a particular use; telephone connection; wire or cable for transmitting electricity; shipping company; railway track; course or direction of movement; prescribed way of thinking; field of interest or activity; row or queue of people; class of goods; row of words ▶ pl words of a theatrical part; school punishment of writing out a sentence a specified number of times ▶ v mark with lines; be or form a border or edge **in line for** likely to receive **in line with** in accordance with ▶ **line dancing** form of dancing performed by rows of people to country and western music ▶ **line-up** n people or things assembled for a particular purpose

line² v give a lining to; cover the inside of

lineage [lin-ee-ij] n descent from an ancestor

lineament n facial feature

linear [lin-ee-er] adj of or in lines

linen n cloth or thread made from flax; sheets, tablecloths, etc.

liner¹ n large passenger ship or aircraft

liner² n something used as a lining

linesman n (in some sports) an official who helps the referee or umpire; person who maintains railway, electricity, or telephone lines

ling¹ n slender food fish

ling² n heather

linger v delay or prolong departure; continue in a weakened state for a long time before dying or disappearing; spend a long time doing something

lingerie [lan-zher-ee] n women's underwear or nightwear

lingo n, pl **-goes** (informal) foreign or unfamiliar language or jargon

lingua franca n language used for communication between people of different mother tongues

lingual adj of the tongue

linguist n person skilled in foreign languages; person who studies linguistics ▶ **linguistic** adj of languages ▶ **linguistics** n scientific study of language

liniment n medicated liquid rubbed on the skin to relieve pain or stiffness

lining n layer of cloth attached to the inside of a garment etc.; inner covering of anything

link n any of the rings forming a chain; person or thing forming a connection; type of communications connection e.g. a radio link ▷ v connect with or as if with links; connect by association ▶ **linkage** n ▶ **link-up** n joining together of two systems or groups

links pl n golf course, esp. one by the sea

linnet n songbird of the finch family

lino n short for **linoleum**

linoleum n floor covering of hessian or jute with a smooth decorative coating of powdered cork

Linotype® n typesetting machine which casts lines of words in one piece

linseed n seed of the flax plant

lint n soft material for dressing a wound

lintel n horizontal beam at the top of a door or window

lion n large animal of the cat family, the male of which has a shaggy mane **the lion's share** the biggest part ▶ **lioness** adj fem ▶ **lion-hearted** adj brave

lip n either of the fleshy edges of the mouth; rim of a jug etc.; (slang) impudence ▶ **lip-reading** n method of understanding speech by interpreting lip movements ▶ **lip service** insincere tribute or respect ▶ **lipstick** n cosmetic in stick form, for colouring the lips

lipo n (informal) short for **liposuction**

liposuction n surgical operation in which body fat is removed

liquefy v **-fying, -fied** make or become liquid ▶ **liquefaction** n

liqueur [lik-cure] n flavoured and sweetened alcoholic spirit

liquid n substance in a physical state which can change shape but not size ▷ adj of or being a liquid; flowing smoothly; (of assets) in the form of money or easily converted into money ▶ **liquidize** v make or become liquid ▶ **liquidizer** n kitchen appliance that liquidizes food ▶ **liquidity** n state of being able to meet financial obligations

liquidate v pay (a debt); dissolve a company and share its assets among creditors; wipe out or kill ▶ **liquidation** n ▶ **liquidator** n official appointed to liquidate a business

liquor n alcoholic drink, esp. spirits; liquid in which food has been cooked

liquorice [lik-ker-iss] n black substance used in medicine and as a sweet

lira n, pl **-re** or **-ras** monetary unit of Turkey and formerly of Italy

lisle [rhymes with **mile**] n strong fine cotton thread or fabric

lisp n speech defect in which s and z are pronounced th ▷ v speak or utter with a lisp

lissom, lissome adj supple, agile

list[1] n item-by-item record of names or things, usu. written one below another ▷ v make a list of; include in a list

list[2] v (of a ship) lean to one side ▷ n leaning to one side

listen v concentrate on hearing something; heed or pay attention to ▶ **listener** n ▶ **listen in** v listen secretly, eavesdrop

listeriosis n dangerous form of food poisoning

listless adj lacking interest or energy ▶ **listlessly** adv

lit v past of **light**[1], **light**[2]

litany n, pl -**nies** prayer with responses from the congregation; any tedious recital

literacy n ability to read and write

literal adj according to the explicit meaning of a word or text, not figurative; (of a translation) word for word; actual, true ▶ **literally** adv

> **USAGE NOTE**
> Note that *literally* means much the same as 'actually'. It is very loosely used to mean 'just about' but this can be ambiguous

literary adj of or knowledgeable about literature; (of a word) formal, not colloquial

literate adj able to read and write; educated ▶ **literati** pl n literary people

literature n written works such as novels, plays, and poetry; books and writings of a particular period, or subject

lithe adj flexible or supple, pliant

lithium n (Chem) chemical element, the lightest known metal

litho n, pl -**thos** short for **lithograph** ▷ adj short for **lithographic**

lithography [lith-og-ra-fee] n method of printing from a metal or stone surface in which the printing areas are made receptive to ink ▶ **lithograph** n print made by lithography ▷ v reproduce by lithography ▶ **lithographer** n ▶ **lithographic** adj

litigant n person involved in a lawsuit

litigation n legal action ▶ **litigate** v bring or contest a lawsuit; engage in legal action ▶ **litigious** [lit-ij-uss] adj frequently going to law

litmus n blue dye turned red by acids and restored to blue by alkalis ▶ **litmus test** something which is regarded as a simple and accurate test of a particular thing

litotes [lie-toe-teez] n ironical understatement used for effect

litre n unit of liquid measure equal to 1000 cubic centimetres or 1.76 pints

litter n untidy rubbish dropped in public places; group of young animals produced at one birth; straw etc. as bedding for an animal; dry material to absorb a cat's excrement; bed or seat on parallel sticks for carrying people ▷ v strew with litter; scatter or be scattered about untidily; give birth to young

little adj small or smaller than average; young ▷ adv not a lot; hardly; not much or often ▷ n small amount, extent, or duration

littoral adj of or by the seashore ▷ n coastal district

liturgy n, pl -**gies** prescribed form of public worship ▶ **liturgical** adj

live[1] v be alive; remain in life or existence; exist in a specified way e.g. *we live well*; reside; continue or last; subsist; enjoy life to the full ▸ **liver** *n* person who lives in a specified way ▸ **live down** *v* wait till people forget a past mistake or misdeed ▸ **live-in** *adj* resident ▸ **live together** *v* (of an unmarried couple) share a house and have a sexual relationship ▸ **live up to** *v* meet (expectations) ▸ **live with** *v* tolerate

live² *adj* living, alive; (of a broadcast) transmitted during the actual performance; (of a performance) done in front of an audience; (of a wire, circuit, etc.) carrying an electric current; causing interest or controversy; capable of exploding; glowing or burning ▸ *adv* in the form of a live performance ▸ **lively** *adj* full of life or vigour; animated; vivid ▸ **liveliness** *n* ▸ **liven up** *v* make (more) lively

livelihood *n* occupation or employment

liver *n* organ secreting bile; animal liver as food ▸ **liverish** *adj* having a disorder of the liver; touchy or irritable

livery *n, pl* **-eries** distinctive dress, esp. of a servant or servants; distinctive design or colours of a company ▸ **liveried** *adj* ▸ **livery stable** stable where horses are kept at a charge or hired out

livestock *n* farm animals

livid *adj* (*informal*) angry or furious; bluish-grey

living *adj* possessing life, not dead or inanimate; currently in use or existing; of everyday life e.g. *living conditions* ▸ *n* condition of being alive; manner of life; financial means ▸ **living room** room in a house used for relaxation and entertainment

lizard *n* four-footed reptile with a long body and tail

llama *n* woolly animal of the camel family used as a beast of burden in S America

LLB Bachelor of Laws

loach *n* carplike freshwater fish

load *n* burden or weight; amount carried; source of worry; amount of electrical energy drawn from a source ▸ *pl* (*informal*) lots ▸ *v* put a load on or into; burden or oppress; cause to be biased; put ammunition into (a weapon); put film into (a camera); transfer (a program) into computer memory ▸ **loaded** *adj* (of a question) containing a hidden trap or implication; (of dice) dishonestly weighted; (*slang*) wealthy

loaf¹ *n, pl* **loaves** shaped mass of baked bread; shaped mass of food; (*slang*) head, esp. as the source of common sense e.g. *use your loaf*

loaf² *v* idle, loiter ▸ **loafer** *n*

loam *n* fertile soil

loan *n* money lent at interest; lending; thing lent ▸ *v* lend ▸ **loan shark** person who lends money at an extremely high interest rate

loath, loth [rhymes with **both**] *adj* unwilling or reluctant (to)

> **USAGE NOTE**
> Distinguish between *loath* 'reluctant' and *loathe* 'be disgusted by'

loathe *v* hate, be disgusted by ▸ **loathing** *n* ▸ **loathsome** *adj*

lob (*Sport*) *n* ball struck or thrown in a high arc ▸ *v* **lobbing, lobbed** strike or throw (a ball) in a high arc

lobby *n, pl* **-bies** corridor into which rooms open; group which tries to influence legislators; hall in a legislative building to which the public has access ▸ *v* try to influence (legislators) in the formulation of policy ▸ **lobbyist** *n*

lobe n rounded projection; soft hanging part of the ear; subdivision of a body organ ▶ **lobed** adj

lobelia n garden plant with blue, red, or white flowers

lobola [law-bawl-a] n (SAfr) (in African custom) price paid by a bridegroom's family to his bride's family

lobotomy n, pl **-mies** surgical incision into a lobe of the brain to treat mental disorders

lobster n shellfish with a long tail and claws, which turns red when boiled; (Aust informal) $20 note

local adj of or existing in a particular place; confined to a particular place ▷ n person belonging to a particular district; (informal) pub close to one's home ▶ **locale** [loh-kahl] n scene of an event ▶ **localism** n policy of devolving power to local bodies ▶ **locality** n neighbourhood or area ▶ **localize** v restrict to a particular place ▶ **locally** adv ▶ **local anaesthetic** anaesthetic which produces loss of feeling in one part of the body ▶ **local authority** governing body of a county, district, or region ▶ **local government** government of towns, counties, and districts by locally elected political bodies

locate v discover the whereabouts of; situate or place ▶ **location** n site or position; act of discovering where something is; site of a film production away from the studio; (SAfr) Black African or coloured township

loch n (Scot) lake; long narrow bay

lock¹ n appliance for fastening a door, case, etc.; section of a canal shut off by gates between which the water level can be altered to aid boats moving from one level to another; extent to which a vehicle's front wheels will turn; interlocking of parts; mechanism for firing a gun;

wrestling hold ▷ v fasten or become fastened securely; become or cause to become fixed or united; become or cause to become immovable; embrace closely ▶ **lockout** n closing of a workplace by an employer to force workers to accept terms ▶ **locksmith** n person who makes and mends locks ▶ **lockup** n prison; garage or storage place away from the main premises

lock² n strand of hair

locker n small cupboard with a lock

locket n small hinged pendant for a portrait etc.

lockjaw n tetanus

locomotive n self-propelled engine for pulling trains ▷ adj of locomotion ▶ **locomotion** n action or power of moving

locum n temporary stand-in for a doctor or clergyman

locus [loh-kuss] n, pl **loci** [loh-sigh] area or place where something happens; (Maths) set of points or lines satisfying one or more specified conditions

locust n destructive African insect that flies in swarms and eats crops

lode n vein of ore ▶ **lodestar** n star used in navigation or astronomy as a point of reference ▶ **lodestone** n magnetic iron ore

lodge n (chiefly Brit) gatekeeper's house; house or cabin used occasionally by hunters, skiers, etc.; porters' room in a university or college; local branch of some societies ▷ v live in another's house at a fixed charge; stick or become stuck (in a place); make (a complaint etc.) formally ▶ **lodger** n ▶ **lodging** n temporary residence ▷ pl rented room or rooms in another person's house

loft n space between the top storey and roof of a building; gallery in a church etc. ▷ v (*Sport*) strike, throw, or kick (a ball) high into the air

lofty adj **loftier, loftiest** of great height; exalted or noble; haughty ▶ **loftily** adv haughtily

log[1] n portion of a felled tree stripped of branches; detailed record of a journey of a ship, aircraft, etc. ▷ v **logging, logged** saw logs from a tree; record in a log ▶ **logging** n work of cutting and transporting logs ▶ **logbook** n book recording the details about a car or a ship's journeys ▶ **log in, log out** v gain entrance to or leave a computer system by keying in a special command

log[2] n short for **logarithm**

loganberry n purplish-red fruit, similar to a raspberry

logarithm n one of a series of arithmetical functions used to make certain calculations easier

loggerheads pl n **at loggerheads** quarrelling, disputing

loggia [loj-ya] n covered gallery at the side of a building

logic n philosophy of reasoning; reasoned thought or argument ▶ **logical** adj of logic; capable of or using clear valid reasoning; reasonable ▶ **logically** adv ▶ **logician** n

logistics n detailed planning and organization of a large, esp. military, operation ▶ **logistical, logistic** adj

logo [loh-go] n, pl **-os** emblem used by a company or other organization

loin n part of the body between the ribs and the hips; cut of meat from this part of an animal ▷ pl hips and inner thighs ▶ **loincloth** n piece of cloth covering the loins only

loiter v stand or wait aimlessly or idly

loll v lounge lazily; hang loosely

lollipop n boiled sweet on a small wooden stick ▶ **lollipop man, lollipop lady** (*Brit informal*) person holding a circular sign on a pole, who controls traffic so that children may cross the road safely

lolly n, pl **-ies** (*informal*) lollipop or ice lolly; (*Aust & NZ informal*) sweet; (*slang*) money ▶ **lolly scramble** (*NZ*) sweets scattered on the ground for children to collect

lone adj solitary ▶ **lonely** adj sad because alone; resulting from being alone; unfrequented ▶ **loneliness** n ▶ **loner** n (*informal*) person who prefers to be alone ▶ **lonesome** adj lonely

long[1] adj having length, esp. great length, in space or time ▷ adv for an extensive period ▶ **long-distance** adj going between places far apart ▶ **long face** glum expression ▶ **longhand** n ordinary writing, not shorthand or typing ▶ **long johns** (*informal*) long underpants ▶ **long-life** adj (of milk, batteries, etc.) lasting longer than the regular kind ▶ **long-lived** adj living or lasting for a long time ▶ **long-range** adj extending into the future; (of vehicles, weapons, etc.) designed to cover great distances ▶ **long shot** competitor, undertaking, or bet with little chance of success ▶ **long-sighted** adj able to see distant objects in focus but not nearby ones ▶ **long-standing** adj existing for a long time ▶ **long-suffering** adj enduring trouble or unhappiness without complaint ▶ **long-term** adj lasting or effective for a long time ▶ **long wave** radio wave with a wavelength of over 1000 metres ▶ **long-winded** adj speaking or writing at tedious length

long² v have a strong desire (for) ▶ **longing** n yearning ▶ **longingly** adv

longevity [lon-jev-it-ee] n long life

longitude n distance east or west from a standard meridian ▶ **longitudinal** adj of length or longitude; lengthways

longshoreman n (US) docker

loo n (Brit informal) toilet

loofah n sponge made from the dried pod of a gourd

look v direct the eyes or attention (towards); have the appearance of being; face in a particular direction; search (for); hope (for) ▶ n instance of looking; (often pl) appearance ▶ **look after** v take care of ▶ **lookalike** n person who is the double of another ▶ **look down on** v treat as inferior or unimportant ▶ **look forward to** v anticipate with pleasure ▶ **look on** v be a spectator; consider or regard ▶ **lookout** n guard; place for watching; (informal) worry or concern; chances or prospect ▶ **look out** v be careful ▶ **look up** v discover or confirm by checking in a book; improve; visit ▶ **look up to** v respect

loom¹ n machine for weaving cloth

loom² v appear dimly; seem ominously close

loony (slang) adj **loonier, looniest** foolish or insane ▶ n, pl **loonies** foolish or insane person

loop n rounded shape made by a curved line or rope crossing itself ▶ v form or fasten with a loop **loop the loop** fly or be flown in a complete vertical circle ▶ **loophole** n means of evading a rule without breaking it

loose adj not tight, fastened, fixed, or tense; vague; dissolute or promiscuous ▶ adv in a loose manner ▶ v free; unfasten; slacken; let fly (an arrow, bullet, etc.) **at a loose end** bored, with nothing to do ▶ **loosely** adv ▶ **looseness** n ▶ **loosen** v make loose ▶ **loosen up** v relax, stop worrying ▶ **looseleaf** adj allowing the addition or removal of pages

loot n, v plunder ▶ n (informal) money ▶ **looter** n ▶ **looting** n

lop v **lopping, lopped** cut away twigs and branches; chop off

lope v run with long easy strides

lop-eared adj having drooping ears

lopsided adj greater in height, weight, or size on one side

loquacious adj talkative ▶ **loquacity** n

lord n person with power over others, such as a monarch or master; male member of the British nobility; (Hist) feudal superior; (L-) God or Jesus; (L-) (in Britain) title given to certain male officials and peers **House of Lords** unelected upper chamber of the British parliament **lord it over** act in a superior manner towards **the Lord's Prayer** prayer taught by Christ to his disciples ▶ **lordly** adj imperious, proud ▶ **Lordship** n (in Britain) title of some male officials and peers

lore n body of traditions on a subject

lorgnette [lor-nyet] n pair of spectacles mounted on a long handle

lorikeet n small brightly coloured Australian parrot

lorry n, pl **-ries** (Brit & SAfr) large vehicle for transporting loads by road

lose v **losing, lost** come to be without, esp. by accident or carelessness; fail to keep or maintain; be deprived of; fail to get or make use of; be defeated in a competition etc.; be or become engrossed e.g. *lost in thought* ▶ **loser** *n* person or thing that loses; (*informal*) person who seems destined to fail

USAGE NOTE
Note the difference in spelling between the verb **lose** and the adjective **loose**

loss *n* losing; that which is lost; damage resulting from losing **at a loss** confused or bewildered; not earning enough to cover costs ▶ **loss leader** item sold at a loss to attract customers

lost *v* past of **lose** ▶ *adj* unable to find one's way; unable to be found

lot *pron* great number ▶ *n* collection of people or things; fate or destiny; one of a set of objects drawn at random to make a selection or choice; item at auction ▶ *pl* (*informal*) great numbers or quantities **a lot** (*informal*) a great deal

loth *adj* same as **loath**

lotion *n* medical or cosmetic liquid for use on the skin

lottery *n*, *pl* **-teries** method of raising money by selling tickets that win prizes by chance; gamble

lotto *n* game of chance like bingo; (**L-**) national lottery

lotus *n* legendary plant whose fruit induces forgetfulness; Egyptian water lily

loud *adj* relatively great in volume; capable of making much noise; insistent and emphatic; unpleasantly patterned or colourful ▶ **loudly** *adv* ▶ **loudness** *n* ▶ **loudspeaker** *n* instrument for converting electrical signals into sound

lough *n* (*Irish*) loch

lounge *n* living room in a private house; more expensive bar in a pub; area for waiting in an airport ▶ *v* sit, lie, or stand in a relaxed manner ▶ **lounge suit** man's suit for daytime wear

lour *v* same as **lower²**

louse *n* (*pl* **lice**) wingless parasitic insect; (*pl* **louses**) unpleasant person ▶ **lousy** *adj* (*slang*) mean or unpleasant; bad, inferior; unwell

lout *n* crude, oafish, or aggressive person ▶ **loutish** *adj*

louvre [**loo-ver**] *n* one of a set of parallel slats slanted to admit air but not rain ▶ **louvred** *adj*

love *v* have a great affection for; feel sexual passion for; enjoy (something) very much ▶ *n* great affection; sexual passion; wholehearted liking for something; beloved person; (*Tennis, squash etc.*) score of nothing **fall in love** become in love **in love (with)** feeling a strong emotional (and sexual) attraction (for) **make love (to)** have sexual intercourse (with) ▶ **lovable, loveable** *adj* ▶ **loveless** *adj* ▶ **lovely** *adj* **-lier, -liest** very attractive; highly enjoyable ▶ **lover** *n* person having a sexual relationship outside marriage; person in love; someone who loves a specified person or thing ▶ **loving** *adj* affectionate, tender ▶ **lovingly** *adv* ▶ **love affair** romantic or sexual relationship between two people who are not married to each other ▶ **lovebird** *n* small parrot ▶ **love child** (*euphemistic*) child of an unmarried couple ▶ **love life** person's romantic or sexual relationship ▶ **lovelorn** *adj* miserable because of unhappiness in love ▶ **lovemaking** *n*

low¹ *adj* not tall, high, or elevated; of little or less than the usual amount, degree, quality, or cost;

coarse or vulgar; dejected; not loud; deep in pitch ▷ *adv* in or to a low position, level, or degree ▷ *n* low position, level, or degree; area of low atmospheric pressure, depression ▶ **lowly** *adj* modest, humble ▶ **lowliness** *n* ▶ **lowbrow** *n*, *adj* (person) with nonintellectual tastes and interests ▶ **Low Church** section of the Anglican Church stressing evangelical beliefs and practices ▶ **lowdown** *n* (*informal*) inside information ▶ **low-down** *adj* (*informal*) mean, underhand, or dishonest ▶ **low-key** *adj* subdued, restrained, not intense ▶ **lowland** *n* low-lying country ▷ *pl* (**L-**) less mountainous parts of Scotland ▶ **low profile** position or attitude avoiding prominence or publicity ▶ **low-spirited** *adj* depressed

low² *n* cry of cattle, moo ▷ *v* moo

lower¹ *adj* below one or more other things; smaller or reduced in amount or value ▷ *v* cause or allow to move down; lessen ▶ **lower case** small, as distinct from capital, letters

lower², **lour** *v* (of the sky or weather) look gloomy or threatening ▶ **lowering** *adj*

loyal *adj* faithful to one's friends, country, or government ▶ **loyally** *adv* ▶ **loyalty** *n* ▶ **loyalty card** swipe card issued by a supermarket or chain store to a customer, used to record credit points awarded for money spent in the store ▶ **loyalist** *n*

lozenge *n* medicated tablet held in the mouth until it dissolves; four-sided diamond-shaped figure

LP *n* record playing approximately 20–25 minutes each side

L-plate *n* (*Brit & Aust*) sign on a car being driven by a learner driver

LSD lysergic acid diethylamide: a hallucinogenic drug

Lt Lieutenant

Ltd (*Brit*) Limited (Liability)

lubricate [loo-brik-ate] *v* oil or grease to lessen friction ▶ **lubricant** *n* lubricating substance, such as oil ▶ **lubrication** *n*

lubricious *adj* (*lit*) lewd

lucerne *n* fodder plant like clover, alfalfa

lucid *adj* clear and easily understood; able to think clearly; bright and clear ▶ **lucidly** *adv* ▶ **lucidity** *n*

Lucifer *n* Satan

luck *n* fortune, good or bad; good fortune ▶ **lucky** *adj* having or bringing good luck ▶ **lucky dip** game in which prizes are picked from a tub at random ▶ **luckily** *adv* fortunately ▶ **luckless** *adj* having bad luck

lucrative *adj* very profitable

lucre [loo-ker] *n* filthy lucre (*facetious*) money

Luddite *n* person opposed to change in industrial methods

luderick *n* Australian fish, usu. black or dark brown in colour

ludicrous *adj* absurd or ridiculous ▶ **ludicrously** *adv*

ludo *n* game played with dice and counters on a board

lug¹ *v* **lugging**, **lugged** carry or drag with great effort

lug² *n* projection serving as a handle; (*Brit informal*) ear

luggage *n* traveller's cases, bags, etc.

lugubrious *adj* mournful, gloomy ▶ **lugubriously** *adv*

lugworm *n* large worm used as bait

lukewarm *adj* moderately warm, tepid; indifferent or half-hearted

lull *v* soothe (someone) by soft sounds or motions; calm (fears or suspicions) by deception ▷ n brief time of quiet in a storm etc.

lullaby n, pl -bies quiet song to send a child to sleep

lumbago [lum-bay-go] n pain in the lower back ▶ **lumbar** adj relating to the lower back

lumber¹ n (Brit) unwanted disused household articles; (chiefly US) sawn timber ▷ v (informal) burden with something unpleasant ▶ **lumberjack** n (US) man who fells trees and prepares logs for transport

lumber² v move heavily and awkwardly ▶ **lumbering** adj

luminous adj reflecting or giving off light ▶ **luminosity** n ▶ **luminary** n famous person; (lit) heavenly body giving off light ▶ **luminescence** n emission of light at low temperatures by any process other than burning ▶ **luminescent** adj

lump¹ n shapeless piece or mass; swelling; (informal) awkward or stupid person ▷ v consider as a single group **lump in one's throat** tight dry feeling in one's throat, usu. caused by great emotion ▶ **lumpy** adj ▶ **lump sum** relatively large sum of money paid at one time

lump² v **lump it** (informal) tolerate or put up with it

lunar adj relating to the moon

lunatic adj foolish and irresponsible ▷ n foolish or annoying person; (old-fashioned) insane person ▶ **lunacy** n

lunch n meal taken in the middle of the day ▷ v eat lunch ▶ **luncheon** n formal lunch ▶ **luncheon meat** tinned ground mixture of meat and cereal ▶ **luncheon voucher** (Brit) voucher for a certain amount, given to an employee and accepted

by some restaurants as payment for a meal

lung n organ that allows an animal or bird to breathe air; humans have two lungs in the chest ▶ **lungfish** n freshwater bony fish with an air-breathing lung of South America and Australia

lunge n sudden forward motion; thrust with a sword ▷ v move with or make a lunge

lupin n garden plant with tall spikes of flowers

lupine adj like a wolf

lurch¹ v tilt or lean suddenly to one side; stagger ▷ n lurching movement

lurch² n **leave someone in the lurch** abandon someone in difficulties

lurcher n crossbred dog trained to hunt silently

lure v tempt or attract by the promise of reward ▷ n person or thing that lures; brightly-coloured artificial angling bait

lurid adj vivid in shocking detail, sensational; glaring in colour ▶ **luridly** adv

lurk v lie hidden or move stealthily, esp. for sinister purposes; be latent

luscious [lush-uss] adj extremely pleasurable to taste or smell; very attractive

lush¹ adj (of grass etc.) growing thickly and healthily; opulent; (slang) very attractive

lush² n (slang) alcoholic

lust n strong sexual desire; any strong desire (for) ▶ **lustful** adj ▶ **lusty** adj vigorous, healthy ▶ **lustily** adv

lustre n gloss, sheen; splendour or glory; metallic pottery glaze ▶ **lustrous** adj shining, luminous

lute n ancient guitar-like musical instrument with a body shaped like a half pear

Lutheran *adj* of Martin Luther (1483–1546), German Reformation leader, his doctrines, or a Church following these doctrines

luxuriant *adj* rich and abundant; very elaborate ▸ **luxuriance** *n* ▸ **luxuriantly** *adv*

luxuriate *v* take self-indulgent pleasure (in); flourish

luxury *n, pl* -**ries** enjoyment of rich, very comfortable living; enjoyable but not essential thing ▷ *adj* of or providing luxury ▸ **luxurious** *adj* full of luxury, sumptuous ▸ **luxuriously** *adv*

lychee [lie-chee] *n* Chinese fruit with a whitish juicy pulp

lych gate *n* roofed gate to a churchyard

Lycra® *n* elastic fabric used for tight-fitting garments, such as swimsuits

lye *n* caustic solution obtained by leaching wood ash

lying *v* present participle of **lie¹**, **lie²**

lymph *n* colourless bodily fluid consisting mainly of white blood cells ▸ **lymphatic** *adj*

lymphocyte *n* type of white blood cell

lynch *v* put to death without a trial

lynx *n* animal of the cat family with tufted ears and a short tail

lyre *n* ancient musical instrument like a U-shaped harp

lyric *adj* (of poetry) expressing personal emotion in song-like style; like a song ▷ *n* short poem in a song-like style ▷ *pl* words of a popular song ▸ **lyrical** *adj* lyric; enthusiastic ▸ **lyricist** *n* person who writes the words of songs or musicals

Mm

M Motorway; Monsieur

m metre(s); mile(s); minute(s)

m. male; married; masculine; meridian; month

MA Master of Arts

ma *n* (*informal*) mother

ma'am *n* madam

mac *n* (*Brit informal*) mackintosh

macabre [mak-**kahb**-ra] *adj* strange and horrible, gruesome

macadam *n* road surface of pressed layers of small broken stones

macadamia [mack-ah-**day**-mee-ah] *n* Australian tree with edible nuts

macaroni *n* pasta in short tube shapes

macaroon *n* small biscuit or cake made with ground almonds

macaw *n* large tropical American parrot

mace¹ *n* ceremonial staff of office; medieval weapon with a spiked metal head

mace² *n* spice made from the dried husk of the nutmeg

macerate [**mass**-er-ate] *v* soften by soaking ▸ **maceration** *n*

machete [mash-**ett**-ee] *n* broad heavy knife used for cutting or as a weapon

Machiavellian [mak-ee-a-**vel**-yan] *adj* unprincipled, crafty, and opportunist

machinations [mak-in-**nay**-shunz] *pl n* cunning plots and ploys

machine *n* apparatus, usu. powered by electricity, designed to perform a particular task; vehicle, such as a car or aircraft; controlling system of an organization ▷ make or produce by machine ▶ **machinery** n machines or machine parts collectively ▶ **machinist** n person who operates a machine ▶ **machine gun** automatic gun that fires rapidly and continuously ▶ **machine-gun** vfire at with such a gun ▶ **machine-readable** adj (of data) in a form suitable for processing by a computer

machismo [mak-**izz**-moh] n exaggerated or strong masculinity

Mach number [mak] n ratio of the speed of a body in a particular medium to the speed of sound in that medium

macho [**match**-oh] adj strongly or exaggeratedly masculine

mackerel n edible sea fish

mackintosh n waterproof raincoat of rubberized cloth

macramé [mak-**rah**-mee] n ornamental work of knotted cord

macrobiotics n dietary system advocating whole grains and vegetables grown without chemical additives ▶ **macrobiotic** adj

macrocosm n the universe; any large complete system

mad adj **madder, maddest** mentally deranged, insane; very foolish; (informal) angry; frantic; (foll by about or on) very enthusiastic (about) **like mad** (informal) with great energy, enthusiasm, or haste ▶ **madly** adv ▶ **madness** n ▶ **madden** v infuriate or irritate ▶ **maddening** adj ▶ **madman, madwoman** n

madam n polite form of address to a woman; (informal) precocious or conceited girl

madame [mad-**dam**] n, pl **mesdames** [may-**dam**] French title equivalent to **Mrs**

madcap adj foolish or reckless

madder n climbing plant; red dye made from its root

made v past of **make**

Madeira [mad-**deer**-a] n fortified white wine ▶ **Madeira cake** rich sponge cake

mademoiselle [mad-mwah-**zel**] n, pl **mesdemoiselles** [maid-mwah-**zel**] French title equivalent to **Miss**

Madiba generation n (S Afr) generation born around 1994, when Nelson Mandela became the first president of a multiracial South Africa

Madonna n the Virgin Mary; picture or statue of her

madrigal n 16th–17th-century part song for unaccompanied voices

maelstrom [**male**-strom] n great whirlpool; turmoil

maestro [**my**-stroh] n, pl **-tri** or **-tros** outstanding musician or conductor; any master of an art

Mafia n international secret criminal organization founded in Sicily ▶ **mafioso** n, pl **-sos** or **-si** member of the Mafia

magazine n periodical publication with articles by different writers; television or radio programme made up of short nonfiction items; appliance for automatically supplying cartridges to a gun or slides to a projector; storehouse for explosives or arms

magenta [maj-**jen**-ta] adj deep purplish-red

maggot n larva of an insect ▶ **maggoty** adj

Magi [**maje**-eye] pl n wise men from the East who came to worship the infant Jesus

magic n supposed art of invoking supernatural powers to influence events; mysterious quality or power ▷ adj (also **magical**) of, using, or like magic; (informal) wonderful, marvellous ▶ **magically** adv ▶ **magician** n conjuror; person with magic powers

magistrate n public officer administering the law; (Brit) Justice of the Peace; (Aust & NZ) former name for **district court judge** ▶ **magisterial** adj commanding or authoritative; of a magistrate

magma n molten rock inside the earth's crust

magnanimous adj noble and generous ▶ **magnanimously** adv ▶ **magnanimity** n

magnate n influential or wealthy person, esp. in industry

magnesia n white tasteless substance used as an antacid and a laxative; magnesium oxide

magnesium n (Chem) silvery-white metallic element

magnet n piece of iron or steel capable of attracting iron and pointing north when suspended ▶ **magnetic** adj having the properties of a magnet; powerfully attractive ▶ **magnetically** adv ▶ **magnetism** n magnetic property; powerful personal charm; science of magnetic properties ▶ **magnetize** v make into a magnet; attract strongly ▶ **magnetic tape** plastic strip coated with a magnetic substance for recording sound or video signals

magneto [mag-nee-toe] n, pl **-tos** apparatus for ignition in an internal-combustion engine

magnificent adj splendid or impressive; excellent ▶ **magnificently** adv ▶ **magnificence** n

magnify v **-fying, -fied** increase in apparent size, as with a lens; exaggerate ▶ **magnification** n

magnitude n relative importance or size

magnolia n shrub or tree with showy white or pink flowers

magnum n large wine bottle holding about 1.5 litres

magpie n black-and-white bird; any of various similar Australian birds, e.g. the butcherbird

maharajah n former title of some Indian princes ▶ **maharani** n fem

mah jong, mah-jongg n Chinese table game for four, played with tiles bearing different designs

mahogany n hard reddish-brown wood of several tropical trees

mahout [ma-howt] n (in India and the East Indies) elephant driver or keeper

maid n (also **maidservant**) female servant; (lit) young unmarried woman

maiden n (lit) young unmarried woman ▷ adj unmarried; first e.g. maiden voyage ▶ **maidenly** adj modest ▶ **maidenhair** n fern with delicate fronds ▶ **maidenhead** n virginity ▶ **maiden name** woman's surname before marriage ▶ **maiden over** (Cricket) over in which no runs are scored

mail n letters and packages transported and delivered by the post office; postal system; single collection or delivery of mail; train, ship, or aircraft carrying mail; same as **e-mail** ▷ v send by mail ▶ **mailbox** n (US, Canadian & Aust) box into which letters and parcels are delivered ▶ **mail order** system of buying goods by post ▶ **mailshot** n (Brit) posting of advertising material to many selected people at once

mail² n flexible armour of interlaced rings or links

maim v cripple or mutilate

main adj chief or principal ▷ n principal pipe or line carrying water, gas, or electricity ▷ pl main distribution network for water, gas, and electricity **in the main** on the whole ▸ **mainly** adv for the most part, chiefly ▸ **mainframe** n, adj (Computing) (denoting) a high-speed general-purpose computer ▸ **mainland** n stretch of land which forms the main part of a country ▸ **mainmast** n chief mast of a ship ▸ **mainsail** n largest sail on a mainmast ▸ **mainspring** n chief cause or motive; chief spring of a watch or clock ▸ **mainstay** n chief support; rope securing a mainmast ▸ **mainstream** n, adj (of) a prevailing cultural trend

maintain v continue or keep in existence; keep up or preserve; support financially; assert ▸ **maintenance** n maintaining; upkeep of a building, car, etc.; provision of money for a separated or divorced spouse

maisonette n (Brit) flat with more than one floor

maître d'hôtel [met-ra dote-**tell**] n (French) head waiter

maize n type of corn with spikes of yellow grains

majesty n, pl -**ties** stateliness or grandeur; supreme power ▸ **majestic** adj ▸ **majestically** adv

major adj greater in number, quality, or extent; significant or serious ▷ n middle-ranking army officer; scale in music; (US, Canadian, S Afr, Aust & NZ) principal field of study at a university etc. ▷ v (foll by in) (US, Canadian, S Afr, Aust & NZ) do one's principal study in (a particular subject) ▸ **major-domo**

n, pl -**domos** chief steward of a great household

majority n, pl -**ties** greater number; number by which the votes on one side exceed those on the other; largest party voting together; state of being legally an adult

make v **making, made** create, construct, or establish; cause to do or be; bring about or produce; perform (an action); serve as or become; amount to; earn ▷ n brand, type, or style **make do** manage with an inferior alternative **make it** (informal) be successful **on the make** (informal) out for profit or conquest ▸ **maker** n ▸ **making** n creation or production ▷ pl necessary requirements or qualities ▸ **make-believe** n fantasy or pretence ▸ **make for** v head towards ▸ **make off with** v steal or abduct ▸ **makeshift** adj serving as a temporary substitute ▸ **make up** v form or constitute; prepare; invent; supply what is lacking, complete; (foll by for) compensate (for); settle a quarrel; apply cosmetics ▸ **make-up** n cosmetics; way something is made; mental or physical constitution ▸ **makeweight** n something unimportant added to make up a lack

mal- combining form bad or badly e.g. malformation

malachite [mal-a-kite] n green mineral

maladjusted adj (Psychol) unable to meet the demands of society ▸ **maladjustment** n

maladministration n inefficient or dishonest administration

maladroit adj clumsy or awkward

malady n, pl -**dies** disease or illness

malaise [mal-**laze**] *n* vague feeling of unease, illness, or depression

malapropism *n* comical misuse of a word by confusion with one which sounds similar, e.g. *I am not under the affluence of alcohol*

malaria *n* infectious disease caused by the bite of some mosquitoes ▶ **malarial** *adj*

Malay *n* member of a people of Malaysia or Indonesia; language of this people ▶ **Malayan** *adj, n*

malcontent *n* discontented person

male *adj* of the sex which can fertilize female reproductive cells ▷ *n* male person or animal

malediction [mal-lid-**dik**-shun] *n* curse

malefactor [mal-if-act-or] *n* criminal or wrongdoer

malevolent [mal-**lev**-a-lent] *adj* wishing evil to others ▶ **malevolently** *adv* ▶ **malevolence** *n*

malfeasance [mal-**fee**-zanss] *n* misconduct, esp. by a public official

malformed *adj* misshapen or deformed ▶ **malformation** *n*

malfunction *v* function imperfectly or fail to function ▷ *n* defective functioning or failure to function

malice [mal-iss] *n* desire to cause harm to others ▶ **malicious** *adj* ▶ **maliciously** *adv*

malign [mal-**line**] *v* slander or defame ▷ *adj* evil in influence or effect ▶ **malignity** *n* evil disposition

malignant [mal-**lig**-nant] *adj* seeking to harm others; (of a tumour) harmful and uncontrollable ▶ **malignancy** *n*

malinger *v* feign illness to avoid work ▶ **malingerer** *n*

mall [mawl] *n* street or shopping area closed to vehicles

mallard *n* wild duck

malleable [mal-lee-a-bl] *adj* capable of being hammered or pressed into shape; easily influenced ▶ **malleability** *n*

mallee *n* (*Aust*) low-growing eucalypt in dry regions

mallet *n* (wooden) hammer; stick with a head like a hammer, used in croquet or polo

mallow *n* plant with pink or purple flowers

malnutrition *n* inadequate nutrition

malodorous [mal-**lode**-or-uss] *adj* bad-smelling

malpractice *n* immoral, illegal, or unethical professional conduct

malt *n* grain, such as barley, prepared for use in making beer or whisky

maltreat *v* treat badly ▶ **maltreatment** *n*

malware *n* computer program, such as a virus, designed to damage or disrupt a system

mama *n* (*old-fashioned*) mother

mamba *n* deadly S African snake

mamma *n* same as **mama**

mammal *n* animal of the type that suckles its young ▶ **mammalian** *adj*

mammary *adj* of the breasts or milk-producing glands

mammon *n* wealth regarded as a source of evil

mammoth *n* extinct elephant-like mammal ▷ *adj* colossal

mampara [mum-**puh**-ruh] *n* (*S Afr slang*) incompetent fool

man *n, pl* **men** adult male; human being or person; mankind; manservant; piece used in chess etc. ▷ *v* **manning, manned** supply with sufficient people for operation or defence ▶ **manhood** *n* ▶ **mankind** *n* human beings collectively ▶ **manly** *adj* (possessing qualities) appropriate

to a man ▸ **manliness** n
▸ **mannish** adj (of a woman) like
a man ▸ **man-hour** n work done
by one person in one hour ▸ **man-made** adj made artificially

mana n (NZ) authority, influence

manacle [man-a-kl] n, v handcuff
or fetter

manage v succeed in doing; be
in charge of, administer; handle
or control; cope with (financial)
difficulties ▸ **manageable** adj
▸ **management** n managers
collectively; administration or
organization

manager, manageress n
person in charge of a business,
institution, actor, sports team, etc.
▸ **managerial** adj

manatee n large tropical plant-
eating aquatic mammal

mandarin n high-ranking
government official; kind of small
orange

Mandarin Chinese, Mandarin n
official language of China

mandate n official or authoritative
command; authorization or
instruction from an electorate to
its representative or government
▸ v give authority to ▸ **mandatory**
adj compulsory

mandible n lower jawbone or
jawlike part

mandolin n musical instrument
with four pairs of strings

mandrake n plant with a forked
root, formerly used as a narcotic

mandrel n shaft on which work is
held in a lathe

mandrill n large blue-faced baboon

mane n long hair on the neck of a
horse, lion, etc.

manful adj determined and brave
▸ **manfully** adv

manganese n (Chem) brittle
greyish-white metallic element

mange n skin disease of domestic
animals

mangelwurzel n variety of beet
used as cattle food

manger n eating trough in a stable
or barn

mangetout [mawnzh-too] n
variety of pea with an edible pod

mangle¹ v destroy by crushing and
twisting; spoil

mangle² n machine with rollers
for squeezing water from washed
clothes ▸ v put through a mangle

mango n, pl -goes or -gos tropical
fruit with sweet juicy yellow flesh

mangrove n tropical tree with
exposed roots, which grows beside
water

mangy adj mangier, mangiest
having mange; scruffy or shabby

manhandle v treat roughly

manhole n hole with a cover,
through which a person can enter a
drain or sewer

mania n extreme enthusiasm;
madness ▸ **maniac** n mad person;
(informal) person who has an
extreme enthusiasm for something
▸ **maniacal** [man-**eye**-a-kl] adj

manic adj affected by mania
▸ **manic-depressive** adj, n
(Psychiatry) (person afflicted with) a
mental disorder that causes mood
swings from extreme euphoria to
deep depression

manicure n cosmetic care of the
fingernails and hands ▸ v care for
(the fingernails and hands) in this
way ▸ **manicurist** n

manifest adj easily noticed,
obvious ▸ v show plainly; be
evidence of ▸ n list of cargo
or passengers for customs
▸ **manifestation** n

manifesto n, pl -tos or -toes
declaration of policy as issued by
a political party

manifold adj numerous and varied ▷ n pipe with several outlets, esp. in an internal-combustion engine

manikin n little man or dwarf; model of the human body

manila, manilla n strong brown paper used for envelopes

manipulate v handle skilfully; control cleverly or deviously ▶ **manipulation** n ▶ **manipulative** adj ▶ **manipulator** n

manna n (Bible) miraculous food which sustained the Israelites in the wilderness; windfall

mannequin n life-size dummy of the human body used to fit or display clothes

manner n way a thing happens or is done; person's bearing or behaviour; type or kind; custom or style ▷ pl (polite) social behaviour ▶ **mannered** adj affected ▶ **mannerism** n person's distinctive habit or trait

mannikin n same as manikin

manoeuvre [man-noo-ver] n skilful movement; contrived, complicated, and possibly deceptive plan or action ▷ pl military or naval exercises ▷ v manipulate or contrive skilfully or cunningly; perform manoeuvres ▶ **manoeuvrable** adj

manor n (Brit) large country house and its lands ▶ **manorial** adj

manpower n available number of workers

manqué [mong-kay] adj would-be e.g. an actor manqué

mansard roof n roof with a break in its slope, the lower part being steeper than the upper

manse n house provided for a minister in some religious denominations

manservant n, pl menservants male servant, esp. a valet

mansion n large house

manslaughter n unlawful but unintentional killing of a person

mantel n structure round a fireplace ▶ **mantelpiece, mantel shelf** n shelf above a fireplace

mantilla n (in Spain) a lace scarf covering a woman's head and shoulders

mantis n, pl **-tises** or **-tes** carnivorous insect like a grasshopper

mantle n loose cloak; covering; responsibilities and duties which go with a particular job or position

mantra n (Hinduism, Buddhism) any sacred word or syllable used as an object of concentration

manual adj of or done with the hands; by human labour rather than automatic means ▷ n handbook; organ keyboard ▶ **manually** adv

manufacture v process or make (goods) on a large scale using machinery; invent or concoct (an excuse etc.) ▷ n process of manufacturing goods

manufacturer n company that manufactures goods

manure n animal excrement used as a fertilizer

manuscript n book or document, orig. one written by hand; copy for printing

Manx adj of the Isle of Man or its inhabitants ▷ n almost extinct language of the Isle of Man ▶ **Manx cat** tailless breed of cat

many adj more, most numerous ▷ n large number

Maoism n form of Marxism advanced by Mao Tse-tung in China ▶ **Maoist** n, adj

Māori n, pl **-ri** or **-ris** member of the indigenous race of New Zealand; language of the Māoris ▷ adj of the Māoris or their language

map n representation of the earth's surface or some part of it, showing geographical features ▷ v **mapping, mapped** make a map of ▶ **map out** v plan

maple n tree with broad leaves, a variety of which (**sugar maple**) yields sugar

mar v **marring, marred** spoil or impair

Mar. March

marabou n large black-and-white African stork; its soft white down, used to trim hats etc.

maraca [mar-**rak**-a] n shaken percussion instrument made from a gourd containing dried seeds etc.

marae n (NZ) enclosed space in front of a Māori meeting house; Māori meeting house and its buildings

maraschino cherry [mar-rass-**kee**-no] n cherry preserved in a cherry liqueur with a taste like bitter almonds

marathon n long-distance race of 26 miles 385 yards (42.195 kilometres); long or arduous task

marauding adj wandering or raiding in search of plunder ▶ **marauder** n

marble n kind of limestone with a mottled appearance, which can be highly polished; slab of or sculpture in this; small glass ball used in playing marbles ▷ pl game of rolling these at one another ▶ **marbled** adj having a mottled appearance like marble

march[1] v walk with a military step; make (a person or group) proceed; progress steadily ▷ n action of marching; steady progress; distance covered by marching; piece of music, as for a march ▶ **marcher** n ▶ **marching girl** (Aust & NZ) girl who does team formation marching as a sport

march[2] n border or frontier

March n third month of the year

marchioness [marsh-on-**ness**] n woman holding the rank of marquis; wife or widow of a marquis

Mardi Gras [mar-dee **grah**] n festival of Shrove Tuesday, celebrated in some cities with great revelry

mare n female horse or zebra ▶ **mare's nest** discovery which proves worthless

margarine n butter substitute made from animal or vegetable fats

marge n (informal) margarine

margin n edge or border; blank space round a printed page; additional amount or one greater than necessary; limit ▶ **marginal** adj insignificant, unimportant; near a limit; (Politics) (of a constituency) won by only a small margin ▷ n (Politics) marginal constituency ▶ **marginalize** v make or treat as insignificant ▶ **marginally** adv

marguerite n large daisy

marigold n plant with yellow or orange flowers

marijuana [mar-ree-**wah**-na] n dried flowers and leaves of the cannabis plant, used as a drug, esp. in cigarettes

marina n harbour for yachts and other pleasure boats

marinade n seasoned liquid in which fish or meat is soaked before cooking ▷ v same as **marinate** ▶ **marinate** v soak in marinade

marine adj of the sea or shipping ▷ n (esp. in Britain and the US) soldier trained for land and sea combat; country's shipping or fleet ▶ **mariner** n sailor

marionette n puppet worked with strings

marital adj relating to marriage

maritime adj relating to shipping; of, near, or living in the sea

marjoram n aromatic herb used for seasoning food and in salads

mark¹ n line, dot, scar, etc. visible on a surface; distinguishing sign or symbol; written or printed symbol; letter or number used to grade academic work; indication of position; indication of some quality; target or goal ▷ v make a mark on; characterize or distinguish; indicate; pay attention to; notice or watch; grade (academic work); stay close to (a sporting opponent) to hamper his or her play ▶ **marked** adj noticeable ▶ **markedly** adv ▶ **marker** n

mark² n same as **Deutschmark**

market n assembly or place for buying and selling; demand for goods ▷ v -**keting, -keted** offer or produce for sale **on the market** for sale ▶ **marketable** adj ▶ **marketing** n part of a business that controls the way that goods or services are sold ▶ **market garden** place where fruit and vegetables are grown for sale ▶ **market maker** (in London Stock Exchange) person who uses a firm's money to create a market for a stock ▶ **marketplace** n market; commercial world ▶ **market research** research into consumers' needs and purchases

marksman n person skilled at shooting ▶ **marksmanship** n

marl n soil formed of clay and lime, used as fertilizer

marlin n large food and game fish of warm and tropical seas, with a very long upper jaw

marlinespike, marlinspike n pointed hook used to separate strands of rope

marmalade n jam made from citrus fruit

marmoreal adj of or like marble

marmoset n small bushy-tailed monkey

marmot n burrowing rodent

maroon¹ adj reddish-purple

maroon² v abandon ashore, esp. on an island; isolate without resources

marquee n large tent used for a party or exhibition

marquess [mar-kwiss] n (Brit) nobleman of the rank below a duke

marquetry n ornamental inlaid work of wood

marquis n (in some European countries) nobleman of the rank above a count

marram grass n grass that grows on sandy shores

marrow n fatty substance inside bones; long thick striped green vegetable with whitish flesh

marry v -**rying, -ried** take as one's partner in marriage; join or give in marriage; unite closely ▶ **marriage** n state of being married; wedding ▶ **marriageable** adj

Mars n Roman god of war; fourth planet from the sun

Marsala [mar-sah-la] n dark sweet wine

marsh n low-lying wet land ▶ **marshy** adj

marshal n officer of the highest rank; official who organizes ceremonies or events; (US) law officer ▷ v -**shalling, -shalled** arrange in order; assemble; conduct with ceremony ▶ **marshalling yard** railway depot for goods trains

marshmallow n spongy pink or white sweet

marsupial [mar-**soop**-ee-al] n animal that carries its young in a pouch, such as a kangaroo

mart n market

Martello tower n round tower for coastal defence, formerly used in Europe

marten n weasel-like animal

martial adj of war, warlike ▶ **martial art** any of various philosophies and techniques of self-defence, orig. Eastern, such as karate ▶ **martial law** law enforced by military authorities in times of danger or emergency

Martian [**marsh**-an] adj of Mars ▷ n supposed inhabitant of Mars

martin n bird with a slightly forked tail

martinet n person who maintains strict discipline

martini n cocktail of vermouth and gin

martyr n person who dies or suffers for his or her beliefs ▷ v make a martyr of **be a martyr to** be constantly suffering from ▶ **martyrdom** n

marvel v **-velling, -velled** be filled with wonder ▷ n wonderful thing ▶ **marvellous** adj amazing; wonderful

Marxism n political philosophy of Karl Marx ▶ **Marxist** n, adj

marzipan n paste of ground almonds, sugar, and egg whites

masala n paste of ground spices, used in Indian cookery

masc. masculine

mascara n cosmetic for darkening the eyelashes

mascot n person, animal, or thing supposed to bring good luck

masculine adj relating to males; manly; (Grammar) of the gender of nouns that includes some male animate things ▶ **masculinity** n

mash n (informal) mashed potatoes; bran or meal mixed with warm water as food for horses etc. ▷ v crush into a soft mass

mask n covering for the face, as a disguise or protection; behaviour that hides one's true feelings ▷ v cover with a mask; hide or disguise

masochism [**mass**-oh-kiz-zum] n condition in which (sexual) pleasure is obtained from feeling pain or from being humiliated ▶ **masochist** n ▶ **masochistic** adj

mason n person who works with stone; (M-) Freemason ▶ **Masonic** adj of Freemasonry ▶ **masonry** n stonework; (M-) Freemasonry

masque [mask] n (Hist) 16th–17th-century form of dramatic entertainment

masquerade [mask-er-**aid**] n deceptive show or pretence; party at which masks and costumes are worn ▷ v pretend to be someone or something else

Mass n service of the Eucharist, esp. in the RC Church

mass n coherent body of matter; large quantity or number; (Physics) amount of matter in a body ▷ adj large-scale; involving many people ▷ v form into a mass **the masses** ordinary people ▶ **massive** adj large and heavy ▶ **mass-market** adj for or appealing to a large number of people ▶ **mass media** means of communication to many people, such as television and newspapers ▶ **mass-produce** v manufacture (standardized goods) in large quantities

massacre [**mass**-a-ker] n indiscriminate killing of large numbers of people ▷ v kill in large numbers

massage [mass-ahzh] n rubbing and kneading of parts of the body to reduce pain or stiffness ▷ v give a massage to ▶ **masseur**, (fem) **masseuse** n person who gives massages

massif [mass-seef] n connected group of mountains

mast¹ n tall pole for supporting something, esp. a ship's sails

mast² n fruit of the beech, oak, etc., used as pig fodder

mastectomy [mass-tek-tom-ee] n, pl -**mies** surgical removal of a breast

master n person in control, such as an employer or an owner of slaves or animals; expert; great artist; original thing from which copies are made; male teacher ▷ adj overall or controlling; main or principal ▷ v acquire knowledge of or skill in; overcome ▶ **masterful** adj domineering; showing great skill ▶ **masterly** adj showing great skill ▶ **mastery** n expertise; control or command ▶ **master key** n key that opens all the locks of a set ▶ **mastermind** v plan and direct (a complex task) ▷ n person who plans and directs a complex task ▶ **masterpiece** n outstanding work of art

mastic n gum obtained from certain trees; putty-like substance used as a filler, adhesive, or seal

masticate v chew ▶ **mastication** n

mastiff n large dog

mastitis n inflammation of a breast or udder

mastodon n extinct elephant-like mammal

mastoid n projection of the bone behind the ear

masturbate v fondle the genitals (of) ▶ **masturbation** n

mat n piece of fabric used as a floor covering or to protect a surface; thick tangled mass ▷ v **matting**, **matted** tangle or become tangled into a dense mass

matador n man who kills the bull in bullfights

match¹ n contest in a game or sport; person or thing exactly like, equal to, or in harmony with another; marriage ▷ v be exactly like, equal to, or in harmony with; put in competition (with); find a match for; join (in marriage) ▶ **matchless** adj unequalled ▶ **matchmaker** n person who schemes to bring about a marriage ▶ **matchmaking** n, adj

match² n small stick with a tip which ignites when scraped on a rough surface ▶ **matchbox** n ▶ **matchstick** n wooden part of a match ▷ adj (of drawn figures) thin and straight ▶ **matchwood** n small splinters

mate¹ n (informal) friend; associate or colleague e.g. team-mate; sexual partner of an animal; officer in a merchant ship; tradesman's assistant ▷ v pair (animals) or (of animals) be paired for reproduction

mate² n, v (Chess) checkmate

material n substance of which a thing is made; cloth; information on which a piece of work may be based ▷ pl things needed for an activity ▷ adj of matter or substance; not spiritual; affecting physical wellbeing; relevant ▶ **materially** adv considerably ▶ **materialism** n excessive interest in or desire for money and possessions; belief that only the material world exists ▶ **materialist** adj, n ▶ **materialistic** adj ▶ **materialize** v actually happen; come into existence or view ▶ **materialization** n

maternal adj of a mother; related through one's mother ▶ **maternity** n motherhood ▷ adj of or for pregnant women

matey adj (Brit informal) friendly or intimate

mathematics n science of number, quantity, shape, and space ▶ **mathematical** adj ▶ **mathematically** adv ▶ **mathematician** n

maths n (informal) mathematics

Matilda n (Aust Hist) swagman's bundle of belongings **waltz Matilda** (Aust) travel about carrying one's bundle of belongings

matinée [mat-in-nay] n afternoon performance in a theatre or cinema

matins pl n early morning service in various Christian Churches

matriarch [mate-ree-ark] n female head of a tribe or family ▶ **matriarchal** adj ▶ **matriarchy** n society governed by a female, in which descent is traced through the female line

matricide n crime of killing one's mother; person who does this

matriculate v enrol or be enrolled in a college or university ▶ **matriculation** n

matrimony n marriage ▶ **matrimonial** adj

matrix [may-trix] n, pl **matrices** substance or situation in which something originates, takes form, or is enclosed; mould for casting; (Maths) rectangular array of numbers or elements

matron n staid or dignified married woman; woman who supervises the domestic or medical arrangements of an institution; former name for **nursing officer** ▶ **matronly** adj

matt adj dull, not shiny

matter n substance of which something is made; physical substance; event, situation, or subject; written material in general; pus ▷ v be of importance **what's the matter?** what is wrong?

mattock n large pick with one of its blade ends flattened for loosening soil

mattress n large stuffed flat case, often with springs, used on or as a bed

mature adj fully developed or grown-up; ripe ▷ v make or become mature; (of a bill or bond) become due for payment ▶ **maturity** n state of being mature ▶ **maturation** n

maudlin adj foolishly or tearfully sentimental

maul v handle roughly; beat or tear

maunder v talk or act aimlessly or idly

mausoleum [maw-so-lee-um] n stately tomb

mauve adj pale purple

maverick n, adj independent and unorthodox (person)

maw n animal's mouth, throat, or stomach

mawkish adj foolishly sentimental

maxim n general truth or principle

maximum adj, n, pl -mums or -ma greatest possible (amount or number) ▶ **maximal** adj ▶ **maximize** v increase to a maximum

May n fifth month of the year; (m-)same as **hawthorn** ▶ **mayfly** n short-lived aquatic insect ▶ **maypole** n pole set up for dancing round on the first day of May to celebrate spring

may v, past tense **might** used as an auxiliary to express possibility, permission, opportunity, etc.

USAGE NOTE
In very careful usage *may* is used in preference to *can* for asking permission. *Might* is used to express a more tentative request: *May/might I ask a favour?*

maybe *adv* perhaps, possibly

Mayday *n* international radio distress signal

mayhem *n* violent destruction or confusion

mayonnaise *n* creamy sauce of egg yolks, oil, and vinegar

SPELLING TIP
There are two *n*s in the middle of *mayonnaise*

mayor *n* head of a municipality ▶ **mayoress** *n* mayor's wife; female mayor ▶ **mayoralty** *n* (term of) office of a mayor

maze *n* complex network of paths or lines designed to puzzle; any confusing network or system

mazurka *n* lively Polish dance; music for this

MB Bachelor of Medicine

MBE (in Britain) Member of the Order of the British Empire

MC Master of Ceremonies

MD Doctor of Medicine

me *pron* objective form of **I**

USAGE NOTE
The use of *it's me* in preference to *it's I* is accepted as quite standard

ME myalgic encephalomyelitis: painful muscles and general weakness sometimes persisting long after a viral illness

mead *n* alcoholic drink made from honey

meadow *n* piece of grassland ▶ **meadowsweet** *n* plant with dense heads of small fragrant flowers

meagre *adj* scanty or insufficient

meal *n* occasion when food is served and eaten; the food itself

meal² *n* grain ground to powder ▶ **mealy** *adj* ▶ **mealy-mouthed** *adj* not outspoken enough

mealie *n* (S Afr) maize

mean¹ *v* meaning, meant intend to convey or express; signify, denote, or portend; intend; have importance as specified ▶ **meaning** *n* sense, significance ▶ **meaningful** *adj* ▶ **meaningless** *adj*

mean² *adj* miserly, ungenerous, or petty; despicable or callous; (*chiefly US informal*) bad-tempered ▶ **meanly** *adv* ▶ **meanness** *n*

mean³ *n* middle point between two extremes; average ▶ *pl* method by which something is done; money ▶ *adj* intermediate in size or quantity; average **by all means** certainly **by no means** in no way ▶ **means test** inquiry into a person's means to decide on eligibility for financial aid

meander [mee-and-er] *v* follow a winding course; wander aimlessly ▶ *n* winding course

meantime *n* intervening period ▶ *adv* meanwhile

meanwhile *adv* during the intervening period; at the same time

measles *n* infectious disease producing red spots ▶ **measly** *adj* (*informal*) meagre

measure *n* size or quantity; graduated scale etc. for measuring size or quantity; unit of size or quantity; extent; action taken; law; poetical rhythm ▶ *v* determine the size or quantity of; be (a specified amount) in size or quantity ▶ **measurable** *adj* ▶ **measured** *adj* slow and steady; carefully considered ▶ **measurement** *n* measuring; size ▶ **measure up to** *v* fulfil (expectations or requirements)

meat n animal flesh as food
▶ **meaty** adj (tasting) of or like meat; brawny; full of significance or interest

Mecca n holy city of Islam; place that attracts visitors

mechanic n person skilled in repairing or operating machinery
▶ **mechanics** n scientific study of motion and force ▶ **mechanical** adj of or done by machines; (of an action) without thought or feeling
▶ **mechanically** adv

mechanism n way a machine works; piece of machinery; process or technique e.g. defence mechanism ▶ **mechanize** v equip with machinery; make mechanical or automatic; (Mil) equip (an army) with armoured vehicles
▶ **mechanization** n

med. medical; medicine; medieval; medium

medal n piece of metal with an inscription etc., given as a reward or memento ▶ **medallion** n disc-shaped ornament worn on a chain round the neck; large medal; circular decorative device in architecture
▶ **medallist** n winner of a medal

meddle v interfere annoyingly
▶ **meddler** n ▶ **meddlesome** adj

media n a plural of **medium**; the mass media collectively

mediaeval adj same as **medieval**

medial adj of or in the middle

median adj, n middle (point or line)

mediate v intervene in a dispute to bring about agreement
▶ **mediation** n ▶ **mediator** n

medic n (informal) doctor or medical student

medical adj of the science of medicine ▶ n (informal) medical examination ▶ **medically** adv ▶ **medicate** v treat with a medicinal substance

▶ **medication** n (treatment with) a medicinal substance

medicine n substance used to treat disease; science of preventing, diagnosing, or curing disease
▶ **medicinal** [med-**diss**-in-al] adj having therapeutic properties
▶ **medicine man** witch doctor

medieval [med-ee-**eve**-al] adj of the Middle Ages

mediocre [mee-dee-**oak**-er] adj average in quality; second-rate
▶ **mediocrity** [mee-dee-**ok**-rit-ee] n

meditate v reflect deeply, esp. on spiritual matters; think about or plan
▶ **meditation** n ▶ **meditative** adj
▶ **meditatively** adv ▶ **meditator** n

medium adj midway between extremes, average ▶ n, pl -**dia** or -**diums** middle state, degree, or condition; intervening substance producing an effect; means of communicating news or information to the public, such as radio or newspapers; person who can supposedly communicate with the dead; surroundings or environment; category of art according to the material used
▶ **medium wave** radio wave with a wavelength between 100 and 1000 metres

medlar n apple-like fruit of a small tree, eaten when it begins to decay

medley n miscellaneous mixture; musical sequence of different tunes

medulla [mid-**dull**-la] n, pl -**las** or -**lae** marrow, pith, or inner tissue

meek adj submissive or humble
▶ **meekly** adv ▶ **meekness** n

meerkat n S African mongoose

meerschaum [meer-**shum**] n white substance like clay; tobacco pipe with a bowl made of this

meet¹ v **meeting, met** come together (with); come into contact (with); be at the place of arrival of; make the acquaintance of; satisfy a need etc.; experience ▷ n meeting, esp. a sports meeting; assembly of a hunt ▶ **meeting** n coming together; assembly

USAGE NOTE
Meet is only followed by with in the context of misfortune: I met his son; I met with an accident

meet² adj (obs) fit or suitable

meg n (Computing informal) short for **megabyte**

mega- combining form denoting one million e.g. megawatt; very great e.g. megastar

megabyte n (Computing) 220 or 1 048 576 bytes

megahertz n, pl **-hertz** one million hertz

megalith n great stone, esp. as part of a prehistoric monument ▶ **megalithic** adj

megalomania n craving for or mental delusions of power ▶ **megalomaniac** adj, n

megaphone n cone-shaped instrument used to amplify the voice

megapixel n (Computing) one million pixels: used to describe the resolution of digital images

megapode n bird of Australia, New Guinea, and adjacent islands

megaton n explosive power equal to that of one million tons of TNT

meh interj expression of indifference or boredom

melaleuca [mel-a-loo-ka] n Australian shrub or tree with a white trunk and black branches

melancholy [mel-an-kol-lee] n sadness or gloom ▷ adj sad or gloomy ▶ **melancholia** [mel-an-kole-lee-a] n state of depression ▶ **melancholic** adj, n

melange [may-lahnzh] n mixture

melanin n dark pigment found in the hair, skin, and eyes of humans and animals

mêlée [mel-lay] n noisy confused fight or crowd

mellifluous [mel-lif-flew-uss] adj (of sound) smooth and sweet

mellow adj soft, not harsh; kind-hearted, esp. through maturity; (of fruit) ripe ▷ v make or become mellow

melodrama n play full of extravagant action and emotion; overdramatic behaviour or emotion ▶ **melodramatic** adj

melody n, pl **-dies** series of musical notes which make a tune; sweet sound ▶ **melodic** [mel-lod-ik] adj of melody; melodious ▶ **melodious** [mel-lode-ee-uss] adj pleasing to the ear; tuneful

melon n large round juicy fruit with a hard rind

melt v (cause to) become liquid by heat; dissolve; disappear; blend (into); soften through emotion ▶ **meltdown** n (in a nuclear reactor) melting of the fuel rods, with the possible release of radiation

member n individual making up a body or society; limb ▶ **membership** n ▶ **Member of Parliament** person elected to parliament

membrane n thin flexible tissue in a plant or animal body ▶ **membranous** adj

meme n video, photo, or story that is viewed by many internet users in a short time

memento n, pl **-tos** or **-toes** thing serving to remind, souvenir

memo n, pl **memos** short for **memorandum**

memoir [mem-wahr] n biography or historical account based on personal knowledge ▷ pl collection of these; autobiography

memorable adj worth remembering, noteworthy ▶ **memorably** adv

memorandum n, pl **-dums** or **-da** written record or communication within a business; note of things to be remembered

memory n, pl **-ries** ability to remember; sum of things remembered; particular recollection; length of time one can remember; commemoration; part of a computer which stores information ▶ **memorize** v commit to memory ▶ **memorial** n something serving to commemorate a person or thing ▶ adj serving as a memorial ▶ **memory card** small removable data storage device, used in mobile phones, digital cameras, etc. ▶ **Memory Stick®** (Computing) standard format for memory cards; (**m-**) (also **USB memory stick**) same as **USB drive**

men n plural of **man**

menace n threat; (informal) nuisance ▶ v threaten, endanger ▶ **menacing** adj

ménage [may-**nahzh**] n household

menagerie [min-**naj**-er-ee] n collection of wild animals for exhibition

mend v repair or patch; recover or heal; make or become better ▶ n mended area **on the mend** regaining health

mendacity n (tendency to) untruthfulness ▶ **mendacious** adj

mendicant adj begging ▶ n beggar

menhir [**men**-hear] n single upright prehistoric stone

menial [**mean**-nee-al] adj involving boring work of low status ▶ n person with a menial job

meningitis [men-in-**jite**-iss] n inflammation of the membranes of the brain

meniscus n curved surface of a liquid; crescent-shaped lens

menopause n time when a woman's menstrual cycle ceases ▶ **menopausal** adj

menstruation n approximately monthly discharge of blood and cellular debris from the womb of a nonpregnant woman ▶ **menstruate** v ▶ **menstrual** adj

mensuration n measuring, esp. in geometry

mental adj of, in, or done by the mind; of or for mental illness; (informal) insane ▶ **mentally** adv ▶ **mentality** n way of thinking

menthol n organic compound found in peppermint, used medicinally

mention v refer to briefly; acknowledge ▶ n brief reference to a person or thing; acknowledgment

mentor n adviser or guide

menu n list of dishes to be served, or from which to order; (Computing) list of options displayed on a screen

MEP Member of the European Parliament

mercantile adj of trade or traders

mercenary adj influenced by greed; working merely for reward ▶ n, pl **-aries** hired soldier

merchandise n commodities

merchant n person engaged in trade, wholesale trader ▶ **merchant bank** bank dealing mainly with businesses and investment ▶ **merchantman** n trading ship ▶ **merchant navy** ships or crew engaged in a nation's commercial shipping

mercury n (Chem) silvery liquid metal; (**M-**) (Roman myth) messenger of the gods; (**M-**) planet nearest the sun ▶ **mercurial** adj lively, changeable

m

mercy n, pl **-cies** compassionate treatment of an offender or enemy who is in one's power; merciful act ► **merciful** adj compassionate; giving relief ► **merciless** adj

mere[1] adj nothing more than e.g. mere chance ► **merely** adv

mere[2] n (Brit obs) lake

meretricious adj superficially or garishly attractive but of no real value

merganser [mer-gan-ser] n large crested diving duck

merge v combine or blend

merger n combination of business firms into one

meridian n imaginary circle of the earth passing through both poles

meringue [mer-rang] n baked mixture of egg whites and sugar; small cake of this

merino n, pl **-nos** breed of sheep with fine soft wool; this wool

merit n excellence or worth ► pl admirable qualities ► v **-iting, -ited** deserve ► **meritorious** adj deserving praise ► **meritocracy** [mer-it-**tok**-rass-ee] n rule by people of superior talent or intellect

merlin n small falcon

mermaid n imaginary sea creature with the upper part of a woman and the lower part of a fish

merry adj **-rier, -riest** cheerful or jolly; (informal) slightly drunk ► **merrily** adv ► **merriment** n ► **merry-go-round** n roundabout ► **merrymaking** n noisy, cheerful celebrations or fun

mesdames n plural of **madame**

mesdemoiselles n plural of **mademoiselle**

mesh n network or net; (open space between) strands forming a network ► v (of gear teeth) engage

mesmerize v hold spellbound; (obs) hypnotize

meson [**mee**-zon] n elementary atomic particle

mess n untidy or dirty confusion; trouble or difficulty; place where service personnel eat; group of service personnel who regularly eat together ► v muddle or dirty; (foll by about) potter about; (foll by with) interfere with; (Brit, Aust & NZ) (of service personnel) eat in a group

message n communication sent; meaning or moral ► **messaging** n sending and receiving of textual communications by mobile phone ► **message board** n internet discussion forum ► **messenger** n bearer of a message

Messiah n Jews' promised deliverer; Christ ► **Messianic** adj

messieurs n plural of **monsieur**

Messrs [**mess**-erz] n plural of **Mr**

messy adj **messier, messiest** dirty, confused, or untidy ► **messily** adv

met v past of **meet**[1]

metabolism [met-**tab**-oh-liz-zum] n chemical processes of a living body ► **metabolic** adj ► **metabolize** v produce or be produced by metabolism

metal n chemical element, such as iron or copper, that is malleable and capable of conducting heat and electricity ► **metallic** adj ► **metallurgy** n scientific study of the structure, properties, extraction, and refining of metals ► **metallurgical** adj ► **metallurgist** n ► **metal road** (NZ) unsealed road covered in gravel

metamorphosis [met-a-**more**-foss-is] n, pl **-phoses** [-foss-eez] change of form or character ► **metamorphic** adj (of rocks) changed in texture or structure by heat and pressure ► **metamorphose** v transform

metaphor n figure of speech in which a term is applied to something it does not literally denote in order to imply a resemblance e.g. *he is a lion in battle* ▶ **metaphorical** adj ▶ **metaphorically** adv

metaphysics n branch of philosophy concerned with being and knowing ▶ **metaphysical** adj

mete v (usually foll by **out**) deal out as punishment

meteor n small fast-moving heavenly body, visible as a streak of incandescence if it enters the earth's atmosphere ▶ **meteoric** [meet-ee-**or**-rik] adj of a meteor; brilliant and very rapid ▶ **meteorite** n meteor that has fallen to earth

meteorology n study of the earth's atmosphere, esp. for weather forecasting ▶ **meteorological** adj ▶ **meteorologist** n

meter n instrument for measuring and recording something, such as the consumption of gas or electricity ▷ v measure by meter

methane n colourless inflammable gas

methanol n colourless poisonous liquid used as a solvent and fuel (also **methyl alcohol**)

methinks v, past tense **methought** (obs) it seems to me

method n way or manner; technique; orderliness ▶ **methodical** adj orderly ▶ **methodically** adv ▶ **methodology** n particular method or procedure

Methodist n member of any of the Protestant Churches originated by John Wesley and his followers ▷ adj of Methodists or their Church ▶ **Methodism** n

meths n (informal) methylated spirits

methyl n (compound containing) a saturated hydrocarbon group of atoms ▶ **methylated spirits** alcohol with methanol added, used as a solvent and for heating

meticulous adj very careful about details ▶ **meticulously** adv

métier [met-ee-ay] n profession or trade; one's strong point

metonymy [mit-on-im-ee] n figure of speech in which one thing is replaced by another associated with it, such as the 'Crown' for 'the queen'

metre n basic unit of length equal to about 1.094 yards (100 centimetres); rhythm of poetry ▶ **metric** adj of the decimal system of weights and measures based on the metre ▶ **metrical** adj of poetic metre ▶ **metrication** n conversion to the metric system

metro n, pl **metros** underground railway system, esp. in Paris

metronome n instrument which marks musical time by means of a ticking pendulum

metropolis [mit-**trop**-oh-liss] n chief city of a country or region ▶ **metropolitan** adj of a metropolis

metrosexual adj, n (Aust) (of) a heterosexual man who is preoccupied with his appearance

mettle n courage or spirit

mew n cry of a cat ▷ v utter this cry

mews n yard or street orig. of stables, now often converted into houses

Mexican adj of Mexico ▷ n person from Mexico

mezzanine [mez-zan-een] n intermediate storey, esp. between the ground and first floor

mezzo-soprano [met-so–] n voice or singer between a soprano and contralto (also **mezzo**)

m

mezzotint [met-so-tint] n method of engraving by scraping the roughened surface of a metal plate; print so made

mg milligram(s)

MHz megahertz

miaow [mee-ow] n, v same as **mew**

miasma [mee-azz-ma] n unwholesome or foreboding atmosphere

mica [my-ka] n glasslike mineral used as an electrical insulator

mice n plural of **mouse**

Michaelmas [mik-kl-mass] n September 29, feast of St Michael the archangel ▶ **Michaelmas daisy** garden plant with small daisy-shaped flowers

mickey n **take the mickey (out of)** (informal) tease

micro n, pl **-cros** short for microcomputer, microprocessor

microbe n minute organism, esp. one causing disease ▶ **microbial** adj

microblog n blog in which there is a limitation on the length of individual postings ▶ **microblogger** n

microchip n small wafer of silicon containing electronic circuits

microcomputer n computer with a central processing unit contained in one or more silicon chips

microcosm n miniature representation of something

microfiche [my-kroh-feesh] n microfilm in sheet form

microfilm n miniaturized recording of books or documents on a roll of film

microlight n very small light private aircraft with large wings

micrometer [my-krom-it-er] n instrument for measuring very small distances or angles

micron [my-kron] n one millionth of a metre

microorganism n organism of microscopic size

microphone n instrument for amplifying or transmitting sounds

microprocessor n integrated circuit acting as the central processing unit in a small computer

microscope n instrument with lens(es) which produces a magnified image of a very small object ▶ **microscopic** adj too small to be seen except with a microscope; very small; of a microscope ▶ **microscopically** adv ▶ **microscopy** n use of a microscope

microsurgery n intricate surgery using a special microscope and miniature precision instruments

microwave n electromagnetic wave with a wavelength of a few centimetres, used in radar and cooking; microwave oven ▷ v cook in a microwave oven ▶ **microwave oven** oven using microwaves to cook food quickly

mid adj intermediate, middle

midday n noon

midden n (Brit & Aust) dunghill or rubbish heap

middle adj equidistant from two extremes; medium, intermediate ▷ n middle point or part ▶ **middle age** period of life between youth and old age ▶ **middle-aged** adj **Middle Ages** period from about 1000 AD to the 15th century ▶ **middle class** social class of business and professional people ▶ **middle-class** adj ▶ **Middle East** area around the eastern Mediterranean up to and including Iran ▶ **middleman** n trader who buys from the producer and sells to the consumer ▶ **middle-of-**

the-road adj politically moderate; (of music) generally popular ▸ **middleweight** n boxer weighing up to 16olb (professional) or 75kg (amateur)

middling adj mediocre; moderate

midge n small mosquito-like insect

midget n very small thing; (offens) very small person

midland n (Brit, Aust & US) middle part of a country ▷ pl (M-) central England

midnight n twelve o'clock at night

midriff n middle part of the body

midshipman n naval officer of the lowest commissioned rank

midst n **in the midst of** surrounded by; at a point during

midsummer n middle of summer; summer solstice ▸ **Midsummer's Day, Midsummer Day** (in Britain and Ireland) June 24th

midway adj, adv halfway

midwife n trained person who assists at childbirth ▸ **midwifery** n

midwinter n middle or depth of winter; winter solstice

mien [meen] n (lit) person's bearing, demeanour, or appearance

miffed adj (informal) offended or upset

might[1] v past tense of **may**

might[2] n power or strength **with might and main** energetically or forcefully ▸ **mighty** adj powerful; important ▷ adv (US & Aust informal) very ▸ **mightily** adv

migraine [mee-grain] n severe headache, often with nausea and visual disturbances

migrate v move from one place to settle in another; (of animals) journey between different habitats at specific seasons ▸ **migration** n ▸ **migrant** n person or animal that moves from one place to another ▷ adj moving from one place to

another ▸ **migratory** adj (of an animal) migrating every year

mike n (informal) microphone

milch adj (chiefly Brit) (of a cow) giving milk

mild adj not strongly flavoured; gentle; calm or temperate ▸ **mildly** adv ▸ **mildness** n

mildew n destructive fungus on plants or things exposed to damp ▸ **mildewed** adj

mile n unit of length equal to 1760 yards or 1.609 kilometres ▸ **mileage** n distance travelled in miles; miles travelled by a motor vehicle per gallon of petrol; (informal) usefulness of something ▸ **mileometer** n (Brit) device that records the number of miles a vehicle has travelled ▸ **milestone** n significant event; stone marker showing the distance to a certain place

milieu [meal-**yer**] n, pl **milieux** or **milieus** environment or surroundings

militant adj aggressive or vigorous in support of a cause ▸ **militancy** n

military adj of or for soldiers, armies, or war ▷ n armed services ▸ **militarism** n belief in the use of military force and methods ▸ **militarist** n ▸ **militarized** adj

militate v (usually foll by against or for) have a strong influence or effect

| USAGE NOTE

Do not confuse militate with mitigate 'make less severe'

militia [mill-**ish**-a] n military force of trained citizens for use in emergency only

milk n white liquid produced by female mammals to feed their young; milk of cows, goats, etc., used by humans as food; fluid in some plants ▷ v draw milk from; exploit (a person or situation) ▶ **milky** adj ▶ **Milky Way** luminous band of stars stretching across the night sky ▶ **milk float** (Brit) small electrically powered vehicle used to deliver milk to houses ▶ **milkmaid** n (esp. in former times) woman who milks cows ▶ **milkman** n (Brit, Aust & NZ) man who delivers milk to people's houses ▶ **milkshake** n frothy flavoured cold milk drink ▶ **milksop** n feeble man ▶ **milk teeth** first set of teeth in young children

mill n factory; machine for grinding, processing, or rolling ▷ v grind, press, or process in or as if in a mill; cut fine grooves across the edges of (coins); (of a crowd) move in a confused manner

millennium n, pl **-nia, -niums** period of a thousand years; future period of peace and happiness

SPELLING TIP

Millennium is often misspelled with only one n: remember there are two ns and two ls

miller n person who works in a mill

millet n type of cereal grass

milli- combining form denoting a thousandth part e.g. millisecond

millibar n unit of atmospheric pressure

millimetre n thousandth part of a metre

milliner n maker or seller of women's hats ▶ **millinery** n

million n one thousand thousands ▶ **millionth** adj, n ▶ **millionaire** n person who owns at least a million pounds, dollars, etc.

SPELLING TIP

Note there is only one n in millionaire

millipede n small animal with a jointed body and many pairs of legs

millstone n flat circular stone for grinding corn

millwheel n waterwheel that drives a mill

milometer n (Brit) same as **mileometer**

milt n sperm of fish

mime n acting without the use of words; performer who does this ▷ v act in mime

mimic v **-icking, -icked** imitate (a person or manner), esp. for satirical effect ▷ n person or animal that is good at mimicking ▶ **mimicry** n

min. minimum; minute(s)

minaret n tall slender tower of a mosque

mince v cut or grind into very small pieces; walk or speak in an affected manner; soften or moderate (one's words) ▷ n minced meat ▶ **mincer** n machine for mincing meat ▶ **mincing** adj affected in manner ▶ **mincemeat** n sweet mixture of dried fruit and spices ▶ **mince pie** pie containing mincemeat

mind n thinking faculties; memory or attention; intention; sanity ▷ v take offence at; pay attention to; take care of; be cautious or careful about (something) ▶ **minded** adj having an inclination as specified e.g. politically minded ▶ **minder** n (informal) aide or bodyguard ▶ **mindful** adj heedful; keeping aware ▶ **mindless** adj stupid; requiring no thought; careless

mine[1] pron belonging to me

mine[2] n deep hole for digging out coal, ores, etc.; bomb placed under the ground or in water; profitable source ▷ v dig for minerals; dig (minerals) from a mine; place explosive mines in or on ▶ **miner** n person who works in a mine ▶ **minefield** n area of land or water containing mines ▶ **minesweeper** n ship for clearing away mines

mineral n naturally occurring inorganic substance, such as metal ▷ adj of, containing, or like minerals ▶ **mineralogy** [min-er-al-a-jee] n study of minerals ▶ **mineral water** water containing dissolved mineral salts or gases

minestrone [min-ness-**strone**-ee] n soup containing vegetables and pasta

minger n (Brit informal) unattractive person ▶ **minging** adj (Brit informal) unattractive or unpleasant

mingle v mix or blend; come into association (with)

mingy adj **-gier, -giest** (informal) miserly

mini n, adj (something) small or miniature; short (skirt)

miniature n small portrait, model, or copy ▷ adj small-scale ▶ **miniaturist** n ▶ **miniaturize** v make to a very small scale

minibar n selection of drinks and confectionery provided in a hotel room

minibus n small bus

minicab n (Brit) ordinary car used as a taxi

minicomputer n computer smaller than a mainframe but more powerful than a microcomputer

minidisc n small recordable compact disc

minim n (Music) note half the length of a semibreve

minimum adj, n, pl **-mums** or **-ma** least possible (amount or number) ▶ **minimal** adj minimum ▶ **minimize** v reduce to a minimum; belittle

minion n servile assistant

miniseries n TV programme shown in several parts, often on consecutive days

minister n head of a government department; diplomatic representative; (in Nonconformist Churches) member of the clergy ▷ v (foll by to) attend to the needs of ▶ **ministerial** adj ▶ **ministration** n giving of help ▶ **ministry** n, pl **-tries** profession or duties of a clergyman; ministers collectively; government department

mink n stoatlike animal; its highly valued fur

minnow n small freshwater fish

minor adj lesser; (Music) (of a scale) having a semitone between the second and third notes ▷ n person regarded legally as a child; (Music) minor scale ▶ **minority** n lesser number; smaller party voting together; group in a minority in any state

minster n (Brit) cathedral or large church

minstrel n medieval singer or musician

mint¹ n plant with aromatic leaves used for seasoning and flavouring; sweet flavoured with this

mint² n place where money is coined ▷ v make (coins)

minuet [min-new-**wet**] n stately dance; music for this

minus prep, adj indicating subtraction ▷ adj less than zero ▷ n sign (-) denoting subtraction or a number less than zero

minuscule [min-niss-skyool] adj very small

> **SPELLING TIP**
> The pronunciation of **minuscule** often leads people to spell it *miniscule*, but it should have only one *i* and two *u*s

minute¹ [min-it] n 60th part of an hour or degree; moment ▷ pl record of the proceedings of a meeting ▷ v record in the minutes

minute² [my-**newt**] *adj* very small; precise ▸ **minutely** *adv* ▸ **minutiae** [my-**new**-shee-eye] *pl n* trifling or precise details

minx *n* bold or flirtatious girl

miracle *n* wonderful supernatural event; marvel ▸ **miraculous** *adj* ▸ **miraculously** *adv* ▸ **miracle play** medieval play based on a sacred subject

mirage [mir-**rahzh**] *n* optical illusion, esp. one caused by hot air

mire *n* swampy ground; mud

mirror *n* coated glass surface for reflecting images ▹ *v* reflect in or as if in a mirror

mirth *n* laughter, merriment, or gaiety ▸ **mirthful** *adj* ▸ **mirthless** *adj*

mis- *prefix* wrong(ly), bad(ly)

misadventure *n* unlucky chance

misanthrope [**miz**-zan-thrope] *n* person who dislikes people in general ▸ **misanthropic** [miz-zan-**throp**-ik] *adj* ▸ **misanthropy** [miz-**zan**-throp-ee] *n*

misapprehend *v* misunderstand ▸ **misapprehension** *n*

misappropriate *v* take and use (money) dishonestly ▸ **misappropriation** *n*

miscarriage *n* spontaneous premature expulsion of a fetus from the womb; failure e.g. *a miscarriage of justice* ▸ **miscarry** *v* have a miscarriage; fail

miscast *v* -**casting**, -**cast** cast (a role or actor) (in a play or film) inappropriately

miscegenation [miss-ij-in-**nay**-shun] *n* interbreeding of races

miscellaneous [miss-sell-**lane**-ee-uss] *adj* mixed or assorted ▸ **miscellany** [miss-**sell**-a-nee] *n* mixed assortment

mischance *n* unlucky event

mischief *n* annoying but not malicious behaviour; inclination to tease; harm ▸ **mischievous** *adj* full of mischief; intended to cause harm ▸ **mischievously** *adv*

miscible [**miss**-sib-bl] *adj* able to be mixed

misconception *n* wrong idea or belief

misconduct *n* immoral or unethical behaviour

miscreant [**miss**-kree-ant] *n* wrongdoer

misdeed *n* wrongful act

misdemeanour *n* minor wrongdoing

miser *n* person who hoards money and hates spending it ▸ **miserly** *adj*

miserable *adj* very unhappy, wretched; causing misery; squalid; mean ▸ **misery** *n*, *pl* -**eries** great unhappiness; (*informal*) complaining person

misfire *v* (of a firearm or engine) fail to fire correctly; (of a plan) fail to turn out as intended

misfit *n* person not suited to his or her social environment

misfortune *n* (piece of) bad luck

misgiving *n* feeling of fear or doubt

misguided *adj* mistaken or unwise

mishandle *v* handle badly or inefficiently

mishap *n* minor accident

misinform *v* give incorrect information to ▸ **misinformation** *n*

misjudge *v* judge wrongly or unfairly ▸ **misjudgment, misjudgement** *n*

mislay *v* lose (something) temporarily

mislead *v* give false or confusing information to ▸ **misleading** *adj*

mismanage *v* organize or run (something) badly ▸ **mismanagement** *n*

misnomer [miss-**no**-mer] *n* incorrect or unsuitable name; use of this

misogyny [miss-**oj**-in-ee] n hatred of women ▶ **misogynist** n

misplace v mislay; put in the wrong place; give (trust or affection) inappropriately

misprint n printing error

misrepresent v represent wrongly or inaccurately

Miss n title of a girl or unmarried woman

miss v fail to notice, hear, hit, reach, find, or catch; not be in time for; notice or regret the absence of; avoid; (of an engine) misfire ▶ n fact or instance of missing ▶ **missing** adj lost or absent

missal n book containing the prayers and rites of the Mass

misshapen adj badly shaped, deformed

missile n object or weapon thrown, shot, or launched at a target

mission n specific task or duty; group of people sent on a mission; building in which missionaries work; (SAfr) long and difficult process ▶ **missionary** n, pl **-aries** person sent abroad to do religious and social work

missive n letter

misspent adj wasted or misused

missus, missis n (informal) one's wife or the wife of the person addressed or referred to

mist n thin fog; fine spray of liquid ▶ **misty** adj full of mist; dim or obscure

mistake n error or blunder ▶ v **-taking, -took, -taken** misunderstand; confuse (a person or thing) with another

Mister n polite form of address to a man

mistletoe n evergreen plant with white berries growing as a parasite on trees

mistral n strong dry northerly wind of S France

mistress n woman who has a continuing sexual relationship with a married man; woman in control of people or animals; female teacher

mistrial n (Law) trial made void because of some error

mistrust v have doubts or suspicions about ▶ n lack of trust ▶ **mistrustful** adj

misunderstand v fail to understand properly ▶ **misunderstanding** n

misuse n incorrect, improper, or careless use ▶ v use wrongly; treat badly

mite n very small spider-like animal; very small thing or amount

mitigate v make less severe ▶ **mitigation** n

USAGE NOTE
Mitigate is often confused with militate 'have an influence on'

mitre [**my**-ter] n bishop's pointed headdress; joint between two pieces of wood bevelled to meet at right angles ▶ v join with a mitre joint

mitt n short for **mitten**; baseball catcher's glove

mitten n glove with one section for the thumb and one for the four fingers together

mix v combine or blend into one mass; form (something) by mixing; be sociable ▶ n mixture ▶ **mixed** adj ▶ **mixer** n ▶ **mixture** n something mixed; combination ▶ **mix up** v confuse; make into a mixture ▶ **mixed up** adj ▶ **mix-up** n

mizzenmast n (on a vessel with three or more masts) third mast from the bow

mm millimetre(s)

mnemonic [nim-**on**-ik] n, adj (something, such as a rhyme) intended to help the memory

m

MO Medical Officer

mo n, pl **mos** (informal) short for **moment**

moa n large extinct flightless New Zealand bird

moan n low cry of pain; (informal) grumble ▷ v make or utter with a moan; (informal) grumble

moat n deep wide ditch, esp. round a castle

mob n disorderly crowd; (slang) gang ▷ v **mobbing, mobbed** surround in a mob to acclaim or attack

mobile adj able to move ▷ n same as **mobile phone**; hanging structure designed to move in air currents ▶ **mobile phone** cordless phone powered by batteries ▶ **mobility** n

mobilize v (of the armed services) prepare for active service; organize for a purpose ▶ **mobilization** n

moccasin n soft leather shoe

> **SPELLING TIP**
>
> Note that **moccasin** has a double c, but only one s

mocha [**mock-a**] n kind of strong dark coffee; flavouring made from coffee and chocolate

mock v make fun of; mimic ▷ adj sham or imitation ▶ **mocks** pl n (informal) (in England and Wales) practice exams taken before public exams **put the mockers on** (Brit, Aust & NZ informal) ruin the chances of success of ▶ **mockery** n derision; inadequate or worthless attempt ▶ **mockingbird** n N American bird which imitates other birds' songs ▶ **mock orange** shrub with white fragrant flowers ▶ **mock-up** n full-scale model for test or study

MOD (in Britain) Ministry of Defence

mod. moderate; modern

mode n method or manner; current fashion

model n (miniature) representation; pattern; person or thing worthy of imitation; person who poses for an artist or photographer; person who wears clothes to display them to prospective buyers ▷ v **-elling, -elled** make a model of; mould; display (clothing) as a model

modem [**mode-em**] n device for connecting two computers by a telephone line

moderate adj not extreme; self-restrained; average ▷ n person of moderate views ▷ v make or become less violent or extreme ▶ **moderately** adv ▶ **moderation** n ▶ **moderator** n (Presbyterian Church) minister appointed to preside over a Church court, general assembly, etc.; person who presides over a public or legislative assembly

modern adj of present or recent times; up-to-date ▶ **modernity** n ▶ **modernism** n (support of) modern tendencies, thoughts, or styles ▶ **modernist** adj, n ▶ **modernize** v bring up to date ▶ **modernization** n

modest adj not vain or boastful; not excessive; not showy; shy ▶ **modestly** adv ▶ **modesty** n

modicum n small quantity

modify v **-fying, -fied** change slightly; tone down; (of a word) qualify (another word) ▶ **modifier** n word that qualifies the sense of another ▶ **modification** n

modish [**mode-ish**] adj in fashion

modulate v vary in tone; adjust; change the key of (music) ▶ **modulation** n ▶ **modulator** n

module n self-contained unit, section, or component with a specific function

modus operandi [**mode-uss op-er-an-die**] n (Latin) method of operating

mogul [moh-gl] n important or powerful person

mohair n fine hair of the Angora goat; yarn or fabric made from this

mohican n punk hairstyle with shaved sides and a stiff central strip of hair, often brightly coloured

moiety [moy-it-ee] n, pl -**ties** half

moist adj slightly wet ▸ **moisten** v make or become moist ▸ **moisture** n liquid diffused as vapour or condensed in drops ▸ **moisturize** v add moisture to (the skin etc.)

mojo n (slang) uncanny power or influence

moke n (Aust & NZ) horse of inferior quality

molar n large back tooth used for grinding

molasses n dark syrup, a by-product of sugar refining

mole¹ n small dark raised spot on the skin

mole² n small burrowing mammal; (informal) spy who has infiltrated and become a trusted member of an organization

mole³ n unit of amount of substance

mole⁴ n breakwater; harbour protected by this

molecule [mol-lik-kyool] n simplest freely existing chemical unit, composed of two or more atoms; very small particle ▸ **molecular** [mol-lek-yew-lar] adj

molest v interfere with sexually; annoy or injure ▸ **molester** n ▸ **molestation** n

moll n (slang) gangster's female accomplice

mollify v -**fying**, -**fied** pacify or soothe

mollusc n soft-bodied, usu. hard-shelled, animal, such as a snail or oyster

mollycoddle v pamper

Molotov cocktail n petrol bomb

molten adj liquefied or melted

molybdenum [mol-lib-din-um] n (Chem) hard silvery-white metallic element

moment n short space of time; (present) point in time ▸ **momentary** adj lasting only a moment ▸ **momentarily** adv

USAGE NOTE
Note that some American speakers use momentarily to mean 'soon' rather than 'for a moment'

momentous [moh-men-tuss] adj of great significance

momentum n impetus of a moving body; product of a body's mass and velocity

monarch n sovereign ruler of a state ▸ **monarchical** adj ▸ **monarchist** n supporter of monarchy ▸ **monarchy** n government by or a state ruled by a sovereign

monastery n, pl -**teries** residence of a community of monks ▸ **monastic** adj of monks, nuns, or monasteries; simple and austere ▸ **monasticism** n

Monday n second day of the week

monetary adj of money or currency ▸ **monetarism** n theory that inflation is caused by an increase in the money supply ▸ **monetarist** n, adj

money n medium of exchange, coins or banknotes ▸ **moneyed, monied** adj rich

mongol n, adj (offens) (person) affected by Down's syndrome ▸ **mongolism** n

Mongolian n person from Mongolia; language of Mongolia ▸ adj of Mongolia or its language

mongoose n, pl -**gooses** stoat-like mammal of Asia and Africa that kills snakes

mongrel n animal, esp. a dog, of mixed breed; something arising from a variety of sources ▷ *adj* of mixed breed or origin

monitor *n* person or device that checks, controls, warns, or keeps a record of something; (Brit, Aust & NZ) pupil assisting a teacher with duties; television set used in a studio to check what is being transmitted; large lizard of Africa, Asia, and Australia ▷ *v* watch and check on

monk *n* member of an all-male religious community bound by vows ▶ **monkish** *adj*

monkey *n* long-tailed primate; mischievous child ▷ *v* (usu. foll by *about* or *around*) meddle or fool ▶ **monkey nut** (Brit) peanut ▶ **monkey puzzle** coniferous tree with sharp stiff leaves ▶ **monkey wrench** wrench with adjustable jaws

mono- *combining form* single e.g. *monosyllable*

monochrome *adj* (Photog) black-and-white; in only one colour

monocle *n* eyeglass for one eye only

monogamy *n* custom of being married to one person at a time

monogram *n* design of combined letters, esp. a person's initials

monograph *n* book or paper on a single subject

monolith *n* large upright block of stone ▶ **monolithic** *adj*

monologue *n* long speech by one person; dramatic piece for one performer

monomania *n* obsession with one thing ▶ **monomaniac** *n, adj*

monoplane *n* aeroplane with one pair of wings

monopoly *n* (pl **-lies**) exclusive possession of or right to do something; (**M-**)® board game for four to six players who deal in

'property' as they move around the board ▶ **monopolize** *v* have or take exclusive possession of

monorail *n* single-rail railway

monotheism *n* belief in only one God ▶ **monotheistic** *adj*

monotone *n* unvaried pitch in speech or sound ▶ **monotonous** *adj* tedious due to lack of variety ▶ **monotonously** *adv* ▶ **monotony** *n*

Monseigneur [mon-sen-**nyur**] *n*, *pl* **Messeigneurs** [may-sen-**nyur**] title of French prelates

monsieur [muss-**syur**] *n, pl* **messieurs** [may-**syur**] French title of address equivalent to *sir* or *Mr*

Monsignor *n* (RC Church) title attached to certain offices

monsoon *n* seasonal wind of SE Asia; rainy season accompanying this

monster *n* imaginary, usu. frightening, beast; huge person, animal, or thing; very wicked person ▷ *adj* huge ▶ **monstrosity** *n* large ugly thing ▶ **monstrous** *adj* unnatural or ugly; outrageous or shocking; huge ▶ **monstrously** *adv*

monstrance *n* (RC Church) container in which the consecrated Host is exposed for adoration

montage [mon-**tahzh**] *n* (making of) a picture composed from pieces of others; method of film editing incorporating several shots to form a single image

month *n* one of the twelve divisions of the calendar year; period of four weeks ▶ **monthly** *adj* happening or payable once a month ▷ *adv* once a month ▷ *n* monthly magazine

monument *n* something, esp. a building or statue, that commemorates something ▶ **monumental** *adj* large, impressive, or lasting; of or being a monument; (*informal*) extreme ▶ **monumentally** *adv*

moo *n* long deep cry of a cow ▷ *v* make this noise

mooch *v* (*slang*) loiter about aimlessly

mood[1] *n* temporary (gloomy) state of mind ▶ **moody** *adj* sullen or gloomy; changeable in mood ▶ **moodily** *adv*

mood[2] *n* (*Grammar*) form of a verb indicating whether it expresses a fact, wish, supposition, or command

moon *n* natural satellite of the earth; natural satellite of any planet ▷ *v* (foll by *about* or *around*) be idle in a listless or dreamy way ▶ **moonlight** *n* light from the moon ▷ *v* (*informal*) work at a secondary job, esp. illegally ▶ **moonshine** *n* (*US & Canadian*) illicitly distilled whisky; nonsense ▶ **moonstone** *n* translucent semiprecious stone ▶ **moonstruck** *adj* slightly mad or odd

moor[1] *n* (*Brit*) tract of open uncultivated ground covered with grass and heather ▶ **moorhen** *n* small black water bird

moor[2] *v* secure (a ship) with ropes etc. ▶ **mooring** *n* place for mooring a ship ▷ *pl* ropes etc. used in mooring a ship

Moor *n* member of a Muslim people of NW Africa who ruled Spain between the 8th and 15th centuries ▶ **Moorish** *adj*

moose *n* large N American deer

moot *adj* debatable e.g. *a moot point* ▷ *v* bring up for discussion

mop *n* long stick with twists of cotton or a sponge on the end, used for cleaning; thick mass of hair ▷ *v* **mopping, mopped** clean or soak up with or as if with a mop

mope *v* be gloomy and apathetic

moped *n* light motorized cycle

mopoke *n* small spotted owl of Australia and New Zealand

moraine *n* accumulated mass of debris deposited by a glacier

moral *adj* concerned with right and wrong conduct; based on a sense of right and wrong; (of support or a victory) psychological rather than practical ▷ *n* lesson to be obtained from a story or event ▷ *pl* principles of behaviour with respect to right and wrong ▶ **morally** *adv* ▶ **moralist** *n* person with a strong sense of right and wrong ▶ **morality** *n* good moral conduct; moral goodness or badness ▶ **morality play** medieval play with a moral lesson ▶ **moralize** *v* make moral pronouncements

morale [mor-**rahl**] *n* degree of confidence or hope of a person or group

morass *n* marsh; mess

moratorium *n, pl* **-ria** *or* **-riums** legally authorized ban or delay

moray *n* large voracious eel

morbid *adj* unduly interested in death or unpleasant events; gruesome

mordant *adj* sarcastic or scathing ▷ *n* substance used to fix dyes

more *adj* greater in amount or degree; comparative of **much**, **many**; additional or further ▷ *adv* to a greater extent; in addition ▷ *pron* greater or additional amount or number ▶ **moreover** *adv* in addition to what has already been said

mores [**more**-rayz] *pl n* customs and conventions embodying the fundamental values of a community

Moreton Bay bug *n* Australian flattish edible shellfish

morganatic marriage *n* marriage of a person of high rank to a lower-ranking person whose status remains unchanged

morgue n mortuary

moribund adj without force or vitality

Mormon n member of a religious sect founded in the USA

morn n (poetic or Aust) morning

morning n part of the day before noon ▸ **morning-glory** n plant with trumpet-shaped flowers which close in the late afternoon

Moroccan adj of Morocco ▸ n person from Morocco

morocco n goatskin leather

moron n (informal) foolish or stupid person; (formerly) person with a low intelligence quotient ▸ **moronic** adj

morose [mor-rohss] adj sullen or moody

morphine, morphia n drug extracted from opium, used as an anaesthetic and sedative

morphology n science of forms and structures of organisms or words ▸ **morphological** adj

morris dance n traditional English folk dance

morrow n (poetic) next day

Morse n former system of signalling in which letters of the alphabet are represented by combinations of short and long signals

morsel n small piece, esp. of food

mortal adj subject to death; causing death ▸ n human being ▸ **mortally** adv ▸ **mortality** n state of being mortal; great loss of life; death rate ▸ **mortal sin** (RC Church) sin meriting damnation

mortar n small cannon with a short range; mixture of lime, sand, and water for holding bricks and stones together; bowl in which substances are pounded ▸ **mortarboard** n black square academic cap

mortgage n conditional pledging of property, esp. a house, as security for the repayment of a loan; the loan itself ▸ v pledge (property) as security thus ▸ **mortgagee** n creditor in a mortgage ▸ **mortgagor** n debtor in a mortgage

mortice, mortise [more-tiss] n hole in a piece of wood or stone shaped to receive a matching projection on another piece ▸ **mortice lock** lock set into a door

mortify v -fying, -fied humiliate; subdue by self-denial; (of flesh) become gangrenous ▸ **mortification** n

mortuary n, pl -aries building where corpses are kept before burial or cremation

Mosaic adj of Moses

mosaic [mow-zay-ik] n design or decoration using small pieces of coloured stone or glass

Moselle n light white German wine

Moslem n, adj same as **Muslim**

mosque n Muslim temple

mosquito n, pl -toes or -tos blood-sucking flying insect

moss n small flowerless plant growing in masses on moist surfaces ▸ **mossy** adj

most n greatest number or degree ▸ adj greatest in number or degree; superlative of **much, many** ▸ adv in the greatest degree ▸ **mostly** adv for the most part, generally

MOT, MOT test n (in Britain) compulsory annual test of the roadworthiness of vehicles over a certain age

motel n roadside hotel for motorists

motet n short sacred choral song

moth n nocturnal insect like a butterfly ▸ **mothball** n small ball of camphor or naphthalene used to repel moths from stored clothes

▷ v store (something operational) for future use; postpone (a project etc.) ▶ **moth-eaten** *adj* decayed or scruffy; eaten or damaged by moth larvae

mother *n* female parent; head of a female religious community ▷ *adj* native or inborn e.g. *mother wit* ▷ v look after as a mother ▶ **motherhood** *n* ▶ **motherless** *adj* ▶ **motherly** *adj* ▶ **mother-in-law** *n* mother of one's husband or wife ▶ **mother of pearl** iridescent lining of certain shells ▶ **mother tongue** one's native language

motif [moh-**teef**] *n* (recurring) theme or design

motion *n* process, action, or way of moving; proposal in a meeting; evacuation of the bowels ▷ v direct (someone) by gesture ▶ **motionless** *adj* not moving ▶ **motion picture** cinema film

motive *n* reason for a course of action ▷ *adj* causing motion ▶ **motivate** v give incentive to ▶ **motivation** *n*

motley *adj* miscellaneous; multicoloured

motocross *n* motorcycle race over a rough course

motor *n* engine, esp. of a vehicle; machine that converts electrical energy into mechanical energy; (*chiefly Brit*) car ▷ v travel by car ▶ **motorist** *n* driver of a car ▶ **motorized** *adj* equipped with a motor or motor transport ▶ **motorbike** *n* ▶ **motorboat** *n* ▶ **motorcar** *n* ▶ **motorcycle** *n* ▶ **motorcyclist** *n* ▶ **motorhome** *n* large motor vehicle designed for living in while travelling ▶ **motor scooter** light motorcycle with small wheels and an enclosed engine ▶ **motorway** *n* main road for fast-moving traffic

mottled *adj* marked with blotches

motto *n*, *pl* **-toes** or **-tos** saying expressing an ideal or rule of conduct; verse or maxim in a paper cracker

mould¹ *n* hollow container in which metal etc. is cast; shape, form, or pattern; nature or character ▷ v shape; influence or direct ▶ **moulding** *n* moulded ornamental edging

mould² *n* fungal growth caused by dampness ▶ **mouldy** *adj* stale or musty; dull or boring

mould³ *n* loose soil ▶ **moulder** v decay into dust

moult v shed feathers, hair, or skin to make way for new growth ▷ *n* process of moulting

mound *n* heap, esp. of earth or stones; small hill

mount v climb or ascend; get up on (a horse etc.); increase or accumulate; fix on a support or backing; organize e.g. *mount a campaign* ▷ *n* backing or support on which something is fixed; horse for riding; hill

mountain *n* hill of great size; large heap ▶ **mountainous** *adj* full of mountains; huge ▶ **mountaineer** *n* person who climbs mountains ▶ **mountaineering** *n* ▶ **mountain bike** bicycle with straight handlebars and heavy-duty tyres, for cycling over rough terrain ▶ **mountain oyster** (*NZ informal*) sheep's testicle eaten as food

mountebank *n* charlatan or fake

Mountie *n* (*informal*) member of the Royal Canadian Mounted Police

mourn v feel or express sorrow for (a dead person or lost thing) ▶ **mournful** *adj* sad or dismal ▶ **mournfully** *adv* ▶ **mourning** *n* grieving; conventional symbols of

m

grief for death, such as the wearing of black

mourner n person attending a funeral

mouse n, pl **mice** small long-tailed rodent; timid person; (Computing) hand-held device for moving the cursor without keying ▶ **mouser** n cat used to catch mice ▶ **mousy** adj like a mouse, esp. in hair colour; meek and shy

mousse n dish of flavoured cream whipped and set

moustache n hair on the upper lip

mouth n opening in the head for eating and issuing sounds; entrance; point where a river enters the sea; opening ▷ v form (words) with the lips without speaking; speak or utter insincerely, esp. in public ▶ **mouthful** n amount of food or drink put into the mouth at any one time when eating or drinking ▶ **mouth organ** same as **harmonica** ▶ **mouthpiece** n part of a telephone into which a person speaks; part of a wind instrument into which the player blows; spokesperson

move v change in place or position; change (one's house etc.); take action; stir the emotions of; incite; suggest (a proposal) formally ▷ n moving; action towards some goal ▶ **movable, moveable** adj ▶ **movement** n action or process of moving; group with a common aim; division of a piece of music; moving parts of a machine

movie n (informal) cinema film

mow v cut (grass or crops) mowed, mowed or mown ▶ **mow down** v kill in large numbers

mower n machine for cutting grass

mozzarella [mot-sa-rel-la] n moist white cheese originally made in Italy from buffalo milk

MP Member of Parliament; Military Police(man)

MP3 (Computing) Motion Picture Expert Group-1, Audio Layer-3: a digital compression format used to compress audio files to a fraction of their original size without loss of sound quality ▶ **MP3 player** small portable digital audio player capable of storing MP3 files downloaded from the internet or transferred from CD

MP4 abbrev MPEG-4 Part 14: software for transmitting data streams to a receiving device in real time

MPEG [em-peg] (Computing) Motion Picture Experts Group: standard compressed file format used for audio and video files; file in this format

mpg miles per gallon

mph miles per hour

Mr Mister

Mrs n title of a married woman

MS manuscript; multiple sclerosis

Ms [mizz] n title used instead of Miss or Mrs

MSc Master of Science

MSP (in Britain) Member of the Scottish Parliament

MSS manuscripts

Mt Mount

much adj **more, most** large amount or degree of ▷ n large amount or degree ▷ adv **more, most** to a great degree; nearly

mucilage [mew-sill-ij] n gum or glue

muck n dirt, filth; manure ▶ **mucky** adj

mucus [mew-kuss] n slimy secretion of the mucous membranes ▶ **mucous**

membrane tissue lining body cavities or passages

mud n wet soft earth ▶ **muddy** adj ▶ **mudguard** n cover over a wheel to prevent mud or water being thrown up by it ▶ **mud pack** cosmetic paste to improve the complexion

muddle v (often foll by up) confuse; mix up ▷ n state of confusion

muesli [mewz-lee] n mixture of grain, nuts, and dried fruit, eaten with milk

muezzin [moo-ezz-in] n official who summons Muslims to prayer

muff[1] n tube-shaped covering to keep the hands warm

muff[2] v bungle (an action)

muffin n light round flat yeast cake

muffle v wrap up for warmth or to deaden sound ▶ **muffler** n (Brit) scarf; device to reduce the noise of an engine exhaust

mufti n civilian clothes worn by a person who usually wears a uniform

mug[1] n large drinking cup

mug[2] n (slang) face; (slang) gullible person ▶ v **mugging, mugged** (informal) attack in order to rob ▶ **mugger** n

mug[3] v **mugging, mugged** (foll by up) (informal) study hard

muggins n (informal) stupid or gullible person

muggy adj **-gier, -giest** (of weather) damp and stifling

mulatto [mew-lat-toe] n, pl **-tos** or **-toes** (offens) child of one Black and one White parent

mulberry n tree whose leaves are used to feed silkworms; purple fruit of this tree

mulch n mixture of wet straw, leaves, etc., used to protect the roots of plants ▷ v cover (land) with mulch

mule[1] n offspring of a horse and a donkey ▶ **mulish** adj obstinate

mule[2] n backless shoe or slipper

mulga n Australian acacia shrub growing in desert regions; (Aust) the outback

mull v think (over) or ponder ▶ **mulled** adj (of wine or ale) flavoured with sugar and spices and served hot

mullah n Muslim scholar, teacher, or religious leader

mullet[1] n edible sea fish

mullet[2] n haircut in which the hair is short at the top and sides and long at the back

mulligatawny n soup made with curry powder

mullion n vertical dividing bar in a window ▶ **mullioned** adj

mulloway n large Australian sea fish, valued for sport and food

multi- combining form many e.g. multicultural; multistorey

multifarious [mull-tee-fare-ee-uss] adj having many various parts

multilateral adj of or involving more than two nations or parties

multiple adj having many parts ▷ n quantity which contains another an exact number of times

multiplex n purpose-built complex containing several cinemas and usu. restaurants and bars ▷ adj having many elements, complex

multiplicity n, pl **-ties** large number or great variety

multiply v **-plying, -plied** (cause to) increase in number, quantity, or degree; add (a number or quantity) to itself a given number of times; increase in number by reproduction ▶ **multiplication** n ▶ **multiplicand** n (Maths) number to be multiplied

multipurpose adj having many uses e.g. a multipurpose tool

▶ **multipurpose vehicle** large vanlike car designed to carry up to eight passengers

multitude n great number; great crowd ▶ **multitudinous** adj very numerous

mum[1] n (informal) mother

mum[2] adj **keep mum** remain silent

mumble v speak indistinctly, mutter

mumbo jumbo n meaningless language; foolish religious ritual or incantation

mummer n actor in a traditional English folk play or mime

mummy[1] n, pl **-mies** body embalmed and wrapped for burial in ancient Egypt ▶ **mummified** adj (of a body) preserved as a mummy

mummy[2] n, pl **-mies** child's word for **mother**

mumps n infectious disease with swelling in the glands of the neck

munch v chew noisily and steadily

mundane adj everyday; earthly

municipal adj relating to a city or town ▶ **municipality** n city or town with local self-government; governing body of this

munificent [mew-niff-fiss-sent] adj very generous ▶ **munificence** n

muniments pl n title deeds or similar documents

munitions pl n military stores

munted adj (NZ slang) destroyed or ruined; abnormal or peculiar

mural n painting on a wall

murder n unlawful intentional killing of a human being ▶ v kill in this way ▶ **murderer**, **murderess** n ▶ **murderous** adj

murky adj dark or gloomy ▶ **murk** n thick darkness

murmur v **-muring**, **-mured** speak or say in a quiet indistinct way; complain ▶ n continuous low indistinct sound

muscle n tissue in the body which produces movement by contracting; strength or power ▶ **muscular** adj with well-developed muscles; of muscles ▶ **muscular dystrophy** disease with wasting of the muscles ▶ **muscle in** v (informal) force one's way in

Muse n (Greek myth) one of nine goddesses, each of whom inspired an art or science; (**m-**) force that inspires a creative artist

muse v ponder quietly

museum n building where natural, artistic, historical, or scientific objects are exhibited and preserved

mush n soft pulpy mass; (informal) cloying sentimentality ▶ **mushy** adj

mushroom n edible fungus with a stem and cap ▶ v grow rapidly

music n art form using a melodious and harmonious combination of notes; written or printed form of this ▶ **musical** adj of or like music; talented in or fond of music; pleasant-sounding ▷ n play or film with songs and dancing ▶ **musically** adv ▶ **musician** n ▶ **musicology** n scientific study of music ▶ **musicologist** n ▶ **music hall** variety theatre

musk n scent obtained from a gland of the musk deer or produced synthetically ▶ **musky** adj ▶ **muskrat** n N American beaver-like rodent; its fur

musket n (Hist) long-barrelled gun ▶ **musketeer** n ▶ **musketry** n (use of) muskets

Muslim n follower of the religion of Islam ▷ adj of or relating to Islam

muslin n fine cotton fabric

mussel n edible shellfish with a dark hinged shell

must[1] v used as an auxiliary to express obligation, certainty, or resolution ▷ *n* essential or necessary thing

must² *n* newly pressed grape juice

mustang *n* wild horse of SW USA

mustard *n* paste made from the powdered seeds of a plant, used as a condiment; the plant ▶ **mustard gas** poisonous gas causing blistering burns and blindness

muster *v* assemble ▷ *n* assembly of military personnel

musty *adj* **mustier, mustiest** smelling mouldy and stale ▶ **mustiness** *n*

mutable [mew-tab-bl] *adj* liable to change ▶ **mutability** *n*

mutation *n* (genetic) change ▶ **mutate** *v* (cause to) undergo mutation ▶ **mutant** *n* mutated animal, plant, etc.

mute *adj* silent; (offens) unable to speak ▷ *n* (offens) person who is unable to speak; (Music) device to soften the tone of an instrument ▶ **muted** *adj* (of sound or colour) softened; (of a reaction) subdued ▶ **mutely** *adv*

muti [moo-ti] *n* (S Afr informal) medicine, esp. herbal medicine

mutilate [mew-till-ate] *v* deprive of a limb or other part; damage (a book or text) ▶ **mutilation** *n*

mutiny [mew-tin-ee] *n, pl* **-nies** rebellion against authority, esp. by soldiers or sailors ▷ *v* **-nying, -nied** commit mutiny ▶ **mutineer** *n* ▶ **mutinous** *adj*

mutt *n* (slang) mongrel dog; stupid person

mutter *v* utter or speak indistinctly; grumble ▷ *n* muttered sound or grumble

mutton *n* flesh of sheep, used as food ▶ **mutton bird** (Aust) sea bird with dark plumage; (NZ) any of a number of migratory sea birds, the young of which are a Māori delicacy

mutual [mew-chew-al] *adj* felt or expressed by each of two people about the other; common to both or all ▶ **mutually** *adv*

> **USAGE NOTE**
> The objection that something *mutual* has to do with two people only is outdated; nowadays *mutual* is equivalent to 'shared, in common'

Muzak® *n* recorded light music played in shops etc.

muzzle *n* animal's mouth and nose; cover for these to prevent biting; open end of a gun ▷ *v* prevent from being heard or noticed; put a muzzle on

muzzy *adj* **-zier, -ziest** confused or muddled; blurred or hazy

MW megawatt(s)

mW milliwatt(s)

my *adj* belonging to me

myall *n* Australian acacia with hard scented wood

mycology *n* study of fungi

myna, mynah, mina *n* Asian bird which can mimic human speech

myopia [my-oh-pee-a] *n* short-sightedness ▶ **myopic** [my-op-ik] *adj*

myriad [mir-ree-ad] *adj* innumerable ▷ *n* large indefinite number

myrrh [mur] *n* aromatic gum used in perfume, incense, and medicine

myrtle [mur-tl] *n* flowering evergreen shrub

myself *pron* emphatic or reflexive form of **I, me**

mystery *n, pl* **-teries** strange or inexplicable event or phenomenon; obscure or secret thing; story or film that arouses suspense ▶ **mysterious** *adj* ▶ **mysteriously** *adv*

mystic *n* person who seeks spiritual knowledge ▷ *adj*

m

mystical ► **mystical** *adj* having a
spiritual or religious significance
beyond human understanding
► **mysticism** *n*
mystify *v* **-fying, -fied** bewilder or
puzzle ► **mystification** *n*
mystique [miss-**steek**] *n* aura of
mystery or power
myth *n* tale with supernatural
characters, usu. of how the world
and mankind began; untrue idea
or explanation; imaginary person
or object ► **mythical, mythic** *adj*
► **mythology** *n* myths collectively;
study of myths ► **mythological** *adj*
myxomatosis [mix-a-mat-oh-siss]
n contagious fatal viral disease
of rabbits

Nn

N (*Chem*) nitrogen; (*Physics*)
newton(s); North(ern)
n. neuter; noun; number
Na (*Chem*) sodium
Naafi *n* (*Brit*) canteen or shop for
military personnel
naan *n* same as **nan bread**
naartjie [nahr-chee] *n* (*SAfr*)
tangerine
nab *v* **nabbing, nabbed** (*informal*)
arrest (someone); catch (someone)
in wrongdoing
nadir *n* point in the sky opposite the
zenith; lowest point
naevus [nee-vuss] *n, pl* **-vi**
birthmark or mole
naff *adj* (*Brit slang*) lacking quality
or taste
nag¹ *v* **nagging, nagged** scold or
find fault constantly; be a constant
source of discomfort or worry to ► *n*
person who nags ► **nagging** *adj, n*
nag² *n* (*informal*) old horse
naiad [nye-ad] *n* (*Greek myth*)
nymph living in a lake or river
nail *n* pointed piece of metal with
a head, hit with a hammer to
join two objects together; hard
covering of the upper tips of
the fingers and toes ► *v* attach
(something) with nails; (*informal*)
catch or arrest **hit the nail on
the head** say something exactly
correct ► **nail file** small metal file
used to smooth or shape the finger
or toe nails ► **nail varnish, nail
polish** cosmetic lacquer applied to
the finger or toe nails

naive [nye-eev] *adj* innocent and gullible; simple and lacking sophistication ▸ **naively** *adv* ▸ **naivety, naïveté** [nye-eev-tee] *n*

naked *adj* without clothes; without any covering **the naked eye** the eye unassisted by any optical instrument ▸ **nakedness** *n*

namby-pamby *adj* (*Brit, Aust & NZ*) sentimental or insipid

name *n* word by which a person or thing is known; reputation, esp. a good one ▸ *v* give a name to; refer to by name; fix or specify **call someone names, call someone a name** insult someone by using rude words to describe him or her ▸ **nameless** *adj* without a name; unspecified; too horrible to be mentioned ▸ **namely** *adv* that is to say ▸ **namesake** *n* person with the same name as another

nan bread *n* slightly leavened Indian bread in a large flat leaf shape

nanny *n, pl* **-nies** woman whose job is looking after young children ▸ **nanny goat** female goat

nano- *combining form* denoting one thousand millionth e.g. *nanosecond*

nap[1] *n* short sleep ▸ *v* **napping, napped** have a short sleep

nap[2] *n* raised fibres of velvet or similar cloth

nap[3] *n* card game similar to whist

napalm *n* highly inflammable jellied petrol, used in bombs

nape *n* back of the neck

naphtha *n* liquid mixture distilled from coal tar or petroleum, used as a solvent and in petrol ▸ **naphthalene** *n* white crystalline product distilled from coal tar or petroleum, used in disinfectants, mothballs, and explosives

napkin *n* piece of cloth or paper for wiping the mouth or protecting the clothes while eating

nappy *n, pl* **-pies** piece of absorbent material fastened round a baby's lower torso to absorb urine and faeces

narcissism *n* exceptional interest in or admiration for oneself ▸ **narcissistic** *adj*

narcissus *n, pl* **-cissi** yellow, orange, or white flower related to the daffodil

narcotic *n, adj* (of) a drug, such as morphine or opium, which produces numbness and drowsiness, used medicinally but addictive ▸ **narcosis** *n* effect of a narcotic

nark (*slang*) *v* annoy ▸ *n* informer or spy; (*Brit*) someone who complains in an irritating manner ▸ **narky** *adj* (*slang*) irritable or complaining

narrate *v* tell (a story); speak the words accompanying and telling what is happening in a film or TV programme ▸ **narration** *n* ▸ **narrator** *n*

narrative *n* account, story

narrow *adj* small in breadth in comparison to length; limited in range, extent, or outlook; with little margin e.g. *a narrow escape* ▸ *v* make or become narrow; (often foll by *down*) limit or restrict ▸ **narrows** *pl n* narrow part of a strait, river, or current ▸ **narrowly** *adv* ▸ **narrowness** *n* ▸ **narrow boat** (*Brit*) long bargelike canal boat ▸ **narrow-minded** *adj* intolerant or bigoted

narwhal *n* arctic whale with a long spiral tusk

NASA (*US*) National Aeronautics and Space Administration

nasal *adj* of the nose; (of a sound) pronounced with air passing through the nose ▸ **nasally** *adv*

nascent *adj* starting to grow or develop

n

NASDAQ (US) National Association of Securities Dealers Automated Quotations (System)

nasturtium n plant with yellow, red, or orange trumpet-shaped flowers

nasty adj **-tier, -tiest** unpleasant; (of an injury) dangerous or painful; spiteful or unkind ▸ **nastily** adv ▸ **nastiness** n

natal adj of or relating to birth

nation n people of one or more cultures or races organized as a single state

national adj characteristic of a particular nation ▸ n citizen of a nation ▸ **nationally** adv ▸ **National Curriculum** curriculum of subjects taught in state schools in England and Wales since 1989 ▸ **National Health Service** (in Britain) system of national medical services financed mainly by taxation ▸ **national insurance** (in Britain) state insurance scheme providing payments to unemployed, sick, and retired people ▸ **national park** area of countryside protected by a government for its natural or environmental importance ▸ **national service** compulsory military service

nationalism n policy of national independence; patriotism, sometimes to an excessive degree ▸ **nationalist** n, adj

nationality n, pl **-ities** fact of being a citizen of a particular nation; group of people of the same race

nationalize v put (an industry or a company) under state control ▸ **nationalization** n

native adj relating to a place where a person was born; born in a specified place; (foll by to) originating (in); inborn ▸ n person born in a specified place;

indigenous animal or plant; member of the original race of a country ▸ **Native American** (person) descended from the original inhabitants of the American continent ▸ **native bear** (Aust) same as **koala** ▸ **native companion** (Aust) same as **brolga** ▸ **native dog** (Aust) dingo

Nativity n (Christianity) birth of Jesus Christ

NATO North Atlantic Treaty Organization

natter (informal) v talk idly or chatter ▸ n long idle chat

natty adj **-tier, -tiest** (informal) smart and spruce

natural adj normal or to be expected; genuine or spontaneous; of, according to, existing in, or produced by nature; not created by human beings; not synthetic ▸ n person with an inborn talent or skill ▸ **naturally** adv of course; in a natural or normal way; instinctively ▸ **natural gas** gas found below the ground, used mainly as a fuel ▸ **natural history** study of animals and plants in the wild ▸ **natural selection** process by which only creatures and plants well adapted to their environment survive

naturalism n movement in art and literature advocating detailed realism ▸ **naturalistic** adj

naturalist n student of natural history

naturalize v give citizenship to (a person born in another country) ▸ **naturalization** n

nature n whole system of the existence, forces, and events of the physical world that are not controlled by human beings; fundamental or essential qualities; kind or sort

naturism n nudism ▸ **naturist** n

naught n (lit) nothing

naughty adj **-tier, -tiest** disobedient or mischievous; mildly indecent
▶ **naughtily** adv ▶ **naughtiness** n

nausea [naw-zee-a] n feeling of being about to vomit ▶ **nauseate** v make (someone) feel sick; disgust
▶ **nauseous** adj as if about to vomit; sickening

nautical adj of the sea or ships
▶ **nautical mile** 1852 metres (6076.12 feet)

nautilus n, pl **-luses** or **-li** shellfish with many tentacles

naval adj see **navy**

nave n long central part of a church

navel n hollow in the middle of the abdomen where the umbilical cord was attached

navigate v direct or plot the path or position of a ship, aircraft, or car; travel over or through
▶ **navigation** n ▶ **navigator** n
▶ **navigable** adj wide, deep, or safe enough to be sailed through; able to be steered

navvy n, pl **-vies** (Brit) labourer employed on a road or a building site

navy n, pl **-vies** branch of a country's armed services comprising warships with their crews and organization; warships of a nation ▷ adj navy-blue ▶ **naval** adj of or relating to a navy or ships
▶ **navy-blue** adj very dark blue

nay interj (obs) no

Nazi n member of the fascist National Socialist Party, which came to power in Germany in 1933 under Adolf Hitler ▷ adj of or relating to the Nazis ▶ **Nazism** n

NB note well

NCO (Mil) noncommissioned officer

NE northeast(ern)

Neanderthal [nee-ann-der-tahl] adj of a type of primitive man that lived in Europe before 12 000 BC

neap tide n tide at the first and last quarters of the moon when there is the smallest rise and fall in tidal level

near prep, adv indicating a place or time not far away ▷ adj almost being the thing specified e.g. a near disaster ▷ v draw close (to)
▶ **nearly** adv almost ▶ **nearness** n ▶ **nearby** adj not far away
▶ **nearside** n side of a vehicle that is nearer the kerb

neat adj tidy and clean; smoothly or competently done; undiluted
▶ **neatly** adv ▶ **neatness** n

nebula n, pl **-lae** (Astronomy) hazy cloud of particles and gases
▶ **nebulous** adj vague and unclear e.g. a nebulous concept

necessary adj needed to obtain the desired result e.g. the necessary skills; certain or unavoidable e.g. the necessary consequences
▶ **necessarily** adv ▶ **necessitate** v compel or require ▶ **necessity** n circumstances that inevitably require a certain result; something needed

SPELLING TIP

The correct spelling of **necessary** has one c and two ss. When you add un- at the beginning, you end up with a double n too: **unnecessary**

neck n part of the body joining the head to the shoulders; part of a garment round the neck; long narrow part of a bottle or violin ▷ v (slang) kiss and cuddle **neck and neck** absolutely level in a race or competition ▶ **neckerchief** n piece of cloth worn tied round the neck ▶ **necklace** n decorative piece of jewellery worn around the neck

necromancy n communication with the dead; sorcery

necropolis [neck-rop-pol-liss] n cemetery

nectar n sweet liquid collected from flowers by bees; drink of the gods

nectarine n smooth-skinned peach

ned n (Scot slang) young working-class male who dresses in casual sports clothes

née [nay] prep indicating the maiden name of a married woman

need v require or be in want of; be obliged (to do something) ▷ n condition of lacking something; requirement or necessity; poverty ▶ **needs** adv (foll by must) necessarily ▶ **needy** adj poor, in need of financial support ▶ **needful** adj necessary or required ▶ **needless** adj unnecessary

needle n thin pointed piece of metal with an eye through which thread is passed for sewing; long pointed rod used in knitting; pointed part of a hypodermic syringe; small pointed part in a record player that touches the record and picks up the sound signals, stylus; pointer on a measuring instrument or compass; long narrow stiff leaf ▷ v (informal) goad or provoke ▶ **needlework** n sewing and embroidery

ne'er adv (lit) never ▶ **ne'er-do-well** n useless or lazy person

nefarious [nif-fair-ee-uss] adj wicked

negate v invalidate; deny the existence of ▶ **negation** n

negative adj expressing a denial or refusal; lacking positive qualities; (of an electrical charge) having the same electrical charge as an electron ▷ n negative word or statement; (Photog) image with a reversal of tones or colours from which positive prints are made

neglect v take no care of; fail (to do something) through carelessness; disregard ▷ n neglecting or being neglected ▶ **neglectful** adj

negligee [neg-lee-zhay] n woman's lightweight usu. lace-trimmed dressing gown

negligence n neglect or carelessness ▶ **negligent** adj ▶ **negligently** adv

negligible adj so small or unimportant as to be not worth considering

negotiate v discuss in order to reach (an agreement); succeed in passing round or over (a place or problem) ▶ **negotiation** n ▶ **negotiator** n ▶ **negotiable** adj

Negro n, pl **-groes** (old-fashioned offens) member of any of the Black peoples originating in Africa ▶ **Negroid** adj of or relating to the Black peoples originating in Africa

neigh n loud high-pitched sound made by a horse ▷ v make this sound

neighbour n person who lives near another ▶ **neighbouring** adj situated nearby ▶ **neighbourhood** n district; surroundings; people of a district ▶ **neighbourly** adj kind, friendly, and helpful

neither adj, pron not one nor the other ▷ conj not both

nemesis [nem-miss-iss] n, pl **-ses** retribution or vengeance

neo- combining form new, recent, or a modern form of e.g. neoclassicism

neocon adj, n short for **neoconservative**

neoconservatism n (in the US) political movement favouring individualism and traditional morality ▶ **neoconservative** adj, n

Neolithic adj of the later Stone Age

neologism [nee-ol-a-jiz-zum] n newly-coined word or an existing word used in a new sense

neon n (Chem) colourless odourless gaseous element used in illuminated signs and lights

neonatal adj relating to the first few weeks of a baby's life

neophyte n beginner or novice; new convert

nephew n son of one's sister or brother

nephritis [nif-**frite**-tiss] n inflammation of a kidney

nepotism [**nep**-a-tiz-zum] n favouritism in business shown to relatives and friends

Neptune n Roman god of the sea; eighth planet from the sun

nerd n (slang) boring person obsessed with a particular subject; stupid and feeble person ▶ **nerdic** n (informal) technical jargon used esp by computing enthusiasts

nerve n cordlike bundle of fibres that conducts impulses between the brain and other parts of the body; bravery and determination; impudence ▶ pl anxiety or tension; ability or inability to remain calm in a difficult situation; **get on someone's nerves** irritate someone **nerve oneself** prepare oneself (to do something difficult or unpleasant) ▶ **nerveless** adj numb, without feeling; fearless ▶ **nervy** adj excitable or nervous ▶ **nerve centre** place from which a system or organization is controlled ▶ **nerve-racking** adj very distressing or harrowing

nervous adj apprehensive or worried; of or relating to the nerves ▶ **nervously** adv ▶ **nervousness** n ▶ **nervous breakdown** mental illness in which the sufferer ceases to function properly

nest n place or structure in which birds or certain animals lay eggs or give birth to young; secluded place;

set of things of graduated sizes designed to fit together ▶ v make or inhabit a nest ▶ **nest egg** fund of money kept in reserve

nestle v snuggle; be in a sheltered position

nestling n bird too young to leave the nest

net' n fabric of meshes of string, thread, or wire with many openings; piece of net used to protect or hold things or to trap animals ▷ v **netting, netted** catch (a fish or animal) in a net **the Net** internet ▶ **netting** n material made of net ▶ **netball** n team game in which a ball has to be thrown through a net hanging from a ring at the top of a pole ▶ **netbook** n type of small laptop computer

net², nett adj left after all deductions; (of weight) excluding the wrapping or container ▷ v **netting, netted** yield or earn as a clear profit

nether adj lower

nettle n plant with stinging hairs on the leaves ▶ **nettled** adj irritated

network n system of intersecting lines, roads, etc.; interconnecting group or system; (in broadcasting) group of stations that all transmit the same programmes simultaneously

neural adj of a nerve or the nervous system

neuralgia n severe pain along a nerve

neuritis [nyoor-**rite**-tiss] n inflammation of a nerve or nerves

neurology n scientific study of the nervous system ▶ **neurologist** n

neurosis n, pl **-ses** mental disorder producing hysteria, anxiety, depression, or obsessive behaviour

▶ **neurotic** adj emotionally unstable; suffering from neurosis ▷ n neurotic person

neuter adj belonging to a particular class of grammatical inflections in some languages ▷ v castrate (an animal)

neutral adj taking neither side in a war or dispute; of or belonging to a neutral party or country; (of a colour) not definite or striking ▷ n neutral person or nation; neutral gear ▶ **neutrality** n ▶ **neutralize** v make ineffective or neutral ▶ **neutral gear** position of the controls of a gearbox that leaves the gears unconnected to the engine

neutrino [new-tree-no] n, pl -nos elementary particle with no mass or electrical charge

neutron n electrically neutral elementary particle of about the same mass as a proton ▶ **neutron bomb** nuclear bomb designed to kill people and animals while leaving buildings virtually undamaged

never adv at no time ▶ **nevertheless** adv in spite of that

> **USAGE NOTE**
> Avoid the use of never with the past tense to mean not: I didn't see her yesterday (not I never saw her yesterday)

never-never n (informal) hire-purchase

new adj not existing before; recently acquired; having lately come into some state; additional; (foll by to) unfamiliar ▷ adv recently ▶ **newness** n ▶ **New Age** philosophy, originating in the late 1980s, characterized by a belief in alternative medicine and spiritualism ▶ **newbie** n (informal) person new to a job, club, etc.

▶ **newborn** adj recently or just born ▶ **newcomer** n recent arrival or participant ▶ **newfangled** adj objectionably or unnecessarily modern ▶ **newlyweds** pl n recently married couple ▶ **new moon** moon when it appears as a narrow crescent at the beginning of its cycle

newel n post at the top or bottom of a flight of stairs that supports the handrail

news n important or interesting new happenings; information about such events reported in the mass media ▶ **newsy** adj full of news ▶ **newsagent** n (Brit) shopkeeper who sells newspapers and magazines ▶ **newsflash** n brief important news item, which interrupts a radio or television programme ▶ **newsletter** n bulletin issued periodically to members of a group ▶ **newspaper** n weekly or daily publication containing news ▶ **newsprint** n inexpensive paper used for newspapers ▶ **newsreader, newscaster** n person who reads the news on the television or radio ▶ **newsreel** n short film giving news ▶ **newsroom** n room where news is received and prepared for publication or broadcasting ▶ **newsworthy** adj sufficiently interesting to be reported as news

newt n small amphibious creature with a long slender body and tail

newton n unit of force

next adj, adv immediately following; nearest ▶ **next-of-kin** n closest relative

nexus n, pl nexus connection or link

NHS (in Britain) National Health Service

nib n writing point of a pen

nibble v take little bites (of) ▷ n little bite

nibs n **his, her nibs** (slang) mock title of respect

nice adj pleasant; kind; good or satisfactory; subtle e.g. a nice distinction ▶ **nicely** adv ▶ **niceness** n

nicety n, pl **-ties** subtle point; refinement or delicacy

niche [neesh] n hollow area in a wall; suitable position for a particular person

nick v make a small cut in; (chiefly Brit slang) steal; (chiefly Brit slang) arrest ▷ n small cut; (slang) prison or police station **in good nick** (informal) in good condition **in the nick of time** just in time

nickel n (Chem) silvery-white metal often used in alloys; US coin worth five cents

nickelodeon n (US) early type of jukebox

nickname n familiar name given to a person or place ▷ v call by a nickname

nicotine n poisonous substance found in tobacco

niece n daughter of one's sister or brother

nifty adj **-tier, -tiest** (informal) neat or smart

niggardly adj stingy ▶ **niggard** n stingy person

nigger n (offens) Black person

niggle v worry slightly; continually find fault (with) ▷ n small worry or doubt

nigh adv, prep (lit) near

night n time of darkness between sunset and sunrise ▶ **nightly** adj, adv (happening) each night ▶ **nightcap** n drink taken just before bedtime; soft cap formerly worn in bed ▶ **nightclub** n establishment for dancing, music, etc., open late at night ▶ **nightdress** n woman's loose dress worn in bed ▶ **nightfall** n approach of darkness ▶ **nightie** (informal) nightdress ▶ **nightingale** n small bird with a musical song usu. heard at night ▶ **nightjar** n nocturnal bird with a harsh cry ▶ **nightlife** n entertainment and social activities available at night in a town or city ▶ **nightmare** n very bad dream; very unpleasant experience ▶ **night school** place where adults can attend educational courses in the evenings ▶ **nightshade** n plant with bell-shaped flowers which are often poisonous ▶ **nightshirt** n long loose shirt worn in bed ▶ **night-time** n time from sunset to sunrise

nihilism [nye-ill-iz-zum] n rejection of all established authority and institutions ▶ **nihilist** n ▶ **nihilistic** adj

nil n nothing, zero

nimble adj agile and quick; mentally alert or acute ▶ **nimbly** adv

nimbus n, pl **-bi** or **-buses** dark grey rain cloud

nincompoop n (informal) stupid person

nine adj, n one more than eight ▶ **ninth** adj, n (of) number nine in a series ▶ **ninepins** n game of skittles

nineteen n, n and ninth ▶ **nineteenth** adj, n

ninety n, n ten times nine ▶ **ninetieth** adj, n

niobium n (Chem) white superconductive metallic element

nip¹ v **nipping, nipped** (informal) hurry; pinch or squeeze; bite lightly ▷ n pinch or light bite; sharp coldness ▶ **nipper** n (Brit, Aust & NZ informal) small child ▶ **nippy** adj frosty or chilly; (informal) quick or nimble

nip² n small alcoholic drink

nipple n projection in the centre of a breast

niqab n veil worn by Muslim women

nirvana [near-vah-na] n (Buddhism, Hinduism) absolute spiritual enlightenment and bliss

nit n egg or larva of a louse; (informal) short for **nitwit** ▶ **nit-picking** adj (informal) overconcerned with insignificant detail, esp. to find fault ▶ **nitwit** n (informal) stupid person

nitrogen [nite-roj-jen] n (Chem) colourless odourless gas that forms four fifths of the air ▶ **nitric, nitrous, nitrogenous** adj of or containing nitrogen ▶ **nitrate** n compound of nitric acid, used as a fertilizer ▶ **nitroglycerine, nitroglycerin** n explosive liquid

nitty-gritty n (informal) basic facts

no interj expresses denial, disagreement, or refusal ▷ adj not any, not a ▷ adv not at all ▷ n, pl **noes** or **nos** answer or vote of 'no'; person who answers or votes 'no' ▶ **no-go area** district barricaded off so that the police or army can enter only by force ▶ **no-man's-land** n land between boundaries, esp. contested land between two opposing forces ▶ **no-one, no one** pron nobody

no. number

nob n (chiefly Brit slang) person of wealth or social distinction

nobble v (Brit slang) attract the attention of (someone) in order to talk to him or her; bribe or threaten

nobelium n (Chem) artificially produced radioactive element

Nobel Prize n prize awarded annually for outstanding achievement in various fields

noble adj showing or having high moral qualities; of the nobility; impressive and magnificent ▷ n member of the nobility ▶ **nobility** n quality of being noble; class of people holding titles and high social rank ▶ **nobly** adv ▶ **nobleman, noblewoman** n

nobody pron no person ▷ n, pl **-bodies** person of no importance

no-brainer n (slang) something that requires little or no mental effort

nocturnal adj of the night; active at night

nocturne n short dreamy piece of music

nod v **nodding, nodded** lower and raise (one's head) briefly in agreement or greeting; let one's head fall forward with sleep ▷ n act of nodding ▶ **nod off** v (informal) fall asleep

noddle n (chiefly Brit informal) the head

node n point on a plant stem from which leaves grow; point at which a curve crosses itself

nodule n small knot or lump; rounded mineral growth on the root of a plant

Noel n Christmas

noggin n (informal) head; small quantity of an alcoholic drink

noise n sound, usu. a loud or disturbing one ▶ **noisy** adj making a lot of noise; full of noise ▶ **noisily** adv ▶ **noiseless** adj

noisome adj (of smells) offensive; harmful or poisonous

nomad n member of a tribe with no fixed dwelling place, wanderer ▶ **nomadic** adj

nom de plume n, pl **noms de plume** pen name

nomenclature n system of names used in a particular subject

nominal adj in name only; very small in comparison with real worth ▶ **nominally** adv

nominate v suggest as a candidate; appoint to an office

or position ▶ **nomination** n
▶ **nominee** n candidate
▶ **nominative** n form of a noun
indicating the subject of a verb
non- prefix indicating negation
e.g. nonexistent; indicating refusal
or failure e.g. noncooperation;
indicating exclusion from a specified
class e.g. nonfiction; indicating lack
or absence e.g. nonevent
nonagenarian n person aged
between ninety and ninety-nine
nonaggression n policy of not
attacking other countries
nonagon n geometric figure with
nine sides
nonalcoholic adj containing no
alcohol
nonaligned adj (of a country) not
part of a major alliance or power bloc
nonce n **for the nonce** for the
present
nonchalant adj casually
unconcerned or indifferent
▶ **nonchalantly** adv
▶ **nonchalance** n
noncombatant n member of the
armed forces whose duties do not
include fighting
noncommissioned officer n (in
the armed forces) a subordinate
officer, risen from the ranks
noncommittal adj not
committing oneself to any
particular opinion
non compos mentis (Latin) adj of
unsound mind
nonconductor n substance
that is a poor conductor of heat,
electricity, or sound
nonconformist n person who
does not conform to generally
accepted patterns of behaviour
or thought; (**N-**) member of a
Protestant group separated
from the Church of England
▷ adj (of behaviour or ideas) not

conforming to accepted patterns
▶ **nonconformity** n
noncontributory adj (Brit)
denoting a pension scheme for
employees, the premiums of which
are paid entirely by the employer
nondescript adj lacking
outstanding features
none pron not any; no-one
▶ **nonetheless** adv despite that,
however
nonentity [non-**enn**-tit-tee] n, pl
-ties insignificant person or thing
nonevent n disappointing or
insignificant occurrence
nonflammable adj not easily
set on fire
nonintervention n refusal to
intervene in the affairs of others
nonpareil [non-par-**rail**] n person
or thing that is unsurpassed
nonpayment n failure to pay
money owed
nonplussed adj perplexed
nonsense n something that
has or makes no sense; absurd
language; foolish behaviour
▶ **nonsensical** adj
non sequitur [**sek**-wit-tur] n
statement with little or no relation
to what preceded it
nonstandard adj denoting
language that is not regarded as
correct by educated native speakers
nonstarter n person or idea that
has little chance of success
nonstick adj coated with a
substance that food will not stick to
when cooked
nonstop adj, adv without a stop
nontoxic adj not poisonous
noob n (slang, chiefly Computing)
same as **newbie**
noodles pl n long thin strips of pasta
nook n sheltered place
noon n twelve o'clock; midday
▶ **noonday** adj happening at noon

n

noose n loop in the end of a rope, tied with a slipknot

nor conj and not

Nordic adj of Scandinavia or its typically tall blond and blue-eyed people

norm n standard that is regarded as normal

normal adj usual, regular, or typical; free from mental or physical disorder ▶ **normally** adv ▶ **normality** n ▶ **normalize** v

Norse n, adj (language) of ancient and medieval Norway

north n direction towards the North Pole, opposite south; area lying in or towards the north ▷ adj to or in the north; (of a wind) from the north ▷ adv in, to, or towards the north ▶ **northerly** adj ▶ **northern** adj ▶ **northerner** n person from the north of a country or area ▶ **northward** adj, adv ▶ **northwards** adv ▶ **North Pole** northernmost point on the earth's axis

nos. numbers

nose n organ of smell, used also in breathing; front part of a vehicle ▷ v move forward slowly and carefully; pry or snoop ▶ **nose dive** sudden drop ▶ **nosegay** n small bunch of flowers ▶ **nosey, nosy** adj (informal) prying or inquisitive ▶ **nosiness** n

nosh (Brit, Aust & NZ slang) n food ▷ v eat

nostalgia n sentimental longing for the past ▶ **nostalgic** adj

nostril n one of the two openings at the end of the nose

nostrum n quack medicine; favourite remedy

not adv expressing negation, refusal, or denial

notable adj worthy of being noted, remarkable ▷ n person of distinction ▶ **notably** adv ▶ **notability** n

notary n, pl -ries person authorized to witness the signing of legal documents

notation n representation of numbers or quantities in a system by a series of symbols; set of such symbols

notch n V-shaped cut; (informal) step or level ▷ v make a notch in; (foll by up) score or achieve

note n short letter; brief comment or record; banknote; (symbol for) a musical sound; hint or mood ▷ v notice, pay attention to; record in writing; remark upon ▶ **noted** adj well-known ▶ **notebook** n book for writing in ▶ **noteworthy** adj worth noting, remarkable

nothing pron not anything; matter of no importance; figure o ▷ adv not at all ▶ **nothingness** n nonexistence; insignificance

> **USAGE NOTE**
> Nothing is usually followed by a singular verb but, if it comes before a plural noun, this can sound odd: Nothing but books was/were on the shelf. A solution is to rephrase the sentence: Only books were...

notice n observation or attention; sign giving warning or an announcement; advance notification of intention to end a contract of employment ▷ v observe, become aware of; point out or remark upon ▶ **noticeable** adj easily seen or detected, appreciable

notify v -fying, -fied inform ▶ **notification** n ▶ **notifiable** adj having to be reported to the authorities

notion n idea or opinion; whim ▶ **notional** adj speculative, imaginary, or unreal

notorious *adj* well known for something bad ▶ **notoriously** *adv* ▶ **notoriety** *n*

notwithstanding *prep* in spite of

nougat *n* chewy sweet containing nuts and fruit

nought *n* figure o; nothing ▶ **noughties** *pl n* (*informal*) decade from 2000 to 2009

noun *n* word that refers to a person, place, or thing

nourish *v* feed; encourage or foster (an idea or feeling) ▶ **nourishment** *n* ▶ **nourishing** *adj* providing the food necessary for life and growth

nouvelle cuisine [noo-vell kwee-**zeen**] *n* style of preparing and presenting food with light sauces and unusual combinations of flavours

Nov November

nova *n*, *pl* **-vae** *or* **-vas** star that suddenly becomes brighter and then gradually decreases to its original brightness

novel¹ *n* long fictitious story in book form ▶ **novelist** *n* writer of novels ▶ **novella** *n*, *pl* **-las** *or* **-lae** short novel

novel² *adj* fresh, new, or original ▶ **novelty** *n* newness; something new or unusual

November *n* eleventh month of the year

novena [no-**vee**-na] *n*, *pl* **-nas** (*RC Church*) set of prayers or services on nine consecutive days

novice *n* beginner; person who has entered a religious order but has not yet taken vows

now *adv* at or for the present time; immediately ▷ *conj* seeing that, since **just now** very recently **now and again, now and then** occasionally ▶ **nowadays** *adv* in these times

nowhere *adv* not anywhere

noxious *adj* poisonous or harmful; extremely unpleasant

nozzle *n* projecting spout through which fluid is discharged

NSPCC (in Britain) National Society for the Prevention of Cruelty to Children

NSW New South Wales

NT (in Britain) National Trust; New Testament; Northern Territory

nuance [**new**-ahnss] *n* subtle difference in colour, meaning, or tone

nub *n* point or gist (of a story etc.)

nubile [**new**-bile] *adj* (of a young woman) sexually attractive; (of a young woman) old enough to get married

nuclear *adj* of nuclear weapons or energy; of a nucleus, esp. the nucleus of an atom ▶ **nuclear energy** energy released as a result of nuclear fission or fusion ▶ **nuclear fission** splitting of an atomic nucleus ▶ **nuclear fusion** combination of two nuclei to form a heavier nucleus with the release of energy ▶ **nuclear power** power produced by a nuclear reactor ▶ **nuclear reaction** change in structure and energy content of an atomic nucleus by interaction with another nucleus or particle ▶ **nuclear reactor** device in which a nuclear reaction is maintained and controlled to produce nuclear energy ▶ **nuclear weapon** weapon whose force is due to uncontrolled nuclear fusion or fission ▶ **nuclear winter** theoretical period of low temperatures and little light after a nuclear war

nucleic acid *n* complex compound, such as DNA or RNA, found in all living cells

nucleus *n*, *pl* **-clei** centre, esp. of an atom or cell; central thing around which others are grouped

n

nude adj naked ▷ n naked figure in painting, sculpture, or photography ▶ **nudity** n practice of not wearing clothes ▶ **nudist** n

nudge v push gently, esp. with the elbow ▷ n gentle push or touch

nugatory [new-gat-tree] adj of little value; not valid

nugget n small lump of gold in its natural state; something small but valuable ▷ v (NZ & S Afr) polish footwear

nuisance n something or someone that causes annoyance or bother

nuke (slang) v attack with nuclear weapons ▷ n nuclear weapon

null adj **null and void** not legally valid ▶ **nullity** v ▶ **nullify** v make ineffective; cancel

nulla-nulla n wooden club used by Australian Aborigines

numb adj without feeling, as through cold, shock, or fear ▷ v make numb ▶ **numbly** adv ▶ **numbness** n ▶ **numbskull** n stupid person

numbat n small Australian marsupial with a long snout and tongue

number n sum or quantity; word or symbol used to express a sum or quantity, numeral; numeral or string of numerals used to identify a person or thing; one of a series, such as a copy of a magazine; song or piece of music; group of people; (Grammar) classification of words depending on how many persons or things are referred to ▷ v count; give a number to; amount to; include in a group ▶ **numberless** adj too many to be counted ▶ **number crunching** (Computing) large-scale processing of numerical data ▶ **number one** n (informal) oneself; bestselling pop record in any one week ▷ adj

first in importance or quality ▶ **numberplate** n plate on a car showing the registration number

numeral n word or symbol used to express a sum or quantity

numerate adj able to do basic arithmetic ▶ **numeracy** n

numeration n act or process of numbering or counting

numerator n (Maths) number above the line in a fraction

numerical adj measured or expressed in numbers ▶ **numerically** adv

numerous adj existing or happening in large numbers

numismatist n coin collector

numskull n same as **numbskull**

nun n female member of a religious order ▶ **nunnery** n convent

nuncio n (RC Church) Pope's ambassador

nuptial adj relating to marriage ▶ **nuptials** pl n wedding

nurse n person employed to look after sick people, usu. in a hospital; woman employed to look after children ▷ v look after (a sick person); breast-feed (a baby); try to cure (an ailment); harbour or foster (a feeling) ▶ **nursing home** private hospital or home for old people ▶ **nursing officer** (in Britain) administrative head of the nursing staff of a hospital

nursery n, pl **-ries** room where children sleep or play; place where children are taken care of while their parents are at work; place where plants are grown for sale ▶ **nurseryman** n person who raises plants for sale ▶ **nursery school** school for children from 3 to 5 years old ▶ **nursery slopes** gentle ski slopes for beginners

nurture n act or process of promoting the development of a child or young plant ▷ v promote or encourage the development of

nut n fruit consisting of a hard shell and a kernel; small piece of metal that screws onto a bolt; (also **nutcase**) (slang) insane or eccentric person; (slang) head ▶ **nutter** n (Brit slang) insane person ▶ **nutty** adj containing or resembling nuts; (slang) insane or eccentric ▶ **nutcracker** n device for cracking the shells of nuts ▶ **nuthatch** n small songbird ▶ **nutmeg** n spice made from the seed of a tropical tree

nutria n fur of the coypu

nutrient n substance that provides nourishment

nutriment n food or nourishment required by all living things to grow and stay healthy

nutrition n process of taking in and absorbing nutrients; process of being nourished ▶ **nutritional** adj ▶ **nutritious, nutritive** adj

nuzzle v push or rub gently with the nose or snout

NW northwest(ern)

nylon n synthetic material used for clothing etc. ▷ pl stockings made of nylon

nymph n mythical spirit of nature, represented as a beautiful young woman; larva of certain insects, resembling the adult form

nymphet n sexually precocious young girl

nymphomaniac n woman with an abnormally intense sexual desire

NZ New Zealand

NZE New Zealand English

NZRFU New Zealand Rugby Football Union

NZSE40 Index New Zealand Share Price 40 Index

Oo

O (chem) oxygen; old; same as **nought** sense 1

oaf n stupid or clumsy person ▶ **oafish** adj

oak n deciduous forest tree; its wood, used for furniture ▶ **oaken** adj ▶ **oak apple** brownish lump found on oak trees

oakum n fibre obtained by unravelling old rope

OAP (in Britain) old-age pensioner

oar n pole with a broad blade, used for rowing a boat

oasis n, pl **-ses** fertile area in a desert

oast n (chiefly Brit) oven for drying hops

oat n hard cereal grown as food ▷ pl grain of this cereal **sow one's wild oats** have many sexual relationships when young ▶ **oatmeal** adj pale brownish-cream

oath n solemn promise, esp. to be truthful in court; swearword

obbligato [ob-lig-**gah**-toe] n, pl **-tos** (Music) essential part or accompaniment

obdurate adj hardhearted or stubborn ▶ **obduracy** n

OBE (in Britain) Officer of the Order of the British Empire

obedient adj obeying or willing to obey ▶ **obedience** n ▶ **obediently** adv

obeisance [oh-**bay**-sanss] n attitude of respect; bow or curtsy

obelisk [**ob**-bill-isk] n four-sided stone column tapering to a pyramid at the top

obese [oh-beess] *adj* very fat
▶ **obesity** *n*

obey *v* carry out instructions or orders

obfuscate *v* make (something) confusing

obituary *n*, *pl* -aries announcement of someone's death, esp. in a newspaper ▶ **obituarist** *n*

object[1] *n* physical thing; focus of thoughts or action; aim or purpose; (*Grammar*) word that a verb or preposition affects **no object** not a hindrance

object[2] *v* express disapproval
▶ **objection** *n* ▶ **objectionable** *adj* unpleasant ▶ **objector** *n*

objective *n* aim or purpose ▷ *adj* not biased; existing in the real world outside the human mind
▶ **objectively** *adv* ▶ **objectivity** *n*

objet d'art [ob-zhay dahr] *n*, *pl* **objets d'art** small object of artistic value

oblation *n* religious offering

oblige *v* compel (someone) morally or by law to do something; do a favour for (someone) ▶ **obliging** *adj* ready to help other people
▶ **obligingly** *adv* ▶ **obligated** *adj* obliged to do something
▶ **obligation** *n* duty ▶ **obligatory** *adj* required by a rule or law

oblique [oh-bleak] *adj* slanting; indirect ▷ *n* the symbol (/)
▶ **obliquely** *adv* ▶ **oblique angle** angle that is not a right angle

obliterate *v* wipe out, destroy
▶ **obliteration** *n*

oblivious *adj* unaware ▶ **oblivion** *n* state of being forgotten; state of being unaware or unconscious

oblong *adj* having two long sides, two short sides, and four right angles ▷ *n* oblong figure

obloquy [ob-lock-wee] *n*, *pl* -quies verbal abuse; discredit

obnoxious *adj* offensive

oboe *n* double-reeded woodwind instrument ▶ **oboist** *n*

obscene *adj* portraying sex offensively; disgusting ▶ **obscenity** *n*

obscure *adj* not well known; hard to understand; indistinct ▷ *v* make (something) obscure ▶ **obscurity** *n*

obsequies [ob-sick-weez] *pl n* funeral rites

obsequious [ob-seek-wee-uss] *adj* overattentive in order to gain favour ▶ **obsequiousness** *n*

observe *v* see or notice; watch (someone or something) carefully; remark; act according to (a law or custom) ▶ **observation** *n* action or habit of observing; remark
▶ **observable** *adj* ▶ **observance** *n* observing of a custom
▶ **observant** *adj* quick to notice things ▶ **observatory** *n* building equipped for studying the weather and the stars

observer *n* person who observes, esp. one who watches someone or something carefully

obsess *v* preoccupy (someone) compulsively ▶ **obsessed** *adj*
▶ **obsessive** *adj* ▶ **obsession** *n*

> **SPELLING TIP**
> Some people get carried away with doubling ss and write *obsession* instead of **obsession**.

obsidian *n* dark glassy volcanic rock

obsolete *adj* no longer in use
▶ **obsolescent** *adj* becoming obsolete ▶ **obsolescence** *n*

obstacle *n* something that makes progress difficult

obstetrics *n* branch of medicine concerned with pregnancy and childbirth ▶ **obstetric** *adj*
▶ **obstetrician** *n*

obstinate *adj* stubborn; difficult to remove or change
▶ **obstinately** *adv* ▶ **obstinacy** *n*

obstreperous adj unruly, noisy

obstruct v block with an obstacle ▸ **obstruction** n ▸ **obstructive** adj

obtain v acquire intentionally; be customary ▸ **obtainable** adj

obtrude v push oneself or one's ideas on others ▸ **obtrusive** adj unpleasantly noticeable ▸ **obtrusively** adv

obtuse adj mentally slow; (Maths) (of an angle) between 90° and 180°; not pointed ▸ **obtuseness** n

obverse n opposite way of looking at an idea; main side of a coin or medal

obviate v make unnecessary

obvious adj easy to see or understand, evident ▸ **obviously** adv

ocarina n small oval wind instrument

occasion n time at which a particular thing happens; reason e.g. *no occasion for complaint*; special event ▹ v cause ▸ **occasional** adj ▸ **occasionally** adv

> **SPELLING TIP**
> *Occassion*, *ocasion*, and *ocassion* are common misspellings of **occasion**. The correct version has two cs and one s

Occident n (lit) the West ▸ **Occidental** adj

occiput [ox-sip-put] n back of the head

occlude v obstruct; close off ▸ **occlusion** n ▸ **occluded front** (Meteorol) front formed when a cold front overtakes a warm front and warm air rises

occult adj relating to the supernatural **the occult** knowledge or study of the supernatural

occupant n person occupying a specified place ▸ **occupancy** n (length of) a person's stay in a specified place

occupation n profession; activity that occupies one's time; control of a country by a foreign military power; being occupied ▸ **occupational** adj ▸ **occupational therapy** purposeful activities, designed to aid recovery from illness etc.

occupy v **-pying, -pied** live or work in (a building); take up the attention of (someone); take up (space or time); take possession of (a place) by force ▸ **occupier** n

occur v **-curring, -curred** happen; exist **occur to** come to the mind of ▸ **occurrence** n something that occurs; fact of occurring

> **SPELLING TIP**
> Note that **occurrence** has a double c and a double r, and in the last syllable an e rather than an a

ocean n vast area of sea between continents ▸ **oceanic** adj ▸ **oceanography** n scientific study of the oceans ▸ **ocean-going** adj able to sail on the open sea

ocelot [oss-ill-lot] n American wild cat with a spotted coat

oche [ok-kee] n (Darts) mark on the floor behind which a player must stand

ochre [oak-er] adj, n brownish-yellow (earth)

o'clock adv used after a number to specify an hour

Oct October

octagon n geometric figure with eight sides ▸ **octagonal** adj

octahedron [ok-ta-heed-ron] n, pl **-drons** or **-dra** three-dimensional geometric figure with eight faces

octane n hydrocarbon found in petrol ▸ **octane rating** measure of petrol quality

octave n (Music) (interval between the first and) eighth note of a scale

octet n group of eight performers; music for such a group

October n tenth month of the year

octogenarian n person aged between eighty and eighty-nine

octopus n, pl **-puses** sea creature with a soft body and eight tentacles

ocular adj relating to the eyes or sight

OD (informal) n overdose ▷ v **OD'ing**, **OD'd** take an overdose

odd adj unusual; occasional; not divisible by two; not part of a set ▷ **odds** pl n (ratio showing) the probability of something happening ▷ **at odds** in conflict ▷ **odds and ends** small miscellaneous items ▷ **oddity** n odd person or thing ▷ **oddness** n quality of being odd ▷ **oddments** pl n things left over

ode n lyric poem, usu. addressed to a particular subject

odium [oh-dee-um] n widespread dislike ▷ **odious** adj offensive

odour n particular smell ▷ **odorous** adj ▷ **odourless** adj

odyssey [odd-iss-ee] n long eventful journey

OE (NZ informal) overseas experience e.g. he's away on his OE.

OECD Organization for Economic Cooperation and Development

oedema [id-deem-a] n, pl **-mata** (Med) abnormal swelling

oesophagus [ee-soff-a-guss] n, pl **-gi** passage between the mouth and stomach

oestrogen [ee-stra-jen] n female hormone that controls the reproductive cycle

of prep belonging to; consisting of; connected with; characteristic of

off prep away from ▷ adv away ▷ adj not operating; cancelled; (of food) gone bad ▷ n (Cricket) side of the field to which the batsman's feet point ▷ **off colour** slightly ill ▷ **offline** adj not connected to the internet ▷ adv while not connected to the internet ▷ **off-message** adj (esp. of a politician) not following the official Party line ▷ **off-road** adj (of a motor vehicle) designed for use away from public roads

USAGE NOTE
Avoid using of after off: He got off the bus (not off of). The use of off to mean from is very informal: They bought milk from (rather than off) a farmer

offal n edible organs of an animal, such as liver or kidneys ▷ **offal pit**, **offal hole** (NZ) place on a farm for the disposal of animal offal

offcut n piece remaining after the required parts have been cut out

offend v hurt the feelings of, insult; commit a crime ▷ **offence** n (cause of) hurt feelings or annoyance; illegal act ▷ **offensive** adj disagreeable; insulting; aggressive ▷ n position or action of attack

offender n person who commits a crime

offer v present (something) for acceptance or rejection; provide; be willing (to do something); propose as payment ▷ n instance of offering something ▷ **offering** n thing offered ▷ **offertory** n (Christianity) offering of the bread and wine for Communion

offhand adj casual, curt ▷ adv without preparation

office n room or building where people work at desks; department of a commercial organization; formal position of responsibility; place where tickets or information can be obtained

officer n person in authority in the armed services; member of the police force; person with special responsibility in an organization

official *adj* of a position of authority; approved or arranged by someone in authority ▷ *n* person who holds a position of authority ▶ **officially** *adv* ▶ **officialdom** *n* officials collectively ▶ **Official Receiver** (*Brit*) person who deals with the affairs of a bankrupt company

officiate *v* act in an official role

officious *adj* interfering unnecessarily

offing *n* area of the sea visible from the shore **in the offing** (*Brit, Aust & NZ*) likely to happen soon

off-licence *n* (*Brit*) shop licensed to sell alcohol for drinking elsewhere

offset *v* cancel out, compensate for

offshoot *n* something developed from something else

offside *adj, adv* (*Sport*) (positioned) illegally ahead of the ball

offspring *n, pl* **offspring** child

often *adv* frequently, much of the time ▶ **oft** *adv* (*poetic*) often

ogle *v* stare at (someone) lustfully

ogre *n* giant that eats human flesh; monstrous or cruel person

oh *interj* exclamation of surprise, pain, etc.

ohm *n* unit of electrical resistance

OHMS (*Brit*) On Her or His Majesty's Service

oil *n* viscous liquid, insoluble in water and usu. flammable; same as **petroleum**; petroleum derivative, used as a fuel or lubricant ▷ *v* lubricate (a machine) with oil ▶ **oily** *adj* ▶ **oilfield** *n* area containing oil reserves ▶ **oil rig** platform constructed for drilling oil wells ▶ **oilskin** *n* (garment made from) waterproof material

ointment *n* greasy substance used for healing skin or as a cosmetic

O.K., okay (*informal*) *interj* expression of approval ▷ *v* approve (something) ▷ *n* approval

okapi [ok-**kah**-pee] *n* African animal related to the giraffe but with a shorter neck

okra *n* tropical plant with edible green pods

old *adj* having lived or existed for a long time; of a specified age e.g. *two years old*; former ▶ **olden** *adj* old e.g. *in the olden days* ▶ **oldie** *n* (*informal*) old but popular song or film ▶ **old-fashioned** *adj* no longer commonly used or valued ▶ **old guard** group of people in an organization who have traditional values ▶ **old hat** boring because so familiar ▶ **old maid** elderly unmarried woman ▶ **old master** European painter or painting from the period 1500–1800 ▶ **Old Nick** (*Brit, Aust & NZ informal*) the Devil ▶ **old school tie** system of mutual help between former pupils of public schools ▶ **Old Testament** part of the Bible recording Hebrew history ▶ **Old World** world as it was known before the discovery of the Americas

oleaginous [ol-lee-**aj**-in-uss] *adj* oily, producing oil

oleander [ol-lee-**ann**-der] *n* Mediterranean flowering evergreen shrub

olfactory *adj* relating to the sense of smell

oligarchy [ol-lee-**gark**-ee] *n, pl* -**chies** government by a small group of people; state governed this way ▶ **oligarchic, oligarchical** *adj*

olive *n* small green or black fruit used as food or pressed for its oil; tree on which this fruit grows ▷ *adj* greyish-green ▶ **olive branch** peace offering

o

Olympian adj of Mount Olympus or the classical Greek gods; majestic or godlike

Olympic Games® pl n four-yearly international sports competition

ombudsman n official who investigates complaints against government organizations

omelette n dish of eggs beaten and fried

> **SPELLING TIP**
> You don't hear it in the pronunciation, but there is an e after the m in **omelette**

omen n happening or object thought to foretell success or misfortune ▶ **ominous** adj worrying, seeming to foretell misfortune

omit v **omitting, omitted** leave out; neglect (to do something) ▶ **omission** n

omnibus n several books or TV or radio programmes made into one; (old-fashioned) bus

omnipotent adj having unlimited power ▶ **omnipotence** n

omnipresent adj present everywhere ▶ **omnipresence** n

omniscient [om-niss-ee-ent] adj knowing everything ▶ **omniscience** n

omnivorous [om-niv-vor-uss] adj eating food obtained from both animals and plants ▶ **omnivore** n omnivorous animal

on prep indicating position above, attachment, closeness, etc. e.g. lying on the ground; a puppet on a string; on the coast ▶ adv in operation; continuing; forwards ▶ adj operating; taking place in (Cricket) side of the field on which the batsman stands ▶ **on line, online** adj relating to the internet e.g. online shopping ▶ adv while connected to the internet

▶ **on-message** adj (esp. of a politician) following the official Party line

once adv on one occasion; formerly ▶ conj as soon as ▶ **at once** immediately; simultaneously ▶ **once-over** n (informal) quick examination

oncogene [on-koh-jean] n gene that can cause cancer when abnormally activated

oncoming adj approaching from the front

one adj single, lone ▶ n number or figure 1; single unit ▶ pron any person ▶ **oneness** n unity ▶ **oneself** pron reflexive form of one ▶ **one-armed bandit** fruit machine operated by a lever on one side ▶ **one-liner** n witty remark ▶ **one-night stand** sexual encounter lasting one night ▶ **one-sided** adj considering only one point of view ▶ **one-way** adj allowing movement in one direction only

> **USAGE NOTE**
> Avoid overuse of the pronoun one as a substitute for I. Many listeners find it affected

onerous [own-er-uss] adj (of a task) difficult to carry out

ongoing adj in progress, continuing

onion n strongly flavoured edible bulb

onlooker n person who watches without taking part

only adj alone of its kind ▶ adv exclusively; merely; no more than ▶ conj but

> **USAGE NOTE**
> In formal use only is placed directly before the words it modifies: The club opens only on Thursdays but in everyday use this becomes: The club only opens on Thursdays

onomatopoeia [on-a-mat-a-pee-a] n use of a word which imitates the sound it represents, such as hiss ▶ **onomatopoeic** adj

onset n beginning

onslaught n violent attack

onto prep to a position on; aware of e.g. she's onto us

ontology n branch of philosophy concerned with existence ▶ **ontological** adj

onus [own-uss] n, pl **onuses** responsibility or burden

onward adj directed or moving forward ▷ adv (also **onwards**) ahead, forward

onyx n type of quartz with coloured layers

oodles pl n (informal) great quantities

ooze[1] v flow slowly ▷ n sluggish flow ▶ **oozy** adj

ooze[2] n soft mud at the bottom of a lake or river

opal n iridescent precious stone ▶ **opalescent** adj iridescent like an opal

opaque adj not able to be seen through, not transparent ▶ **opacity** n

op. cit. [op sit] in the work cited

OPEC Organization of Petroleum-Exporting Countries

open adj not closed; not covered; unfolded; ready for business; free from obstruction, accessible; frank ▷ v (cause to) become open; begin ▷ n (Sport) competition which all may enter **in the open** outdoors ▶ **openly** adv without concealment ▶ **opening** n opportunity; hole ▷ adj first ▶ **opencast mining** mining at the surface and not underground ▶ **open day** day on which a school or college is open to the public ▶ **open-handed** adj generous

▶ **open-hearted** adj generous; frank ▶ **open-heart surgery** surgery on the heart during which the blood circulation is maintained by machine ▶ **open house** hospitality to visitors at any time ▶ **open letter** letter to an individual that the writer makes public in a newspaper or magazine ▶ **open-minded** adj receptive to new ideas ▶ **open-plan** adj (of a house or office) having few interior walls ▶ **open prison** prison with minimal security ▶ **open source** (Computing) intellectual property, esp computer source code, that is made freely available to the public by its creators ▶ **open-standard** adj (of computer programs, codes, etc) freely available to all users ▶ **open verdict** coroner's verdict not stating the cause of death

opera[1] n drama in which the text is sung to an orchestral accompaniment ▶ **operatic** adj ▶ **operetta** n light-hearted comic opera

opera[2] n a plural of **opus**

operate v (cause to) work; direct; perform an operation ▶ **operator** n ▶ **operation** n method or procedure of working; medical procedure in which the body is worked on to repair a damaged part ▶ **operational** adj in working order; relating to an operation ▶ **operative** adj working ▷ n worker with a special skill

ophthalmic adj relating to the eye ▶ **ophthalmology** n study of the eye and its diseases ▶ **ophthalmologist** n

opiate n narcotic drug containing opium

opinion n personal belief or judgment ▶ **opinionated** adj having strong opinions ▶ **opine** v (old-fashioned) express an opinion ▸ **opinion poll** see **poll**

opium n addictive narcotic drug made from poppy seeds

opossum n small marsupial of America or Australasia

opponent n person one is working against in a contest, battle, or argument

opportunity n, pl **-ties** favourable time or condition; good chance ▸ **opportunity shop** (Aust & NZ) shop selling second-hand clothes, sometimes for charity (also **op-shop**) ▸ **opportune** adj happening at a suitable time ▸ **opportunist** n, adj (person) doing whatever is advantageous without regard for principles ▸ **opportunism** n

> **SPELLING TIP**
>
> Lots of people forget that **opportunity**, which is a very common word, has two ps

oppose v work against **be opposed to** disagree with or disapprove of ▸ **opposition** n obstruction or hostility; group opposing another; political party not in power

opposite adj situated on the other side; facing; completely different ▸ n person or thing that is opposite ▸ prep facing ▸ adv on the other side

oppress v control by cruelty or force; depress ▸ **oppression** n ▸ **oppressor** n ▸ **oppressive** adj tyrannical; (of weather) hot and humid ▸ **oppressively** adv

opprobrium [op-probe-ree-um] n state of being criticized severely for wrong one has done

opt v show a preference, choose ▸ **opt out** v choose not to be part (of)

optic adj relating to the eyes or sight ▸ **optics** n science of sight and light ▸ **optical** adj ▸ **optical character reader** device that electronically reads and stores text ▸ **optical fibre** fine glass-fibre tube used to transmit information

optician n (also **ophthalmic optician**) person qualified to prescribe glasses; (also **dispensing optician**) person who supplies and fits glasses

optimism n tendency to take the most hopeful view ▸ **optimist** n ▸ **optimistic** adj ▸ **optimistically** adv

optimum n, pl **-ma** or **-mums** best possible conditions ▸ adj most favourable ▸ **optimal** adj ▸ **optimize** v make the most of

option n choice; thing chosen; right to buy or sell something at a specified price within a given time ▸ **optional** adj possible but not compulsory

optometrist n person qualified to prescribe glasses ▸ **optometry** n

opulent [op-pew-lent] adj having or indicating wealth ▸ **opulence** n

opus n, pl **opuses** or **opera** artistic creation, esp. a musical work

or conj used to join alternatives e.g. tea or coffee

oracle n shrine of an ancient god; prophecy, often obscure, revealed at a shrine; person believed to make infallible predictions ▸ **oracular** adj

oral adj spoken; (of a drug) to be taken by mouth ▸ n spoken examination ▸ **orally** adv

orange n reddish-yellow citrus fruit ▸ adj reddish-yellow ▸ **orangeade** n (Brit) orange-flavoured, usu. fizzy drink ▸ **orangery** n greenhouse for growing orange trees

orang-utan, orang-utang n large reddish-brown ape with long arms

orator [or-rat-tor] n skilful public speaker ▸ **oration** n formal speech

oratorio [or-rat-**tor**-ee-oh] *n, pl* **-rios** musical composition for choir and orchestra, usu. with a religious theme

oratory[1] [or-rat-tree] *n* art of making speeches ▶ **oratorical** *adj*

oratory[2] *n, pl* **-ries** small private chapel

orb *n* ceremonial decorated sphere with a cross on top, carried by a monarch

orbit *n* curved path of a planet, satellite, or spacecraft around another body; sphere of influence ▷ *v* **orbiting, orbited** move in an orbit around; put (a satellite or spacecraft) into orbit ▶ **orbital** *adj*

orca (*Zool*) same as **killer whale**

orchard *n* area where fruit trees are grown

orchestra *n* large group of musicians, esp. playing a variety of instruments; (also **orchestra pit**) area of a theatre in front of the stage, reserved for the musicians ▶ **orchestral** *adj* ▶ **orchestrate** *v* arrange (music) for orchestra; organize (something) to produce a particular result ▶ **orchestration** *n*

orchid *n* plant with flowers that have unusual lip-shaped petals

ordain *v* make (someone) a member of the clergy; order or establish with authority

ordeal *n* painful or difficult experience

order *n* instruction to be carried out; methodical arrangement or sequence; established social system; condition of a law-abiding society; request for goods to be supplied; kind, sort; religious society of monks or nuns ▷ *v* give an instruction to; request (something) to be supplied **in order** so that it is possible ▶ **orderly** *adj* well-organized; well-

behaved ▷ *n, pl* **-lies** male hospital attendant ▶ **orderliness** *n*

ordinal number *n* number showing a position in a series e.g. *first; second*

ordinance *n* official rule or order

ordinary *adj* usual or normal; dull or commonplace ▶ **ordinarily** *adv*

ordination *n* act of making someone a member of the clergy

ordnance *n* weapons and military supplies ▶ **Ordnance Survey** official organization making maps of Britain

ordure *n* excrement

ore *n* (rock containing) a mineral which yields metal

oregano [or-rig-**gah**-no] *n* sweet-smelling herb used in cooking

organ *n* part of an animal or plant that has a particular function, such as the heart or lungs; musical keyboard instrument in which notes are produced by forcing air through pipes; means of conveying information, esp. a newspaper ▶ **organist** *n* organ player

organdie *n* fine cotton fabric

organic *adj* of or produced from animals or plants; grown without artificial fertilizers or pesticides; (*Chem*) relating to compounds of carbon; organized systematically ▶ **organically** *adv* ▶ **organism** *n* any living animal or plant

organize *v* make arrangements for; arrange systematically ▶ **organization** *n* group of people working together; act of organizing ▶ **organizational** *adj* ▶ **organizer** *n*

orgasm *n* most intense point of sexual pleasure ▶ **orgasmic** *adj*

orgy *n, pl* **-gies** party involving promiscuous sexual activity; unrestrained indulgence e.g. *an orgy of destruction* ▶ **orgiastic** *adj*

oriel window n upper window built out from a wall

Orient n the Orient (lit) East Asia ▶ **Oriental** adj ▶ **Orientalist** n specialist in the languages and history of the Far East

orient, orientate v position (oneself) according to one's surroundings; position (a map) in relation to the points of the compass ▶ **orientation** n ▶ **orienteering** n sport in which competitors hike over a course using a compass and a map

orifice [or-rif-fiss] n opening or hole

origami [or-rig-gah-mee] n Japanese decorative art of paper folding

origin n point from which something develops; ancestry ▶ **original** adj first or earliest; new, not copied or based on something else; able to think up new ideas ▷ n first version, from which others are copied ▶ **original sin** human imperfection and mortality as a result of Adam's disobedience ▶ **originality** n ▶ **originally** adv ▶ **originate** v come or bring into existence ▶ **origination** n ▶ **originator** n

oriole n tropical or American songbird

ormolu n gold-coloured alloy used for decoration

ornament n decorative object ▷ v decorate ▶ **ornamental** adj ▶ **ornamentation** n

ornate adj highly decorated, elaborate

ornithology n study of birds ▶ **ornithological** adj ▶ **ornithologist** n

orphan n child whose parents are dead ▶ **orphanage** n children's home for orphans ▶ **orphaned** adj having no living parents

orrery n, pl -ries mechanical model of the solar system

orris n kind of iris; (also **orris root**) fragrant root used for perfume

orthodontics n branch of dentistry concerned with correcting irregular teeth ▶ **orthodontist** n

orthodox adj conforming to established views ▶ **orthodoxy** n ▶ **Orthodox Church** dominant Christian Church in Eastern Europe

orthography n correct spelling

orthopaedics n branch of medicine concerned with disorders of the bones or joints ▶ **orthopaedic** adj

oryx n large African antelope

Oscar n award in the form of a statuette given for achievements in films

oscillate [oss-ill-late] v swing back and forth ▶ **oscillation** n ▶ **oscillator** n ▶ **oscilloscope** [oss-sill-oh-scope] n instrument that shows the shape of a wave on a cathode-ray tube

osier [oh-zee-er] n willow tree

osmium n (Chem) heaviest known metallic element

osmosis n movement of a liquid through a membrane from a lower to a higher concentration; process of subtle influence ▶ **osmotic** adj

osprey n large fish-eating bird of prey

ossify v -fying, -fied (cause to) become bone, harden; become inflexible ▶ **ossification** n

ostensible adj apparent, seeming ▶ **ostensibly** adv

ostentation n pretentious display ▶ **ostentatious** adj ▶ **ostentatiously** adv

osteopathy n medical treatment involving manipulation of the joints ▶ **osteopath** n

osteoporosis n brittleness of the bones, caused by lack of calcium

ostracize v exclude (a person) from a group ▶ **ostracism** n

ostrich n large African bird that runs fast but cannot fly

OT Old Testament

other adj remaining in a group of which one or some have been specified; different from the ones specified or understood; additional ▶ n other person or thing ▶ **otherwise** conj or else, if not ▶ adv differently, in another way ▶ **otherworldly** adj concerned with spiritual rather than practical matters

otiose [oh-tee-oze] adj not useful e.g. otiose language

otter n small brown freshwater mammal that eats fish

ottoman n, pl **-mans** storage chest with a padded lid for use as a seat ▶ **Ottoman** n, adj (Hist) (member) of the former Turkish empire

oubliette [oo-blee-ett] n dungeon entered only by a trapdoor

ouch interj exclamation of sudden pain

ought v used as an auxiliary to express obligation e.g. you ought to pay; used as an auxiliary to express advisability e.g. you ought to diet; used as an auxiliary to express probability e.g. you ought to know by then

> **USAGE NOTE**
> In standard English did and had are not used with ought: ought not to (not didn't/hadn't ought to)

Ouija board® n lettered board on which supposed messages from the dead are spelt out

ounce n unit of weight equal to one sixteenth of a pound (28.4 grams)

our adj belonging to us ▶ **ours** pron thing(s) belonging to us ▶ **ourselves** pron emphatic and reflexive form of **we, us**

ousel n see **dipper**

oust v force (someone) out, expel

out adv, adj denoting movement or distance away from, a state of being used up or extinguished, public availability, etc. e.g. oil was pouring out; turn the light out; her new book is out ▶ v (informal) name (a public figure) as being homosexual **out of** at or to a point outside ▶ **out-of-date** adj old-fashioned ▶ **outer** adj on the outside ▶ **outermost** adj furthest out ▶ **outer space** space beyond the earth's atmosphere ▶ **outing** n leisure trip ▶ **outward** adj apparent ▶ adv (also **outwards**) away from somewhere ▶ **outwardly** adv

out- prefix surpassing e.g. outlive; outdistance

outback n remote bush country of Australia

outbid v offer a higher price than

outboard motor n engine externally attached to the stern of a boat

outbox n (on a computer) folder in a mailbox in which outgoing messages are stored and displayed

outbreak n sudden occurrence (of something unpleasant)

outburst n sudden expression of emotion

outcast n person rejected by a particular group

outclass v surpass in quality

outcome n result

outcrop n part of a rock formation that sticks out of the earth

outcry n, pl **-cries** vehement or widespread protest

outdated adj old-fashioned

outdo v surpass in performance

outdoors adv in(to) the open air ▶ n the open air ▶ **outdoor** adj

outface v subdue or disconcert (someone) by staring

outfield n (Cricket) area far from the pitch

outfit n matching set of clothes; (informal) group of people working together ▶ **outfitter** n supplier of men's clothes

outflank v get round the side of (an enemy army); outdo (someone)

outgoing adj leaving; sociable ▶ **outgoings** pl n expenses

outgrow v become too large or too old for ▶ **outgrowth** n natural development

outhouse n building near a main building

outlandish adj extremely unconventional

outlaw n (Hist) criminal deprived of legal protection, bandit ▶ v make illegal; (Hist) make (someone) an outlaw

outlay n expenditure

outlet n means of expressing emotion; market for a product; place where a product is sold; opening or way out

outline n short general explanation; line defining the shape of something ▶ v summarize; show the general shape of

outlook n attitude; probable outcome

outlying adj distant from the main area

outmanoeuvre v get an advantage over

outmoded adj no longer fashionable or accepted

outnumber v exceed in number

outpatient n patient who does not stay in hospital overnight

outpost n outlying settlement

outpouring n passionate outburst

output n amount produced; power, voltage, or current delivered by an electrical circuit; (Computing) data produced ▶ v (Computing) produce (data) at the end of a process

outrage n great moral indignation; gross violation of morality ▶ v offend morally ▶ **outrageous** adj shocking; offensive ▶ **outrageously** adv

outré [oo-tray] adj shockingly eccentric

outrider n motorcyclist acting as an escort

outrigger n stabilizing frame projecting from a boat

outright adj, adv absolute(ly); open(ly) and direct(ly)

outrun v run faster than; exceed

outset n beginning

outshine v surpass (someone) in excellence

outside prep, adj, adv indicating movement to or position on the exterior ▶ adj unlikely e.g. an outside chance; coming from outside ▶ n external area or surface ▶ **outsider** n person outside a specific group; contestant thought unlikely to win

> **USAGE NOTE**
> Outside is not followed by of in standard English

outsize, outsized adj larger than normal

outskirts pl n outer areas, esp. of a town

outsmart v (informal) outwit

outspan n (SAfr) relax

outspoken adj tending to say what one thinks; said openly

outstanding adj excellent; still to be dealt with or paid

outstrip v surpass; go faster than

outtake n unreleased take from a recording session, film, or TV programme

outweigh v be more important, significant, or influential than

outwit v -witting, -witted get the better of (someone) by cunning

ouzel [ooze-el] n see **dipper**

ova n plural of **ovum**

oval adj egg-shaped ▷ n anything that is oval in shape

ovary n, pl **-ries** female egg-producing organ ► **ovarian** adj

ovation n enthusiastic round of applause

oven n heated compartment or container for cooking or for drying or firing ceramics

over prep, adv indicating position on the top of, movement to the other side of, amount greater than, etc. e.g. a room over the garage; climbing over the fence; over fifty pounds ▷ n (Cricket) series of six balls bowled from one end ► **overly** adv excessively

over- prefix too much or e.g. overeat; above e.g. overlord; on top e.g. overshoe

overall adj, adv in total ▷ n coat-shaped protective garment ▷ pl protective garment consisting of trousers with a jacket or bib and braces attached

overarm adj, adv (thrown) with the arm above the shoulder

overawe v affect (someone) with an overpowering sense of awe

overbalance v lose balance

overbearing adj unpleasantly forceful

overblown adj excessive

overboard adv from a boat into the water **go overboard** go to extremes, esp. in enthusiasm

overcast adj (of the sky) covered by clouds

overcoat n heavy coat

overcome v gain control over after an effort; (of an emotion) affect strongly

overcrowded adj containing more people or things than is desirable

overdo v do to excess; exaggerate (something) **overdo it** do something to a greater degree than is advisable

overdose n excessive dose of a drug ▷ v take an overdose

overdraft n overdrawing; amount overdrawn

overdraw v withdraw more money than is in (one's bank account)

overdrawn adj having overdrawn one's account; (of an account) in debit

overdrive n very high gear in a motor vehicle

overdue adj still due after the time allowed

overgrown adj thickly covered with plants and weeds

overhaul v examine and repair ▷ n examination and repair

overhead adj, adv above one's head ► **overheads** pl n general cost of maintaining a business

overhear v hear (a speaker or remark) unintentionally or without the speaker's knowledge

overjoyed adj extremely pleased

overkill n treatment that is greater than required

overland adj, adv by land

overlap v share part of the same space or period of time (as) ▷ n area overlapping

overleaf adv on the back of the current page

overlook v fail to notice; ignore; look at from above

overnight adj, adv (taking place) during one night; (happening) very quickly

overpower v subdue or overcome (someone)

overreach v **overreach oneself** fail by trying to be too clever

override v overrule; replace

o

overrule v reverse the decision of (a person with less power); reverse (someone else's decision)

overrun v spread over (a place) rapidly; extend beyond a set limit

overseas adv, adj to, of, or from a distant country

oversee v watch over from a position of authority ▸ **overseer** n

overshadow v reduce the significance of (a person or thing) by comparison; sadden the atmosphere of

oversight n mistake caused by not noticing something

overspill n (Brit) rehousing of people from crowded cities in smaller towns

overstay v **overstay one's welcome** stay longer than one's host or hostess would like ▸ **overstayer** n (NZ) person who remains in New Zealand after their permit has expired

overt adj open, not hidden ▸ **overtly** adv

overtake v move past (a vehicle or person) travelling in the same direction

overthrow v defeat and replace ▷ n downfall, destruction

overtime n, adv (paid work done) in addition to one's normal working hours

overtone n additional meaning

overture n (Music) orchestral introduction ▷ pl opening moves in a new relationship

overturn v turn upside down; overrule (a legal decision); overthrow (a government)

overweight adj weighing more than is healthy

overwhelm v overpower, esp. emotionally; defeat by force ▸ **overwhelming** adj ▸ **overwhelmingly** adv

overwrought adj nervous and agitated

ovoid [oh-void] adj egg-shaped

ovulate [ov-yew-late] v produce or release an egg cell from an ovary ▸ **ovulation** n

ovum [oh-vum] n, pl **ova** unfertilized egg cell

owe v be obliged to pay (a sum of money) to (a person) **owing to** as a result of

owl n night bird of prey ▸ **owlish** adj

own adj used to emphasize possession e.g. my own idea ▷ v possess ▸ **owner** n ▸ **ownership** n ▸ **own up** v confess

ox n, pl **oxen** castrated bull

Oxbridge n (Brit) universities of Oxford and Cambridge considered together

Oxfam Oxford Committee for Famine Relief

oxide n compound of oxygen and one other element ▸ **oxidize** v combine chemically with oxygen, as in burning or rusting

oxygen n (Chem) gaseous element essential to life and combustion ▸ **oxygenate** v add oxygen to

oxymoron [ox-see-more-on] n figure of speech that combines two apparently contradictory ideas e.g. cruel kindness

oyez interj (Hist) shouted three times by a public crier, listen

oyster n edible shellfish ▸ **oystercatcher** n wading bird with black-and-white feathers

Oz n (slang) Australia

oz. ounce

ozone n strong-smelling form of oxygen ▸ **ozone layer** layer of ozone in the upper atmosphere that filters out ultraviolet radiation

Pp

P parking

p (*Brit, Aust & NZ*) penny; (*Brit*) pence

p. (*pl* **pp.**) page

PA personal assistant; public-address system

pa *n* (*NZ*) (formerly) a fortified Māori settlement

p.a. per annum: each year

pace *n* single step in walking; length of a step; rate of progress ▷ *v* walk up and down, esp. in anxiety; (foll by *out*) cross or measure with steps ▶ **pacemaker** *n* electronic device surgically implanted in a person with heart disease to regulate the heartbeat; person who, by taking the lead early in a race, sets the pace for the rest of the competitors

pachyderm [pak-ee-durm] *n* thick-skinned animal such as an elephant

pacifist *n* person who refuses on principle to take part in war ▶ **pacifism** *n*

pacify *v* **-fying, -fied** soothe, calm ▶ **pacification** *n*

pack *v* put (clothes etc.) together in a suitcase or bag; put (goods) into containers or parcels; fill with people or things ▷ *n* bag carried on a person's or animal's back; (*chiefly US*)same as **packet**; set of playing cards; group of dogs or wolves that hunt together ▶ **pack ice** mass of floating ice in the sea ▶ **pack in** (*informal*) stop doing ▶ **pack off** *v* (*informal*) send away

package *n* small parcel; (*also* **package deal**) deal in which separate items are presented together as a unit ▷ *v* put into a package ▶ **packaging** *n* ▶ **package holiday** holiday in which everything is arranged by one company for a fixed price

packet *n* small container (and contents); small parcel; (*slang*) large sum of money

packhorse *n* horse used for carrying goods

pact *n* formal agreement

pad *n* piece of soft material used for protection, support, absorption of liquid, etc.; number of sheets of paper fastened at the edge; fleshy underpart of an animal's paw; place for launching rockets; (*slang*) home ▷ *v* **padding, padded** protect or fill with soft material; walk with soft steps ▶ **padding** *n* soft material used to pad something; unnecessary words put into a speech or written work to make it longer

paddle¹ *n* short oar with a broad blade at one or each end ▷ *v* move (a canoe etc.) with a paddle ▶ **paddle steamer** ship propelled by paddle wheels ▶ **paddle wheel** wheel with crosswise blades that strike the water successively to propel a ship

paddle² *v* walk barefoot in shallow water

paddock *n* small field or enclosure for horses

paddy *n* (*Brit informal*) fit of temper

paddy field *n* field where rice is grown (*also* **paddy**)

pademelon, paddymelon [pad-ee-mel-an] *n* small Australian wallaby

padlock *n* detachable lock with a hinged hoop fastened over a ring on the object to be secured

padre [pah-dray] *n* chaplain to the armed forces

paean [pee-an] *n* song of triumph or thanksgiving

paediatrics *n* branch of medicine concerned with diseases of children ▶ **paediatrician** *n*

paedophilia *n* condition of being sexually attracted to children ▶ **paedophile** *n* person who is sexually attracted to children

paella [pie-**ell**-a] *n* Spanish dish of rice, chicken, shellfish, and vegetables

pagan *n, adj* (person) not belonging to one of the world's main religions

page¹ *n* (one side of) a sheet of paper forming a book etc.; screenful of information from a website or teletext service

page² *n* (also **pageboy**) small boy who attends a bride at her wedding; (*Hist*) boy in training for knighthood ▷ *v* summon (someone) by bleeper or loudspeaker, in order to pass on a message

pageant *n* parade or display of people in costume, usu. illustrating a scene from history ▶ **pageantry** *n*

pagination *n* numbering of the pages of a book etc.

pagoda *n* pyramid-shaped Asian temple or tower

paid *v* past of **pay**; **put paid to** (*informal*) end or destroy

pail *n* (contents of) a bucket

pain *n* physical or mental suffering ▷ *pl* trouble, effort **on pain of** subject to the penalty of ▶ **painful** *adj* ▶ **painfully** *adv* ▶ **painless** *adj* ▶ **painlessly** *adv* ▶ **painkiller** *n* drug that relieves pain

painstaking *adj* extremely thorough and careful

paint *n* coloured substance, spread on a surface with a brush or roller ▷ *v* colour or coat with paint; use paint to make a picture of ▶ **painter** *n* ▶ **painting** *n*

painter *n* rope at the front of a boat, for tying it up

pair *n* set of two things matched for use together ▷ *v* group or be grouped in twos

> **USAGE NOTE**
>
> *Pair* is followed by a singular verb if it refers to a unit: *A pair of shoes was on the floor*, and by a plural verb if it refers to two individuals: *That pair are good friends*

paisley pattern *n* pattern of small curving shapes, used in fabric

Pakeha [pah-kee-ha] *n* (*NZ*) New Zealander who is not of Māori descent

Pakistani *n, adj* (person) from Pakistan

pal *n* (*informal*) friend

palace *n* residence of a king, bishop, etc.; large grand building

palaeography [pal-ee-og-ra-fee] *n* study of ancient manuscripts

Palaeolithic [pal-ee-oh-lith-ik] *adj* of the Old Stone Age

palaeontology [pal-ee-on-tol-a-jee] *n* study of past geological periods and fossils

Palagi [pa-lang-ee] *n, pl -gis* (*NZ*) Samoan name for a Pakeha

palatable *adj* pleasant to taste

palate *n* roof of the mouth; sense of taste

palatial *adj* like a palace, magnificent

palaver [pal-lah-ver] *n* time-wasting fuss

pale¹ *adj* light, whitish; whitish in the face, esp. through illness or shock ▷ *v* become pale

pale² *n* wooden or metal post used in fences **beyond the pale** outside the limits of social convention

palette n artist's flat board for mixing colours on

palindrome n word or phrase that reads the same backwards as forwards

paling n wooden or metal post used in fences

palisade n fence made of wooden posts driven into the ground

pall[1] n cloth spread over a coffin; dark cloud (of smoke); depressing oppressive atmosphere ▶ **pallbearer** n person who helps to carry the coffin at a funeral

pall[2] v become boring

palladium n (Chem) silvery-white element of the platinum metal group

pallet[1] n portable platform for storing and moving goods

pallet[2] n straw-filled mattress or bed

palliate v lessen the severity of (something) without curing it

palliative adj giving temporary or partial relief ▶ n something, for example a drug, that palliates

pallid adj pale, esp. because ill or weak ▶ **pallor** n

pally adj **-lier, -liest** (informal) on friendly terms

palm[1] n inner surface of the hand ▶ **palm off** v get rid of (an unwanted thing or person), esp. by deceit

palm[2] n tropical tree with long pointed leaves growing out of the top of a straight trunk ▶ **Palm Sunday** Sunday before Easter

palmistry n fortune-telling from lines on the palm of the hand ▶ **palmist** n

palmtop adj (of a computer) small enough to be held in the hand ▶ n computer small enough to be held in the hand

palomino n, pl **-nos** gold-coloured horse with a white mane and tail

palpable adj obvious e.g. a palpable hit; so intense as to seem capable of being touched e.g. the tension is almost palpable ▶ **palpably** adv

palpate v (Med) examine (an area of the body) by touching

palpitate v (of the heart) beat rapidly; flutter or tremble ▶ **palpitation** n

palsy [pawl-zee] n paralysis ▶ **palsied** adj affected with palsy

paltry adj **-trier, -triest** insignificant

pampas pl n vast grassy plains in S America ▶ **pampas grass** tall grass with feathery ornamental flower branches

pamper v treat (someone) with great indulgence, spoil

pamphlet n thin paper-covered booklet ▶ **pamphleteer** n writer of pamphlets

pan[1] n wide long-handled metal container used in cooking; bowl of a toilet ▶ v **panning, panned** sift gravel from (a river) in a pan to search for gold; (informal) criticize harshly ▶ **pan out** v result

pan[2] v **panning, panned** (of a film camera) be moved slowly so as to cover a whole scene or follow a moving object

pan- combining form all e.g. pan-American

panacea [pan-a-see-a] n remedy for all diseases or problems

panache [pan-ash] n confident elegant style

panama hat n straw hat

panatella n long slender cigar

pancake n thin flat circle of fried batter

panchromatic adj (Photog) sensitive to light of all colours

pancreas [pang-kree-ass] n large gland behind the stomach that produces insulin and helps digestion ▶ **pancreatic** adj

panda n large black-and-white bearlike mammal from China ▸ **panda car** (Brit) police patrol car

pandemic n, adj (a disease) occurring over a wide area

pandemonium n wild confusion, uproar

pander¹ v (foll by to) indulge (a person in his or her desires)

pander² n (old-fashioned) person who procures a sexual partner for someone

p & p postage and packing

pane n sheet of glass in a window or door

panegyric [pan-ih-jir-ik] n formal speech or piece of writing in praise of someone or something

panel n flat distinct section of a larger surface, for example in a door; group of people as a team in a quiz etc.; list of jurors, doctors, etc.; board or surface containing switches and controls to operate equipment ▸ v **-elling, -elled** cover or decorate with panels ▸ **panelling** n panels collectively, esp. on a wall ▸ **panellist** n member of a panel ▸ **panel beater** person who repairs damage to car bodies

pang n sudden sharp feeling of pain or sadness

pangolin n animal of tropical countries with a scaly body and a long snout for eating ants and termites (also **scaly anteater**)

panic n sudden overwhelming fear, often affecting a whole group of people ▸ v **-icking, -icked** feel or cause to feel panic ▸ **panicky** adj ▸ **panic-stricken** adj

panini [pa-nee-nee] n, pl **-ni** or **-nis** type of Italian sandwich, usu. served grilled

pannier n bag fixed on the back of a cycle; basket carried by a beast of burden

panoply n magnificent array

panorama n wide unbroken view of a scene ▸ **panoramic** adj

pansy n, pl **-sies** small garden flower with velvety purple, yellow, or white petals; (offens) effeminate or homosexual man

pant v breathe quickly and noisily during or after exertion

pantaloons pl n baggy trousers gathered at the ankles

pantechnicon n large van for furniture removals

pantheism n belief that God is present in everything ▸ **pantheist** n ▸ **pantheistic** adj

pantheon n (in ancient Greece and Rome) temple built to honour all the gods

panther n leopard, esp. a black one

panties pl n women's underpants

pantile n roofing tile with an S-shaped cross section

pantomime n play based on a fairy tale, performed at Christmas time

pantry n, pl **-tries** small room or cupboard for storing food

pants pl n undergarment for the lower part of the body; (US, Canadian, Aust & NZ) trousers

pap n soft food for babies or invalids; worthless entertainment or information

papacy [pay-pa-see] n, pl **-cies** position or term of office of a pope ▸ **papal** adj of the pope

paparazzo [pap-a-rat-so] n, pl **-razzi** photographer specializing in candid photographs of famous people

papaya [pa-pie-ya] n large sweet West Indian fruit

paper n material made in sheets from wood pulp or other fibres; printed sheet of this; newspaper; set of examination questions; article or essay ▸ pl personal documents ▷ v cover (walls) with wallpaper ▶ **paperback** n book with covers made of flexible card ▶ **paperweight** n heavy decorative object placed on top of loose papers ▶ **paperwork** n clerical work, such as writing reports and letters

papier-mâché [pap-yay **mash**-ay] n material made from paper mixed with paste and moulded when moist

papist n, adj (offens) Roman Catholic

papoose n Native American child

paprika n mild powdered seasoning made from red peppers

papyrus [pap-**ire**-uss] n, pl **-ri** or **-ruses** tall water plant; (manuscript written on) a kind of paper made from this plant

par n usual or average condition e.g. feeling under par; (Golf) expected standard score; face value of stocks and shares **on a par with** equal to

parable n story that illustrates a religious teaching

parabola [par-**ab**-bol-a] n regular curve resembling the course of an object thrown forward and up ▶ **parabolic** adj

paracetamol n mild pain-relieving drug

parachute n large fabric canopy that slows the descent of a person or object from an aircraft ▷ v land or drop by parachute ▶ **parachutist** n

parade n procession or march; street or promenade ▷ v display or flaunt; march in procession

paradigm [par-a-dime] n example or model

paradise n heaven; place or situation that is near-perfect

paradox n statement that seems self-contradictory but

may be true ▶ **paradoxical** adj ▶ **paradoxically** adv

paraffin n (Brit & S Afr) liquid mixture distilled from petroleum and used as a fuel or solvent

> **SPELLING TIP**
> Remember that **paraffin** has a single r and double f

paragliding n cross-country gliding wearing a parachute shaped like wings

paragon n model of perfection

paragraph n section of a piece of writing starting on a new line

parakeet n small long-tailed parrot

parallax n apparent change in an object's position due to a change in the observer's position

parallel adj separated by an equal distance at every point; exactly corresponding ▷ n line separated from another by an equal distance at every point; thing with similar features to another; line of latitude ▷ v correspond to

parallelogram n four-sided geometric figure with opposite sides parallel

paralysis n inability to move or feel, because of damage to the nervous system ▶ **paralyse** v affect with paralysis; make temporarily unable to move or take action ▶ **paralytic** n, adj (person) affected with paralysis

paramedic n person working in support of the medical profession ▶ **paramedical** adj

parameter [par-**am**-it-er] n limiting factor, boundary

paramilitary adj organized on military lines

paramount adj of the greatest importance

paramour n (old-fashioned) lover, esp. of a person married to someone else

P

paranoia n mental illness causing delusions of grandeur or persecution; (informal) intense fear or suspicion ▶ **paranoid, paranoiac** adj, n

paranormal adj beyond scientific explanation

parapet n low wall or railing along the edge of a balcony or roof

paraphernalia n personal belongings or bits of equipment

paraphrase v put (a statement or text) into other words

paraplegia [par-a-pleej-ya] n paralysis of the lower half of the body ▶ **paraplegic** adj, n

parapsychology n study of mental phenomena such as telepathy

Paraquat® n extremely poisonous weedkiller

parasite n animal or plant living in or on another; person who lives at the expense of others ▶ **parasitic** adj

parasol n umbrella-like sunshade

paratrooper n soldier trained to be dropped by parachute into a battle area ▶ **paratroops** pl n

parboil v boil until partly cooked

parcel n something wrapped up, package ▷ v **-celling, -celled** (often foll by up) wrap up ▶ **parcel out** v divide into parts

parch v make very hot and dry; make thirsty

parchment n thick smooth writing material made from animal skin

pardon v forgive, excuse ▷ n forgiveness; official release from punishment for a crime ▶ **pardonable** adj

parentage n ancestry or family

parenting n activity of bringing up children

parenthesis [par-en-thiss-iss] n, pl **-ses** word or sentence inserted into a passage, marked off by brackets or dashes ▷ pl round brackets, ( ) ▶ **parenthetical** adj

pariah [par-rye-a] n social outcast

parietal [par-rye-it-al] adj of the walls of a body cavity such as the skull

parish n area that has its own church and a priest or pastor ▶ **parishioner** n inhabitant of a parish

parity n equality or equivalence

park n area of open land for recreational use by the public; area containing a number of related enterprises e.g. a business park; (Brit) area of private land around a large country house ▷ v stop and leave (a vehicle) temporarily

parka n large waterproof jacket with a hood

Parkinson's disease n progressive disorder of the central nervous system which causes impaired muscular coordination and tremor (also **Parkinsonism**)

parky adj **parkier, parkiest** (Brit informal) (of the weather) chilly

parlance n particular way of speaking, idiom

parley n meeting between leaders or representatives of opposing forces to discuss terms ▷ v have a parley

parliament n law-making assembly of a country ▶ **parliamentary** adj

parlour n (old-fashioned) living room for receiving visitors

parlous adj (old-fashioned) dire; dangerously bad

Parmesan n hard strong-flavoured Italian cheese, used grated on pasta dishes and soups

parent n father or mother ▶ **parental** adj ▶ **parenthood** n

p

parochial adj narrow in outlook; of a parish ▸ **parochialism** n

parody n, pl **-dies** exaggerated and amusing imitation of someone else's style ▸ v **-dying, -died** make a parody of

parole n early freeing of a prisoner on condition that he or she behaves well ▸ v put on parole **on parole** (of a prisoner) released on condition that he or she behaves well

paroxysm n uncontrollable outburst of rage, delight, etc.; spasm or convulsion of coughing, pain, etc.

parquet [par-kay] n floor covering made of wooden blocks arranged in a geometric pattern ▸ **parquetry** n

parricide n crime of killing either of one's parents; person who does this

parrot n tropical bird with a short hooked beak and an ability to imitate human speech ▸ v **-roting, -roted** repeat (someone else's words) without thinking

parry v **-rying, -ried** ward off (an attack); cleverly avoid (an awkward question)

parse [parz] v analyse (a sentence) in terms of grammar

parsimony n extreme caution in spending money ▸ **parsimonious** adj

parsley n herb used for seasoning and decorating food

parsnip n long tapering cream-coloured root vegetable

parson n Anglican parish priest; any member of the clergy ▸ **parsonage** n parson's house

part n one of the pieces that make up a whole; one of several equal divisions; actor's role; (often pl) region, area; component of a vehicle or machine ▸ v divide or separate; (of people) leave each other **take someone's part** support someone in an argument etc. **take (something) in good part** respond to (teasing or criticism) with good humour ▸ **parting** n occasion when one person leaves another; line of scalp between sections of hair combed in opposite directions; dividing or separating ▸ **partly** adv not completely ▸ **part of speech** particular grammatical class of words, such as noun or verb ▸ **part-time** adj occupying or working less than the full working week ▸ **part with** v give away, hand over

partake v **-taking, -took, -taken** (foll by of) take (food or drink); (foll by in) take part in

partial adj not complete; prejudiced **partial to** having a liking for ▸ **partiality** n ▸ **partially** adv

participate v become actively involved ▸ **participant** n ▸ **participation** n

participle n form of a verb used in compound tenses or as an adjective e.g. worried; worrying

particle n extremely small piece or amount; (Physics) minute piece of matter, such as a proton or electron

particular adj relating to one person or thing, not general; exceptional or special; very exact; difficult to please, fastidious ▸ n item of information, detail ▸ **particularly** adv ▸ **particularize** v give details about

partisan n strong supporter of a party or group; guerrilla, member of a resistance movement ▸ adj prejudiced or one-sided

partition n screen or thin wall that divides a room; division of a country into independent parts ▷ v divide with a partition

partner n either member of a couple in a relationship or activity; member of a business partnership ▷ v be the partner of ▶ **partnership** n joint business venture between two or more people

partridge n game bird of the grouse family

parturition n act of giving birth

party n, pl **-ties** social gathering for pleasure; group of people travelling or working together; political organization whose members have common aims and beliefs; person or people forming one side in a lawsuit or contest ▶ **party line** official view of a political party; telephone line shared by two or more subscribers ▶ **party wall** common wall separating adjoining buildings

parvenu [par-ven-new] n person newly risen to a position of power or wealth

pascal n unit of pressure

pash (Aust & NZ slang) v kiss and cuddle ▷ n act of kissing and cuddling

pashmina [pash-mee-na] n shawl or scarf made from fine soft goat's wool

paspalum [pass-pale-um] n (Aust & NZ) type of grass with wide leaves

pass v go by, past, or through; be successful in (a test or examination); spend (time) or (of time) go by; give, hand; be inherited by; (Sport) hit, kick, or throw (the ball) to another player; (of a law-making body) agree to (a law); exceed ▷ n successful result in a test or examination; permit or licence **make a pass at** (informal) make sexual advances to ▶ **passable** adj (just) acceptable; (of a road) capable of being travelled along

▶ **passing** adj brief or transitory; cursory or casual ▶ **pass away** v die

pass out v (informal) faint ▶ **pass up** v (informal) fail to take advantage of (something)

passage n channel or opening providing a way through; hall or corridor; section of a book etc.; journey by sea; right or freedom to pass ▶ **passageway** n passage or corridor

passbook n book issued by a bank or building society for keeping a record of deposits and withdrawals; (SAfr) formerly, an official identity document

passé [pas-say] adj old-fashioned

passenger n person travelling in a vehicle driven by someone else; member of a team who does not pull his or her weight

passer-by n, pl passers-by person who is walking past something or someone

passim adv (Latin) everywhere, throughout

passion n intense sexual love; any strong emotion; great enthusiasm; (P-) (Christianity) the suffering of Christ ▶ **passionate** adj ▶ **passionflower** n tropical American plant ▶ **passion fruit** edible fruit of the passionflower ▶ **Passion play** play about Christ's suffering

passive adj not playing an active part; submissive and receptive to outside forces; (Grammar) (of a verb) in a form indicating that the subject receives the action, e.g. was jeered in he was jeered by the crowd ▶ **passivity** n ▶ **passive resistance** resistance to a government, law, etc. by nonviolent acts ▶ **passive smoking** inhalation of smoke from others' cigarettes by a nonsmoker

Passover n Jewish festival commemorating the sparing of the Jews in Egypt

passport n official document of nationality granting permission to travel abroad

password n secret word or phrase that allows access

past adj of the time before the present; ended, gone by; (Grammar) (of a verb tense) indicating that the action took place earlier ▷ n period of time before the present; person's earlier life, esp. a disreputable period; (Grammar) past tense ▷ adv by, along ▷ prep beyond **past it** (informal) unable to do the things one could do when younger ▶ **past master** person with great talent or experience in a particular subject

pasta n type of food, such as spaghetti, that is made in different shapes from flour and water

paste n moist soft mixture, such as toothpaste; adhesive, esp. for paper; (Brit) pastry dough; shiny glass used to make imitation jewellery ▷ v fasten with paste ▶ **pasting** n (informal) heavy defeat; strong criticism ▶ **pasteboard** n stiff thick paper

pastel n coloured chalk crayon for drawing; picture drawn in pastels; pale delicate colour ▷ adj pale and delicate in colour

pasteurize v sterilize by heating ▶ **pasteurization** n

pastiche [pass-**teesh**] n work of art that mixes styles or copies the style of another artist

pastille n small fruit-flavoured and sometimes medicated sweet

pastime n activity that makes time pass pleasantly

pastor n member of the clergy in charge of a congregation

▶ **pastoral** adj of or depicting country life; of a member of the clergy or his or her duties

pastrami n highly seasoned smoked beef

pastry n, pl-**ries** baking dough made of flour, fat, and water; cake or pie

pasture n grassy land for farm animals to graze on

pasty¹ [**pay**-stee] adj pastier, pastiest (of a complexion) pale and unhealthy

pasty² [pass-tee] n, pl **pasties** round of pastry folded over a savoury filling

pat¹ v patting, patted tap lightly ▷ n gentle tap or stroke; small shaped mass of butter etc.

pat² adj quick, ready, or glib **off pat** learned thoroughly

patch n piece of material sewn on a garment; small contrasting section; plot of ground; protective pad for the eye ▷ v mend with a patch ▶ **patchy** adj of uneven quality or intensity ▶ **patch up** v repair clumsily; make up (a quarrel) ▶ **patchwork** n needlework made of pieces of different materials sewn together

pate n (old-fashioned) head

pâté [pat-ay] n spread of finely minced liver etc.

patella n, pl-**lae** kneecap

patent n document giving the exclusive right to make or sell an invention ▷ adj open to public inspection e.g. letters patent; obvious; protected by a patent ▷ v obtain a patent for ▶ **patently** adv obviously ▶ **patent leather** leather processed to give a hard glossy surface

paternal adj fatherly; related through one's father ▶ **paternity** n fact or state of being a father ▶ **paternalism** n authority exercised in a way that limits individual responsibility ▶ **paternalistic** adj

path n surfaced walk or track; course of action ▶ **pathname** n (Computing) file name listing the sequence of directories leading to a particular file or directory

pathetic adj causing feelings of pity or sadness; distressingly inadequate ▶ **pathetically** adv

pathogen n thing that causes disease ▶ **pathogenic** adj

pathology n scientific study of diseases ▶ **pathological** adj of pathology; (informal) compulsively motivated ▶ **pathologist** n

pathos n power of arousing pity or sadness

patient adj enduring difficulties or delays calmly ▶ n person receiving medical treatment ▶ **patience** n quality of being patient; card game for one

patina n fine layer on a surface; sheen of age on woodwork

patio n, pl **-tios** paved area adjoining a house

patois [pat-wah] n, pl **patois** [pat-wahz] regional dialect, esp. of French

patriarch n male head of a family or tribe; highest-ranking bishop in Orthodox Churches ▶ **patriarchal** adj ▶ **patriarchy** n, pl **-chies** society in which men have most of the power

patrician n member of the nobility ▶ adj of noble birth

patricide n crime of killing one's father; person who does this

patrimony n, pl **-nies** property inherited from ancestors

patriot n person who loves his or her country and supports its interests ▶ **patriotic** adj ▶ **patriotism** n

patrol n regular circuit by a guard; person or small group patrolling; unit of Scouts or Guides ▶ v **-trolling, -trolled** go round on guard, or reconnoitring

patron n person who gives financial support to charities, artists, etc.; regular customer of a shop, pub, etc. ▶ **patronage** n support given by a patron ▶ **patronize** v treat in a condescending way; be a patron of ▶ **patron saint** saint regarded as the guardian of a country or group

patronymic n name derived from one's father or a male ancestor

patter[1] v make repeated soft tapping sounds ▶ n quick succession of taps

patter[2] n glib rapid speech

pattern n arrangement of repeated parts or decorative designs; regular way that something is done; diagram or shape used as a guide to make something ▶ **patterned** adj decorated with a pattern

patty n, pl **-ties** small flattened cake of minced food

paucity n scarcity; smallness of amount or number

paunch n protruding belly

pauper n very poor person

pause v stop for a time ▶ n stop or rest in speech or action

pave v form (a surface) with stone or brick ▶ **pavement** n (Brit) paved path for pedestrians

pavilion n building on a playing field etc.; building for housing an exhibition etc.

paw n animal's foot with claws and pads ▶ v scrape with the paw or hoof; (informal) touch in a rough or overfamiliar way

pawn[1] v deposit (an article) as security for money borrowed **in pawn** deposited as security with a pawnbroker ▶ **pawnbroker**

n lender of money on goods deposited

pawn² *n* chessman of the lowest value; person manipulated by someone else

pay *v* **paying, paid** give money etc. in return for goods or services; settle a debt or obligation; compensate (for); give; be profitable to ▷ *n* wages or salary ▶ **payment** *n* act of paying; money paid ▶ **payable** *adj* due to be paid ▶ **payee** *n* person to whom money is paid or due ▶ **payday** *n* day on which wages or salary is paid ▶ **paying guest** lodger or boarder ▶ **pay off** *v* pay (debt) in full; turn out successfully ▶ **pay out** *v* spend; release (a rope) bit by bit ▶ **payroll** *n* list of employees who receive regular pay ▶ **paywall** *n* system preventing access to part of a website unless a fee is paid

PAYE pay as you earn: system by which income tax is paid by an employer straight to the government

payload *n* passengers or cargo of an aircraft; explosive power of a missile etc.

payola *n* (*chiefly US informal*) bribe to get special treatment, esp. to promote a commercial product

PC personal computer; (in Britain) Police Constable; politically correct; (in Britain) Privy Councillor

pc per cent

PDA personal digital assistant: a handheld computer used for personal information such as contact details

PDF (*Computing*) portable document format: a format in which documents may be viewed

PE physical education

pea *n* climbing plant with seeds growing in pods; its seed, eaten as a vegetable

peace *n* calm, quietness; absence of anxiety; freedom from war; harmony between people ▶ **peaceable** *adj* inclined towards peace ▶ **peaceably** *adv* ▶ **peaceful** *adj* ▶ **peacefully** *adv*

peach *n* soft juicy fruit with a stone and a downy skin; (*informal*) very pleasing person or thing ▷ *adj* pinkish-orange

peacock *n* large male bird with a brilliantly coloured fanlike tail ▶ **peahen** *n fem*

peak *n* pointed top, esp. of a mountain; point of greatest development etc.; projecting piece on the front of a cap ▷ *v* form or reach a peak ▷ *adj* of or at the point of greatest demand ▶ **peaked** *adj* ▶ **peaky** *adj* pale and sickly

peal *n* long loud echoing sound, esp. of bells or thunder ▷ *v* sound with a peal or peals

peanut *n* pea-shaped nut that ripens underground ▷ *pl* (*informal*) trifling amounts of money

pear *n* sweet juicy fruit with a narrow top and rounded base

pearl *n* hard round shiny object found inside some oyster shells and used as a jewel ▶ **pearly** *adj*

peasant *n* person working on the land, esp. in poorer countries or in the past ▶ **peasantry** *n* peasants collectively

peat *n* decayed vegetable material found in bogs, used as fertilizer or fuel

pebble *n* small roundish stone ▶ **pebbly** *adj* ▶ **pebble dash** coating for exterior walls consisting of small stones set in plaster

pecan [pee-kan] *n* edible nut of a N American tree

peccadillo *n, pl* **-loes** *or* **-los** trivial misdeed

peck *v* strike or pick up with the beak; (*informal*) kiss quickly ▷ *n* pecking movement ▸ **peckish** adj (informal) slightly hungry ▸ **peck at** v nibble, eat reluctantly

pecs pl n (informal) pectoral muscles

pectin n substance in fruit that makes jam set

pectoral adj of the chest or thorax ▷ n pectoral muscle or fin

peculiar adj strange; distinct, special; belonging exclusively to ▸ **peculiarity** n, pl **-ties** oddity, eccentricity; distinguishing trait

pecuniary adj relating to, or consisting of, money

pedagogue n schoolteacher, esp. a pedantic one

pedal n foot-operated lever used to control a vehicle or machine, or to modify the tone of a musical instrument ▷ v **-alling, -alled** propel (a bicycle) by using its pedals

pedant n person who is excessively concerned with details and rules, esp. in academic work ▸ **pedantic** adj ▸ **pedantry** n

peddle v sell (goods) from door to door

peddler n person who sells illegal drugs

pederast n man who has homosexual relations with boys ▸ **pederasty** n

pedestal n base supporting a column, statue, etc.

pedestrian n person who walks ▷ adj dull, uninspiring ▸ **pedestrian crossing** place marked where pedestrians may cross a road ▸ **pedestrian precinct** (Brit) (shopping) area for pedestrians only

pedicure n medical or cosmetic treatment of the feet

pedigree n register of ancestors, esp. a purebred animal

pediment n triangular part over a door etc.

pedlar n person who sells goods from door to door

pee (informal) ▷ v **peeing, peed** urinate ▷ n urine; act of urinating

peek v, n peep or glance

peel v remove the skin or rind of (a vegetable or fruit); (of skin or a surface) come off in flakes ▷ n rind or skin ▸ **peelings** pl n

peep¹ v look slyly or quickly ▷ n quick or furtive look ▸ **Peeping Tom** man who furtively watches women undressing

peep² v make a small shrill noise ▷ n small shrill noise

peer¹ n (fem **peeress**) (in Britain) member of the nobility; person of the same status, age, etc. ▸ **peerage** n (Brit) whole body of peers; rank of a peer ▸ **peerless** adj unequalled, unsurpassed ▸ **peer group** group of people of similar age, status, etc. ▸ **peer pressure** influence from one's peer group

peer² v look closely and intently

peeved adj (informal) annoyed

peevish adj fretful or irritable ▸ **peevishly** adv

peewee n black-and-white Australian bird

peewit n same as **lapwing**

peg n pin or clip for joining, fastening, marking, etc.; hook or knob for hanging things on ▷ v **pegging, pegged** fasten with pegs; stabilize (prices) **off the peg** (of clothes) ready-to-wear, not tailor-made

peggy square n (NZ) small hand-knitted square

peignoir [pay-nwahr] n woman's light dressing gown

pejorative [pij-jor-a-tiv] adj (of words etc.) with an insulting or critical meaning

Pekingese, Pekinese n, pl **-ese** small dog with a short wrinkled muzzle

pelargonium n plant with red, white, purple, or pink flowers

pelican n large water bird with a pouch beneath its bill for storing fish ▶ **pelican crossing** (in Britain) road crossing with pedestrian-operated traffic lights

pellagra n disease caused by lack of vitamin B

pellet n small ball of something

pell-mell adv in utter confusion, headlong

pellucid adj very clear

pelmet n ornamental drapery or board, concealing a curtain rail

pelt[1] v throw missiles at; run fast, rush; rain heavily **at full pelt** at top speed

pelt[2] n skin of a fur-bearing animal

pelvis n framework of bones at the base of the spine, to which the hips are attached ▶ **pelvic** adj

pen[1] n instrument for writing in ink ▷ v **penning, penned** write or compose ▶ **pen friend** friend with whom a person corresponds without meeting ▶ **penknife** n small knife with blade(s) that fold into the handle ▶ **pen name** name used by a writer instead of his or her real name

pen[2] n small enclosure for domestic animals ▷ v **penning, penned** put or keep in a pen

pen[3] n female swan

penal [pee-nal] adj of or used in punishment ▶ **penalize** v impose a penalty on; handicap, hinder

penalty n, pl -**ties** punishment for a crime or offence; (Sport) handicap or disadvantage imposed for breaking a rule

penance n voluntary self-punishment to make amends for wrongdoing

pence n (Brit) a plural of **penny**

penchant [pon-shon] n inclination or liking

pencil n thin cylindrical instrument containing graphite, for writing or drawing ▷ v -**cilling, -cilled** draw, write, or mark with a pencil

pendant n ornament worn on a chain round the neck

pendent adj hanging

pending prep while waiting for ▷ adj not yet decided or settled

pendulous adj hanging, swinging

pendulum n suspended weight swinging to and fro, esp. as a regulator for a clock

penetrate v find or force a way into or through; arrive at the meaning of ▶ **penetrable** adj capable of being penetrated ▶ **penetrating** adj (of a sound) loud and unpleasant; quick to understand ▶ **penetration** n

penguin n flightless black-and-white sea bird of the southern hemisphere

penicillin n antibiotic drug effective against a wide range of diseases and infections

peninsula n strip of land nearly surrounded by water ▶ **peninsular** adj

penis n organ of copulation and urination in male mammals

penitent adj feeling sorry for having done wrong ▷ n someone who is penitent ▶ **penitence** n ▶ **penitentiary** n, pl -**ries** (US) prison ▷ adj (also **penitential**) relating to penance

pennant n long narrow flag

penny n, pl **pence** or **pennies** British bronze coin worth one hundredth of a pound; former British and Australian coin worth one twelfth of a shilling ▶ **penniless** adj very poor

pension[1] n regular payment to people above a certain age, retired employees, widows, etc.

▶ **pensionable** adj ▶ **pensioner**
n person receiving a pension
▶ **pension off** v force (someone)
to retire from a job and pay him or
her a pension
pension² [pon-syon] n boarding
house in Europe
pensive adj deeply thoughtful,
often with a tinge of sadness
pentagon n geometric figure with
five sides; (**P-**) headquarters of the
US military ▶ **pentagonal** adj
pentameter [pen-tam-it-er] n line
of poetry in five metrical feet
Pentateuch [pent-a-tyuke] n first
five books of the Old Testament
Pentecost n Christian festival
celebrating the descent of the Holy
Spirit to the apostles, Whitsuntide
penthouse n flat built on the roof
or top floor of a building
pent-up adj (of an emotion) not
released, repressed
penultimate adj second last
penumbra n, pl **-brae** or **-bras** (in
an eclipse) the partially shadowed
region which surrounds the full
shadow; partial shadow
penury n extreme poverty
▶ **penurious** adj
peony n, pl **-nies** garden plant with
showy red, pink, or white flowers
people pl n persons generally;
the community; one's family ▶ n
race or nation ▶ v provide with
inhabitants ▶ **people mover** (Brit,
Aust & NZ) same as **multipurpose
vehicle**
pep n (informal) high spirits,
energy, or enthusiasm ▶ **pep talk**
(informal) talk designed to increase
confidence and enthusiasm ▶ **pep
up** v **pepping, pepped** stimulate,
invigorate
pepper n sharp hot condiment
made from the fruit of an East
Indian climbing plant; colourful

tropical fruit used as a vegetable,
capsicum ▶ v season with
pepper; sprinkle, dot; pelt with
missiles ▶ **peppery** adj tasting of
pepper; irritable ▶ **peppercorn**
n dried berry of the pepper plant
▶ **peppercorn rent** (chiefly Brit) low
or nominal rent ▶ **pepper spray**
aerosol spray causing temporary
blindness and breathing difficulty,
used esp. for self-defence
peppermint n plant that yields
an oil with a strong sharp flavour;
sweet flavoured with this
peptic adj relating to digestion or
the digestive juices
per prep for each as **per** in
accordance with
perambulate v (old-fashioned)
walk through or about (a
place) ▶ **perambulation** n
▶ **perambulator** n pram
per annum adv (Latin) in each year
per capita adj, adv (Latin) of or for
each person
perceive v become aware of
(something) through the senses;
understand
per cent adv in each hundred
▶ **percentage** n proportion or rate
per hundred
perceptible adj discernible,
recognizable
perception n act of
perceiving; intuitive judgment
▶ **perceptive** adj
perch¹ n resting place for a bird ▶ v
alight, rest, or place on or as if on
a perch
perch² n any of various edible fishes
perchance adv (old-fashioned)
perhaps
percipient adj quick to notice
things, observant
percolate v pass or filter
through small holes; spread
gradually; make (coffee) or (of

coffee) be made in a percolator ▶ **percolation** n ▶ **percolator** n coffee pot in which boiling water is forced through a tube and filters down through coffee

percussion n striking of one thing against another ▶ **percussion instrument** musical instrument played by being struck, such as drums or cymbals

perdition n (Christianity) spiritual ruin

peregrination n (obs) travels, roaming

peregrine falcon n falcon with dark upper parts and a light underside

peremptory adj authoritative, imperious

perennial adj lasting through many years ▷ n plant lasting more than two years ▶ **perennially** adv

perfect adj having all the essential elements; faultless; correct, precise; utter or absolute; excellent ▷ n (Grammar) perfect tense ▷ v improve; make fully correct ▶ **perfectly** adv ▶ **perfection** n state of being perfect ▶ **perfectionist** n person who demands the highest standards of excellence ▶ **perfectionism** n

perfidious adj (lit) treacherous, disloyal ▶ **perfidy** n

perforate v make holes in ▶ **perforation** n

perforce (old-fashioned) adv of necessity

perform v carry out (an action); act, sing, or present a play before an audience; fulfil (a request etc.) ▶ **performance** n ▶ **performer** n

perfume n liquid cosmetic worn for its pleasant smell; fragrance ▷ v give a pleasant smell to ▶ **perfumery** n perfumes in general

perfunctory adj done only as a matter of routine; superficial ▶ **perfunctorily** adv

pergola n arch or framework of trellis supporting climbing plants

perhaps adv possibly, maybe

pericardium n, pl -**dia** membrane enclosing the heart

perihelion n, pl -**lia** point in the orbit of a planet or comet that is nearest to the sun

peril n great danger ▶ **perilous** adj ▶ **perilously** adv

perimeter [per-**rim**-it-er] n (length of) the outer edge of an area

perinatal adj of or in the weeks shortly before or after birth

period n particular portion of time; single occurrence of menstruation; division of time at school etc. when a particular subject is taught; (US) full stop ▷ adj (of furniture, dress, a play, etc.) dating from or in the style of an earlier time ▶ **periodic** adj recurring at intervals ▶ **periodic table** (Chem) chart of the elements, arranged to show their relationship to each other ▶ **periodical** n magazine issued at regular intervals ▷ adj periodic

peripatetic [per-rip-a-**tet**-ik] adj travelling about from place to place

periphery [per-**if**-er-ee] n, pl -**eries** boundary or edge; fringes of a field of activity ▶ **peripheral** [per-**if**-er-al] adj unimportant, not central; of or on the periphery

periscope n instrument used, esp. in submarines, to give a view of objects on a different level

perish v be destroyed or die; decay, rot ▶ **perishable** adj liable to rot quickly ▶ **perishing** adj (informal) very cold

peritoneum [per-rit-toe-**nee**-um] n, pl -**nea** or -**neums** membrane lining the internal surface of the abdomen ▶ **peritonitis** [per-rit-tone-**ite**-iss] n inflammation of the peritoneum

periwinkle¹ n small edible shellfish, the winkle

periwinkle² n plant with trailing stems and blue flowers

perjury n, pl **-juries** act or crime of lying while under oath in a court ▶ **perjure oneself** commit perjury

perk n (informal) incidental benefit gained from a job, such as a company car

perk up v cheer up ▶ **perky** adj lively or cheerful

perlemoen n (S Afr) edible sea creature with a shell lined with mother of pearl

perm n long-lasting curly hairstyle produced by treating the hair with chemicals ▷ v give (hair) a perm

permafrost n permanently frozen ground

permanent adj lasting forever ▶ **permanence** n ▶ **permanently** adv

permeate v pervade or pass through the whole of (something) ▶ **permeable** adj able to be permeated, esp. by liquid

permit v -mitting, -mitted give permission, allow ▷ n document giving permission to do something ▶ **permission** n authorization to do something ▶ **permissible** adj ▶ **permissive** adj (excessively) tolerant, esp. in sexual matters

permutation n any of the ways a number of things can be arranged or combined

pernicious adj wicked; extremely harmful, deadly

pernickety adj (informal) (excessively) fussy about details

peroration n concluding part of a speech, usu. summing up the main points

peroxide n hydrogen peroxide used as a hair bleach; oxide containing a high proportion of oxygen

perp n (US & Canadian informal) person who has committed a crime

perpendicular adj at right angles to a line or surface; upright or vertical ▷ n line or plane at right angles to another

perpetrate v commit or be responsible for (a wrongdoing) ▶ **perpetration** n ▶ **perpetrator** n

perpetual adj lasting forever; continually repeated ▶ **perpetually** adv ▶ **perpetuate** v cause to continue or be remembered ▶ **perpetuation** n in perpetuity forever

perplex v puzzle, bewilder ▶ **perplexity** n, pl **-ties**

perquisite n (formal) same as **perk**

perry n, pl **-ries** alcoholic drink made from fermented pears

per se [per say] adv (Latin) in itself

persecute v treat cruelly because of race, religion, etc.; subject to persistent harassment ▶ **persecution** n ▶ **persecutor** n

persevere v keep making an effort despite difficulties ▶ **perseverance** n

Persian adj of ancient Persia or modern Iran, their people, or their languages ▷ n person from modern Iran, Iranian ▶ **Persian carpet, Persian rug** hand-made carpet or rug with flowing or geometric designs in rich colours ▶ **Persian cat** long-haired domestic cat

persimmon n sweet red tropical fruit

persist v continue to be or happen, last; continue in spite of obstacles or objections ▶ **persistent** adj ▶ **persistently** adv ▶ **persistence** n

person n human being; body of a human being; (Grammar) form of pronouns and verbs that shows if a person is speaking, spoken to, or spoken of **in person** actually present

persona [per-**soh**-na] n, pl -**nae** [-nee] someone's personality as presented to others

personable adj pleasant in appearance and personality

personage n important person

personal adj individual or private; of the body e.g. personal hygiene; (of a remark etc.) offensive ▶ **personally** adv directly, not by delegation to others; in one's own opinion ▶ **personal computer** small computer used for word processing or computer games ▶ **personal pronoun** pronoun like I or she that stands for a definite person ▶ **personal stereo** very small portable cassette player with headphones

personality n, pl -**ties** person's distinctive characteristics; celebrity ▷ pl personal remarks e.g. the discussion degenerated into personalities

personify v -fying, -fied give human characteristics to; be an example of, typify ▶ **personification** n

personnel n people employed in an organization; department in an organization that appoints or keeps records of employees

perspective n view of the relative importance of situations or facts; method of drawing that gives the effect of solidity and relative distances and sizes

Perspex® n transparent acrylic substitute for glass

perspicacious adj having quick mental insight ▶ **perspicacity** n

perspire v sweat ▶ **perspiration** n

persuade v make (someone) do something by argument, charm,

etc.; convince ▶ **persuasion** n act of persuading; way of thinking or belief ▶ **persuasive** adj

pert adj saucy and cheeky

pertain v belong to or be relevant (to)

pertinacious adj (formal) very persistent and determined ▶ **pertinacity** n

pertinent adj relevant ▶ **pertinence** n

perturb v disturb greatly ▶ **perturbation** n

peruse v read in a careful or leisurely manner ▶ **perusal** n

pervade v spread right through (something) ▶ **pervasive** adj

perverse adj deliberately doing something different from what is thought normal or proper ▶ **perversely** adv ▶ **perversity** n

pervert v use or alter for a wrong purpose; lead into abnormal (sexual) behaviour ▷ n person who practises sexual perversion ▶ **perversion** n sexual act or desire considered abnormal; act of perverting

pervious adj able to be penetrated, permeable

peseta [pa-**say**-ta] n former monetary unit of Spain

pessary n, pl -**ries** appliance worn in the vagina, either to prevent conception or to support the womb; vaginal suppository

pessimism n tendency to expect the worst in all things ▶ **pessimist** n ▶ **pessimistic** adj ▶ **pessimistically** adv

pest n annoying person; insect or animal that damages crops ▶ **pesticide** n chemical for killing insect pests

pester v annoy or nag continually

pestilence n deadly epidemic disease ▶ **pestilent** adj annoying, troublesome; deadly ▶ **pestilential** adj

P

pestle n club-shaped implement for grinding things to powder in a mortar

pet n animal kept for pleasure and companionship; person favoured or indulged ▷ adj particularly cherished ▷ v **petting, petted** treat as a pet; pat or stroke affectionately; (old-fashioned) kiss and caress erotically

petal n one of the brightly coloured outer parts of a flower ▶ **petalled** adj

petard n hoist with one's own petard being the victim of one's own schemes

peter out v gradually come to an end

petite adj (of a woman) small and dainty

petition n formal request, esp. one signed by many people and presented to parliament ▷ v present a petition to ▶ **petitioner** n

petrel n sea bird with a hooked bill and tubular nostrils

petrify v **-fying, -fied** frighten severely; turn to stone ▶ **petrification** n

petrochemical n substance, such as acetone, obtained from petroleum

petrol n flammable liquid obtained from petroleum, used as fuel in internal-combustion engines ▶ **petrol bomb** home-made incendiary device consisting of a bottle filled with petrol

petroleum n thick dark oil found underground

petticoat n woman's skirt-shaped undergarment

pettifogging adj excessively concerned with unimportant detail

petty adj **-tier, -tiest** unimportant, trivial; small-minded; on a small

scale e.g. petty crime ▶ **pettiness** n ▶ **petty cash** cash kept by a firm to pay minor expenses ▶ **petty officer** noncommissioned officer in the navy

petulant adj childishly irritable or peevish ▶ **petulance** n ▶ **petulantly** adv

petunia n garden plant with funnel-shaped flowers

pew n fixed benchlike seat in a church; (informal) chair, seat

pewter n greyish metal made of tin and lead

pH (Chem) n measure of the acidity of a solution

phalanger n long-tailed Australian tree-dwelling marsupial

phalanx n, pl **phalanxes** or **phalanges** closely grouped mass of people

phallus n, pl **-luses** or **-li** penis, esp. as a symbol of reproductive power in primitive rites ▶ **phallic** adj

phantasm n unreal vision, illusion ▶ **phantasmal** adj

phantasmagoria n shifting medley of dreamlike figures

phantom n ghost; unreal vision

Pharaoh [fare-oh] n title of the ancient Egyptian kings

pharmaceutical adj of pharmacy

pharmacology n study of drugs ▶ **pharmacological** adj ▶ **pharmacologist** n

pharmacopoeia [far-ma-koh-pee-a] n book with a list of and directions for the use of drugs

pharmacy n, pl **-cies** preparation and dispensing of drugs and medicines; pharmacist's shop ▶ **pharmacist** n person qualified to prepare and sell drugs and medicines

pharynx [far-rinks] n, pl **pharynges** or **pharynxes** cavity forming the back part of the mouth

▶ **pharyngitis** [far-rin-**jite**-iss] *n* inflammation of the pharynx

phase *n* any distinct or characteristic stage in a development or chain of events ▷ *v* arrange or carry out in stages or to coincide with something else ▶ **phase in, phase out** *v* introduce or discontinue gradually

PhD Doctor of Philosophy

pheasant *n* game bird with bright plumage

phenobarbitone *n* drug inducing sleep or relaxation

phenol *n* chemical used in disinfectants and antiseptics

phenomenon *n, pl* **-ena** anything appearing or observed; remarkable person or thing ▶ **phenomenal** *adj* extraordinary, outstanding ▶ **phenomenally** *adv*

phial *n* small bottle for medicine etc.

philadelphus *n* shrub with sweet-scented flowers

philanderer *n* man who flirts or has many casual love affairs ▶ **philandering** *adj, n*

philanthropy *n* practice of helping people less well-off than oneself ▶ **philanthropic** *adj* ▶ **philanthropist** *n*

philately [fill-**lat**-a-lee] *n* stamp collecting ▶ **philatelist** *n*

philharmonic *adj* (in names of orchestras etc.) music-loving

philistine *adj, n* boorishly uncultivated (person) ▶ **philistinism** *n*

philology *n* science of the structure and development of languages ▶ **philological** *adj* ▶ **philologist** *n*

philosopher *n* person who studies philosophy

philosophy *n, pl* **-phies** study of the meaning of life, knowledge, thought, etc.; theory or set of ideas held by a particular philosopher; person's outlook on life ▶ **philosophical, philosophic** *adj* of philosophy; calm in the face of difficulties or disappointments ▶ **philosophically** *adv* ▶ **philosophize** *v* discuss in a philosophical manner

philtre *n* magic drink supposed to arouse love in the person who drinks it

phishing [**fish**-ing] *n* practice of tricking computer users into revealing their financial data in order to defraud them

phlebitis [fleb-**bite**-iss] *n* inflammation of a vein

phlegm [**flem**] *n* thick yellowish substance formed in the nose and throat during a cold ▶ **phlegmatic** [fleg-**mat**-ik] *adj* not easily excited, unemotional ▶ **phlegmatically** *adv*

phlox *n, pl* **phlox** or **phloxes** flowering garden plant

phobia *n* intense and unreasoning fear or dislike

phoenix *n* legendary bird said to set fire to itself and rise anew from its ashes

phone *n, v* telephone ▶ **phonecard** *n* card used to operate certain public telephones ▶ **phone in** make a telephone call to deliver information (esp to a broadcasting studio or place of work); (*informal*) deliver (a performance) in a perfunctory manner ▶ **phone-in** *n* (*Brit, Aust & SAfr*) broadcast in which telephone comments or questions from the public are transmitted live

phonetic *adj* of speech sounds; (of spelling) written as it is sounded ▶ **phonetics** *n* science of speech sounds ▶ **phonetically** *adv*

phoney, phony (*informal*) *adj* **phonier, phoniest** not genuine;

insincere ▷ n, pl **phoneys** or **phonies** phoney person or thing

phonograph n (US, old-fashioned) record player

phosphorescence n faint glow in the dark ▶ **phosphorescent** adj

phosphorus n (Chem) toxic flammable nonmetallic element which appears luminous in the dark ▶ **phosphate** n compound of phosphorus; fertilizer containing phosphorus

photo n, pl **photos** short for **photograph** ▶ **photo finish** finish of a race in which the contestants are so close that a photograph is needed to decide the result ▶ **Photoshop**® n software application for managing and editing digital images ▷ v (informal) alter (a digital image) using Photoshop or a similar application

photocopy n, pl **-copies** photographic reproduction ▷ v **-copying, -copied** make a photocopy of ▶ **photocopier** n

photoelectric adj using or worked by electricity produced by the action of light

photogenic adj always looking attractive in photographs

photograph n picture made by the chemical action of light on sensitive film ▷ v take a photograph of ▶ **photographic** adj ▶ **photography** n art of taking photographs

photographer n person who takes photographs, esp. professionally

photostat n copy made by photocopying machine

photosynthesis n process by which a green plant uses sunlight to build up carbohydrate reserves

phrase n group of words forming a unit of meaning, esp. within

a sentence; short effective expression ▷ v express in words ▶ **phrasal verb** phrase consisting of a verb and an adverb or preposition, with a meaning different from the parts, such as take in meaning deceive

phraseology n, pl **-gies** way in which words are used

physical adj of the body, as contrasted with the mind or spirit; of material things or nature; of physics ▶ **physically** adv ▶ **physical education** training and practice in sports and gymnastics

physician n doctor of medicine

physics n science of the properties of matter and energy ▶ **physicist** n person skilled in or studying physics

physiognomy [fiz-ee-on-om-ee] n face

physiology n science of the normal function of living things ▶ **physiological** adj ▶ **physiologist** n

physiotherapy n treatment of disease or injury by physical means such as massage, rather than by drugs ▶ **physiotherapist** n

physique n person's bodily build and muscular development

pi n (Maths) ratio of a circle's circumference to its diameter

pianissimo adv (Music) very quietly

piano adv (Music) quietly

Pianola® n mechanically played piano

piazza n square or marketplace, esp. in Italy

pic n, pl **pics** or **pix** (informal) photograph or illustration

picador n mounted bullfighter with a lance

picaresque adj denoting a type of fiction in which the hero, a rogue, has a series of adventures

piccalilli *n* pickle of vegetables in mustard sauce

piccolo *n*, *pl* **-los** small flute

pick[1] *v* choose; remove (flowers or fruit) from a plant; take hold of and move with the fingers; provoke (a fight etc.) deliberately; open (a lock) by means other than a key ▷ *n* choice; best part ▶ **pick-me-up** *n* (*informal*) stimulating drink, tonic ▶ **pick on** *v* continually treat unfairly ▶ **pick out** *v* recognize, distinguish ▶ **pick up** *v* raise, lift; collect; improve, get better; become acquainted with for a sexual purpose ▶ **pick-up** *n* small truck; casual acquaintance made for a sexual purpose

pick[2] *n* tool with a curved iron crossbar and wooden shaft, for breaking up hard ground or rocks

pickaxe *n* large pick

picket *n* person or group standing outside a workplace to deter would-be workers during a strike; sentry or sentries posted to give warning of an attack; pointed stick used as part of a fence ▷ *v* form a picket outside (a workplace) ▶ **picket line** line of people acting as pickets

pickings *pl n* money easily acquired

pickle *n* food preserved in vinegar or salt water; (*informal*) awkward situation ▷ *v* preserve in vinegar or salt water ▶ **pickled** *adj* (of food) preserved; (*informal*) drunk

pickpocket *n* thief who steals from someone's pocket

picnic *n* informal meal out of doors ▷ *v*-**nicking, -nicked** have a picnic

Pict *n* member of an ancient race of N Britain ▶ **Pictish** *adj*

pictorial *adj* of or in painting or pictures

picture *n* drawing or painting; photograph; mental image;

beautiful or picturesque object; image on a TV screen ▷ *pl* cinema ▷ *v* visualize, imagine; represent in a picture ▶ **picturesque** *adj* (of a place or view) pleasant to look at; (of language) forceful, vivid ▶ **picture window** large window made of a single sheet of glass

piddle *v* (*informal*) urinate

pidgin *n* language, not a mother tongue, made up of elements of two or more other languages

pie *n* dish of meat, fruit, etc. baked in pastry ▶ **pie chart** circular diagram with sectors representing quantities

piebald *n*, *adj* (horse) with irregular black-and-white markings

piece *n* separate bit or part; instance e.g. *a piece of luck*; example, specimen; literary or musical composition; coin; small object used in draughts, chess, etc. ▶ **piece together** *v* assemble bit by bit

pièce de résistance [pyess de ray-**ziss**-tonss] *n* (*French*) most impressive item

piecemeal *adv* bit by bit

piecework *n* work paid for according to the quantity produced

pied *adj* having markings of two or more colours

pied-à-terre [pyay da **tair**] *n*, *pl* **pieds-à-terre** [pyay da **tair**] small flat or house for occasional use

pier *n* platform on stilts sticking out into the sea; pillar, esp. one supporting a bridge

pierce *v* make a hole in or through with a sharp instrument; make a way through ▶ **piercing** *adj* (of a sound) shrill and high-pitched

Pierrot [**pier**-roe] *n* pantomime clown with a whitened face

piety *n*, *pl* -**ties** deep devotion to God and religion

p

piffle n (informal) nonsense

pig n animal kept and killed for pork, ham, and bacon; (informal) greedy, dirty, or rude person; (offens slang) policeman ▶ **piggish, piggy** adj (informal) dirty; greedy; stubborn ▶ **piggery** n place for keeping and breeding pigs ▶ **pig-headed** adj obstinate ▶ **pig iron** crude iron produced in a blast furnace ▶ **pigsty** n pen for pigs; (Brit, Aust & NZ) dirty or untidy place

pigeon[1] n bird with a heavy body and short legs, sometimes trained to carry messages ▶ **pigeonhole** n compartment for papers in a desk etc. ▶ v classify; put aside and do nothing about ▶ **pigeon-toed** adj with the feet or toes turned inwards

pigeon[2] n (informal) concern or responsibility

piggyback n ride on someone's shoulders ▶ adv carried on someone's shoulders

pigment n colouring matter, paint or dye ▶ **pigmentation** n

Pigmy n, pl **-mies** same as **Pygmy**

pigtail n plait of hair hanging from the back or either side of the head

pike[1] n large predatory freshwater fish

pike[2] n (Hist) long-handled spear

pikelet n (Aust & NZ) small thick pancake

piker n (Aust & NZ slang) shirker

pikey n (Brit offens slang) vagrant; person considered to be lower class

pilaster n square column, usu. set in a wall

pilau, pilaf, pilaff n Middle Eastern dish of meat, fish, or poultry boiled with rice, spices, etc.

pilchard n small edible sea fish of the herring family

pile[1] n number of things lying on top of each other; (informal) large amount; large building ▶ v collect into a pile; (foll by in or out) move in a group ▶ **pile-up** n (informal) traffic accident involving several vehicles

pile[2] n beam driven into the ground, esp. as a foundation for building

pile[3] n fibres of a carpet or a fabric, esp. velvet, that stand up from the weave

piles pl n swollen veins in the rectum, haemorrhoids

pilfer v steal in small quantities

pilgrim n person who journeys to a holy place ▶ **pilgrimage** n

pill n small ball of medicine swallowed whole **the pill** pill taken by a woman to prevent pregnancy

pillage v steal property by violence in war ▶ n violent seizure of goods, esp. in war

pillar n upright post, usu. supporting a roof; strong supporter ▶ **pillar box** (in Britain) red pillar-shaped letter box in the street

pillion n seat for a passenger behind the rider of a motorcycle

pillory n, pl **-ries** (Hist) frame with holes for the head and hands in which an offender was locked and exposed to public abuse ▶ v **-rying, -ried** ridicule publicly

pillow n stuffed cloth bag for supporting the head in bed ▶ v rest as if on a pillow ▶ **pillowcase, pillowslip** n removable cover for a pillow

pilot n person qualified to fly an aircraft or spacecraft; person employed to steer a ship entering or leaving a harbour ▶ adj experimental and preliminary ▶ v act as the pilot of; guide, steer ▶ **pilot light** small flame lighting the main one in a gas appliance

pimento n, pl **-tos** mild-tasting red pepper

pimp *n* man who gets customers for a prostitute in return for a share of his or her earnings ▷ *v* act as a pimp ▶ **pimp up, pimp out** *v* decorate (someone or something) with flashy accessories

pimpernel *n* wild plant with small star-shaped flowers

pimple *n* small pus-filled spot on the skin ▶ **pimply** *adj*

PIN personal identification number: number used with a credit or debit card to withdraw money, confirm a purchase, etc. ▶ **PIN pad** small pad into which a customer keys his or her PIN to confirm a purchase

pin *n* short thin piece of stiff wire with a point and head, for fastening things; wooden or metal peg or stake ▷ *v* **pinning, pinned** fasten with a pin; seize and hold fast ▶ **pin down** *v* force (someone) to make a decision, take action, etc.; define clearly ▶ **pin money** small amount earned to buy small luxuries ▶ **pin-up** *n* picture of a sexually attractive person, esp. (partly) naked

pinafore *n* apron; dress with a bib top

pinball *n* electrically operated table game in which a small ball is shot through various hazards

pince-nez [panss-**nay**] *n*, *pl* **pince-nez** glasses kept in place only by a clip on the bridge of the nose

pincers *pl n* tool consisting of two hinged arms, for gripping; claws of a lobster etc.

pinch *v* squeeze between finger and thumb; cause pain by being too tight; (*informal*) steal ▷ *n* act of pinching; as much as can be taken up between the finger and thumb ▶ **at a pinch** if absolutely necessary ▶ **feel the pinch** have to economize

pinchbeck *n* alloy of zinc and copper, used as imitation gold

pine¹ *n* evergreen coniferous tree; its wood ▶ **pine cone** woody seed case of the pine tree ▶ **pine marten** wild mammal of the coniferous forests of Europe and Asia

pine² *v* (foll by *for*) feel great longing (for); become thin and ill through grief etc.

pineal gland *n* small cone-shaped gland at the base of the brain

pineapple *n* large tropical fruit with juicy yellow flesh and a hard skin

ping *v*, *n* (make) a short high-pitched sound

Ping-Pong® *n* table tennis

pinion¹ *n* bird's wing ▷ *v* immobilize (someone) by tying or holding his or her arms

pinion² *n* small cogwheel

pink *n* pale reddish colour; fragrant garden plant ▷ *adj* of the colour pink ▷ *v* (of an engine) make a metallic noise because not working properly, knock ▶ **in the pink** in good health

pinking shears *pl n* scissors with a serrated blade or blades that give a wavy edge to material to prevent fraying

pinnacle *n* highest point of fame or success; mountain peak; small slender spire

pinotage [pin-no-**tajj**] *n* blended red wine of S Africa

pinpoint *v* locate or identify exactly

pinstripe *n* very narrow stripe in fabric; the fabric itself

pint *n* liquid measure, 1/8 gallon (.568 litre)

pioneer *n* explorer or early settler of a new country; originator or developer of something new ▷ *v* be the pioneer or leader of

pious *adj* deeply religious, devout

pip¹ *n* small seed in a fruit

P

pip² n high-pitched sound used as a time signal on radio; (*informal*) star on a junior army officer's shoulder showing rank

pip³ n **give someone the pip** (*Brit, NZ & S Afr slang*) annoy

pipe n tube for conveying liquid or gas; tube with a small bowl at the end for smoking tobacco; tubular musical instrument ▷ *pl* bagpipes ▷ v play on a pipe; utter in a shrill tone; convey by pipe; decorate with piping ▶ **piper** n player on a pipe or bagpipes ▶ **piping** n system of pipes; decoration of icing on a cake etc.; fancy edging on clothes etc. ▶ **piped music** recorded music played as background music in public places ▶ **pipe down** v (*informal*) stop talking ▶ **pipe dream** fanciful impossible plan ▶ **pipeline** n long pipe for transporting oil, water, etc.; means of communication **in the pipeline** in preparation ▶ **pipe up** v speak suddenly or shrilly

pipette n slender glass tube used to transfer or measure fluids

pipi n (*Aust & NZ*) edible mollusc often used as bait

pipit n small brownish songbird

pippin n type of eating apple

piquant [pee-kant] adj having a pleasant spicy taste; mentally stimulating ▶ **piquancy** n

pique [peek] n feeling of hurt pride, baffled curiosity, or resentment ▷ v hurt the pride of; arouse (curiosity)

piqué [pee-kay] n stiff ribbed cotton fabric

piquet [pik-ket] n card game for two

piranha n small fierce freshwater fish of tropical America

pirate n sea robber; person who illegally publishes or sells work owned by someone else; person or company that broadcasts illegally ▷ v sell or reproduce (artistic work etc.) illegally ▶ **piracy** n ▶ **piratical** adj

pirouette v, n (make) a spinning turn balanced on the toes of one foot

piss (*vulgar slang*) v urinate ▷ n act of urinating; urine

pistachio n, pl -chios edible nut of a Mediterranean tree

piste [peest] n ski slope

pistil n seed-bearing part of a flower

pistol n short-barrelled handgun

piston n cylindrical part in an engine that slides to and fro in a cylinder

pit n deep hole in the ground; coal mine; dent or depression; servicing and refuelling area on a motor-racing track; same as **orchestra** ▷ v **pitting, pitted** mark with small dents or scars **pit one's wits against** compete against in a test or contest ▶ **pit bull terrier** strong muscular terrier with a short coat

pitch¹ v throw, hurl; set up (a tent); fall headlong; (of a ship or plane) move with the front and back going up and down alternately; set the level or tone of ▷ n area marked out for playing sport; degree or angle of slope; degree of highness or lowness of a (musical) sound; place where a street or market trader regularly sells; (*informal*) persuasive sales talk ▶ **pitch in** v join in enthusiastically ▶ **pitch into** v (*informal*) attack

pitch² n dark sticky substance obtained from tar ▶ **pitch-black, pitch-dark** adj very dark

pitchblende n mineral composed largely of uranium oxide, yielding radium

pitcher n large jug with a narrow neck

pitchfork n large long-handled fork for lifting hay ▷ v thrust abruptly or violently

pitfall n hidden difficulty or danger

pith n soft white lining of the rind of oranges etc.; essential part; soft tissue in the stems of certain plants ▶ **pithy** adj short and full of meaning

piton [peet-on] n metal spike used in climbing to secure a rope

pittance n very small amount of money

pituitary n, pl **-taries** gland at the base of the brain that helps to control growth (also **pituitary gland**)

pity n, pl **pities** sympathy or sorrow for others' suffering; regrettable fact ▷ v **pitying, pitied** feel pity for ▶ **piteous, pitiable** adj arousing pity ▶ **pitiful** adj arousing pity; woeful, contemptible ▶ **pitifully** adv ▶ **pitiless** adj feeling no pity or mercy ▶ **pitilessly** adv

pivot n central shaft on which something turns ▷ v provide with or turn on a pivot ▶ **pivotal** adj of crucial importance

pix n (informal) a plural of **pic**

pixel n smallest constituent unit of an image, as on a computer screen

pixie n (in folklore) fairy

pizza n flat disc of dough covered with a wide variety of savoury toppings and baked

pizzazz n (informal) attractive combination of energy and style

pizzicato [pit-see-kah-toe] adj (Music) played by plucking the string of a violin etc. with the finger

placard n notice that is carried or displayed in public

placate v make (someone) stop feeling angry or upset ▶ **placatory** adj

place n particular part of an area or space; particular town, building,

etc.; position or point reached; seat or space; duty or right; position of employment; usual position ▷ v put in a particular place; identify, put in context; make (an order, bet, etc.) **be placed** (of a competitor in a race) be among the first three **take place** happen, occur

placebo [plas-see-bo] n, pl **-bos** or **-boes** sugar pill etc. given to an unsuspecting patient instead of an active drug

placenta [plass-ent-a] n, pl **-tas** or **-tae** organ formed in the womb during pregnancy, providing nutrients for the fetus ▶ **placental** adj

placid adj not easily excited or upset, calm ▶ **placidity** n

plagiarize [play-jer-ize] v steal ideas, passages, etc. from (someone else's work) and present them as one's own ▶ **plagiarism** n

plague n fast-spreading fatal disease; (Hist) bubonic plague; widespread infestation ▷ v plaguing, plagued trouble or annoy continually

plaice n edible European flatfish

plaid n long piece of tartan cloth worn as part of Highland dress; tartan cloth or pattern

plain adj easy to see or understand; expressed honestly and clearly; without decoration or pattern; not beautiful; simple, ordinary ▷ n large stretch of level country ▶ **plainly** adv ▶ **plainness** n ▶ **plain clothes** ordinary clothes, as opposed to uniform ▶ **plain sailing** easy progress ▶ **plain speaking** saying exactly what one thinks

plainsong n unaccompanied singing, esp. in a medieval church

plaintiff n person who sues in a court of law

plaintive adj sad, mournful ▷ **plaintively** adv

plait [platt] n intertwined length of hair ▷ v intertwine separate strands in a pattern

plan n way thought out to do or achieve something; diagram showing the layout or design of something ▷ v **planning, planned** arrange beforehand; make a diagram of ▷ **planner** n

plane¹ n aeroplane; (Maths) flat surface; level of attainment etc. ▷ adj perfectly flat or level ▷ v glide or skim

plane² n tool for smoothing wood ▷ v smooth (wood) with a plane

plane³ n tree with broad leaves

planet n large body in space that revolves round the sun or another star ▷ **planetary** adj

planetarium n, pl **-iums** or **-ia** building where the movements of the stars, planets, etc. are shown by projecting lights on the inside of a dome

plangent adj (of sounds) mournful and resounding

plank n long flat piece of sawn timber

plankton n minute animals and plants floating in the surface water of a sea or lake

plant n living organism that grows in the ground and has no power to move; equipment or machinery used in industrial processes; factory or other industrial premises ▷ v put in the ground to grow; place firmly in position; (informal) put (a person) secretly in an organization to spy; (informal) hide (stolen goods etc.) on a person to make him or her seem guilty ▷ **planter** n owner of a plantation

plantain¹ n low-growing wild plant with broad leaves

plantain² n tropical fruit like a green banana

plantation n estate for the cultivation of tea, coffee, etc.; wood of cultivated trees

plaque n inscribed commemorative stone or metal plate; filmy deposit on teeth that causes decay

plasma n clear liquid part of blood

plasma screen n type of flat screen on a television or visual display unit

plaster n mixture of lime, sand, etc. for coating walls; adhesive strip of material for dressing cuts etc. ▷ v cover with plaster; coat thickly ▷ **plastered** adj (slang) drunk ▷ **plaster of Paris** white powder which dries to form a hard solid when mixed with water, used for sculptures and casts for broken limbs

plastic n synthetic material that can be moulded when soft but sets in a hard long-lasting shape; credit cards etc. as opposed to cash ▷ adj made of plastic; easily moulded, pliant ▷ **plasticity** n ability to be moulded ▷ **plastic bullet** solid PVC cylinder fired by police in riot control ▷ **plastic surgery** repair or reconstruction of missing or malformed parts of the body

Plasticine® n soft coloured modelling material used esp. by children

plate n shallow dish for holding food; flat thin sheet of metal, glass, etc.; thin coating of metal on another metal; dishes or cutlery made of gold or silver; illustration, usu. on fine quality paper, in a book; (informal) set of false teeth ▷ v cover with a thin coating of gold, silver, or other metal ▷ **plateful** n ▷ **plate glass** glass in thin sheets, used for mirrors and windows ▷ **plate**

tectonics study of the structure of the earth's crust, esp. the movement of layers of rocks

plateau n, pl **-teaus** or **-teaux** area of level high land; stage when there is no change or development

platen n roller of a typewriter, against which the paper is held

platform n raised floor; raised area in a station from which passengers board trains; structure in the sea which holds machinery, stores, etc. for drilling an oil well; programme of a political party

platinum n (Chem) valuable silvery-white metal ▶ **platinum blonde** woman with silvery-blonde hair

platitude n remark that is true but not interesting or original ▶ **platitudinous** adj

platonic adj (of a relationship) friendly or affectionate but not sexual

platoon n smaller unit within a company of soldiers

platteland n (S Afr) rural district

platter n large dish

platypus n Australian egg-laying amphibious mammal, with dense fur, webbed feet, and a ducklike bill (also **duck-billed platypus**)

plaudits pl n expressions of approval

plausible adj apparently true or reasonable; persuasive but insincere ▶ **plausibly** adv ▶ **plausibility** n

play v occupy oneself in (a game or recreation); compete against in a game or sport; behave carelessly; act (a part) on the stage; perform on (a musical instrument); cause (a radio, record player, etc.) to give out sound; move lightly or irregularly, flicker ▶ n story performed on stage or broadcast; activities children take part in for amusement; playing of a game; conduct e.g. fair play;

(scope for) freedom of movement ▶ **playful** adj lively ▶ **play back** v listen to or watch (something recorded) ▶ **playcentre** n (NZ & S Afr) centre for preschool children run by parents ▶ **play down** v minimize the importance of ▶ **playgroup** n regular meeting of very young children for supervised play ▶ **playhouse** n theatre ▶ **playing card** one of a set of 52 cards used in card games ▶ **playing field** extensive piece of ground for sport ▶ **play-lunch** n (Aust & NZ) child's mid-morning snack at school ▶ **play off** v set (two people) against each other for one's own ends ▶ **play on** v exploit or encourage (someone's sympathy or weakness) ▶ **playschool** n nursery group for young children ▶ **PlayStation®** n type of video games console ▶ **plaything** n toy; person regarded or treated as a toy ▶ **play up** v give prominence to; cause trouble ▶ **playwright** n author of plays

playboy n rich man who lives only for pleasure

player n person who plays a game or sport; actor or actress; person who plays a musical instrument

plaza n open space or square; modern shopping complex

PLC, plc (in Britain) Public Limited Company

plea n serious or urgent request, entreaty; statement of a prisoner or defendant; excuse

plead v ask urgently or with deep feeling; give as an excuse; (Law) declare oneself to be guilty or innocent of a charge made against one

pleasant adj pleasing, enjoyable ▶ **pleasantly** adv ▶ **pleasantry** n, pl **-tries** polite or joking remark

please v give pleasure or satisfaction to ▷ adv polite word of request **please oneself** do as one likes ▶ **pleased** adj ▶ **pleasing** adj

pleasure n feeling of happiness and satisfaction; something that causes this ▶ **pleasurable** adj ▶ **pleasurably** adv

pleat n fold made by doubling material back on itself ▷ v arrange (material) in pleats

plebeian [pleb-**ee**-an] adj of the lower social classes; vulgar or rough ▷ n (also **pleb**) member of the lower social classes

plebiscite [pleb-**iss**-ite] n decision by direct voting of the people of a country

plectrum n, pl **-trums** or **-tra** small implement for plucking the strings of a guitar etc.

pledge n solemn promise; something valuable given as a guarantee that a promise will be kept or a debt paid ▷ v promise solemnly; bind by or as if by a pledge

plenary adj (of a meeting) attended by all members

plenipotentiary adj having full powers ▷ n, pl **-aries** diplomat or representative having full powers

plenitude n completeness, abundance

plenteous adj plentiful

plenty n large amount or number; quite enough ▶ **plentiful** adj existing in large amounts or numbers ▶ **plentifully** adv

pleonasm n use of more words than necessary

plethora n excess

pleurisy n inflammation of the membrane covering the lungs

pliable adj easily bent; easily influenced ▶ **pliability** n

pliant adj pliable ▶ **pliancy** n

pliers pl n tool with hinged arms and jaws for gripping

plight¹ n difficult or dangerous situation

plight² v **plight one's troth** (old-fashioned) promise to marry

Plimsoll line n mark on a ship showing the level water should reach when the ship is fully loaded

plimsolls pl n (Brit) rubber-soled canvas shoes

plinth n slab forming the base of a statue, column, etc.

PLO Palestine Liberation Organization

plod v **plodding, plodded** walk with slow heavy steps; work slowly but determinedly ▶ **plodder** n

plonk¹ v put (something) down heavily and carelessly

plonk² n (informal) cheap inferior wine

plop n sound of an object falling into water without a splash ▷ v **plopping, plopped** make this sound

plot¹ n secret plan to do something illegal or wrong; story of a film, novel, etc. ▷ v **plotting, plotted** plan secretly, conspire; mark the position or course of (a ship or aircraft) on a map; mark and join up (points on a graph)

plot² n small piece of land

plough n agricultural tool for turning over soil ▷ v turn over (earth) with a plough; move or work through slowly and laboriously ▶ **ploughman** n ▶ **ploughshare** n blade of a plough

plover n shore bird with a straight bill and long pointed wings

ploy n manoeuvre designed to gain an advantage

pluck v pull or pick off; pull out the feathers of (a bird for cooking); sound the strings of (a guitar etc.) with the fingers or a plectrum

▷ *n* courage ▶ **plucky** *adj* brave
▶ **pluckily** *adv* ▶ **pluck up** *v*
summon up (courage)

plug *n* thing fitting into and filling a hole; device connecting an appliance to an electricity supply; (*informal*) favourable mention of a product etc., to encourage people to buy it ▷ *v* **plugging, plugged** block or seal (a hole or gap) with a plug; (*informal*) advertise (a product etc.) by constant repetition ▶ **plug away** *v* (*informal*) work steadily ▶ **plug in** *v* connect (an electrical appliance) to a power source by pushing a plug into a socket

plum *n* oval usu. dark red fruit with a stone in the middle ▷ *adj* dark purplish-red; very desirable

plumage *n* bird's feathers

plumb *v* understand (something obscure); test with a plumb line ▷ *adv* exactly **plumb the depths of** experience the worst extremes of (an unpleasant quality or emotion) ▶ **plumbing** *n* pipes and fixtures used in water and drainage systems ▶ **plumb in** *v* connect (an appliance such as a washing machine) to a water supply ▶ **plumb line** string with a weight at the end, used to test the depth of water or to test whether something is vertical

plumber *n* person who fits and repairs pipes and fixtures for water and drainage systems

plume *n* feather, esp. one worn as an ornament

plummet *v* **-meting, -meted** plunge downward

plump¹ *adj* moderately or attractively fat ▶ **plumpness** *n* ▶ **plump up** *v* make (a pillow, cushion, etc) fuller or rounded

plump² *v* sit or fall heavily and suddenly ▶ **plump for** *v* choose, vote for

plunder *v* take by force, esp. in time of war ▷ *n* things plundered, spoils

plunge *v* put or throw forcibly or suddenly (into); descend steeply ▷ *n* plunging, dive **take the plunge** (*informal*) embark on a risky enterprise ▶ **plunger** *n* rubber suction cup used to clear blocked pipes ▶ **plunge into** *v* become deeply involved in

Plunket baby *n* (*NZ*) baby brought up on the diet recommended by the Plunket Society ▶ **Plunket nurse** (*NZ*) nurse working for the Plunket Society

pluperfect *n, adj* (*Grammar*) (tense) expressing an action completed before a past time, e.g. *had gone* in *his wife had gone already*

plural *adj* of or consisting of more than one ▷ *n* word indicating more than one

pluralism *n* existence and toleration of a variety of peoples, opinions, etc. in a society ▶ **pluralist** *n* ▶ **pluralistic** *adj*

plus *prep, adj* indicating addition ▷ *adj* more than zero; positive; advantageous ▷ *n* sign (+) denoting addition; advantage

> **USAGE NOTE**
> Avoid using *plus* to mean 'additionally' except in very informal contexts

plus fours *pl n* trousers gathered in just below the knee

plush *n* fabric with long velvety pile ▷ *adj* (also **plushy**) luxurious

plus-one *n* (*informal*) person who accompanies an invited person to a social function

Pluto *n* Greek god of the underworld; farthest planet from the sun

plutocrat *n* person who is powerful because of being very rich ▶ **plutocratic** *adj*

P

plutonium n (Chem) radioactive metallic element used esp. in nuclear reactors and weapons

ply¹ v **plying, plied** work at (a job or trade); use (a tool); (of a ship) travel regularly along or between ▸ **ply with** v supply with or subject to persistently

ply² n thickness of wool, fabric, etc.

plywood n board made of thin layers of wood glued together

PM prime minister

p.m. after noon; postmortem

PMS premenstrual syndrome

PMT premenstrual tension

pneumatic adj worked by or inflated with wind or air

pneumonia n inflammation of the lungs

PO (Brit) postal order; Post Office

poach¹ v catch (animals) illegally on someone else's land; encroach on or steal something belonging to someone else

poach² v simmer (food) gently in liquid

poacher n person who catches animals illegally on someone else's land

POC proof of concept

pocket n small bag sewn into clothing for carrying things; pouchlike container, esp. for catching balls at the edge of a snooker table; isolated or distinct group or area ▷ v **pocketing, pocketed** put into one's pocket; take secretly or dishonestly ▷ adj small **out of pocket** having made a loss ▸ **pocket money** small regular allowance given to children by parents; money for small personal expenses

pockmarked adj (of the skin) marked with hollow scars where diseased spots have been

pod n long narrow seed case of peas, beans, etc.

podcast n audio file similar to a radio broadcast which can be downloaded to a computer, iPod™, etc. ▷ v create such files and make them available for downloading

podgy adj **podgier, podgiest** short and fat

podiatrist [poe-die-a-trist] n same as **chiropodist** ▸ **podiatry** n

podium n, pl **-diums** or **-dia** small raised platform for a conductor or speaker

poem n imaginative piece of writing in rhythmic lines

poep n (SAfr slang) emission of gas from the anus

poesy n (obs)

poet n writer of poems ▸ **poetry** n poems; art of writing poems; beautiful or pleasing quality ▸ **poetic, poetical** adj of or like poetry ▸ **poetically** adv ▸ **poetic justice** suitable reward or punishment for someone's past actions ▸ **poet laureate** poet appointed by the British sovereign to write poems on important occasions

pogrom n organized persecution and massacre

poignant adj sharply painful to the feelings ▸ **poignancy** n

poinsettia n Central American shrub widely grown for its clusters of scarlet leaves, which resemble petals

point n main idea in a discussion, argument, etc.; aim or purpose; detail or item; characteristic; particular position, stage, or time; dot indicating decimals; sharp end; unit for recording a value or score; one of the direction marks of a compass; electrical socket ▷ v show the direction or position of something or draw attention to it by extending a finger or other pointed object towards it; direct or face towards **on the point of** very shortly going to ▶ **pointed** *adj* having a sharp end; (of a remark) obviously directed at a particular person ▶ **pointer** *n* helpful hint; indicator on a measuring instrument; breed of gun dog ▶ **pointless** *adj* meaningless, irrelevant ▶ **point-blank** *adj* fired at a very close target; (of a remark or question) direct, blunt ▶ *adv* directly or bluntly ▶ **point duty** control of traffic by a policeman at a road junction ▶ **point of view** way of considering something ▶ **point-to-point** *n* (*Brit*) horse race across open country

poise *n* calm dignified manner ▶ **poised** *adj* absolutely ready; behaving with or showing poise

poison *n* substance that kills or injures when swallowed or absorbed ▶ *v* give poison to; have a harmful or evil effect on, spoil ▶ **poisoner** *n* ▶ **poisonous** *adj* ▶ **poison-pen letter** malicious anonymous letter

poke *v* jab or prod with one's finger, a stick, etc.; thrust forward or out ▶ *n* poking ▶ **poky** *adj* small and cramped

poker[1] *n* metal rod for stirring a fire

poker[2] *n* card game in which players bet on the hands dealt ▶ **poker-faced** *adj* expressionless

polar *adj* of or near either of the earth's poles ▶ **polar bear** white bear that lives in the regions around the North Pole

polarize *v* form or cause to form into groups with directly opposite views; (*Physics*) restrict (light waves) to certain directions of vibration ▶ **polarization** *n*

Polaroid® *n* plastic which polarizes light and so reduces glare; camera that develops a print very quickly inside itself

polder *n* land reclaimed from the sea, esp. in the Netherlands

pole[1] *n* long rounded piece of wood ▶ **pole dancing** entertainment in which a woman dances erotically using a vertical fixed pole

pole[2] *n* point furthest north or south on the earth's axis of rotation; either of the opposite ends of a magnet or electric cell ▶ **Pole Star** star nearest to the North Pole in the northern hemisphere

poleaxe *v* hit with a heavy blow

polecat *n* small animal of the weasel family

polemic [pol-**em**-ik] *n* fierce attack on or defence of a particular opinion, belief, etc. ▶ **polemical** *adj*

police *n* organized force in a state which keeps law and order ▶ *v* control or watch over with police or a similar body ▶ **policeman, policewoman** *n* member of a police force

policy[1] *n*, *pl* **-cies** plan of action adopted by a person, group, or state

policy[2] *n*, *pl* **-cies** document containing an insurance contract

polio *n* disease affecting the spinal cord, which often causes paralysis (also **poliomyelitis**)

polish *v* make smooth and shiny by rubbing; make more nearly perfect ▶ *n* substance used for polishing; pleasing elegant style ▶ **polished** *adj* accomplished; done or performed well or professionally ▶ **polish off** *v* (*informal*) finish completely, dispose of

Polish *adj* of Poland, its people, or their language ▶ *n* official language of Poland

p

polite adj showing consideration for others in one's manners, speech, etc.; socially correct or refined ▶ **politely** adv ▶ **politeness** n

politic adj wise and likely to prove advantageous

politics n winning and using of power to govern society; (study of) the art of government; person's beliefs about how a country should be governed ▶ **political** adj of the state, government, or public administration ▶ **politically** adv ▶ **politically correct** (of language) intended to avoid any implied prejudice ▶ **political prisoner** person imprisoned because of his or her political beliefs ▶ **politician** n person actively engaged in politics, esp. a member of parliament

polka n lively 19th-century dance; music for this ▶ **polka dots** pattern of bold spots on fabric

poll n (also **opinion poll**) questioning of a random sample of people to find out general opinion; voting; number of votes recorded ▷ v receive (votes); question in an opinion poll ▶ **pollster** n person who conducts opinion polls ▶ **polling station** building where people vote in an election

pollarded adj (of a tree) growing very bushy because its top branches have been cut short

pollen n fine dust produced by flowers to fertilize other flowers ▶ **pollinate** v fertilize with pollen ▶ **pollen count** measure of the amount of pollen in the air, esp. as a warning to people with hay fever

pollute v contaminate with something poisonous or harmful ▶ **pollution** n ▶ **pollutant** n something that pollutes

polly n, pl **-lies** (Aust informal) politician

polo n game like hockey played by teams of players on horseback ▶ **polo neck** sweater with tight turned-over collar

polonaise n old stately dance; music for this

polonium n (Chem) radioactive element that occurs in trace amounts in uranium ores

poltergeist n spirit believed to move furniture and throw objects around

poltroon n (obs) utter coward

poly- combining form many, much

polyandry n practice of having more than one husband at the same time

polyanthus n garden primrose

polychromatic adj many-coloured

polyester n synthetic material used to make plastics and textile fibres

polygamy [pol-ig-a-mee] n practice of having more than one husband or wife at the same time ▶ **polygamous** adj ▶ **polygamist** n

polyglot n, adj (person) able to speak or write several languages

polygon n geometrical figure with three or more angles and sides ▶ **polygonal** adj

polyhedron n, pl **-drons** or **-dra** solid figure with four or more sides

polymer n chemical compound with large molecules made of simple molecules of the same kind ▶ **polymerize** v form into polymers ▶ **polymerization** n

polyp n small simple sea creature with a hollow cylindrical body; small growth on a mucous membrane

polyphonic adj (Music) consisting of several melodies played simultaneously

polystyrene n synthetic material used esp. as white rigid foam for packing and insulation

polytechnic n (in New Zealand and formerly in Britain) college offering courses in many subjects at and below degree level

polytheism n belief in many gods ▶ **polytheistic** adj

polythene n light plastic used for bags etc.

polyunsaturated adj of a group of fats that do not form cholesterol in the blood

polyurethane n synthetic material used esp. in paints

pom n (Aust & NZ slang) person from England (also **pommy**)

pomander n (container for) a mixture of sweet-smelling petals, herbs, etc.

pomegranate n round tropical fruit with a thick rind containing many seeds in a red pulp

Pomeranian n small dog with long straight hair

pommel n raised part on the front of a saddle; knob at the top of a sword hilt

pomp n stately display or ceremony

pompom n decorative ball of tufted wool, silk, etc.

pompous adj foolishly serious and grand, self-important ▶ **pompously** adv ▶ **pomposity** n

ponce n (offens) effeminate man; pimp ▶ **ponce around** v (Brit, Aust & NZ) behave in a ridiculous or posturing way

poncho n, pl **-chos** loose circular cloak with a hole for the head

pond n small area of still water

ponder v think thoroughly or deeply (about)

ponderous adj serious and dull; heavy and unwieldy; (of movement) slow and clumsy ▶ **ponderously** adv

pong v, n (informal) (give off) a strong unpleasant smell

pontiff n the Pope ▶ **pontificate** v state one's opinions as if they were the only possible correct ones ▷ n period of office of a Pope

pontoon[1] n floating platform supporting a temporary bridge

pontoon[2] n gambling card game

pony n, pl **ponies** small horse ▶ **ponytail** n long hair tied in one bunch at the back of the head

Ponzi scheme n fraudulent investment operation that pays quick returns to initial contributors using money from subsequent contributors rather than profit

poodle n dog with curly hair often clipped fancifully

poof, poofter n (Brit, Aust & NZ offens) homosexual man

pool[1] n small body of still water; puddle of spilt liquid; swimming pool

pool[2] n shared fund or group of workers or resources; game similar to snooker ▷ pl (Brit) short for **football pools** ▷ v put in a common fund

poop n raised part at the back of a sailing ship

poor adj having little money and few possessions; less, smaller, or weaker than is needed or expected; inferior; unlucky, pitiable ▶ **poorly** adv in a poor manner ▷ adj not in good health

pop[1] v **popping, popped** make or cause to make a small explosive sound; (informal) (often foll by in or out) go, put, or come unexpectedly or suddenly ▷ n small explosive sound; (Brit) nonalcoholic fizzy drink ▶ **popcorn** n grains of maize heated until they puff up and burst ▶ **pop-up** n (Computing) image that appears above the open window on a computer screen

pop[2] n music of general appeal, esp. to young people

pop³ n (informal) father

Pope n head of the Roman Catholic Church ▸ **popish** adj (offens) Roman Catholic

poplar n tall slender tree

poplin n ribbed cotton material

poppadom n thin round crisp Indian bread

poppy n, pl **-pies** plant with a large red flower

populace n the ordinary people

popular adj widely liked and admired; of or for the public in general ▸ **popularly** adv ▸ **popularity** n ▸ **popularize** v make popular; make (something technical or specialist) easily understood

populate v live in, inhabit; fill with inhabitants ▸ **populous** adj densely populated

population n all the people who live in a particular place; the number of people living in a particular place

populist n, adj (person) appealing to the interests or prejudices of ordinary people ▸ **populism** n

porbeagle n kind of shark

porcelain n fine china; objects made of it

porch n covered approach to the entrance of a building

porcine adj of or like a pig

porcupine n animal covered with long pointed quills

pore n tiny opening in the skin or in the surface of a plant

pork n pig meat ▸ **porker** n pig raised for food

porn, porno n, adj (informal) short for **pornography, pornographic**

pornography n writing, films, or pictures designed to be sexually exciting ▸ **pornographer** n producer of pornography ▸ **pornographic** adj

porous adj allowing liquid to pass through gradually ▸ **porosity** n

porphyry [por-fir-ee] n reddish rock with large crystals in it

porpoise n fishlike sea mammal

porridge n breakfast food made of oatmeal cooked in water or milk; (chiefly Brit slang) term in prison

port¹ n (town with) a harbour

port² n left side of a ship or aircraft when facing the front of it

port³ n strong sweet wine, usu. red

port⁴ n opening in the side of a ship; porthole

portable adj easily carried ▸ **portability** n

portal n large imposing doorway or gate; (Computing) internet site providing links to other sites

portcullis n grating suspended above a castle gateway, that can be lowered to block the entrance

portend v be a sign of

portent n sign of a future event ▸ **portentous** adj of great or ominous significance; pompous, self-important

porter¹ n man who carries luggage; hospital worker who transfers patients between rooms etc.

porter² n doorman or gatekeeper of a building

portfolio n, pl **-os** (flat case for carrying) examples of an artist's work; area of responsibility of a government minister; list of investments held by an investor

porthole n small round window in a ship or aircraft

portico n, pl **-coes** or **-cos** porch or covered walkway with columns supporting the roof

portion n part or share; helping of food for one person; destiny or fate ▸ **portion out** v divide into shares

portly adj **-lier, -liest** rather fat

portmanteau n, pl -**teaus** or -**teaux** (old-fashioned) large suitcase that opens into two compartments ▷ adj combining aspects of different things

portrait n picture of a person; lifelike description

portray v describe or represent by artistic means, as in writing or film ▶ **portrayal** n

Portuguese adj of Portugal, its people, or their language ▷ n person from Portugal; language of Portugal and Brazil ▶ **Portuguese man-of-war** sea creature resembling a jellyfish, with stinging tentacles

pose v place in or take up a particular position to be photographed or drawn; raise (a problem); ask (a question) ▷ n position while posing; behaviour adopted for effect **pose as** pretend to be ▶ **poser** n puzzling question; poseur ▶ **poseur** n person who behaves in an affected way to impress others

posh adj (informal) smart, luxurious; affectedly upper-class

posit [pozz-it] v lay down as a basis for argument

position n place; usual or expected place; way in which something is placed or arranged; attitude, point of view; social standing; job ▷ v place

positive adj feeling no doubts, certain; confident, hopeful; helpful, providing encouragement; absolute, downright; (Maths) greater than zero; (of an electrical charge) having a deficiency of electrons ▶ **positively** adv ▶ **positive discrimination** provision of special opportunities for a disadvantaged group

positron n (Physics) particle with same mass as an electron but a positive charge

posse [poss-ee] n (US) group of men organized to maintain law and order; (Brit & Aust informal) group of friends or associates

possess v have as one's property; (of a feeling, belief, etc.) have complete control of, dominate ▶ **possessor** n ▶ **possession** n state of possessing, ownership ▷ pl things a person possesses ▶ **possessive** adj wanting all the attention or love of another person; (of a word) indicating the person or thing that something belongs to ▶ **possessiveness** n

possible adj able to exist, happen, or be done; worthy of consideration ▷ n person or thing that might be suitable or chosen ▶ **possibility** n, pl -**ties** ▶ **possibly** adv perhaps, not necessarily

possum n same as opossum; (Aust & NZ) same as phalanger; **play possum** pretend to be dead or asleep to deceive an opponent

post[1] n official system of delivering letters and parcels; (single collection or delivery of) letters and parcels sent by this system ▷ v send by post **keep someone posted** supply someone regularly with the latest information ▶ **postage** n charge for sending a letter or parcel by post ▶ **postal** adj ▶ **postal order** (Brit) written money order sent by post and cashed at a post office by the person who receives it ▶ **postbag** n postman's bag; post received by a magazine, famous person, etc. ▶ **postcode** n system of letters and numbers used to aid the sorting of mail ▶ **postie** n (Scot, Aust & NZ informal) postman ▶ **postman, postwoman** n person who collects and delivers post ▶ **postmark** n official mark stamped on letters showing place and date of posting ▶ **postmaster, postmistress** n (in some countries) official in charge of a post office ▶ **post office** place where postal business is conducted ▶ **post shop** (NZ) shop providing postal services

post² n length of wood, concrete, etc. fixed upright to support or mark something ▷ v put up (a notice) in a public place

post³ n job; position to which someone, esp. a soldier, is assigned for duty; military establishment ▷ v send (a person) to a new place to work; put (a guard etc.) on duty

post- prefix after, later than e.g. postwar

postcard n card for sending a message by post without an envelope

postdate v write a date on (a cheque) that is later than the actual date

poster n large picture or notice stuck on a wall

posterior n buttocks ▷ adj behind, at the back of

posterity n future generations, descendants

postern n small back door or gate

postgraduate n person with a degree who is studying for a more advanced qualification

posthaste adv with great speed

posthumous [poss-tume-uss] adj occurring after one's death ▶ **posthumously** adv

postilion, postillion n (Hist) person riding one of a pair of horses drawing a carriage

postmortem n medical examination of a body to establish the cause of death

postnatal adj occurring after childbirth

postpone v put off to a later time ▶ **postponement** n

postscript n passage added at the end of a letter

postulant n candidate for admission to a religious order

postulate v assume to be true as the basis of an argument or theory

posture n position or way in which someone stands, walks, etc. ▷ v behave in an exaggerated way to get attention

posy n, pl -sies small bunch of flowers

pot¹ n deep round container; teapot ▷ pl (informal) large amount ▷ v potting, potted plant in a pot; (Snooker) hit (a ball) into a pocket ▶ **potted** adj grown in a pot; (of meat or fish) cooked or preserved in a pot; (informal) abridged ▶ **pot shot** shot taken without aiming carefully ▶ **potting shed** shed where plants are potted

pot² n (slang) cannabis

potable [pote-a-bl] adj drinkable

potash n white powdery substance obtained from ashes and used as fertilizer

potassium n (Chem) silvery metallic element

potato n, pl -toes roundish starchy vegetable that grows underground

poteen n (in Ireland) illegally made alcoholic drink

potent adj having great power or influence; (of a male) capable of having sexual intercourse ▶ **potency** n

potentate n ruler or monarch

potential adj possible but not yet actual ▷ n ability or talent not yet fully used; (Electricity) level of electric pressure ▶ **potentially** adv ▶ **potentiality** n, pl -ties

pothole n hole in the surface of a road; deep hole in a limestone area ▶ **potholing** n sport of exploring underground caves ▶ **potholer** n

potion n dose of medicine or poison

potjie n (S Afr) three-legged iron pot used for cooking over a wood fire

potluck n **take potluck** accept whatever happens to be available

potoroo n, pl -**roos** Australian leaping rodent

potpourri [po-poor-ee] n fragrant mixture of dried flower petals; assortment or medley

pottage n (old-fashioned) thick soup or stew

potter¹ n person who makes pottery

potter² v be busy in a pleasant but aimless way

pottery n, pl -**ries** articles made from baked clay; place where they are made

potty¹ adj -**tier, -tiest** (informal) crazy or silly

potty² n, pl -**ties** bowl used by a small child as a toilet

pouch n small bag; baglike pocket of skin on an animal

pouf, pouffe [poof] n large solid cushion used as a seat

poulterer n (Brit) person who sells poultry

poultice [pole-tiss] n moist dressing, often heated, applied to inflamed skin

poultry n domestic fowls

pounce v spring upon suddenly to attack or capture ▷ n pouncing

pound¹ n monetary unit of Britain and some other countries; unit of weight equal to 0.454 kg

pound² v hit heavily and repeatedly; crush to pieces or powder; (of the heart) throb heavily; run heavily

pound³ n enclosure for stray animals or officially removed vehicles

pour v flow or cause to flow out in a stream; rain heavily; come or go in large numbers

pout v thrust out one's lips, look sulky ▷ n pouting look

poverty n state of being without enough food or money; lack of, scarcity

POW prisoner of war

powder n substance in the form of tiny loose particles; medicine or cosmetic in this form ▷ v apply powder to ▶ **powdered** adj in the form of a powder e.g. powdered milk ▶ **powdery** adj ▶ **powder room** (old-fashioned) ladies' toilet

power n ability to do or act; strength; position of authority or control; (Maths) product from continuous multiplication of a number by itself; (Physics) rate at which work is done; electricity supply; particular form of energy e.g. nuclear power ▶ **powered** adj having or operated by mechanical or electrical power ▶ **powerful** adj ▶ **powerless** adj ▶ **power cut** temporary interruption in the supply of electricity ▶ **power point** socket on a wall for plugging in electrical appliances ▶ **power station** installation for generating and distributing electric power

powwow n (informal) talk or conference

pox n disease in which skin pustules form; (informal) syphilis

pp (in signing a document) for and on behalf of

pp. pages

PPTA (in New Zealand) Post Primary Teachers Association

PR proportional representation; public relations

practicable adj capable of being done successfully; usable ▶ **practicability** n

practical adj involving experience or actual use rather than theory; sensible, useful, and effective;

good at making or doing things; in effect though not in name ▷ n examination in which something has to be done or made ▶ **practically** adv ▶ **practical joke** trick intended to make someone look foolish

practice n something done regularly or habitually; repetition of something so as to gain skill; doctor's or lawyer's place of work **in practice** in reality: referring to what actually happens as distinct from what is supposed to happen **put into practice** carry out, do

> SPELLING TIP
> Note the -ice ending for the noun, with -ise for the verb (practise)

practise v do repeatedly so as to gain skill; take part in, follow (a religion etc.); work at e.g. practise medicine; do habitually

practitioner n person who practises a profession

pragmatic adj concerned with practical consequences rather than theory ▶ **pragmatism** n ▶ **pragmatist** n

prairie n large treeless area of grassland, esp. in N America ▶ **prairie dog** rodent that lives in burrows in the N American prairies

praise v express approval or admiration of (someone or something); express honour and thanks to (one's God) ▷ n something said or written to show approval or admiration **sing someone's praises** praise someone highly ▶ **praiseworthy** adj

praline [prah-leen] n sweet made of nuts and caramelized sugar

pram n four-wheeled carriage for a baby, pushed by hand

prance v walk with exaggerated bouncing steps

prang v, n (slang) (have) a crash in a car or aircraft

prank n mischievous trick

prat n (Brit, Aust & NZ informal) stupid person

prattle v chatter in a childish or foolish way ▷ n childish or foolish talk

prawn n edible shellfish like a large shrimp

praxis n practice as opposed to theory

pray v say prayers; ask earnestly, entreat

prayer n thanks or appeal addressed to one's God; set form of words used in praying; earnest request

pre- prefix before, beforehand e.g. prenatal; prerecorded; preshrunk

preach v give a talk on a religious theme as part of a church service; speak in support of (an idea, principle, etc.)

preacher n person who preaches, esp. in church

preamble n introductory part to something said or written

prearranged adj arranged beforehand

prebendary n, pl -daries clergyman who is a member of the chapter of a cathedral

precarious adj insecure, unsafe, likely to fall or collapse ▶ **precariously** adv

precaution n action taken in advance to prevent something bad happening ▶ **precautionary** adj

precede v go or be before ▶ **precedence** [press-ee-denss] n formal order of rank or position **take precedence over** be more important than ▶ **precedent** n previous case or occurrence regarded as an example to be followed

precentor n person who leads the singing in a church

precept n rule of behaviour ▶ **preceptive** adj

precinct n (Brit, Aust & S Afr) area in a town closed to traffic; (Brit, Aust & S Afr) enclosed area round a building; (US) administrative area of a city ▷ pl surrounding region

precious adj of great value and importance; loved and treasured; (of behaviour) affected, unnatural ▶ **precious metal** gold, silver, or platinum ▶ **precious stone** rare mineral, such as a ruby, valued as a gem

precipice n very steep face of cliff or rockface ▶ **precipitous** adj sheer

precipitate v cause to happen suddenly; (Chem) cause to be deposited in solid form from a solution; throw headlong ▷ adj done rashly or hastily ▷ n (Chem) substance precipitated from a solution ▶ **precipitately** adv ▶ **precipitation** n precipitating; rain, snow, etc.

précis [pray-see] n, pl **précis** short written summary of a longer piece ▷ v make a précis of

precise adj exact, accurate in every detail; strict in observing rules or standards ▶ **precisely** adv ▶ **precision** n

preclude v make impossible to happen

precocious adj having developed or matured early or too soon ▶ **precocity** n

precognition n alleged ability to foretell the future

preconceived adj (of an idea) formed without real experience or reliable information ▶ **preconception** n

precondition n something that must happen or exist before something else can

precursor n something that precedes and is a signal of something else; predecessor

predate v occur at an earlier date than; write a date on (a document) that is earlier than the actual date

predatory [pred-a-tree] adj habitually hunting and killing other animals for food ▶ **predator** n predatory animal

predecease v die before (someone else)

predecessor n person who precedes another in an office or position; ancestor

predestination n (Theology) belief that future events have already been decided by God or fate ▶ **predestined** adj

predetermined adj decided in advance

predicament n embarrassing or difficult situation

predicate n (Grammar) part of a sentence in which something is said about the subject, e.g. went home in I went home ▷ v declare or assert

predict v tell about in advance, prophesy ▶ **predictable** adj ▶ **prediction** n ▶ **predictive** adj relating to or able to make predictions; (of a word processor or mobile phone) able to complete words after only part of a word has been keyed

predilection n (formal) preference or liking

predispose v influence (someone) in favour of something; make (someone) susceptible to something ▶ **predisposition** n

predominate v be the main or controlling

P

element ▸ **predominance** n
▸ **predominant** adj
▸ **predominantly** adv

pre-eminent adj excelling all others, outstanding ▸ **pre-eminence** n

pre-empt v prevent an action by doing something which makes it pointless or impossible ▸ **pre-emption** n ▸ **pre-emptive** adj

preen v (of a bird) clean or trim (feathers) with the beak **preen oneself** smarten oneself; show self-satisfaction

prefab n prefabricated house

prefabricated adj (of a building) manufactured in shaped sections for rapid assembly on site

preface [pref-iss] n introduction to a book ▹ v serve as an introduction to (a book, speech, etc.)
▸ **prefatory** adj

prefect n senior pupil in a school, with limited power over others; senior administrative officer in some countries ▸ **prefecture** n office or area of authority of a prefect

prefer v **-ferring, -ferred** like better; (Law) bring (charges) before a court ▸ **preferable** adj more desirable ▸ **preferably** adv ▸ **preference** n ▸ **preferential** adj showing preference ▸ **preferment** n promotion or advancement

prefigure v represent or suggest in advance

prefix n letter or group of letters put at the beginning of a word to make a new word, such as un- in unhappy ▹ v put as an introduction or prefix (to)

pregnant adj carrying a fetus in the womb; full of meaning or significance e.g. a pregnant pause ▸ **pregnancy** n, pl **-cies**

prehensile adj capable of grasping

prehistoric adj of the period before written history begins
▸ **prehistory** n

prejudice n unreasonable or unfair dislike or preference ▹ v cause (someone) to have a prejudice; harm, cause disadvantage to ▸ **prejudicial** adj disadvantageous, harmful

> **SPELLING TIP**
> There are examples in the Bank of English of **prejudice** being misspelt as prejudice, with an extra d. Although d often combines with g in English, it is not necessary before j except in words beginning with the prefix ad-

prelate [prel-it] n bishop or other churchman of high rank

preliminary adj happening before and in preparation, introductory ▹ n, pl **-naries** preliminary remark, contest, etc.

prelude n introductory movement in music; event preceding and introducing something else

premarital adj occurring before marriage

premature adj happening or done before the normal or expected time; (of a baby) born before the end of the normal period of pregnancy ▸ **prematurely** adv

premeditated adj planned in advance ▸ **premeditation** n

premenstrual adj occurring or experienced before a menstrual period e.g. premenstrual tension

premier n prime minister ▹ adj chief, leading ▸ **premiership** n

première n first performance or showing of a play, film, etc.

premise, premiss n statement assumed to be true and used as the basis of reasoning

premises pl n house or other building and its land

premium n additional sum of money, as on a wage or charge; (regular) sum paid for insurance ▶ **at a premium** in great demand because scarce ▶ **premium bonds** (in Britain) savings certificates issued by the government, on which no interest is paid but cash prizes can be won

premonition n feeling that something unpleasant is going to happen; foreboding ▶ **premonitory** adj

prenatal adj before birth, during pregnancy

preoccupy v -pying, -pied fill the thoughts or attention of (someone) to the exclusion of other things ▶ **preoccupation** n

preordained adj decreed or determined in advance

prep. preparatory; preposition

prepacked adj sold already wrapped

prepaid adj paid for in advance

prepare v make or get ready ▶ **prepared** adj willing; ready ▶ **preparation** n preparing; something done in readiness for something else; mixture prepared for use as a cosmetic, medicine, etc. ▶ **preparatory** [prip-**par**-a-tree] adj preparing ▶ **preparatory school** (Brit & SAfr) private school for children between 7 and 13

preponderance n greater force, amount, or influence ▶ **preponderant** adj

preposition n word placed before a noun or pronoun to show its relationship with other words, such as by in go by bus ▶ **prepositional** adj

prepossessing adj making a favourable impression, attractive

preposterous adj utterly absurd

prep school n short for **preparatory school**

prepuce [pree-pyewss] n retractable fold of skin covering the tip of the penis, foreskin

prerecorded adj recorded in advance to be played or broadcast later

prerequisite n, adj (something) required before something else is possible

prerogative n special power or privilege

> **SPELLING TIP**
>
> The way **prerogative** is often pronounced is presumably the reason why *perogative* is a common way of misspelling it

presage [press-ij] v be a sign or warning of

Presbyterian n, adj (member) of a Protestant church governed by lay elders ▶ **Presbyterianism** n

presbytery n, pl -teries (Presbyterian Church) local church court; (RC Church) priest's house

prescience [press-ee-enss] n knowledge of events before they happen ▶ **prescient** adj

prescribe v recommend the use of (a medicine); lay down as a rule ▶ **prescription** n written instructions from a doctor for the making up and use of a medicine ▶ **prescriptive** adj laying down rules

presence n fact of being in a specified place; impressive dignified appearance ▶ **presence of mind** ability to act sensibly in a crisis

present¹ adj being in a specified place; existing or happening now; (Grammar) (of a verb tense) indicating that the action specified is taking place now ▷ n present

p

time or tense ▸ **presently** adv
soon; (US & Scot) now

present² n something given
to bring pleasure to another
person ▹ v introduce formally or
publicly; introduce and compere
(a TV or radio show); cause e.g.
present a difficulty; give, award
▸ **presentation** n ▸ **presentable**
adj attractive, neat, fit for people
to see ▸ **presenter** n person
introducing a TV or radio show

presentiment [priz-zen-tim-ent]
n sense of something unpleasant
about to happen

preserve v keep from being
damaged, changed, or ended;
treat (food) to prevent it
decaying ▹ n area of interest
restricted to a particular person
or group; fruit preserved by
cooking in sugar; area where
game is kept for private hunting
or fishing ▸ **preservation** n
▸ **preservative** n chemical that
prevents decay

preshrunk v (of fabric or a
garment) having been shrunk
during manufacture so that further
shrinkage will not occur when
washed

preside v be in charge, esp. of a
meeting

president n head of state in many
countries; head of a society,
institution, etc. ▸ **presidential** adj
▸ **presidency** n, pl -cies

press¹ v apply force or weight to;
squeeze; smooth by applying
pressure or heat; urge insistently;
crowd, push ▹ n printing machine
pressed for short of ▸ **pressing**
adj urgent ▸ **press box** room
at a sports ground reserved for
reporters ▸ **press conference**
interview for reporters given
by a celebrity

press² v **press into service** force to
be involved or used ▸ **press gang**
(Hist) group of men used to capture
men and boys and force them to
join the navy

pressure n force produced
by pressing; urgent claims or
demands; (Physics) force applied
to a surface per unit of area
▸ **pressure cooker** airtight pot
which cooks food quickly by steam
under pressure ▸ **pressure group**
group that tries to influence
policies, public opinion, etc.

prestidigitation n skilful quickness
with the hands, conjuring

prestige n high status or respect
resulting from success or
achievements ▸ **prestigious** adj

presto adv (Music) very quickly

prestressed adj (of concrete)
containing stretched steel wires to
strengthen it

presume v suppose to be the case;
dare (to) ▸ **presumably** adv one
supposes (that) ▸ **presumption**
n bold insolent behaviour; strong
probability ▸ **presumptive**
adj assumed to be true or valid
until the contrary is proved
▸ **presumptuous** adj doing things
one has no right to do

presuppose v need as a previous
condition in order to be true
▸ **presupposition** n

pretend v claim or give the
appearance of (something untrue)
to deceive or in play ▸ **pretender**
n person who makes a false or
disputed claim to a position of
power ▸ **pretence** n behaviour
intended to deceive, pretending
▸ **pretentious** adj making
(unjustified) claims to special merit
or importance ▸ **pretension** n

preternatural adj beyond what is
natural, supernatural

pretext n false reason given to hide the real one

pretty adj -tier, -tiest pleasing to look at ▷ adv fairly, moderately e.g. I'm pretty certain ▶ **prettily** adv ▶ **prettiness** n

pretzel n brittle salted biscuit

prevail v gain mastery; be generally established ▶ **prevailing** adj widespread; predominant ▶ **prevalence** n ▶ **prevalent** adj widespread, common

prevaricate v avoid giving a direct or truthful answer ▶ **prevarication** n

prevent v keep from happening or doing ▶ **preventable** adj ▶ **prevention** n ▶ **preventive** adj, n

preview n advance showing of a film or exhibition before it is shown to the public

previous adj coming or happening before ▶ **previously** adv

prey n animal hunted and killed for food by another; victim **bird of prey** bird that kills and eats other birds or animals ▶ **prey on** v hunt and kill for food; worry, obsess

price n amount of money for which a thing is bought or sold; unpleasant thing that must be endured to get something desirable ▷ v fix or ask the price of ▶ **priceless** adj very valuable; (informal) very funny ▶ **pricey** adj **pricier**, **priciest** (informal) expensive

prick v pierce lightly with a sharp point; cause to feel mental pain; (of an animal) make (the ears) stand erect ▷ n sudden sharp pain caused by pricking; mark made by pricking; remorse **prick up one's ears** listen intently

prickle n thorn or spike on a plant ▷ v have a tingling or pricking sensation ▶ **prickly** adj ▶ **prickly**

heat itchy rash occurring in hot moist weather

pride n feeling of pleasure and satisfaction when one has done well; too high an opinion of oneself; sense of dignity and self-respect; something that causes one to feel pride; group of lions **pride of place** most important position **pride oneself on** feel pride about

priest n (in the Christian church) a person who can administer the sacraments and preach; (in some other religions) an official who performs religious ceremonies ▶ **priestess** n fem ▶ **priesthood** n ▶ **priestly** adj

prig n self-righteous person who acts as if superior to others ▶ **priggish** adj ▶ **priggishness** n

prim adj primmer, primmest formal, proper, and rather prudish ▶ **primly** adv

prima ballerina n leading female ballet dancer

primacy n, pl -cies state of being first in rank, grade, etc.; office of an archbishop

prima donna n leading female opera singer; (informal) temperamental person

primaeval adj same as **primeval**

prima facie [prime-a-fay-shee] adv (Latin) as it seems at first

primal adj of basic causes or origins

primary adj chief, most important; being the first stage, elementary ▶ **primarily** adv ▶ **primary colours** (in physics) red, green, and blue or (in art) red, yellow, and blue, from which all other colours can be produced by mixing ▶ **primary school** school for children from five to eleven years or (in New Zealand) from five to thirteen years

p

primate[1] n member of an order of mammals including monkeys and humans

primate[2] n archbishop

prime adj main, most important; of the highest quality ▷ n time when someone is at his or her best or most vigorous ▷ v give (someone) information in advance to prepare them for something; prepare (a surface) for painting; prepare (a gun, pump, etc.) for use ▶ **primer** n special paint applied to bare wood etc. before the main paint ▶ **Prime Minister** leader of a government ▶ **prime number** number that can be divided exactly only by itself and one

primer n beginners' school book or manual

primeval [prime-ee-val] adj of the earliest age of the world

primitive adj of an early simple stage of development; basic, crude

primogeniture n system under which the eldest son inherits all his parents' property

primordial adj existing at or from the beginning

primrose n pale yellow spring flower

primula n type of primrose with brightly coloured flowers

Primus® n portable cooking stove used esp. by campers

prince n male member of a royal family, esp. the son of the king or queen; male ruler of a small country ▶ **princely** adj of or like a prince; generous, lavish, or magnificent ▶ **prince consort** husband of a reigning queen ▶ **Prince of Wales** eldest son of the British sovereign ▶ **princess** n female member of a royal family, esp. the daughter of the king or queen ▶ **Princess Royal** title

sometimes given to the eldest daughter of the British sovereign

principal adj main, most important ▷ n head of a school or college; person taking a leading part in something; sum of money lent on which interest is paid ▶ **principally** adv ▶ **principal boy** (Brit) leading male role in pantomime, played by a woman

USAGE NOTE
Do not confuse principal and principle

principality n, pl -ties territory ruled by a prince

principle n moral rule guiding behaviour; general or basic truth; scientific law concerning the working of something **in principle** in theory but not always in practice **on principle** because of one's beliefs

print v reproduce (a newspaper, book, etc.) in large quantities by mechanical or electronic means; reproduce (text or pictures) by pressing ink onto paper etc.; write in letters that are not joined up; stamp (fabric) with a design; (Photog) produce (pictures) from negatives ▷ n printed words etc.; printed copy of a painting; printed lettering; photograph; printed fabric; mark left on a surface by something that has pressed against it **out of print** no longer available from a publisher ▶ **printer** n person or company engaged in printing; machine that prints ▶ **printing** n ▶ **printed circuit** electronic circuit with wiring printed on an insulating base ▶ **print-out** n printed information from a computer

prior[1] adj earlier **prior to** before

prior[2] n head monk in a priory ▶ **prioress** n deputy head nun in a convent ▶ **priory** n, pl **-ries** place where certain orders of monks or nuns live

priority n, pl **-ties** most important thing that must be dealt with first; right to be or go before others

prise v force open by levering

prism n transparent block usu. with triangular ends and rectangular sides, used to disperse light into a spectrum or refract it in optical instruments ▶ **prismatic** adj of or shaped like a prism; (of colour) as if produced by refraction through a prism, rainbow-like

prison n building where criminals and accused people are held

prisoner n person held captive ▶ **prisoner of war** serviceman captured by an enemy in wartime

prissy adj **-sier, -siest** prim, correct, and easily shocked ▶ **prissily** adv

pristine adj clean, new, and unused

private adj for the use of one person or group only; secret; personal, unconnected with one's work; owned or paid for by individuals rather than by the government; quiet, not likely to be disturbed ▷ n soldier of the lowest rank ▶ **privately** adv ▶ **privacy** n

privateer n (Hist) privately owned armed vessel authorized by the government to take part in a war; captain of such a ship

privation n loss or lack of the necessities of life

privatize v sell (a publicly owned company) to individuals or a private company ▶ **privatization** n

privet n bushy evergreen shrub used for hedges

privilege n advantage or favour that only some people have ▶ **privileged** adj enjoying a special right or immunity

SPELLING TIP

Note that in **privilege** there is an i in each of the first two syllables, and that there is no d in the last: *priviledge* is a common misspelling

privy adj sharing knowledge of something secret ▷ n, pl **privies** (obs) toilet, esp. an outside one ▶ **Privy Council** private council of the British monarch

prize¹ n reward given for success in a competition etc. ▷ adj winning or likely to win a prize ▶ **prizefighter** n boxer who fights for money

prize² v value highly

prize³ v same as **prise**

pro¹ adv, prep in favour of ▶ **pros and cons** arguments for and against

pro² n, pl **pros** (informal) professional; prostitute

pro- prefix in favour of e.g. *pro-Russian*; instead of e.g. *pronoun*

probable adj likely to happen or be true ▶ **probability** n, pl **-ties**

probably adv in all likelihood

probate n process of proving the validity of a will; certificate stating that a will is genuine

probation n system of dealing with law-breakers, esp. juvenile ones, by placing them under supervision; period when someone is assessed for suitability for a job etc. ▶ **probationer** n person on probation

probe v search into or examine closely ▷ n surgical instrument used to examine a wound, cavity, etc.

probiotic adj, n (of) a bacterium that protects the body from harmful bacteria e.g. *probiotic yogurts*

probity n honesty, integrity

problem n something difficult to deal with or solve; question

or puzzle set for solution
▸ **problematic**, **problematical** adj

proboscis [pro-boss-iss] n long trunk or snout; elongated mouth of some insects

procedure n way of doing something, esp. the correct or usual one ▸ **procedural** adj

proceed v start or continue doing; (formal) walk, go; start a legal action; arise from ▸ **proceeds** pl n money obtained from an event or activity ▸ **proceedings** pl n organized or related series of events; minutes of a meeting; legal action

process n series of actions or changes; method of doing or producing something ▸ v handle or prepare by a special method of manufacture ▸ **processed** adj (of food) treated to prevent it decaying ▸ **processor** n

procession n line of people or vehicles moving forward together in order

proclaim v declare publicly ▸ **proclamation** n

proclivity n, pl -**ties** inclination, tendency

procrastinate v put off taking action, delay ▸ **procrastination** n

procreate v (formal) produce offspring ▸ **procreation** n

procurator fiscal n (in Scotland) law officer who acts as public prosecutor and coroner

procure v get, provide; obtain (people) to act as prostitutes ▸ **procurement** n ▸ **procurer**, **procuress** n person who obtains people to act as prostitutes

prod v **prodding**, **prodded** poke with something pointed; goad (someone) into action ▸ n prodding

prodigal adj recklessly extravagant, wasteful ▸ **prodigality** n

prodigy n, pl -**gies** person with some marvellous talent; wonderful thing ▸ **prodigious** adj very large, immense; wonderful ▸ **prodigiously** adv

produce v bring into existence; present to view, show; make, manufacture; present on stage, film, or television ▸ n food grown for sale ▸ **producer** n person with control over the making of a film, record, etc.; person or company that produces something

product n something produced; number resulting from multiplication ▸ **production** n producing; things produced; presentation of a play, opera, etc. ▸ **productive** adj producing large quantities; useful, profitable ▸ **productivity** n

profane adj showing disrespect for religion or holy things; (of language) coarse, blasphemous ▸ v treat (something sacred) irreverently, desecrate ▸ **profanation** n act of profaning ▸ **profanity** n, pl -**ties** profane talk or behaviour; blasphemy

profess v state or claim (something as true), sometimes falsely; have as one's belief or religion ▸ **professed** adj supposed

profession n type of work, such as being a doctor, that needs special training; all the people employed in a profession e.g. the legal profession; declaration of a belief or feeling ▸ **professional** adj working in a profession; taking part in an activity, such as sport or music, for money; very competent ▸ n person who works in a profession; person paid to take part in sport, music, etc. ▸ **professionally** adv ▸ **professionalism** n

professor n teacher of the highest rank in a university ▶ **professorial** adj ▶ **professorship** n

proffer v offer

proficient adj skilled, expert ▶ **proficiency** n

profile n outline, esp. of the face, as seen from the side; brief biographical sketch ▶ **profiling** n practice of categorizing and predicting the behaviour of people according to certain characteristics e.g. racial profiling

profit n money gained; benefit obtained ▷ v gain or benefit ▶ **profitable** adj making profit ▶ **profitably** adv ▶ **profitability** n ▶ **profiteer** n person who makes excessive profits at the expense of the public ▶ **profiteering** n

profligate adj recklessly extravagant; shamelessly immoral ▷ n profligate person ▶ **profligacy** n

pro forma adj (Latin) prescribing a set form

profound adj showing or needing great knowledge; strongly felt, intense ▶ **profundity** n, pl **-ties**

profuse adj plentiful ▶ **profusion** n

progeny [proj-in-ee] n, pl **-nies** children ▶ **progenitor** [pro-jen-it-er] n ancestor

progesterone n hormone which prepares the womb for pregnancy and prevents further ovulation

prognosis n, pl **-noses** doctor's forecast about the progress of an illness; any forecast

prognostication n forecast or prediction

program n sequence of coded instructions for a computer ▷ v **-gramming, -grammed** arrange (data) so that it can be processed by a computer; feed a program into (a computer) ▶ **programmer** n ▶ **programmable** adj

programme n planned series of events; broadcast on radio or television; list of items or performers in an entertainment

progress n improvement, development; movement forward ▷ v become more advanced or skilful; move forward **in progress** taking place ▶ **progression** n ▶ **progressive** adj favouring political or social reform; happening gradually ▶ **progressively** adv

prohibit v forbid or prevent from happening ▶ **prohibition** n act of forbidding; ban on the sale or drinking of alcohol ▶ **prohibitive** adj (of prices) too high to be affordable ▶ **prohibitively** adv

project n planned scheme to do or examine something over a period ▷ v make a forecast based on known data; make (a film or slide) appear on a screen; communicate (an impression); stick out beyond a surface or edge ▶ **projector** n apparatus for projecting photographic images, films, or slides on a screen ▶ **projection** n ▶ **projectionist** n person who operates a projector

projectile n object thrown as a weapon or fired from a gun

prolapse n slipping down of an internal organ of the body from its normal position

prole adj, n (chiefly Brit slang) proletarian

proletariat [pro-lit-air-ee-at] n working class ▶ **proletarian** adj, n

proliferate v grow or reproduce rapidly ▶ **proliferation** n

prolific adj very productive ▶ **prolifically** adv

prolix adj (of speech or a piece of writing) overlong and boring

prologue n introduction to a play or book

prolong v make (something) last longer ▶ **prolongation** n

prom n short for **promenade**, **promenade concert**

promenade n (chiefly Brit) paved walkway along the seafront at a holiday resort ▷ v, n (old-fashioned) (take) a leisurely walk ▶ **promenade concert** (Brit) concert at which part of the audience stands rather than sits

prominent adj very noticeable; famous, widely known ▶ **prominently** adv ▶ **prominence** n

promiscuous adj having many casual sexual relationships ▶ **promiscuity** n

promise v say that one will definitely do or not do something; show signs of, seem likely ▷ n undertaking to do or not to do something; indication of future success ▶ **promising** adj likely to succeed or turn out well

promo n, pl -**mos** (informal) short film to promote a product

promontory n, pl -**ries** point of high land jutting out into the sea

promote v help to make (something) happen or increase; raise to a higher rank or position; encourage the sale of by advertising ▶ **promoter** n person who organizes or finances an event etc. ▶ **promotion** n ▶ **promotional** adj

prompt v cause (an action); remind (an actor or speaker) of words that he or she has forgotten ▷ adj done without delay ▷ adv exactly e.g. six o'clock prompt ▶ **promptly** adv immediately, without delay ▶ **promptness** n ▶ **prompter**, **prompt** n person offstage who prompts actors

promulgate v put (a law etc.) into effect by announcing it

officially; make widely known ▶ **promulgation** n

prone adj (foll by to) likely to do or be affected by (something); lying face downwards

prong n one spike of a fork or similar instrument ▶ **pronged** adj

pronoun n word, such as she or it, used to replace a noun

pronounce v form the sounds of (words or letters), esp. clearly or in a particular way; declare formally or officially ▶ **pronounceable** adj ▶ **pronounced** adj very noticeable ▶ **pronouncement** n formal announcement ▶ **pronunciation** n way in which a word or language is pronounced

USAGE NOTE
Note the difference in spelling between pronounce and pronunciation. The pronunciation also changes from 'nown' to 'nun'.

pronto adv (informal) at once

proof n evidence that shows that something is true or has happened; copy of something printed, such as the pages of a book, for checking before final production ▷ adj able to withstand e.g. proof against criticism; denoting the strength of an alcoholic drink e.g. seventy proof ▶ **proofread** v read and correct (printer's proofs) ▶ **proofreader** n

prop¹ v **propping, propped** support (something) so that it stays upright or in place ▷ n pole, beam, etc. used as a support

prop² n movable object used on the set of a film or play

prop³ n (informal) propeller

propaganda n (organized promotion of) information to assist or damage the cause of a government or movement ▶ **propagandist** n

propagate v spread (information and ideas); reproduce, breed, or grow ▶ **propagation** n

propane n flammable gas found in petroleum and used as a fuel

propel v -**pelling, -pelled** cause to move forward ▶ **propellant** n something that provides or causes propulsion; gas used in an aerosol spray ▶ **propulsion** n method by which something is propelled; act of propelling or state of being propelled

propeller n revolving shaft with blades for driving a ship or aircraft

propensity n, pl -**ties** natural tendency

proper adj real or genuine; suited to a particular purpose; correct in behaviour; excessively moral; (Brit, Aust & NZ informal) complete ▶ **properly** adv

property n, pl -**ties** something owned; possessions collectively; land or buildings owned by somebody; quality or attribute

prophet n person supposedly chosen by God to spread His word; person who predicts the future ▶ **prophetic** adj ▶ **prophetically** adv ▶ **prophecy** n, pl -**cies** prediction; message revealing God's will ▶ **prophesy** v -**sying, -sied** foretell

SPELLING TIP
Note the difference in spelling between the noun (**prophecy**) and the verb (**prophesy**)

prophylactic n, adj (drug) used to prevent disease

propitiate v appease, win the favour of ▶ **propitiation** n ▶ **propitious** adj favourable or auspicious

proponent n person who argues in favour of something

proportion n relative size or extent; correct relation between connected parts; part considered with respect to the whole ▷ pl dimensions or size ▷ v adjust in relative amount or size **in proportion** comparable in size, rate of increase, etc.; without exaggerating ▶ **proportional, proportionate** adj being in proportion ▶ **proportionally, proportionately** adv

propose v put forward for consideration; nominate; intend or plan (to do); make an offer of marriage ▶ **proposal** n ▶ **proposition** n offer; statement or assertion; (Maths) theorem; (informal) thing to be dealt with ▷ v (informal) ask (someone) to have sexual intercourse

propound v put forward for consideration

proprietor n owner of a business establishment ▶ **proprietress** n fem ▶ **proprietary** adj made and distributed under a trade name; denoting or suggesting ownership

propriety n, pl -**ties** correct conduct

propulsion n see **propel**

pro rata adv, adj (Latin) in proportion

prorogue v suspend (parliament) without dissolving it ▶ **prorogation** n

prosaic [pro-**zay**-ik] adj lacking imagination, dull ▶ **prosaically** adv

proscenium n, pl -**nia** or -**niums** arch in a theatre separating the stage from the auditorium

proscribe v prohibit, outlaw ▶ **proscription** n ▶ **proscriptive** adj

prose n ordinary speech or writing as opposed to poetry

prosecute v bring a criminal charge against; continue to do ▶ **prosecution** n ▶ **prosecutor** n

proselyte [pross-ill-ite] n recent convert

proselytize [pross-ill-it-ize] v attempt to convert

prospect n something anticipated; (old-fashioned) view from a place ▷ pl probability of future success ▷ v explore, esp. for gold
▶ **prospective** adj future; expected
▶ **prospector** n ▶ **prospectus** n booklet giving details of a university, company, etc.

prosper v be successful
▶ **prosperity** n success and wealth
▶ **prosperous** adj

prostate n gland in male mammals that surrounds the neck of the bladder

prosthesis [pross-**theess**-iss] n, pl -**ses** [-seez] artificial body part, such as a limb or breast
▶ **prosthetic** adj

prostitute n person who offers sexual intercourse in return for payment ▷ v make a prostitute of; offer (oneself or one's talents) for unworthy purposes
▶ **prostitution** n

prostrate adj lying face downwards; physically or emotionally exhausted ▷ v lie face downwards; exhaust physically or emotionally ▶ **prostration** n

protagonist n supporter of a cause; leading character in a play or a story

protea [pro-tee-a] n African shrub with showy flowers

protean [pro-**tee**-an] adj constantly changing

protect v defend from trouble, harm, or loss ▶ **protection** n ▶ **protectionism** n policy of protecting industries by taxing competing imports
▶ **protectionist** n, adj
▶ **protective** adj giving protection

e.g. protective clothing; tending or wishing to protect someone
▶ **protector** n person or thing that protects; regent ▶ **protectorate** n territory largely controlled by a stronger state; (period of) rule of a regent

protégé, (fem) **protégée** [pro-ti-zhay] n person who is protected and helped by another

protein n any of a group of complex organic compounds that are essential for life

pro tempore adv, adj for the time being (also **pro tem**)

protest n declaration or demonstration of objection
▷ v object, disagree; assert formally ▶ **protestation** n strong declaration

Protestant n follower of any of the Christian churches that split from the Roman Catholic Church in the sixteenth century ▷ adj of or relating to such a church
▶ **Protestantism** n

proto- combining form first e.g. protohuman

protocol n rules of behaviour for formal occasions; (Computing) set of rules for transfer of data, esp. between different systems

proton n positively charged particle in the nucleus of an atom

protoplasm n substance forming the living contents of a cell

prototype n original or model to be copied or developed

protozoan [pro-toe-**zoe**-an] n, pl -**zoa** microscopic one-celled creature

protracted adj lengthened or extended

protractor n instrument for measuring angles

protrude v stick out, project
▶ **protrusion** n

protuberant adj swelling out, bulging ▸ **protuberance** n

proud adj feeling pleasure and satisfaction; feeling honoured; thinking oneself superior to other people; dignified ▸ **proudly** adv

prove v **proving, proved, proved** or **proven** establish the validity of; demonstrate, test; be found to be ▸ **proven** adj known from experience to work

provenance [prov-in-anss] n place of origin

provender n (old-fashioned) fodder

proverb n short saying that expresses a truth or gives a warning ▸ **proverbial** adj

provide v make available **provided that, providing** on condition that ▸ **provider** n ▸ **provide for** v take precautions (against); support financially

USAGE NOTE
The usage is either providing on its own or provided that (not providing that)

providence n God or nature seen as a protective force that arranges people's lives ▸ **provident** adj thrifty; showing foresight ▸ **providential** adj lucky

province n area governed as a unit of a country or empire; area of learning, activity, etc. ▷ pl parts of a country outside the capital ▸ **provincial** adj of a province or the provinces; unsophisticated and narrow-minded ▷ n unsophisticated person; person from a province or the provinces ▸ **provincialism** n narrow-mindedness and lack of sophistication

provision n act of supplying something; something supplied; (Law) condition incorporated in a document ▷ pl food ▷ v supply with

food ▸ **provisional** adj temporary or conditional ▸ **provisionally** adv

proviso [pro-vize-oh] n, pl **-sos** or **-soes** condition, stipulation

provoke v deliberately anger; cause (an adverse reaction) ▸ **provocation** n ▸ **provocative** adj

provost n head of certain university colleges in Britain; chief councillor of a Scottish town

prow n bow of a vessel

prowess n superior skill or ability; bravery, fearlessness

prowl v move stealthily around a place as if in search of prey or plunder ▷ n prowling

prowler n person who moves stealthily around a place as if in search of prey or plunder

proximity n nearness in space, time, or in a series ▸ **proximate** adj

proxy n, pl **proxies** person authorized to act on behalf of someone else; authority to act on behalf of someone else

prude n person who is excessively modest, prim, or proper ▸ **prudish** adj ▸ **prudery** n

prudent adj cautious, discreet, and sensible ▸ **prudence** n ▸ **prudential** adj (old-fashioned) prudent

prune[1] n dried plum

prune[2] v cut off dead parts or excessive branches from (a tree or plant); shorten, reduce

prurient adj excessively interested in sexual matters ▸ **prurience** n

pry v **prying, pried** make an impertinent or uninvited inquiry into a private matter

PS postscript

PSA (in New Zealand) Public Service Association

psalm n sacred song ▸ **psalmist** n writer of psalms

Psalter n book containing (a version of) psalms from the Bible ▸ **psaltery** n, pl **-ries** ancient instrument played by plucking strings

PSBR (in Britain) public sector borrowing requirement

psephology [sef-**fol**-a-jee] n statistical study of elections

pseud n (informal) pretentious person

pseudo- combining form false, pretending, or unauthentic e.g. pseudoclassical

pseudonym n fictitious name adopted esp. by an author ▸ **pseudonymous** adj

psittacosis n disease of parrots that can be transmitted to humans

psoriasis [so-**rye**-a-siss] n skin disease with reddish spots and patches covered with silvery scales

psyche [**sye**-kee] n human mind or soul

psychedelic adj denoting a drug that causes hallucinations; having vivid colours and complex patterns similar to those experienced during hallucinations

> SPELLING TIP
> The main problem with **psychedelic** is which vowel follows the ch; it should be e

psychiatry n branch of medicine concerned with mental disorders ▸ **psychiatric** adj ▸ **psychiatrist** n

psychic adj (also **psychical**) having mental powers which cannot be explained by natural laws; relating to the mind ▸ n person with psychic powers

psycho n, pl **-chos** (informal) psychopath

psychoanalysis n method of treating mental and emotional disorders by discussion and analysis of one's thoughts and feelings ▸ **psychoanalyse** v ▸ **psychoanalyst** n

psychology n, pl **-gies** study of human and animal behaviour; (informal) person's mental make-up ▸ **psychologist** n ▸ **psychological** adj of or affecting the mind; of psychology ▸ **psychologically** adv

psychopath n person afflicted with a personality disorder causing him or her to commit anti-social or violent acts ▸ **psychopathic** adj

psychosis n, pl **-ses** severe mental disorder in which the sufferer's contact with reality becomes distorted ▸ **psychotic** adj

psychosomatic adj (of a physical disorder) thought to have psychological causes

psychotherapy n treatment of nervous disorders by psychological methods ▸ **psychotherapeutic** adj ▸ **psychotherapist** n

psych up v prepare (oneself) mentally for a contest or task

PT (old-fashioned) physical training

pt part; point

pt. pint

PTA Parent-Teacher Association

ptarmigan [**tar**-mig-an] n bird of the grouse family which turns white in winter

pterodactyl [terr-roe-**dak**-til] n extinct flying reptile with batlike wings

PTO please turn over

ptomaine [**toe**-main] n any of a group of poisonous alkaloids found in decaying matter

Pty (Aust, NZ & SAfr) Proprietary

pub n building with a bar licensed to sell alcoholic drinks

puberty n beginning of sexual maturity ▸ **pubertal** adj

pubescent adj reaching or having reached puberty; covered with fine short hairs or down, as some plants and animals ▸ **pubescence** n

pubic adj of the lower abdomen e.g. pubic hair

public adj of or concerning the people as a whole; for use by everyone; well-known; performed or made openly ▸ in the community, people in general ▸ **publicly** adv ▸ **public house** pub ▸ **public relations** promotion of a favourable opinion towards an organization among the public ▸ **public school** private fee-paying school in Britain ▸ **public-spirited** adj having or showing an active interest in the good of the community

publican n (Brit, Aust & NZ) person who owns or runs a pub

publicity n process or information used to arouse public attention; public interest so aroused ▸ **publicist** n person, esp. a press agent or journalist, who publicizes something ▸ **publicize** v bring to public attention

publish v produce and issue (printed matter) for sale; announce formally or in public ▸ **publication** n ▸ **publisher** n

puce adj purplish-brown

puck¹ n small rubber disc used in ice hockey

puck² n mischievous or evil spirit ▸ **puckish** adj

pucker v gather into wrinkles ▸ n wrinkle or crease

pudding n dessert, esp. a cooked one served hot; savoury dish with pastry or batter e.g. steak-and-kidney pudding; sausage-like mass of meat e.g. black pudding

puddle n small pool of water, esp. of rain

puerile adj silly and childish

puerperal [pew-**er**-per-al] adj concerning the period following childbirth

puff n (sound of) a short blast of breath, wind, etc.; act of inhaling cigarette smoke ▸ v blow or breathe in short quick draughts; take draws at (a cigarette); send out in small clouds; swell out ▸ **puff out** of breath ▸ **puffy** adj ▸ **puffball** n ball-shaped fungus ▸ **puff pastry** light flaky pastry

puffin n black-and-white sea bird with a brightly-coloured beak

pug n small snub-nosed dog ▸ **pug nose** short stubby upturned nose

pugilist [pew-jil-ist] n boxer ▸ **pugilism** n ▸ **pugilistic** adj

pugnacious adj ready and eager to fight ▸ **pugnacity** n

puissance [pwee-**sonce**] n showjumping competition that tests a horse's ability to jump large obstacles

puke (slang) v vomit ▸ n act of vomiting; vomited matter

pulchritude n (lit) beauty

pull v exert force on (an object) to move it towards the source of the force; strain or stretch; remove or extract; attract ▸ n act of pulling; force used in pulling; act of taking in drink or smoke; (informal) power, influence ▸ **pull in** v (of a vehicle or driver) draw in to the side of the road or stop; reach a destination; attract in large numbers; (Brit, Aust & NZ slang) arrest ▸ **pull off** v (informal) succeed in performing ▸ **pull out** v (of a vehicle or driver) move away from the side of the road or move out to overtake; (of a train) depart; withdraw; remove by pulling ▸ **pull up** v (of a vehicle or driver) stop; remove by the roots; reprimand

pullet n young hen

pulley n wheel with a grooved rim in which a belt, chain, or piece of rope runs in order to lift weights by a downward pull

Pullman n, pl **-mans** luxurious railway coach

pullover n sweater that is pulled on over the head

pulmonary adj of the lungs

pulp n soft wet substance made from crushed or beaten matter; flesh of a fruit; poor-quality books and magazines ▷ v reduce to pulp

pulpit n raised platform for a preacher

pulsar n small dense star which emits regular bursts of radio waves

pulse¹ n regular beating of blood through the arteries at each heartbeat; any regular beat or vibration ▶ **pulsate** v throb, quiver ▶ **pulsation** n

pulse² n edible seed of a pod-bearing plant such as a bean or pea

pulverize v reduce to fine pieces; destroy completely

puma n large American wild cat with a greyish-brown coat

pumice [pumm-iss] n light porous stone used for scouring

pummel v **-melling, -melled** strike repeatedly with or as if with the fists

pump¹ n machine used to force a liquid or gas to move in a particular direction ▷ v raise or drive with a pump; supply in large amounts; operate or work in the manner of a pump; extract information from

pump² n light flat-soled shoe

pumpkin n large round fruit with an orange rind, soft flesh, and many seeds, eaten as a vegetable

pun n use of words to exploit double meanings for humorous effect ▷ v **punning, punned** make puns

punch¹ v strike with a clenched fist ▷ n blow with a clenched fist; (informal) effectiveness or vigour ▶ **punchy** adj forceful ▶ **punch-drunk** adj dazed by or as if by repeated blows to the head

punch² n tool or machine for shaping, piercing, or engraving ▷ v pierce, cut, stamp, shape, or drive with a punch

punch³ n drink made from a mixture of wine, spirits, fruit, sugar, and spices

punctilious adj paying great attention to correctness in etiquette; careful about small details

punctual adj arriving or taking place at the correct time ▶ **punctuality** n ▶ **punctually** adv

punctuate v put punctuation marks in; interrupt at frequent intervals ▶ **punctuation** n (use of) marks such as commas, colons, etc. in writing, to assist in making the sense clear

puncture n small hole made by a sharp object, esp. in a tyre ▷ v pierce a hole in

pundit n expert who speaks publicly on a subject

pungent adj having a strong sharp bitter flavour ▶ **pungency** n

punish v cause (someone) to suffer or undergo a penalty for some wrongdoing ▶ **punishing** adj harsh or difficult ▶ **punishment** n ▶ **punitive** [pew-nit-tiv] adj relating to punishment

punk n anti-Establishment youth movement and style of rock music of the late 1970s; follower of this music; worthless person

punnet n small basket for fruit

punt¹ n open flat-bottomed boat propelled by a pole ▷ v travel in a punt

punt² (*Sport*) *n* kick of a ball before it touches the ground when dropped from the hands ▷ *v* kick (a ball) in this way

punt³ *n* former monetary unit of the Irish Republic

punter *n* who bets; (*Brit, Aust & NZ*) any member of the public

puny *adj* **-nier, -niest** small and feeble

pup *n* young of certain animals, such as dogs and seals

pupa *n, pl* **-pae** *or* **-pas** insect at the stage of development between a larva and an adult

pupil¹ *n* person who is taught by a teacher

pupil² *n* round dark opening in the centre of the eye

puppet *n* small doll or figure moved by strings or by the operator's hand; person or country controlled by another ▶ **puppeteer** *n*

puppy *n, pl* **-pies** young dog

purchase *v* obtain by payment ▷ *n* thing that is bought; act of buying; leverage, grip ▶ **purchaser** *n*

purdah *n* Muslim and Hindu custom of keeping women in seclusion, with clothing that conceals them completely when they go out

pure *adj* unmixed, untainted; innocent; complete e.g. *pure delight*; concerned with theory only e.g. *pure mathematics* ▶ **purely** *adv* ▶ **purity** *n* ▶ **purify** *v* **-fying, -fied** make or become pure ▶ **purification** *n* ▶ **purist** *n* person concerned with strict obedience to the traditions of a subject

purée [**pure-ray**] *n* pulp of cooked food ▷ *v* **-réeing, -réed** make into a purée

purgatory *n* place or state of temporary suffering; (**P-**) (RC Church) place where souls of the dead undergo punishment for their sins before being admitted to Heaven ▶ **purgatorial** *adj*

purge *v* rid (a thing or place) of (unwanted things or people) ▷ *n* purging ▶ **purgative** *n, adj* (medicine) designed to cause defecation

Puritan *n* (*Hist*) member of the English Protestant group who wanted simpler church ceremonies; (**p-**) person with strict moral and religious principles ▶ **puritanical** *adj* ▶ **puritanism** *n*

purl *n* stitch made by knitting a plain stitch backwards ▷ *v* knit in purl

purlieus [**per-lyooz**] *pl n* (*lit*) outskirts

purloin *v* steal

purple *adj, n* (of) a colour between red and blue

purport *v* claim (to be or do something) ▷ *n* apparent meaning, significance

purpose *n* reason for which something is done or exists; determination; practical advantage or use e.g. *use the time to good purpose* ▶ **purposely** *adv* intentionally (also **on purpose**)

purr *v* (of cats) make low vibrant sound, usu. when pleased ▷ *n* this sound

purse *n* small bag for money; (*US & NZ*) handbag; financial resources; prize money ▷ *v* draw (one's lips) together into a small round shape ▶ **purser** *n* ship's officer who keeps the accounts

pursue *v* chase; follow (a goal); engage in; continue to discuss or ask about (something) ▶ **pursuer** *n* ▶ **pursuit** *n* pursuing; occupation or pastime

SPELLING TIP

This family of words should be spelt with a *u* in each of the first two syllables, as in **pursuing** and **pursued**

purulent [pure-yoo-lent] *adj* of or containing pus

purvey *v* supply (provisions) ▸ **purveyor** *n*

purview *n* scope or range of activity or outlook

pus *n* yellowish matter produced by infected tissue

push *v* move or try to move by steady force; drive or spur (oneself or another person) to do something; (*informal*) sell (drugs) illegally ▸ *n* act of pushing; special effort **the push** (*slang*) dismissal from a job or relationship ▸ **pusher** *n* person who sells illegal drugs ▸ **pushy** *adj* too assertive or ambitious ▸ **pushchair** *n* (*Brit*) folding chair on wheels for a baby

pusillanimous *adj* timid and cowardly ▸ **pusillanimity** *n*

puss, pussy *n, pl* **pusses** or **pussies** (*informal*) cat

pussyfoot *v* (*informal*) behave too cautiously

pustule *n* pimple containing pus

put *v* **putting, put** cause to be (in a position, state, or place); express; throw (the shot) in the shot put ▸ *n* throw in putting the shot ▸ **put across** *v* express successfully ▸ **put off** *v* postpone; disconcert; repel ▸ **put up** *v* erect; accommodate; nominate ▸ **put-upon** *adj* taken advantage of

putative *adj* reputed, supposed

putrid *adj* rotten and foul-smelling ▸ **putrefy** *v* **-fying, -fied** rot and produce an offensive smell ▸ **putrefaction** *n* ▸ **putrescent** *adj* rotting

putsch *n* sudden violent attempt to remove a government from power

putt (*Golf*) *n* stroke on the putting green to roll the ball into or near the hole ▸ *v* strike (the ball) in this way ▸ **putter** ▸ *n* golf club for putting

putty *n* adhesive used to fix glass into frames and fill cracks in woodwork

puzzle *v* perplex and confuse or be perplexed or confused ▸ *n* problem that cannot be easily solved; toy, game, or question that requires skill or ingenuity to solve ▸ **puzzlement** *n* ▸ **puzzling** *adj*

PVC polyvinyl chloride: plastic material used in clothes etc.

Pygmy *n, pl* **-mies** member of one of the very short peoples of Equatorial Africa ▸ *adj* (**p-**) very small

pyjamas *pl n* loose-fitting trousers and top worn in bed

pylon *n* steel tower-like structure supporting electrical cables

pyramid *n* solid figure with a flat base and triangular sides sloping upwards to a point; building of this shape, esp. an ancient Egyptian one ▸ **pyramidal** *adj*

pyre *n* pile of wood for burning a corpse on

Pyrex® *n* heat-resistant glassware

pyromania *n* uncontrollable urge to set things on fire ▸ **pyromaniac** *n*

pyrotechnics *n* art of making fireworks; firework display ▸ **pyrotechnic** *adj*

Pyrrhic victory [pir-ik] *n* victory in which the victor's losses are as great as those of the defeated

python *n* large nonpoisonous snake that crushes its prey

Qq

QC Queen's Counsel
QED which was to be shown or proved
Qld Queensland
QM Quartermaster
qr. quarter; quire
qt. quart
qua [kwah] *prep* in the capacity of
quack¹ *v* (of a duck) utter a harsh guttural sound ▷ *n* sound made by a duck
quack² *n* unqualified person who claims medical knowledge
quad *n* see **quadrangle**; (*informal*) quadruplet ▷ *adj* short for **quadraphonic ▶ quad bike, quad** vehicle like a small motorcycle with four large wheels, designed for agricultural and sporting uses
quadrangle *n* (also **quad**) rectangular courtyard with buildings on all four sides; geometric figure consisting of four points connected by four lines ▶ **quadrangular** *adj*
quadrant *n* quarter of a circle; quarter of a circle's circumference; instrument for measuring the altitude of the stars
quadraphonic *adj* using four independent channels to reproduce or record sound
quadratic (*Maths*) *n* equation in which the variable is raised to the power of two, but nowhere raised to a higher power ▷ *adj* of the second power

quadrennial *adj* occurring every four years; lasting four years
quadri- *combining form* four e.g. *quadrilateral*
quadrilateral *adj* having four sides ▷ *n* polygon with four sides
quadrille *n* square dance for four couples
quadriplegia *n* paralysis of all four limbs
quadruped [kwod-roo-ped] *n* any animal with four legs
quadruple *v* multiply by four ▷ *adj* four times as much or as many; consisting of four parts
quadruplet *n* one of four offspring born at one birth
quaff [kwoff] *v* drink heartily or in one draught
quagmire [kwog-mire] *n* soft wet area of land
quail¹ *n* small game bird of the partridge family
quail² *v* shrink back with fear
quaint *adj* attractively unusual, esp. in an old-fashioned style ▶ **quaintly** *adv*
quake *v* shake or tremble with or as if with fear ▷ *n* (*informal*) earthquake
Quaker *n* member of a Christian sect, the Society of Friends ▶ **Quakerism** *n*
qualify *v* -**fying, -fied** provide or be provided with the abilities necessary for a task, office, or duty; moderate or restrict (a statement) ▶ **qualified** *adj* ▶ **qualification** *n* official record of achievement in a course or examination; quality or skill needed for a particular activity; condition that modifies or limits; act of qualifying
quality *n, pl* -**ties** degree or standard of excellence; distinguishing characteristic or attribute; basic character or nature

of something ▷ *adj* excellent or superior ▸ **qualitative** *adj* of or relating to quality ▸ **quality assurance** system in which the quality of a product or service is compared with that required

qualm [kwahm] *n* pang of conscience; sudden sensation of misgiving

quandary *n*, *pl* **-ries** difficult situation or dilemma

quandong [kwon-dong] *n* small Australian tree with edible fruit and nuts used in preservesAustralian tree with pale timber

quango *n*, *pl* **-gos** (*chiefly Brit*) quasi-autonomous nongovernmental organization: any partly independent official body set up by a government

quanta *n* plural of **quantum**

quantify *v* **-fying, -fied** discover or express the quantity of ▸ **quantifiable** *adj* ▸ **quantification** *n*

quantity *n*, *pl* **-ties** specified or definite amount or number; aspect of anything that can be measured, weighed, or counted ▸ **quantitative** *adj* of or relating to quantity ▸ **quantitative easing** practice of increasing the supply of money in order to stimulate economic activity ▸ **quantity surveyor** person who estimates the cost of the materials and labour necessary for a construction job

quantum *n*, *pl* **-ta** desired or required amount, esp. a very small one ▸ **quantum leap, quantum jump** (*informal*) sudden large change, increase, or advance ▸ **quantum theory** physics theory based on the idea that energy of electrons is discharged in discrete quanta

quarantine *n* period of isolation of people or animals to prevent the spread of disease ▷ *v* isolate in or as if in quarantine

quark *n* (*Physics*) subatomic particle thought to be the fundamental unit of matter

quarrel *n* angry disagreement; cause of dispute ▷ *v* **-relling, -relled** have a disagreement or dispute ▸ **quarrelsome** *adj*

quarry[1] *n*, *pl* **-ries** place where stone is dug from the surface of the earth ▷ *v* **-rying, -ried** extract (stone) from a quarry

quarry[2] *n*, *pl* **-ries** person or animal that is being hunted

quart *n* unit of liquid measure equal to two pints (1.136 litres)

quarter *n* one of four equal parts of something; fourth part of a year; (*informal*) unit of weight equal to 4 ounces; region or district of a town or city; (*US*) 25-cent piece; mercy or pity, as shown towards a defeated opponent ▷ *pl* lodgings ▷ *v* divide into four equal parts; billet or be billeted in lodgings ▸ **quarterly** *adj* occurring, due, or issued at intervals of three months ▷ *n* magazine issued every three months ▷ *adv* once every three months ▸ **quarter day** (*Brit*) any of the four days in the year when certain payments become due ▸ **quarterdeck** *n* (*Naut*) rear part of the upper deck of a ship ▸ **quarterfinal** *n* round before the semifinal in a competition ▸ **quartermaster** *n* military officer responsible for accommodation, food, and equipment

quartet *n* group of four performers; music for such a group

quarto *n*, *pl* **-tos** book size in which the sheets are folded into four leaves

quartz n hard glossy mineral

quasar [kway-zar] n extremely distant starlike object that emits powerful radio waves

quash v annul or make void; subdue forcefully and completely

quasi- [kway-zie] combining form almost but not really e.g. quasi-religious; a quasi-scholar

quatrain n stanza or poem of four lines

quaver v (of a voice) quiver or tremble ▷ n (Music) note half the length of a crotchet; tremulous sound or note

quay [kee] n wharf built parallel to the shore

queasy adj -sier, -siest having the feeling that one is about to vomit; feeling or causing uneasiness ▶ **queasiness** n

queen n female sovereign who is the official ruler or head of state; wife of a king; woman, place, or thing considered to be the best of her or its kind; (slang) effeminate male homosexual; only fertile female in a colony of bees, wasps, or ants; the most powerful piece in chess ▶ **queenly** adj ▶ **Queen's Counsel** barrister or advocate appointed Counsel to the Crown

queer adj not normal or usual; (Brit) faint, giddy, or queasy; (offens) homosexual ▷ n (offens) homosexual **queer someone's pitch** (informal) spoil someone's chances of something

> **USAGE NOTE**
> Although the term queer meaning homosexual is still considered derogatory when used by non-homosexuals, it is now being used by homosexuals of themselves as a positive term: queer politics, queer cinema

quell v suppress; overcome

quench v satisfy (one's thirst); put out or extinguish

quern n stone hand mill for grinding corn

querulous [kwer-yoo-luss] adj complaining or whining ▶ **querulously** adv

query n, pl -ries question, esp. one raising doubt; question mark ▷ v -rying, -ried express uncertainty, doubt, or an objection concerning (something)

quest n long and difficult search ▷ v (foll by for or after) go in search of

question n form of words addressed to a person in order to obtain an answer; point at issue; difficulty or uncertainty ▷ v put a question or questions to (a person); express uncertainty about **in question** under discussion **out of the question** impossible ▶ **questionable** adj of disputable value or authority ▶ **questionably** adv ▶ **questionnaire** n set of questions on a form, used to collect information from people ▶ **question mark** punctuation mark (?) written at the end of questions

> **SPELLING TIP**
> Remember that questionnaire has two ns

queue n line of people or vehicles waiting for something ▷ v queuing or queueing, queued (often foll by up) form or remain in a line while waiting

quibble v make trivial objections ▷ n trivial objection

quiche [keesh] n savoury flan with an egg custard filling to which vegetables etc. are added

quick adj speedy, fast; lasting or taking a short time; alert and responsive; easily excited or aroused ▷ n area of sensitive

q

flesh under a nail ▷ *adv* (*informal*) in a rapid manner **cut someone to the quick** hurt someone's feelings deeply ▶ **quickly** *adv* ▶ **quicken** *v* make or become faster; make or become more lively ▶ **quicklime** *n* white solid used in the manufacture of glass and steel ▶ **quicksand** *n* deep mass of loose wet sand that sucks anything on top of it into it ▶ **quicksilver** *n* mercury ▶ **quickstep** *n* fast ballroom dance

quid *n*, *pl* **quid** (*Brit slang*) pound (sterling)

quid pro quo *n*, *pl* **quid pro quos** one thing, esp. an advantage or object, given in exchange for another

quiescent [kwee-**ess**-ent] *adj* quiet, inactive, or dormant ▶ **quiescence** *n*

quiet *adj* with little noise; calm or tranquil; untroubled ▷ *n* quietness ▷ *v* make or become quiet **on the quiet** without other people knowing, secretly ▶ **quietly** *adv* ▶ **quietness** *n* ▶ **quieten** *v* (often foll by *down*) make or become quiet ▶ **quietude** *n* quietness, peace, or tranquillity

quietism *n* passivity and calmness of mind towards external events

quiff *n* tuft of hair brushed up above the forehead

quill *n* pen made from the feather of a bird's wing or tail; stiff hollow spine of a hedgehog or porcupine

quilt *n* padded covering for a bed ▶ **quilted** *adj* consisting of two layers of fabric with a layer of soft material between them

quin *n* short for **quintuplet**

quince *n* acid-tasting pear-shaped fruit

quinine *n* bitter drug used as a tonic and formerly to treat malaria

quinquennial *adj* occurring every five years; lasting five years

quinsy *n* inflammation of the throat or tonsils

quintessence *n* most perfect representation of a quality or state ▶ **quintessential** *adj*

quintet *n* group of five performers; music for such a group

quintuplet *n* one of five offspring born at one birth

quip *n* witty saying ▷ *v* **quipping**, **quipped** make a quip

quire *n* set of 24 or 25 sheets of paper

quirk *n* peculiarity of character; unexpected twist or turn e.g. *a quirk of fate* ▶ **quirky** *adj*

quisling *n* traitor who aids an occupying enemy force

quit *v* **quitting**, **quit** stop (something); give up (a job); depart from ▶ **quitter** *n* person who lacks perseverance ▶ **quits** *adj* (*informal*) on an equal footing

quite *adv* somewhat e.g. *she's quite pretty*; absolutely e.g. *you're quite right*; in actuality, truly ▷ *interj* expression of agreement

> **USAGE NOTE**
> Note that because *quite* can mean 'extremely': *quite amazing*, or can express a reservation: *quite friendly*, it should be used carefully

quiver¹ *v* shake with a tremulous movement ▷ *n* shaking or trembling

quiver² *n* case for arrows

quixotic [kwik-**sot**-ik] *adj* romantic and unrealistic ▶ **quixotically** *adv*

quiz *n*, *pl* **quizzes** entertainment in which the knowledge of the players is tested by a series of questions ▷ *v* **quizzing**, **quizzed** investigate by close questioning ▶ **quizzical** *adj* questioning and mocking e.g. *a quizzical look* ▶ **quizzically** *adv*

quod *n* (*Brit slang*) jail

quoit *n* large ring used in the game of quoits ▷ *pl* game in which quoits are tossed at a stake in the ground in attempts to encircle it

quokka *n* small Australian wallaby

quorum *n* minimum number of people required to be present at a meeting before any transactions can take place

quota *n* share that is due from, due to, or allocated to a group or person; prescribed number or quantity allowed, required, or admitted

quote *v* repeat (words) exactly from (an earlier work, speech, or conversation); state (a price) for goods or a piece of work ▷ *n* (*informal*) quotation ▶ **quotable** *adj* ▶ **quotation** *n* written or spoken passage repeated exactly in a later work, speech, or conversation; act of quoting; estimate of costs submitted by a contractor to a prospective client ▶ **quotation marks** raised commas used in writing to mark the beginning and end of a quotation or passage of speech

quoth *v* (*obs*) said

quotidian *adj* daily; commonplace

quotient *n* result of the division of one number or quantity by another

q.v. which see: used to refer a reader to another item in the same book

Rr

R King; Queen; River

r radius; ratio; right

RA (in Britain) Royal Academy; (in Britain) Royal Artillery

RAAF Royal Australian Air Force

rabbi [rab-bye] *n, pl* -**bis** Jewish spiritual leader ▶ **rabbinical** *adj*

rabbit *n* small burrowing mammal with long ears ▶ **rabbit on** *v* **rabbiting, rabbited** (*Brit informal*) talk too much

rabble *n* disorderly crowd of noisy people

rabid *adj* fanatical; having rabies ▶ **rabidly** *adv*

rabies [ray-beez] *n* usu. fatal viral disease transmitted by dogs and certain other animals

RAC (in Britain) Royal Automobile Club

raccoon *n* small N American mammal with a long striped tail

race[1] *n* contest of speed ▷ *pl* meeting for horse racing ▷ *v* compete (with) in a race; run swiftly; (of an engine) run faster than normal ▶ **racer** *n* ▶ **racecourse** *n* ▶ **racehorse** *n* ▶ **racetrack** *n*

race[2] *n* group of people of common ancestry with distinguishing physical features, such as skin colour ▶ **racial** *adj* ▶ **racism, racialism** *n* hostile attitude or behaviour to members of other races, based on a belief in the innate superiority of one's own race ▶ **racist, racialist** *adj, n*

raceme [rass-**eem**] n cluster of flowers along a central stem, as in the foxglove

rack[1] n framework for holding particular articles, such as coats or luggage; (Hist) instrument of torture that stretched the victim's body ▷ v cause great suffering to **rack one's brains** try very hard to remember

rack[2] n **go to rack and ruin** be destroyed

racket[1] n noisy disturbance; occupation by which money is made illegally

racket[2], **racquet** n bat with strings stretched in an oval frame, used in tennis etc. ▶ **rackets** n ball game played in a paved walled court

racketeer n person making illegal profits

raconteur [rak-on-**tur**] n skilled storyteller

racy adj **racier**, **raciest** slightly shocking; spirited or lively

radar n device for tracking distant objects by bouncing high-frequency radio pulses off them

radial adj spreading out from a common central point; of a radius; (also **radial-ply**) (of a tyre) having flexible sides strengthened with radial cords

radiant adj looking happy; shining; emitting radiation ▶ **radiance** n

radiate v spread out from a centre; emit or be emitted as radiation ▶ **radiator** n (Brit) arrangement of pipes containing hot water or steam to heat a room; tubes containing water as cooling apparatus for a car engine; (Aust & NZ) electric fire

radiation n transmission of energy from one body to another; particles or waves emitted in nuclear decay; process of radiating

radical adj fundamental; thorough; advocating fundamental change ▷ n person advocating fundamental (political) change; number expressed as the root of another ▶ **radically** adv ▶ **radicalism** n

radicle n small or developing root

radii n a plural of **radius**

radio n, pl **-dios** use of electromagnetic waves for broadcasting, communication, etc.; device for receiving and amplifying radio signals; sound broadcasting ▷ v transmit (a message) by radio

radio- combining form of radio, radiation, or radioactivity

radioactive adj emitting radiation as a result of nuclear decay ▶ **radioactivity** n

radiography [ray-dee-**og**-ra-fee] n production of an image on a film or plate by radiation ▶ **radiographer** n

radiology [ray-dee-**ol**-a-jee] n science of using X-rays in medicine ▶ **radiologist** n

radiotherapy n treatment of disease, esp. cancer, by radiation ▶ **radiotherapist** n

radish n small hot-flavoured root vegetable eaten raw in salads

radium n (Chem) radioactive metallic element

radius n, pl **radii** or **radiuses** (length of) a straight line from the centre to the circumference of a circle; outer of two bones in the forearm

radon [**ray**-don] n (Chem) radioactive gaseous element

RAF (in Britain) Royal Air Force

raffia n prepared palm fibre for weaving mats etc.

raffish adj slightly disreputable

raffle n lottery with goods as prizes ▷ v offer as a prize in a raffle

raft n floating platform of logs, planks, etc.

rafter n one of the main beams of a roof

rag¹ n fragment of cloth; (Brit, Aust & NZ informal) newspaper ▷ pl tattered clothing ▶ **ragged** [rag-gid] adj dressed in shabby or torn clothes; torn; lacking smoothness

rag² (Brit) v **ragging, ragged** tease ▷ adj, n (of) events organized by students to raise money for charities

ragamuffin n ragged dirty child

rage n violent anger or passion ▷ v speak or act with fury; proceed violently and without check **all the rage** very popular

raglan adj (of a sleeve) joined to a garment by diagonal seams from the neck to the underarm

ragout [rag-goo] n richly seasoned stew of meat and vegetables

ragtime n style of jazz piano music

raid n sudden surprise attack or search ▷ v make a raid on ▶ **raider** n

rail¹ n horizontal bar, esp. as part of a fence or track; railway ▶ **railing** n fence made of rails supported by posts ▶ **railway** n track of iron rails on which trains run; company operating a railway

rail² v (foll by at or against) complain bitterly or loudly ▶ **raillery** n teasing or joking

rail³ n small marsh bird

raiment n (obs) clothing

rain n water falling in drops from the clouds ▷ v fall or pour down as rain; fall rapidly and in large quantities ▶ **rainy** adj ▶ **rainbow** n arch of colours in the sky ▶ **rainbow nation** South African nation ▶ **raincoat** n water-resistant overcoat ▶ **rainfall** n amount of rain ▶ **rainforest** n dense forest in tropical and temperate areas

raise v lift up; set upright; increase in amount or intensity; collect or levy; bring up (a family); put forward for consideration

raisin n dried grape

raison d'être [ray-zon det-ra] n, pl **raisons d'être** (French) reason or justification for existence

Raj n **the Raj** former British rule in India

raja, rajah n (Hist) Indian prince or ruler

rake¹ n tool with a long handle and a crosspiece with teeth, used for smoothing earth or gathering leaves, hay, etc. ▷ v gather or smooth with a rake; search (through); sweep (with gunfire) **rake it in** (informal) make a large amount of money ▶ **rake-off** n (slang) share of profits, esp. illegal ▶ **rake up** revive memories of (a forgotten unpleasant event)

rake² n dissolute or immoral man ▶ **rakish** adj

rakish adj dashing or jaunty

rally n, pl **-lies** large gathering of people for a meeting; marked recovery of strength; (Tennis etc.) lively exchange of strokes; car-driving competition on public roads ▶ **v -lying, -lied** bring or come together after dispersal or for a common cause; regain health or strength, revive

RAM (Computing) random access memory

ram n male sheep; hydraulic machine ▷ v **ramming, rammed** strike against with force; force or drive; cram or stuff

Ramadan n 9th Muslim month; strict fasting from dawn to dusk observed during this time

ramble v walk without a definite route; talk incoherently ▷ n walk, esp. in the country

rambler n person who rambles; climbing rose

ramekin [ram-ik-in] n small ovenproof dish for a single serving of food

ramifications pl n consequences resulting from an action

ramp n slope joining two level surfaces

rampage v dash about violently **on the rampage** behaving violently or destructively

rampant adj growing or spreading uncontrollably; (of a heraldic beast) on its hind legs

rampart n mound or wall for defence

ramshackle adj tumbledown, rickety, or makeshift

ran v past tense of **run**

ranch n large cattle farm in the American West ▶ **rancher** n

rancid adj (of butter, bacon, etc.) stale and having an offensive smell ▶ **rancidity** n

rancour n deep bitter hate ▶ **rancorous** adj

rand n monetary unit of S Africa

R & D research and development

random adj made or done by chance or without plan **at random** haphazardly

randy adj **randier, randiest** (informal) sexually aroused

rang v past tense of **ring¹**

range n limits of effectiveness or variation; distance that a missile or plane can travel; distance of a mark shot at; whole set of related things; chain of mountains; place for shooting practice or rocket testing; kitchen stove ▷ v vary between one point and another; cover or extend over; roam ▶ **ranger** n official in charge of a nature reserve etc.; **(R-)** member of the senior branch of Guides ▶ **rangefinder**

n instrument for finding how far away an object is

rangy [rain-jee] adj **rangier, rangiest** having long slender limbs

rank¹ n relative place or position; status; social class; row or line ▷ v have a specific rank or position; arrange in rows or lines **the ranks** common soldiers ▶ **rank and file** ordinary people or members

rank² adj complete or absolute e.g. rank favouritism; smelling offensively strong; growing too thickly

rankle v continue to cause resentment or bitterness

ransack v search thoroughly; pillage, plunder

ransom n money demanded in return for the release of someone who has been kidnapped

rant v talk in a loud and excited way ▶ **ranter** n

rap v **rapping, rapped** hit with a sharp quick blow; utter (a command) abruptly; perform a rhythmic monologue with musical backing ▷ n quick sharp blow; rhythmic monologue performed to music **take the rap** (slang) suffer punishment for something whether guilty or not ▶ **rapper** n

rapacious adj greedy or grasping ▶ **rapacity** n

rape¹ v force to submit to sexual intercourse ▷ n act of raping; any violation or abuse ▶ **rapist** n

rape² n plant with oil-yielding seeds, also used as fodder

rapid adj quick, swift ▶ **rapids** pl n part of a river with a fast turbulent current ▶ **rapidly** adv ▶ **rapidity** n

rapier [ray-pyer] n fine-bladed sword

rapport [rap-pore] n harmony or agreement

rapprochement [rap-prosh-mong] n re-establishment of

friendly relations, esp. between nations

rapt adj engrossed or spellbound ► **rapture** n ecstasy ► **rapturous** adj

rare¹ adj uncommon; infrequent; of uncommonly high quality; (of air at high altitudes) having low density, thin ► **rarely** adv seldom ► **rarity** n

rare² adj (of meat) lightly cooked

rarebit n see **Welsh rarebit**

rarefied [rare-if-ide] adj highly specialized, exalted; (of air) thin

raring adj **raring to** enthusiastic, willing, or ready to

rascal n rogue; naughty (young) person ► **rascally** adj

rash¹ adj hasty, reckless, or incautious ► **rashly** adv

rash² n eruption of spots or patches on the skin; outbreak of (unpleasant) occurrences

rasher n thin slice of bacon

rasp n harsh grating noise; coarse file ► v speak in a grating voice; make a scraping noise

raspberry n red juicy edible berry; (informal) spluttering noise made with the tongue and lips, to show contempt

Rastafarian n, adj (member) of a religion originating in Jamaica and regarding Haile Selassie as God (also **Rasta**)

rat n small rodent; (informal) contemptible person, esp. a deserter or informer ► v **ratting, ratted** (informal) inform (on); hunt rats ► **ratty** adj (Brit & NZ informal) bad-tempered, irritable ► **rat race** continual hectic competitive activity

ratafia [rat-a-fee-a] n liqueur made from fruit; (chiefly Brit) almond-flavoured biscuit

ratatouille [rat-a-twee] n vegetable casserole of tomatoes, aubergines, etc.

ratchet n set of teeth on a bar or wheel allowing motion in one direction only

rate n degree of speed or progress; proportion between two things; charge ► pl local tax on business ► v consider or value; estimate the value of **at any rate** in any case ► **rateable** adj able to be rated; (of property) liable to payment of rates ► **ratepayer** n

rather adv to some extent; more truly or appropriately; more willingly

ratify v **-fying, -fied** give formal approval to ► **ratification** n

rating n valuation or assessment; classification; noncommissioned sailor ► pl size of the audience for a TV programme

ratio n, pl **-tios** relationship between two numbers or quantities expressed as a proportion

ration n fixed allowance of food etc. ► v limit to a certain amount per person

rational adj reasonable, sensible; capable of reasoning ► **rationally** adv ► **rationality** n ► **rationale** [rash-a-nahl] n reason for an action or decision ► **rationalism** n philosophy that regards reason as the only basis for beliefs or actions ► **rationalist** n ► **rationalize** v justify by plausible reasoning; reorganize to improve efficiency or profitability ► **rationalization** n

rattan n climbing palm with jointed stems used for canes

rattle v give out a succession of short sharp sounds; shake briskly causing sharp sounds; (informal) confuse or fluster ► n short sharp sound; instrument for making such a sound ► **rattlesnake** n poisonous snake with loose horny segments on the tail that make a rattling sound

raucous adj hoarse or harsh

raunchy adj **-chier, -chiest** (slang) earthy, sexy

ravage v cause extensive damage to ▸ **ravages** pl n damaging effects

rave v talk wildly or with enthusiasm ▸ n large-scale party with electronic dance music ▸ **raving** adj delirious; (informal) exceptional e.g. a raving beauty

ravel v **-elling, -elled** tangle or become entangled

raven n black bird like a large crow ▸ adj (of hair) shiny black

ravenous adj very hungry

ravine [rav-veen] n narrow steep-sided valley worn by a stream

ravioli pl n small squares of pasta with a savoury filling

ravish v enrapture; (lit) rape ▸ **ravishing** adj lovely or entrancing

raw adj uncooked; not manufactured or refined; inexperienced; chilly ▸ **raw deal** unfair or dishonest treatment ▸ **rawhide** n untanned hide

ray[1] n single line or narrow beam of light

ray[2] n large sea fish with a flat body and a whiplike tail

rayon n (fabric made of) a synthetic fibre

raze v destroy (buildings or a town) completely

razor n sharp instrument for shaving ▸ **razorbill** n sea bird of the North Atlantic with a stout bill flattened at the sides

razzle-dazzle, razzmatazz n (slang) showy activity

RC Roman Catholic; Red Cross

Rd Road

re prep with reference to, concerning

RE (in Britain) religious education

re- prefix again e.g. re-enter; retrial

reach v arrive at; make a movement in order to grasp or touch; succeed in touching; make contact or communication with; extend as far as ▸ n distance that one can reach; range of influence ▸ **reaches** pl n stretch of a river ▸ **reachable** adj

react v act in response (to); (foll by against) act in an opposing or contrary manner ▸ **reaction** n physical or emotional response to a stimulus; any action resisting another; opposition to change; chemical or nuclear change, combination, or decomposition ▸ **reactionary** n ▸ adj (person) opposed to change, esp. in politics ▸ **reactance** n (Electricity) resistance to the flow of an alternating current caused by the inductance or capacitance of the circuit ▸ **reactive** adj chemically active ▸ **reactor** n apparatus in which a nuclear reaction is maintained and controlled to produce nuclear energy

read v **reading, read** look at and understand or take in (written or printed matter); look at and say aloud; interpret the significance or meaning of; (of an instrument) register; study ▸ n matter suitable for reading e.g. a good read ▸ **readable** adj enjoyable to read; legible ▸ **reading** n

reader n person who reads; textbook; (chiefly Brit) senior university lecturer ▸ **readership** n readers of a publication collectively

readjust v adapt to a new situation ▸ **readjustment** n

ready adj **readier, readiest** prepared for use or action; willing; prompt ▸ **readily** adv ▸ **readiness** n ▸ **ready-made** adj for immediate use by any customer

reagent [ree-age-ent] n chemical substance that reacts with another, used to detect the presence of the other

real adj existing in fact; actual; genuine ▸ **really** adv very; truly ▸ interj exclamation of dismay, doubt, or surprise ▸ **reality** n state of things as they are ▸ **reality TV** television programmes focusing on members of the public living in conditions created especially by the programme makers ▸ **real ale** (chiefly Brit) beer allowed to ferment in the barrel ▸ **real estate** property consisting of land and houses

> **USAGE NOTE**
> To intensify an adjective, use the adverb form really, not real: He's really strong

realistic adj seeing and accepting things as they really are, practical ▸ **realistically** adv ▸ **realism** n ▸ **realist** n

realize v become aware or grasp the significance of; achieve (a plan, hopes, etc.); convert into money ▸ **realization** n

realm n kingdom; sphere of interest

ream n twenty quires of paper, generally 500 sheets ▸ pl (informal) large quantity (of written matter)

reap v cut and gather (a harvest); receive as the result of a previous activity ▸ **reaper** n

reappear v appear again ▸ **reappearance** n

rear¹ n back part; part of an army, procession, etc. behind the others **bring up the rear** come last ▸ **rearmost** adj ▸ **rear admiral** high-ranking naval officer ▸ **rearguard** n troops protecting the rear of an army

rear² v care for and educate (children); breed (animals); (of a horse) rise on its hind feet

rearrange v organize differently, alter ▸ **rearrangement** n

reason n cause or motive; faculty of rational thought; sanity ▸ v think logically in drawing conclusions **reason with** persuade by logical argument into doing something ▸ **reasonable** adj sensible; not excessive; logical ▸ **reasonably** adv

reassess v reconsider the value or importance of

reassure v restore confidence to ▸ **reassurance** n

rebate n discount or refund

rebel v -belling, -belled revolt against the ruling power; reject accepted conventions ▸ n person who rebels ▸ **rebellion** n organized open resistance to authority; rejection of conventions ▸ **rebellious** adj

rebirth n revival or renaissance ▸ **reborn** adj active again after a period of inactivity

rebore, reboring n boring of a cylinder to restore its true shape

rebound v spring back; misfire so as to hurt the perpetrator of a plan or deed **on the rebound** (informal) while recovering from rejection

rebuff v reject or snub ▸ n blunt refusal, snub

rebuke v scold sternly ▸ n stern scolding

rebus n, pl -buses puzzle consisting of pictures and symbols representing words or syllables

rebut v -butting, -butted prove that (a claim) is untrue ▸ **rebuttal** n

recalcitrant adj wilfully disobedient ▸ **recalcitrance** n

recall v recollect or remember; order to return; annul or cancel ▸ n ability to remember; order to return

recant v withdraw (a statement or belief) publicly ▸ **recantation** n

recap (*informal*) v **-capping,
-capped** recapitulate ▷ n
recapitulation

recapitulate v state again briefly,
repeat ▶ **recapitulation** n

recapture v experience again;
capture again

recce (*chiefly Brit slang*) v **-ceing,
-ced** or **-ceed** reconnoitre ▷ n
reconnaissance

recede v move to a more distant
place; (of the hair) stop growing
at the front

receipt n written acknowledgment
of money or goods received;
receiving or being received

receive v take, accept, or get;
experience; greet (guests)
▶ **received** adj generally accepted
▶ **receiver** n part of telephone
that is held to the ear; equipment
in a telephone, radio, or television
that converts electrical signals
into sound; person appointed by a
court to manage the property of a
bankrupt ▶ **receivership** n state of
being administered by a receiver

recent adj having happened lately;
new ▶ **recently** adv

receptacle n object used to
contain something

reception n area for receiving
guests, clients, etc.; formal party;
manner of receiving; welcome; (in
broadcasting) quality of signals
received ▶ **receptionist** n person
who receives guests, clients, etc.

receptive adj willing to accept
new ideas, suggestions, etc.
▶ **receptivity** n

recess n niche or alcove; holiday
between sessions of work; secret
hidden place ▶ **recessed** adj hidden
or placed in a recess

recession n period of economic
difficulty when little is being bought
or sold ▶ **recessive** adj receding

recherché [rish-**air**-shay] adj
refined or elegant; known only
to experts

recidivism n habitual relapse into
crime ▶ **recidivist** n

recipe n directions for cooking a dish;
method for achieving something

recipient n person who receives
something

reciprocal [ris-**sip**-pro-kl] adj
mutual; given or done in return
▶ **reciprocally** adv ▶ **reciprocate**
v give or feel in return; (of a
machine part) move backwards
and forwards ▶ **reciprocation** n
▶ **reciprocity** n

recite v repeat (a poem etc.)
aloud to an audience ▶ **recital**
n musical performance by a
soloist or soloists; act of reciting
▶ **recitation** n recital, usu. from
memory, of poetry or prose
▶ **recitative** [ress-it-a-**teev**] n
speechlike style of singing, used
esp. for narrative passages in opera

reckless adj heedless of danger
▶ **recklessly** adv ▶ **recklessness** n

reckon v consider or think; make
calculations, count; expect
▶ **reckoning** n

reclaim v regain possession
of; make fit for cultivation
▶ **reclamation** n

recline v rest in a leaning position
▶ **reclining** adj

recluse n person who avoids other
people ▶ **reclusive** adj

recognize v identify as (a person
or thing) already known; accept
as true or existing; treat as valid;
notice, show appreciation of
▶ **recognition** n ▶ **recognizable**
adj ▶ **recognizance** [rik-og-nizz-
anss] n undertaking before a court
to observe some condition

recoil v jerk or spring back; draw
back in horror; (of an action) go

wrong so as to hurt the doer ▷ n backward jerk; recoiling

recollect v call back to mind, remember ▶ **recollection** n

recommend v advise or counsel; praise or commend; make acceptable ▶ **recommendation** n

> **SPELLING TIP**
> Most people who misspell these words use single letters throughout (*recomend* and *recomendation*); they should, of course, double the *m*, as in **recommendation**

recompense v pay or reward; compensate or make up for ▷ n compensation; reward or remuneration

reconcile v harmonize (conflicting beliefs etc.); bring into friendship; accept or cause to accept (an unpleasant situation) ▶ **reconciliation** n

recondite adj difficult to understand

recondition v restore to good condition or working order

reconnaissance [rik-**kon**-iss-anss] n survey for military or engineering purposes

> **SPELLING TIP**
> The Collins Corpus shows that the most common way to misspell **reconnaissance** is to miss out an s, although there are examples where an n has been missed out instead. Remember, there are two ns in the middle and two ss

reconnoitre [rek-a-**noy**-ter] v make a reconnaissance of

reconsider v think about again, consider changing

reconstitute v reorganize; restore (dried food) to its former state by adding water ▶ **reconstitution** n

reconstruct v rebuild; use evidence to re-create ▶ **reconstruction** n

record n [**rek**-ord] document or other thing that preserves information; disc with indentations which a record player transforms into sound; best recorded achievement; known facts about a person's past ▷ v [rik-**kord**] put in writing; preserve (sound, TV programmes, etc.) for reproduction on a playback device; show or register **off the record** not for publication ▶ **recorder** n person or machine that records sound or images; type of flute, held vertically; judge in certain courts ▶ **recording** n ▶ **record player** instrument for reproducing sound on records

recount v tell in detail

re-count v count again ▷ n second or further count, esp. of votes

recoup [rik-**koop**] v regain or make good (a loss); recompense or compensate

recourse n source of help **have recourse to** turn to a source of help or course of action

recover v become healthy again; regain a former condition; find again; get back (a loss or expense) ▶ **recovery** n ▶ **recoverable** adj

re-create v make happen or exist again

recreation n agreeable or refreshing occupation, relaxation, or amusement ▶ **recreational** adj

recrimination n mutual blame ▶ **recriminatory** adj

recruit v enlist (new soldiers, members, etc.) ▷ n newly enlisted soldier; new member or supporter ▶ **recruitment** n

rectangle n oblong four-sided figure with four right angles ▶ **rectangular** adj

rectify v **-fying, -fied** put right, correct; (Chem) purify by distillation; (Electricity) convert (alternating current) into direct current ▶ **rectification** n ▶ **rectifier** n

rectilinear adj in a straight line; characterized by straight lines

rectitude n moral correctness

recto n, pl **-tos** right-hand page of a book

rector n clergyman in charge of a parish; head of certain academic institutions ▶ **rectory** n rector's house

rectum n, pl **-tums** or **-ta** final section of the large intestine

recumbent adj lying down

recuperate v recover from illness ▶ **recuperation** n ▶ **recuperative** adj

recur v **-curring, -curred** happen again ▶ **recurrence** n repetition ▶ **recurrent** adj

recycle v reprocess (used materials) for further use ▶ **recyclable** adj

red adj **redder, reddest** of a colour varying from crimson to orange and seen in blood, fire, etc.; flushed in the face from anger, shame, etc. ▶ n red colour; (R-) (informal) communist **in the red** (informal) in debt **see red** (informal) be angry ▶ **redness** n ▶ **redden** v make or become red ▶ **reddish** adj ▶ **redback spider** small venomous Australian spider with a red stripe on the back of the abdomen ▶ **red-blooded** adj (informal) vigorous or virile ▶ **redbrick** adj (of a university in Britain) founded in the late 19th or early 20th century ▶ **red card** (Soccer) piece of red pasteboard shown by a referee to indicate that a player has been sent off ▶ **red carpet** very special welcome for an important guest ▶ **redcoat** n (Hist) British soldier ▶ **Red Cross** international organization providing help for victims of war or natural disasters ▶ **redcurrant** n small round edible red berry ▶ **red-handed** adj (informal) (caught) in the act of doing something wrong or illegal ▶ **red herring** something which diverts attention from the main issue ▶ **red-hot** adj glowing red; extremely hot; very keen ▶ **Red Indian** (offens) Native American ▶ **red light** traffic signal to stop; danger signal ▶ **red meat** dark meat, esp. beef or lamb ▶ **red tape** excessive adherence to official rules

redeem v make up for; reinstate (oneself) in someone's good opinion; free from sin; buy back; pay off (a loan or debt) **the Redeemer** Jesus Christ ▶ **redeemable** adj ▶ **redemption** n ▶ **redemptive** adj

redeploy v assign to a new position or task ▶ **redeployment** n

redevelop v rebuild or renovate (an area or building) ▶ **redevelopment** n

redolent adj reminiscent (of); smelling strongly (of)

redouble v increase, multiply, or intensify

redoubt n small fort defending a hilltop or pass

redoubtable adj formidable

redound v cause advantage or disadvantage (to)

redox n chemical reaction in which one substance is reduced and the other is oxidized

redress v make amends for ▶ n compensation or amends

reduce v bring down, lower; lessen, weaken; bring by force or necessity to some state or action; slim; simplify; make (a sauce) more concentrated ▶ **reducible** adj ▶ **reduction** n

redundant adj (of a worker) no longer needed; superfluous **be made redundant** be laid off ▶ **redundancy** n

reed n tall grass that grows in swamps and shallow water; tall straight stem of this plant; (Music) vibrating cane or metal strip in certain wind instruments ▶ **reedy** adj harsh and thin in tone

reef¹ n ridge of rock or coral near the surface of the sea; vein of ore

reef² n part of a sail which can be rolled up to reduce its area ▷ v take in a reef ▶ **reefer** n short thick jacket worn esp. by sailors; (old-fashioned slang) hand-rolled cigarette containing cannabis ▶ **reef knot** two simple knots turned opposite ways

reek v smell strongly ▷ n strong unpleasant smell **reek of** give a strong suggestion of

reel¹ n cylindrical object on which film, tape, thread, or wire is wound; winding apparatus, as of a fishing rod ▶ **reel in** draw in by means of a reel ▶ **reel off** v recite or write fluently or quickly

reel² v stagger, sway, or whirl

reel³ n lively Scottish dance

ref n (informal) referee in sport

refectory n, pl -tories room for meals in a college etc.

refer v -ferring, -ferred (foll by to) allude (to); be relevant (to); send (to) for information; submit (to) for decision ▶ **referral** n act of referring; citation or direction in a book; written testimonial regarding character or capabilities **with reference to** concerning

> **USAGE NOTE**
> Avoid using refer back; refer includes the sense 'back' in its meaning

referee n umpire in sports, esp. soccer or boxing; person willing to testify to someone's character etc.; arbitrator ▷ v -eeing, -eed act as referee

referendum n, pl -dums or -da direct vote of the electorate on an important question

refill v fill again ▷ n second or subsequent filling; replacement supply of something in a permanent container

refine v purify; improve ▶ **refined** adj cultured or polite; purified ▶ **refinement** n improvement or elaboration; fineness of taste or manners; subtlety ▶ **refinery** n place where sugar, oil, etc. is refined

reflation n increase in the supply of money and credit designed to encourage economic activity ▶ **reflate** v ▶ **reflationary** adj

reflect v throw back, esp. rays of light, heat, etc.; form an image of; show; consider at length; bring credit or discredit upon ▶ **reflection** n act of reflecting; return of rays of heat, light, etc. from a surface; image of an object given back by a mirror etc.; conscious thought or meditation; attribution of discredit or blame ▶ **reflective** adj quiet, contemplative; capable of reflecting images ▶ **reflector** n polished surface for reflecting light etc.

reflex n involuntary response to a stimulus or situation ▷ adj (of a muscular action) involuntary; reflected; (of an angle) more than 180° ▶ **reflexive** adj (Grammar) denoting a verb whose subject is the same as its object e.g. dress oneself

reflexology n foot massage as a therapy in alternative medicine

reform n improvement ▷ v improve; abandon evil practices ▶ **reformer** n ▶ **reformation** n act or instance of something being reformed; (**R-**) religious movement in 16th-century Europe that resulted in the establishment of the Protestant Churches ▶ **reformatory** n (formerly) institution for reforming young offenders

refract v change the course of (light etc.) passing from one medium to another ▶ **refraction** n ▶ **refractive** adj ▶ **refractor** n

refractory adj unmanageable or rebellious; (Med) resistant to treatment; resistant to heat

refrain[1] v **refrain from** keep oneself from doing

refrain[2] n frequently repeated part of a song

refresh v revive or reinvigorate, as through food, drink, or rest; stimulate (the memory) ▶ **refresher** n ▶ **refreshing** adj having a reviving effect; pleasantly different or new ▶ **refreshment** n something that refreshes, esp. food or drink

refrigerate v cool or freeze in order to preserve ▶ **refrigeration** n ▶ **refrigerator** n full name for **fridge**

refuge n (source of) shelter or protection ▶ **refugee** n person who seeks refuge, esp. in a foreign country

refulgent adj shining, radiant

refund v pay back ▷ n return of money; amount returned

refurbish v renovate and brighten up

refuse[1] v decline, deny, or reject ▶ **refusal** n denial of anything demanded or offered

refuse[2] n rubbish or useless matter

refute v disprove ▶ **refutation** n

> **USAGE NOTE**
> *Refute* is not the same as *deny*. It means 'show evidence to disprove something', while *deny* means only 'say something is not true'

regain v get back or recover; reach again

regal adj of or like a king or queen ▶ **regally** adv ▶ **regalia** pl n ceremonial emblems of royalty or high office

regale v entertain (someone) with stories etc.

regard v consider; look at; heed ▷ n respect or esteem; attention; look ▷ pl expression of goodwill **as regards, regarding** in respect of, concerning ▶ **regardless** adj heedless ▷ adv in spite of everything

regatta n meeting for yacht or boat races

regenerate v (cause to) undergo renewal, moral, or physical renewal; reproduce or re-create ▶ **regeneration** n ▶ **regenerative** adj

regent n ruler of a kingdom during the absence, childhood, or illness of its monarch ▷ adj ruling as a regent e.g. *prince regent* ▶ **regency** n status or period of office of a regent

reggae n style of Jamaican popular music with a strong beat

regicide n killing of a king; person who kills a king

regime [ray-zheem] n system of government; particular administration

regimen n prescribed system of diet etc.

regiment n organized body of troops as a unit of the army ▶ **regimental** adj

▶ **regimentation** n ▶ **regimented** adj very strictly controlled

region n administrative division of a country; area considered as a unit but with no definite boundaries; part of the body ▶ **regional** adj

register n (book containing) an official list or record of things; range of a voice or instrument ▷ v enter in a register or set down in writing; show or be shown on a meter or the face ▶ **registration** n ▶ **registration number** numbers and letters displayed on a vehicle to identify it ▶ **registrar** n keeper of official records; senior hospital doctor, junior to a consultant ▶ **register office, registry office** place where births, marriages, and deaths are recorded

Regius professor [reej-yuss] n (in Britain) professor appointed by the Crown to a university chair founded by a royal patron

regress v revert to a former worse condition ▶ **regression** n act of regressing; (Psychol) use of an earlier (inappropriate) mode of behaviour ▶ **regressive** adj

regret v -gretting, -gretted feel sorry about; express apology or distress ▷ n feeling of repentance, guilt, or sorrow ▶ **regretful** adj ▶ **regrettable** adj

regular adj normal, customary, or usual; symmetrical or even; done or occurring according to a rule; periodical; employed continuously in the armed forces ▷ n regular soldier; (informal) frequent customer ▶ **regularity** n ▶ **regularize** v ▶ **regularly** adv

regulate v control, esp. by rules; adjust slightly ▶ **regulation** n rule; regulating ▶ **regulator** n device that automatically controls pressure, temperature, etc.

regurgitate v vomit; (of some birds and animals) bring back (partly digested food) into the mouth; reproduce (ideas, facts, etc.) without understanding them ▶ **regurgitation** n

rehabilitate v help (a person) to readjust to society after illness, imprisonment, etc.; restore to a former position or rank; restore the good reputation of ▶ **rehabilitation** n

rehash v rework or reuse ▷ n old ideas presented in a new form

rehearse v practise (a play, concert, etc.); repeat aloud ▶ **rehearsal** n

rehouse v provide with a new (and better) home

reign n period of a sovereign's rule ▷ v rule (a country); be supreme

reimburse v refund, pay back ▶ **reimbursement** n

rein v check or manage with reins; control or limit ▶ **reins** pl n narrow straps attached to a bit to guide a horse; means of control

reincarnation n rebirth of a soul in successive bodies; one of a series of such transmigrations ▶ **reincarnate** v

reindeer n, pl -deer or -deers deer of arctic regions with large branched antlers

reinforce v strengthen with new support, material, or force; strengthen with additional troops, ships, etc. ▶ **reinforcement** n ▶ **reinforced concrete** concrete strengthened by having steel mesh or bars embedded in it

reinstate v restore to a former position ▶ **reinstatement** n

reiterate v repeat again and again ▶ **reiteration** n

reject v refuse to accept or believe; rebuff (a person); discard as useless

▷ *n* person or thing rejected as not up to standard ▶ **rejection** *n*

rejig *v* **-jigging, -jigged** re-equip (a factory or plant); rearrange

rejoice *v* feel or express great happiness

rejoin[1] *v* join again

rejoin[2] *v* reply ▶ **rejoinder** *n* answer, retort

rejuvenate *v* restore youth or vitality to ▶ **rejuvenation** *n*

relapse *v* fall back into bad habits, illness, etc. ▷ *n* return of bad habits, illness, etc.

relate *v* establish a relation between; have reference or relation to; have an understanding of (people or ideas); tell (a story) or describe (an event) ▶ **related** *adj*

relation *n* connection between things; relative; connection by blood or marriage; act of relating (a story) ▷ *pl* social or political dealings; family ▶ **relationship** *n* dealings and feelings between people or countries; emotional or sexual affair; connection between two things; association by blood or marriage, kinship

relative *adj* dependent on relation to something else, not absolute; having reference or relation (to); (*Grammar*) referring to a word or clause earlier in the sentence ▷ *n* person connected by blood or marriage ▶ **relatively** *adv* ▶ **relativity** *n* subject of two theories of Albert Einstein, dealing with relationships of space, time, and motion, and acceleration and gravity; state of being relative

relax *v* make or become looser, less tense, or less rigid; ease up from effort or attention, rest; be less strict about; become more friendly ▶ **relaxing** *adj* ▶ **relaxation** *n*

relay *n* fresh set of people or animals relieving others; (*Electricity*) device for making or breaking a local circuit; broadcasting station receiving and retransmitting programmes ▷ *v* **-laying, -layed** pass on (a message) ▶ **relay race** race between teams in which each runner races part of the distance

release *v* set free; let go or fall; issue (a record, film, etc.) for sale or public showing; emit heat, energy, etc. ▷ *n* setting free; statement to the press; act of issuing for sale or publication; newly issued film, record, etc.

relegate *v* put in a less important position; demote (a sports team) to a lower league ▶ **relegation** *n*

relent *v* give up a harsh intention, become less severe ▶ **relentless** *adj* unremitting; merciless

relevant *adj* to do with the matter in hand ▶ **relevance** *n*

> **SPELLING TIP**
> A common mistake in English, **relevant** is not always spelt correctly. The final syllable is the problem and sometimes appears incorrectly in the Collins Corpus as *-ent*

reliable *adj* able to be trusted, dependable ▶ **reliably** *adv* ▶ **reliability** *n*

reliance *n* dependence, confidence, or trust ▶ **reliant** *adj*

relic *n* something that has survived from the past; body or possession of a saint, regarded as holy ▷ *pl* remains or traces ▶ **relict** *n* (*obs*) widow

relief *n* gladness at the end or removal of pain, distress, etc.; release from monotony or duty; money or food given to victims of disaster, poverty, etc.; freeing of a besieged city etc.; person who replaces another; projection of a carved design from the surface; any vivid effect resulting from contrast e.g. *comic relief* ▶ **relieve** *v* bring relief to ▶ **relieve oneself** urinate or defecate ▶ **relief map** map showing the shape and height of land by shading

religion *n* system of belief in and worship of a supernatural power or god ▶ **religious** *adj* of religion; pious or devout; scrupulous or conscientious ▶ **religiously** *adv*

relinquish *v* give up or abandon

reliquary *n, pl* **-quaries** case or shrine for holy relics

relish *v* enjoy, like very much ▷ *n* liking or enjoyment; appetizing savoury food, such as pickle; zestful quality or flavour

relocate *v* move to a new place to live or work ▶ **relocation** *n*

reluctant *adj* unwilling or disinclined ▶ **reluctantly** *adv* ▶ **reluctance** *n*

rely *v* **-lying, -lied** depend (on); trust

remain *v* continue; stay, be left behind; be left (over); be left to be done, said, etc. ▶ **remains** *pl n* relics, esp. of ancient buildings; dead body ▶ **remainder** *n* part which is left; amount left over after subtraction or division ▷ *v* offer (copies of a poorly selling book) at reduced prices

remand *v* send back into custody or put on bail before trial **on remand** in custody or on bail before trial ▶ **remand centre** (in Britain) place where accused people are detained awaiting trial

remark *v* make a casual comment (on); say; observe or notice ▷ *n* observation or comment ▶ **remarkable** *adj* worthy of note or attention; striking or unusual ▶ **remarkably** *adv*

remedy *n, pl* **-edies** means of curing pain or disease; means of solving a problem ▷ *v* **-edying, -edied** put right ▶ **remedial** *adj* intended to correct a specific disability etc.

remember *v* retain in or recall to one's memory; keep in mind ▶ **remembrance** *n* memory; token or souvenir; honouring of the memory of a person or event

remind *v* cause to remember; put in mind (of) ▶ **reminder** *n* something that recalls the past; note to remind a person of something not done

reminisce *v* talk or write of past times, experiences, etc. ▶ **reminiscence** *n* remembering; thing recollected ▷ *pl* memoirs ▶ **reminiscent** *adj* reminding or suggestive (of)

remiss *adj* negligent or careless

remission *n* reduction in the length of a prison term; easing of intensity, as of an illness

remit *v* [rim-**mitt**] **-mitting, -mitted** send (money) for goods, services, etc., esp. by post; cancel (a punishment or debt); refer (a decision) to a higher authority or later date ▷ *n* [**ree**-mitt] area of competence or authority ▶ **remittance** *n* money sent as payment

remnant *n* small piece, esp. of fabric, left over; surviving trace

remonstrate *v* argue in protest ▶ **remonstrance** *n*

remorse *n* feeling of sorrow and regret for something one did ▶ **remorseful** *adj* ▶ **remorseless** *adj* pitiless; persistent ▶ **remorselessly** *adv*

remote *adj* far away, distant; aloof; slight or faint ▶ **remotely** *adv* ▶ **remote control** control of an apparatus from a distance by an electrical device

remould v (Brit) renovate (a worn tyre) ▷ n (Brit) renovated tyre

remove v take away or off; get rid of; dismiss from office ▷ n degree of difference ▶ **removable** adj ▶ **removal** n removing, esp. changing residence

Remuera tractor n (NZ informal) four-wheel drive vehicle

remunerate v reward or pay ▶ **remunerative** adj

remuneration n reward or payment

renaissance n revival or rebirth; (R-) revival of learning in the 14th–16th centuries

renal [ree-nal] adj of the kidneys

renascent adj becoming active or vigorous again

rend v rending, rent tear or wrench apart; (of a sound) break (the silence) violently

render v cause to become; give or provide (aid, a service, etc.); submit or present (a bill); portray or represent; cover with plaster; melt down (fat)

rendezvous [ron-day-voo] n, pl **-vous** appointment; meeting place ▷ v meet as arranged

rendition n performance; translation

renegade n person who deserts a cause

renege [rin-nayg] v go back (on a promise etc.)

renew v begin again; make valid again; grow again; restore to a former state; replace (a worn part); restate or reaffirm ▶ **renewable** adj able to be renewed; (of energy or an energy source) inexhaustible or capable of being perpetually replenished ▶ **renewables** pl n renewable energy sources ▶ **renewal** n

rennet n substance for curdling milk to make cheese

renounce v give up (a belief, habit, etc.) voluntarily; give up (a title or claim) formally ▶ **renunciation** n

renovate v restore to good condition ▶ **renovation** n

renown n widespread good reputation

renowned adj famous

rent[1] v give or have use of in return for regular payments ▷ n regular payment for use of land, a building, machine, etc. ▶ **rental** n sum payable as rent

rent[2] n tear or fissure ▷ v past of **rend**

renunciation n see **renounce**

reoffend v commit another offence

reorganize v organize in a new and more efficient way ▶ **reorganization** n

rep[1] n short for **repertory company**

rep[2] n short for **representative**

repair[1] v restore to good condition, mend ▷ n act of repairing; repaired part; state or condition e.g. in good repair ▶ **reparation** n something done or given as compensation

repair[2] v go (to)

repartee n interchange of witty retorts; witty retort

repast n meal

repatriate v send (someone) back to his or her own country ▶ **repatriation** n

repay v repaying, repaid pay back, refund; do something in return for e.g. repay hospitality ▶ **repayable** adj ▶ **repayment** n

repeal v cancel (a law) officially ▷ n act of repealing

repeat v say or do again; happen again, recur ▷ n act or instance of repeating; programme broadcast again ▶ **repeatedly** adv

▶ **repeater** n firearm that may be discharged many times without reloading

repel v **-pelling, -pelled** be disgusting to; drive back, ward off; resist ▶ **repellent** adj distasteful; resisting water etc. ▷ n something that repels, esp. a chemical to repel insects

repent v feel regret for (a deed or omission) ▶ **repentance** n ▶ **repentant** adj

repercussions pl n indirect effects, often unpleasant

repertoire n stock of plays, songs, etc. that a player or company can give

repertory n, pl **-ries** repertoire ▶ **repertory company** permanent theatre company producing a succession of plays

repetition n act of repeating; thing repeated ▶ **repetitive, repetitious** adj full of repetition

rephrase v express in different words

repine v fret or complain

replace v substitute for; put back ▶ **replacement** n

replay n (also **action replay**) immediate reshowing on TV of an incident in sport, esp. in slow motion; second sports match, esp. one following an earlier draw ▷ v play (a match, recording, etc.) again

replenish v fill up again, resupply ▶ **replenishment** n

replete adj filled or gorged

replica n exact copy ▶ **replicate** v make or be a copy of

reply v **-plying, -plied** answer or respond ▷ n, pl **-plies** answer or response

report v give an account of; make a report (on); make a formal complaint about; present oneself (to); be responsible (to) ▷ n account or statement; rumour; written statement of a child's progress at school; bang ▶ **reportedly** adv according to rumour ▶ **reporter** n person who gathers news for a newspaper, TV, etc.

repose n peace; composure; sleep ▷ v lie or lay at rest

repository n, pl **-ries** place where valuables are deposited for safekeeping, store

repossess v (of a lender) take back property from a customer who is behind with payments ▶ **repossession** n

reprehensible adj open to criticism, unworthy

represent v act as a delegate or substitute for; stand for; symbolize; make out to be; portray, as in art ▶ **representation** n ▶ **representative** n person chosen to stand for a group; (travelling) salesperson ▷ adj typical

repress v keep (feelings) in check; restrict the freedom of ▶ **repression** n ▶ **repressive** adj

reprieve v postpone the execution of (a condemned person); give temporary relief to ▷ n (document granting) postponement or cancellation of a punishment; temporary relief

reprimand v blame (someone) officially for a fault ▷ n official blame

reprint v print further copies of (a book) ▷ n reprinted copy

reprisal n retaliation

reproach n, v blame, rebuke ▶ **reproachful** adj ▶ **reproachfully** adv

reprobate adj, n depraved or disreputable (person)

reproduce v produce a copy of; bring new

individuals into existence; re-create ▶ **reproducible** adj ▶ **reproduction** n process of reproducing; facsimile, as of a painting etc.; quality of sound from an audio system ▶ **reproductive** adj

reprove v speak severely to (someone) about a fault ▶ **reproof** n severe blaming of someone for a fault

reptile n cold-blooded egg-laying vertebrate with horny scales or plates, such as a snake or tortoise ▶ **reptilian** adj

republic n form of government in which the people or their elected representatives possess the supreme power; country in which a president is the head of state ▶ **Republican** n, adj (member or supporter) of the Republican Party, the more conservative of the two main political parties in the US ▶ **Republicanism** n

repudiate [rip-pew-dee-ate] v reject the authority or validity of; disown ▶ **repudiation** n

repugnant adj offensive or distasteful ▶ **repugnance** n

repulse v be disgusting to; drive (an army) back; rebuff or reject ▶ n driving back; rejection or rebuff ▶ **repulsion** n distaste or aversion; (Physics) force separating two objects ▶ **repulsive** adj loathsome, disgusting

reputation n estimation in which a person is held ▶ **reputable** adj of good reputation, respectable ▶ **repute** n reputation ▶ **reputed** adj supposed ▶ **reputedly** adv

request v ask ▶ n asking; thing asked for

Requiem [rek-wee-em] n Mass for the dead; music for this

require v want or need; demand ▶ **requirement** n essential condition; specific need or want

USAGE NOTE
Require suggests a demand imposed by some regulation. Need is usually something that comes from a person

requisite [rek-wizz-it] adj necessary, essential ▷ n essential thing

requisition v demand (supplies) ▷ n formal demand, such as for materials or supplies

requite v return to someone (the same treatment or feeling as received)

reredos [rear-doss] n ornamental screen behind an altar

rescind v annul or repeal

rescue v -cuing, -cued deliver from danger or trouble, save ▷ n rescuing ▶ **rescuer** n

research n systematic investigation to discover facts or collect information ▷ v carry out investigations ▶ **researcher** n

resemble v be or look like ▶ **resemblance** n

resent v feel bitter about ▶ **resentful** adj ▶ **resentment** n

reservation n doubt; exception or limitation; seat, room, etc. that has been reserved; area of land reserved for use by a particular group; (also **central reservation**) (Brit) strip of ground separating the two carriageways of a dual carriageway or motorway

reserve v set aside, keep for future use; obtain by arranging beforehand, book; retain ▷ n something, esp. money or troops, kept for emergencies; area of land reserved for a particular purpose; (Sport) substitute; concealment of feelings or friendliness ▶ **reserved**

adj not showing one's feelings, lacking friendliness; set aside for use by a particular person ▸ **reservist** *n* member of a military reserve

reservoir *n* natural or artificial lake storing water for community supplies; store or supply of something

reshuffle *n* reorganization ▷ *v* reorganize

reside *v* dwell permanently

resident *n* person who lives in a place ▷ *adj* living in a place ▸ **residence** *n* home or house ▸ **residential** *adj* (of part of a town) consisting mainly of houses; providing living accommodation

residue *n* what is left, remainder ▸ **residual** *adj*

resign *v* give up office, a job, etc.; reconcile (oneself) to ▸ **resigned** *adj* content to endure ▸ **resignation** *n* resigning; passive endurance of difficulties

resilient *adj* (of a person) recovering quickly from a shock etc.; able to return to normal shape after stretching etc. ▸ **resilience** *n*

resin [rezz-in] *n* sticky substance from plants, esp. pines; similar synthetic substance ▸ **resinous** *adj*

resist *v* withstand or oppose; refrain from despite temptation; be proof against ▸ **resistance** *n* act of resisting; capacity to withstand something; (Electricity) opposition offered by a circuit to the passage of a current through it ▸ **resistant** *adj* ▸ **resistible** *adj* ▸ **resistor** *n* component of an electrical circuit producing resistance

resit *v* take (an exam) again ▷ *n* exam that has to be retaken again

resolute *adj* firm in purpose ▸ **resolutely** *adv*

resolution *n* firmness of conduct or character; thing resolved upon;

decision of a court or vote of an assembly; act of resolving; ability of a television, microscope, etc. to show fine detail

resolve *v* decide with an effort of will; form (a resolution) by a vote; separate the component parts of; make clear, settle ▸ **resolved** *adj* determined

resonance *n* echoing, esp. with a deep sound; sound produced in one object by sound waves coming from another object ▸ **resonant** *adj* ▸ **resonate** *v*

resort *v* have recourse (to) for help etc. ▷ *n* place for holidays; recourse

resound [riz-zownd] *v* echo or ring with sound ▸ **resounding** *adj* echoing; clear and emphatic

resource *n* thing resorted to for support; ingenuity; means of achieving something ▷ *pl* sources of economic wealth; stock that can be drawn on, funds ▸ **resourceful** *adj* ▸ **resourcefulness** *n*

respect *n* consideration; deference or esteem; point or aspect; reference or relation e.g. *with respect to* ▷ *v* treat with esteem; show consideration for ▸ **respecter** *n* ▸ **respectful** *adj* ▸ **respecting** *prep* concerning

respectable *adj* worthy of respect; fairly good ▸ **respectably** *adv* ▸ **respectability** *n*

respective *adj* relating separately to each of those in question ▸ **respectively** *adv*

respiration [ress-per-ray-shun] *n* breathing ▸ **respirator** *n* apparatus worn over the mouth and breathed through as protection against dust, poison gas, etc., or to provide artificial respiration ▸ **respiratory** *adj* ▸ **respire** *v* breathe

r

respite n pause, interval of rest; delay

resplendent adj brilliant or splendid; shining ▶ **resplendence** n

respond v answer; act in answer to any stimulus; react favourably to a stimulus ▶ **respondent** n (Law) defendant ▶ **response** n answer; reaction to a stimulus ▶ **responsive** adj readily reacting to some influence ▶ **responsiveness** n

responsible adj having control and authority; reporting or accountable (to); sensible and dependable; involving responsibility ▶ **responsibly** adv ▶ **responsibility** n, pl -**ties** state of being responsible; person or thing for which one is responsible

rest¹ n freedom from exertion etc.; repose; pause, esp. in music; object used for support ▷ v take a rest; give a rest (to); be supported; place on a support ▶ **restful** adj ▶ **restless** adj unable to relax or concentrate

rest² n what is left; others ▷ v remain, continue to be

restaurant n commercial establishment serving meals ▶ **restaurateur** [rest-er-a-**tur**] n person who owns or runs a restaurant

restitution n giving back; reparation or compensation

restive adj restless or impatient

restore v return (a building, painting, etc.) to its original condition; cause to recover health or spirits; give back, return; re-establish ▶ **restoration** n ▶ **restorative** adj restoring ▷ n food or medicine to strengthen etc. ▶ **restorer** n

restrain v hold (someone) back from action; control or restrict

▶ **restrained** adj not displaying emotion ▶ **restraint** n control, esp. self-control; restraining ▶ **restraining order** (Aust, NZ, Canadian & US) temporary court order imposing restrictions on a company or a person

restrict v confine to certain limits ▶ **restriction** n ▶ **restrictive** adj

restructure v organize in a different way

result n outcome or consequence; score; number obtained from a calculation; exam mark or grade ▷ v (foll by from) be the outcome or consequence (of); (foll by in) end (in) ▶ **resultant** adj

resume v begin again; occupy or take again ▶ **resumption** n

résumé [rezz-yew-may] n summary; (US & Canadian) outline of someone's educational and professional history, prepared for job applications.

resurgence n rising again to vigour ▶ **resurgent** adj

resurrect v restore to life; use once more (something discarded etc.), revive ▶ **resurrection** n rising again (esp. from the dead); revival

resuscitate [ris-**suss**-it-tate] v restore to consciousness ▶ **resuscitation** n

> **SPELLING TIP**
> There is a silent c in **resuscitate**, but only one: it comes after the second s

retail n selling of goods individually or in small amounts to the public ▷ adv by retail ▷ v sell or be sold retail; recount in detail

retailer n person or company that sells goods to the public

retain v keep in one's possession; engage the services of ▶ **retainer** n fee to retain someone's services; old-established servant of a family

retaliate v repay an injury or wrong in kind ▸ **retaliation** n ▸ **retaliatory** adj

retard v delay or slow (progress or development) ▸ **retarded** adj (offens) underdeveloped, esp. mentally ▸ **retardation** n

retch v try to vomit

retention n retaining; ability to remember; abnormal holding of something, esp. fluid, in the body ▸ **retentive** adj capable of retaining or remembering

rethink v consider again, esp. with a view to changing one's tactics

reticent adj uncommunicative, reserved ▸ **reticence** n

retina n, pl **-nas** or **-nae** light-sensitive membrane at the back of the eye

retinue n band of attendants

retire v (cause to) give up office or work, esp. through age; go away or withdraw; go to bed ▸ **retired** adj having retired from work etc. ▸ **retirement** n ▸ **retiring** adj shy

retort[1] v reply quickly, wittily, or angrily ▸ n quick, witty, or angry reply

retort[2] n glass container with a bent neck used for distilling

retouch v restore or improve by new touches, esp. of paint

retrace v go back over (a route etc.) again

retract v withdraw (a statement etc.); draw in or back ▸ **retractable, retractile** adj able to be retracted ▸ **retraction** n

retread v, n same as **remould**

retreat v move back from a position, withdraw ▸ n act of or military signal for retiring or withdrawal; place to which anyone retires, refuge

retrench v reduce expenditure, cut back ▸ **retrenchment** n

retrial n second trial of a case or defendant in a court of law

retribution n punishment or vengeance for evil deeds ▸ **retributive** adj

retrieve v fetch back again; restore to a better state; recover (information) from a computer ▸ **retrievable** adj ▸ **retrieval** n ▸ **retriever** n dog trained to retrieve shot game

retro adj associated with or revived from the past e.g. retro fashion

retroactive adj effective from a date in the past

retrograde adj tending towards an earlier worse condition

retrogressive adj going back to an earlier worse condition ▸ **retrogression** n

retrorocket n small rocket engine used to slow a spacecraft

retrospect n in retrospect when looking back on the past ▸ **retrospective** adj looking back in time; applying from a date in the past ▸ n exhibition of an artist's life's work

retroussé [rit-troo-say] adj (of a nose) turned upwards

retsina n Greek wine flavoured with resin

return v go or come back; give, put, or send back; reply; elect ▸ n returning; (thing) being returned; profit; official report, as of taxable income; return ticket ▸ **returnable** adj ▸ **returning officer** person in charge of an election ▸ **return ticket** ticket allowing a passenger to travel to a place and back

reunion n meeting of people who have been apart ▸ **reunite** v bring or come together again after a separation

reuse v use again ▸ **reusable** adj

r

rev (informal) n revolution (of an engine) ▷ v **revving, revved** (foll by up) increase the speed of revolution of (an engine)

Rev., Revd. Reverend

revalue v adjust the exchange value of (a currency) upwards ▶ **revaluation** n

revamp v renovate or restore

reveal v make known; expose or show ▶ **revelation** n

reveille [riv-val-ee] n morning bugle call to waken soldiers

revel v -elling, -elled take pleasure (in); make merry ▶ **revels** pl n merrymaking ▶ **reveller** n ▶ **revelry** n festivity

revenge n retaliation for wrong done ▷ v make retaliation for; avenge (oneself or another) ▶ **revengeful** adj

revenue n income, esp. of a state

reverberate v echo or resound ▶ **reverberation** n

revere v be in awe of and respect greatly ▶ **reverence** n awe mingled with respect and esteem ▶ **reverent** adj showing reverence ▶ **reverently** adv ▶ **reverential** adj marked by reverence

Reverend adj title of respect for a clergyman

reverie n absent-minded daydream

revers [riv-veer] n turned back part of a garment, such as the lapel

reverse v turn upside down or the other way round; change completely; move (a vehicle) backwards ▷ n opposite; back side; change for the worse; reverse gear ▷ adj opposite or contrary ▶ **reversal** n ▶ **reversible** adj ▶ **reverse gear** mechanism enabling a vehicle to move backwards

revert v return to a former state; come back to a subject; (of

property) return to its former owner ▶ **reversion** n

review n critical assessment of a book, concert, etc.; publication with critical articles; general survey; formal inspection ▷ v hold or write a review of; examine, reconsider, or look back on; inspect formally ▶ **reviewer** n writer of reviews

revile v be abusively scornful of

revise v change or alter; restudy (work) in preparation for an examination ▶ **revision** n

revive v bring or come back to life, vigour, use, etc. ▶ **revival** n reviving or renewal; movement seeking to restore religious faith ▶ **revivalism** n ▶ **revivalist** n

revoke v cancel (a will, agreement, etc.) ▶ **revocation** n

revolt n uprising against authority ▷ v rise in rebellion; cause to feel disgust ▶ **revolting** adj disgusting, horrible

revolution n overthrow of a government by the governed; great change; spinning round; complete rotation ▶ **revolutionary** adj advocating or engaged in revolution; radically new or different ▷ n, pl -aries person advocating or engaged in revolution ▶ **revolutionize** v change considerably

revolve v turn round, rotate **revolve around** be centred on

revolver n repeating pistol

revue n theatrical entertainment with topical sketches and songs

revulsion n strong disgust

reward n something given in return for a service; sum of money offered for finding a criminal or missing property ▷ v pay or give something to (someone) for a service, information, etc. ▶ **rewarding**

adj giving personal satisfaction, worthwhile

rewind v run (a tape or film) back to an earlier point in order to replay

rewire v provide (a house, engine, etc.) with new wiring

rewrite v write again in a different way ▷ n something rewritten

rhapsody n, pl -dies freely structured emotional piece of music; expression of ecstatic enthusiasm ▶ **rhapsodic** adj ▶ **rhapsodize** v speak or write with extravagant enthusiasm

rhea [ree-a] n S American three-toed ostrich

rhenium n (Chem) silvery-white metallic element with a high melting point

rheostat n instrument for varying the resistance of an electrical circuit

rhesus [ree-suss] n small long-tailed monkey of S Asia ▶ **rhesus factor, Rh factor** antigen commonly found in human blood

rhetoric n art of effective speaking or writing; artificial or exaggerated language ▶ **rhetorical** adj (of a question) not requiring an answer ▶ **rhetorically** adv

rheumatism n painful inflammation of joints or muscles ▶ **rheumatic** n, adj (person) affected by rheumatism ▶ **rheumatoid** adj of or like rheumatism

Rh factor n see **rhesus**

rhinestone n imitation diamond

rhino n short for **rhinoceros**

rhinoceros n, pl -oses or -os large thick-skinned animal with one or two horns on its nose

SPELLING TIP
The pronunciation of **rhinoceros** probably misleads some people into making the mistake of adding a u before the final s (*rhinocerous*)

rhizome n thick underground stem producing new plants

rhodium n (Chem) hard metallic element

rhododendron n evergreen flowering shrub

rhombus n, pl -buses or -bi parallelogram with sides of equal length but no right angles, diamond-shaped figure ▶ **rhomboid** n parallelogram with adjacent sides of unequal length

rhubarb n garden plant of which the fleshy stalks are cooked as fruit

rhyme n sameness of the final sounds at the ends of lines of verse, or in words; word identical in sound to another in its final sounds; verse marked by rhyme ▷ v make a rhyme

rhythm n any regular movement or beat; arrangement of the durations of and stress on the notes of a piece of music, usu. grouped into a regular pattern; (in poetry) arrangement of words to form a regular pattern of stresses ▶ **rhythmic, rhythmical** adj ▶ **rhythmically** adv ▶ **rhythm and blues** popular music, orig. Black American, influenced by the blues

SPELLING TIP
The second letter of **rhythm** is a silent h, which people often forget in writing

rib[1] n one of the curved bones forming the framework of the upper part of the body; cut of meat including the rib(s); curved supporting part, as in the hull of a boat; raised series of rows in knitting ▷ v **ribbing, ribbed** provide or mark with ribs; knit to form a rib pattern ▶ **ribbed** adj ▶ **ribbing** n ▶ **ribcage** n bony structure of ribs enclosing the lungs

rib² v **ribbing, ribbed** (*informal*) tease or ridicule ▸ **ribbing** n

ribald *adj* humorously or mockingly rude or obscene ▸ **ribaldry** n

ribbon n narrow band of fabric used for trimming, tying, etc.; any long strip, for example of inked tape in a typewriter

riboflavin [rye-boe-**flay**-vin] n form of vitamin B

rice n cereal plant grown on wet ground in warm countries; its seeds as food

rich *adj* owning a lot of money or property, wealthy; abounding; fertile; (of food) containing much fat or sugar; mellow; amusing ▸ **riches** pl n wealth ▸ **richly** adv elaborately; fully ▸ **richness** n

Richter scale n scale for measuring the intensity of earthquakes

rick¹ n stack of hay etc.

rick² v, n sprain or wrench

rickets n disease of children marked by softening of the bones, bow legs, etc., caused by vitamin D deficiency

rickety *adj* shaky or unstable

rickshaw n light two-wheeled man-drawn Asian vehicle

ricochet [rik-osh-ay] v (of a bullet) rebound from a solid surface ▸ n such a rebound

rid v **ridding, rid** clear or relieve (of) **get rid of** free oneself of (something undesirable) **good riddance** relief at getting rid of something or someone

ridden v past participle of **ride** ▸ adj afflicted or affected by the thing specified e.g. disease-ridden

riddle¹ n question made puzzling to test one's ingenuity; puzzling person or thing

riddle² v pierce with many holes ▸ n coarse sieve for gravel etc. **riddled with** full of

ride v **riding, rode, ridden** sit on and control or propel (a horse, bicycle, etc.); go on horseback or in a vehicle; travel over; be carried on or across; lie at anchor ▸ n journey on a horse etc., or in a vehicle; type of movement experienced in a vehicle ▸ **ride up** v (of a garment) move up from the proper position

rider n person who rides; supplementary clause added to a document

ridge n long narrow hill; long narrow raised part on a surface; line where two sloping surfaces meet; (*Meteorol*) elongated area of high pressure ▸ **ridged** adj

ridiculous *adj* deserving to be laughed at, absurd ▸ **ridicule** n treatment of a person or thing as ridiculous ▸ v laugh at, make fun of

Riding n former administrative district of Yorkshire

riesling n type of white wine

rife *adj* widespread or common **rife with** full of

riff n (*jazz, rock*) short repeated melodic figure

riffle v flick through (pages etc.) quickly

riffraff n rabble, disreputable people

rifle¹ n firearm with a long barrel

rifle² v search and rob; steal

rift n break in friendly relations; crack, split, or cleft ▸ **rift valley** long narrow valley resulting from subsidence between faults

rig v **rigging, rigged** arrange in a dishonest way; equip, esp. a ship ▸ n apparatus for drilling for oil and gas; way a ship's masts and sails are arranged; (*informal*) outfit of clothes ▸ **rigging** n ship's spars

and ropes ► **rig up** v set up or build
temporarily

right adj just; true or correct;
proper; in a satisfactory condition;
of the side that faces east when
the front is turned to the north;
of the outer side of a fabric ► n adv
properly; straight or directly;
on or to the right side ► n claim,
title, etc. allowed or due; what
is just or due; (**R-**) conservative
political party or group ► v bring
or come back to a normal or
correct state; bring or come back
to a vertical position **in the right**
morally or legally correct ► **right
away** immediately ► **rightly** adv
► **rightful** adj ► **rightfully** adv
► **rightist** n, adj (person) on the
political right ► **right angle** angle
of 90° ► **right-handed** adj using
or for the right hand ► **right-hand
man** person's most valuable
assistant ► **right of way** right of
one vehicle to go before another;
legal right to pass over someone's
land ► **right-wing** adj conservative
or reactionary; belonging to
the more conservative part of a
political party

righteous [rye-chuss] adj upright,
godly, or virtuous; morally justified
► **righteousness** n

rigid adj inflexible or strict;
unyielding or stiff ► **rigidly** adv
► **rigidity** n

rigmarole n long complicated
procedure

rigor mortis n stiffening of the
body after death

rigour n harshness, severity, or
strictness; hardship ► **rigorous** adj
harsh, severe, or stern

rile v anger or annoy

rill n small stream

rim n edge or border; outer ring of a
wheel ► **rimmed** adj

rime n (lit) hoarfrost

rimu n (NZ) New Zealand tree
whose wood is used for building
and furniture

rind n tough outer coating of fruits,
cheese, or bacon

ring¹ v **ringing, rang, rung** give out
a clear resonant sound, as a bell;
cause (a bell) to sound; telephone;
resound ► n ringing; telephone
call ► **ring off** v end a telephone
call ► **ringtone** n tune played by
a mobile phone when it receives a
call ► **ring up** v telephone; record
on a cash register

> **USAGE NOTE**
> The simple past is *rang*: *He
> rang the bell*. Avoid the use of
> the past participle *rung* for the
> simple past

ring² n circle of gold etc., esp. for
a finger; any circular band, coil,
or rim; circle of people; enclosed
area, esp. a circle for a circus or
a roped-in square for boxing;
group operating (illegal) control
of a market ► v put a ring round;
mark (a bird) with a ring; kill (a
tree) by cutting the bark round the
trunk ► **ringer** n (Brit, Aust & NZ
slang) person or thing apparently
identical to another (also **dead
ringer**) ► **ringlet** n curly lock of
hair ► **ringleader** n instigator
of a mutiny, riot, etc. ► **ring
road** (Brit, Aust & S Afr) main road
that bypasses a town (centre)
► **ringside** n row of seats nearest
a boxing or circus ring ► **ringtail**
n (Aust) possum with a curling
tail used to grip branches while
climbing ► **ringworm** n fungal skin
disease in circular patches

rink n sheet of ice for skating or
curling; floor for roller-skating

rinkhals n S African cobra that
can spit venom

rinse v remove soap from (washed clothes, hair, etc.) by applying clean water; wash lightly ▷ n rinsing; liquid to tint hair

riot n disorderly unruly disturbance; (Brit, Aust & NZ) loud revelry; profusion; (slang) very amusing person or thing ▷ v take part in a riot **read the riot act** reprimand severely **run riot** behave without restraint; grow profusely ▸ **riotous** adj unrestrained; unruly or rebellious

RIP rest in peace

rip v **ripping, ripped** tear violently; tear away; (informal) rush ▷ n split or tear **let rip** speak without restraint ▸ **ripcord** n cord pulled to open a parachute ▸ **rip off** v (slang) cheat by overcharging ▸ **rip-off** n (slang) cheat or swindle ▸ **rip-roaring** adj (informal) boisterous and exciting

riparian [rip-pair-ee-an] adj of or on the banks of a river

ripe adj ready to be reaped, eaten, etc.; matured; ready or suitable ▸ **ripen** v grow ripe; mature

riposte [rip-posst] n verbal retort; counterattack, esp. in fencing ▷ v make a riposte

ripple n slight wave or ruffling of a surface; sound like ripples of water ▷ v flow or form into little waves (on); (of sounds) rise and fall gently

rise v **rising, rose, risen** get up from a lying, sitting, or kneeling position; move upwards; (of the sun or moon) appear above the horizon; reach a higher level; (of an amount or price) increase; rebel; (of a court) adjourn ▷ n rising; upward slope; increase, esp. of wages **give rise to** cause ▸ **riser** n person who rises, esp. from bed; vertical part of a step ▸ **rising** n revolt ▷ adj increasing in rank or maturity

risible [riz-zib-bl] adj causing laughter, ridiculous

risk n chance of disaster or loss; person or thing considered as a potential hazard ▷ v act in spite of the possibility of (injury or loss); expose to danger or loss ▸ **risky** adj full of risk, dangerous

risotto n, pl **-tos** dish of rice cooked in stock with vegetables, meat, etc.

risqué [risk-ay] adj bordering on indecency

rissole n cake of minced meat, coated with breadcrumbs and fried

rite n formal custom, esp. religious

ritual n prescribed order of rites; regular repeated action or behaviour ▷ adj concerning rites ▸ **ritually** adv ▸ **ritualistic** adj like a ritual

ritzy adj **ritzier, ritziest** (slang) luxurious or elegant

rival n person or thing that competes with or equals another for favour, success, etc. ▷ adj in the position of a rival ▷ v **-valling, -valled** (try to) equal ▸ **rivalry** n keen competition

riven adj split apart

river n large natural stream of water; plentiful flow

rivet [riv-vit] n bolt for fastening metal plates, the end being put through holes and then beaten flat ▷ v **riveting, riveted** fasten with rivets; cause to be fixed, as in fascination ▸ **riveting** adj very interesting and exciting

rivulet n small stream

RME (in Scotland) religious and moral education

RN (in Britain) Royal Navy

RNA ribonucleic acid: substance in living cells essential for the synthesis of protein

RNZ Radio New Zealand

RNZAF Royal New Zealand Air Force

RNZN Royal New Zealand Navy

roach n Eurasian freshwater fish

road n way prepared for passengers, vehicles, etc.; route in a town or city with houses along it; way or course e.g. *the road to fame* **on the road** travelling ▶ **roadblock** n barricade across a road to stop traffic for inspection etc. ▶ **road hog** (*informal*) selfish aggressive driver ▶ **roadhouse** n (*Brit, Aust & S Afr*) pub or restaurant on a country road ▶ **road map** map for drivers; plan or guide for future actions ▶ **roadie** n (*Brit, Aust & NZ informal*) person who transports and sets up equipment for a band ▶ **roadside** n, adj ▶ **road test** test of a vehicle etc. in actual use ▶ **roadway** n the part of a road used by vehicles ▶ **roadworks** pl n repairs to a road, esp. blocking part of the road ▶ **roadworthy** adj (of a vehicle) mechanically sound

roam v wander about

roan adj (of a horse) having a brown or black coat sprinkled with white hairs ▷ n roan horse

roar v make or utter a loud deep hoarse sound like that of a lion; shout (something) as in anger; laugh loudly ▷ n such a sound **a roaring trade** (*informal*) brisk and profitable business **roaring drunk** noisily drunk

roast v cook by dry heat, as in an oven; make or be very hot ▷ n roasted joint of meat ▷ adj roasted ▶ **roasting** (*informal*) adj extremely hot ▷ n severe criticism or scolding

rob v **robbing, robbed** steal from; deprive ▶ **robber** n ▶ **robbery** n

robe n long loose outer garment ▷ v put a robe on

robin n small brown bird with a red breast

robot n automated machine, esp. one performing functions in a human manner; person of machine-like efficiency; (*SAfr*) set of coloured lights at a junction to control the traffic flow ▶ **robotic** adj ▶ **robotics** n science of designing and using robots

robust adj very strong and healthy ▶ **robustly** adv ▶ **robustness** n

roc n monstrous bird of Arabian mythology

rock¹ n hard mineral substance that makes up part of the earth's crust, stone; large rugged mass of stone; (*Brit*) hard sweet in sticks **on the rocks** (of a marriage) about to end; (of an alcoholic drink) served with ice ▶ **rocky** adj having many rocks ▶ **rockery** n mound of stones in a garden for rock plants ▶ **rock bottom** lowest possible level ▶ **rock cake** small fruit cake with a rough surface

rock² v (cause to) sway to and fro; (*slang*) be very good ▷ n (also **rock music**) style of pop music with a heavy beat ▶ **rocky** adj shaky or unstable ▶ **rock and roll, rock'n'roll** style of pop music blending rhythm and blues and country music ▶ **rocking chair** chair allowing the sitter to rock backwards and forwards

rocker n rocking chair; curved piece of wood etc. on which something may rock **off one's rocker** (*informal*) insane

rocket n self-propelling device powered by the burning of explosive contents (used as a firework, weapon, etc.); vehicle propelled by a rocket engine, as a weapon or carrying a spacecraft ▷ v -**eting, -eted** move fast, esp. upwards, like a rocket

r

rock melon n (US, Aust & NZ) kind of melon with sweet orange flesh

rococo [rok-koe-koe] adj (of furniture, architecture, etc.) having much elaborate decoration in an early 18th-century style

rod n slender straight bar, stick; cane

rode v past tense of **ride**

rodent n animal with teeth specialized for gnawing, such as a rat, mouse, or squirrel

rodeo [roh-**dee**-oh] n, pl -**deos** display of skill by cowboys, such as bareback riding

roe[1] n mass of eggs in a fish, sometimes eaten as food

roe[2] n small species of deer

roentgen [ront-gan] n unit measuring a radiation dose

rogue n dishonest or unprincipled person; mischief-loving person ▸ adj (of a wild beast) having a savage temper and living apart from the herd ▸ **roguish** adj

roister v make merry noisily or boisterously

role, rôle n task or function; actor's part

roll v move by turning over and over; move or sweep along; wind round; undulate; smooth out with a roller; (of a ship or aircraft) turn from side to side about a line from nose to tail ▸ n act of rolling over or from side to side; piece of paper etc. rolled up; small round individually baked piece of bread; list or register; continuous sound, as of drums, thunder, etc.; swaying unsteady movement or gait ▸ **roll call** calling out of a list of names, as in a school or the army, to check who is present ▸ **rolled gold** metal coated with a thin layer of gold ▸ **rolling pin** cylindrical roller for flattening pastry ▸ **rolling stock** locomotives and coaches of a railway ▸ **rolling stone** restless wandering person ▸ **roll-on/roll-off** adj (Brit, Aust & NZ) denoting a ship allowing vehicles to be driven straight on and off ▸ **roll-top** adj (of a desk) having a flexible lid sliding in grooves ▸ **roll up** v (informal) appear or arrive ▸ **roll-up** n (Brit informal) cigarette made by the smoker from loose tobacco and cigarette paper

roller n rotating cylinder used for smoothing or supporting a thing to be moved, spreading paint, etc.; long wave of the sea ▸ **Rollerblade**® n roller skate with the wheels set in one straight line ▸ **roller coaster** (at a funfair) narrow railway with steep slopes ▸ **roller skate** skate with wheels

rollicking adj boisterously carefree

roly-poly adj round or plump

ROM (Computing) read only memory

Roman adj of Rome or the Roman Catholic Church ▸ **Roman Catholic** (member) of that section of the Christian Church that acknowledges the supremacy of the Pope ▸ **Roman numerals** the letters I, V, X, L, C, D, M, used to represent numbers ▸ **roman type** plain upright letters in printing

Romance adj (of a language) developed from Latin, such as French or Spanish

romance n love affair; mysterious or exciting quality; novel or film dealing with love, esp. sentimentally; story with scenes remote from ordinary life

romantic adj of or dealing with love; idealistic but impractical; (of literature, music, etc.) displaying passion and imagination rather than order and form ▷ n romantic person or artist ▶ **romantically** adv ▶ **romanticism** n ▶ **romanticize** v describe or regard in an idealized and unrealistic way

Romany n, pl **-nies** ▷ adj Gypsy

romp v play wildly and joyfully ▷ n boisterous activity **romp home** win easily ▶ **rompers** pl n child's overalls

rondavel n (SAfr) circular building, often thatched

rondo n, pl **-dos** piece of music with a leading theme continually returned to

roo n (Aust informal) kangaroo

rood n (Christianity) the Cross; crucifix ▶ **rood screen** (in a church) screen separating the nave from the choir

roof n, pl **roofs** outside upper covering of a building, car, etc. ▷ v put a roof on

rooibos [roy-boss] n (SAfr) tea prepared from the dried leaves of an African plant

rook¹ n Eurasian bird of the crow family ▶ **rookery** n, pl **-eries** colony of rooks, penguins, or seals

rook² n chess piece shaped like a castle

rookie n (informal) new recruit

room n enclosed area in a building; unoccupied space; scope or opportunity ▷ pl lodgings ▶ **roomy** adj spacious

roost n perch for fowls ▷ v perch

rooster n domestic cock

root¹ n part of a plant that grows down into the earth obtaining nourishment; plant with an edible root, such as a carrot; part of a tooth, hair, etc. below the skin; source or origin; form of a word from which other words and forms are derived; (Maths) factor of a quantity which, when multiplied

by itself the number of times indicated, gives the quantity ▷ pl person's sense of belonging ▷ v establish a root and start to grow ▶ **rootless** adj having no sense of belonging ▶ **root for** v (informal) cheer on ▶ **root out** v get rid of completely

root² v dig or burrow

rope n thick cord **know the ropes** be thoroughly familiar with an activity ▶ **rope in** v persuade to join in

ropey, ropy adj **ropier, ropiest** (Brit informal) inferior or inadequate; not well

rorqual n toothless whale with a dorsal fin

rort (Aust informal) n dishonest scheme ▷ v take unfair advantage of something

rosary n, pl **-saries** series of prayers; string of beads for counting these prayers

rose¹ n shrub or climbing plant with prickly stems and fragrant flowers; flower of this plant; perforated flat nozzle for a hose; pink colour ▷ adj pink ▶ **roseate** [roe-zee-ate] adj rose-coloured ▶ **rose window** circular window with spokes branching from the centre ▶ **rosewood** n fragrant wood used to make furniture

rose² v past tense of **rise**

rosé [roe-zay] n pink wine

rosehip n berry-like fruit of a rose plant

rosella n type of Australian parrot

rosemary n fragrant flowering shrub; its leaves as a herb

rosette n rose-shaped ornament, esp. a circular bunch of ribbons

rosin [rozz-in] n resin used for treating the bows of violins etc.

roster n list of people and their turns of duty

rostrum n, pl **-trums** or **-tra** platform or stage

rosy adj **rosier, rosiest** pink-coloured; hopeful or promising

rot v **rotting, rotted** decompose or decay; slowly deteriorate physically or mentally ▷ n decay; (informal) nonsense

rota n list of people who take it in turn to do a particular task

rotary adj revolving; operated by rotation

rotate v (cause to) move round a centre or on a pivot; (cause to) follow a set sequence ▶ **rotation** n

rote n mechanical repetition **by rote** by memory

rotisserie n rotating spit for cooking meat

rotor n revolving portion of a dynamo, motor, or turbine; rotating device with long blades that provides thrust to lift a helicopter

rotten adj decaying; (informal) very bad; corrupt

rotter n (chiefly Brit slang) despicable person

Rottweiler [rot-vile-er] n large sturdy dog with a smooth black and tan coat and usu. a docked tail

rotund [roe-tund] adj round and plump; sonorous ▶ **rotundity** n

rotunda n circular building or room, esp. with a dome

rouble [roo-bl] n monetary unit of Russia, Belarus and Tajikistan

roué [roo-ay] n man given to immoral living

rouge n red cosmetic used to colour the cheeks

rough adj uneven or irregular; not careful or gentle; difficult or unpleasant; approximate; violent, stormy, or boisterous; in preliminary form; lacking refinement ▷ v make rough ▷ n rough state or area **rough it** live without the usual comforts etc. ▶ **roughen** v ▶ **roughly** adv ▶ **roughness** n ▶ **roughage** n indigestible constituents of food which aid digestion ▶ **rough-and-ready** adj hastily prepared but adequate ▶ **rough-and-tumble** n playful fight ▶ **rough-hewn** adj roughly shaped ▶ **roughhouse** n (chiefly US slang) fight ▶ **rough out** v prepare (a sketch or report) in preliminary form

roughcast n mixture of plaster and small stones for outside walls ▷ v coat with this

roughshod adv **ride roughshod over** act with total disregard for

roulette n gambling game played with a revolving wheel and a ball

round adj spherical, cylindrical, circular, or curved ▷ adv, prep indicating an encircling movement, presence on all sides, etc. e.g. tied round the waist; books scattered round the room ▷ v move round ▷ n customary course, as of a milkman; game (of golf); stage in a competition; one of several periods in a boxing match etc.; number of drinks bought at one time; bullet or shell for a gun ▶ **roundly** adv thoroughly ▶ **rounders** n bat-and-ball team game ▶ **round robin** petition signed with names in a circle to conceal the order; tournament in which each player plays against every other player ▶ **round-the-clock** adj throughout the day and night ▶ **round trip** journey out and back again ▶ **round up** v gather (people or animals) together ▶ **roundup** n

roundabout n road junction at which traffic passes round a central island; revolving circular platform on which people ride for amusement ▷ *adj* not straightforward

roundel *n* small disc

roundelay *n* simple song with a refrain

Roundhead *n* (*Hist*) supporter of Parliament against Charles I in the English Civil War

rouse¹ [rhymes with **cows**] *v* wake up; provoke or excite ▶ **rousing** *adj* lively, vigorous

rouse² [rhymes with **mouse**] *v* (foll by *on*) (*Aust*) scold or rebuke

rouseabout *n* (*Aust & NZ*) labourer in a shearing shed

roustabout *n* labourer on an oil rig

rout *n* overwhelming defeat; disorderly retreat ▷ *v* defeat and put to flight

route *n* roads taken to reach a destination; chosen way ▶ **route march** long military training march

routine *n* usual or regular method of procedure; set sequence ▷ *adj* ordinary or regular

roux [roo] *n* fat and flour cooked together as a basis for sauces

rove *v* wander

rover *n* wanderer, traveller

row¹ [rhymes with **go**] *n* straight line of people or things **in a row** in succession

row² [rhymes with **go**] *v* propel (a boat) by oars ▷ *n* spell of rowing ▶ **rowing boat** boat propelled by oars

row³ [rhymes with **now**] (*informal*) *n* dispute; disturbance; reprimand ▷ *v* quarrel noisily

rowan *n* tree producing bright red berries, mountain ash

rowdy *adj* **-dier, -diest** disorderly, noisy, and rough ▷ *n*, *pl* **-dies** person like this

rowel [rhymes with **towel**] *n* small spiked wheel on a spur

rowlock [rol-luk] *n* device on a boat that holds an oar in place

royal *adj* of, befitting, or supported by a king or queen; splendid ▷ *n* (*informal*) member of a royal family ▶ **royally** *adv* ▶ **royalist** *n* supporter of monarchy ▶ **royalty** *n* royal people; rank or power of a monarch; (*pl* **-ties**) payment from an author, musician, inventor, etc. ▶ **royal blue** deep blue

RPI (in Britain) retail price index: measure of change in the average level of prices

rpm revolutions per minute

RSA Republic of South Africa; (in New Zealand) Returned Services Association

RSI repetitive strain injury

RSPCA (in Britain) Royal Society for the Prevention of Cruelty to Animals

RSS Really Simple Syndication: a way of allowing web users to receive updates on their browsers from selected websites

RSVP please reply

rub *v* **rubbing, rubbed** apply pressure and friction to (something) with a circular or backwards-and-forwards movement; clean, polish, or dry by rubbing; chafe or fray through rubbing ▷ *n* act of rubbing **rub it in** emphasize an unpleasant fact ▶ **rub out** *v* remove with a rubber

rubato *adv*, *n* (*Music*) (with) expressive flexibility of tempo

rubber¹ *n* strong waterproof elastic material, orig. made from the dried sap of a tropical tree, now usu. synthetic; piece of rubber used for erasing writing ▷ *adj* made of or producing rubber ▶ **rubbery** *adj* ▶ **rubberneck** *v* stare with unthinking curiosity ▶ **rubber stamp** device for imprinting the date, a name, etc.; automatic authorization

rubber² n match consisting of three games of bridge, whist, etc.; series of matches

rubbish n waste matter; anything worthless; nonsense ▶ **rubbishy** adj

rubble n fragments of broken stone, brick, etc.

rubella n same as **German measles**

rubicund adj ruddy

rubidium n (Chem) soft highly reactive radioactive element

rubric n heading or explanation inserted in a text

ruby n, pl **-bies** red precious gemstone ▷ adj deep red

ruck¹ n rough crowd of common people; (Rugby) loose scrummage

ruck² n, v wrinkle or crease

rucksack n (Brit, Aust & S Afr) large pack carried on the back

ructions pl n (informal) noisy uproar

rudder n vertical hinged piece at the stern of a boat or at the rear of an aircraft, for steering

ruddy adj **-dier, -diest** of a fresh healthy red colour

rude adj impolite or insulting; coarse, vulgar, or obscene; unexpected and unpleasant; roughly made; robust ▶ **rudely** adv ▶ **rudeness** n

rudiments pl n simplest and most basic stages of a subject ▶ **rudimentary** adj basic, elementary

rue¹ v **ruing, rued** feel regret for ▶ **rueful** adj regretful or sorry ▶ **ruefully** adv

rue² n plant with evergreen bitter leaves

ruff n starched and frilled collar; natural collar of feathers, fur, etc. on certain birds and animals

ruffian n violent lawless person

ruffle v disturb the calm of; annoy, irritate ▷ n frill or pleat

rug n small carpet; thick woollen blanket ▶ **rug up** (Aust & NZ) put on warm clothing

rugby n form of football played with an oval ball which may be handled by the players

rugged [rug-gid] adj rocky or steep; uneven and jagged; strong-featured; tough and sturdy

rugger n (chiefly Brit informal) rugby

ruin v destroy or spoil completely; impoverish ▷ n destruction or decay; loss of wealth, position, etc.; broken-down unused building ▶ **ruination** n act of ruining; state of being ruined; cause of ruin ▶ **ruinous** adj causing ruin; more expensive than can be afforded ▶ **ruinously** adv

rule n statement of what is allowed, for example in a game or procedure; what is usual; government, authority, or control; measuring device with a straight edge ▷ v govern; be pre-eminent; give a formal decision; mark with straight line(s); restrain **as a rule** usually ▶ **ruler** n person who governs; measuring device with a straight edge ▶ **ruling** n formal decision ▶ **rule of thumb** practical but imprecise approach ▶ **rule out** v dismiss from consideration

rum¹ n alcoholic drink distilled from sugar cane

rum² adj (Brit informal) odd, strange

rumba n lively ballroom dance of Cuban origin

rumble v make a low continuous noise; (Brit informal) discover the (disreputable) truth about ▷ n deep resonant sound

rumbustious adj boisterous or unruly

ruminate v chew the cud; ponder or meditate ▶ **ruminant** adj, n cud-chewing (animal, such as a cow, sheep, or deer) ▶ **rumination** n quiet meditation and reflection ▶ **ruminative** adj

rummage v search untidily and at length ▷ n untidy search through a collection of things

rummy n card game in which players try to collect sets or sequences

rumour n unproved statement; gossip or common talk ▶ **rumoured** adj suggested by rumour

rump n buttocks; rear of an animal

rumple v make untidy, crumpled, or dishevelled

rumpus n, pl **-puses** noisy commotion

run v **running, ran, run** move with a more rapid gait than walking; compete in a race, election, etc.; travel according to schedule; function; manage; continue in a particular direction for a specified period; expose oneself to (a risk); flow; spread; (of stitches) unravel ▷ n act or spell of running; ride in a car; continuous period; series of unravelled stitches, ladder ▶ **run away** v make one's escape, flee ▶ **run down** v be rude about; reduce in number or size; stop working ▶ **rundown** n ▶ **run-down** adj exhausted ▶ **run into** v meet ▶ **run-of-the-mill** adj ordinary ▶ **run out** v be completely used up ▶ **run over** v knock down (a person) with a moving vehicle ▶ **run up** v incur (a debt)

rune n any character of the earliest Germanic alphabet ▶ **runic** adj

rung[1] n crossbar on a ladder

rung[2] v past participle of **ring**[1]

runnel n small brook

runner n competitor in a race; messenger; part underneath an ice skate etc., on which it slides; slender horizontal stem of a plant, such as a strawberry, running along the ground and forming new roots at intervals; long strip of carpet or decorative cloth ▶ **runner-up** n person who comes second in a competition

running adj continuous; consecutive; (of water) flowing ▷ n act of moving or flowing quickly; management of a business etc. **in the running, out of the running** having or not having a good chance in a competition

runny adj **-nier, -niest** tending to flow; exuding moisture

runt n smallest animal in a litter; (derogatory) undersized person

runway n hard level roadway where aircraft take off and land

rupee n monetary unit of India and Pakistan

rupture n breaking, breach; hernia ▷ v break, burst, or sever

rural adj in or of the countryside

ruse [rooz] n stratagem or trick

rush[1] v move or do very quickly; force (someone) to act hastily; make a sudden attack upon (a person or place) ▷ n sudden quick or violent movement ▷ pl first unedited prints of a scene for a film ▷ adj done with speed, hasty ▶ **rush hour** period at the beginning and end of the working day, when many people are travelling to or from work

rush[2] n marsh plant with a slender pithy stem ▶ **rushy** adj full of rushes

rusk n hard brown crisp biscuit, used esp. for feeding babies

russet adj reddish-brown ▷ n apple with rough reddish-brown skin

Ss

Russian *adj* of Russia ▷ *n* person from Russia; official language of Russia and, formerly, of the Soviet Union ▶ **Russian roulette** act of bravado in which a person spins the cylinder of a revolver loaded with only one cartridge and presses the trigger with the barrel against his or her own head

rust *n* reddish-brown coating formed on iron etc. that has been exposed to moisture; disease of plants which produces rust-coloured spots ▷ *adj* reddish-brown ▷ *v* become coated with rust ▶ **rusty** *adj* coated with rust; of a rust colour; out of practice

rustic *adj* of or resembling country people; rural; crude, awkward, or uncouth; (of furniture) made of untrimmed branches ▷ *n* person from the country

rustle[1] *v*, *n* (make) a low whispering sound

rustle[2] *v* (US) steal (cattle) ▶ **rustler** *n* (US) cattle thief ▶ **rustle up** *v* prepare at short notice

rut[1] *n* furrow made by wheels; dull settled habits or way of living

rut[2] *n* recurrent period of sexual excitability in male deer ▷ *v* **rutting, rutted** be in a period of sexual excitability

ruthenium *n* (Chem) rare hard brittle white element

ruthless *adj* pitiless, merciless ▶ **ruthlessly** *adv* ▶ **ruthlessness** *n*

rye *n* kind of grain used for fodder and bread; (US) whiskey made from rye

rye-grass *n* any of several grasses cultivated for fodder

S South(ern)

s second(s)

SA Salvation Army; South Africa; South Australia

SAA South African Airways

Sabbath *n* day of worship and rest: Saturday for Jews, Sunday for Christians ▶ **sabbatical** *adj*, *n* (denoting) leave for study

SABC South African Broadcasting Corporation

sable *n* dark fur from a small weasel-like Arctic animal ▷ *adj* black

sabot [sab-oh] *n* wooden shoe traditionally worn by peasants in France

sabotage *n* intentional damage done to machinery, systems, etc. ▷ *v* damage intentionally ▶ **saboteur** *n* person who commits sabotage

sabre *n* curved cavalry sword

sac *n* pouchlike structure in an animal or plant

saccharin *n* artificial sweetener ▶ **saccharine** *adj* excessively sweet

sacerdotal *adj* of priests

sachet *n* small envelope or bag containing a single portion

sack[1] *n* large bag made of coarse material; (informal) dismissal; (slang) bed ▷ *v* (informal) dismiss ▶ **sackcloth** *n* coarse fabric used for sacks, formerly worn as a penance

sack[2] *n* plundering of a captured town ▷ *v* plunder (a captured town)

sacrament n ceremony of the Christian Church, esp. Communion ▶ **sacramental** adj

sacred adj holy; connected with religion; set apart, reserved

sacrifice n giving something up; thing given up; making of an offering to a god; thing offered ▷ v offer as a sacrifice; give (something) up ▶ **sacrificial** adj

sacrilege n misuse or desecration of something sacred ▶ **sacrilegious** adj

> **SPELLING TIP**
> One could be forgiven for thinking that **sacrilegious** had something to do with the word 'religious', which might explain why the most common misspelling of the word in the Collins Corpus is *sacreligious*

sacristan n person in charge of the contents of a church ▶ **sacristy** n, pl **-ties** room in a church where sacred objects are kept

sacrosanct adj regarded as sacred, inviolable

sacrum [say-krum] n, pl **-cra** wedge-shaped bone at the base of the spine

sad adj **sadder, saddest** sorrowful, unhappy; deplorably bad ▶ **sadden** v make sad ▶ **saddo** n, pl **-dos** or **-does** (Brit informal) socially inadequate or pathetic person ▶ **sadly** adv ▶ **sadness** n

saddle n rider's seat on a horse or bicycle; joint of meat ▷ v put a saddle on (a horse); burden (with a responsibility) ▶ **saddler** n maker or seller of saddles

sadism [say-dizz-um] n gaining of (sexual) pleasure from inflicting pain ▶ **sadist** n ▶ **sadistic** adj ▶ **sadistically** adv

sadomasochism n combination of sadism and masochism ▶ **sadomasochist** n

s.a.e. (Brit, Aust & NZ) stamped addressed envelope

safari n, pl **-ris** expedition to hunt or observe wild animals, esp. in Africa ▶ **safari park** park where lions, elephants, etc. are kept uncaged so that people can see them from cars

safe adj secure, protected; uninjured, out of danger; not involving risk ▷ n strong lockable container ▶ **safely** adv ▶ **safe-conduct** n permit allowing travel through a dangerous area ▶ **safekeeping** n protection

safeguard v protect ▷ n protection

safety n, pl **-ties** state of being safe ▶ **safety net** net to catch performers on a trapeze or high wire if they fall ▶ **safety pin** pin with a spring fastening and a guard over the point when closed ▶ **safety valve** valve that allows steam etc. to escape if pressure becomes excessive

saffron n orange-coloured spice obtained from a crocus, used for flavouring and colouring ▷ adj orange

sag v **sagging, sagged** sink in the middle; tire; (of clothes) hang loosely ▷ n droop

saga [sah-ga] n legend of Norse heroes; any long story or series of events

sagacious adj wise ▶ **sagacity** n

sage[1] n very wise man ▷ adj (lit) wise ▶ **sagely** adv

sage[2] n aromatic herb with grey-green leaves

sago n starchy cereal from the powdered pith of the sago palm tree

said v past of **say**

sail n sheet of fabric stretched to catch the wind for propelling a sailing boat; arm of a windmill ▷ v travel by water; begin a voyage; move smoothly ▶ **sailor** n member of a ship's crew ▸ **sailboard** n board with a mast and single sail, used for windsurfing

saint n (Christianity) person venerated after death as specially holy; exceptionally good person ▸ **saintly** adj ▸ **saintliness** n

sake[1] n benefit; purpose **for the sake of** for the purpose of; to please or benefit (someone)

sake[2], **saki** [sah-kee] n Japanese alcoholic drink made from fermented rice

salaam [sal-ahm] n low bow of greeting among Muslims

salacious adj excessively concerned with sex

salad n dish of raw vegetables, eaten as a meal or part of a meal

salamander n amphibian which looks like a lizard

salami n highly spiced sausage

salary n, pl -**ries** fixed regular payment, usu. monthly, to an employee ▸ **salaried** adj

sale n exchange of goods for money; selling of goods at unusually low prices; auction ▸ **saleable** adj fit or likely to be sold ▸ **salesman, saleswoman, salesperson** n person who sells goods ▸ **salesmanship** n skill in selling

salient [say-lee-ent] adj prominent, noticeable ▸ n (Mil) projecting part of a front line

saline [say-line] adj containing salt ▸ **salinity** n

saliva n liquid that forms in the mouth, spittle ▸ **salivary** adj ▸ **salivate** v produce saliva

sallee n (Aust) (also **snow gum**) SE Australian eucalyptus with a pale grey barkacacia tree

sallow adj of an unhealthy pale or yellowish colour

sally n, pl -**lies** witty remark; sudden brief attack by troops

▸ v -**lying, -lied** (foll by forth) rush out; go out

salmon n large fish with orange-pink flesh valued as food ▸ adj orange-pink

salmonella n, pl -**lae** bacterium causing food poisoning

salon n commercial premises of a hairdresser, beautician, etc.; elegant reception room for guests

saloon n two-door or four-door car with body closed off from rear luggage area; large public room, as on a ship; (US) bar serving alcoholic drinks ▸ **saloon bar** more expensive bar in a pub

salt n white crystalline substance used to season food; chemical compound of acid and metal ▸ v season or preserve with salt **old salt** experienced sailor **with a pinch of salt** allowing for exaggeration **worth one's salt** efficient ▸ **salty** adj ▸ **saltbush** n shrub that grows in alkaline desert regions ▸ **salt cellar** small container for salt at table

saltire n (Heraldry) diagonal cross on a shield

saltpetre n compound used in gunpowder and as a preservative

salubrious adj favourable to health

Saluki n tall hound with a silky coat

salutary adj producing a beneficial result

salute n motion of the arm as a formal military sign of respect; firing of guns as a military greeting of honour ▸ v greet with a salute; make a salute; acknowledge with praise ▸ **salutation** n greeting by words or actions

salvage n saving of a ship or other property from destruction; property so saved ▸ v save from destruction or waste

salvation n fact or state of being saved from harm or the consequences of sin

salve n healing or soothing ointment ▷ v soothe or appease

salver n (silver) tray on which something is presented

salvia n plant with blue or red flowers

salvo n, pl -vos or -voes simultaneous discharge of guns etc.; burst of applause or questions

sal volatile [sal vol-at-ill-ee] n preparation of ammonia, used to revive a person who feels faint

SAM surface-to-air missile

Samaritan n person who helps people in distress

samba n lively Brazilian dance

same adj identical, not different, unchanged; just mentioned ▷ **sameness** n

samovar n Russian tea urn

Samoyed n dog with a thick white coat and tightly curled tail

sampan n small boat with oars used in China

samphire n plant found on rocks by the seashore

sample n part taken as representative of a whole; (Music) short extract from an existing recording mixed into a backing track to produce a new recording ▷ v take and test a sample of; (Music) take a short extract from (one recording) and mix it into a backing track; record (a sound) and feed it into a computerized synthesizer so that it can be reproduced at any pitch ▷ **sampler** n piece of embroidery showing the embroiderer's skill; (Music) piece of electronic equipment used for sampling ▷ **sampling** n

samurai n, pl -rai member of an ancient Japanese warrior caste

sanatorium n, pl -riums or -ria institution for invalids or convalescents; room for sick pupils at a boarding school

sanctify v -fying, -fied make holy

sanctimonious adj pretending to be religious and virtuous

sanction n permission, authorization; coercive measure or penalty ▷ v allow, authorize

sanctity n sacredness, inviolability

sanctuary n, pl -aries holy place; part of a church nearest the altar; place of safety for a fugitive; place where animals or birds can live undisturbed

sanctum n, pl -tums or -ta sacred place; person's private room

sand n substance consisting of small grains of rock, esp. on a beach or in a desert ▷ pl stretches of sand forming a beach or desert ▷ v smooth with sandpaper ▷ **sandy** adj covered with sand; (of hair) reddish-fair ▷ **sandbag** n bag filled with sand, used as protection against gunfire or flood water ▷ **sandblast** v, n (clean with) a jet of sand blown from a nozzle under pressure ▷ **sandpaper** n paper coated with sand for smoothing a surface ▷ **sandpiper** n shore bird with a long bill and slender legs ▷ **sandstone** n rock composed of sand ▷ **sandstorm** n desert wind that whips up clouds of sand

sandal n light shoe consisting of a sole attached by straps

sandalwood n sweet-scented wood

sander n power tool for smoothing surfaces

S & M sadomasochism

sandwich n two slices of bread with a layer of food between ▷ v insert between two other things ▷ **sandwich board** pair of boards hung over a person's shoulders to display advertisements in front and behind

sane adj of sound mind; sensible, rational ▸ **sanity** n

sang v past tense of **sing**

sang-froid [sahng-frwah] n composure and calmness in a difficult situation

sangoma n (S Afr) witch doctor or herbalist

sanguinary adj accompanied by bloodshed; bloodthirsty

sanguine adj cheerful, optimistic

sanitary adj promoting health by getting rid of dirt and germs ▸ **sanitation** n sanitary measures, esp. drainage or sewerage

sank v past tense of **sink**

Sanskrit n ancient language of India

sap¹ n moisture that circulates in plants; (informal) gullible person

sap² v **sapping**, **sapped** undermine; weaken ▸ **sapper** n soldier in an engineering unit

SAP South African Police

sapient [say-pee-ent] adj (lit) wise, shrewd

sapling n young tree

sapphire n blue precious stone ▷ adj deep blue

sarabande, **saraband** n slow stately Spanish dance

Saracen n (Hist) Arab or Muslim who opposed the Crusades

sarcasm n (use of) bitter or wounding ironic language ▸ **sarcastic** adj ▸ **sarcastically** adv

sarcophagus n, pl **-gi** or **-guses** stone coffin

sardine n small fish of the herring family, usu. preserved tightly packed in tins

sardonic adj mocking or scornful ▸ **sardonically** adv

sargassum, **sargasso** n type of floating seaweed

sari, **saree** n long piece of cloth draped around the body and over one shoulder, worn by Hindu women

sarmie n (S Afr slang) sandwich

sarong n long piece of cloth tucked around the waist or under the armpits, worn esp. in Malaysia

sarsaparilla n soft drink, orig. made from the root of a tropical American plant

sartorial adj of men's clothes or tailoring

SARU South African Rugby Union

SAS (in Britain) Special Air Service

sash¹ n decorative strip of cloth worn round the waist or over one shoulder

sash² n wooden frame containing the panes of a window ▸ **sash window** window consisting of two sashes that can be opened by sliding one over the other

sassafras n American tree with aromatic bark used medicinally

Sassenach n (Scot) English person

sat v past of **sit**

Satan n the Devil ▸ **the Great Satan** radical Islamic term for the United States ▸ **satanic** adj of Satan; supremely evil ▸ **Satanism** n worship of Satan

satay, **saté** [sat-ay] n Indonesian and Malaysian dish consisting of pieces of chicken, pork, etc., grilled on skewers and served with peanut sauce

satchel n bag, usu. with a shoulder strap, for carrying books

sate v satisfy (a desire or appetite) fully

satellite n man-made device orbiting in space; heavenly body that orbits another; country that is dependent on a more powerful one ▷ adj of or used in the transmission of television signals from a satellite to the home

satiate [say-she-ate] v provide with more than enough, so as to disgust ▸ **satiety** [sat-**tie**-a-tee] n feeling of having had too much

satin n silky fabric with a glossy surface on one side ▸ **satiny** adj of or like satin ▸ **satinwood** n tropical tree yielding hard wood

satire n use of ridicule to expose vice or folly; poem or other work that does this ▸ **satirical** adj ▸ **satirist** n ▸ **satirize** v ridicule by means of satire

satisfy v **-fying, -fied** please, content; provide amply for (a need or desire); convince, persuade ▸ **satisfaction** n ▸ **satisfactory** adj

satnav n (Motoring informal) satellite navigation

satsuma n kind of small orange

saturate v soak thoroughly; cause to absorb the maximum amount of something ▸ **saturation** n

Saturday n seventh day of the week

Saturn n Roman god of agriculture; sixth planet from the sun ▸ **saturnine** adj gloomy in temperament or appearance ▸ **saturnalia** n wild party or orgy

satyr n woodland god, part man, part goat; lustful man

sauce n liquid added to food to enhance flavour; (chiefly Brit informal) impudence ▸ **saucy** adj impudent; pert, jaunty ▸ **saucily** adv ▸ **saucepan** n cooking pot with a long handle

saucer n small round dish put under a cup

sauerkraut n shredded cabbage fermented in brine

sauna n Finnish-style steam bath

saunter v walk in a leisurely manner, stroll ▷ n leisurely walk

sausage n minced meat in an edible tube-shaped skin

▸ **sausage roll** skinless sausage covered in pastry

sauté [so-tay] v **-téing** or **-téeing, -téed** fry quickly in a little fat

savage adj wild, untamed; cruel and violent; uncivilized, primitive ▷ n uncivilized person ▷ v attack ferociously ▸ **savagely** adv ▸ **savagery** n

savannah, savanna n extensive open grassy plain in Africa

savant n learned person

save v rescue or preserve from harm, protect; keep for the future; set aside (money); (Sport) prevent the scoring of (a goal) ▷ n (Sport) act of preventing a goal ▸ **saver** n ▸ **saving** n economy ▷ pl money put by for future use

saveloy n (Brit, Aust & NZ) spicy smoked sausage

saviour n person who rescues another; (**S-**) Christ

savoir-faire [sav-wahr-**fair**] n (French) ability to do and say the right thing in any situation

savory n aromatic herb used in cooking

savour v enjoy, relish; (foll by of) have a flavour or suggestion of ▷ n characteristic taste or odour; slight but distinctive quality ▸ **savoury** adj salty or spicy ▷ n, pl **-vouries** savoury dish served before or after a meal

savoy n variety of cabbage

savvy (slang) v **-vying, -vied** understand ▷ n understanding, intelligence

saw[1] n cutting tool with a toothed metal blade ▷ v **sawing, sawed, sawed** or **sawn** cut with a saw; move (something) back and forth ▸ **sawyer** n person who saws timber for a living ▸ **sawdust** n fine wood fragments made in sawing ▸ **sawfish** n fish with a long toothed snout ▶ **sawmill** n mill where timber is sawn into planks

saw² v past tense of **see¹**

saw³ n wise saying, proverb

sax n (informal) short for **saxophone**

saxifrage n alpine rock plant with small flowers

Saxon n member of the W Germanic people who settled widely in Europe in the early Middle Ages ▷ adj of the Saxons

saxophone n brass wind instrument with keys and a curved body ▶ **saxophonist** n

say v **saying, said** speak or utter; express (an idea) in words; give as one's opinion; suppose as an example or possibility ▷ n right or chance to speak; share in a decision ▶ **saying** n maxim, proverb

scab n crust formed over a wound; (offens) blackleg ▶ **scabby** adj covered with scabs; (informal) despicable

scabbard n sheath for a sword or dagger

scabies [skay-beez] n itchy skin disease

scabrous [skay-bruss] adj rough and scaly; indecent

scaffold n temporary platform for workmen; gallows ▶ **scaffolding** n (materials for building) scaffolds

scalar n, adj (variable quantity) having magnitude but no direction

scald v burn with hot liquid or steam; sterilize with boiling water; heat (liquid) almost to boiling point ▷ n injury by scalding

scale¹ n one of the thin overlapping plates covering fishes and reptiles; thin flake; coating which forms in kettles etc. due to hard water; tartar formed on the teeth ▷ v remove scales from; come off in scales ▶ **scaly** adj

scale² n (often pl) weighing instrument

scale³ n graduated table or sequence of marks at regular intervals, used as a reference in making measurements; ratio of size between a thing and a representation of it; relative degree or extent; fixed series of notes in music ▷ v climb ▶ **scale down** v decrease proportionately in size ▶ **scale up** v increase proportionately in size

scalene adj (of a triangle) with three unequal sides

scallop n edible shellfish with two fan-shaped shells; one of a series of small curves along an edge ▶ **scalloped** adj decorated with small curves along the edge

scallywag n (informal) scamp, rascal

scalp n skin and hair on top of the head ▷ v cut off the scalp of

scalpel n small surgical knife

scam n (informal) dishonest scheme

scamp n mischievous child

scamper v run about hurriedly or in play ▷ n scampering

scampi pl n large prawns

scan v **scanning, scanned** scrutinize carefully; glance over quickly; examine or search (an area) by passing a radar or sonar beam over it; (of verse) conform to metrical rules ▷ n scanning ▶ **scanner** n electronic device used for scanning ▶ **scansion** n metrical scanning of verse

scandal n disgraceful action or event; malicious gossip ▶ **scandalize** v shock by scandal ▶ **scandalous** adj

Scandinavian n, adj (inhabitant or language) of Scandinavia (Norway, Denmark, Sweden, Finland, and Iceland)

S

scandium n (Chem) rare silvery-white metallic element

scant adj barely sufficient, meagre

scanty adj **scantier, scantiest** barely sufficient or not sufficient ▶ **scantily** adv

scapegoat n person made to bear the blame for others

scapula n, pl **-lae** or **-las** shoulder blade ▶ **scapular** adj

scar n mark left by a healed wound; permanent emotional damage left by an unpleasant experience ▷ v **scarring, scarred** mark or become marked with a scar

scarab n sacred beetle of ancient Egypt

scarce adj insufficient to meet demand; not common, rarely found **make oneself scarce** (informal) go away ▶ **scarcely** adv hardly at all; definitely or probably not ▶ **scarcity** n

scare v frighten or be frightened ▷ n fright, sudden panic ▶ **scary** adj (informal) frightening ▶ **scarecrow** n figure dressed in old clothes, set up to scare birds away from crops; raggedly dressed person ▶ **scaremonger** n person who spreads alarming rumours

scarf n, pl **scarves** or **scarfs** piece of material worn round the neck, head, or shoulders

scarf n joint between two pieces of timber made by notching the ends and fastening them together ▷ v join in this way

scarify v **-fying, -fied** scratch or cut slightly all over; break up and loosen (topsoil); criticize mercilessly ▶ **scarification** n

scarlatina n scarlet fever

scarlet adj, n brilliant red ▶ **scarlet fever** infectious fever with a scarlet rash

scarp n steep slope

scarper v (Brit slang) run away

scat v **scatting, scatted** (informal) go away

scat n jazz singing using improvised vocal sounds instead of words

scathing adj harshly critical

scatological adj preoccupied with obscenity, esp. with references to excrement ▶ **scatology** n

scatter v throw about in various directions; disperse ▶ **scatterbrain** n empty-headed person

scatty adj **-tier, -tiest** (informal) empty-headed

scavenge v search for (anything usable) among discarded material

scavenger n person who scavenges; animal that feeds on decaying matter

scenario n, pl **-rios** summary of the plot of a play or film; imagined sequence of future events

scene n place of action of a real or imaginary event; subdivision of a play or film in which the action is continuous; view of a place; display of emotion; (informal) specific activity or interest e.g. the fashion scene **behind the scenes** backstage; in secret ▶ **scenery** n natural features of a landscape; painted backcloths or screens used on stage to represent the scene of action ▶ **scenic** adj picturesque

scent n pleasant smell; smell left in passing, by which an animal can be traced; series of clues; perfume ▷ v detect by smell; suspect; fill with fragrance

sceptic [skep-tik] n person who habitually doubts generally accepted beliefs ▶ **sceptical** adj ▶ **sceptically** adv ▶ **scepticism** n

sceptre n ornamental rod symbolizing royal power

schedule n plan of procedure for a project; list; timetable ▷ v plan to occur at a certain time

schema n, pl **-mata** overall plan or diagram ▶ **schematic** adj presented as a plan or diagram

scheme n systematic plan; secret plot ▷ v plan in an underhand manner ▶ **scheming** adj, n

scherzo [skairt-so] n, pl **-zos** or **-zi** brisk lively piece of music

schism [skizz-um] n (group resulting from) division in an organization ▶ **schismatic** adj, n

schist [shist] n crystalline rock which splits into layers

schizoid adj abnormally introverted; (offens) contradictory ▷ n (offens) schizoid person

schizophrenia n mental disorder involving deterioration of or confusion about the personality; (informal) contradictory behaviour or attitudes ▶ **schizophrenic** adj, n

schmaltz n excessive sentimentality ▶ **schmaltzy** adj

schnapps n strong alcoholic spirit

schnitzel n thin slice of meat, esp. veal

scholar n learned person; student receiving a scholarship; pupil ▶ **scholarly** adj learned ▶ **scholarship** n learning; financial aid given to a student because of academic merit ▶ **scholastic** adj of schools or scholars

school¹ n place where children are taught or instruction is given in a subject; group of artists, thinkers, etc. with shared principles or methods ▷ v educate or train ▶ **schoolie** (Aust) schoolteacher or high-school student ▶ **schoolies week** (Aust informal) week of post-exam celebrations for students who have just completed their final year of high school

school² n shoal of fish, whales, etc.

schooner n sailing ship rigged fore-and-aft; large glass

sciatica n severe pain in the large nerve in the back of the leg ▶ **sciatic** adj of the hip; of or afflicted with sciatica

science n systematic study and knowledge of natural or physical phenomena ▶ **scientific** adj of science; systematic ▶ **scientifically** adv ▶ **scientist** n person who studies or practises a science ▶ **science fiction** stories making imaginative use of scientific knowledge ▶ **science park** area where scientific research and commercial development are carried on in cooperation

sci-fi n short for **science fiction**

scimitar n curved oriental sword

scintillate v give off sparks

scintillating adj very lively and amusing

scion [sy-on] n descendant or heir; shoot of a plant for grafting

scissors pl n cutting instrument with two crossed pivoted blades

sclerosis n, pl **-ses** abnormal hardening of body tissues

scoff¹ v express derision

scoff² v (informal) eat rapidly

scold v find fault with, reprimand ▷ n person who scolds ▶ **scolding** n

sconce n bracket on a wall for holding candles or lights

scone n small plain cake baked in an oven or on a griddle

scoop n shovel-like tool for ladling or hollowing out; news story reported in one newspaper before all its rivals ▷ v take up or hollow out with or as if with a scoop; beat (rival newspapers) in reporting a news item

scoot v (slang) leave or move quickly

scooter n child's vehicle propelled by pushing on the ground with one foot; light motorcycle

scope n opportunity for using abilities; range of activity

scorch v burn on the surface; parch or shrivel from heat ▷ n slight burn ▶ **scorcher** n (informal) very hot day

score n points gained in a game or competition; twenty; written version of a piece of music showing parts for each musician; mark or cut; grievance e.g. settle old scores ▷ pl lots ▷ v gain (points) in a game; keep a record of points; mark or cut; (foll by out) cross out; arrange music (for); achieve a success

scorn n open contempt ▷ v reject with contempt ▶ **scornful** adj ▶ **scornfully** adv

scorpion n small lobster-shaped animal with a sting at the end of a jointed tail

Scot n person from Scotland ▶ **Scottish** adj of Scotland, its people, or their languages ▶ **Scotch** n whisky distilled in Scotland ▶ **Scotch broth** thick soup of beef or lamb and vegetables ▶ **Scots** adj Scottish ▶ **Scotsman, Scotswoman** n

> **USAGE NOTE**
> Scotch is used only in certain fixed expressions like Scotch bonnet. The use of Scotch for Scots or Scottish is otherwise felt to be incorrect, esp. when applied to people

scotch v put an end to

scot-free adj without harm or punishment

scoundrel n (old-fashioned) cheat or deceiver

scour¹ v clean or polish by rubbing with something rough; clear or flush out ▶ **scourer** n small rough nylon pad used for cleaning pots and pans

scour² v search thoroughly and energetically

scourge n person or thing causing severe suffering; whip ▷ v cause severe suffering to; whip

scout n person sent out to reconnoitre; (**S-**) member of the Scout Association, an organization for young people which aims to develop character and promotes outdoor activities ▷ v act as a scout; reconnoitre

scowl v, n (have) an angry or sullen expression

scrabble v scrape at with the hands, feet, or claws

scrag n thin end of a neck of mutton ▶ **scraggy** adj thin, bony

scram v scramming, scrammed (informal) go away quickly

scramble v climb or crawl hastily or awkwardly; struggle with others (for); mix up; cook (eggs beaten up with milk); (of an aircraft or aircrew) take off hurriedly in an emergency; make (transmitted speech) unintelligible by the use of an electronic device ▷ n scrambling; rough climb; disorderly struggle; motorcycle race over rough ground ▶ **scrambler** n electronic device that makes transmitted speech unintelligible

scrap¹ n small piece; waste metal collected for reprocessing ▷ pl leftover food ▷ v scrapping, scrapped discard as useless ▶ **scrappy** adj fragmentary, disjointed ▶ **scrapbook** n book with blank pages in which newspaper cuttings or pictures are stuck

scrap² n, v scrapping, scrapped (informal) fight or quarrel

scrape v rub with something rough or sharp; clean or smooth thus;

S

rub with a harsh noise; economize ▷ *n* act or sound of scraping; mark or wound caused by scraping; (*informal*) awkward situation ▶ **scraper** *n* ▶ **scrape through** *v* succeed in or obtain with difficulty

scratch *v* mark or cut with claws, nails, or anything rough or sharp; scrape (skin) with nails or claws to relieve itching; withdraw from a race or competition ▷ *n* wound, mark, or sound made by scratching ▷ *adj* put together at short notice **from scratch** from the very beginning **up to scratch** up to standard ▶ **scratchy** *adj* ▶ **scratchcard** *n* ticket that reveals whether or not the holder has won a prize when the surface is removed by scratching

scrawl *v* write carelessly or hastily ▷ *n* scribbled writing

scrawny *adj* **scrawnier, scrawniest** thin and bony

scream *v* utter a piercing cry, esp. of fear or pain; utter with a scream ▷ *n* shrill piercing cry; (*informal*) very funny person or thing

scree *n* slope of loose shifting stones

screech *v*, *n* (utter) a shrill cry

screed *n* long tedious piece of writing

screen *n* surface of a television set, VDU, etc., on which an image is formed; white surface on which films or slides are projected; movable structure used to shelter, divide, or conceal something ▷ *v* shelter or conceal with or as if with a screen; examine (a person or group) to determine suitability for a task or to detect the presence of disease or weapons; show (a film) **the screen** cinema generally ▶ **screen saver** (*Computing*) software that produces changing

images on a monitor when the computer is operative but idle ▶ **screenshot** *n* (*Computing*) copied image of a computer screen at a particular moment

screw *n* metal pin with a spiral ridge along its length, twisted into materials to fasten them together; (*slang*) prison guard ▷ *v* turn (a screw); twist; fasten with screw(s); (*informal*) extort ▶ **screwy** *adj* (*informal*) crazy or eccentric ▶ **screwdriver** *n* tool for turning screws ▶ **screw up** *v* (*informal*) bungle; distort

scribble *v* write hastily or illegibly; make meaningless or illegible marks ▷ *n* something scribbled

scribe *n* person who copied manuscripts before the invention of printing; (*Bible*) scholar of the Jewish Law

scrimmage *n* rough or disorderly struggle

scrimp *v* be very economical

scrip *n* certificate representing a claim to stocks or shares

script *n* text of a film, play, or TV programme; a particular system of writing e.g. *Arabic script*; handwriting

scripture *n* sacred writings of a religion ▶ **scriptural** *adj*

scrofula *n* tuberculosis of the lymphatic glands ▶ **scrofulous** *adj*

scroggin *n* (*NZ*) mixture of nuts and dried fruits

scroll *n* roll of parchment or paper; ornamental carving shaped like a scroll ▷ *v* move (text) up or down on a VDU screen

scrotum *n*, *pl* **-ta** or **-tums** pouch of skin containing the testicles

scrounge *v* (*informal*) get by cadging or begging ▶ **scrounger** *n*

scrub[1] *v* **scrubbing, scrubbed** clean by rubbing, often with a hard brush and water; (informal) delete or cancel ▷ n act or instance of scrubbing ▷ pl (Med) hygienic clothing worn by operating theatre staff

scrub² n stunted trees; area of land covered with scrub ▶ **scrubby** adj covered with scrub; stunted; (informal) shabby

scruff¹ n nape (of the neck)

scruff² n (informal) untidy person ▶ **scruffy** adj unkempt or shabby

scrum, scrummage n (Rugby) restarting of play in which opposing packs of forwards push against each other to gain possession of the ball; disorderly struggle

scrumptious adj (informal) delicious

scrunch v crumple or crunch or be crumpled or crunched ▷ n act or sound of scrunching

scruple n doubt produced by one's conscience or morals ▷ v have doubts on moral grounds ▶ **scrupulous** adj very conscientious; very careful or precise ▶ **scrupulously** adv

scrutiny n, pl -**nies** close examination ▶ **scrutinize** v examine closely

scuba diving n sport of swimming under water using cylinders containing compressed air attached to breathing apparatus

scud v **scudding, scudded** move along swiftly

scuff v drag (the feet) while walking; scrape (one's shoes) by doing so ▷ n mark caused by scuffing

scuffle v fight in a disorderly manner ▷ v in disorderly struggle; scuffling sound

scull n small oar ▷ v row (a boat) using sculls

scullery n, pl -**leries** small room where washing-up and other kitchen work is done

sculpture n art of making figures or designs in wood, stone, etc.; product of this art ▷ v (also **sculpt**) represent in sculpture ▶ **sculptor, sculptress** ▶ **sculptural** adj

scum n impure or waste matter on the surface of a liquid; worthless people ▶ **scummy** adj

scungy adj -**ier, -iest** (Aust & NZ informal) sordid or dirty

scupper v (informal) defeat or ruin

scurf n flaky skin on the scalp

scurrilous adj untrue and defamatory

scurry v -**rying, -ried** move hastily ▷ n act or sound of scurrying

scurvy n disease caused by lack of vitamin C

scut n short tail of the hare, rabbit, or deer

scuttle¹ n fireside container for coal

scuttle² v run with short quick steps ▷ n hurried run

scuttle³ v make a hole in (a ship) to sink it

scythe n long-handled tool with a curved blade for cutting grass ▷ v cut with a scythe

SE southeast(ern)

sea n mass of salt water covering three quarters of the earth's surface; particular area of this; vast expanse **at sea** in a ship on the ocean; confused or bewildered ▶ **sea anemone** sea animal with suckers like petals ▶ **seaboard** n coast ▶ **sea dog** experienced sailor ▶ **seafaring** adj working or travelling by sea ▶ **seafood** n edible saltwater fish or shellfish ▶ **seagull** n gull ▶ **sea horse** small sea fish with a plated body and horselike head ▶ **sea level** average level of the sea's surface in relation

s

to the land ▸ **sea lion** kind of large seal ▸ **seaman** n sailor ▸ **seaplane** n aircraft designed to take off from and land on water ▸ **seasick** adj suffering from nausea caused by the motion of a ship ▸ **seasickness** n ▸ **seaside** n area, esp. a holiday resort, on the coast ▸ **sea urchin** sea animal with a round spiky shell ▸ **seaweed** n plant growing in the sea ▸ **seaworthy** adj (of a ship) in fit condition for a sea voyage

seal[1] n piece of wax, lead, etc. with a special design impressed upon it, attached to a letter or document as a mark of authentication; device or material used to close an opening tightly ▷ v close with or as if with a seal; make airtight or watertight; affix a seal to or stamp with a seal; decide (one's fate) irrevocably ▸ **sealant** n any substance used for sealing ▸ **seal off** v enclose or isolate (a place) completely

seal[2] n amphibious mammal with flippers as limbs ▸ **sealskin** n

seam n line where two edges are joined, as by stitching; thin layer of coal or ore ▷ v mark with furrows or wrinkles ▸ **seamless** adj

seamstress n woman who sews, esp. professionally

seamy adj sordid

seance [say-anss] n meeting at which spiritualists attempt to communicate with the dead

sear v scorch, burn the surface of ▸ **searing** adj (of pain) very sharp; highly critical

search v examine closely in order to find something ▷ n searching ▸ **searching** adj keen or thorough ▸ **search engine** (Computing) internet service enabling users to search for items of interest ▸ **searchlight** n powerful light

with a beam that can be shone in any direction

season n one of four divisions of the year, each of which has characteristic weather conditions; period during which a thing happens or is plentiful; fitting or proper time ▷ v flavour with salt, herbs, etc.; dry (timber) till ready for use ▸ **seasonable** adj appropriate for the season; timely or opportune ▸ **seasonal** adj depending on or varying with the seasons ▸ **seasoned** adj experienced ▸ **seasoning** n salt, herbs, etc. added to food to enhance flavour ▸ **season ticket** ticket for a series of journeys or events within a specified period

seat n thing designed or used for sitting on; place to sit in a theatre, esp. one that requires a ticket; buttocks; (Brit) country house; membership of a legislative or administrative body ▷ v cause to sit; provide seating for ▸ **seat belt** belt worn in a car or aircraft to prevent a person being thrown forward in a crash

sebaceous adj of, like, or secreting fat or oil

secateurs pl n small pruning shears

secede v withdraw formally from a political alliance or federation ▸ **secession** n

seclude v keep (a person) from contact with others ▸ **secluded** adj private, sheltered ▸ **seclusion** n

second[1] adj coming directly after the first; alternate, additional; inferior ▷ n person or thing coming second; attendant in a duel or boxing match ▷ pl inferior goods ▷ v express formal support for (a motion proposed in a meeting) ▸ **secondly** adv ▸ **second-class**

inferior; cheaper, slower, or less comfortable than first-class ▸ **second-hand** adj bought after use by another ▸ **second nature** something so habitual that it seems part of one's character ▸ **second sight** supposed ability to predict events ▸ **second thoughts** revised opinion on a matter already considered ▸ **second wind** renewed ability to continue effort

second² n sixtieth part of a minute of an angle or time; moment

second³ [si-**kond**] v transfer (a person) temporarily to another job ▸ **secondment** n

secondary adj of less importance; coming after or derived from what is primary or first; relating to the education of people between the ages of 11 and 18 or, in New Zealand, between 13 and 18

secret adj kept from the knowledge of others ▹ n something kept secret; mystery; underlying explanation e.g. the secret of my success **in secret** without other people knowing ▸ **secretly** adv ▸ **secrecy** n ▸ **secretive** adj inclined to keep things secret ▸ **secretiveness** n

secretariat n administrative office or staff of a legislative body

secretary n, pl -ries person who deals with correspondence and general clerical work; (S-) head of a state department e.g. Home Secretary ▸ **secretarial** adj ▸ **Secretary of State** head of a major government department

secrete¹ v (of an organ, gland, etc.) produce and release (a substance) ▸ **secretion** n ▸ **secretory** [sek-**reet-or-ee**] adj

secrete² v hide or conceal

sect n subdivision of a religious or political group, esp. one with extreme beliefs ▸ **sectarian** adj of a sect or religious group; narrow-minded

section n part cut off; part or subdivision of something; distinct part of a country or community; cutting; drawing of something as if cut through ▹ v cut or divide into sections ▸ **sectional** adj

sector n part or subdivision; part of a circle enclosed by two radii and the arc which they cut off

secular adj worldly, as opposed to sacred; not connected with religion or the church

secure adj free from danger; free from anxiety; firmly fixed; reliable ▹ v obtain; make safe; make firm; guarantee payment of (a loan) by giving something as security ▸ **securely** adv ▸ **security** n, pl -ties precautions against theft, espionage, or other danger; state of being secure; certificate of ownership of a share, stock, or bond; something given or pledged to guarantee payment of a loan

sedan n (US, Aust & NZ) two-door or four-door car with the body closed off from the rear luggage area ▸ **sedan chair** (Hist) enclosed chair for one person, carried on poles by two bearers

sedate¹ adj calm and dignified; slow or unhurried ▸ **sedately** adv

sedate² v give a sedative drug to ▸ **sedation** n ▸ **sedative** adj having a soothing or calming effect ▹ n sedative drug

sedentary adj done sitting down, involving little exercise

sedge n coarse grasslike plant growing on wet ground

sediment n matter which settles to the bottom of a liquid; material deposited by water, ice, or wind ▸ **sedimentary** adj

S

sedition n speech or action encouraging rebellion against the government ▸ **seditious** adj

seduce v persuade into sexual intercourse; tempt into wrongdoing ▸ **seducer, seductress** n ▸ **seduction** n ▸ **seductive** adj

sedulous adj diligent or persevering ▸ **sedulously** adv

see[1] v **seeing, saw, seen** perceive with the eyes or mind; understand; watch; find out; make sure (of something); consider or decide; have experience of; meet or visit; accompany ▸ **seeing** conj (often foll by as or that) in view of the fact that

see[2] n diocese of a bishop

seed n mature fertilized grain of a plant; such grains used for sowing; origin; (obs) offspring; (Sport) player ranked according to his or her ability ▸ v sow with seed; remove seeds from; arrange (the draw of a sports tournament) so that the outstanding competitors will not meet in the early rounds **go to seed, run to seed** (of plants) produce or shed seeds after flowering; lose vigour or usefulness ▸ **seedling** n young plant raised from a seed ▸ **seedy** adj shabby

seek v **seeking, sought** try to find or obtain; try (to do something)

seem v appear to be ▸ **seeming** adj apparent but not real ▸ **seemingly** adv

seemly adj proper or fitting

seen v past participle of **see**[1]

seep v trickle through slowly, ooze ▸ **seepage** n

seer n prophet

seersucker n light cotton fabric with a slightly crinkled surface

seesaw n plank balanced in the middle so that two people seated on either end ride up and down alternately ▸ v move up and down

seethe v **seething, seethed** be very agitated; (of a liquid) boil or foam

segment n one of several sections into which something may be divided ▸ v divide into segments ▸ **segmentation** n

segregate v set apart ▸ **segregation** n

seine [sane] n large fishing net that hangs vertically from floats

seismic adj relating to earthquakes ▸ **seismology** n study of earthquakes ▸ **seismological** adj ▸ **seismologist** n ▸ **seismograph, seismometer** n instrument that records the strength of earthquakes

seize v take hold of forcibly or quickly; take immediate advantage of; (usu. foll by up) (of mechanical parts) stick tightly through overheating ▸ **seizure** n sudden violent attack of an illness; seizing or being seized

seldom adv not often, rarely

select v pick out or choose ▸ adj chosen in preference to others; restricted to a particular group, exclusive ▸ **selection** n selecting; things that have been selected; range from which something may be selected ▸ **selective** adj chosen or choosing carefully ▸ **selectively** adv ▸ **selectivity** n ▸ **selector** n

selenium n (Chem) nonmetallic element with photoelectric properties

self n, pl **selves** distinct individuality or identity of a person or thing; one's basic nature; one's own welfare or interests ▸ **selfish** adj caring too much about oneself and not enough about others

▶ **selfishly** adv ▶ **selfishness** n
▶ **selfless** adj unselfish ▶ **selfie** n photograph taken by pointing camera at oneself ▶ **selfie stick** extendable rod to which a camera may be attached to take a photograph of oneself

self- prefix denoting of oneself or itself; denoting by, to, in, due to, for, or from the self; denoting automatic(ally) ▶ **self-assured** adj confident ▶ **self-catering** adj (of accommodation) for people who provide their own food ▶ **self-coloured** adj having only a single colour ▶ **self-conscious** adj embarrassed at being the object of others' attention ▶ **self-contained** adj containing everything needed, complete; (of a flat) having its own facilities ▶ **self-determination** n the right of a nation to decide its own form of government ▶ **self-evident** adj obvious without proof ▶ **self-help** n use of one's own abilities to solve problems; practice of solving one's problems within a group of people with similar problems ▶ **self-interest** n one's own advantage ▶ **self-made** adj having achieved wealth or status by one's own efforts ▶ **self-possessed** adj having control of one's emotions, calm ▶ **self-raising** adj (of flour) containing a raising agent ▶ **self-righteous** adj thinking oneself more virtuous than others ▶ **selfsame** adj the very same ▶ **self-seeking** adj, n seeking to promote only one's own interests ▶ **self-service** adj denoting a shop, café, or garage where customers serve themselves and then pay a cashier ▶ **self-styled** adj using a title or name that one has taken without right ▶ **self-sufficient** adj able to provide for oneself without help ▶ **self-willed** adj stubbornly determined to get one's own way

sell v **selling**, **sold** exchange (something) for money; stock, deal in; (of goods) be sold; (foll by for) have a specified price; (informal) persuade (someone) to accept (something) ▷ n manner of selling ▶ **seller** n ▶ **sell-by date** (Brit) date on packaged food after which it should not be sold ▶ **sell out** v dispose of (something) completely by selling; (informal) betray ▶ **sellout** n performance of a show etc. for which all the tickets are sold; (informal) betrayal

Sellotape® n type of adhesive tape ▷ v stick with Sellotape

selvage, **selvedge** n edge of cloth, woven so as to prevent unravelling

selves n plural of **self**

semantic adj relating to the meaning of words ▶ **semantics** n study of linguistic meaning

Semantic Web n proposed development of the World Wide Web in which computers can interpret and act on natural language

semaphore n system of signalling by holding two flags in different positions to represent letters of the alphabet

semblance n outward or superficial appearance

semen n sperm-carrying fluid produced by male animals

semester n either of two divisions of the academic year

semi n (Brit & S Afr informal) semidetached house

semi- prefix indicating half e.g. semicircle; indicating partly or almost e.g. semiprofessional

semibreve n musical note four beats long

semicolon n the punctuation mark (;)

semiconductor n substance with an electrical conductivity that increases with temperature

semidetached adj (of a house) joined to another on one side

semifinal n match or round before the final ▸ **semifinalist** n

seminal adj original and influential; capable of developing; of semen or seed

seminar n meeting of a group of students for discussion

seminary n, pl -**ries** college for priests

semiprecious adj (of gemstones) having less value than precious stones

semiquaver n musical note half the length of a quaver

Semite n member of the group of peoples including Jews and Arabs

Semitic adj of the group of peoples including Jews and Arabs

semitone n smallest interval between two notes in Western music

semitrailer n (Aust) large truck in two separate sections joined by a pivoted bar (also **semi**)

semolina n hard grains of wheat left after the milling of flour, used to make puddings and pasta

Senate n upper house of some parliaments; governing body of some universities ▸ **senator** n member of a Senate ▸ **senatorial** adj

send v **sending**, **sent** cause (a person or thing) to go to or be taken or transmitted to a place; bring into a specified state or condition ▸ **send off** n demonstration of good wishes at a person's departure ▸ **send up** v (informal) make fun of by imitating ▸ **send-up** n (informal) imitation

senile adj mentally or physically weak because of old age ▸ **senility** n

senior adj superior in rank or standing; older; of or for older pupils ▸ n senior person ▸ **seniority** n

senna n tropical plant; its dried leaves or pods used as a laxative

señor [sen-**nyor**] n, pl -**ores** Spanish term of address equivalent to sir or Mr ▸ **señora** [sen-**nyor-a**] n Spanish term of address equivalent to madam or Mrs ▸ **señorita** [sen-nyor-**ee-ta**] n Spanish term of address equivalent to madam or Miss

sensation n ability to feel things physically; physical feeling; general feeling or awareness; state of excitement; exciting person or thing ▸ **sensational** adj causing intense shock, anger, or excitement; (informal) very good ▸ **sensationalism** n deliberate use of sensational language or subject matter ▸ **sensationalist** adj, n

sense n any of the faculties of perception or feeling (sight, hearing, touch, taste, or smell); ability to perceive; feeling perceived through one of the senses; awareness; (sometimes pl) sound practical judgment or intelligence; specific meaning ▸ v perceive ▸ **senseless** adj

sensible adj having or showing good sense; practical e.g. sensible shoes; (foll by of) aware ▸ **sensibly** adv ▸ **sensibility** n ability to experience deep feelings

sensitive adj easily hurt or offended; responsive to external stimuli; (of a subject) liable to arouse controversy or strong feelings; (of an instrument) responsive to slight changes ▸ **sensitively** adv ▸ **sensitivity** n ▸ **sensitize** v make sensitive

sensor n device that detects or measures the presence of something, such as radiation

sensory adj of the senses or sensation

sensual adj giving pleasure to the body and senses rather than the mind; having a strong liking for physical pleasures ▸ **sensually** adv ▸ **sensuality** n ▸ **sensualist** n

sensuous adj pleasing to the senses ▸ **sensuously** adv

sent v past of **send**

sentence n sequence of words capable of standing alone as a statement, question, or command; punishment given to a criminal ▷ v pass sentence on (a convicted person)

sententious adj trying to sound wise; pompously moralizing

sentient [sen-tee-ent] adj capable of feeling ▸ **sentience** n

sentiment n thought, opinion, or attitude; feeling expressed in words; exaggerated or mawkish emotion ▸ **sentimental** adj excessively romantic or nostalgic ▸ **sentimentalism** n ▸ **sentimentality** n ▸ **sentimentalize** v make sentimental

sentinel n sentry

sentry n, pl **-tries** soldier on watch

sepal n leaflike division of the calyx of a flower

separate v act as a barrier between; distinguish between; divide up into parts; (of a couple) stop living together ▷ adj not the same, different; set apart; not shared, individual ▸ **separately** adv ▸ **separation** n separating or being separated; (Law) living apart of a married couple without divorce ▸ **separable** adj ▸ **separatist** n person who advocates the separation of a group from an organization or country ▸ **separatism** n

SPELLING TIP
There are 101 examples of *seperate* in the Collins Corpus, which makes it the most popular misspelling of **separate**.

sepia adj, n reddish-brown (pigment)

sepoy n (formerly) Indian soldier in the service of the British

sepsis n poisoning caused by pus-forming bacteria

Sept September

September n ninth month of the year

septet n group of seven performers; music for such a group

septic adj (of a wound) infected; of or caused by harmful bacteria ▸ **septic tank** tank in which sewage is decomposed by the action of bacteria

septicaemia [sep-tis-**see**-mee-a] n infection of the blood

septuagenarian n person aged between seventy and seventy-nine

sepulchre [**sep**-pull-ker] n tomb or burial vault ▸ **sepulchral** [sip-**pulk**-ral] adj gloomy

sequel n novel, play, or film that continues the story of an earlier one; consequence

sequence n arrangement of two or more things in successive order; the successive order of two or more things; section of a film showing a single uninterrupted episode ▸ **sequential** adj

sequester v seclude; sequestrate

sequestrate v confiscate (property) until its owner's debts are paid or a court order is complied with ▸ **sequestration** n

sequin n small ornamental metal disc on a garment ▸ **sequined** adj

sequoia n giant Californian coniferous tree

seraglio [sir-ah-lee-oh] *n, pl* **-raglios** harem of a Muslim palace; Turkish sultan's palace

seraph *n, pl* **-aphs** *or* **-aphim** member of the highest order of angels ▶ **seraphic** *adj*

Serbian, Serb *adj* of Serbia ▷ *n* person from Serbia ▶ **Serbo-Croat, Serbo-Croatian** *adj*, *n* (of) the chief official language of Serbia and Croatia

serenade *n* music played or sung to a woman by a lover ▷ *v* sing or play a serenade to (someone)

serendipity *n* gift of making fortunate discoveries by accident

serene *adj* calm, peaceful ▶ **serenely** *adv* ▶ **serenity** *n*

serf *n* medieval farm labourer who could not leave the land he worked on ▶ **serfdom** *n*

serge *n* strong woollen fabric

sergeant *n* noncommissioned officer in the army; police officer ranking between constable and inspector ▶ **sergeant at arms** parliamentary or court officer with ceremonial duties ▶ **sergeant major** highest rank of noncommissioned officer in the army

serial *n* story or play produced in successive instalments ▷ *adj* of or forming a series; published or presented as a serial ▷ *v* publish or present as a serial ▶ **serialize** *v* publish or present as a serial ▶ **serial killer** person who commits a series of murders

series *n, pl* **-ries** group or succession of related things, usu. arranged in order; set of radio or TV programmes about the same subject or characters

serious *adj* giving cause for concern; concerned with important matters; not cheerful, grave; sincere, not joking ▶ **seriously** *adv* ▶ **seriousness** *n*

sermon *n* speech on a religious or moral subject by a clergyman in a church service; long moralizing speech ▶ **sermonize** *v* make a long moralizing speech

serotonin [ser-a-tone-in] *n* chemical compound that occurs in the brain, thought to be beneficial to mental health

serpent *n* (*lit*) snake ▶ **serpentine** *adj* twisting like a snake

serrated *adj* having a notched or sawlike edge

serried *adj* in close formation

serum [seer-um] *n* watery fluid left after blood has clotted; this fluid from the blood of immunized animals used for inoculation or vaccination

servant *n* person employed to do household work for another

serve *v* work for (a person, community, or cause); perform official duties; attend to (customers); provide (someone) with (food or drink); provide with a service; be a member of the armed forces; spend (time) in prison; be useful or suitable; (*Tennis etc.*) put (the ball) into play ▷ *n* (*Tennis etc.*) act of serving the ball

server *n* player who serves in racket games; (*Computing*) computer or program that supplies data to other machines on a network

service *n* system that provides something needed by the public; department of public employment and its employees; availability for use; overhaul of a machine or vehicle; formal religious ceremony; (*Tennis etc.*) act, manner, or right of serving the ball ▷ *pl* armed forces ▷ *v* overhaul (a machine or vehicle) ▶ **serviceable** *adj* useful or helpful; able or ready to be used ▶ **service area** area beside a motorway with garage, restaurant, and toilet facilities ▶ **serviceman,**

servicewoman n member of the armed forces ▶ **service road** narrow road giving access to houses and shops ▶ **service station** garage selling fuel for motor vehicles

serviette n table napkin

servile adj too eager to obey people, fawning; suitable for a slave ▶ **servility** n

servitude n bondage or slavery

servo n (Aust informal) service station

sesame [sess-am-ee] n plant cultivated for its seeds and oil, which are used in cooking

session n period spent in an activity; meeting of a court, parliament, or council; series or period of such meetings; academic term or year

set¹ v **setting, set** put in a specified position or state; make ready; make or become firm or rigid; establish, arrange; prescribe; assign; (of the sun) go down ▷ n scenery used in a play or film ▷ adj fixed or established beforehand; rigid or inflexible; determined (to do something) ▶ **setback** n anything that delays progress ▶ **set square** flat right-angled triangular instrument used for drawing angles ▶ **set up** v arrange or establish ▶ **setup** n way in which anything is organized or arranged

set² n number of things or people grouped or belonging together; (Maths) group of numbers or objects that satisfy a given condition or share a property; television or radio receiver; (Sport) group of games in a match ▶ **set-top box** device that enables digital television broadcasts to be viewed on a standard television set

sett, set n badger's burrow

settee n couch

setter n long-haired gun dog

setting n background or surroundings; time and place where a film, book, etc. is supposed to have taken place; music written for the words of a text; decorative metalwork in which a gem is set; plates and cutlery for a single place at a table; position or level to which the controls of a machine can be adjusted

settle¹ v arrange or put in order; come to rest; establish or become established as a resident; make quiet, calm, or stable; pay (a bill); bestow (property) legally ▶ **settlement** n act of settling; place newly colonized; subsidence (of a building); property bestowed legally ▶ **settler** n colonist

settle² n long wooden bench with high back and arms

seven adj, n one more than six ▶ **seventh** adj, n (of) number seven in a series ▶ **seventeen** adj, n ten and seven ▶ **seventeenth** adj, n ▶ **seventy** adj, n ten times seven ▶ **seventieth** adj, n

sever v cut through or off; break off (a relationship) ▶ **severance** n ▶ **severance pay** compensation paid by a firm to an employee who leaves because the job he or she was appointed to do no longer exists

several adj some, a few; various, separate ▶ **severally** adv separately

severe adj strict or harsh; very intense or unpleasant; strictly restrained in appearance ▶ **severely** adv ▶ **severity** n

sew v **sewing, sewed, sewn** or **sewed** join with thread repeatedly passed through with a needle; make or fasten by sewing

sewage n waste matter or excrement carried away in sewers

S

sewer n drain to remove waste water and sewage ▶ **sewerage** n system of sewers

sewn v a past participle of **sew**

sex n state of being male or female; male or female category; sexual intercourse; sexual feelings or behaviour ▷ v find out the sex of ▶ **sexy** adj sexually exciting or attractive; (informal) exciting or trendy ▶ **sexism** n discrimination on the basis of a person's sex ▶ **sexist** adj, n ▶ **sexual** adj ▶ **sexually** adv ▶ **sexuality** n ▶ **sexual intercourse** sexual act in which the male's penis is inserted into the female's vagina ▶ **sex up** v (informal) make (something) more exciting

sexagenarian n person aged between sixty and sixty-nine

sextant n navigator's instrument for measuring angles, as between the sun and horizon, to calculate one's position

sextet n group of six performers; music for such a group

sexton n official in charge of a church and churchyard

SF science fiction

Sgt. Sergeant

shabby adj -bier, -biest worn or dilapidated in appearance; mean or unworthy e.g. shabby treatment ▶ **shabbily** adv ▶ **shabbiness** n

shack n rough hut ▶ **shack up with** v (slang) live with (one's lover)

shackle n one of a pair of metal rings joined by a chain, for securing a person's wrists or ankles ▷ v fasten with shackles

shad n herring-like fish

shade n relative darkness; place sheltered from sun; screen or cover used to protect from a direct source of light; depth of colour; slight amount; (lit) ghost ▷ pl (informal) sunglasses ▷ v screen from light; darken; represent (darker areas) in drawing; change slightly or by degrees ▶ **shady** adj situated in or giving shade; of doubtful honesty or legality

shadow n dark shape cast on a surface when something stands between a light and the surface; patch of shade; slight trace; threatening influence; inseparable companion ▷ v cast a shadow over; follow secretly ▶ **shadowy** adj ▶ **shadow-boxing** n boxing against an imaginary opponent for practice ▶ **Shadow Cabinet** members of the main opposition party in Parliament who would be ministers if their party were in power

shaft n long narrow straight handle of a tool or weapon; ray of light; revolving rod that transmits power in a machine; vertical passageway, as for a lift or mine; one of the bars between which an animal is harnessed to a vehicle

shag¹ n coarse shredded tobacco ▷ adj (of a carpet) having a long pile ▶ **shaggy** adj covered with rough hair or wool; tousled, unkempt ▶ **shaggy-dog story** long anecdote with a humorous twist at the end

shag² n kind of cormorant

shagreen n sharkskin; rough grainy untanned leather

shah n formerly, ruler of Iran

shake v **shaking, shook, shaken** move quickly up and down or back and forth; make unsteady; tremble; grasp (someone's hand) in greeting or agreement; shock or upset ▷ n shaking; vibration; (informal) short period of time ▶ **shaky** adj unsteady; uncertain or questionable ▶ **shakily** adv

shale *n* flaky sedimentary rock

shall *v, past tense* **should** used as an auxiliary to make the future tense or to indicate intention, obligation, or inevitability

> **USAGE NOTE**
> The use of *shall* with I and *we* is a matter of preference, not rule

shallot *n* kind of small onion

shallow *adj* not deep; lacking depth of character or intellect ▷ **shallows** *pl n* area of shallow water ▷ **shallowness** *n*

sham *n* thing or person that is not genuine ▷ *adj* not genuine ▷ *v* shamming, shammed fake, feign

shamble *v* walk in a shuffling awkward way

shambles *n* disorderly event or place

shame *n* painful emotion caused by awareness of having done something dishonourable or foolish; capacity to feel shame; cause of shame; cause for regret ▷ *v* cause to feel shame; disgrace; compel by shame ▷ *interj* (*S Afr informal*) exclamation of sympathy or endearment ▷ **shameful** *adj* causing or deserving shame ▷ **shamefully** *adv* ▷ **shameless** *adj* with no sense of shame ▷ **shamefaced** *adj* looking ashamed

shammy *n, pl* -mies (*informal*) piece of chamois leather

shampoo *n* liquid soap for washing hair, carpets, or upholstery; process of shampooing ▷ *v* wash with shampoo

shamrock *n* clover leaf, esp. as the Irish emblem

shandy *n, pl* -dies drink made of beer and lemonade

shanghai *v* -haiing, -haied force or trick (someone) into doing something ▷ *n* (*Aust & NZ*) catapult

shank *n* lower leg; shaft or stem

shan't shall not

shantung *n* soft Chinese silk with a knobbly surface

shanty[1] *n, pl* -ties shack or crude dwelling ▷ **shantytown** *n* slum consisting of shanties

shanty[2] *n, pl* -ties sailor's traditional song

shape *n* outward form of an object; way in which something is organized; pattern or mould; condition or state ▷ *v* form or mould; devise or develop ▷ **shapeless** *adj* ▷ **shapely** *adj* having an attractive shape

shard *n* broken piece of pottery or glass

share[1] *n* part of something that belongs to or is contributed by a person; one of the equal parts into which the capital stock of a public company is divided ▷ *v* give or take a share of (something); join with others in doing or using (something) ▷ **shareholder** *n* ▷ **sharemilker** *n* (*NZ*) person who works on a dairy farm belonging to someone else

share[2] *n* blade of a plough

shark *n* large usu. predatory sea fish; person who cheats others

shark biscuit *n* (*Aust facetious*) bodyboard; young or inexperienced surfer

sharkskin *n* stiff glossy fabric

sharp *adj* having a keen cutting edge or fine point; not gradual; clearly defined; mentally acute; shrill; bitter or sour in taste; (*Music*) above the true pitch ▷ *adv* promptly; (*Music*) too high in pitch ▷ *n* (*Music*) symbol raising a note one semitone above natural pitch ▷ **sharply** *adv* ▷ **sharpness** *n* ▷ **sharpen** *v* make or become sharp or sharper ▷ **sharpener** *n* ▷ **sharpshooter** *n* marksman

S

shatter v break into pieces; destroy completely ▸ **shattered** adj (informal) completely exhausted; badly upset

shave v **shaving, shaved, shaved** or **shaven** remove (hair) from (the face, head, or body) with a razor or shaver; pare away; touch lightly in passing ▸ n shaving **close shave** (informal) narrow escape ▸ **shaver** n electric razor ▸ **shavings** pl n parings

shawl n piece of cloth worn over a woman's head or shoulders or wrapped around a baby

she pron referring to female person or animal previously mentioned; referring to something regarded as female, such as a car, ship, or nation

sheaf n, pl **sheaves** bundle of papers; tied bundle of reaped corn

shear v **shearing, sheared, sheared** or **shorn** clip hair or wool from; cut through ▸ **shears** pl n large scissors or a cutting tool shaped like these ▸ **shearer** n ▸ **shearing shed** (Aust & NZ) farm building with equipment for shearing sheep

shearwater n medium-sized sea bird

sheath n close-fitting cover, esp. for a knife or sword; (Brit, Aust & NZ) condom ▸ **sheathe** v put into a sheath

shebeen n (Scot, Irish & S Afr) place where alcohol is sold illegally

shed¹ n building used for storage or shelter or as a workshop

shed² v **shedding, shed** pour forth (tears); cast off (skin, hair, or leaves)

sheen n glistening brightness on the surface of something

sheep n, pl **sheep** ruminant animal bred for wool and meat ▸ **sheep-dip** n liquid disinfectant in which sheep are immersed ▸ **sheepdog** n dog used for herding sheep ▸ **sheepskin** n skin of a sheep with the fleece still on, used for clothing or rugs

sheepish adj embarrassed because of feeling foolish ▸ **sheepishly** adv

sheer¹ adj absolute, complete e.g. sheer folly; perpendicular, steep; (of material) so fine as to be transparent

sheer² v change course suddenly

sheet¹ n large piece of cloth used as an inner bed cover; broad thin piece of any material; large expanse

sheet² n rope for controlling the position of a sail ▸ **sheet anchor** strong anchor for use in an emergency; person or thing relied on

sheikh, sheik [shake] n Arab chief ▸ **sheikhdom, sheikdom** n

sheila n (Aust & NZ slang) girl or woman

shekel n monetary unit of Israel ▸ pl (informal) money

shelf n, pl **shelves** board fixed horizontally for holding things; ledge ▸ **shelf life** time a packaged product will remain fresh

shell n hard outer covering of an egg, nut, or certain animals; external frame of something; explosive projectile fired from a large gun ▸ v take the shell from; fire at with artillery shells ▸ **shellfish** n sea-living animal, esp. one that can be eaten, with a shell ▸ **shell out** v (informal) pay out or hand over (money) ▸ **shell shock** nervous disorder caused by exposure to battle conditions ▸ **shell suit** (Brit) lightweight tracksuit made of a waterproof nylon layer over a cotton layer

shellac n resin used in varnishes ▸ v -**lacking, -lacked** coat with shellac

shelter n structure providing protection from danger or the weather; protection ▷ v give shelter to; take shelter

shelve¹ v put aside or postpone; provide with shelves ▶ **shelving** n (material for) shelves

shelve² v slope

shenanigans pl n (informal) mischief or nonsense; trickery

shepherd n person who tends sheep ▷ v guide or watch over (people) ▶ **shepherdess** n fem ▶ **shepherd's pie** baked dish of mince covered with mashed potato

sherbet n (Brit, Aust & NZ) fruit-flavoured fizzy powder; (US, Canadian & SAfr) flavoured water ice

sheriff n (in the US) chief law enforcement officer of a county; (in England and Wales) chief executive officer of the Crown in a county; (in Scotland) chief judge of a district; (in Australia) officer of the Supreme Court

Sherpa n member of a people of Tibet and Nepal

sherry n, pl **-ries** pale or dark brown fortified wine

shibboleth n slogan or principle, usu. considered outworn, characteristic of a particular group

shield n piece of armour carried on the arm to protect the body from blows or missiles; anything that protects; sports trophy in the shape of a shield ▷ v protect

shift v move; transfer (blame or responsibility); remove or be removed ▷ n shifting; group of workers who work during a specified period; period of time during which they work; loose-fitting straight underskirt or dress ▶ **shiftless** adj lacking in ambition or initiative ▶ **shifty** adj evasive or untrustworthy ▶ **shiftiness** n

shillelagh [shil-**lay**-lee] n (in Ireland) a cudgel

shilling n former British coin, replaced by the 5p piece; former Australian coin, worth one twentieth of a pound

shillyshally v **-lying, -lied** (informal) be indecisive

shimmer v, n (shine with) a faint unsteady light

shin n front of the lower leg ▷ v **shinning, shinned** climb by using the hands or arms and legs ▶ **shinbone** n tibia

shindig n (informal) noisy party; brawl

shine v **shining, shone** give out or reflect light; aim (a light); polish; excel ▷ n brightness or lustre **take a shine to** (informal) take a liking to (someone) ▶ **shiny** adj **-nier, -niest** ▶ **shiner** n (informal) black eye

shingle¹ n wooden roof tile ▷ v cover (a roof) with shingles

shingle² n coarse gravel found on beaches ▶ **shingle slide** (NZ) loose stones on a steep slope

shingles n disease causing a rash of small blisters along a nerve

Shinto n Japanese religion in which ancestors and nature spirits are worshipped ▶ **Shintoism** n

shinty n game like hockey

ship n large seagoing vessel ▷ v **shipping, shipped** send or transport by carrier, esp. a ship; bring or go aboard a ship ▶ **shipment** n act of shipping cargo; consignment of goods shipped ▶ **shipping** n freight transport business; ships collectively ▶ **shipshape** adj orderly or neat ▶ **shipwreck** n destruction of a ship through storm or collision ▷ v cause to undergo shipwreck ▶ **shipyard** n place where ships are built

shire n (Brit) county; (Aust) rural area with an elected council

shire horse n large powerful breed of horse

shirk v avoid (duty or work) ▸ **shirker** n

shirt n garment for the upper part of the body

shirty adj **-tier, -tiest** (chiefly Brit slang) bad-tempered or annoyed

shish kebab n meat and vegetable dish cooked on a skewer

shiver[1] v tremble, as from cold or fear ▸ n shivering

shiver[2] v splinter into pieces

shoal[1] n large number of fish swimming together

shoal[2] n stretch of shallow water; sandbank

shock[1] v horrify, disgust, or astonish ▸ n sudden violent emotional disturbance; sudden violent blow or impact; something causing this; state of bodily collapse caused by physical or mental shock; pain and muscular spasm caused by an electric current passing through the body ▸ **shocker** n ▸ **shocking** adj causing horror, disgust, or astonishment; (informal) very bad

shock[2] n bushy mass (of hair)

shod v past of **shoe**

shoddy adj **-dier, -diest** made or done badly

shoe n outer covering for the foot, ending below the ankle; horseshoe ▸ v **shoeing, shod** fit with a shoe or shoes ▸ **shoehorn** n smooth curved implement inserted at the heel of a shoe to ease the foot into it ▸ **shoestring** n on a shoestring using a very small amount of money

shone v past of **shine**

shonky adj **-kier, -kiest** (Aust & NZ informal) unreliable or unsound

shoo interj go away! ▸ v drive away as by saying 'shoo'

shook v past tense of **shake**

shoot v **shooting, shot** hit, wound, or kill with a missile fired from a weapon; fire (a missile from) a weapon; hunt; send out or move rapidly; (of a plant) sprout; photograph or film; (Sport) take a shot at goal ▸ n new branch or sprout of a plant; hunting expedition ▸ **shooting star** meteor ▸ **shooting stick** stick with a spike at one end and a folding seat at the other

shop n place for sale of goods and services; workshop ▸ v **shopping, shopped** visit a shop or shops to buy goods; (Brit, Aust & NZ slang) inform against (someone) **talk shop** discuss one's work, esp. on a social occasion ▸ **shop around** v visit various shops to compare goods and prices ▸ **shop floor** production area of a factory; workers in a factory ▸ **shoplifter** n person who steals from a shop ▸ **shop-soiled** adj soiled or faded from being displayed in a shop ▸ **shop steward** (in some countries) trade-union official elected to represent his or her fellow workers

shore[1] n edge of a sea or lake

shore[2] v (foll by up) prop or support

shorn v a past participle of **shear**

short adj not long; not tall; not lasting long, brief; deficient e.g. short of cash; abrupt, rude; (of a drink) consisting chiefly of a spirit; (of pastry) crumbly ▸ adv abruptly ▸ n drink of spirits; short film; (informal) short circuit ▸ pl short trousers ▸ **shortage** n deficiency ▸ **shorten** v make or become shorter ▸ **shortly** adv soon; rudely ▸ **shortbread, shortcake** n crumbly biscuit made with butter ▸ **short-change**

v give (someone) less than the correct amount of change; (*slang*) swindle ▸ **short circuit** faulty or accidental connection in a circuit, which deflects current through a path of low resistance ▸ **shortcoming** *n* failing or defect ▸ **short cut** quicker route or method ▸ **shortfall** *n* deficit ▸ **shorthand** *n* system of rapid writing using symbols to represent words ▸ **short-handed** *adj* not having enough workers ▸ **short list** selected list of candidates for a job or prize, from which the final choice will be made ▸ **short-list** *v* put on a short list ▸ **short shrift** brief and unsympathetic treatment ▸ **short-sighted** *adj* unable to see distant things clearly; lacking in foresight ▸ **short wave** radio wave with a wavelength of less than 60 metres

shot[1] *n* shooting; small lead pellets used in a shotgun; person with specified skill in shooting; (*slang*) attempt; (*Sport*) act or instance of hitting, kicking, or throwing the ball; photograph; uninterrupted film sequence; (*informal*) injection ▸ **shotgun** *n* gun for firing a charge of shot at short range

shot[2] *v* past of **shoot** ▸ *adj* woven to show changing colours

shot put *n* athletic event in which contestants hurl a heavy metal ball as far as possible ▸ **shot-putter** *n*

should *v* past tense of **shall** used as an auxiliary to make the subjunctive mood or to indicate obligation or possibility

shoulder *n* part of the body to which an arm, foreleg, or wing is attached; cut of meat including the upper foreleg; side of a road ▸ *v* bear (a burden or responsibility); push with one's shoulder; put on

one's shoulder ▸ **shoulder blade** large flat triangular bone at the shoulder

shouldn't should not

shout *n* loud cry; (*informal*) person's turn to buy a round of drinks ▸ *v* cry out loudly; (*Aust & NZ informal*) treat (someone) to (something, such as a drink) ▸ **shout down** *v* silence (someone) by shouting

shove *v* push roughly; (*informal*) put ▸ *n* rough push ▸ **shove off** *v* (*informal*) go away

shovel *n* tool for lifting or moving loose material ▸ *v* **-elling, -elled** lift or move as with a shovel

show *v* **showing, showed, shown** or **showing** make, be, or become noticeable or visible; exhibit or display; indicate; instruct by demonstration; prove; guide; reveal or display (an emotion) ▸ *n* public exhibition; theatrical or other entertainment; mere display or pretence ▸ **showy** *adj* gaudy; ostentatious ▸ **showily** *adv* ▸ **show business** the entertainment industry ▸ **showcase** *n* situation in which something is displayed to best advantage; glass case used to display objects ▸ **showdown** *n* confrontation that settles a dispute ▸ **showjumping** *n* competitive sport of riding horses to demonstrate skill in jumping ▸ **showman** *n* man skilled at presenting anything spectacularly ▸ **showmanship** *n* ▸ **show off** *v* exhibit to invite admiration; (*informal*) behave flamboyantly in order to attract attention ▸ **show-off** *n* (*informal*) person who shows off ▸ **showpiece** *n* excellent specimen shown for display or as an example ▸ **showroom** *n* room in which goods for sale are on

s

display ▸ **show up** v reveal or be revealed clearly; expose the faults or defects of; (informal) embarrass; (informal) arrive

shower n kind of bath in which a person stands while being sprayed with water; wash in this; short period of rain, hail, or snow; sudden abundant fall of objects ▷ v wash in a shower; bestow (things) or present (someone) with things liberally ▸ **showery** adj

shown v a past participle of **show**

shrank v a past tense of **shrink**

shrapnel n artillery shell filled with pellets which scatter on explosion; fragments from this

shred n long narrow strip torn from something; small amount ▷ v **shredding**, **shredded** or **shred** tear to shreds

shrew n small mouselike animal; bad-tempered nagging woman ▸ **shrewish** adj

shrewd adj clever and perceptive ▸ **shrewdly** adv ▸ **shrewdness** n

shriek n shrill cry ▷ v utter (with) a shriek

shrike n songbird with a heavy hooked bill

shrill adj (of a sound) sharp and high-pitched ▸ **shrillness** n ▸ **shrilly** adv

shrimp n small edible shellfish; (informal) small person ▸ **shrimping** n fishing for shrimps

shrine n place of worship associated with a sacred person or object

shrink v **shrinking**, **shrank** or **shrunk**, **shrunk** or **shrunken** become or make smaller; recoil or withdraw ▷ n (slang) psychiatrist ▸ **shrinkage** n decrease in size, value, or weight

shrivel v **-elling**, **-elled** shrink and wither

shroud n piece of cloth used to wrap a dead body; anything which conceals ▷ v conceal

Shrove Tuesday n day before Ash Wednesday

shrub n woody plant smaller than a tree ▸ **shrubbery** n, pl **-beries** area planted with shrubs

shrug v **shrugging**, **shrugged** raise and then drop (the shoulders) as a sign of indifference, ignorance, or doubt ▷ n shrugging ▸ **shrug off** v dismiss as unimportant

shrunk v a past of **shrink**

shrunken v a past participle of **shrink**

shudder v shake or tremble violently, esp. with horror ▷ n shaking or trembling

shuffle v walk without lifting the feet; jumble together; rearrange ▷ n shuffling; rearrangement

shun v **shunning**, **shunned** avoid

shunt v move (objects or people) to a different position; move (a train) from one track to another

shush interj be quiet!

shut v **shutting**, **shut** bring together or fold, close; prevent access to; (of a shop etc.) stop operating for the day ▸ **shutter** n hinged doorlike cover for closing off a window; device in a camera letting in the light required to expose a film ▸ **shut down** v close or stop (a factory, machine, or business) ▸ **shutdown** n

shuttle n vehicle going to and fro over a short distance; instrument which passes the weft thread between the warp threads in weaving ▷ v travel by or as if by shuttle

shuttlecock n small light cone with feathers stuck in one end, struck to and fro in badminton

shy¹ adj not at ease in company; timid; (foll by of) cautious or wary

▷ v **shying, shied** start back in fear; (foll by *away from*) avoid (doing something) through fear or lack of confidence ▸ **shyly** adv ▸ **shyness** n

shy² v **shying, shied** throw ▷ n, pl **shies** throw

SI (*French*) Système International (d'Unités): international metric system of units of measurement

Siamese adj of Siam, former name of Thailand ▸ **Siamese cat** breed of cat with cream fur, dark ears and face, and blue eyes ▸ **Siamese twins** twins born joined to each other at some part of the body

sibilant adj hissing ▷ n consonant pronounced with a hissing sound

sibling n brother or sister

sibyl n (in ancient Greece and Rome) prophetess

sic (*Latin*) adv thus: used to indicate that an odd spelling or reading is in fact accurate

sick adj vomiting or likely to vomit; physically or mentally unwell; (*informal*) amused or fascinated by something sadistic or morbid; (foll by *of*) (*informal*) disgusted (by) or weary (of) ▸ **sickness** n ▸ **sicken** v make nauseated or disgusted; become ill ▸ **sickly** adj unhealthy, weak; causing revulsion or nausea ▸ **sick bay** place for sick people, such as that on a ship

sickle n tool with a curved blade for cutting grass or grain

side n line or surface that borders anything; either of two halves into which something can be divided; either surface of a flat object; area immediately next to a person or thing; aspect or part; one of two opposing groups or teams ▷ adj at or on the side; subordinate **on the side** as an extra; unofficially ▸ **siding** n short stretch of railway track on which trains or wagons

are shunted from the main line ▸ **sidebar** n short article placed alongside a longer one on a website ▸ **sideboard** n piece of furniture for holding plates, cutlery, etc. in a dining room ▸ **sideburns, sideboards** pl n man's side whiskers ▸ **side effect** additional undesirable effect ▸ **sidekick** n (*informal*) close friend or associate ▸ **sidelight** n either of two small lights on the front of a vehicle ▸ **sideline** n subsidiary interest or source of income; (*Sport*) line marking the boundary of a playing area ▸ **sidelong** adj sideways ▷ adv obliquely ▸ **side-saddle** n saddle designed to allow a woman rider to sit with both legs on the same side of the horse ▸ **sidestep** v dodge (an issue); avoid by stepping sideways ▸ **sidetrack** v divert from the main topic ▸ **sidewalk** n (*US*) paved path for pedestrians, at the side of a road ▸ **sideways** adv to or from the side; obliquely ▸ **side with** v support (one side in a dispute)

sidereal [side-eer-ee-al] adj of or determined with reference to the stars

sidle v walk in a furtive manner

SIDS sudden infant death syndrome: cot death

siege n surrounding and blockading of a place

sienna n reddish- or yellowish-brown pigment made from natural earth

sierra n range of mountains in Spain or America with jagged peaks

siesta n afternoon nap, taken in hot countries

sieve [siv] n utensil with mesh through which a substance is sifted or strained ▷ v sift or strain through a sieve

sift v remove the coarser particles from a substance with a sieve;

examine (information or evidence) to select what is important

sigh n long audible breath expressing sadness, tiredness, relief, or longing ▸ v utter a sigh

sight n ability to see; instance of seeing; range of vision; device for guiding the eye while using a gun or optical instrument; thing worth seeing; (informal) a lot ▸ v catch sight of ▸ **sightless** adj blind ▸ **sight-read** v play or sing printed music without previous preparation ▸ **sightseeing** n visiting places of interest ▸ **sightseer** n

sign n indication of something not immediately or outwardly observable; gesture, mark, or symbol conveying a meaning; notice displayed to advertise, inform, or warn; omen ▸ v write (one's name) on (a document or letter) to show its authenticity or one's agreement; communicate using sign language; make a sign or gesture ▸ **sign language** system of communication by gestures, as used by deaf people (also **signing**) ▸ **sign on** v register as unemployed; sign a document committing oneself to a job, course, etc. ▸ **signpost** n post bearing a sign that shows the way

signal n sign or gesture to convey information; sequence of electrical impulses or radio waves transmitted or received ▸ adj (formal) very important ▸ v -**nalling**, -**nalled** convey (information) by signal ▸ **signally** adv ▸ **signal box** building from which railway signals are operated ▸ **signalman** n railwayman in charge of signals and points

signatory n, pl -**ries** one of the parties who sign a document

signature n person's name written by himself or herself in signing something; sign at the start of a piece of music to show the key or tempo ▸ **signature tune** tune used to introduce a particular television or radio programme

signet n small seal used to authenticate documents ▸ **signet ring** finger ring bearing a signet

significant adj important; having or expressing a meaning ▸ **significantly** adv ▸ **significance** n

signify v -**fying**, -**fied** indicate or suggest; be a symbol or sign for; be important ▸ **signification** n

signor [see-nyor] n Italian term of address equivalent to sir or Mr ▸ **signora** [see-nyor-a] n Italian term of address equivalent to madam or Mrs ▸ **signorina** [see-nyor-ee-na] n Italian term of address equivalent to madam or Miss

Sikh [seek] n member of an Indian religion having only one God

silage [sile-ij] n fodder crop harvested while green and partially fermented in a silo or plastic bags

silence n absence of noise or speech ▸ v make silent; put a stop to ▸ **silent** adj ▸ **silently** adv ▸ **silencer** n device to reduce the noise of an engine exhaust or gun

silhouette n outline of a dark shape seen against a light background ▸ v show in silhouette

silica n hard glossy mineral found as quartz and in sandstone ▸ **silicosis** n lung disease caused by inhaling silica dust

silicon n (Chem) brittle nonmetallic element widely used in chemistry and industry ▸ **silicon chip** tiny wafer of silicon processed to form an integrated circuit

silicone n tough synthetic substance made from silicon and used in lubricants, paints, and resins

silk n fibre made by the larva (**silkworm**) of a certain moth; thread or fabric made from this ► **silky, silken** adj or like silk

sill n ledge at the bottom of a window or door

silly adj **-lier, -liest** foolish ► **silliness** n

silo n, pl **-los** pit or airtight tower for storing silage or grains; underground structure in which nuclear missiles are kept ready for launching

silt n mud deposited by moving water ► v (foll by up) fill or be choked with silt

silvan adj same as **sylvan**

silver n white precious metal; coins or articles made of silver ► adj made of or of the colour of silver ► **silverbeet** n (Aust & NZ) leafy green vegetable with white stalks ► **silver birch** tree with silvery-white bark ► **silver fern** (NZ) sporting symbol of New Zealand ► **silverfish** n small wingless silver-coloured insect ► **silverside** n cut of beef from below the rump and above the leg ► **silver wedding** twenty-fifth wedding anniversary

sim n computer game that simulates an activity such as flying or playing a sport

simian adj, n (of or like) a monkey or ape

similar adj alike but not identical ► **similarity** n ► **similarly** adv

simile [sim-ill-ee] n figure of speech comparing one thing to another, using 'as' or 'like' e.g. as blind as a bat

similitude n similarity, likeness

simmer v cook gently at just below boiling point; be in a state of

suppressed rage ► **simmer down** v (informal) calm down

simnel cake n (Brit) fruit cake with marzipan

simper v smile in a silly or affected way; utter (something) with a simper ► n simpering smile

simple adj easy to understand or do; plain or unpretentious; not combined or complex; sincere or frank; feeble-minded ► **simply** adv ► **simplicity** n ► **simplify** v make less complicated ► **simplification** n ► **simplistic** adj too simple or naive ► **simpleton** n foolish or half-witted person

simulate v make a pretence of; imitate the conditions of (a particular situation); have the appearance of ► **simulation** n ► **simulator** n

simultaneous adj occurring at the same time ► **simultaneously** adv

sin[1] n breaking of a religious or moral law; offence against a principle or standard ► v **sinning, sinned** commit a sin ► **sinful** adj guilty of sin; being a sin ► **sinfully** adv ► **sinner** n

sin[2] (Maths) sine

since prep during the period of time after ► conj from the time when; for the reason that ► adv from that time

sincere adj without pretence or deceit ► **sincerely** adv ► **sincerity** n

sine n (in trigonometry) ratio of the length of the opposite side to that of the hypotenuse in a right-angled triangle

sinecure [sin-ee-cure] n paid job with minimal duties

sine die [sin-ay dee-ay] adv (Latin) with no date fixed for future action

sine qua non [sin-ay kwah non] n (Latin) essential requirement

sinew n tough fibrous tissue joining muscle to bone; muscles or strength ▷ **sinewy** adj

sing v **singing, sang, sung** make musical sounds with the voice; perform (a song); make a humming or whistling sound ▷ **singing telegram** service in which a messenger brings greetings to a person by singing ▷ **singsong** n informal singing session ▷ adj (of the voice) repeatedly rising and falling in pitch

> **USAGE NOTE**
> The simple past of *sing* is *sang*: *He sang the chorus*. Avoid the use of the past participle *sung* for the simple past

singe v **singeing, singed** burn the surface of ▷ n superficial burn

singer n person who sings, esp. professionally

single adj one only; distinct from others of the same kind; unmarried; designed for one user; formed of only one part; (of a ticket) valid for an outward journey only ▷ n single thing; thing intended for one person; record with one short song or tune on each side; single ticket ▷ pl game between two players ▷ v (foll by **out**) pick out from others ▷ **singly** adv ▷ **single file** (of people or things) arranged in one line ▷ **single-handed** adj without assistance ▷ **single-minded** adj having one aim only ▷ **single parent** parent bringing up a child or children alone

singlet n sleeveless vest

singular adj (of a word or form) denoting one person or thing; remarkable, unusual ▷ n singular form of a word ▷ **singularity** n ▷ **singularly** adv

sinister adj threatening or suggesting evil or harm

sink v **sinking, sank, sunk** or **sunken** submerge (in liquid); descend or cause to descend; decline in value or amount; become weaker in health; dig or drill (a hole or shaft); invest (money); (Golf, snooker) hit (a ball) into a hole or pocket ▷ n fixed basin with a water supply and drainage pipe ▷ **sinker** n weight for a fishing line ▷ **sink in** v penetrate the mind ▷ **sinking fund** money set aside regularly to repay a long-term debt

Sino- combining form Chinese

sinuous adj curving; lithe ▷ **sinuously** adv

sinus [sine-uss] n hollow space in a bone, esp. an air passage opening into the nose

sip v **sipping, sipped** drink in small mouthfuls ▷ n amount sipped

siphon n bent tube which uses air pressure to draw liquid from a container ▷ v draw off thus; redirect (resources)

sir n polite term of address for a man; (**S-**) title of a knight or baronet

sire n male parent of a horse or other domestic animal; respectful term of address to a king ▷ v father

siren n device making a loud wailing noise as a warning; dangerously alluring woman

sirloin n prime cut of loin of beef

sirocco n, pl **-cos** hot wind blowing from N Africa into S Europe

sis interj (S Afr informal) exclamation of disgust

sisal [size-al] n (fibre of) plant used in making ropes

siskin n yellow-and-black finch

sissy adj, n, pl **-sies** weak or cowardly (person)

sister n girl or woman with the same parents as another person; female fellow-member of a group;

senior nurse; nun ▷ *adj* closely related, similar ▸ **sisterhood** *n* state of being a sister; group of women united by common aims or beliefs ▸ **sisterly** *adj* ▸ **sister-in-law** *n, pl* **sisters-in-law** sister of one's husband or wife; wife of one's sibling

sit *v* **sitting, sat** rest one's body upright on the buttocks; cause to sit; perch; occupy an official position; (of an official body) hold a session; take (an examination) ▸ **sitting room** room in a house where people sit and relax ▸ **sit-in** *n* protest in which demonstrators occupy a place and refuse to move

sitar *n* Indian stringed musical instrument

sitcom *n* (*informal*) situation comedy

site *n* place where something is, was, or is intended to be located; same as **website** ▷ *v* provide with a site

situate *v* place

situation *n* state of affairs; location and surroundings; position of employment ▸ **situation comedy** radio or television series involving the same characters in various situations

six *adj, n* one more than five ▸ **sixth** *adj, n* (of) number six in a series ▸ **sixteen** *adj, n* six and ten ▸ **sixteenth** *adj, n* ▸ **sixty** *adj, n* six times ten ▸ **sixtieth** *adj, n*

size¹ *n* dimensions, bigness; one of a series of standard measurements of goods ▷ *v* arrange according to size ▸ **sizeable, sizable** *adj* quite large ▸ **size up** *v* (*informal*) assess

size² *n* gluey substance used as a protective coating

sizzle *v* make a hissing sound like frying fat

skanky *adj* (*slang*) dirty or unattractive; promiscuous

skate¹ *n* boot with a steel blade or sets of wheels attached to the sole for gliding over ice or a hard surface ▷ *v* glide on or as if on skates ▸ **skateboard** *n* board mounted on small wheels for riding on while standing up ▸ **skateboarding** *n* ▸ **skate over, skate round** *v* avoid discussing or dealing with (a matter) fully

skate² *n* large marine flatfish

skedaddle *v* (*informal*) run off

skein [skayn] *n* yarn wound in a loose coil; flock of geese in flight

skeleton *n* framework of bones inside a person's or animal's body; essential framework of a structure; small steel-frame racing sledge ▷ *adj* reduced to a minimum ▸ **skeletal** *adj* ▸ **skeleton key** key which can open many different locks

sketch *n* rough drawing; brief description; short humorous play ▷ *v* make a sketch (of) ▸ **sketchy** *adj* incomplete or inadequate

skew *v* make slanting or crooked ▷ *adj* slanting or crooked ▸ **skew-whiff** *adj* (*Brit informal*) slanting or crooked

skewer *n* pin to hold meat together during cooking ▷ *v* fasten with a skewer

ski *n* one of a pair of long runners fastened to boots for gliding over snow or water ▷ *v* **skiing, skied** or **ski'd** travel on skis ▸ **skier** *n*

skid *v* **skidding, skidded** (of a moving vehicle) slide sideways uncontrollably ▷ *n* skidding

skiff *n* small boat

skill *n* special ability or expertise; something requiring special training or expertise ▸ **skilful** *adj* having or showing skill ▸ **skilfully** *adv* ▸ **skilled** *adj*

SPELLING TIP
When you make an adjective

S

from **skill**, you should drop an *l* to make **skilful**. American English keeps the double *l* in this case

skillet *n* small frying pan or shallow cooking pot

skim *v* **skimming, skimmed** remove floating matter from the surface of (a liquid); glide smoothly over; read quickly ▶ **skimmed milk, skim milk** milk from which the cream has been removed

skimp *v* not invest enough time, money, material, etc. ▶ **skimpy** *adj* scanty or insufficient

skin *n* outer covering of the body; complexion; outer layer or covering; film on a liquid; animal skin used as a material or container ▷ *v* **skinning, skinned** remove the skin of ▶ **skinless** *adj* ▶ **skinny** *adj* thin ▶ **skin-deep** *adj* superficial ▶ **skin diving** underwater swimming using flippers and light breathing apparatus ▶ **skin-diver** *n* ▶ **skinflint** *n* miser ▶ **skinhead** *n* youth with very short hair

skint *adj* (*Brit slang*) having no money

skip¹ *v* **skipping, skipped** leap lightly from one foot to the other; jump over a rope as it is swung under one; (*informal*) pass over, omit ▷ *n* skipping

skip² *n* large open container for builders' rubbish

skipper *n, v* captain

skirl *n* sound of bagpipes

skirmish *n* brief or minor fight or argument ▷ *v* take part in a skirmish

skirt *n* woman's garment hanging from the waist; part of a dress or coat below the waist; cut of beef from the flank ▷ *v* border; go round; avoid dealing with (an issue) ▶ **skirting board** narrow

board round the bottom of an interior wall

skit *n* brief satirical sketch

skite *v, n* (*Aust & NZ*) boast

skittish *adj* playful or lively

skittle *n* bottle-shaped object used as a target in some games ▷ *pl* game in which players try to knock over skittles by rolling a ball at them

skive *v* (*Brit informal*) evade work or responsibility

skivvy *n, pl* **-vies** (*Brit*) female servant who does menial work

skua *n* large predatory gull

skulduggery *n* (*informal*) trickery

skulk *v* move stealthily; lurk

skull *n* bony framework of the head ▶ **skullcap** *n* close-fitting brimless cap

skunk *n* small black-and-white N American mammal which emits a foul-smelling fluid when attacked; (*slang*) despicable person

sky *n, pl* **skies** upper atmosphere as seen from the earth ▶ **skydiving** *n* sport of jumping from an aircraft and performing manoeuvres before opening one's parachute ▶ **skylark** *n* lark that sings while soaring at a great height ▶ **skylight** *n* window in a roof or ceiling ▶ **skyscraper** *n* very tall building

Skype® *n* software application that allows users to make voice and video calls over the internet

slab *n* broad flat piece

slack *adj* not tight; negligent; not busy ▷ *n* slack part ▷ *pl informal* trousers ▷ *v* neglect one's work or duty ▶ **slackness** *n* ▶ **slacken** *v* make or become slack ▶ **slacker** *n* lazy person

slag *n* waste left after metal is smelted ▷ *v* **slagging, slagged** (foll by *off*) (*Brit, Aust & NZ slang*) criticize

slain v past participle of **slay**

slake v satisfy (thirst or desire); combine (quicklime) with water

slalom n skiing or canoeing race over a winding course

slam v **slamming, slammed** shut, put down, or hit violently and noisily; (informal) criticize harshly ▷ n act or sound of slamming **grand slam** see **grand**

slander n false and malicious statement about a person; crime of making such a statement ▷ v utter slander about ▶ **slanderous** adj

slang n very informal language ▶ **slangy** adj ▶ **slanging match** abusive argument

slant v lean at an angle, slope; present (information) in a biased way ▷ n slope; point of view, esp. a biased one ▶ **slanting** adj

slap n blow with the open hand or a flat object ▷ v **slapping, slapped** strike with the open hand or a flat object; (informal) place forcefully or carelessly ▶ **slapdash** adj careless and hasty ▶ **slap-happy** adj (informal) cheerfully careless ▶ **slapstick** n boisterous knockabout comedy ▶ **slap-up** adj (of a meal) large and luxurious

slash v cut with a sweeping stroke; gash; reduce drastically ▷ n sweeping stroke; gash

slat n narrow strip of wood or metal

slate¹ n rock which splits easily into thin layers; piece of this for covering a roof or, formerly, for writing on

slate² v (informal) criticize harshly

slattern n (old-fashioned) slovenly woman ▶ **slatternly** adj

slaughter v kill (animals) for food; kill (people) savagely or indiscriminately ▷ n slaughtering ▶ **slaughterhouse** n place where animals are killed for food

Slav n member of any of the peoples of E Europe or the former Soviet Union who speak a Slavonic language ▶ **Slavonic** n language group including Russian, Polish, and Czech ▷ adj of this language group

slave n person owned by another for whom he or she has to work; person dominated by another or by a habit; drudge ▷ v work like a slave ▶ **slaver** n ship or person engaged in the slave trade ▶ **slavery** n state or condition of being a slave; practice of owning slaves ▶ **slavish** adj of or like a slave; imitative ▶ **slave-driver** n person who makes others work very hard

slaver [slav-ver] v dribble saliva from the mouth

slay v **slaying, slew, slain** kill

sleazy adj -**zier**, -**ziest** run-down or sordid ▶ **sleaze** n

sledge¹, sled n carriage on runners for sliding on snow; light wooden frame for sliding over snow ▷ v travel by sledge

sledge², sledgehammer n heavy hammer with a long handle

sleek adj glossy, smooth, and shiny

sleep n state of rest characterized by unconsciousness; period of this ▷ v **sleeping, slept** be in or as if in a state of sleep; have sleeping accommodation for (a specified number) ▶ **sleeper** n railway car fitted for sleeping in; beam supporting the rails of a railway; ring worn in a pierced ear to stop the hole from closing up; person who sleeps ▶ **sleepy** adj ▶ **sleepily** adv ▶ **sleepiness** n ▶ **sleepless** adj ▶ **sleeping bag** padded bag for sleeping in ▶ **sleeping sickness** African disease spread by the tsetse fly ▶ **sleepout** n (NZ) small building for sleeping in ▶ **sleepover** n occasion when a person stays overnight at a friend's house ▸ **sleep with, sleep together** v have sexual intercourse (with)

sleet n rain and snow or hail falling together

sleeve n part of a garment which covers the arm; tubelike cover; gramophone record cover **up one's sleeve** secretly ready ▸ **sleeveless** adj

sleigh n, v sledge

sleight of hand [**slite**] n skilful use of the hands when performing conjuring tricks

slender adj slim; small in amount

slept v past of **sleep**

sleuth [**slooth**] n detective

slew[1] v past tense of **slay**

slew[2] v twist or swing round

slice n thin flat piece cut from something; share; kitchen tool with a broad flat blade; (Sport) hitting of a ball so that it travels obliquely ▸ v cut into slices; (Sport) hit (a ball) with a slice

slick adj persuasive and glib; skilfully devised or carried out; well-made and attractive, but superficial ▸ n patch of oil on water ▸ v make smooth or sleek

slide v **sliding, slid** slip smoothly along (a surface); pass unobtrusively ▸ n sliding; piece of glass holding an object to be viewed under a microscope; photographic transparency; surface or structure for sliding on or down; ornamental hair clip ▸ **slide rule** mathematical instrument formerly used for rapid calculations ▸ **sliding scale** variable scale according to which things such as wages alter in response to changes in other factors

slight adj small in quantity or extent; not important; slim and delicate ▸ v, n snub ▸ **slightly** adv

slim adj **slimmer, slimmest** not heavy or stout; thin; slight ▸ v **slimming, slimmed** make or become slim by diet and exercise ▸ **slimmer** n

slime n unpleasant thick slippery substance ▸ **slimy** adj of, like, or covered with slime; ingratiating

sling[1] n bandage hung from the neck to support an injured hand or arm; rope or strap for lifting something; strap with a string at each end for throwing a stone ▸ v **slinging, slung** throw; carry, hang, or throw with or as if with a sling

sling[2] n sweetened drink with a spirit base e.g. *gin sling*

slink v **slinking, slunk** move furtively or guiltily ▸ **slinky** adj (of clothes) figure-hugging

slip[1] v **slipping, slipped** lose balance by sliding; move smoothly, easily, or quietly; (foll by on or off) put on or take off easily or quickly; pass out of (the mind) ▸ n slipping; mistake; petticoat **give someone the slip** escape from someone ▸ **slippy** adj (informal) slippery ▸ **slipknot** n knot tied so that it will slip along the rope round which it is made ▸ **slipped disc** painful condition in which one of the discs connecting the bones of the spine becomes displaced ▸ **slip road** narrow road giving access to a motorway ▸ **slipshod** adj (of an action) careless ▸ **slipstream** n stream of air forced backwards by a fast-moving object ▸ **slip up** v make a mistake ▸ **slipway** n launching slope on which ships are built or repaired

slip[2] n small piece (of paper)

slip[3] n clay mixed with water used for decorating pottery

slipper n light shoe for indoor wear

slippery *adj* so smooth or wet as to cause slipping or be difficult to hold; (of a person) untrustworthy

slit *n* long narrow cut or opening ▷ *v* **slitting, slit** make a long straight cut in

slither *v* slide unsteadily

sliver [sliv-ver] *n* small thin piece

slob *n* (*informal*) lazy and untidy person ▶ **slobbish** *adj*

slobber *v* dribble or drool ▶ **slobbery** *adj*

sloe *n* sour blue-black fruit

slog *v* **slogging, slogged** work hard and steadily; make one's way with difficulty; hit hard ▷ *n* long and exhausting work or walk

slogan *n* catchword or phrase used in politics or advertising

sloop *n* small single-masted ship

slop *v* **slopping, slopped** splash or spill ▷ *n* spilt liquid; liquid food ▷ *pl* liquid refuse and waste food used to feed animals ▶ **sloppy** *adj* careless or untidy; gushingly sentimental

slope *v* slant ▷ *n* sloping surface; degree of inclination ▷ *pl* hills ▶ **slope off** *v* (*informal*) go furtively

slosh *v* splash carelessly; (*slang*) hit hard ▷ *n* splashing sound ▶ **sloshed** *adj* (*slang*) drunk

slot *n* narrow opening for inserting something; (*informal*) place in a series or scheme ▷ *v* **slotting, slotted** make a slot or slots in; fit into a slot ▶ **slot machine** automatic machine worked by placing a coin in a slot

sloth [rhymes with **both**] *n* slow-moving animal of tropical America; laziness ▶ **slothful** *adj* lazy or idle

slouch *v* sit, stand, or move with a drooping posture ▷ *n* drooping posture **be no slouch** (*informal*) be very good or talented

slough¹ [rhymes with **now**] *n* bog

slough² [sluff] *v* (of a snake) shed (its skin) or (of a skin) be shed ▶ **slough off** *v* get rid of (something unwanted or unnecessary)

sloven *n* habitually dirty or untidy person ▶ **slovenly** *adj* dirty or untidy; careless

Slovene *adj*, *n* (also **Slovenian**) (person) from Slovenia ▷ *n* language of Slovenia

slow *adj* taking a longer time than is usual or expected; not fast; (of a clock or watch) showing a time earlier than the correct time; stupid ▷ *v* reduce the speed (of) ▷ *adj*, *adv* ▶ **slowness** *n* ▶ **slowcoach** *n* (*informal*) person who moves or works slowly

slowworm *n* small legless lizard

sludge *n* thick mud; sewage

slug¹ *n* land snail with no shell ▶ **sluggish** *adj* slow-moving, lacking energy ▶ **sluggishly** *adv* ▶ **sluggishness** *n* ▶ **sluggard** *n* lazy person

slug² *n* bullet; (*informal*) mouthful of an alcoholic drink

slug³ *v* **slugging, slugged** hit hard ▷ *n* heavy blow

sluice *n* channel carrying off water; sliding gate used to control the flow of water in this; water controlled by a sluice ▷ *v* pour a stream of water over or through

slum *n* squalid overcrowded house or area ▶ *v* **slumming, slummed** temporarily and deliberately experience poorer places or conditions than usual

slumber *v*, *n* (*lit*) sleep

slump *v* (of prices or demand) decline suddenly; sink or fall heavily ▷ *n* sudden decline in prices or demand; time of substantial unemployment

slung *v* past of **sling¹**

slunk *v* past of **slink**

S

slur v **slurring, slurred** pronounce or utter (words) indistinctly; (*Music*) sing or play (notes) smoothly without a break ▷ n slurring of words; remark intended to discredit someone; (*Music*) slurring of notes; curved line indicating notes to be slurred

slurp (*informal*) v eat or drink noisily ▷ n slurping sound

slurry n, pl **-ries** muddy liquid mixture

slush n watery muddy substance; sloppy sentimental talk or writing ▶ **slushy** adj ▶ **slush fund** fund for financing bribery or corruption

slut n (*offens*) dirty or immoral woman ▶ **sluttish** adj

sly adj **slyer, slyest** or **slier, sliest** crafty; secretive and cunning; roguish **on the sly** secretly ▶ **slyly** adv ▶ **slyness** n

smack v slap sharply; open and close (the lips) loudly in enjoyment or anticipation ▷ n sharp slap; loud kiss; slapping sound ▷ adv (*informal*) squarely or directly e.g. *smack in the middle* ▶ **smacker** n (*slang*) loud kiss

smack n slight flavour or trace; (*slang*) heroin ▷ v have a slight flavour or trace (of)

smack n small single-masted fishing boat

small adj not large in size, number, or amount; unimportant; mean or petty ▷ n narrow part of the lower back ▷ pl (*informal*) underwear ▶ **smallness** n ▶ **smallholding** n small area of farming land ▶ **small hours** hours just after midnight ▶ **small-minded** adj intolerant, petty ▶ **smallpox** n contagious disease with blisters that leave scars ▶ **small talk** light social conversation ▶ **small-time** adj insignificant or minor

smarmy adj **smarmier, smarmiest** (*informal*) unpleasantly suave or flattering

smart adj well-kept and neat; astute; witty; fashionable; brisk; (of a system or machine) using computer technology ▷ v feel or cause stinging pain ▷ n stinging pain ▶ **smartly** adv ▶ **smartness** n ▶ **smarten** v make or become smart ▶ **smart aleck** (*informal*) irritatingly clever person ▶ **smartphone** n mobile phone allowing access to the internet

smash v break violently and noisily; throw (against) violently; collide forcefully; destroy ▷ n act or sound of smashing; violent collision of vehicles; (*informal*) popular success; (*Sport*) powerful overhead shot ▶ **smasher** n (*informal*) attractive person or thing ▶ **smashing** adj (*informal*) excellent

smattering n slight knowledge

smear v spread with a greasy or sticky substance; rub so as to produce a dirty mark or smudge; slander ▷ n dirty mark or smudge; slander; (*Med*) sample of a secretion smeared on to a slide for examination under a microscope

smell v **smelling, smelt** or **smelled** perceive (a scent or odour) by means of the nose; have or give off a smell; have an unpleasant smell; detect by instinct ▷ n ability to perceive odours by the nose; odour or scent; smelling ▶ **smelly** adj having a nasty smell ▶ **smelling salts** preparation of ammonia used to revive a person who feels faint

smelt v extract (a metal) from (an ore) by heating

smelt n small fish of the salmon family

smelt v a past of **smell**

smelter n industrial plant where smelting is carried out

smile n turning up of the corners of the mouth to show pleasure, amusement, or friendliness ▷ v give a smile ▶ **smiley** n symbol depicting a smile or other facial expression, used in e-mail ▶ **smile on, smile upon** v regard favourably

smirch v, n stain

smirk n smug smile ▷ v give a smirk

smite v **smiting, smote, smitten** (old-fashioned) strike hard; affect severely

smith n worker in metal ▶ **smithy** n blacksmith's workshop

smithereens pl n shattered fragments

smitten v past participle of **smite**

smock n loose garment overall; woman's loose blouselike garment ▷ v gather (material) by sewing in a honeycomb pattern ▶ **smocking** n

smog n mixture of smoke and fog

smoke n cloudy mass that rises from something burning; act of smoking tobacco ▷ v give off smoke; inhale and expel smoke of (a cigar, cigarette, or pipe); do this habitually; cure (meat, fish, or cheese) by treating with smoke ▶ **smokeless** adj ▶ **smoker** n ▶ **smoky** adj ▶ **smoke screen** something said or done to hide the truth

smooch (informal) v kiss and cuddle ▷ n smooching

smooth adj even in surface, texture, or consistency; without obstructions or difficulties; charming and polite but possibly insincere; free from jolts; not harsh in taste ▷ v make smooth; calm ▶ **smoothie** n (informal) charming but possibly insincere man; thick drink made from puréed fresh fruit ▶ **smoothly** adv

smorgasbord n buffet meal of assorted dishes

smote v past tense of **smite**

smother v suffocate or stifle; suppress; cover thickly

smoulder v burn slowly with smoke but no flame; (of feelings) exist in a suppressed state

SMS short message system: used for sending data to mobile phones

smudge v make or become smeared or soiled ▷ n dirty mark; blurred form ▶ **smudgy** adj

smug adj **smugger, smuggest** self-satisfied ▶ **smugly** adv ▶ **smugness** n

smuggle v import or export (goods) secretly and illegally; take somewhere secretly ▶ **smuggler** n

smut n obscene jokes, pictures, etc.; speck of soot or dark mark left by soot ▶ **smutty** adj

snack n light quick meal ▶ **snack bar** place where snacks are sold

snaffle n jointed bit for a horse ▷ v (Brit, Aust & NZ slang) steal

snag n difficulty or disadvantage; sharp projecting point; hole in fabric caused by a sharp object ▷ v **snagging, snagged** catch or tear on a point

snail n slow-moving mollusc with a spiral shell ▶ **snail mail** (informal) conventional post, as opposed to e-mail ▶ **snail's pace** very slow speed

snake n long thin scaly limbless reptile ▷ v move in a winding course like a snake **snake in the grass** treacherous person ▶ **snaky** adj twisted or winding

snap v **snapping, snapped** break suddenly; (cause to) make a sharp cracking sound; move suddenly; bite (at) suddenly; speak sharply and angrily; take a snapshot of ▷ n act or sound of snapping; (informal) snapshot; sudden brief spell of cold weather; card game in which the word 'snap' is called when two similar cards are put down ▷ *adj* made on the spur of the moment ▶ **snappy** *adj* (also **snappish**) irritable; (*slang*) quick; (*slang*) smart and fashionable ▶ **snapdragon** *n* plant with flowers that can open and shut like a mouth ▶ **snapper** *n* food fish of Australia and New Zealand with a pinkish body covered with blue spots ▶ **snapshot** *n* informal photograph ▶ **snap up** *v* take eagerly and quickly

snare *n* trap with a noose ▷ *v* catch in or as if in a snare

snarl¹ *v* (of an animal) growl with bared teeth; speak or utter fiercely ▷ *n* act or sound of snarling

snarl² *n* tangled mess ▷ *v* make tangled ▶ **snarl-up** *n* (*informal*) confused situation such as a traffic jam

snatch *v* seize or try to seize suddenly; take (food, rest, etc.) hurriedly ▷ *n* snatching; fragment

snazzy *adj* **-zier, -ziest** (*informal*) stylish and flashy

sneak *v* move furtively; bring, take, or put furtively; (*informal*) tell tales ▷ *n* cowardly or underhand person ▶ **sneaking** *adj* slight but persistent; secret ▶ **sneaky** *adj*

sneakers *pl n* canvas shoes with rubber soles

sneer *n* contemptuous expression or remark ▷ *v* show contempt by a sneer

sneeze *v* expel air from the nose suddenly, involuntarily, and noisily ▷ *n* act or sound of sneezing

snib *n* (*Scot & NZ*) catch of a door or window

snicker *n*, *v* same as **snigger**

snide *adj* critical in an unfair and nasty way

sniff *v* inhale through the nose in short audible breaths; smell by

sniffing ▷ *n* act or sound of sniffing ▶ **sniffle** *v* sniff repeatedly, as when suffering from a cold ▷ *n* slight cold ▶ **sniff at** *v* express contempt for ▶ **sniffer dog** police dog trained to detect drugs or explosives by smell

snifter *n* (*informal*) small quantity of alcoholic drink

snigger *n* sly disrespectful laugh, esp. one partly stifled ▷ *v* utter a snigger

snip *v* **snipping, snipped** cut in small quick strokes with scissors or shears ▷ *n* (*informal*) bargain; act or sound of snipping ▶ **snippet** *n* small piece

snipe *n* wading bird with a long straight bill ▷ *v* (foll by *at*) shoot at (a person) from cover; make critical remarks about

sniper *n* person who shoots at someone from cover

snitch (*informal*) *v* act as an informer; steal ▷ *n* informer

snivel *v* **-elling, -elled** cry in a whining way

snob *n* person who judges others by social rank; person who feels smugly superior in his or her tastes or interests ▶ **snobbery** *n* ▶ **snobbish** *adj*

snoek *n* (*SAfr*) edible marine fish

snog *v* **snogging, snogged** (*Brit, NZ & S Afr informal*) kiss and cuddle

snood *n* pouch, often of net, loosely holding a woman's hair at the back

snook *n* **cock a snook at** show contempt for

snooker *n* game played on a billiard table ▷ *v* leave (a snooker opponent) in a position such that another ball blocks the target ball; (*informal*) put (someone) in a position where he or she can do nothing

snoop (*informal*) *v* pry ▷ *n* snooping ▶ **snooper** *n*

snooty adj **snootier, snootiest** (informal) haughty

snooze (informal) v take a brief light sleep ▷ n brief light sleep

snore v make snorting sounds while sleeping ▷ n sound of snoring

snorkel n tube allowing a swimmer to breathe while face down on the surface of the water ▷ v **-kelling, -kelled** swim using a snorkel

snort v exhale noisily through the nostrils; express contempt or anger by snorting ▷ n act or sound of snorting

snot n (slang) mucus from the nose

snout n animal's projecting nose and jaws

snow n frozen vapour falling from the sky in flakes; (slang) cocaine ▷ v fall as or like snow **be snowed under** be overwhelmed, esp. with paperwork **snowy** adj ▶ **snowball** n snow pressed into a ball for throwing ▷ v increase rapidly ▶ **snowboard** n board on which a person stands to slide across the snow ▶ **snowboarding** n ▶ **snowdrift** n bank of deep snow ▶ **snowdrop** n small white bell-shaped spring flower ▶ **snowflake** n single crystal of snow ▶ **snow gum** same as **sallee** ▶ **snow line** (on a mountain) height above which there is permanent snow ▶ **snowman** n figure shaped out of snow ▶ **snowplough** n vehicle for clearing away snow ▶ **snowshoes** pl n racket-shaped shoes for walking on deep snow

snub v **snubbing, snubbed** insult deliberately ▷ n deliberate insult ▷ adj (of a nose) short and blunt ▶ **snub-nosed** adj

snuff[1] n powdered tobacco for sniffing up the nostrils

snuff[2] v extinguish (a candle) **snuff it** (informal) die

snuffle v breathe noisily or with difficulty

snug adj **snugger, snuggest** warm and comfortable; comfortably close-fitting ▷ n (in Britain and Ireland) small room in a pub ▶ **snugly** adv

snuggle v nestle into a person or thing for warmth or from affection

so adv to such an extent; in such a manner; very; also; thereupon ▷ conj in order that; with the result that; therefore ▷ interj exclamation of surprise, triumph, or realization **so that** in order that ▶ **so-and-so** n (informal) person whose name is not specified; unpleasant person or thing ▶ **so-called** adj called (in the speaker's opinion, wrongly) by that name ▶ **so long** goodbye

soak v make wet; put or lie in liquid so as to become thoroughly wet; (of liquid) penetrate ▷ n soaking; (slang) drunkard ▶ **soaking** n, adj ▶ **soak up** v absorb

soap n compound of alkali and fat, used with water as a cleaning agent; (informal) soap opera ▷ v apply soap to ▶ **soapy** adj ▶ **soap opera** radio or television serial dealing with domestic themes

soar v rise or fly upwards; increase suddenly

sob v **sobbing, sobbed** weep with convulsive gasps; utter with sobs ▷ n act or spasm of sobbing ▶ **sob story** tale of personal distress told to arouse sympathy

sober adj not drunk; serious; (of colours) plain and dull ▷ v make or become sober ▶ **soberly** adv ▶ **sobriety** n state of being sober

sobriquet [so-brik-ay] n nickname

soccer n football played by two teams of eleven kicking a spherical ball

sociable adj friendly or companionable; (of an occasion)

S

providing companionship
▸ **sociability** n ▸ **sociably** adv

social adj living in a community; of society or its organization; sociable ▹ n informal gathering ▸ **socially** adv ▸ **socialite** n member of fashionable society ▸ **socialize** v meet others socially ▸ **social media** websites and applications that allow users to interact ▸ **social networking site** website that allows subscribers to interact, esp. by forming online communities based around shared interests, experiences, etc. ▸ **social security** state provision for unemployed, elderly, or sick people ▸ **social services** welfare services provided by local authorities or the state ▸ **social work** work which involves helping or advising people with serious financial or family problems

socialism n political system which advocates public ownership of industries, resources, and transport ▸ **socialist** n, adj

society n, pl **-ties** human beings considered as a group; organized community; structure and institutions of such a community; organized group with common aims and interests; upper-class or fashionable people collectively; companionship

sociology n study of human societies ▸ **sociological** adj ▸ **sociologist** n

sock¹ n knitted covering for the foot

sock² (slang) v hit hard ▹ n hard blow

socket n hole or recess into which something fits

sod¹ n (piece of) turf

sod² n (slang) obnoxious person

soda n compound of sodium; soda water ▸ **soda water** fizzy drink made from water charged with carbon dioxide

sodden adj soaked

sodium n (Chem) silvery-white metallic element ▸ **sodium bicarbonate** white soluble compound used in baking powder ▸ **sodium chloride** common table salt ▸ **sodium hydroxide** white alkaline substance used in making paper and soap

sodomy n anal intercourse ▸ **sodomite** n person who practises sodomy

sofa n couch

soft adj easy to shape or cut; not hard, rough, or harsh; (of a breeze or climate) mild; (too) lenient; easily influenced or imposed upon; (of drugs) not liable to cause addiction ▸ **softly** adv ▸ **soften** v make or become soft or softer ▸ **soft drink** nonalcoholic drink ▸ **soft furnishings** curtains, rugs, lampshades, and furniture covers ▸ **soft option** easiest alternative ▸ **soft-pedal** v deliberately avoid emphasizing something ▸ **soft-soap** v (informal) flatter ▸ **software** n computer programs ▸ **softwood** n wood of a coniferous tree

soggy adj **-gier, -giest** soaked; moist and heavy ▸ **sogginess** n

soigné, (fem) soignée [swah-nyay] adj well-groomed, elegant

soil¹ n top layer of earth; country or territory

soil² v make or become dirty; disgrace

soiree [swah-ray] n evening party or gathering

sojourn [soj-urn] n temporary stay ▹ v stay temporarily

solace [sol-iss] n, v comfort in distress

solar adj of the sun; using the energy of the sun ▸ **solar plexus** network of nerves at the pit of the stomach; this part of the stomach

▶ **solar system** the sun and the heavenly bodies that go round it

solarium n, pl **-lariums** or **-laria** place with beds and ultraviolet lights used for acquiring an artificial suntan

sold v past of **sell**

solder n soft alloy used to join two metal surfaces ▷ v join with solder ▶ **soldering iron** tool for melting and applying solder

soldier n member of an army ▷ v serve in an army ▶ **soldierly** adj ▶ **soldier on** v persist doggedly

sole¹ adj one and only; not shared, exclusive ▶ **solely** adv only, completely; alone ▶ **sole charge school** (NZ) country school with only one teacher

sole² n underside of the foot; underside of a shoe ▷ v provide (a shoe) with a sole

sole³ n small edible flatfish

solecism [sol-iss-izz-um] n minor grammatical mistake; breach of etiquette

solemn adj serious, deeply sincere; formal ▶ **solemnly** adv ▶ **solemnity** n

solenoid [sole-in-oid] n coil of wire magnetized by passing a current through it

sol-fa n system of syllables used as names for the notes of a scale

solicit v **-iting, -ited** request; (of a prostitute) offer (a person) sex for money ▶ **solicitation** n

solicitor n (Brit, Aust & NZ) lawyer who advises clients and prepares documents and cases

solicitous adj anxious about someone's welfare ▶ **solicitude** n

solid adj (of a substance) keeping its shape; not liquid or gas; not hollow; of the same substance throughout; strong or substantial; sound or reliable; having three dimensions ▷ n three-dimensional shape; solid substance ▶ **solidly** adv ▶ **solidify** v make or become solid or firm ▶ **solidity** n

solidarity n agreement in aims or interests, total unity

soliloquy n, pl **-quies** speech made by a person while alone, esp. in a play

solipsism n doctrine that the self is the only thing known to exist ▶ **solipsist** n

solitaire n game for one person played with pegs set in a board; gem set by itself

solitary adj alone, single; (of a place) lonely ▶ **solitude** n state of being alone

solo n, pl **-los** music for one performer; any act done without assistance ▷ adj done alone ▷ adv by oneself, alone ▶ **soloist** n ▶ **solo parent** (NZ) parent bringing up a child or children alone

solstice n either the shortest (in winter) or longest (in summer) day of the year

soluble adj able to be dissolved; able to be solved ▶ **solubility** n

solution n answer to a problem; act of solving a problem; liquid with something dissolved in it; process of dissolving

solve v find the answer to (a problem) ▶ **solvable** adj

solvent adj having enough money to pay one's debts ▷ n liquid capable of dissolving other substances ▶ **solvency** n ▶ **solvent abuse** deliberate inhaling of intoxicating fumes from certain solvents

sombre adj dark, gloomy

sombrero n, pl **-ros** wide-brimmed Mexican hat

some adj unknown or unspecified; unknown or unspecified quantity or number of; considerable number or amount of; (*informal*) remarkable ▷ *pron* certain unknown or unspecified people or things; unknown or unspecified number or quantity ▶ **somebody** *pron* some person ▷ *n* important person ▶ **somehow** *adv* in some unspecified way ▶ **someone** *pron* somebody ▶ **something** *pron* unknown or unspecified thing or amount; impressive or important thing ▶ **sometime** *adv* at some unspecified time ▷ *adj* former ▶ **sometimes** *adv* from time to time, now and then ▶ **somewhat** *adv* to some extent, rather ▶ **somewhere** *adv* in, to, or at some unspecified or unknown place

somersault *n* leap or roll in which the trunk and legs are turned over the head ▷ *v* perform a somersault

somnambulist *n* person who walks in his or her sleep ▶ **somnambulism** *n*

somnolent *adj* drowsy

son *n* male offspring ▶ **son-in-law** *n*, *pl* **sons-in-law** husband of one's child

sonar *n* device for detecting underwater objects by the reflection of sound waves

sonata *n* piece of music in several movements for one instrument with or without piano

son et lumière [sawn eh loo-mee-er] *n* (*French*) night-time entertainment with lighting and sound effects, telling the story of the place where it is staged

song *n* music for the voice; tuneful sound made by certain birds; singing **for a song** very cheaply ▶ **songster, songstress** *n* singer ▶ **songbird** *n* any bird with a musical call

sonic *adj* of or producing sound ▶ **sonic boom** loud bang caused by an aircraft flying faster than the speed of sound

sonnet *n* fourteen-line poem with a fixed rhyme scheme

sonorous *adj* (of sound) deep or resonant ▶ **sonorously** *adv* ▶ **sonority** *n*

soon *adv* in a short time

sooner *adv* rather e.g. *I'd sooner go alone*; **sooner or later** eventually

soot *n* black powder formed by the incomplete burning of an organic substance ▶ **sooty** *adj*

soothe *v* make calm; relieve (pain etc.)

soothsayer *n* seer or prophet

sop *n* concession to pacify someone ▷ *v* **sopping, sopped** mop up or absorb (liquid) ▶ **sopping** *adj* completely soaked ▶ **soppy** *adj* (*informal*) oversentimental

sophist *n* person who uses clever but invalid arguments

sophisticate *v* make less natural or innocent; make more complex or refined ▷ *n* sophisticated person

sophisticated *adj* having or appealing to refined or cultured tastes and habits; complex and refined ▶ **sophistication** *n*

sophistry, sophism *n* clever but invalid argument

sophomore *n* (*US*) student in second year at college

soporific *adj* causing sleep ▷ *n* drug that causes sleep

soprano *n, pl* **-pranos** (singer with) the highest female or boy's voice; highest pitched of a family of instruments

sorbet *n* flavoured water ice

sorcerer *n* magician ▶ **sorceress** *n fem* ▶ **sorcery** *n* witchcraft or magic

sordid *adj* dirty, squalid; base, vile; selfish and grasping ▶ **sordidly** *adv* ▶ **sordidness** *n*

sore adj painful; causing annoyance; resentful; (of need) urgent ▷ n painful area on the body ▷ adv (obs) greatly ▶ **sorely** adv greatly ▶ **soreness** n

sorghum n kind of grass cultivated for grain

sorrel n bitter-tasting plant

sorrow n grief or sadness; cause of sorrow ▷ v grieve ▶ **sorrowful** adj ▶ **sorrowfully** adv

sorry adj **-rier, -riest** feeling pity or regret; pitiful or wretched

sort n group all sharing certain qualities or characteristics; (informal) type of character ▷ v arrange according to kind; mend or fix **out of sorts** slightly unwell or bad-tempered

> **USAGE NOTE**
> Note the singular/plural usage: this (or that) sort of thing; these (or those) sorts of thing. In the second, plural example you can also say these sorts of things

sortie n relatively short return trip; operational flight made by military aircraft

SOS n international code signal of distress; call for help

so-so adj (informal) mediocre

sot n habitual drunkard

sotto voce [sot-toe voe-chay] adv in an undertone

soubriquet [so-brik-ay] n same as **sobriquet**

soufflé [soo-flay] n light fluffy dish made with beaten egg whites and other ingredients

sough [rhymes with **now**] v (of the wind) make a sighing sound

sought [sawt] v past of **seek**

souk [sook] n marketplace in Muslim countries, often open-air

soul n spiritual and immortal part of a human being; essential part or fundamental nature; deep and

sincere feelings; person regarded as typifying some quality; person; type of Black music combining blues, pop, and gospel ▶ **soulful** adj full of emotion ▶ **soulless** adj lacking human qualities, mechanical; (of a person) lacking sensitivity

sound[1] n something heard, noise ▷ v make or cause to make a sound; seem to be as specified; pronounce ▶ **sound barrier** (informal) sudden increase in air resistance against an object as it approaches the speed of sound ▶ **sound bite** short pithy sentence or phrase extracted from a longer speech, esp. by a politician, for use on television or radio ▶ **soundproof** adj not penetrable by sound ▷ v make soundproof ▶ **soundtrack** n recorded sound accompaniment to a film

sound[2] adj in good condition; firm, substantial; financially reliable; showing good judgment; ethically correct; (of sleep) deep; thorough ▶ **soundly** adv

sound[3] v find the depth of (water etc.); examine (the body) by tapping or with a stethoscope; ascertain the views of ▶ **soundings** pl n measurements of depth taken by sounding ▶ **sounding board** person or group used to test a new idea

sound[4] n channel or strait

soup n liquid food made from meat, vegetables, etc. ▶ **soupy** adj ▶ **soup kitchen** place where food and drink is served to needy people ▶ **souped-up** adj (of an engine) adjusted so as to be more powerful than normal

soupçon [soop-sonn] n small amount

sour adj sharp-tasting; (of milk) gone bad; (of a person's temperament) sullen ▷ v make or become sour ▶ **sourly** adv ▶ **sourness** n

source n origin or starting point; person, book, etc. providing information; spring where a river or stream begins

souse v plunge (something) into liquid; drench; pickle

soutane [soo-**tan**] n Roman Catholic priest's cassock

south n direction towards the South Pole, opposite north; area lying in or towards the south ▷ adj to or in the south; (of a wind) from the south ▷ adv in, to, or towards the south ▶ **southerly** adj ▶ **southern** adj ▶ **southerner** n person from the south of a country or area ▶ **southward** adj, adv ▶ **southwards** adv ▶ **southpaw** n (informal) left-handed person, esp. a boxer ▶ **South Pole** southernmost point on the earth's axis

souvenir n keepsake, memento

sou'wester n seaman's waterproof hat covering the head and back of the neck

sovereign n king or queen; former British gold coin worth one pound ▷ adj (of a state) independent; supreme in rank or authority; excellent ▶ **sovereignty** n

soviet n formerly, elected council at various levels of government in the USSR ▷ adj (**S-**) of the former USSR

sow¹ [rhymes with **know**] v **sowing, sowed, sown** or **sowed** scatter or plant (seed) in or on (the ground); implant or introduce

sow² [rhymes with **cow**] n female adult pig

soya n plant whose edible bean (**soya bean**) is used for food and as a source of oil ▶ **soy sauce** sauce made from fermented soya beans,

used in Chinese and Japanese cookery

sozzled adj (Brit, Aust & NZ slang) drunk

spa n resort with a mineral-water spring

space n unlimited expanse in which all objects exist and move; interval; blank portion; unoccupied area; the universe beyond the earth's atmosphere ▷ v place at intervals ▶ **spacious** adj having a large capacity or area ▶ **spacecraft, spaceship** n vehicle for travel beyond the earth's atmosphere ▶ **space shuttle** manned reusable vehicle for repeated space flights ▶ **spacesuit** n sealed pressurized suit worn by an astronaut

spade¹ n tool for digging ▶ **spadework** n hard preparatory work

spade² n playing card of the suit marked with black leaf-shaped symbols

spaghetti n pasta in the form of long strings

spam v **spamming, spammed** send unsolicited e-mail or text messages to multiple recipients ▷ n unsolicited e-mail or text messages sent in this way

span n space between two points; complete extent; distance from thumb to little finger of the expanded hand ▷ v **spanning, spanned** stretch or extend across

spangle n small shiny metallic ornament ▷ v decorate with spangles

Spaniard n person from Spain

spaniel n dog with long ears and silky hair

Spanish n official language of Spain and most countries of S and Central America ▷ adj of Spain or its language or people

spank v slap with the open hand, on the buttocks or legs ▷ n such a slap ▶ **spanking** n

spanking adj (informal) outstandingly fine or smart; quick

spanner n tool for gripping and turning a nut or bolt

spar¹ n pole used as a ship's mast, boom, or yard

spar² v **sparring, sparred** box or fight using light blows for practice; argue (with someone)

spare adj extra; in reserve; (of a person) thin ▷ n duplicate kept in case of damage or loss ▷ v refrain from punishing or harming; protect (someone) from (something unpleasant); afford to give **to spare** in addition to what is needed ▶ **sparing** adj economical ▶ **spare ribs** pork ribs with most of the meat trimmed off

spark n fiery particle thrown out from a fire or caused by friction; flash of light produced by an electrical discharge; trace or hint (of a particular quality) ▷ v give off sparks; initiate ▶ **sparkie** n (NZ informal) electrician ▶ **spark plug** device in an engine that ignites the fuel by producing an electric spark

sparkle v glitter with many points of light; be vivacious or witty ▷ n sparkling points of light; vivacity or wit ▶ **sparkler** n hand-held firework that emits sparks ▶ **sparkling** adj (of wine or mineral water) slightly fizzy

sparrow n small brownish bird ▶ **sparrowhawk** n small hawk

sparse adj thinly scattered ▶ **sparsely** adv ▶ **sparseness** n

spartan adj strict and austere

spasm n involuntary muscular contraction; sudden burst of activity or feeling ▶ **spasmodic** adj occurring in spasms ▶ **spasmodically** adv

spastic n (offens) person with cerebral palsy ▷ adj (offens) suffering from cerebral palsy; affected by spasms

spat¹ n slight quarrel

spat² v past of **spit¹**

spate n large number of things happening within a period of time; sudden outpouring or flood

spatial adj of or in space

spats pl n coverings formerly worn over the ankle and instep

spatter v scatter or be scattered in drops over (something) ▷ n spattering sound; something spattered

spatula n utensil with a broad flat blade for spreading or stirring

spawn n jelly-like mass of eggs of fish, frogs, or molluscs ▷ v (of fish, frogs, or molluscs) lay eggs; generate

spay v remove the ovaries from (a female animal)

spaza shop n (S Afr slang) small informal shop in a township

speak v **speaking, spoke, spoken** say words, talk; communicate or express in words; give a speech or lecture; know how to talk in (a specified language) ▶ **speaker** n person who speaks, esp. at a formal occasion; loudspeaker; (**S-**) official chairman of a body ▶ **speakerphone** n telephone with a microphone and loudspeaker allowing more than one person to participate in a call

spear¹ n weapon consisting of a long shaft with a sharp point ▷ v pierce with or as if with a spear ▶ **spearhead** v lead (an attack or campaign) ▷ n leading force in an attack or campaign

spear² n slender shoot

spearmint n type of mint

spec n **on spec** (informal) as a risk or gamble

s

special *adj* distinguished from others of its kind; for a specific purpose; exceptional; particular ▶ **specially** *adv* ▶ **specialist** *n* expert in a particular activity or subject ▶ **speciality** *n* special interest or skill; product specialized in ▶ **specialize** *v* be a specialist ▶ **specialization** *n*

specie *n* coins as distinct from paper money

species *n, pl* **-cies** group of plants or animals that are related closely enough to interbreed naturally

specific *adj* particular, definite ▷ *n* drug used to treat a particular disease ▷ *pl* particular details ▶ **specifically** *adv* ▶ **specification** *n* detailed description of something to be made or done ▶ **specify** *v* refer to or state specifically ▶ **specific gravity** ratio of the density of a substance to that of water

specimen *n* individual or part typifying a whole; sample of blood etc. taken for analysis

specious [spee-shuss] *adj* apparently true, but actually false

speck *n* small spot or particle ▶ **speckle** *n* small spot ▷ *v* mark with speckles

specs *pl n* (*informal*) short for **spectacles**

spectacle *n* strange, interesting, or ridiculous sight; impressive public show ▷ *pl* pair of glasses for correcting faulty vision ▶ **spectacular** *adj* impressive ▷ *n* spectacular public show ▶ **spectacularly** *adv*

spectator *n* person viewing anything, onlooker ▶ **spectate** *v* watch

spectre *n* ghost; menacing mental image ▶ **spectral** *adj*

spectroscope *n* instrument for producing or examining spectra

spectrum *n, pl* **-tra** range of different colours, radio waves, etc. in order of their wavelengths; entire range of anything

speculate *v* guess, conjecture; buy property, shares, etc. in the hope of selling them at a profit ▶ **speculation** *n* ▶ **speculative** *adj* ▶ **speculator** *n*

sped *v* a past of **speed**

speech *n* act, power, or manner of speaking; talk given to an audience; language or dialect ▶ **speechless** *adj* unable to speak because of great emotion

speed *n* swiftness; rate at which something moves or acts; (*slang*) amphetamine ▷ *v* **speeding, sped** or **speeded** go quickly; drive faster than the legal limit ▶ **speedy** *adj* prompt; rapid ▶ **speedily** *adv* ▶ **speedboat** *n* light fast motorboat ▶ **speed camera** *n* (*Brit, Aust & NZ*) camera for photographing vehicles breaking the speed limit ▶ **speed dating** dating method in which each participant engages in a timed chat with all the others in turn ▶ **speedometer** *n* instrument to show the speed of a vehicle ▶ **speed up** *v* accelerate ▶ **speedway** *n* track for motorcycle racing; (*US, Canadian & NZ*) track for motor racing ▶ **speedwell** *n* plant with small blue flowers

speleology *n* study and exploration of caves

spell¹ *v* **spelling, spelt** or **spelled** give in correct order the letters that form (a word); (of letters) make up (a word); indicate ▶ **spelling** *n* way a word is spelt; person's ability to spell ▶ **spellchecker** *n* (*Computing*) program that highlights wrongly spelled words in a word-processed document ▶ **spell out** *v* make explicit

spell² n formula of words supposed to have magic power; effect of a spell; fascination ▶ **spellbound** adj entranced

spell³ n period of time of weather or activity; (Scot, Aust & NZ) period of rest

spelt v a past of **spell¹**

spend v **spending, spent** pay out (money); use or pass (time); use up completely ▶ **spendthrift** n person who spends money wastefully

sperm n, pl **sperms** or **sperm** male reproductive cell; semen ▶ **spermicide** n substance that kills sperm ▶ **sperm whale** large toothed whale

spermaceti [sper-ma-**set**-ee] n waxy solid obtained from the sperm whale

spermatozoon [sper-ma-toe-**zoe**-on] n, pl -**zoa** sperm

spew v vomit; send out in a stream

sphagnum n moss found in bogs

sphere n perfectly round solid object; field of activity ▶ **spherical** adj

sphincter n ring of muscle which controls the opening and closing of a hollow organ

Sphinx n statue in Egypt with a lion's body and human head; (**s-**) enigmatic person

spice n aromatic substance used as flavouring; something that adds zest or interest ▶ v flavour with spices ▶ **spicy** adj flavoured with spices; (informal) slightly scandalous

spick-and-span adj neat and clean

spider n small eight-legged creature which spins a web to catch insects for food ▶ **spidery** adj

spiel n speech made to persuade someone to do something

spigot n stopper for, or tap fitted to, a cask

spike n sharp point; sharp pointed metal object ▶ pl sports shoes with spikes for greater grip ▶ v put spikes on; pierce or fasten with a spike; add alcohol to (a drink) **spike someone's guns** thwart someone ▶ **spiky** adj

spill¹ v **spilling, spilt** or **spilled** pour from or as if from a container ▶ n fall; amount spilt **spill the beans** (informal) give away a secret ▶ **spillage** n

spill² n thin strip of wood or paper for lighting pipes or fires

spin v **spinning, spun** revolve or cause to revolve rapidly; draw out and twist (fibres) into thread; (informal) present information in a way that creates a favourable impression ▶ n revolving motion; continuous spiral descent of an aircraft; (informal) short drive for pleasure; (informal) presenting of information in a way that creates a favourable impression **spin a yarn** tell an improbable story ▶ **spinner** n ▶ **spin doctor** (informal) person who provides a favourable slant to a news item or policy on behalf of a politician or a political party ▶ **spin-dry** v dry (clothes) in a spin-dryer ▶ **spin-dryer** n machine in which washed clothes are spun in a perforated drum to remove excess water ▶ **spin-off** n incidental benefit ▶ **spin out** v prolong

spina bifida n condition in which part of the spinal cord protrudes through a gap in the backbone, often causing paralysis

spinach n dark green leafy vegetable

spindle n rotating rod that acts as an axle; weighted rod rotated for spinning thread by hand ▶ **spindly** adj long, slender, and frail

spindrift n spray blown up from the sea

spine n backbone; edge of a book on which the title is printed; sharp point on an animal or plant ▶ **spinal** adj of the spine ▶ **spineless** adj lacking courage ▶ **spiny** adj covered with spines

spinet n small harpsichord

spinifex n coarse spiny Australian grass

spinnaker n large sail on a racing yacht

spinney n (chiefly Brit) small wood

spinster n unmarried woman

spiral n continuous curve formed by a point winding about a central axis at an ever-increasing distance from it; same as **helix**; steadily accelerating increase or decrease ▶ v **-ralling, -ralled** move in a spiral; increase or decrease with steady acceleration ▶ adj having the form of a spiral

spire n pointed part of a steeple

spirit¹ n nonphysical aspect of a person concerned with profound thoughts; nonphysical part of a person believed to live on after death; courage and liveliness; essential meaning as opposed to literal interpretation; ghost ▶ pl emotional state ▶ v **-iting, -ited** carry away mysteriously ▶ **spirited** adj lively

spirit² n liquid obtained by distillation ▶ **spirit level** glass tube containing a bubble in liquid, used to check whether a surface is level

spiritual adj relating to the spirit; relating to sacred things ▶ n type of religious folk song originating among Black slaves in America ▶ **spiritually** adv ▶ **spirituality** n

spiritualism n belief that the spirits of the dead can communicate with the living ▶ **spiritualist** n

spit¹ v **spitting, spat** eject (saliva or food) from the mouth; throw

out particles explosively; rain slightly; utter (words) in a violent manner ▶ n saliva ▶ **spitting image** (informal) person who looks very like another ▶ **spittle** n fluid produced in the mouth, saliva ▶ **spittoon** n bowl to spit into

spit² n sharp rod on which meat is skewered for roasting; long narrow strip of land jutting out into the sea

spite n deliberate nastiness ▶ v annoy or hurt from spite **in spite of** in defiance of ▶ **spiteful** adj ▶ **spitefully** adv

spitfire n person with a fiery temper

spiv n (Brit, Aust & NZ slang) smartly dressed man who makes a living by shady dealings

splash v scatter liquid on (something); scatter (liquid) or (of liquid) be scattered in drops; print (a story or photograph) prominently in a newspaper ▶ n splashing sound; patch (of colour or light); extravagant display; small amount of liquid added to a drink ▶ **splash out** v (informal) spend extravagantly

splatter v, n splash

splay v spread out, with ends spreading in different directions

spleen n abdominal organ which filters bacteria from the blood; bad temper ▶ **splenetic** adj spiteful or irritable

splendid adj excellent; brilliant in appearance ▶ **splendidly** adv ▶ **splendour** n

splice v join by interweaving or overlapping ends **get spliced** (slang) get married

splint n rigid support for a broken bone

splinter n thin sharp piece broken off, esp. from wood ▶ v break into fragments ▶ **splinter group** group that has broken away from an organization

split v **splitting, split** break into separate pieces; separate; share ▷ n crack or division caused by splitting ▷ pl act of sitting with the legs outstretched in opposite directions ▶ **split second** very short period of time

splotch, splodge n, v splash, daub

splurge v spend money extravagantly ▷ n bout of extravagance

splutter v utter with spitting or choking sounds; make hissing spitting sounds ▷ n spluttering

spoil v **spoiling, spoilt** or **spoiled** damage; harm the character of (a child) by giving it all it wants; rot, go bad **spoiling for** eager for ▶ **spoils** pl n booty ▶ **spoilsport** n person who spoils the enjoyment of others

spoke¹ v past tense of **speak**

spoke² n bar joining the hub of a wheel to the rim

spoken v past participle of **speak**

spokesman, spokeswoman, spokesperson n person chosen to speak on behalf of a group

spoliation n plundering

sponge n sea animal with a porous absorbent skeleton; skeleton of a sponge, or a substance like it, used for cleaning; type of light cake ▷ v wipe with a sponge; live at the expense of others ▶ **sponger** n (slang) person who sponges on others ▶ **spongy** adj

sponsor n person who promotes something; person who agrees to give money to a charity on completion of a specified activity by another; godparent ▷ v act as a sponsor for ▶ **sponsorship** n

spontaneous adj not planned or arranged; occurring through

natural processes without outside influence ▶ **spontaneously** adv ▶ **spontaneity** n

spoof n mildly satirical parody

spook n (informal) ghost ▶ **spooky** adj

spool n cylinder round which something can be wound

spoon n shallow bowl attached to a handle for eating, stirring, or serving food ▷ v lift with a spoon ▶ **spoonful** n ▶ **spoonbill** n wading bird of warm regions with a long flat bill ▶ **spoon-feed** v feed with a spoon; give (someone) too much help

spoonerism n accidental changing over of the initial letters of a pair of words, such as half-warmed fish for half-formed wish

spoor n trail of an animal

sporadic adj intermittent, scattered ▶ **sporadically** adv

spore n minute reproductive body of some plants

sporran n pouch worn in front of a kilt

sport n activity for pleasure, competition, or exercise; such activities collectively; enjoyment; playful joking; person who reacts cheerfully ▷ v wear proudly ▶ **sporting** adj of sport; behaving in a fair and decent way **sporting chance** reasonable chance of success ▶ **sporty** adj ▶ **sportive** adj playful ▶ **sports car** fast low-built car, usu. open-topped ▶ **sports jacket** man's casual jacket ▶ **sportsman, sportswoman** n person who plays sports; person who plays fair and is good-humoured when losing ▶ **sportsmanlike** adj ▶ **sportsmanship** n

spot n small mark on a surface; pimple; location; (informal) small

S

quantity; (*informal*) awkward situation ▷ v **spotting, spotted** notice; mark with spots; watch for and take note of **on the spot** at the place in question; immediately; in an awkward predicament ▶ **spotless** *adj* absolutely clean ▶ **spotlessly** *adv* ▶ **spotty** *adj* with spots ▶ **spot check** random examination ▶ **spotlight** *n* powerful light illuminating a small area; centre of attention ▶ **spot-on** *adj* (*informal*) absolutely accurate

spouse *n* husband or wife

spout *v* pour out in a stream or jet; (*slang*) utter (a stream of words) lengthily ▷ *n* projecting tube or lip for pouring liquids; stream or jet of liquid

sprain *v* injure (a joint) by a sudden twist ▷ *n* such an injury

sprang *v* a past tense of **spring**

sprat *n* small sea fish

sprawl *v* lie or sit with the limbs spread out; spread out in a straggling manner ▷ *n* part of a city that has spread untidily over a large area

spray[1] *n* (device for producing) fine drops of liquid ▷ *v* scatter in fine drops; cover with a spray ▶ **spray gun** device for spraying paint etc.

spray[2] *n* branch with buds, leaves, flowers, or berries; ornament like this

spread *v* **spreading, spread** open out or be displayed to the fullest extent; extend over a larger expanse; apply as a coating; send or be sent in all directions ▷ *n* spreading; extent; (*informal*) large meal; soft food which can be spread ▶ **spread-eagled** *adj* with arms and legs outstretched ▶ **spreadsheet** *n* computer program for manipulating figures

spree *n* session of overindulgence, usu. in drinking or spending money

sprig *n* twig or shoot; (NZ) stud on the sole of a soccer or rugby boot

sprightly *adj* **-lier, -liest** lively and brisk ▶ **sprightliness** *n*

spring *v* **springing, sprang** or **sprung, sprung** move suddenly upwards or forwards in a single motion, jump; develop unexpectedly; originate (from); (*informal*) arrange the escape of (someone) from prison ▷ *n* season between winter and summer; jump; coil which can be compressed, stretched, or bent and returns to its original shape when released; natural pool forming the source of a stream; elasticity ▶ **springy** *adj* elastic ▶ **springboard** *n* flexible board used to gain height or momentum in diving or gymnastics ▶ **spring-clean** *v* clean (a house) thoroughly ▶ **spring tide** high tide at new or full moon

springbok *n* S African antelope

springer *n* small spaniel

sprinkle *v* scatter (liquid or powder) in tiny drops or particles over (something) ▶ **sprinkler** *n* ▶ **sprinkling** *n* small quantity or number

sprint *n* short race run at top speed; fast run ▷ *v* run a short distance at top speed ▶ **sprinter** *n*

sprite *n* elf

sprocket *n* wheel with teeth on the rim, that drives or is driven by a chain

sprout *v* put forth shoots; begin to grow or develop ▷ *n* shoot; short for **Brussels sprout**

spruce[1] *n* kind of fir

spruce[2] *adj* neat and smart ▶ **spruce up** *v* make neat and smart

sprung v a past of **spring**

spry adj **spryer, spryest** or **sprier, spriest** active or nimble

spud n (informal) potato

spume n, v froth

spun v past of **spin**

spunk n (informal) courage, spirit ▷ **spunky** adj

spur n stimulus or incentive; spiked wheel on the heel of a rider's boot used to urge on a horse; projection ▷ v **spurring, spurred** urge on, incite (someone) **on the spur of the moment** on impulse

spurge n plant with milky sap

spurious adj not genuine

spurn v reject with scorn

spurt v gush or cause to gush out in a jet ▷ n short sudden burst of activity or speed; sudden gush

sputnik n early Soviet artificial satellite

sputter v, n splutter

sputum n, pl **-ta** spittle, usu. mixed with mucus

spy n, pl **spies** person employed to obtain secret information; person who secretly watches others ▷ v **spying, spied** act as a spy; catch sight of ▷ **spyware** n (Computing) software installed surreptitiously on a computer to collect and transmit data from that computer to another

Sq. Square

squabble v, n (engage in) a petty or noisy quarrel

squad n small group of people working or training together

squadron n division of an air force, fleet, or cavalry regiment

squalid adj dirty and unpleasant; morally sordid ▷ **squalor** n disgusting dirt and filth

squall[1] n sudden strong wind

squall[2] v cry noisily, yell ▷ n harsh cry

squander v waste (money or resources)

square n geometric figure with four equal sides and four right angles; open area in a town in this shape; product of a number multiplied by itself ▷ adj square in shape; denoting a measure of area; straight or level; fair and honest; with all accounts or debts settled ▷ v multiply (a number) by itself; make square; be or cause to be consistent ▷ adv squarely, directly ▷ **squarely** adv in a direct way; in an honest and frank manner ▷ **square dance** formation dance in which the couples form squares ▷ **square meal** substantial meal ▷ **square root** number of which a given number is the square ▷ **square up to** v prepare to confront (a person or problem)

squash[1] v crush flat; suppress; push into a confined space; humiliate with a crushing retort ▷ n sweet fruit drink diluted with water; crowd of people in a confined space; game played in an enclosed court with a rubber ball and long-handled rackets ▷ **squashy** adj

squash[2] n marrow-like vegetable

squat v **squatting, squatted** crouch with the knees bent and the weight on the feet; occupy unused premises to which one has no legal right ▷ n place where squatters live ▷ adj short and broad

squatter n illegal occupier of unused premises

squaw n (offens) Native American woman

squawk n loud harsh cry ▷ v utter a squawk

squeak n short shrill cry or sound ▷ v make or utter a squeak ▷ **squeaky** adj

squeal n long shrill cry or sound ▷ v make or utter a squeal; (slang) inform on someone to the police

s

squeamish adj easily sickened or shocked

squeegee n tool with a rubber blade for clearing water from a surface

squeeze v grip or press firmly; crush or press to extract liquid; push into a confined space; hug; obtain (something) by force or great effort ▷ n squeezing; amount extracted by squeezing; hug; crush of people in a confined space; restriction on borrowing

squelch v make a wet sucking sound, as by walking through mud ▷ n squelching sound

squib n small firework that hisses before exploding

squid n sea creature with a long soft body and ten tentacles

squiggle n wavy line ▶ **squiggly** adj

squint v have eyes which face in different directions; glance sideways ▷ n squinting condition of the eye; (informal) glance ▷ adj crooked

squire n country gentleman, usu. the main landowner in a community; (Hist) knight's apprentice

squirm v wriggle, writhe; feel embarrassed ▷ n wriggling movement

squirrel n small bushy-tailed tree-living animal

squirt v force (a liquid) or (of a liquid) be forced out of a narrow opening; squirt liquid at ▷ n jet of liquid; (informal) small or insignificant person

squish v, n (make) a soft squelching sound ▶ **squishy** adj

Sr Senior; Señor

SS Schutzstaffel: Nazi paramilitary security force; steamship

SSL (Computing) secure sockets layer: protocol for encrypting

and transmitting sensitive data securely over the internet

St Saint; Street

st. stone (weight)

stab v stabbing, stabbed pierce with something pointed; jab (at) ▷ n stabbing; sudden unpleasant sensation; (informal) attempt

stabilize v make or become stable ▶ **stabilization** ▶ **stabilizer** n device for stabilizing a child's bicycle, an aircraft, or a ship

stable[1] n building in which horses are kept; establishment that breeds and trains racehorses; establishment that manages or trains several entertainers or athletes ▷ v put or keep (a horse) in a stable

stable[2] adj firmly fixed or established; firm in character; (Science) not subject to decay or decomposition ▶ **stability** n

staccato [stak-ah-toe] adj, adv (Music) with the notes sharply separated ▷ adj consisting of short abrupt sounds

stack n ordered pile; large amount; chimney ▷ v pile in a stack; control (aircraft waiting to land) so that they fly at different altitudes

stadium n, pl -diums or -dia sports arena with tiered seats for spectators

staff[1] n people employed in an organization; stick used as a weapon, support, etc. ▷ v supply with personnel

staff[2] n, pl staves set of five horizontal lines on which music is written

stag n adult male deer ▶ **stag beetle** beetle with large branched jaws ▶ **stagette** n (informal) young unmarried professional woman ▶ **stag night, stag party** party for men only

stage n step or period of development; platform in a theatre where actors perform; portion of a journey ▷ v put (a play) on stage; organize and carry out (an event) **the stage** theatre as a profession ▶ **stagey** adj overtheatrical ▶ **stagecoach** n large horse-drawn vehicle formerly used to carry passengers and mail ▶ **stage fright** nervousness felt by a person about to face an audience ▶ **stage whisper** loud whisper intended to be heard by an audience

stagger v walk unsteadily; astound; set apart to avoid congestion ▷ n staggering

stagnant adj (of water or air) stale from not moving; not growing or developing ▶ **stagnate** v be stagnant ▷ n **stagnation** n

staid adj sedate, serious, and rather dull

stain v discolour, mark; colour with a penetrating pigment ▷ n discoloration or mark; moral blemish or slur; penetrating liquid used to colour things ▶ **stainless** adj ▶ **stainless steel** steel alloy that does not rust

stairs pl n flight of steps between floors, usu. indoors ▶ **staircase**, **stairway** n flight of stairs with a handrail or banisters

stake¹ n pointed stick or post driven into the ground as a support or marker ▷ v support or mark out with stakes **stake a claim to** claim a right to ▶ **stake out** (slang) (of police) keep (a place) under surveillance

stake² n money wagered; interest, usu. financial, held in something ▷ v wager, risk; support financially ▶ **stakeholder** n person who has a concern or interest in something, esp. a business

stalactite n lime deposit hanging from the roof of a cave

stalagmite n lime deposit sticking up from the floor of a cave

stale adj not fresh; lacking energy or ideas through overwork or monotony; uninteresting from overuse ▶ **staleness** n

stalemate n (Chess) position in which any of a player's moves would put his or her king in check, resulting in a draw; deadlock, impasse

stalk¹ n plant's stem

stalk² n follow or approach stealthily; pursue persistently and, sometimes, attack (a person with whom one is obsessed); walk in a stiff or haughty manner ▶ **stalker** n person who follows or stealthily approaches a person or an animal; person who persistently pursues and, sometimes, attacks someone with whom he or she is obsessed ▶ **stalking-horse** n pretext

stall¹ n small stand for the display and sale of goods; compartment in a stable; small room or compartment ▷ pl ground-floor seats in a theatre or cinema; row of seats in a church for the choir or clergy ▷ v stop (a motor vehicle or engine) or (of a motor vehicle or engine) stop accidentally

stall² n employ delaying tactics

stallion n uncastrated male horse

stalwart [stawl-wart] adj strong and sturdy; dependable ▷ n stalwart person

stamen n pollen-producing part of a flower

stamina n enduring energy and strength

stammer v speak or say with involuntary pauses or repetition of syllables ▷ n tendency to stammer

stamp n (also **postage stamp**) piece of gummed paper stuck to

s

an envelope or parcel to show that the postage has been paid; act of stamping; instrument for stamping a pattern or mark; pattern or mark stamped; characteristic feature ▷ v bring (one's foot) down forcefully; walk with heavy footsteps; characterize; impress (a pattern or mark) on; stick a postage stamp on ▶ **stamping ground** favourite meeting place ▶ **stamp out** v suppress by force

stampede n sudden rush of frightened animals or of a crowd ▷ v (cause to) take part in a stampede

stance n attitude; manner of standing

stanch v same as **staunch²**

stanchion n upright bar used as a support

stand v **standing, stood** be in, rise to, or place in an upright position; be situated; be in a specified state or position; remain unchanged or valid; tolerate; offer oneself as a candidate; (informal) treat to ▷ n stall for the sale of goods; structure for spectators at a sports ground; firmly held opinion; (US & Aust) witness box; rack or piece of furniture on which things may be placed ▶ **standing** adj permanent, lasting ▷ n reputation or status; duration ▶ **stand for** v represent or mean; (informal) tolerate ▶ **stand in** v act as a substitute ▶ **stand-in** n substitute ▶ **standoffish** adj reserved or haughty ▶ **stand up for** v support or defend

standard n level of quality; example against which others are judged or measured; moral principle; distinctive flag; upright pole ▷ adj usual, regular, or average; of recognized authority;

accepted as correct ▶ **standardize** v cause to conform to a standard ▶ **standardization** n ▶ **standard lamp** lamp attached to an upright pole on a base

standpipe n tap attached to a water main to provide a public water supply

standpoint n point of view

standstill n complete halt

stank v a past tense of **stink**

stanza n verse of a poem

staple¹ n U-shaped piece of metal used to fasten papers or secure things ▷ v fasten with staples ▶ **stapler** n small device for fastening papers together

staple² adj of prime importance, principal ▷ n main constituent of anything

star n hot gaseous mass in space, visible in the night sky as a point of light; star-shaped mark used to indicate excellence; asterisk; celebrity in the entertainment or sports world ▷ pl astrological forecast, horoscope ▷ v **starring, starred** feature or be featured as a star; mark with a star or stars ▷ adj leading, famous ▶ **stardom** n status of a star in the entertainment or sports world ▶ **starry** adj full of or like stars ▶ **starry-eyed** adj full of naive optimism ▶ **starfish** n star-shaped sea creature

starboard n right-hand side of a ship, when facing forward ▷ adj of or on this side

starch n carbohydrate forming the main food element in bread, potatoes, etc., and used mixed with water for stiffening fabric ▷ v stiffen (fabric) with starch ▶ **starchy** adj containing starch; stiff and formal

stare v look or gaze fixedly (at) ▷ n fixed gaze

stark adj harsh, unpleasant, and plain; desolate, bare; absolute ▷ adv completely

starling n songbird with glossy black speckled feathers

start v take the first step, begin; set or be set in motion; make a sudden involuntary movement from fright; establish or set up ▷ n first part of something; place or time of starting; advantage or lead in a competitive activity; sudden movement made from fright ▶ **starter** n first course of a meal; device for starting a car's engine; person who signals the start of a race ▶ **start-up** n recently launched project or business enterprise ▷ adj recently launched e.g. start-up grants

startle v slightly surprise or frighten

starve v die or suffer or cause to die or suffer from hunger; deprive of something needed ▶ **starvation** n

stash (informal) v store in a secret place ▷ n secret store

state n condition of a person or thing; sovereign political power or its territory; (**S-**) the government; (informal) excited or agitated condition; pomp ▷ adj of or concerning the State; involving ceremony ▷ v express in words ▶ **stately** adj dignified or grand ▶ **statehouse** n (NZ) publicly-owned house rented to a low-income tenant ▶ **statement** n something stated; printed financial account ▶ **stateroom** n private cabin on a ship; large room in a palace, used for ceremonial occasions ▶ **statesman, stateswoman** n experienced and respected political leader ▶ **statesmanship** n

static adj stationary or inactive; (of a force) acting but producing no movement ▷ n crackling sound or speckled picture caused by interference in radio or television reception; (also **static electricity**) electric sparks produced by friction

statin n (Med) drug used to lower cholesterol

station n place where trains stop for passengers; headquarters or local offices of the police or a fire brigade; building with special equipment for a particular purpose e.g. power station; television or radio channel; position in society; large Australian sheep or cattle property ▷ v assign (someone) to a particular place ▶ **station wagon** (US & Aust) car with a rear door and luggage space behind the rear seats

stationary adj not moving

USAGE NOTE
The words **stationary** and **stationery** are completely different in meaning and should not be confused

stationery n writing materials such as paper and pens ▶ **stationer** n dealer in stationery

statistic n numerical fact collected and classified systematically ▶ **statistics** n science of classifying and interpreting numerical information ▶ **statistical** adj ▶ **statistically** adv ▶ **statistician** n person who compiles and studies statistics

statue n large sculpture of a human or animal figure ▶ **statuary** n statues collectively ▶ **statuesque** adj (of a woman) tall and well-proportioned ▶ **statuette** n small statue

stature n person's height; reputation of a person or their achievements

status n social position; prestige; person's legal standing ▶ **status quo** existing state of affairs

s

statute n written law ▸ **statutory** adj required or authorized by law

staunch[1] adj loyal, firm

staunch[2] , **stanch** v stop (a flow of blood)

stave n one of the strips of wood forming a barrel; (Music) same as **staff**[2] ▸ **stave in** v **staving, stove** burst a hole in ▸ **stave off** v **staving, staved** ward off

stay[1] v remain in a place or condition; reside temporarily; endure; (Scot & SAfr) live permanently e.g. where do you stay? ▸ n period of staying in a place; postponement ▸ **staying power** stamina

stay[2] n prop or buttress ▸ pl corset

stay[3] n rope or wire supporting a ship's mast

STD sexually transmitted disease; (Brit, Aust & SAfr) subscriber trunk dialling; (NZ) subscriber toll dialling

stead n in **someone's stead** in someone's place **stand someone in good stead** be useful to someone

steadfast adj firm, determined ▸ **steadfastly** adv

steady adj **steadier, steadiest** not shaky or wavering; regular or continuous; sensible and dependable ▸ v **steadying, steadied** make steady ▸ adv in a steady manner ▸ **steadily** adv ▸ **steadiness** n

steak n thick slice of meat, esp. beef; slice of fish

steal v **stealing, stole, stolen** take unlawfully or without permission; move stealthily

stealth n secret or underhand behaviour ▸ adj (of technology) able to render an aircraft almost invisible to radar; disguised or hidden e.g. stealth taxes ▸ **stealthy** adj ▸ **stealthily** adv

steam n vapour into which water changes when boiled; power,

energy, or speed ▸ v give off steam; (of a vehicle) move by steam power; cook or treat with steam ▸ **steamer** n steam-propelled ship; container used to cook food in steam ▸ **steam engine** engine worked by steam ▸ **steamroller** n steam-powered vehicle with heavy rollers, used to level road surfaces ▸ v use overpowering force to make (someone) do what one wants

steed n (lit) horse

steel n hard malleable alloy of iron and carbon; steel rod used for sharpening knives; hardness of character or attitude ▸ v prepare (oneself) for something unpleasant ▸ **steely** adj

steep[1] adj sloping sharply; (informal) (of a price) unreasonably high ▸ **steeply** adv ▸ **steepness** n

steep[2] v soak or be soaked in liquid **steeped in** filled with

steeple n church tower with a spire ▸ **steeplejack** n person who repairs steeples and chimneys

steeplechase n horse race with obstacles to jump; track race with hurdles and a water jump

steer[1] v direct the course of (a vehicle or ship); direct (one's course) ▸ **steerage** n cheapest accommodation on a passenger ship ▸ **steering wheel** wheel turned by the driver of a vehicle in order to steer it

steer[2] n castrated male ox

stein [stine] n earthenware beer mug

stellar adj of stars

stem[1] n long thin central part of a plant; long slender part, as of a wineglass; part of a word to which inflections are added ▸ v **stemming, stemmed** ▸ **stem from** originate from

stem² v **stemming, stemmed** stop (the flow of something)

stench n foul smell

stencil n thin sheet with cut-out pattern through which ink or paint passes to form the pattern on the surface below; pattern made thus ▷ v **-cilling, -cilled** make (a pattern) with a stencil

stenographer n shorthand typist

stent n surgical implant used to keep an artery open

stentorian adj (of a voice) very loud

step v **stepping, stepped** move and set down the foot, as when walking; walk a short distance ▷ n stepping; distance covered by a step; sound made by stepping; foot movement in a dance; one of a sequence of actions taken in order to achieve a goal; degree in a series or scale; flat surface for placing the foot on when going up or down ▷ pl stepladder ▸ **step in** v intervene ▸ **stepladder** n folding portable ladder with supporting frame ▸ **stepping stone** one of a series of stones for stepping on in crossing a stream; means of progress towards a goal ▸ **step up** v increase (something) by stages; take responsibility for something

step- prefix denoting a relationship created by the remarriage of a parent e.g. stepmother

steppes pl n wide grassy treeless plains in Russia and Ukraine

stereo adj short for **stereophonic** ▷ n stereophonic record player; stereophonic sound

stereophonic adj using two separate loudspeakers to give the effect of naturally distributed sound

stereotype n standardized idea of a type of person or thing ▷ v form a stereotype of

sterile adj free from germs; unable to produce offspring or seeds; lacking inspiration or vitality ▸ **sterility** n ▸ **sterilize** v make sterile ▸ **sterilization** n

sterling n British money system ▷ adj genuine and reliable

stern¹ adj severe, strict ▸ **sternly** adv ▸ **sternness** n

stern² n rear part of a ship

sternum n, pl **-na** or **-nums** same as **breastbone**

steroid n organic compound containing a carbon ring system, such as many hormones

stethoscope n medical instrument for listening to sounds made inside the body

Stetson® n tall broad-brimmed hat, worn mainly by cowboys

stevedore n person who loads and unloads ships

stew n food cooked slowly in a closed pot; (informal) troubled or worried state ▷ v cook slowly in a closed pot

steward n person who looks after passengers on a ship or aircraft; official who helps at a public event such as a race; person who administers another's property ▸ **stewardess** n fem

stick¹ n long thin piece of wood; such a piece of wood shaped for a special purpose e.g. hockey stick; something like a stick e.g. stick of celery; (slang) verbal abuse, criticism

stick² v **sticking, stuck** push (a pointed object) into (something); fasten or be fastened by or as if by pins or glue; (foll by out) extend beyond something else, protrude; (informal) remain for a long time ▸ **sticker** n adhesive label or sign ▸ **sticky** adj covered with an adhesive substance; (informal) difficult, unpleasant; (of weather) warm and humid ▶ **stick-in-the-mud** n person who does not like anything new ▶ **stick-up** n (slang) robbery at gunpoint ▶ **stick up for** v (informal) support or defend

stickleback n small fish with sharp spines on its back

stickler n person who insists on something e.g. stickler for detail

stiff adj not easily bent or moved; severe e.g. a stiff punishment; unrelaxed or awkward; firm in consistency; strong e.g. a stiff drink ▶ n (slang) corpse ▶ **stiffly** adv ▶ **stiffness** n ▶ **stiffen** v make or become stiff ▶ **stiff-necked** adj haughtily stubborn

stifle v suppress; suffocate

stigma n, pl **-mas** or **-mata** mark of social disgrace; part of a plant that receives pollen ▶ **stigmata** pl n marks resembling the wounds of the crucified Christ ▶ **stigmatize** v mark as being shameful

stile n set of steps allowing people to climb a fence

stiletto n, pl **-tos** high narrow heel on a woman's shoe; small slender dagger

still[1] adv now or in the future as before; up to this or that time; even or yet e.g. still more insults; quietly or without movement ▶ adj motionless; silent and calm, undisturbed; (of a drink) not fizzy ▶ n photograph from a film scene ▶ v make still ▶ **stillness** n ▶ **stillborn** adj born dead ▶ **still life** painting of inanimate objects

still[2] n apparatus for distilling alcoholic drinks

stilted adj stiff and formal in manner

stilts pl n pair of poles with footrests for walking raised from the ground;

long posts supporting a building above ground level

stimulus n, pl **-li** something that rouses a person or thing to activity ▶ **stimulant** n something, such as a drug, that acts as a stimulus ▶ **stimulate** v act as a stimulus (on) ▶ **stimulation** n

sting v **stinging, stung** (of certain animals or plants) wound by injecting with poison; feel or cause to feel sharp physical or mental pain; (slang) cheat (someone) by overcharging ▶ n wound or pain caused by or as if by stinging; mental pain; sharp pointed organ of certain animals or plants by which poison can be injected

stingy [stin-jee] adj **-gier, -giest** mean or miserly ▶ **stinginess** n

stink n strong unpleasant smell; (slang) unpleasant fuss ▶ v **stinking, stank** or **stunk, stunk** give off a strong unpleasant smell; (slang) be very unpleasant

stint v (foll by on) be miserly with (something) ▶ n allotted amount of work

stipend [sty-pend] n regular allowance or salary, esp. that paid to a clergyman ▶ **stipendiary** adj receiving a stipend

stipple v paint, draw, or engrave using dots

stipulate v specify as a condition of an agreement ▶ **stipulation** n

stir v **stirring, stirred** mix up (a liquid) by moving a spoon etc. around in it; move; excite or stimulate (a person) emotionally ▶ n a stirring; strong reaction, usu. of excitement ▶ **stir-fry** v **-fries, -frying, -fried** cook (food) quickly by stirring it in a pan over a high heat ▶ n, pl **-fries** dish cooked in this way

stirrup n metal loop attached to a saddle for supporting a rider's foot

S

stitch n link made by drawing thread through material with a needle; loop of yarn formed round a needle or hook in knitting or crochet; sharp pain in the side ▷ v sew **in stitches** (informal) laughing uncontrollably **not a stitch** (informal) no clothes at all

stoat n small mammal of the weasel family, with brown fur that turns white in winter

stock n total amount of goods available for sale in a shop; supply stored for future use; financial shares in, or capital of, a company; liquid produced by boiling meat, fish, bones, or vegetables ▷ pl (Hist) instrument of punishment consisting of a wooden frame with holes into which the hands and feet of the victim were locked ▷ adj kept in stock, standard; hackneyed ▷ v keep for sale or future use; supply (a farm) with livestock or (a lake etc.) with fish ▶ **stockist** n dealer who stocks a particular product ▶ **stocky** adj (of a person) broad and sturdy ▶ **stockbroker** n person who buys and sells stocks and shares for customers ▶ **stock car** car modified for a form of racing in which the cars often collide ▶ **stock exchange, stock market** institution for the buying and selling of shares ▶ **stockpile** v store a large quantity of (something) for future use ▷ n accumulated store ▶ **stock-still** adj motionless ▶ **stocktaking** n counting and valuing of the goods in a shop

stockade n enclosure or barrier made of stakes

stocking n close-fitting covering for the foot and leg

stodgy adj **stodgier, stodgiest** (of food) heavy and starchy; (of a person) serious and boring

▶ **stodge** (Brit, Aust & NZ) heavy starchy food

stoep [stoop] n (S Afr) verandah

stoic [stow-ik] n person who suffers hardship without showing his or her feelings ▷ adj (also **stoical**) suffering hardship without showing one's feelings ▶ **stoically** adv ▶ **stoicism** [stow-iss-izz-um] n

stoke v feed and tend (a fire or furnace) ▶ **stoker** n

stole¹ v past tense of **steal**

stole² n long scarf or shawl

stolen v past participle of **steal**

stolid adj showing little emotion or interest ▶ **stolidly** adv

stomach n organ in the body which digests food; front of the body around the waist; desire or inclination ▷ v put up with

stomp v (informal) tread heavily

stone n material of which rocks are made; piece of this; gem; hard central part of a fruit; unit of weight equal to 14 pounds or 6.350 kilograms; hard deposit formed in the kidney or bladder ▷ v throw stones at; remove stones from (a fruit) ▶ **stoned** adj (slang) under the influence of alcohol or drugs ▶ **stony** adj of or like stone; unfeeling or hard ▶ **stony-broke** adj (slang) completely penniless ▶ **stonily** adv ▶ **Stone Age** prehistoric period when tools were made of stone ▶ **stone-cold** adj completely cold ▶ **stone-deaf** adj completely deaf ▶ **stonewall** v obstruct or hinder discussion ▶ **stoneware** n hard kind of pottery fired at a very high temperature

stood v past of **stand**

stooge n actor who feeds lines to a comedian or acts as the butt of his or her jokes; (slang) person taken advantage of by a superior

stool n chair without arms or back; piece of excrement

stool pigeon n informer for the police

stoop v bend (the body) forward and downward; carry oneself habitually in this way; degrade oneself ▷ n stooping posture

stop v stopping, stopped cease or cause to cease from doing (something); bring to or come to a halt; prevent or restrain; withhold; block or plug; stay or rest ▷ n stopping or being stopped; place where something stops; full stop; knob on an organ that is pulled out to allow a set of pipes to sound ▶ stoppage n ▶ stoppage time same as injury time ▶ stopper n plug for closing a bottle etc. ▶ stopcock n device to control or stop the flow of fluid in a pipe ▶ stopgap n temporary substitute ▶ stopover n short break in a journey ▶ stop press news item put into a newspaper after printing has been started ▶ stopwatch n watch which can be stopped instantly for exact timing of a sporting event

stop-work, stop-work meeting n (Aust) temporary stoppage of work as a form of protest

store v collect and keep (things) for future use; put (furniture etc.) in a warehouse for safekeeping; stock (goods); (Computing) enter or retain (data) ▷ n shop; supply kept for future use; storage place, such as a warehouse ▷ pl stock of provisions **in store** about to happen **set great store by** value greatly ▶ storage n storing; space for storing ▶ storage heater electric device that can accumulate and radiate heat generated by off-peak electricity

storey n floor or level of a building

stork n large wading bird

storm n violent weather with wind, rain, or snow; strongly expressed reaction ▷ v attack or capture (a place) suddenly; shout angrily; rush violently or angrily ▶ stormy adj characterized by storms; involving violent emotions

story n, pl -ries description of a series of events told or written for entertainment; plot of a book or film; news report; (informal) lie

stoup [stoop] n small basin for holy water

stout adj fat; thick and strong; brave and resolute ▷ n strong dark beer ▶ stoutly adv

stove[1] n apparatus for cooking or heating

stove[2] v a past of stave

stow v pack or store ▶ stowaway n person who hides on a ship or aircraft in order to travel free ▶ stow away v hide as a stowaway

straddle v have one leg or part on each side of (something)

strafe v attack (an enemy) with machine guns from the air

straggle v go or spread in a rambling or irregular way ▶ straggler n ▶ straggly adj

straight adj not curved or crooked; level or upright; honest or frank; (of spirits) undiluted; (slang) heterosexual ▷ adv in a straight line; immediately; in a level or upright position ▷ n straight part, esp. of a racetrack; (slang) heterosexual person **go straight** (informal) reform after being a criminal ▶ straighten v ▶ straightaway adv immediately ▶ straight face serious facial expression concealing a desire to laugh ▶ straightforward adj honest, frank; (of a task) easy

strain v cause (something) to be used or tested beyond its limits; make an intense effort; injure by overexertion; sieve ▷ n tension or tiredness; force exerted by straining; injury from overexertion; great demand on strength or resources; melody or theme ▶ **strained** adj not natural, forced; not relaxed, tense ▶ **strainer** n sieve

strain n breed or race; trace or streak

strait n narrow channel connecting two areas of sea ▷ pl position of acute difficulty ▶ **straitjacket** n strong jacket with long sleeves used to bind the arms of a violent person ▶ **strait-laced, straight-laced** adj prudish or puritanical

straitened adj **in straitened circumstances** not having much money

strand v run aground; leave in difficulties ▷ n (poetic) shore

strand n single thread of string, wire, etc.

strange adj odd or unusual; not familiar; inexperienced (in) or unaccustomed (to) ▶ **strangely** adv ▶ **strangeness** n

stranger n person who is not known or is new to a place or experience

strangle v kill by squeezing the throat; prevent the development of ▶ **strangler** n ▶ **strangulation** n strangling ▶ **stranglehold** n strangling grip in wrestling; powerful control

strap n strip of flexible material for lifting, fastening, or holding in place ▶ **strapping, strapped** fasten with a strap or straps ▶ **strapping** adj tall and sturdy

strata n plural of **stratum**

stratagem n clever plan, trick

strategy n, pl **-gies** overall plan; art of planning in war ▶ **strategic**

[strat-**ee**-jik] adj advantageous; (of weapons) aimed at an enemy's homeland ▶ **strategically** adv ▶ **strategist** n

strathspey n Scottish dance with gliding steps

stratosphere n atmospheric layer between about 15 and 50 kilometres above the earth

stratum [**strah**-tum] n, pl **strata** layer, esp. of rock; social class ▶ **stratified** adj divided into strata ▶ **stratification** n

straw n dried stalks of grain; single stalk of straw; long thin tube used to suck up liquid into the mouth ▶ **straw poll** unofficial poll taken to determine general opinion

strawberry n sweet fleshy red fruit with small seeds on the outside ▶ **strawberry mark** red birthmark

stray v wander; digress; deviate from strayed moral standards ▷ adj having strayed; scattered, random ▷ n stray animal

streak n long band of contrasting colour or substance; quality or characteristic; short stretch (of good or bad luck) ▷ v mark with streaks; move rapidly; (informal) run naked in public ▶ **streaker** n ▶ **streaky** adj

stream n small river; steady flow, as of liquid, speech, or people; schoolchildren grouped together because of similar ability ▷ v flow steadily; move in unbroken succession; (Computing) send video or audio material over the internet so that the receiving system can play it almost simultaneously; float in the air; group (pupils) in streams ▶ **streamer** n strip of coloured paper that unrolls when tossed; long narrow flag ▶ **streaming** n

streamline v make more efficient by simplifying; give (a car, plane,

s

etc.) a smooth even shape to offer least resistance to the flow of air or water

street n public road, usu. lined with buildings ▸ **streetcar** n (US) tram ▸ **streetwise** adj knowing how to survive in big cities

strength n quality of being strong; quality or ability considered an advantage; degree of intensity; total number of people in a group **on the strength of** on the basis of ▸ **strengthen** v

strenuous adj requiring great energy or effort ▸ **strenuously** adv

streptococcus [strep-toe-kok-uss] n, pl -cocci bacterium occurring in chains, many species of which cause disease

stress n tension or strain; emphasis; stronger sound in saying a word or syllable; (Physics) force producing strain ▸ v emphasize; put stress on (a word or syllable) ▸ **stressed-out** adj (informal) suffering from tension

stretch v extend or be extended; be able to be stretched; extend the limbs or body; strain (resources or abilities) to the utmost ▸ n stretching; continuous expanse; period; (informal) term of imprisonment ▸ **stretchy** adj

stretcher n frame covered with canvas, on which an injured person is carried

strew v **strewing, strewed, strewed** or **strewn** scatter (things) over a surface

striated adj having a pattern of scratches or grooves

stricken adj seriously affected by disease, grief, pain, etc.

strict adj stern or severe; adhering closely to specified rules; complete, absolute ▸ **strictly** adv ▸ **strictness** n

stricture n severe criticism

stride v **striding, strode, stridden** walk with long steps ▸ n long step; regular pace ▸ pl progress

strident adj loud and harsh ▸ **stridently** adv ▸ **stridency** n

strife n conflict, quarrelling

strike v **striking, struck** cease work as a protest; hit; attack suddenly; ignite (a match) by friction; (of a clock) indicate (a time) by sounding a bell; enter the mind of; afflict; discover (gold, oil, etc.); agree (a bargain) ▸ n stoppage of work as a protest ▸ **striking** adj impressive; noteworthy **strike camp** dismantle and pack up tents **strike home** have the desired effect ▸ **strike off, strike out** v cross out ▸ **strike up** v begin (a conversation or friendship); begin to play music

striker n striking worker; attacking player at soccer

string n thin cord used for tying; set of objects threaded on a string; series of things or events; stretched wire or cord in a musical instrument that produces sound when vibrated ▸ pl restrictions or conditions; section of an orchestra consisting of stringed instruments ▸ v **stringing, strung** provide with a string or strings; thread on a string **pull strings** use one's influence ▸ **stringed** adj (of a musical instrument) having strings that are plucked or played with a bow ▸ **stringy** adj like string; (of meat) fibrous ▸ **string along** v deceive over a period of time ▸ **string up** v (informal) kill by hanging ▸ **stringybark** n Australian eucalyptus with a fibrous bark

stringent [strin-jent] adj strictly controlled or enforced ▸ **stringently** adv ▸ **stringency** n

strip[1] v **stripping, stripped** take (the covering or clothes) off; take a title or possession away from (someone); dismantle (an engine) ▸ **stripper** n person who performs a striptease ▸ **striptease** n entertainment in which a performer undresses to music

strip[2] n long narrow piece; (Brit, Aust & NZ) clothes a sports team plays in ▸ **strip cartoon** sequence of drawings telling a story

stripe n long narrow band of contrasting colour or substance; chevron or band worn on a uniform to indicate rank ▸ **striped, stripy, stripey** adj

stripling n youth

strive v **striving, strove, striven** make a great effort

strobe n short for **stroboscope**

stroboscope n instrument producing a very bright flashing light

strode v past tense of **stride**

stroke v touch or caress lightly with the hand ▸ n light touch or caress with the hand; rupture of a blood vessel in the brain; blow; action or occurrence of the kind specified e.g. *a stroke of luck*; chime of a clock; mark made by a pen or paintbrush; style or method of swimming

stroll v walk in a leisurely manner ▸ n leisurely walk

strong adj having physical power; not easily broken; great in degree or intensity; having moral force; having a specified number e.g. *twenty strong* ▸ **strongly** adv ▸ **stronghold** n area of predominance of a particular belief; fortress ▸ **strongroom** n room designed for the safekeeping of valuables

strontium n (Chem) silvery-white metallic element

strop n leather strap for sharpening razors

stroppy adj **-pier, -piest** (slang) angry or awkward

strove v past tense of **strive**

struck v past of **strike**

structure n complex construction; manner or basis of construction or organization ▸ v give a structure to ▸ **structural** adj ▸ **structuralism** n approach to literature, social sciences, etc., which sees changes in the subject as caused and organized by a hidden set of universal rules ▸ **structuralist** n, adj

strudel n thin sheet of filled dough rolled up and baked, usu. with an apple filling

struggle v work, strive, or make one's way with difficulty; move about violently in an attempt to get free; fight (with someone) ▸ n striving; fight

strum v **strumming, strummed** play (a guitar or banjo) by sweeping the thumb or a plectrum across the strings

strumpet n (old-fashioned) prostitute

strung v past of **string**

strut v **strutting, strutted** walk pompously; swagger ▸ n bar supporting a structure

strychnine [strik-neen] n very poisonous drug used in small quantities as a stimulant

stub n short piece left after use; counterfoil of a cheque or ticket ▸ v **stubbing, stubbed** strike (the toe) painfully against an object; put out (a cigarette) by pressing the end against a surface ▸ **stubby** adj short and broad

stubble n short stalks of grain left in a field after reaping; short growth of hair on the chin of a man who has not shaved recently ▸ **stubbly** adj

stubborn adj refusing to agree or give in; difficult to deal with ▸ **stubbornly** adv ▸ **stubbornness** n

stucco n plaster used for coating or decorating walls

stuck v past of **stick²** ▸ **stuck-up** adj (informal) conceited or snobbish

stud¹ n small piece of metal attached to a surface for decoration; disc-like removable fastener for clothes; one of several small round objects fixed to the sole of a football boot to give better grip ▷ v **studding, studded** set with studs

stud² n male animal, esp. a stallion, kept for breeding; (also **stud farm**) place where horses are bred; (slang) virile or sexually active man

student n person who studies a subject, esp. at university

studio n, pl **-dios** workroom of an artist or photographer; room or building in which television or radio programmes, records, or films are made ▸ **studio flat** (Brit) one-room flat with a small kitchen and bathroom

study v **studying, studied** be engaged in learning (a subject); investigate by observation and research; scrutinize ▷ n, pl **studies** act or process of studying; room for studying in; book or paper produced as a result of study; sketch done as practice or preparation; musical composition designed to improve playing technique ▸ **studied** adj carefully practised or planned ▸ **studious** adj fond of study; careful and deliberate ▸ **studiously** adv

stuff n substance or material; collection of unnamed things ▷ v

pack, cram, or fill completely; fill (food) with a seasoned mixture; fill (an animal's skin) with material to restore the shape of the live animal ▸ **stuffing** n seasoned mixture with which food is stuffed; padding

stuffy adj **stuffier, stuffiest** lacking fresh air; (informal) dull or conventional

stultifying adj very boring and repetitive

stumble v trip and nearly fall; walk in an unsure way; make frequent mistakes in speech ▷ n **stumbling** ▸ **stumble across** v discover accidentally ▸ **stumbling block** obstacle or difficulty

stump n base of a tree left when the main trunk has been cut down; part of a thing left after a larger part has been removed; (Cricket) one of the three upright sticks forming the wicket ▷ v baffle; (Cricket) dismiss (a batsman) by breaking the wicket with the ball; walk with heavy steps ▸ **stumpy** adj short and thick ▸ **stump up** v (informal) give (the money required)

stun v **stunning, stunned** shock or overwhelm; knock senseless ▸ **stunning** adj very attractive or impressive

stung v past of **sting**

stunk v past of **stink**

stunt¹ v prevent or impede the growth of ▸ **stunted** adj

stunt² n acrobatic or dangerous action; anything spectacular done to gain publicity

stupefy v **-fying, -fied** make insensitive or lethargic; astound ▸ **stupefaction** n

stupendous adj very large or impressive ▸ **stupendously** adv

stupid adj lacking intelligence; silly; in a stupor ▸ **stupidity** n ▸ **stupidly** adv

stupor n dazed or unconscious state

sturdy adj **-dier, -diest** healthy and robust; strongly built ▸ **sturdily** adv

sturgeon n fish from which caviar is obtained

stutter v speak with repetition of initial consonants ▸ n tendency to stutter

sty n, pl **sties** pen for pigs

stye, sty n, pl **styes** or **sties** inflammation at the base of an eyelash

style n shape or design; manner of writing, speaking, or doing something; elegance, refinement; prevailing fashion ▸ v shape or design; name or call ▸ **stylish** adj smart, elegant, and fashionable ▸ **stylishly** adv ▸ **stylist** n hairdresser; person who writes or performs with great attention to style ▸ **stylistic** adj of literary or artistic style ▸ **stylize** v cause to conform to an established stylistic form

stylus n needle-like device on a record player that rests in the groove of the record and picks up the sound signals

stymie v **-mieing, -mied** hinder or thwart

styptic n, adj (drug) used to stop bleeding

suave [swahv] adj smooth and sophisticated in manner ▸ **suavely** adv

sub n subeditor; submarine; subscription; substitute; (Brit informal) advance payment of wages or salary ▸ v **subbing, subbed** act as a substitute; grant advance payment to

sub- prefix indicating under or beneath e.g. submarine; indicating subordinate e.g. sublieutenant;

indicating falling short of e.g. subnormal; indicating forming a subdivision e.g. subheading

subaltern n British army officer below the rank of captain

subatomic adj of or being one of the particles which make up an atom

subcommittee n small committee formed from some members of a larger committee

subconscious adj happening or existing without one's awareness ▸ n (Psychoanalysis) that part of the mind of which one is not aware but which can influence one's behaviour ▸ **subconsciously** adv

subcontinent n large land mass that is a distinct part of a continent

subcontract n secondary contract by which the main contractor for a job puts work out to others ▸ v put out (work) on a subcontract ▸ **subcontractor** n

subcutaneous [sub-cute-**ayn**-ee-uss] adj under the skin

subdivide v divide (a part of something) into smaller parts ▸ **subdivision** n

subdue v **-duing, -dued** overcome; make less intense

subeditor n person who checks and edits text for a newspaper or magazine

subject n person or thing being dealt with or studied; (Grammar) word or phrase that represents the person or thing performing the action of the verb in a sentence; person under the rule of a monarch or government ▸ adj being under the rule of a monarch or government ▸ v (foll by to) cause to undergo **subject to** liable to; conditional upon ▸ **subjection** n ▸ **subjective** adj based on personal feelings ▸ **subjectively** adv

sub judice [sub joo-diss-ee] *adj* (Latin) before a court of law and therefore prohibited from public discussion

subjugate *v* bring (a group of people) under one's control ▸ **subjugation** *n*

subjunctive (*Grammar*) *n* mood of verbs used when the content of the clause is doubted, supposed, or wished ▸ *adj* in or of that mood

sublet *v* **-letting, -let** rent out (property rented from someone else)

sublimate *v* (*Psychol*) direct the energy of (a strong desire, esp. a sexual one) into socially acceptable activities ▸ **sublimation** *n*

sublime *adj* of high moral, intellectual, or spiritual value; unparalleled, supreme ▸ *v* (*Chem*) change from a solid to a vapour without first melting ▸ **sublimely** *adv*

subliminal *adj* relating to mental processes of which the individual is not aware

sub-machine gun *n* portable machine gun with a short barrel

submarine *n* vessel which can operate below the surface of the sea ▸ *adj* below the surface of the sea

submerge *v* put or go below the surface of water or other liquid ▸ **submersion** *n*

submit *v* **-mitting, -mitted** surrender; put forward for consideration; be (voluntarily) subjected to a process or treatment ▸ **submission** *n* submitting; something submitted for consideration; state of being submissive ▸ **submissive** *adj* meek and obedient

subordinate *adj* of lesser rank or importance ▸ *n* subordinate person or thing ▸ *v* make or treat as subordinate ▸ **subordination** *n*

suborn *v* (*formal*) bribe or incite (a person) to commit a wrongful act

subpoena [sub-pee-na] *n* writ requiring a person to appear before a lawcourt ▸ *v* summon (someone) with a subpoena

subprime *adj* (of a loan) made to a borrower with a poor credit rating e.g. *subprime mortgage* ▸ *n* such a loan

subscribe *v* pay (a subscription); give support or approval (to) ▸ **subscriber** *n* ▸ **subscription** *n* payment for issues of a publication over a period; money contributed to a charity etc.; membership fees paid to a society

subsection *n* division of a section

subsequent *adj* occurring after, succeeding ▸ **subsequently** *adv*

subservient *adj* submissive, servile ▸ **subservience** *n*

subside *v* become less intense; sink to a lower level ▸ **subsidence** *n* act or process of subsiding

subsidiary *adj* of lesser importance ▸ *n, pl* **-aries** subsidiary person or thing

subsidize *v* help financially ▸ **subsidy** *n, pl* **-dies** financial aid

subsist *v* manage to live ▸ **subsistence** *n*

subsonic *adj* moving at a speed less than that of sound

substance *n* physical composition of something; solid, powder, liquid, or paste; essential meaning of something; solid or meaningful quality; wealth ▸ **substantial** *adj* of considerable size or value; (of food or a meal) sufficient and nourishing; solid or strong; real ▸ **substantially** *adv* ▸ **substantiate** *v* support (a story) with evidence ▸ **substantiation** *n*

▶ **substantive** n noun ▷ adj of or being the essential element of a thing
substitute v take the place of or put in place of another ▷ n person or thing taking the place of (another) ▶ **substitution** n
subsume v include (an idea, case, etc.) under a larger classification or group
subterfuge n trick used to achieve an objective
subterranean adj underground
subtitle n secondary title of a book ▷ pl printed translation at the bottom of the picture in a film with foreign dialogue ▷ v provide with a subtitle or subtitles
subtle adj not immediately obvious; having or requiring ingenuity ▶ **subtly** adv ▶ **subtlety** n
subtract v take (one number) from another ▶ **subtraction** n
subtropical adj of the regions bordering on the tropics
suburb n residential area on the outskirts of a city ▶ **suburban** adj of or inhabiting a suburb; narrow or unadventurous in outlook ▶ **suburbia** n suburbs and their inhabitants
subvention n (formal) subsidy
subvert v overthrow the authority of ▶ **subversion** n ▶ **subversive** adj, n
subway n passage under a road or railway; underground railway
succeed v accomplish an aim; turn out satisfactorily; come next in order after (something); take over a position from (someone) ▶ **success** n achievement of something attempted; attainment of wealth, fame, or position; successful person or thing ▶ **successful** adj having success

▶ **successfully** adv ▶ **succession** n series of people or things following one another in order; act or right by which one person succeeds another in a position ▶ **successive** adj consecutive ▶ **successively** adv ▶ **successor** n person who succeeds someone in a position

> **SPELLING TIP**
> The Collins Corpus evidence shows that people are able to remember the double s at the end of **success** more easily than the double c in the middle

succinct adj brief and clear ▶ **succinctly** adv
succour v, n help in distress
succulent adj juicy and delicious; (of a plant) having thick fleshy leaves ▷ n succulent plant ▶ **succulence** n
succumb v (foll by to) give way (to something overpowering); die of (an illness)
such adj of the kind specified; so great, so much ▷ pron such things ▶ **such-and-such** adj specific, but not known or named ▶ **suchlike** pron such or similar things
suck v draw (liquid or air) into the mouth; take (something) into the mouth and moisten, dissolve, or roll it around with the tongue; (foll by in) draw in by irresistible force ▷ n sucking ▶ **sucker** n (slang) person who is easily deceived or swindled; organ or device which adheres by suction; shoot coming from a plant's root or the base of its main stem ▶ **suckhole** n (Aust slang) sycophant, toady ▶ **suck up to** v (informal) flatter (someone) for one's own profit
suckle v feed at the breast ▶ **suckling** n unweaned baby or young animal

sucrose [soo-kroze] n chemical name for sugar

suction n sucking; force produced by drawing air out of a space to make a vacuum that will suck in a substance from another space

sudden adj done or occurring quickly and unexpectedly **all of a sudden** quickly and unexpectedly ▸ **suddenly** adv ▸ **suddenness** n ▸ **sudden death** (Sport) period of extra time in which the first competitor to score wins

sudoku [soo-doe-koo] n logic puzzle involving the insertion of each of the numbers 1 to 9 into each row, column, and individual grid of a larger square made up of 9 3x3 grids

sudorific [syoo-dor-if-ik] n, adj (drug) causing sweating

suds pl n froth of soap and water

sue v suing, sued start legal proceedings against

suede n leather with a velvety finish on one side

suet n hard fat obtained from sheep and cattle, used in cooking

suffer v undergo or be subjected to; tolerate ▸ **sufferer** n ▸ **suffering** n ▸ **sufferance** n **on sufferance** tolerated with reluctance

suffice [suf-fice] v be enough for a purpose

sufficient adj enough, adequate ▸ **sufficiency** n adequate amount ▸ **sufficiently** adv

suffix n letter or letters added to the end of a word to form another word, such as -s and -ness in dogs and softness

suffocate v kill or be killed by deprivation of oxygen; feel uncomfortable from heat and lack of air ▸ **suffocation** n

suffragan n bishop appointed to assist an archbishop

suffrage n right to vote in public elections ▸ **suffragette** n (in Britain in the early 20th century) a woman who campaigned militantly for the right to vote

suffuse v spread through or over (something) ▸ **suffusion** n

sugar n sweet crystalline carbohydrate found in many plants and used to sweeten food and drinks ▷ v sweeten or cover with sugar ▸ **sugary** adj ▸ **sugar beet** beet grown for the sugar obtained from its roots ▸ **sugar cane** tropical grass grown for the sugar obtained from its canes ▸ **sugar daddy** (slang) older man who gives a young person money and gifts in return for sexual favours ▸ **sugar glider** common Australian phalanger that glides from tree to tree feeding on insects and nectar

suggest v put forward (an idea) for consideration; bring to mind by the association of ideas; give a hint of ▸ **suggestible** adj easily influenced ▸ **suggestion** n thing suggested; hint or indication ▸ **suggestive** adj suggesting something indecent; conveying a hint (of) ▸ **suggestively** adv

suicide n killing oneself intentionally; person who kills himself or herself intentionally; self-inflicted ruin of one's own prospects or interests ▸ **suicidal** adj liable to commit suicide ▸ **suicidally** adv

suit n set of clothes designed to be worn together; outfit worn for a specific purpose; one of the four sets into which a pack of cards is divided; lawsuit ▷ v be appropriate for; be acceptable to ▸ **suitable** adj appropriate or proper ▸ **suitably** adv ▸ **suitability** n ▸ **suitcase** n portable travelling case for clothing

suite n set of connected rooms in a hotel; matching set of furniture; set of musical pieces in the same key

suitor n (old-fashioned) man who is courting a woman

sulk v be silent and sullen because of bad temper ▷ n resentful or sullen mood ▸ **sulky** adj ▸ **sulkily** adv

sullen adj unwilling to talk ▸ **sullenly** adv ▸ **sullenness** n

sully v -lying, -lied ruin (someone's reputation); make dirty

sulphate n salt or ester of sulphuric acid

sulphide n compound of sulphur with another element

sulphite n salt or ester of sulphurous acid

sulphonamide [sulf-on-a-mide] n any of a class of drugs that prevent the growth of bacteria

sulphur n (Chem) pale yellow nonmetallic element ▸ **sulphuric**, **sulphurous** adj of or containing sulphur

sultan n sovereign of a Muslim country ▸ **sultana** n kind of raisin; sultan's wife, mother, or daughter ▸ **sultanate** n territory of a sultan

sultry adj -trier, -triest (of weather or climate) hot and humid; passionate, sensual

sum n result of addition, total; problem in arithmetic; quantity of money ▸ **sum total** complete or final total ▸ **sum up** v summing, summed summarize; form a quick opinion of

summary n, pl -ries brief account giving the main points of something ▷ adj done quickly, without formalities ▸ **summarily** adv ▸ **summarize** v make or be a summary of (something) ▸ **summation** n summary; adding up

summer n warmest season of the year, between spring and autumn ▸ **summery** adj ▸ **summerhouse** n small building in a garden ▸ **summertime** n period or season of summer

summit n top of a mountain or hill; highest point; conference between heads of state or other high officials

summon v order (someone) to come; call upon (someone) to do something; gather (one's courage, strength, etc.) ▸ **summons** n command summoning someone; order requiring someone to appear in court ▷ v order (someone) to appear in court

sumo n Japanese style of wrestling

sump n container in an internal-combustion engine into which oil can drain; hollow into which liquid drains

sumptuous adj lavish, magnificent ▸ **sumptuously** adv

sun n star around which the earth and other planets revolve; any star around which planets revolve; heat and light from the sun ▷ v **sunning, sunned** expose (oneself) to the sun's rays ▸ **sunless** adj ▸ **sunny** adj full of or exposed to sunlight; cheerful ▸ **sunbathe** v lie in the sunshine in order to get a suntan ▸ **sunbeam** n ray of sun ▸ **sunburn** n painful reddening of the skin caused by overexposure to the sun ▸ **sunburnt, sunburned** adj ▸ **sundial** n device showing the time by means of a pointer that casts a shadow on a marked dial ▸ **sundown** n sunset ▸ **sunflower** n tall plant with large golden flowers ▸ **sunrise** n daily appearance of the sun above the horizon; time of this ▸ **sunset** n daily disappearance of the sun

S

below the horizon; time of this ▶ **sunshine** n light and warmth from the sun ▶ **sunspot** n dark patch appearing temporarily on the sun's surface; (Aust) small area of skin damage caused by exposure to the sun ▶ **sunstroke** n illness caused by prolonged exposure to intensely hot sunlight ▶ **suntan** n browning of the skin caused by exposure to the sun

sundae n ice cream topped with fruit etc.

Sunday n first day of the week and the Christian day of worship ▶ **Sunday school** school for teaching children about Christianity

sundry adj several, various ▶ **sundries** pl n several things of various sorts **all and sundry** everybody

sung v past participle of **sing**

sunk v a past participle of **sink**

sunken v a past participle of **sink**

sup v **supping, supped** take (liquid) by sips ▷ n **sip**

super adj (informal) excellent

super- prefix indicating above or over e.g. superimpose; indicating outstanding e.g. superstar; indicating greater size or extent e.g. supermarket

superannuation n regular payment by an employee into a pension fund; pension paid from this ▶ **superannuated** adj discharged with a pension, owing to old age or illness

superb adj excellent, impressive, or splendid ▶ **superbly** adv

superbug n (informal) bacterium resistant to antibiotics

supercharged adj (of an engine) having a supercharger ▶ **supercharger** n device that increases the power of an internal-combustion engine by forcing extra air into it

supercilious adj showing arrogant pride or scorn

superconductor n substance which has almost no electrical resistance at very low temperatures ▶ **superconductivity** n

superficial adj not careful or thorough; (of a person) without depth of character, shallow; of or on the surface ▶ **superficially** adv ▶ **superficiality** n

superfluous [soo-per-flew-uss] adj more than is needed ▶ **superfluity** n

superfood n highly nutritious foodstuff

superhuman adj beyond normal human ability or experience

superimpose v place (something) on or over something else

superintendent n senior police officer; supervisor ▶ **superintend** v supervise (a person or activity)

superior adj greater in quality, quantity, or merit; higher in position or rank; believing oneself to be better than others ▷ n person of greater rank or status ▶ **superiority** n

superlative [soo-per-lat-iv] adj of outstanding quality; (Grammar) denoting the form of an adjective or adverb indicating most ▷ n (Grammar) superlative form of a word

superman n man with great physical or mental powers

supermarket n large self-service store selling food and household goods

supermodel n famous and highly-paid fashion model

supernatural adj of or relating to things beyond the laws of nature **the supernatural** supernatural forces, occurrences, and beings collectively

supernova n, pl **-vae** or **-vas** star that explodes and briefly becomes exceptionally bright

supernumerary adj exceeding the required or regular number ▷ n, pl **-ries** supernumerary person or thing

superpower n extremely powerful nation

superscript n, adj (character) printed above the line

supersede v replace, supplant

> **SPELLING TIP**
> Although there is a word 'cede', spelt with a c, the word **supersede** must have an s in the middle

supersize, supersized adj larger than standard size

supersonic adj of or travelling at a speed greater than the speed of sound

superstition n belief in omens, ghosts, etc.; idea or practice based on this ▶ **superstitious** adj

superstore n large supermarket

superstructure n structure erected on something else; part of a ship above the main deck

supertax n extra tax on incomes above a certain level

supervene v occur as an unexpected development

supervise v watch over to direct or check ▶ **supervision** n ▶ **supervisor** n ▶ **supervisory** adj

supine adj lying flat on one's back

supper n light evening meal

supplant v take the place of, oust

supple adj (of a person) moving and bending easily and gracefully; bending easily without damage ▶ **suppleness** n

supplement n thing added to complete something or make up for a lack; magazine inserted into a newspaper; section added to

a publication to supply further information ▷ v provide or be a supplement to (something) ▶ **supplementary** adj

supplication n humble request ▶ **supplicant** n person who makes a humble request

supply v **-plying, -plied** provide with something required ▷ n, pl **-plies** supplying; amount available; (Economics) willingness and ability to provide goods and services ▷ pl food or equipment ▶ **supplier** n

support v bear the weight of; provide the necessities of life for; give practical or emotional help to; take an active interest in (a sports team, political principle, etc.); help to prove (a theory etc.); speak in favour of ▷ n supporting; means of support ▶ **supporter** n person who supports a team, principle, etc. ▶ **supportive** adj

suppose v presume to be true; consider as a proposal for the sake of discussion ▶ **supposed** adj presumed to be true without proof, doubtful **supposed to** expected or required to e.g. you were supposed to phone me; permitted to e.g. we're not supposed to swim here ▶ **supposedly** adv ▶ **supposition** n supposing; something supposed

suppository n, pl **-ries** solid medication inserted into the rectum or vagina and left to melt

suppress v put an end to; prevent publication of (information); restrain (an emotion or response) ▶ **suppression** n

suppurate v (of a wound etc.) produce pus

supreme adj highest in authority, rank, or degree ▶ **supremely** adv extremely ▶ **supremacy** n supreme power; state of being

S

supreme ▶ **supremo** n (informal) person in overall authority

surcharge n additional charge

surd n (Maths) number that cannot be expressed in whole numbers

sure adj free from uncertainty or doubt; reliable; inevitable ▷ adv, interj (informal) certainly ▶ **surely** adv it must be true that ▶ **sure-footed** adj unlikely to slip or stumble

surety n, pl **-ties** person who takes responsibility, or thing given as a guarantee, for the fulfilment of another's obligation

surf n foam caused by waves breaking on the shore ▷ v take part in surfing; move quickly through a medium such as the internet ▶ **surfing** n sport of riding towards the shore on a surfboard on the crest of a wave ▶ **surfer** n ▶ **surfboard** n long smooth board used in surfing

surface n outside or top of an object; material covering the surface of an object; superficial appearance ▷ v rise to the surface; put a surface on

surfeit n excessive amount

surge n sudden powerful increase; strong rolling movement, esp. of the sea ▷ v increase suddenly; move forward strongly

surgeon n doctor who specializes in surgery ▶ **surgery** n, pl **-geries** treatment in which the patient's body is cut open in order to treat the affected part; place where, or time when, a doctor, dentist, etc. can be consulted; (Brit) occasion when an elected politician can be consulted ▶ **surgical** adj ▶ **surgically** adv

surly adj **-lier, -liest** ill-tempered and rude ▶ **surliness** n

surmise v, n guess, conjecture

surmount v overcome (a problem); be on top of (something) ▶ **surmountable** adj

surname n family name

surpass v be greater than or superior to

surplice n loose white robe worn by clergymen and choristers

surplus n amount left over in excess of what is required

surprise n unexpected event; amazement and wonder ▷ v cause to feel amazement or wonder; come upon, attack, or catch suddenly and unexpectedly

surrealism n movement in art and literature involving the combination of incongruous images, as in a dream ▶ **surreal** adj bizarre ▶ **surrealist** n, adj ▶ **surrealistic** adj

surrender v give oneself up; give (something) up to another; yield (to a temptation or influence) ▷ n surrendering

surreptitious adj done secretly or stealthily ▶ **surreptitiously** adv

surrogate n substitute ▶ **surrogate mother** woman who gives birth to a child on behalf of a couple who cannot have children

surround v be, come, or place all around (a person or thing) ▷ n border or edging ▶ **surroundings** pl n area or environment around a person, place, or thing

surveillance n close observation

survey v view or consider in a general way; make a map of (an area); inspect (a building) to assess its condition and value; find out the incomes, opinions, etc. of (a group of people) ▷ n surveying; report produced by a survey ▶ **surveyor** n

survive v continue to live or exist after (a difficult experience); live after the death of (another)

▶ **survival** n condition of having survived ▶ **survivor** n

susceptible adj liable to be influenced or affected by ▶ **susceptibility** n

sushi [soo-shee] n Japanese dish of small cakes of cold rice with a topping of raw fish

suspect v believe (someone) to be guilty without having any proof; think (something) to be false or questionable; believe (something) to be the case ▷ adj not to be trusted ▷ n person who is suspected

suspend v hang from a high place; cause to remain floating or hanging; cause to cease temporarily; remove (someone) temporarily from a job or team ▶ **suspenders** pl n straps for holding up stockings; (US) braces

suspense n state of uncertainty while awaiting news, an event, etc.

suspension n suspending or being suspended; system of springs and shock absorbers supporting the body of a vehicle; mixture of fine particles of a solid in a fluid ▶ **suspension bridge** bridge hanging from cables attached to towers at each end

suspicion n feeling of not trusting a person or thing; belief that something is true without definite proof; slight trace ▶ **suspicious** adj feeling or causing suspicion ▶ **suspiciously** adv

suss out v (Brit slang) work out using one's intuition

sustain v maintain or prolong; keep up the vitality or strength of; suffer (an injury or loss); support ▶ **sustenance** n food

suture [soo-cher] n stitch joining the edges of a wound

SUV sport (or sports) utility vehicle: a powerful car with four-wheel drive, designed for road and off-road use

suzerain n state or sovereign with limited authority over another self-governing state ▶ **suzerainty** n

svelte adj attractively or gracefully slim

SW southwest(ern)

swab n small piece of cotton wool used to apply medication, clean a wound, etc. ▷ v **swabbing, swabbed** clean (a wound) with a swab; clean (the deck of a ship) with a mop

swaddle v wrap (a baby) in swaddling clothes ▶ **swaddling clothes** long strips of cloth formerly wrapped round a newborn baby

swag n (slang) stolen property ▶ **swagman** n (Aust Hist) tramp who carries his belongings in a bundle on his back

swagger v walk or behave arrogantly ▷ n arrogant walk or manner

swain n (poetic) suitor; country youth

swallow¹ v cause to pass down one's throat; make a gulping movement in the throat, as when nervous; (informal) believe (something) gullibly; refrain from showing (a feeling); engulf or absorb ▷ n swallowing; amount swallowed

swallow² n small migratory bird with long pointed wings and a forked tail

swam v past tense of **swim**

swamp n watery area of land, bog ▷ v cause (a boat) to fill with water and sink; overwhelm ▶ **swampy** adj

swan n large usu. white water bird with a long graceful neck ▷ v **swanning, swanned** (informal) wander aimlessly ▸ **swan song** person's last performance before retirement or death

swank (slang) v show off or boast ▸ n showing off or boasting ▸ **swanky** ▸ adj (slang) expensive and showy, stylish

swanndri® [swan-dry] n (NZ) weatherproof woollen shirt or jacket (also **swannie**)

swap v **swapping, swapped** exchange (something) for something else ▸ n exchange

sward n stretch of short grass

swarm¹ n large group of bees or other insects; large crowd ▸ v move in a swarm; (of a place) be crowded or overrun

swarm² v (foll by up) climb (a ladder or rope) by gripping with the hands and feet

swarthy adj **-thier, -thiest** dark-complexioned

swashbuckling adj having the exciting behaviour of pirates, esp. those depicted in films ▸ **swashbuckler** n

swastika n symbol in the shape of a cross with the arms bent at right angles, used as the emblem of Nazi Germany

swat v **swatting, swatted** hit sharply ▸ n sharp blow

swatch n sample of cloth

swath [swawth] n see **swathe**

swathe v wrap in bandages or layers of cloth ▸ n long strip of cloth wrapped around something; (also **swath**) the width of one sweep of a scythe or mower

sway v swing to and fro or from side to side; waver or cause to waver in opinion ▸ n power or influence; swaying motion

swear v **swearing, swore, sworn** use obscene or blasphemous language; state or promise on

oath; state earnestly ▸ **swear by** v have complete confidence in ▸ **swear in** v cause to take an oath ▸ **swearword** n word considered obscene or blasphemous

sweat n salty liquid given off through the pores of the skin; (slang) drudgery or hard labour ▸ v have sweat coming through the pores; be anxious ▸ **sweaty** adj ▸ **sweatband** n strip of cloth tied around the forehead or wrist to absorb sweat ▸ **sweatshirt** n long-sleeved cotton jersey ▸ **sweatshop** n place where employees work long hours in poor conditions for low pay

sweater n (woollen) garment for the upper part of the body

swede n kind of turnip; (**S-**) person from Sweden ▸ **Swedish** n, adj (language) of Sweden

sweep v **sweeping, swept** remove dirt from (a floor) with a broom; move smoothly and quickly; spread rapidly; move majestically; carry away suddenly or forcefully; stretch in a long wide curve ▸ n sweeping; sweeping motion; wide expanse; sweepstake; chimney sweep ▸ **sweeping** adj wide-ranging; indiscriminate ▸ **sweepstake** n lottery in which the stakes of the participants make up the prize

sweet adj tasting of or like sugar; kind and charming; agreeable to the senses or mind; (of wine) with a high sugar content ▸ n shaped piece of food consisting mainly of sugar; dessert ▸ **sweetly** adv ▸ **sweetness** n ▸ **sweeten** v ▸ **sweetener** n sweetening agent that does not contain sugar; (Brit, Aust & NZ slang) bribe ▸ **sweetbread** n animal's pancreas used as food

▶ **sweet corn** type of maize with sweet yellow kernels, eaten as a vegetable ▶ **sweetheart** n lover ▶ **sweetmeat** n (old-fashioned) sweet delicacy such as a small cake ▶ **sweet pea** climbing plant with bright fragrant flowers ▶ **sweet potato** tropical root vegetable with yellow flesh ▶ **sweet-talk** v (informal) coax or flatter ▶ **sweet tooth** strong liking for sweet foods

swell v **swelling, swelled, swollen** or **swelled** expand or increase; (of a sound) become gradually louder ▷ n swelling or being swollen; movement of waves in the sea; (old-fashioned slang) fashionable person ▷ adj (US slang) excellent or fine ▶ **swelling** n enlargement of part of the body, caused by injury or infection

swelter v feel uncomfortably hot
sweltering adj uncomfortably hot
swept v past of **sweep**
swerve v turn aside from a course sharply or suddenly ▷ n swerving
swift adj moving or able to move quickly ▷ n fast-flying bird with pointed wings ▶ **swiftly** adv ▶ **swiftness** n

swig n large mouthful of drink ▷ v **swigging, swigged** drink in large mouthfuls

swill v drink greedily; rinse (something) in large amounts of water ▷ n sloppy mixture containing waste food, fed to pigs; deep drink

swim v **swimming, swam, swum** move along in water by movements of the limbs; be covered or flooded with liquid; reel e.g. *her head was swimming* ▷ n act or period of swimming ▶ **swimmer** n ▶ **swimmingly** adv successfully and effortlessly ▶ **swimming pool** (building containing) an artificial pond for swimming in

swindle v cheat (someone) out of money ▷ n instance of swindling ▶ **swindler** n

swine n contemptible person; pig ▶ **swine flu** n influenza which affects pigs, or human influenza caused by a related virus

swing v **swinging, swung** move to and fro, sway; move in a curve; (of an opinion or mood) change sharply; hit out with a sweeping motion; (slang) be hanged ▷ n swinging; suspended seat on which a child can swing to and fro; sudden or extreme change ▶ **swing by** v (informal) go somewhere to pay a visit

swingeing [swin-jing] adj punishing, severe

swipe v strike (at) with a sweeping blow; (slang) steal; pass (a credit card or debit card) through a machine that electronically reads information stored in the card; activate by moving one's finger across (an item on a screen) ▷ n hard blow ▶ **swipe card** credit or debit card that is passed through a machine that electronically reads information stored in the card

swirl v turn with a whirling motion ▷ n whirling motion; twisting shape

swish v move with a whistling or hissing sound ▷ n whistling or hissing sound ▷ adj (informal) fashionable, smart

Swiss adj of Switzerland or its people ▷ n, pl **Swiss** person from Switzerland ▶ **swiss roll** sponge cake spread with jam or cream and rolled up

switch n device for opening and closing an electric circuit; abrupt change; exchange or swap; flexible rod or twig ▷ v change abruptly; exchange or swap ▶ **switchback** n road or railway with many sharp bends or hills ▶ **switchboard**

S

n installation in a telephone exchange or office where telephone calls are connected ▸ **switch on, switch off** *v* turn (a device) on or off by means of a switch

swivel *v* -elling, -elled turn on a central point ▷ *n* coupling device that allows an attached object to turn freely

swizzle stick *n* small stick used to stir cocktails

swollen *v* a past participle of **swell**

swoon *v, n* faint

swoop *v* sweep down or pounce on suddenly ▸ *n* swooping

swop *v* swopping, swopped ▸ *n* same as **swap**

sword *n* weapon with a long sharp blade ▸ **swordfish** *n* large fish with a very long upper jaw ▸ **swordsman** *n* person skilled in the use of a sword

swore *v* past tense of **swear**

sworn *v* past participle of **swear** ▷ *adj* bound by or as if by an oath e.g. *sworn enemies*

swot (*informal*) *v* swotting, swotted study hard ▷ *n* person who studies hard

swum *v* past participle of **swim**

swung *v* past tense and past participle of **swing**

sybarite [sib-bar-ite] *n* lover of luxury ▸ **sybaritic** *adj*

sycamore *n* tree with five-pointed leaves and two-winged fruits

sycophant *n* person who uses flattery to win favour from people with power or influence ▸ **sycophantic** *adj* ▸ **sycophancy** *n*

syllable *n* part of a word pronounced as a unit ▸ **syllabic** *adj*

syllabub *n* dessert of beaten cream, sugar, and wine

syllabus *n, pl* -buses *or* -bi list of subjects for a course of study

syllogism *n* form of logical reasoning consisting of two premises and a conclusion

sylph *n* slender graceful girl or woman; imaginary being supposed to inhabit the air ▸ **sylphlike** *adj*

sylvan *adj* (*lit*) relating to woods and trees

symbiosis *n* close association of two species living together to their mutual benefit ▸ **symbiotic** *adj*

symbol *n* sign or thing that stands for something else ▸ **symbolic** *adj* ▸ **symbolically** *adv* ▸ **symbolism** *n* representation of something by symbols; movement in art and literature using symbols to express abstract and mystical ideas ▸ **symbolist** *n, adj* ▸ **symbolize** *v* be a symbol of; represent with a symbol

symmetry *n* state of having two halves that are mirror images of each other ▸ **symmetrical** *adj* ▸ **symmetrically** *adv*

sympathy *n, pl* -thies compassion for someone's pain or distress; agreement with someone's feelings or interests ▸ **sympathetic** *adj* feeling or showing sympathy; likeable or appealing ▸ **sympathetically** *adv* ▸ **sympathize** *v* feel or express sympathy ▸ **sympathizer** *n*

symphony *n, pl* -nies composition for orchestra, with several movements ▸ **symphonic** *adj*

symposium *n, pl* -siums *or* -sia conference for discussion of a particular topic

symptom *n* sign indicating the presence of an illness; sign that something is wrong ▸ **symptomatic** *adj*

synagogue *n* Jewish place of worship and religious instruction

sync, synch (*informal*) *n* synchronization ▷ *v* synchronize

synchromesh adj (of a gearbox) having a device that synchronizes the speeds of gears before they engage

synchronize v (of two or more people) perform (an action) at the same time; set (watches) to show the same time; match (the soundtrack and action of a film) precisely ▸ **synchronization** n ▸ **synchronous** adj happening or existing at the same time

syncopate v (Music) stress the weak beats in (a rhythm) instead of the strong ones ▸ **syncopation** n

syncope [sing-kop-ee] n (Med) a faint

syndicate n group of people or firms undertaking a joint business project; agency that sells material to several newspapers; association of individuals who control organized crime ▷ v publish (material) in several newspapers; form a syndicate ▸ **syndication** n

syndrome n combination of symptoms indicating a particular disease; set of characteristics indicating a particular problem

synergy n potential ability for people or groups to be more successful working together than on their own

synod n church council

synonym n word with the same meaning as another ▸ **synonymous** adj

synopsis n, pl **-ses** summary or outline

syntax n (Grammar) way in which words are arranged to form phrases and sentences ▸ **syntactic** adj

synthesis n, pl **-ses** combination of objects or ideas into a whole; artificial production of a substance

▸ **synthesize** v produce by synthesis ▸ **synthesizer** n electronic musical instrument producing a range of sounds ▸ **synthetic** adj (of a substance) made artificially; not genuine, insincere ▸ **synthetically** adv

syphilis n serious sexually transmitted disease ▸ **syphilitic** adj

syphon n, v same as **siphon**

Syrian adj of Syria, its people, or their dialect of Arabic ▷ n person from Syria

syringe n device for withdrawing or injecting fluids, consisting of a hollow cylinder, a piston, and a hollow needle ▷ v wash out or inject with a syringe

syrup n solution of sugar in water; thick sweet liquid ▸ **syrupy** adj

system n method or set of methods; scheme of classification or arrangement; network or assembly of parts that form a whole ▸ **systematic** adj ▸ **systematically** adv ▸ **systematize** v organize using a system ▸ **systematization** n ▸ **systemic** adj affecting the entire animal or body

systole [siss-tol-ee] n regular contraction of the heart as it pumps blood ▸ **systolic** adj

S

Tt

T *n* **to a T** in every detail; perfectly

t tonne

t. ton

TA (in Britain) Territorial Army

ta *interj* (informal) thank you

TAB (in New Zealand) Totalisator Agency Board

tab *n* small flap or projecting label **keep tabs on** (informal) watch closely

tabard *n* short sleeveless tunic decorated with a coat of arms, worn in medieval times

Tabasco® *n* very hot red pepper sauce

tabby *n, pl* **-bies** ▷ *adj* (cat) with dark stripes on a lighter background

tabernacle *n* portable shrine of the Israelites; Christian place of worship not called a church; (*RC Church*) receptacle for the consecrated Host

tabla *n, pl* **-bla** *or* **-blas** one of a pair of Indian drums played with the hands

table *n* piece of furniture with a flat top supported by legs; arrangement of information in columns ▷ *v* submit (a motion) for discussion by a meeting; (*US*) suspend discussion of (a proposal) ▶ **table football** *n* game like soccer played on a table with sets of miniature figures on rods allowing them to be moved to hit a ball ▶ **tableland** *n* high plateau ▶ **tablespoon** *n* large spoon for serving food ▶ **table tennis** game

like tennis played on a table with small bats and a light ball

tableau [tab-loh] *n, pl* **-leaux** silent motionless group arranged to represent some scene

table d'hôte [tah-bla **dote**] *n, pl* **tables d'hôte** ▷ *adj* (meal) having a set number of dishes at a fixed price

tablet *n* pill of compressed medicinal substance; inscribed slab of stone etc.; handheld computer operated by touching a screen

tabloid *n* small-sized newspaper with many photographs and a concise, usu. sensational style

taboo *n, pl* **-boos** prohibition resulting from religious or social conventions ▷ *adj* forbidden by a taboo

tabular *adj* arranged in a table ▶ **tabulate** *v* arrange (information) in a table ▶ **tabulation** *n*

tachograph *n* device for recording the speed and distance travelled by a motor vehicle

tachometer *n* device for measuring speed, esp. that of a revolving shaft

tacit [tass-it] *adj* implied but not spoken ▶ **tacitly** *adv*

taciturn [tass-it-turn] *adj* habitually uncommunicative ▶ **taciturnity** *n*

tack¹ *n* short nail with a large head; long loose stitch ▷ *v* fasten with tacks; stitch with tacks ▶ **tack on** *v* append

tack² *n* course of a ship sailing obliquely into the wind; course of action ▷ *v* sail into the wind on a zigzag course

tack³ *n* riding harness for horses

tackies, takkies *pl n, sing* **tacky** (*S Afr informal*) tennis shoes or plimsolls

tackle *v* deal with (a task); confront (an opponent); (*Sport*) attempt to get the ball from (an opposing player) ▷ n (Sport) act of tackling an opposing player; equipment for a particular activity; set of ropes and pulleys for lifting heavy weights

tacky¹ adj **tackier, tackiest** slightly sticky

tacky² adj **tackier, tackiest** (informal) vulgar and tasteless; shabby

taco [tah-koh] n, pl **tacos** (Mexican cookery) tortilla fried until crisp, served with a filling

tact n skill in avoiding giving offence ▶ **tactful** adj ▶ **tactfully** adv ▶ **tactless** adj ▶ **tactlessly** adv

tactics n art of directing military forces in battle ▶ **tactic** n method or plan to achieve an end ▶ **tactical** adj ▶ **tactician** n

tactile adj of or having the sense of touch

tadpole n limbless tailed larva of a frog or toad

TAFE (Aust) Technical and Further Education

taffeta n shiny silk or rayon fabric

tag¹ n label bearing information; pointed end of a cord or lace; trite quotation ▷ v **tagging, tagged** attach a tag to ▶ **tag along** v accompany someone, esp. if uninvited

tag² n children's game where the person being chased becomes the chaser upon being touched ▷ v **tagging, tagged** touch and catch in this game

tagliatelle n pasta in long narrow strips

tail n rear part of an animal's body, usu. forming a flexible appendage; rear or last part or parts of something; (informal) person employed to follow and spy on another ▷ pl (informal) tail coat ▷ adj at the rear ▷ v (informal)

follow (someone) secretly **turn tail** run away ▶ **tailless** adj ▶ **tails** adv with the side of a coin uppermost that does not have a portrait of a head on it ▶ **tailback** n (Brit) queue of traffic stretching back from an obstruction ▶ **tailboard** n removable or hinged rear board on a truck etc. ▶ **tail coat** man's coat with a long back split into two below the waist ▶ **tail off, tail away** v diminish gradually ▶ **tailplane** n small stabilizing wing at the rear of an aircraft ▶ **tailspin** n uncontrolled spinning dive of an aircraft ▶ **tailwind** n wind coming from the rear

tailor n person who makes men's clothes ▷ v adapt to suit a purpose ▶ **tailor-made** adj made by a tailor; perfect for a purpose

taint v spoil with a small amount of decay, contamination, or other bad quality ▷ n something that taints

taipan n large poisonous Australian snake

take v **taking, took, taken** remove from a place; carry or accompany; use; get possession of, esp. dishonestly; capture; require (time, resources, or ability); assume; accept ▷ n one of a series of recordings from which the best will be used **take place** happen ▶ **taking** adj charming ▶ **takings** pl n money received by a shop ▶ **take after** v look or behave like (a parent etc.) ▶ **take away** v remove or subtract ▶ **takeaway** n shop or restaurant selling meals for eating elsewhere; meal bought at a takeaway ▶ **take in** v understand; deceive or swindle; make (clothing) smaller ▶ **take off** v (of an aircraft) leave the ground; (informal) depart; (informal) parody ▶ **takeoff** n ▶ **takeover** n act of taking control

t

of a company by buying a large number of its shares ▸ **take up** occupy or fill (space or time); adopt the study or activity of; shorten (a garment); accept (an offer)

talc n talcum powder; soft mineral of magnesium silicate ▸ **talcum powder** powder, usu. scented, used to dry or perfume the body

tale n story; malicious piece of gossip

talent n natural ability; ancient unit of weight or money ▸ **talented** adj

talisman n, pl -**mans** object believed to have magic power ▸ **talismanic** adj

talk v express ideas or feelings by means of speech; utter; discuss e.g. *let's talk business*; reveal information; (be able to) speak in a specified language ▷ n speech or lecture ▸ **talker** n ▸ **talkative** adj fond of talking ▸ **talk back** v answer impudently ▸ **talkback** n (NZ) broadcast in which telephone comments or questions from the public are transmitted live ▸ **talking-to** n (informal) telling-off

tall adj higher than average; of a specified height ▸ **tall order** difficult task ▸ **tall story** unlikely and probably untrue tale

tallboy n high chest of drawers

tallow n hard animal fat used to make candles

tally v -**lying, -lied** (of two things) correspond ▷ n, pl -**lies** record of a debt or score

tally-ho interj huntsman's cry when the quarry is sighted

Talmud n body of Jewish law ▸ **Talmudic** adj

talon n bird's hooked claw

tamarind n tropical tree; its acid fruit

tamarisk n evergreen shrub with slender branches and feathery flower clusters

tambourine n percussion instrument like a small drum with jingling metal discs attached

tame adj (of animals) brought under human control; (of animals) not afraid of people; meek or submissive; uninteresting ▷ v make tame ▸ **tamely** adv

tamer n person who tames wild animals

Tamil n member of a people of Sri Lanka and S India; their language

tam-o'-shanter n brimless wool cap with a bobble in the centre

tamp v pack down by repeated taps

tamper v (foll by with) interfere

tampon n absorbent plug of cotton wool inserted into the vagina during menstruation

tan¹ n brown coloration of the skin from exposure to sunlight ▷ v **tanning, tanned** (of skin) go brown from exposure to sunlight; convert (a hide) into leather ▷ adj yellowish-brown ▸ **tannery** n place where hides are tanned

tan² (maths) tangent

tandem n bicycle for two riders, one behind the other **in tandem** together

tandoori adj (of food) cooked in an Indian clay oven

tang n strong taste or smell; trace or hint ▸ **tangy** adj

tangata whenua [tang-ah-tah fen-noo-ah] pl n (NZ) original Polynesian settlers in New Zealand

tangent n line that touches a curve without intersecting it; (in trigonometry) ratio of the length of the opposite side to that of the adjacent side of a right-angled triangle **go off at a tangent** suddenly take a completely different line of thought or action ▸ **tangential** adj of superficial relevance only; of a tangent ▶ **tangentially** adv

tangerine n small orange-like fruit of an Asian citrus tree

tangible adj able to be touched; clear and definite ▶ **tangibly** adv

tangle n confused mass or situation ▷ v twist together in a tangle; (often foll by with) come into conflict

tango n, pl **-gos** S American dance ▷ v dance a tango

taniwha [tun-ee-fah] n (NZ) mythical Māori monster that lives in water

tank n container for liquids or gases; armoured fighting vehicle moving on tracks ▶ **tanker** n ship or truck for carrying liquid in bulk

tankard n large beer-mug, often with a hinged lid

tannin, tannic acid n vegetable substance used in tanning

Tannoy® n (Brit) type of public-address system

tansy n, pl **-sies** yellow-flowered plant

tantalize v torment by showing but withholding something desired ▶ **tantalizing** adj ▶ **tantalizingly** adv

tantalum n (Chem) hard greyish-white metallic element

tantamount adj **tantamount to** equivalent in effect to

tantrum n childish outburst of temper

tap[1] v **tapping, tapped** knock lightly and usu. repeatedly ▷ n light knock ▶ **tap dancing** style of dancing in which the feet beat out an elaborate rhythm

tap[2] n valve to control the flow of liquid from a pipe or cask ▷ v **tapping, tapped** listen in on (a telephone call) secretly by making an illegal connection; draw off with or as if with a tap **on tap** (informal)

readily available; (of beer etc.) drawn from a cask

tape n narrow long strip of material; (recording made on) a cassette containing magnetic tape; string stretched across a race track to mark the finish ▷ v record on magnetic tape; bind or fasten with tape ▶ **tape measure** tape marked off in centimetres or inches for measuring ▶ **tape recorder** device for recording and reproducing sound on magnetic tape ▶ **tapeworm** n long flat parasitic worm living in the intestines of vertebrates

taper v become narrower towards one end ▷ n long thin candle ▶ **taper off** v become gradually less

tapestry n, pl **-tries** fabric decorated with coloured woven designs

tapioca n beadlike starch made from cassava root, used in puddings

tapir [tape-er] n piglike mammal of tropical America and SE Asia, with a long snout

tappet n short steel rod in an engine, transferring motion from one part to another

taproot n main root of a plant, growing straight down

tar n thick black liquid distilled from coal etc. ▷ v **tarring, tarred** coat with tar ▶ **tar-seal** n (NZ) tarred road surface

taramasalata n creamy pink pâté made from fish roe

tarantella n lively Italian dance; music for this

tarantula n large hairy spider with a poisonous bite

tardy adj **tardier, tardiest** slow or late ▶ **tardily** adv ▶ **tardiness** n

tare n type of vetch plant; (Bible) weed

target n object or person a missile is aimed at; goal or objective; object of criticism ▷ v **-geting, -geted** aim or direct

> **SPELLING TIP**
> Note that **targeting** and **targeted** have only one t in the middle

tariff n tax levied on imports; list of fixed prices

Tarmac® n mixture of tar, bitumen, and crushed stones used for roads etc.; **(t-)** airport runway

tarn n small mountain lake

tarnish v make or become stained or less bright; damage or taint ▷ n discoloration or blemish

tarot [**tarr-oh**] n special pack of cards used mainly in fortune-telling ▶ **tarot card** card in a tarot pack

tarpaulin n (sheet of) heavy waterproof fabric

tarragon n aromatic herb

tarry v **-rying, -ried** (old-fashioned) linger or delay; stay briefly

tarsus n, pl **-si** bones of the heel and ankle collectively

tart¹ n pie or flan with a sweet filling

tart² adj sharp or bitter ▶ **tartly** adv ▶ **tartness** n

tart³ n (informal) sexually provocative or promiscuous woman ▶ **tart up** v (informal) dress or decorate in a smart or flashy way

tartan n design of straight lines crossing at right angles, esp. one associated with a Scottish clan; cloth with such a pattern

tartar¹ n hard deposit on the teeth; deposit formed during the fermentation of wine

tartar² n fearsome or formidable person

tartare sauce n mayonnaise sauce mixed with chopped herbs and capers, served with seafood

tartrazine [tar-**traz**-zeen] n artificial yellow dye used in food etc.

TAS Tasmania

task n (difficult or unpleasant) piece of work to be done **take to task** criticize or scold ▶ **task force** (military) group formed to carry out a specific task ▶ **taskmaster** n person who enforces hard work

Tasmanian n, adj (person) from Tasmania ▶ **Tasmanian devil** small carnivorous Tasmanian marsupial ▶ **Tasmanian tiger** same as **thylacine**

tassel n decorative fringed knot of threads

taste n sense by which the flavour of a substance is distinguished in the mouth; distinctive flavour; small amount tasted; brief experience of something; liking; ability to appreciate what is beautiful or excellent ▷ v distinguish the taste of (a substance); take a small amount of (something) into the mouth; have a specific taste; experience briefly ▶ **tasteful** adj having or showing good taste ▶ **tastefully** adv ▶ **tasteless** adj bland or insipid; showing bad taste ▶ **tastelessly** adv ▶ **tasty** adj pleasantly flavoured ▶ **taste bud** small organ on the tongue which perceives flavours

tat n (Brit) tatty or tasteless article(s)

tattered adj ragged or torn **in tatters** in ragged pieces

tattle v, n (Brit, Aust & NZ) gossip or chatter

tattoo¹ n pattern made on the body by pricking the skin and staining it with indelible inks ▷ v **-tooing, -tooed** make such a pattern on the skin ▶ **tattooist** n

tattoo² n military display or pageant; drumming or tapping

tatty adj **-tier, -tiest** shabby or worn out

taught v past of **teach**

taunt v tease with jeers ▷ n jeering remark

taupe adj brownish-grey

taut adj drawn tight; showing nervous strain ▶ **tauten** v make or become taut

tautology n, pl **-gies** use of words which merely repeat something already stated ▶ **tautological** adj

tavern n (old-fashioned) pub

tawdry adj **-drier, -driest** cheap, showy, and of poor quality

tawny adj **-nier, -niest** yellowish-brown

tax n compulsory payment levied by a government on income, property, etc. to raise revenue ▷ v levy a tax on; make heavy demands on ▶ **taxable** adj ▶ **taxation** n levying of taxes ▶ **tax-free** adj (of goods, services and income) not taxed ▶ **taxpayer** n person who pays income tax ▶ **tax relief** reduction in the amount of tax a person or company has to pay ▶ **tax return** statement of personal income for tax purposes

taxi n (also **taxicab**) car with a driver that may be hired to take people to any specified destination ▷ v **taxiing, taxied** (of an aircraft) run along the ground before taking off or after landing ▶ **taxi meter** meter in a taxi that registers the fare ▶ **taxi rank** place where taxis wait to be hired

taxidermy n art of stuffing and mounting animal skins to give them a lifelike appearance ▶ **taxidermist** n

taxonomy n classification of plants and animals into groups ▶ **taxonomic** adj ▶ **taxonomist** n

TB tuberculosis

tba, TBA to be arranged

tbc to be confirmed

tbs., tbsp. tablespoon(ful)

tea n drink made from infusing the dried leaves of an Asian bush in boiling water; leaves used to make this drink; (Brit, Aust & NZ) main evening meal; (chiefly Brit) light afternoon meal of tea, cakes, etc.; drink like tea, made from other plants ▶ **tea bag** small porous bag of tea leaves ▶ **tea cosy** covering for a teapot to keep the tea warm ▶ **teapot** n container with a lid, spout, and handle for making and serving tea ▶ **teaspoon** n small spoon for stirring tea ▶ **tea towel, tea cloth** towel for drying dishes ▶ **tea tree** tree of Australia and New Zealand that yields an oil used as an antiseptic

teach v **teaching, taught** tell or show (someone) how to do something; give lessons in (a subject); cause to learn or understand ▶ **teaching** n

teacher n person who teaches, esp. in a school

teak n very hard wood of an E Indian tree

teal n kind of small duck

team n group of people forming one side in a game; group of people or animals working together ▶ **teamster** n (US) commercial vehicle driver ▶ **team up** v make or join a team ▶ **teamwork** n cooperative work by a team

tear¹, teardrop n drop of fluid appearing in and falling from the eye **in tears** weeping ▶ **tearful** adj weeping or about to weep ▶ **tear gas** gas that stings the eyes and causes temporary blindness ▶ **tear-jerker** n (informal) excessively sentimental film or book

tear² v **tearing, tore, torn** rip a hole in; rip apart; rush ▷ n hole or split ▶ **tearaway** n wild or unruly person

tease v make fun of (someone) in a provoking or playful way ▷ n person who teases ▸ **teasing** adj, ▸ **tease out** v remove tangles from (hair etc.) by combing

teasel, teazel, teazle n plant with prickly leaves and flowers

teat n nipple of a breast or udder; rubber nipple of a feeding bottle

tech n (informal) technical college

techie (informal) n person who is skilled in the use of technology ▷ adj relating to or skilled in the use of technology

technetium [tek-neesh-ee-um] n (Chem) artificially produced silvery-grey metallic element

technical adj of or specializing in industrial, practical, or mechanical arts and applied sciences; skilled in technical subjects; relating to a particular field; according to the letter of the law; showing technique e.g. technical brilliance ▸ **technically** adv ▸ **technicality** n petty point based on a strict application of rules ▸ **technician** n person skilled in a particular technical field ▸ **technical college** higher educational institution with courses in art and technical subjects

Technicolor® n system of colour photography used for the cinema

technique n method or skill used for a particular task; technical proficiency

techno n type of electronic dance music with a very fast beat

technocracy n, pl -**cies** government by technical experts ▸ **technocrat** n

technology n application of practical or mechanical sciences to industry or commerce; scientific methods used in a particular field ▸ **technological** adj ▸ **technologist** n

tectonics n study of the earth's crust and the forces affecting it

teddy n, pl -**dies** teddy bear; combined camisole and knickers ▸ **teddy bear** soft toy bear

tedious adj causing fatigue or boredom ▸ **tediously** adv ▸ **tedium** n monotony

tee n small peg from which a golf ball can be played at the start of each hole; area of a golf course from which the first stroke of a hole is made ▸ **tee off** v make the first stroke of a hole in golf

teem¹ v be full of

teem² v rain heavily

teenager n person aged between 13 and 19 ▸ **teenage** adj

teens pl n period of being a teenager

teepee n same as **tepee**

tee-shirt n same as **T-shirt**

teeter v wobble or move unsteadily

teeth n plural of **tooth**

teethe v (of a baby) grow his or her first teeth ▸ **teething troubles** problems during the early stages of something

teetotal adj drinking no alcohol ▸ **teetotaller** n

TEFL Teaching of English as a Foreign Language

Teflon® n substance used for nonstick coatings on saucepans etc.

tele- combining form distance e.g. telecommunications; telephone or television e.g. teleconference

telecommunications n communications using telephone, radio, television, etc.

telegram n formerly, a message sent by telegraph

telegraph n formerly, a system for sending messages over a distance along a cable ▷ v communicate by telegraph ▸ **telegraphic** adj

▶**telegraphist** n ▶**telegraphy** n science or use of a telegraph

telekinesis n movement of objects by thought or willpower

telemetry n use of electronic devices to record or measure a distant event and transmit the data to a receiver

teleology n belief that all things have a predetermined purpose ▶**teleological** adj

telepathy n direct communication between minds ▶**telepathic** adj ▶**telepathically** adv

telephone n device for transmitting sound over a distance along wires ▷ v call or talk to (a person) by telephone ▶**telephony** n ▶**telephonic** adj ▶**telephonist** n person operating a telephone switchboard

telephoto lens n camera lens producing a magnified image of a distant object

teleprinter n (Brit) apparatus like a typewriter for sending and receiving typed messages by wire

telesales n selling of a product or service by telephone

telescope n optical instrument for magnifying distant objects ▷ v shorten ▶**telescopic** adj

television n system of producing a moving image and accompanying sound on a distant screen; device for receiving broadcast signals and converting them into sound and pictures; content of television programmes ▶**televise** v broadcast on television ▶**televisual** adj

telex n international communication service using teleprinters; message sent by telex ▷ v transmit by telex

tell v telling, told make known in words; order or instruct; give an

account of; discern or distinguish; have an effect; (informal) reveal secrets ▶**teller** n narrator; bank cashier; person who counts votes ▶**telling** adj having a marked effect ▶**tell off** v reprimand ▶**telling-off** n ▶**telltale** n person who reveals secrets ▷ adj revealing

tellurium n (Chem) brittle silvery-white nonmetallic element

telly n, pl -lies (informal) television

temerity [tim-merr-it-tee] n boldness or audacity

temp (Brit informal) n temporary employee, esp. a secretary ▷ v work as a temp

temp. temperature; temporary

temper n outburst of anger; tendency to become angry; calm mental condition e.g. I lost my temper; frame of mind ▷ v make less extreme; strengthen or toughen (metal)

tempera n painting medium for powdered pigments

temperament n person's character or disposition ▶**temperamental** adj having changeable moods; (informal) erratic and unreliable ▶**temperamentally** adv

temperate adj (of climate) not extreme; self-restrained or moderate ▶**temperance** n moderation; abstinence from alcohol

temperature n degree of heat or cold; (informal) abnormally high body temperature

tempest n violent storm ▶**tempestuous** adj violent or stormy

template n pattern used to cut out shapes accurately

temple¹ n building for worship

temple² n region on either side of the forehead ▶**temporal** adj

tempo n, pl **-pi** or **-pos** rate or pace; speed of a piece of music

temporal adj of time; worldly rather than spiritual

temporary adj lasting only for a short time ▸ **temporarily** adv

temporize v gain time by negotiation or evasiveness; adapt to circumstances

tempt v entice (a person) to do something wrong **tempt fate** take foolish or unnecessary risks ▸ **tempter,** (fem) **temptress** n ▸ **temptation** n tempting; tempting thing ▸ **tempting** adj attractive or inviting

ten adj, n one more than nine ▸ **tenth** adj, n (of) number ten in a series

tenable adj able to be upheld or maintained

tenacious adj holding fast; stubborn ▸ **tenaciously** adv ▸ **tenacity** n

tenant n person who rents land or a building ▸ **tenancy** n

tench n, pl **tench** freshwater game fish of the carp family

tend¹ v be inclined; go in the direction of ▸ **tendency** n inclination to act in a certain way ▸ **tendentious** adj biased, not impartial

tend² v take care of

tender¹ adj not tough; gentle and affectionate; vulnerable or sensitive ▸ **tenderly** adv ▸ **tenderness** n ▸ **tenderize** v soften (meat) by pounding or treatment with a special substance

tender² v offer; make a formal offer to supply goods or services at a stated cost ▸ n such an offer **legal tender** currency that must, by law, be accepted as payment

tender³ n small boat that brings supplies to a larger ship in a port; carriage for fuel and water attached to a steam locomotive

tendon n strong tissue attaching a muscle to a bone

tendril n slender stem by which a climbing plant clings

tenement n (esp. in Scotland and the US) building divided into several flats

tenet [ten-nit] n doctrine or belief

tenner n (Brit informal) ten-pound note

tennis n game in which players use rackets to hit a ball back and forth over a net

tenon n projecting end on a piece of wood fitting into a slot in another

tenor n (singer with) the second highest male voice; general meaning ▸ adj (of a voice or instrument) between alto and baritone

tenpin bowling n game in which players try to knock over ten skittles by rolling a ball at them

tense¹ adj emotionally strained; stretched tight ▷ v make or become tense

tense² n (Grammar) form of a verb showing the time of action

tensile adj of tension

tension n hostility or suspense; emotional strain; degree of stretching

tent n portable canvas shelter

tentacle n flexible organ of many invertebrates, used for grasping, feeding, etc.

tentative adj provisional or experimental; cautious or hesitant ▸ **tentatively** adv

tenterhooks pl n **on tenterhooks** in anxious suspense

tenuous adj slight or flimsy ▸ **tenuously** adv

tenure n (period of) the holding of an office or position

tepee [tee-pee] n cone-shaped tent, formerly used by Native Americans

tepid adj slightly warm; half-hearted

tequila n Mexican alcoholic drink

tercentenary adj, n, pl **-naries** (of) a three hundredth anniversary

term n word or expression; fixed period; period of the year when a school etc. is open or a lawcourt holds sessions ▷ pl conditions; mutual relationship ▷ v name or designate

terminal adj (of an illness) ending in death; at or being an end ▷ n place where people or vehicles begin or end a journey; point where current enters or leaves an electrical device; keyboard and VDU having input and output links with a computer ▸ **terminally** adv

terminate v bring or come to an end ▸ **termination** n bringing to an end; abortion

terminology n technical terms relating to a subject

terminus n, pl **-ni** or **-nuses** railway or bus station at the end of a line

termite n white ant-like insect that destroys timber

tern n gull-like sea bird with a forked tail and pointed wings

ternary adj consisting of three parts

Terpsichorean adj of dancing

terrace n row of houses built as one block; paved area next to a building; level tier cut out of a hill ▷ pl (also **terracing**) tiered area in a stadium where spectators stand ▷ v form into or provide with a terrace

terracotta adj, n (made of) brownish-red unglazed pottery ▷ adj brownish-red

terra firma n (Latin) dry land or solid ground

terrain n area of ground, esp. with reference to its physical character

terrapin n small turtle-like reptile

terrarium n, pl **-raria** or **-rariums** enclosed container for small plants or animals

terrazzo n, pl **-zos** floor of marble chips set in mortar and polished

terrestrial adj of the earth; of or living on land

terrible adj very serious; (informal) very bad; causing fear ▸ **terribly** adv

terrier n any of various breeds of small active dog

terrific adj great or intense; (informal) excellent

terrify v **-fying, -fied** fill with fear ▸ **terrified** adj ▸ **terrifying** adj

terrine [terr-reen] n earthenware dish with a lid; pâté or similar food

territory n, pl **-ries** district; area under the control of a particular government; area inhabited and defended by an animal; area of knowledge ▸ **territorial** adj ▸ **Territorial Army** (in Britain) reserve army

terror n great fear; terrifying person or thing; (Brit, Aust & NZ informal) troublesome person or thing ▸ **terrorism** n use of violence and intimidation to achieve political ends ▸ **terrorist** n, adj ▸ **terrorize** v force or oppress by fear or violence

terry n fabric with small loops covering both sides, used esp. for making towels

terse adj neat and concise; curt ▸ **tersely** adv

tertiary [tur-shar-ee] adj third in degree, order, etc.

Terylene® n synthetic polyester yarn or fabric

tessellated adj paved or inlaid with a mosaic of small tiles

test v try out to ascertain the worth, capability, or endurance of; carry out an examination on ▷ *n* critical examination; Test match ▶ **testing** *adj* ▶ **test case** lawsuit that establishes a precedent ▶ **Test match** one of a series of international cricket or rugby matches ▶ **test tube** narrow round-bottomed glass tube used in scientific experiments ▶ **test-tube baby** baby conceived outside the mother's body

testament *n* proof or tribute; (*Law*) will; (**T-**) one of the two main divisions of the Bible

testator [test-**tay**-tor], (*fem*) **testatrix** [test-**tay**-triks] *n* maker of a will

testicle *n* either of the two male reproductive glands

testify v -**fying**, -**fied** give evidence under oath ▶ **testify to** be evidence of

testimony *n*, *pl* -**nies** declaration of truth or fact; evidence given under oath ▶ **testimonial** *n* recommendation of the worth of a person or thing; tribute for services or achievement

testis *n*, *pl* -**tes** testicle

testosterone *n* male sex hormone secreted by the testes

testy *adj* -**tier**, -**tiest** irritable or touchy ▶ **testily** *adv* ▶ **testiness** *n*

tetanus *n* acute infectious disease producing muscular spasms and convulsions

tête-à-tête *n*, *pl* -**têtes** or -**tête** private conversation

tether *n* rope or chain for tying an animal to a spot ▷ *v* tie up with rope **at the end of one's tether** at the limit of one's endurance

tetrahedron [tet-ra-**heed**-ron] *n*, *pl* -**drons** or -**dra** solid figure with four faces

tetralogy *n*, *pl* -**gies** series of four related works

Teutonic [tew-**tonn**-ik] *adj* of or like the (ancient) Germans

text *n* main body of a book as distinct from illustrations etc.; passage of the Bible as the subject of a sermon; novel or play studied for a course; text message ▷ *v* send a text message to (someone) ▶ **textual** *adj* ▶ **textbook** *n* standard book on a particular subject ▷ *adj* perfect e.g. *a textbook landing* ▶ **text message** message sent in text form, esp. by means of a mobile phone

textile *n* fabric or cloth, esp. woven

texture *n* structure, feel, or consistency ▶ **textured** *adj* ▶ **textural** *adj*

Thai *adj* of Thailand ▷ *n* (*pl* **Thais** or **Thai**) person from Thailand; language of Thailand

thalidomide [thal-**lid**-oh-mide] *n* drug formerly used as a sedative, but found to cause abnormalities in developing fetuses

thallium *n* (*Chem*) highly toxic metallic element

than *conj*, *prep* used to introduce the second element of a comparison

thane *n* (*Hist*) Anglo-Saxon or medieval Scottish nobleman

thank v express gratitude to; hold responsible ▶ **thanks** *pl n* words of gratitude ▷ *interj* (also **thank you**) polite expression of gratitude **thanks to** because of ▶ **thankful** *adj* grateful ▶ **thankless** *adj* unrewarding or unappreciated ▶ **Thanksgiving (Day)** autumn public holiday in Canada and the US

that *adj*, *pron* used to refer to something already mentioned or familiar, or further away ▷ *conj* used to introduce a clause ▷ *pron* used to introduce a relative clause

thatch n roofing material of reeds or straw ▷ v roof (a house) with reeds or straw

thaw v make or become unfrozen; become more relaxed or friendly ▷ n thawing; weather causing snow or ice to melt

the adj the definite article, used before a noun

theatre n place where plays etc. are performed; hospital operating room; drama and acting in general ▶ **theatrical** adj of the theatre; exaggerated or affected ▶ **theatricals** pl n (amateur) dramatic performances ▶ **theatrically** adv ▶ **theatricality** n

thee pron (obs) objective form of **thou**

theft n act or an instance of stealing

their adj of or associated with them ▶ **theirs** pron (thing or person) belonging to them

USAGE NOTE
Note the difference between their and there. Their is used for possession: their new baby. There indicates place and has a similar '-ere' spelling pattern to here and where. Also, do not confuse their and theirs, which do not have apostrophes, with they're and there's, which do have them (because letters have been missed out where two words have been joined together).

theism [thee-iz-zum] n belief in a God or gods ▶ **theist** n, adj ▶ **theistic** adj

them pron refers to people or things other than the speaker or those addressed ▶ **themselves** pron emphatic and reflexive form of **they, them**

USAGE NOTE
Them may be used after a singular to avoid the clumsy him or her: If you see a person looking lost, help them

theme n main idea or subject being discussed; recurring melodic figure in music ▶ **thematic** adj ▶ **theme park** leisure area in which all the activities and displays are based on a single theme

then adv at that time; after that; that being so

thence adv (old-fashioned) from that place or time; therefore

theocracy n, pl -cies government by a god or priests ▶ **theocratic** adj

theodolite [thee-odd-oh-lite] n surveying instrument for measuring angles

theology n, pl -gies study of religions and religious beliefs ▶ **theologian** n ▶ **theological** adj ▶ **theologically** adv

theorem n proposition that can be proved by reasoning

theory n, pl -ries set of ideas to explain something; abstract knowledge or reasoning; idea or opinion **in theory** in an ideal or hypothetical situation ▶ **theoretical** adj based on theory rather than practice or fact ▶ **theoretically** adv ▶ **theorist** n ▶ **theorize** v form theories, speculate

theosophy n religious or philosophical system claiming to be based on intuitive insight into the divine nature ▶ **theosophical** adj

therapy n, pl -pies curing treatment ▶ **therapist** n ▶ **therapeutic** [ther-rap-pew-tik] adj curing ▶ **therapeutics** n art of curing

there adv in or to that place; in that respect ▶ **thereby** adv by that means ▶ **therefore** adv

t

consequently, that being so ► **thereupon** *adv* immediately after that

therm *n* unit of measurement of heat

thermal *adj* of heat; hot or warm; (of clothing) retaining heat ▷ *n* rising current of warm air

thermodynamics *n* scientific study of the relationship between heat and other forms of energy

thermometer *n* instrument for measuring temperature

thermonuclear *adj* involving nuclear fusion

thermoplastic *adj* (of a plastic) softening when heated and resetting on cooling

Thermos® *n* vacuum flask

thermosetting *adj* (of a plastic) remaining hard when heated

thermostat *n* device for automatically regulating temperature ► **thermostatic** *adj* ► **thermostatically** *adv*

thesaurus [thiss-sore-uss] *n, pl* **-ruses** book containing lists of synonyms and related words

these *adj, pron* plural of **this**

thesis *n, pl* **theses** written work submitted for a degree; opinion supported by reasoned argument

Thespian *n* actor or actress ▷ *adj* of the theatre

they *pron* refers to people or things other than the speaker or people addressed; refers to people in general; (informal) refers to he or she

If a person is born gloomy, they cannot help it

thiamine *n* vitamin found in the outer coat of rice and other grains

thick *adj* of great or specified extent from one side to the other; having a dense consistency; (informal) stupid or insensitive; (Brit, Aust & NZ informal) friendly **a bit thick** (informal) unfair or unreasonable **the thick** busiest or most intense part **thick with** full of ► **thicken** *v* make or become thick or thicker ► **thickly** *adv* ► **thickness** *n* state of being thick; dimension through an object; layer ► **thickset** *adj* stocky in build

thicket *n* dense growth of small trees

thief *n, pl* **thieves** person who steals ► **thieve** *v* steal ► **thieving** *adj*

thigh *n* upper part of the human leg

thimble *n* cap used to protect the end of the finger when sewing

thin *adj* **thinner, thinnest** not thick; slim or lean; sparse or meagre; of low density; poor or unconvincing ▷ *v* **thinning, thinned** make or become thin ► **thinly** *adv* ► **thinness** *n*

thine *pron, adj* (obs) (something) of or associated with you (thou)

thing *n* material object; object, fact, or idea considered as a separate entity; (informal) obsession ▷ *pl* possessions, clothes, etc.

think *v* **thinking, thought** consider, judge, or believe; make use of the mind; be considerate enough or remember to do something ► **thinker** *n* ► **thinking** *adj, n* ► **think-tank** *n* group of experts studying specific problems ► **think up** *v* invent or devise

third adj of number three in a series; rated or graded below the second level ▷ n one of three equal parts ▶ **third degree** violent interrogation ▶ **third party** (applying to) a person involved by chance or only incidentally in legal proceedings, an accident, etc. ▶ **Third World** developing countries of Africa, Asia, and Latin America

thirst n desire to drink; craving or yearning ▷ v feel thirst ▶ **thirsty** adj ▶ **thirstily** adv

thirteen adj, n three plus ten ▶ **thirteenth** adj, n

thirty adj, n three times ten ▶ **thirtieth** adj, n

this adj, pron used to refer to a thing or person nearby, just mentioned, or about to be mentioned ▷ adj used to refer to the present time e.g. this morning

thistle n prickly plant with dense flower heads

thither adv (obs) to or towards that place

thong n thin strip of leather etc.; skimpy article of underwear or beachwear that covers the genitals while leaving the buttocks bare

thorax n, pl **thoraxes** or **thoraces** part of the body between the neck and the abdomen ▶ **thoracic** adj

thorn n prickle on a plant; bush with thorns **thorn in one's side**, **thorn in one's flesh** source of irritation ▶ **thorny** adj

thorough adj complete; careful or methodical ▶ **thoroughly** adv ▶ **thoroughness** n ▶ **thoroughbred** n, adj (animal) of pure breed ▶ **thoroughfare** n way through from one place to another

those adj, pron plural of **that**

thou pron (obs) singular form of **you**

though conj despite the fact that ▷ adv nevertheless

thought v past of **think** ▷ n thinking; concept or idea; ideas typical of a time or place; consideration; intention or expectation ▶ **thoughtful** adj considerate; showing careful thought; pensive or reflective ▶ **thoughtless** adj inconsiderate

thousand adj, n ten hundred; large but unspecified number ▶ **thousandth** adj, n (of) number one thousand in a series

thrall n state of being in the power of another person

thrash v beat, esp. with a stick or whip; defeat soundly; move about wildly; thresh ▶ **thrashing** n severe beating ▶ **thrash out** v solve by thorough argument

thread n fine strand or yarn; unifying theme; spiral ridge on a screw, nut, or bolt ▷ v pass thread through; pick (one's way etc.) ▶ **threadbare** adj (of fabric) with the nap worn off; hackneyed; shabby

threat n declaration of intent to harm; dangerous person or thing ▶ **threaten** v make or be a threat to; be a menacing indication of

three adj, n one more than two ▶ **threesome** n group of three ▶ **three-dimensional, 3-D** adj having three dimensions

threnody n, pl **-dies** lament for the dead

thresh v beat (wheat etc.) to separate the grain from the husks and straw ▶ **thresh about** move about wildly

threshold n bar forming the bottom of a doorway; entrance; starting point; point at which something begins to take effect

threw v past tense of **throw**

t

thrice adv (lit) three times

thrift n wisdom and caution with money; low-growing plant with pink flowers ▶ **thrifty** adj

thrill n sudden feeling of excitement ▷ v (cause to) feel a thrill ▶ **thrilling** adj

thriller n book, film, etc. with an atmosphere of mystery or suspense

thrive v **thriving, thrived** or **throve, thrived** or **thriven** flourish or prosper; grow well

throat n passage from the mouth and nose to the stomach and lungs; front of the neck ▶ **throaty** adj (of the voice) hoarse

throb v **throbbing, throbbed** pulsate repeatedly; vibrate rhythmically ▶ n **throbbing**

throes pl n violent pangs or pains **in the throes of** struggling to cope with

thrombosis n, pl -**ses** forming of a clot in a blood vessel or the heart

throne n ceremonial seat of a monarch or bishop; sovereign power

throng n, v crowd

throstle n song thrush

throttle n device controlling the amount of fuel entering an engine ▷ v strangle

through prep from end to end or side to side of; because of; during ▷ adj finished; (of transport) going directly to a place **through and through** completely ▶ **throughout** prep, adv in every part (of) ▶ **throughput** n amount of material processed

throve v a past tense of **thrive**

throw v **throwing, threw, thrown** hurl through the air; move or put suddenly or carelessly; bring into a specified state, esp. suddenly; give (a party); (informal) baffle or disconcert ▷ n throwing; distance

thrown ▶ **throwaway** adj done or said casually; designed to be discarded after use ▶ **throwback** n person or thing that reverts to an earlier type ▶ **throw up** v vomit

thrush¹ n brown songbird

thrush² n fungal disease of the mouth or vagina

thrust v **thrusting, thrust** push forcefully ▷ n forceful stab; force or power; intellectual or emotional drive

thud n dull heavy sound ▷ v **thudding, thudded** make such a sound

thug n violent man, esp. a criminal ▶ **thuggery** n ▶ **thuggish** adj

thumb n short thick finger set apart from the others ▷ v touch or handle with the thumb; signal with the thumb for a lift in a vehicle ▶ **thumb through** flick through (a book or magazine)

thump n (sound of) a dull heavy blow ▷ v strike heavily

thunder n loud noise accompanying lightning ▷ v rumble with thunder; shout; move fast, heavily, and noisily ▶ **thunderous** adj ▶ **thundery** adj ▶ **thunderbolt** n lightning flash; something sudden and unexpected ▶ **thunderclap** n peal of thunder ▶ **thunderstruck** adj amazed

Thursday n fifth day of the week

thus adv therefore; in this way

thwack v, n whack

thwart v foil or frustrate ▷ n seat across a boat

thy adj (obs) of or associated with you (thou) ▶ **thyself** pron (obs) emphatic form of **thou**

thylacine n extinct doglike Tasmanian marsupial

thyme [time] n aromatic herb

thymus n, pl -**muses** or -**mi** small gland at the base of the neck

thyroid adj, n (of) a gland in the neck controlling body growth

tiara n semicircular jewelled headdress

tibia n, pl **tibiae** or **tibias** inner bone of the lower leg ▶ **tibial** adj

tic n spasmodic muscular twitch

tick¹ n mark (✓) used to check off or indicate the correctness of something; recurrent tapping sound, as of a clock; (informal) moment ▷ v mark with a tick; make a ticking sound ▶ **tick off** v mark with a tick; reprimand ▶ **tick over** v (of an engine) idle; function smoothly ▶ **ticktack** n (Brit) bookmakers' sign language

tick² n tiny bloodsucking parasitic animal

tick³ n (informal) credit or account

ticket n card or paper entitling the holder to admission, travel, etc.; label, esp. showing price; official notification of a parking or traffic offence; (chiefly US & NZ) declared policy of a political party ▷ v -**eting**, -**eted** attach or issue a ticket to

ticking n strong material for mattress covers

tickle v touch or stroke (a person) to produce laughter; itch or tingle; please or amuse ▷ n tickling ▶ **ticklish** adj sensitive to tickling; requiring care or tact

tiddler n (informal) very small fish

tiddly¹ adj -**dlier**, -**dliest** tiny

tiddly² adj -**dlier**, -**dliest** (informal) slightly drunk

tiddlywinks n game in which players try to flip small plastic discs into a cup

tide n rise and fall of the sea caused by the gravitational pull of the sun and moon; current caused by this; widespread feeling or tendency ▶ **tidal** adj ▶ **tidal wave** large

destructive wave ▶ **tide over** v help (someone) temporarily

tidings pl n news

tidy adj -**dier**, -**diest** neat and orderly; (Brit, Aust & NZ informal) considerable ▷ v -**dying**, -**died** put in order ▶ **tidily** adv ▶ **tidiness** n

tie v **tying**, **tied** fasten or be fastened with string, rope, etc.; make (a knot or bow) in (something); restrict or limit; score the same as another competitor ▷ n long narrow piece of material worn knotted round the neck; bond or fastening; drawn game or contest ▶ **tied** adj (Brit) (of a cottage etc.) rented to the tenant only as long as he or she is employed by the owner

tier n one of a set of rows placed one above and behind the other

tiff n petty quarrel

tiger n large yellow-and-black striped Asian cat ▶ **tiger snake** n highly venomous brown-and-yellow Australian snake ▶ **tigress** n female tiger; (informal) fierce woman

tight adj stretched or drawn taut; closely fitting; secure or firm; cramped; (Brit, Aust & NZ informal) not generous; (of a match or game) very close; (informal) drunk ▶ **tights** pl n one-piece clinging garment covering the body from the waist to the feet ▶ **tightly** adv ▶ **tighten** v make or become tight or tighter ▶ **tightrope** n rope stretched taut on which acrobats perform

tiki n (NZ) small carving of a grotesque person worn as a pendant

tikka adj (Indian cookery) marinated in spices and dry-roasted e.g. chicken tikka

tilde n mark (~) used in Spanish to indicate that the letter 'n' is to be pronounced in a particular way

t

tile n flat piece of ceramic, plastic, etc. used to cover a roof, floor, or wall ▷ v cover with tiles ▸ **tiled** adj ▸ **tiling** n tiles collectively

till¹ conj, prep until

till² v cultivate (land) ▸ **tillage** n

till³ n drawer for money, usu. in a cash register

tiller n lever to move a rudder of a boat

tilt v slant at an angle; (Hist) compete against in a jousting contest ▷ n slope; (Hist) jousting contest; attempt **at full tilt** at full speed or force

timber n wood as a building material; trees collectively; wooden beam in the frame of a house, boat, etc. ▸ **timbered** adj ▸ **timber line** limit beyond which trees will not grow

timbre [**tam**-bra] n distinctive quality of sound of a voice or instrument

time n past, present, and future as a continuous whole; specific point in time; unspecified interval; instance or occasion; period with specific features; musical tempo; (Brit, Aust & NZ slang) imprisonment ▷ v note the time taken by; choose a time for ▸ **timeless** adj unaffected by time; eternal ▸ **timely** adj at the appropriate time ▸ **time-honoured** adj sanctioned by custom ▸ **time-lag** n period between cause and effect ▸ **timepiece** n watch or clock ▸ **time-poor** adj having little free time ▸ **timeserver** n person who changes his or her views to gain support or favour ▸ **time sharing** system of part ownership of a holiday property for a specified period each year ▸ **timetable** n plan showing the times when something takes place, e.g. departure and arrival times of trains or buses, etc.

timid adj easily frightened; shy, not bold ▸ **timidly** adv ▸ **timidity** n ▸ **timorous** adj timid

timpani [**tim**-pan-ee] pl n set of kettledrums ▸ **timpanist** n

tin n soft metallic element; (airtight) metal container ▸ **tinned** adj (of food) preserved by being sealed in a tin ▸ **tinny** adj (of sound) thin and metallic ▸ **tinpot** adj (informal) worthless or unimportant

tincture n medicinal extract in a solution of alcohol

tinder n dry easily-burning material used to start a fire ▸ **tinderbox** n formerly, small box for tinder, esp. one fitted with a flint and steel

tine n prong of a fork or antler

ting n high metallic sound, as of a small bell

tinge n slight tint; trace ▷ v tingeing, tinged give a slight tint or trace to

tingle v, n (feel) a prickling or stinging sensation

tinker n travelling mender of pots and pans; (Scot & Irish derogatory) Gypsy ▷ v fiddle with (an engine etc.) in an attempt to repair it

tinkle v ring with a high tinny sound like a small bell ▷ n this sound or action

tinsel n decorative metallic strips or threads

tint n (pale) shade of a colour; dye for the hair ▷ v give a tint to

tiny adj tinier, tiniest very small

tip¹ n narrow or pointed end of anything; small piece forming an end ▷ v tipping, tipped put a tip on

tip² n money given in return for service; helpful hint or warning; piece of inside information ▷ v

tipping, tipped give a tip to
▶ **tipster** n person who sells tips about races

tip³ v **tipping, tipped** tilt or overturn; dump (rubbish) ▷ n rubbish dump ▶ **tipping point** moment or event that marks a decisive change

tipple v drink alcohol habitually, esp. in small quantities ▷ n alcoholic drink ▶ **tippler** n

tipsy adj **-sier, -siest** slightly drunk

tiptoe v **-toeing, -toed** walk quietly with the heels off the ground

tiptop adj of the highest quality or condition

tirade n long angry speech

tire v reduce the energy of, as by exertion; weary or bore ▶ **tired** adj exhausted; hackneyed or stale ▶ **tiring** adj ▶ **tireless** adj energetic and determined ▶ **tiresome** adj boring and irritating

tissue n substance of an animal body or plant; piece of thin soft paper used as a handkerchief etc.; interwoven threads

tit¹ n any of various small songbirds

tit² n (slang) female breast

titanic adj huge or very important

titanium n (Chem) strong light metallic element used to make alloys

titbit n tasty piece of food; pleasing scrap of scandal

tit-for-tat adj done in retaliation

tithe n esp. formerly, one tenth of one's income or produce paid to the church as a tax

Titian [tish-an] adj (of hair) reddish-gold

titillate v excite or stimulate pleasurably ▶ **titillating** adj ▶ **titillation** n

titivate v smarten up

title n name of a book, film, etc.; name signifying rank or position;

formal designation, such as Mrs; (Sport) championship; (Law) legal right of possession ▶ **titled** adj aristocratic ▶ **title deed** legal document of ownership

titter v laugh in a suppressed way ▷ n suppressed laugh

tittle-tattle n, v gossip

titular adj in name only; of a title

tizzy n, pl **-zies** (informal) confused or agitated state

TNT n trinitrotoluene, a powerful explosive

to prep indicating movement towards, equality or comparison, etc. e.g. walking to school; forty miles to the gallon; used to mark the indirect object or infinitive of a verb ▷ adv to a closed position e.g. pull the door to ▶ **to and fro** back and forth

toad n animal like a large frog

toad-in-the-hole n (Brit) sausages baked in batter

toadstool n poisonous fungus like a mushroom

toady n, pl **toadies** ingratiating person ▷ v **toadying, toadied** be ingratiating

toast¹ n sliced bread browned by heat ▷ v brown (bread) by heat; warm or be warmed ▶ **toaster** n electrical device for toasting bread

toast² n tribute or proposal of health or success marked by people raising glasses and drinking together; person or thing so honoured ▷ v drink a toast to

tobacco n, pl **-cos or -coes** plant with large leaves dried for smoking ▶ **tobacconist** n person or shop selling tobacco, cigarettes, etc.

toboggan n narrow sledge for sliding over snow ▷ v **-ganing, -ganed** ride a toboggan

toby jug n (chiefly Brit) mug in the form of a stout seated man

toccata [tok-kah-ta] n rapid piece of music for a keyboard instrument

today n this day; the present age ▷ adv on this day; nowadays

toddler n child beginning to walk ▶ **toddle** v walk with short unsteady steps

toddy n, pl -**dies** sweetened drink of spirits and hot water

to-do n, pl -**dos** (Brit, Aust & NZ) fuss or commotion

toe n digit of the foot; part of a shoe or sock covering the toes ▷ v **toeing, toed** touch or kick with the toe **toe the line** conform

toff n (Brit slang) well-dressed or upper-class person

toffee n chewy sweet made of boiled sugar

tofu n soft food made from soya-bean curd

tog n unit for measuring the insulating power of duvets

toga [toe-ga] n garment worn by citizens of ancient Rome

together adv in company; simultaneously ▷ adj (informal) organized

> **USAGE NOTE**
> Two nouns linked by together with do not make a plural subject so the following verb must be singular: Jones, together with his partner, has had great success

toggle n small bar-shaped button inserted through a loop for fastening; switch used to turn a machine or computer function on or off

toil n hard work ▷ v work hard; progress with difficulty

toilet n (room with) a bowl connected to a drain for receiving and disposing of urine and faeces; washing and dressing ▶ **toiletry** n, pl -**ries** object or cosmetic used to clean or groom oneself ▶ **toilet water** light perfume

token n sign or symbol; voucher exchangeable for goods of a specified value; disc used as money in a slot machine ▷ adj nominal or slight ▶ **tokenism** n policy of making only a token effort, esp. to comply with a law

told v past of **tell**

tolerate v allow to exist or happen; endure patiently ▶ **tolerable** adj bearable; (informal) quite good ▶ **tolerably** adv ▶ **tolerance** n acceptance of other people's rights to their own opinions or actions; ability to endure something ▶ **tolerant** adj ▶ **tolerantly** adv ▶ **toleration** n

toll v ring (a bell) slowly and regularly, esp. to announce a death ▷ n tolling

toll n charge for the use of a bridge or road; total loss or damage from a disaster

tom n male cat

tomahawk n fighting axe of the Native Americans

tomato n, pl -**toes** red fruit used in salads and as a vegetable

tomb n grave; monument over a grave ▶ **tombstone** n gravestone

tombola n lottery with tickets drawn from a revolving drum

tomboy n girl who acts or dresses like a boy

tome n large heavy book

tomfoolery n foolish behaviour

Tommy gun n light sub-machine-gun

tomorrow adv, n (on) the day after today; (in) the future

tom-tom n drum beaten with the hands

ton n unit of weight equal to 2240 pounds or 1016 kilograms (**long ton**) or, in the US, 2000 pounds or 907 kilograms (**short ton**) ▸ **tonnage** n weight capacity of a ship

tone n sound with reference to its pitch, volume, etc.; (US) musical note; (Music) (also **whole tone**) interval of two semitones; quality of a sound or colour; general character; healthy bodily condition ▸ v harmonize (with); give tone to; give more firmness or strength to (the body or a part of the body) ▸ **tonal** adj (Music) written in a key ▸ **tonality** n ▸ **toneless** adj ▸ **tone-deaf** adj unable to perceive subtle differences in pitch ▸ **tone down** v make or become more moderate

tongs pl n pincers for picking up (pieces of coal, sugar, etc)

tongue n muscular organ in the mouth, used in speaking and tasting; language; animal tongue as food; thin projecting strip; flap of leather on a shoe

tonic n medicine to improve body tone ▸ adj invigorating ▸ **tonic water** mineral water containing quinine

tonight adv, n (in or during) the night or evening of this day

tonne [tunn] n unit of weight equal to 1000 kilograms

tonsil n small gland in the throat ▸ **tonsillectomy** n surgical removal of the tonsils ▸ **tonsillitis** n inflammation of the tonsils

tonsure n shaving of all or the top of the head as a religious practice; shaved part of the head ▸ **tonsured** adj

too adv also, as well; to excess; extremely

took v past tense of **take**

tool n implement used by hand; person used by another to perform unpleasant or dishonourable tasks ▸ **toolbar** n (Computing) row or column of selectable buttons on a computer screen, each allowing the user to access a particular function ▸ **toolie** n (Aust slang) older adult who gate-crashes the schoolies week celebrations

Toorak tractor n (Aust derogatory) luxury four-wheel drive or sport utility vehicle

toot n short hooting sound ▸ v (cause to) make such a sound

tooth n, pl **teeth** bonelike projection in the jaws of most vertebrates for biting and chewing; toothlike prong or point **sweet tooth** strong liking for sweet food ▸ **toothless** adj ▸ **toothpaste** n paste used to clean the teeth ▸ **toothpick** n small stick for removing scraps of food from between the teeth

top[1] n highest point or part; lid or cap; highest rank; garment for the upper part of the body ▸ adj at or of the top ▸ v **topping**, **topped** form a top on; be at the top of; exceed or surpass ▸ **topping** n sauce or garnish for food ▸ **topless** adj (of a costume or woman) with no covering for the breasts ▸ **topmost** adj highest or best ▸ **top brass** most important officers or leaders ▸ **top hat** man's tall cylindrical hat ▸ **top-heavy** adj unstable through being overloaded at the top ▸ **top-notch** adj excellent, first-class ▸ **topsoil** n surface layer of soil

top[2] n toy which spins on a pointed base

topaz [toe-pazz] n semiprecious stone in various colours

topee, topi [toe-pee] n lightweight hat worn in tropical countries

topiary [tope-yar-ee] n art of trimming trees and bushes into decorative shapes

topic n subject of a conversation, book, etc. ▶ **topical** adj relating to current events ▶ **topicality** n

topography n, pl **-phies** (science of describing) the surface features of a place ▶ **topographer** n ▶ **topographical** adj

topology n geometry of the properties of the shape which are unaffected by continuous distortion ▶ **topological** adj

topple v (cause to) fall over; overthrow (a government etc.)

topsy-turvy adj upside down; in confusion

toque [toke] n small round hat

tor n high rocky hill

Torah n body of traditional Jewish teaching

torch n small portable battery-powered lamp; wooden shaft dipped in wax and set alight ▷ v (informal) deliberately set (a building) on fire

tore v past tense of **tear²**

toreador [torr-ee-a-dor] n bullfighter

torment v cause (someone) great suffering; tease cruelly ▷ n great suffering; source of suffering ▶ **tormentor** n

torn v past participle of **tear²**

tornado n, pl **-dos** or **-does** violent whirlwind

torpedo n, pl **-does** self-propelled underwater missile ▷ v **-doing, -doed** attack or destroy with or as if with torpedoes

torpid adj sluggish and inactive ▶ **torpor** n torpid state

torque [tork] n force causing rotation; Celtic necklace or armband of twisted metal

torrent n rushing stream; rapid flow of questions, abuse, etc.; (Computing) file that controls the transfer of data in a BitTorrent system ▶ **torrential** adj (of rain) very heavy

torrid adj very hot and dry; highly emotional

torsion n twisting of a part by equal forces being applied at both ends but in opposite directions

torso n, pl **-sos** trunk of the human body; statue of a nude human trunk

tort n (Law) civil wrong or injury for which damages may be claimed

tortilla n thin Mexican pancake

tortoise n slow-moving land reptile with a dome-shaped shell ▶ **tortoiseshell** n mottled brown shell of a turtle, used for making ornaments ▷ adj having brown, orange, and black markings

tortuous adj winding or twisting; not straightforward

torture v cause (someone) severe pain or mental anguish ▷ n severe physical or mental pain; torturing ▶ **torturer** n

Tory n, pl **Tories** member of the Conservative Party in Great Britain or Canada ▷ adj of Tories ▶ **Toryism** n

toss v throw lightly; fling or be flung about; coat (food) by gentle stirring or mixing; (of a horse) throw (its rider); throw up (a coin) to decide between alternatives by guessing which side will land uppermost ▷ n tossing ▶ **toss up** v toss a coin ▶ **toss-up** n even chance or risk

tot¹ n small child; small drink of spirits

tot² v **totting, totted** ▶ **tot up** add (numbers) together

total n whole, esp. a sum of parts ▷ adj complete; of or being a total ▷ v **-talling, -talled** amount to; add up ▶ **totally** adv ▶ **totality** n ▶ **totalizator** n betting system in which money is paid out in proportion to the winners' stakes

totalitarian adj of a dictatorial one-party government ▸ **totalitarianism** n

tote¹ v carry (a gun etc.)

tote² n short for **totalizator**

totem n tribal badge or emblem ▸ **totem pole** post carved or painted with totems by Native Americans

totter v move unsteadily; be about to fall

toucan n tropical American bird with a large bill

touch v come into contact with; tap, feel, or stroke; affect; move emotionally; eat or drink; equal or match; (Brit, Aust & NZ slang) ask for money ▸ n sense by which an object's qualities are perceived when they come into contact with part of the body; gentle tap, push, or caress; small amount; characteristic style; detail ▸ adj of a non-contact version of particular sport e.g. touch rugby ▸ **touch and go** risky or critical ▸ **touched** emotionally moved; slightly mad ▸ **touching** adj emotionally moving ▸ **touchy** adj easily offended ▸ **touchless** adj (of a device) able to be controlled without touching a keypad or screen ▸ **touch down** v (of an aircraft) land ▸ **touchline** n side line of the pitch in some games ▸ **touch on** v refer to in passing ▸ **touch-type** v type without looking at the keyboard

touché [too-**shay**] interj acknowledgment of the striking home of a remark or witty reply

touchstone n standard by which a judgment is made

tough adj strong or resilient; difficult to chew or cut; firm and determined; rough and violent; difficult; (informal) unlucky or unfair

▸ n (informal) rough violent person ▸ **toughness** n ▸ **toughen** v make or become tough or tougher

toupee [too-**pay**] n small wig

tour n journey visiting places of interest along the way; trip to perform or play in different places ▸ v make a tour (of) ▸ **tourism** n tourist travel as an industry ▸ **tourist** n person travelling for pleasure ▸ **touristy** adj (informal often derogatory) full of tourists or tourist attractions

tour de force n, pl **tours de force** (French) brilliant achievement

tournament n sporting competition with several stages to decide the overall winner; (Hist) contest between knights on horseback

tourniquet [**toor**-nick-kay] n something twisted round a limb to stop bleeding

tousled adj ruffled and untidy

tout [rhymes with **shout**] v seek business in a persistent manner; recommend (a person or thing) ▸ n person who sells tickets for a popular event at inflated prices

tow¹ v drag, esp. by means of a rope ▸ n tow rope ▸ **on tow** being towed closely behind ▸ **towbar** n metal bar on a car for towing vehicles ▸ **towpath** n path beside a canal or river, originally for horses towing boats

tow² n fibre of hemp or flax

towards, toward prep in the direction of; with regard to; as a contribution to

towel n cloth for drying things ▸ **towelling** n material used for making towels

tower n tall structure, often forming part of a larger building ▸ **tower of strength** person who supports or comforts ▸ **tower over** v be much taller than

t

town n group of buildings larger than a village; central part of this; people of a town ▶ **town hall** large building used for council meetings, etc. ▶ **township** n small town; (in S Africa) urban settlement of Black or Coloured people

toxaemia [tox-seem-ya] n blood poisoning; high blood pressure in pregnancy

toxic adj poisonous; caused by poison ▶ **toxicity** n ▶ **toxicology** n study of poisons ▶ **toxin** n poison of bacterial origin

toy n something designed to be played with ▷ adj (of a dog) of a variety much smaller than is normal for that breed ▶ **toy with** v play or fiddle with

toy-toy (S Afr) n dance of political protest ▷ v perform this dance

trace v track down and find; follow the course of; copy exactly by drawing on a thin sheet of transparent paper set on top of the original ▷ n track left by something; minute quantity; indication ▶ **traceable** adj ▶ **tracer** n projectile which leaves a visible trail ▶ **tracery** n pattern of interlacing lines ▶ **tracing** n traced copy ▶ **trace element** chemical element occurring in very small amounts in soil etc.

traces pl n strap by which a horse pulls a vehicle **kick over the traces** escape or defy control

trachea [track-kee-a] n, pl **tracheae** windpipe

tracheotomy [track-ee-ot-a-mee] n surgical incision into the trachea

track n rough road or path; mark or trail left by the passage of anything; railway line; course for racing; separate section on a record, tape, or CD; course of action or thought; endless

band round the wheels of a tank, bulldozer, etc. ▷ v follow the trail or path of ▶ **track down** v hunt for and find ▶ **track event** athletic sport held on a running track ▶ **track record** past accomplishments of a person or organization ▶ **tracksuit** n warm loose-fitting suit worn by athletes etc., esp. during training

tract¹ n wide area; (Anat) system of organs with a particular function

tract² n pamphlet, esp. a religious one

tractable adj easy to manage or control

traction n pulling, esp. by engine power; (Med) application of a steady pull on an injured limb by weights and pulleys; grip of the wheels of a vehicle on the ground ▶ **traction engine** old-fashioned steam-powered vehicle for pulling heavy loads

tractor n motor vehicle with large rear wheels for pulling farm machinery

trade n buying, selling, or exchange of goods; person's job or craft; (people engaged in) a particular industry or business ▷ v buy and sell; exchange; engage in trade ▶ **trader** n ▶ **trading** n ▶ **trade-in** n used article given in part payment for a new one ▶ **trademark** n (legally registered) name or symbol used by a firm to distinguish its goods ▶ **trade-off** n exchange made as a compromise ▶ **tradesman** n skilled worker; shopkeeper ▶ **trade union** society of workers formed to protect their interests ▶ **trade wind** wind blowing steadily towards the equator

tradition n body of beliefs, customs, etc. handed down

from generation to generation; custom or practice of long standing ▶ **traditional** adj ▶ **traditionally** adv

traduce v slander

traffic n vehicles coming and going on a road; (illicit) trade ▷ v **-ficking, -ficked** trade, usu. illicitly ▶ **trafficker** n ▶ **traffic lights** set of coloured lights at a junction to control the traffic flow ▶ **traffic warden** (Brit) person employed to control the movement and parking of traffic

tragedy n, pl **-dies** shocking or sad event; serious play, film, etc. in which the hero is destroyed by a personal failing in adverse circumstances ▶ **tragedian** [traj-jee-dee-an], **tragedienne** [traj-jee-dee-**enn**](fem) n person who acts in or writes tragedies ▶ **tragic** adj of or like a tragedy ▶ **tragically** adv ▶ **tragicomedy** n play with both tragic and comic elements

trail n path, track, or road; tracks left by a person, animal, or object ▷ v drag along the ground; lag behind; follow the tracks of

trailer n vehicle designed to be towed by another vehicle; extract from a film or programme used to advertise it

train v instruct in a skill; learn the skills needed to do a particular job or activity; prepare for a sports event etc.; aim (a gun etc.); cause (an animal) to perform or (a plant) to grow in a particular way ▷ n line of railway coaches or wagons drawn by an engine; sequence or series; long trailing back section of a dress ▶ **trainer** n person who trains an athlete or sportsman; sports shoe ▶ **trainee** n person being trained

traipse v (informal) walk wearily

trait n characteristic feature

traitor n person guilty of treason or treachery ▶ **traitorous** adj

trajectory n, pl **-ries** line of flight, esp. of a projectile

tram n public transport vehicle powered by an overhead wire and running on rails laid in the road ▶ **tramlines** pl n track for trams

tramp v travel on foot, hike; walk heavily ▷ n homeless person who travels on foot; hike; sound of tramping; cargo ship available for hire; (US, Aust & NZ slang) promiscuous woman

trample v tread on and crush

trampoline n tough canvas sheet attached to a frame by springs, used by acrobats etc. ▷ v bounce on a trampoline

trance n unconscious or dazed state

tranche n portion of something large, esp. a sum of money

tranquil adj calm and quiet ▶ **tranquillity** adv

tranquillize v make calm ▶ **tranquillizer** n drug which reduces anxiety or tension

trans- prefix across, through, or beyond

transact v conduct or negotiate (a business deal)

transaction n business deal transacted

transatlantic adj on, from, or to the other side of the Atlantic

transceiver n transmitter and receiver of radio or electronic signals

transcend v rise above; be superior to ▶ **transcendence** n ▶ **transcendent** adj ▶ **transcendental** adj based on intuition rather than experience; supernatural or mystical ▶ **transcendentalism** n

t

transcribe v write down (something said); record for a later broadcast; arrange (music) for a different instrument ▶ **transcript** n copy

transducer n device that converts one form of energy to another

transept n either of the two shorter wings of a cross-shaped church

transfer v **-ferring, -ferred** move or send from one person or place to another ▷ n transferring; design which can be transferred from one surface to another ▶ **transferable** adj ▶ **transference** n transferring ▶ **transfer station** (NZ) depot where rubbish is sorted for recycling

transfigure v change in appearance ▶ **transfiguration** n

transfix v astound or stun; pierce through

transform v change the shape or character of ▶ **transformation** n ▶ **transformer** n device for changing the voltage of an alternating current

transfusion n injection of blood into the blood vessels of a patient ▶ **transfuse** v give a transfusion to; permeate or infuse

transgress v break (a moral law) ▶ **transgression** n ▶ **transgressor** n

transient adj lasting only for a short time ▶ **transience** n

transistor n semiconducting device used to amplify electric currents; portable radio using transistors

transit n movement from one place to another ▶ **transition** n change from one state to another ▶ **transitional** adj ▶ **transitive** adj (Grammar) (of a verb) requiring a direct object ▶ **transitory** adj not lasting long

translate v turn from one language into another ▶ **translation** n ▶ **translator** n

transliterate v convert to the letters of a different alphabet ▶ **transliteration** n

translucent adj letting light pass through, but not transparent ▶ **translucency, translucence** n

transmigrate v (of a soul) pass into another body ▶ **transmigration** n

transmit v **-mitting, -mitted** pass (something) from one person or place to another; send out (signals) by radio waves; broadcast (a radio or television programme) ▶ **transmission** n transmitting; shafts and gears through which power passes from a vehicle's engine to its wheels ▶ **transmittable** adj ▶ **transmitter** n

transmogrify v **-fying, -fied** (informal) change completely

transmute v change the form or nature of ▶ **transmutation** n

transom n horizontal bar across a window; bar separating a door from the window over it

transparent adj able to be seen through, clear; easily understood or recognized ▶ **transparently** adv ▶ **transparency** n transparent quality; colour photograph on transparent film that can be viewed by means of a projector

transpire v become known; (informal) happen; give off water vapour through pores ▶ **transpiration** n

transplant v transfer (an organ or tissue) surgically from one part or body to another; remove and transfer (a plant) to another place ▷ n surgical transplanting; thing transplanted ▶ **transplantation** n

transport v convey from one place to another; (*Hist*) exile (a criminal) to a penal colony; enrapture ▷ n business or system of transporting; vehicle used in transport; ecstasy or rapture ▶ **transportation** n ▶ **transporter** n large goods vehicle

transpose v interchange two things; put (music) into a different key ▶ **transposition** n

transsexual, transexual n person of one sex who believes his or her true identity is the opposite sex

transubstantiation n (*Christianity*) doctrine that the bread and wine consecrated in Communion changes into the substance of Christ's body and blood

transuranic [tranz-yoor-**ran**-ik] adj (of an element) having an atomic number greater than that of uranium

transverse adj crossing from side to side

transvestite n person who seeks sexual pleasure by wearing the clothes of the opposite sex ▶ **transvestism** n

trap n device for catching animals; plan for tricking or catching a person; bend in a pipe containing liquid to prevent the escape of gas; stall in which greyhounds are enclosed before a race; two-wheeled carriage; (*Brit, Aust & NZ slang*) mouth ▷ v **trapping, trapped** catch; trick ▶ **trapper** n person who traps animals for their fur ▶ **trapdoor** n door in floor or roof ▶ **trap-door spider** spider that builds a silk-lined hole in the ground closed by a hinged door of earth and silk

trapeze n horizontal bar suspended from two ropes, used by circus acrobats

trapezium n, pl -**ziums** or -**zia** quadrilateral with two parallel sides of unequal length ▶ **trapezoid** [**trap**-piz-zoid] n quadrilateral with no sides parallel; (*chiefly US*) trapezium

trappings pl n accessories that symbolize an office or position

Trappist n member of an order of Christian monks who observe strict silence

trash n anything worthless; (*US & S Afr*) rubbish ▶ **trashy** adj

trauma [**traw**-ma] n emotional shock; injury or wound ▶ **traumatic** adj ▶ **traumatize** v

travail n (*lit*) labour or toil

travel v -**elling, -elled** go from one place to another, through an area, or for a specified distance ▷ n travelling, esp. as a tourist ▷ pl (account of) travelling ▶ **traveller** n ▶ **travelogue** n film or talk about someone's travels

traverse v move over or back and forth over

travesty n, pl -**ties** grotesque imitation or mockery

trawl n net dragged at deep levels behind a fishing boat ▷ v fish with such a net ▶ **trawler** n trawling boat

tray n flat board, usu. with a rim, for carrying things; open receptacle for office correspondence

treachery n, pl -**eries** wilful betrayal ▶ **treacherous** adj disloyal; unreliable or dangerous ▶ **treacherously** adv

treacle n thick dark syrup produced when sugar is refined ▶ **treacly** adj

tread v **treading, trod, trodden** or **trod** set one's foot on; crush by walking on ▷ n way of walking or dancing; upper surface of a step; part of a tyre or shoe that touches the ground ▶ **treadmill** n (*Hist*) cylinder turned by treading on steps projecting from it; dreary routine

treadle [tred-dl] n lever worked by the foot to turn a wheel

treason n betrayal of one's sovereign or country; treachery or disloyalty ▶ **treasonable** adj

treasure n collection of wealth, esp. gold or jewels; valued person or thing ▶ prize or cherish ▶ **treasurer** n official in charge of funds ▶ **treasure-trove** n treasure found with no evidence of ownership ▶ **treasury** n storage place for treasure; (**T-**) government department in charge of finance

treat v deal with or regard in a certain manner; give medical treatment to; subject to a chemical or industrial process; provide (someone) with (something) as a treat ▶ n pleasure, entertainment, etc. given or paid for by someone else ▶ **treatment** n medical care; way of treating a person or thing

treatise [treat-izz] n formal piece of writing on a particular subject

treaty n, pl **-ties** signed contract between states

treble adj triple; (Music) high-pitched ▶ n (singer with or part for) a soprano voice ▶ v increase three times ▶ **trebly** adv

tree n large perennial plant with a woody trunk ▶ **treeless** adj ▶ **tree kangaroo** tree-living kangaroo of New Guinea and N Australia ▶ **tree surgery** treatment of damaged trees ▶ **tree surgeon**

trefoil [tref-foil] n plant, such as clover, with a three-lobed leaf; carved ornament like this

trek n long difficult journey, esp. on foot; (SAfr) migration by ox wagon ▶ v **trekking, trekked** make such a journey

trellis n framework of horizontal and vertical strips of wood

tremble v shake or quiver; feel fear or anxiety ▶ n trembling ▶ **trembling** adj

tremendous adj huge; (informal) great in quality or amount ▶ **tremendously** adv

tremolo n, pl **-los** (Music) quivering effect in singing or playing

tremor n involuntary shaking; minor earthquake

tremulous adj trembling, as from fear or excitement

trench n long narrow ditch, esp. one used as a shelter in war ▶ **trench coat** double-breasted waterproof coat

trenchant adj incisive; effective

trencher n (Hist) wooden plate for serving food ▶ **trencherman** n hearty eater

trend n general tendency or direction; fashion ▶ v be widely discussed on a social media site ▶ **trendy** adj, n (informal) consciously fashionable (person) ▶ **trendiness** n

trepidation n fear or anxiety

trespass v go onto another's property without permission ▶ n trespassing; (old-fashioned) sin or wrongdoing ▶ **trespasser** n ▶ **trespass on** take unfair advantage of (someone's friendship, patience, etc.)

tresses pl n long flowing hair

trestle n board fixed on pairs of spreading legs, used as a support

trevally n, pl **-lies** (Aust & NZ) any of various food and game fishes

trews pl n close-fitting tartan trousers

tri- combining form three

triad n group of three; (**T-**) Chinese criminal secret society

trial n (Law) investigation of a case before a judge; trying or testing; thing or person straining endurance or patience ▷ pl sporting competition for individuals

triangle n geometric figure with three sides; triangular percussion instrument; situation involving three people ▶ **triangular** adj

tribe n group of clans or families believed to have a common ancestor ▶ **tribal** adj ▶ **tribalism** n loyalty to a tribe

tribulation n great distress

tribunal n board appointed to inquire into a specific matter; lawcourt

tribune n people's representative, esp. in ancient Rome

tributary n, pl **-taries** stream or river flowing into a larger one ▷ adj (of a stream or river) flowing into a larger one

tribute n sign of respect or admiration; tax paid by one state to another

trice n in a trice instantly

triceps n muscle at the back of the upper arm

trichology [trick-ol-a-jee] n study and treatment of hair and its diseases ▶ **trichologist** n

trick n deceitful or cunning action or plan; joke or prank; feat of skill or cunning; mannerism; cards played in one round ▷ v cheat or deceive ▶ **trickery** n ▶ **trickster** n ▶ **tricky** adj difficult, needing careful handling; crafty

trickle v (cause to) flow in a thin stream or drops; move gradually ▷ n gradual flow

tricolour [trick-kol-or] n three-coloured striped flag

tricycle n three-wheeled cycle

trident n three-pronged spear

triennial adj happening every three years

trifle n insignificant thing or amount; dessert of sponge cake, fruit, custard, and cream ▶ **trifling** adj insignificant ▶ **trifle with** v toy with

trigger n small lever releasing a catch on a gun or machine; action that sets off a course of events ▷ v (usu. foll by off) set (an action or process) in motion ▶ **trigger-happy** adj too quick to fire guns

trigonometry n branch of mathematics dealing with relations of the sides and angles of triangles

trike n (informal) tricycle

trilateral adj having three sides

trilby n, pl **-bies** man's soft felt hat

trill n (Music) rapid alternation between two notes; shrill warbling sound made by some birds ▷ v play or sing a trill

trillion n one million million, 10¹²; (formerly) one million million million, 10¹⁸

trilobite [trile-oh-bite] n small prehistoric sea animal

trilogy n, pl **-gies** series of three related books, plays, etc.

trim adj **trimmer, trimmest** neat and smart; slender ▷ v **trimming, trimmed** cut or prune into good shape; decorate with lace, ribbons, etc.; adjust the balance of (a ship or aircraft) by shifting the cargo etc. ▷ n decoration; upholstery and decorative facings in a car; trim state; haircut that neatens the existing style ▶ **trimming** n decoration ▷ pl usual accompaniments

trimaran [trime-a-ran] n three-hulled boat

trinitrotoluene n full name for **TNT**

trinity n, pl **-ties** group of three; (**T-**) (Christianity) union of three persons, Father, Son, and Holy Spirit, in one God

trinket n small or worthless ornament or piece of jewellery

trio n, pl **trios** group of three; piece of music for three performers

trip n journey to a place and back, esp. for pleasure; stumble; (informal) hallucinogenic drug experience; switch on a mechanism ▷ v **tripping, tripped** (cause to) stumble; (often foll by up) catch (someone) in a mistake; move or tread lightly; (informal) experience the hallucinogenic effects of a drug ▶ **tripper** n tourist

tripe n stomach of a cow used as food; (Brit, Aust & NZ informal) nonsense

triple adj having three parts; three times as great or as many ▷ v increase three times ▶ **triplet** n one of three babies born at one birth ▶ **triple jump** athletic event in which competitors make a hop, a step, and a jump as a continuous movement

triplicate adj triple **in triplicate** in three copies

tripod [tripe-pod] n three-legged stand, stool, etc.

tripos [tripe-poss] n final examinations for an honours degree at Cambridge University

triptych [trip-tick] n painting or carving on three hinged panels, often forming an altarpiece

trite adj (of a remark or idea) commonplace and unoriginal

tritium n radioactive isotope of hydrogen

triumph n (happiness caused by) victory or success ▷ v be victorious or successful; rejoice over a victory ▶ **triumphal** adj celebrating a triumph ▶ **triumphant** adj feeling or showing triumph

triumvirate [try-umm-vir-rit] n group of three people in joint control

trivet [triv-vit] n metal stand for a pot or kettle

trivial adj of little importance ▶ **trivially** adv ▶ **trivia** pl n trivial things or details ▶ **triviality** n ▶ **trivialize** v make (something) seem less important or complex than it is

trod v past tense and a past participle of **tread**

trodden v a past participle of **tread**

troglodyte n cave dweller

troika n Russian vehicle drawn by three horses abreast; group of three people in authority

Trojan adj of ancient Troy ▷ n person from ancient Troy; hard-working person; destructive computer program ▶ **Trojan Horse** trap intended to undermine an enemy

troll n giant or dwarf in Scandinavian folklore

trolley n small wheeled table for food and drink; wheeled cart for moving goods ▶ **trolley bus** bus powered by electricity from an overhead wire but not running on rails

trollop n (old-fashioned) promiscuous or slovenly woman

trombone n brass musical instrument with a sliding tube ▶ **trombonist** n

troop n large group; artillery or cavalry unit; Scout company ▷ pl soldiers ▷ v move in a crowd ▶ **trooper** n cavalry soldier

trope n figure of speech

trophy n, pl **-phies** cup, shield, etc. given as a prize; memento of success

tropic n either of two lines of latitude at 23½°N (**tropic of Cancer**) or 23½°S (**tropic of Capricorn**) ▷ pl part of the earth's surface between these lines ▶ **tropical** adj of or in the tropics; (of climate) very hot

trot v **trotting, trotted** (of a horse) move at a medium pace, lifting the feet in diagonal pairs; (of a person) move at a steady brisk pace ▷ n

trotting ▸ **trotter** n pig's foot
▸ **trot out** v repeat (old ideas etc.)
without fresh thought

troth [rhymes with **growth**] n (obs)
pledge of devotion, esp. a betrothal

troubadour [troo-bad-oor] n
medieval travelling poet and singer

trouble n (cause of) distress or
anxiety; disease or malfunctioning;
state of disorder or unrest; care or
effort ▷ v (cause to) worry; exert
oneself; cause inconvenience to
▸ **troubled** adj ▸ **troublesome**
adj ▸ **troubleshooter** n person
employed to locate and deal with
faults or problems

trough [troff] n long open
container, esp. for animals' food or
water; narrow channel between
two waves or ridges; (Meteorol)
area of low pressure

trounce v defeat utterly

troupe [troop] n company of
performers ▸ **trouper** n

trousers pl n two-legged outer
garment with legs reaching usu. to
the ankles ▸ **trouser** adj of trousers

trousseau [troo-so] n, pl -seaux
or -seaus bride's collection of
clothing etc. for her marriage

trout n game fish related to the
salmon

trowel n hand tool with a wide
blade for spreading mortar, lifting
plants, etc.

troy weight, troy n system of
weights used for gold, silver, and
jewels

truant n pupil who stays away
from school without permission
play truant stay away from school
without permission ▸ **truancy** n

truce n temporary agreement to
stop fighting

truck¹ n railway goods wagon; large
vehicle for transporting loads by
road ▸ **trucker** n truck driver

truck² n **have no truck with** refuse
to be involved with

truculent [truck-yew-lent] adj
aggressively defiant ▸ **truculence** n

trudge v walk heavily or wearily ▷ n
long tiring walk

true adj **truer, truest** in accordance
with facts; genuine; faithful; exact
▸ **truly** adv ▸ **truism** n self-evident
truth ▸ **truth** n state of being true;
something true ▸ **truthful** adj
honest; exact ▸ **truthfully** adv

truffle n edible underground
fungus; sweet made with
chocolate

trug n (Brit) long shallow basket
used by gardeners

trump¹ n, adj (card) of the suit
outranking the others ▷ v
play a trump card on (another
card) **trumped up** invented
or concocted **turn up trumps**
achieve an unexpected success

trump² n (lit) (sound of) a trumpet

trumpet n valved brass instrument
with a flared tube ▷ v -**peting**,
-**peted** proclaim loudly; (of an
elephant) cry loudly ▸ **trumpeter** n

truncate v cut short

truncheon n club formerly carried
by a policeman

trundle v move heavily on wheels

trunk n main stem of a tree;
large case or box for clothes etc.;
person's body excluding the head
and limbs; elephant's long nose;
(US) car boot ▷ pl man's swimming
shorts ▸ **trunk call** (chiefly Brit)
long-distance telephone call
▸ **trunk road** main road

truss v tie or bind up ▷ n device
for holding a hernia in place;
framework supporting a roof,
bridge, etc.

trust v believe in and rely on; allow
to someone's care; expect or hope
▷ n confidence in the truth, reliability,

etc. of a person or thing; obligation arising from responsibility; arrangement in which one person administers property, money, etc. on another's behalf; property held for another; (Brit) self-governing hospital or group of hospitals within the National Health Service; group of companies joined to control a market ▶ **trustee** n person holding property on another's behalf ▶ **trustful, trusting** adj inclined to trust others ▶ **trustworthy** adj reliable or honest ▶ **trusty** adj faithful or reliable

truth n see **true**

try v **trying, tried** make an effort or attempt; test or sample; put strain on ie he tries my patience; investigate (a case); examine (a person) in a lawcourt ▷ n, pl **tries** attempt or effort; (Rugby) score gained by touching the ball down over the opponent's goal line **try it on** (informal) try to deceive or fool someone ▶ **trying** adj (informal) difficult or annoying

> **USAGE NOTE**
> The idiom **try** to can be used at any time. The alternative **try** and is less formal and often signals a 'dare': Just **try** and stop me!

tryst n arrangement to meet

tsar, czar [zahr] n (Hist) Russian emperor

tsetse fly [tset-see] n bloodsucking African fly whose bite transmits disease, esp. sleeping sickness

T-shirt n short-sleeved casual shirt or top

tsp. teaspoon

T-square n T-shaped ruler

tsunami n, pl **-mis** or **-mi** tidal wave, usu. caused by an earthquake under the sea

TT teetotal

tuatara n large lizard-like New Zealand reptile

tub n open, usu. round container; bath ▶ **tubby** adj (of a person) short and fat

tuba [tube-a] n valved low-pitched brass instrument

tube n hollow cylinder; flexible cylinder with a cap to hold pastes **the tube** underground railway, esp. the one in London ▶ **tubing** n length of tube; system of tubes ▶ **tubular** [tube-yew-lar] adj of or shaped like a tube

tuber [tube-er] n fleshy underground root of a plant such as a potato ▶ **tuberous** adj

tubercle [tube-er-kl] n small rounded swelling

tuberculosis [tube-berk-yew-lohss-iss] n infectious disease causing tubercles, esp. in the lungs ▶ **tubercular** adj ▶ **tuberculin** n extract from a bacillus used to test for tuberculosis

TUC (in Britain and S Africa) Trades Union Congress

tuck v push or fold into a small space; stitch in folds ▷ n stitched fold; (Brit informal) food ▶ **tuck away** v eat (a large amount of food)

tucker n (Aust & NZ informal) food

Tudor adj of the English royal house ruling from 1485–1603

Tuesday n third day of the week

tufa [tew-fa] n porous rock formed as a deposit from springs

tuffet n small mound or seat

tuft n bunch of feathers, grass, hair, etc. held or growing together at the base

tug v **tugging, tugged** pull hard ▷ n hard pull; (also **tugboat**) small ship used to tow other vessels ▶ **tug of war** contest in which two teams pull against one another on a rope

tuition n instruction, esp. received individually or in a small group

tulip n plant with bright cup-shaped flowers

tulle [tewl] n fine net fabric of silk etc.

tumble v (cause to) fall, esp. awkwardly or violently; roll or twist, esp. in play; rumple ▷ n fall; somersault ▶ **tumbler** n stemless drinking glass; acrobat; spring catch in a lock ▶ **tumbledown** adj dilapidated ▶ **tumble dryer, tumble drier** machine that dries laundry by rotating it in warm air ▶ **tumble to** v (informal) realize, understand

tumbril, tumbrel n farm cart used during the French Revolution to take prisoners to the guillotine

tumescent [tew-mess-ent] adj swollen or becoming swollen

tummy n, pl -mies (informal) stomach

tumour [tew-mer] n abnormal growth in or on the body

tumult n uproar or commotion ▶ **tumultuous** [tew-mull-tew-uss] adj

tumulus n, pl -li burial mound

tun n large beer cask

tuna n large marine food fish

tundra n vast treeless Arctic region with permanently frozen subsoil

tune n (pleasing) sequence of musical notes; correct musical pitch e.g. she sang out of tune ▷ v adjust (a musical instrument) so that it is in tune; adjust (a machine) to obtain the desired performance ▶ **tuneful** adj ▶ **tunefully** adv ▶ **tuneless** adj ▶ **tuner** n ▶ **tune in** v (adjust (a radio or television) to receive (a station or programme)

tungsten n (Chem) greyish-white metal

tunic n close-fitting jacket forming part of some uniforms; loose knee-length garment

tunnel n underground passage ▷ v **-nelling, -nelled** make a tunnel (through)

tunny n, pl -nies or -ny same as **tuna**

tup n male sheep

turban n Muslim, Hindu, or Sikh man's head covering, made by winding cloth round the head

turbid adj muddy, not clear

turbine n machine or generator driven by gas, water, etc. turning blades

turbot n large European edible flatfish

turbulence n confusion, movement, or agitation; atmospheric instability causing gusty air currents ▶ **turbulent** adj

tureen n serving dish for soup

turf n, pl turfs or turves short thick even grass; square of this with roots and soil attached ▷ v cover with turf **the turf** racecourse; horse racing ▶ **turf accountant** bookmaker ▶ **turf out** v (informal) throw out

turgid [tur-jid] adj (of language) pompous; swollen and thick

turkey n large bird bred for food

Turkish adj of Turkey, its people, or their language ▷ n Turkish language ▶ **Turkish bath** steam bath ▶ **Turkish delight** jelly-like sweet coated with icing sugar

turmeric n yellow spice obtained from the root of an Asian plant

turmoil n agitation or confusion

turn v change the position or direction (of); move around an axis, rotate; (usu. foll by into) change in nature or character; reach or pass in age, time, etc. e.g. she has just turned twenty; shape on a

t

lathe; become sour ▷ *n* turning; opportunity to do something as part of an ongoing succession; direction or drift; period or spell; short theatrical performance **good turn, bad turn** helpful or unhelpful act ▶ **turner** *n* ▶ **turning** *n* road or path leading off a main route ▶ **turncoat** *n* person who deserts one party or cause to join another ▶ **turn down** *v* reduce the volume or brightness (of); refuse or reject ▶ **turn in** *v* go to bed; hand in ▶ **turning point** moment when a decisive change occurs ▶ **turn off** *v* stop (something) working by using a knob etc. ▶ **turn on** *v* start (something) working by using a knob etc.; become aggressive towards; (*informal*) excite, esp. sexually ▶ **turnout** *n* number of people appearing at a gathering ▶ **turnover** *n* total sales made by a business over a certain period; small pastry; rate at which staff leave and are replaced ▶ **turnpike** *n* (*Brit Hist*) road where a toll is collected at barriers ▶ **turnstile** *n* revolving gate for admitting one person at a time ▶ **turntable** *n* revolving platform ▶ **turn up** *v* arrive or appear; find or be found; increase the volume or brightness (of) ▶ **turn-up** *n* turned-up fold at the bottom of a trouser leg; (*informal*) unexpected event

turnip *n* root vegetable with orange or white flesh

turpentine *n* (oil made from) the resin of certain trees ▶ **turps** *n* turpentine oil

turpitude *n* wickedness

turquoise *adj* blue-green ▷ *n* blue-green precious stone

turret *n* small tower; revolving gun tower on a warship or tank

turtle *n* sea tortoise **turn turtle** capsize ▶ **turtledove** *n* small wild

dove ▶ **turtleneck** *n* (sweater with) a round high close-fitting neck

tusk *n* long pointed tooth of an elephant, walrus, etc.

tussle *n*, *v* fight or scuffle

tussock *n* tuft of grass

tutelage [tew-till-lij] *n* instruction or guidance, esp. by a tutor; state of being supervised by a guardian or tutor ▶ **tutelary** [tew-till-lar-ee] *adj*

tutor *n* person teaching individuals or small groups ▷ *v* act as a tutor to ▶ **tutorial** *n* period of instruction with a tutor

tutu *n* short stiff skirt worn by ballerinas

tuxedo *n*, *pl* **-dos** (*US & Aust*) dinner jacket

TV television

twaddle *n* silly or pretentious talk or writing

twain *n* (*obs*) two

twang *n* sharp ringing sound; nasal speech ▷ *v* (cause to) make a twang

tweak *v* pinch or twist sharply; (*informal*) alter slightly ▷ *n* tweaking

twee *adj* (*informal*) too sentimental, sweet, or pretty

tweed *n* thick woollen cloth ▷ *pl* suit of tweed ▶ **tweedy** *adj*

tweet *n* chirp; short message posted on the Twitter website ▷ *v* chirp; post a short message on the Twitter website

tweeter *n* loudspeaker reproducing high-frequency sounds

tweezers *pl n* small pincer-like tool

twelve *adj*, *n* two more than ten ▶ **twelfth** *adj*, *n* (of) number twelve in a series

twenty *adj*, *n* two times ten ▶ **twentieth** *adj*, *n*

twenty-four-seven, 24/7 *adv* (*informal*) all the time

twerp n (informal) silly person

twice adv two times

twiddle v fiddle or twirl in an idle way **twiddle one's thumbs** be bored, have nothing to do

twig¹ n small branch or shoot

twig² v **twigging, twigged** (informal) realize or understand

twilight n soft dim light just after sunset

twill n fabric woven to produce parallel ridges

twin n one of a pair, esp. of two children born at one birth ▷ v **twinning, twinned** pair or be paired

twine n string or cord ▷ v twist or coil round

twinge n sudden sharp pain or emotional pang

twinkle v shine brightly but intermittently ▷ n flickering brightness

twirl v turn or spin around quickly; twist or wind, esp. idly

twist v turn out of the natural position; distort or pervert; wind or twine ▷ n twisting; twisted thing; unexpected development in the plot of a film, book, etc.; bend or curve; distortion ▶ **twisted** adj (of a person) cruel or perverted ▶ **twister** n (Brit informal) swindler

twit¹ v **twitting, twitted** poke fun at (someone)

twit² n (informal) foolish person

twitch v move spasmodically; pull sharply ▷ n nervous muscular spasm; sharp pull

twitter v (of birds) utter chirping sounds ▷ n act or sound of twittering; (T-)® website where people can post short messages about their current thoughts and activities

two adj, n one more than one ▶ **two-edged** adj (of a remark)

having both a favourable and an unfavourable interpretation ▶ **two-faced** adj deceitful, hypocritical ▶ **two-time** v (informal) deceive (a lover) by having an affair with someone else

tycoon n powerful wealthy businessman

tyke n (Brit, Aust & NZ informal) small cheeky child

type n class or category; (informal) person, esp. of a specified kind; block with a raised character used for printing; printed text ▷ v write with a typewriter or word processor; typify; classify ▶ **typist** n person who types with a typewriter or word processor ▶ **typecast** v continually cast (an actor or actress) in similar roles ▶ **typewriter** n machine which prints a character when the appropriate key is pressed

typhoid fever n acute infectious feverish disease

typhoon n violent tropical storm

typhus n infectious feverish disease

typical adj true to type, characteristic ▶ **typically** adv ▶ **typify** v **-fying, -fied** be typical of

typography n art or style of printing ▶ **typographical** adj ▶ **typographer** n

tyrannosaurus [tirr-ran-oh-sore-uss] n large two-footed flesh-eating dinosaur

tyrant n oppressive or cruel ruler; person who exercises authority oppressively ▶ **tyrannical** adj like a tyrant, oppressive ▶ **tyrannize** v exert power (over) oppressively or cruelly ▶ **tyrannous** adj ▶ **tyranny** n tyrannical rule

tyre n rubber ring, usu. inflated, over the rim of a vehicle's wheel to grip the road

tyro n, pl **-ros** novice or beginner

Uu

ubiquitous [yew-bik-wit-uss] *adj* being or seeming to be everywhere at once ▶ **ubiquity** *n*

udder *n* large baglike milk-producing gland of cows, sheep, or goats

UEFA Union of European Football Associations

UFO unidentified flying object

ugly *adj* **uglier, ugliest** of unpleasant appearance; ominous or menacing ▶ **ugliness** *n*

UHF ultrahigh frequency

UHT (of milk or cream) ultra-heat-treated

UK United Kingdom

ukulele, ukelele [yew-kal-**lay**-lee] *n* small guitar with four strings

ulcer *n* open sore on the surface of the skin or mucous membrane. ▶ **ulcerated** *adj* made or becoming ulcerous ▶ **ulceration** *n* ▶ **ulcerous** *adj* of, like, or characterized by ulcers

ulna *n*, *pl* **-nae** or **-nas** inner and longer of the two bones of the human forearm

ulterior *adj* (of an aim, reason, etc.) concealed or hidden

ultimate *adj* final in a series or process; highest or supreme ▶ **ultimately** *adv*

ultimatum [ult-im-**may**-tum] *n* final warning stating that action will be taken unless certain conditions are met

ultra- *prefix* beyond a specified extent, range, or limit e.g. *ultrasonic*; extremely e.g. *ultramodern*

ultrahigh frequency *n* radio frequency between 300 and 3000 megahertz

ultramarine *adj* vivid blue

ultrasonic *adj* of or producing sound waves with a higher frequency than the human ear can hear

ultrasound *n* ultrasonic waves, used in medical diagnosis and therapy and in echo sounding

ultraviolet *adj*, *n* (of) light beyond the limit of visibility at the violet end of the spectrum

ululate [yewl-yew-late] *v* howl or wail ▶ **ululation** *n*

umber *adj* dark brown to reddish-brown

umbilical *adj* of the navel ▶ **umbilical cord** long flexible tube of blood vessels that connects a fetus with the placenta

umbrage *n* **take umbrage** feel offended or upset

umbrella *n* portable device used for protection against rain, consisting of a folding frame covered in material attached to a central rod; single organization, idea, etc. that contains or covers many different organizations, ideas, etc.

umpire *n* official who rules on the playing of a game ▷ *v* act as umpire in (a game)

umpteen *adj* (informal) very many ▶ **umpteenth** *n*, *adj*

UN United Nations

un- *prefix* not e.g. *unidentified*; denoting reversal of an action e.g. *untie*; denoting removal from e.g. *unthrone*

unable *adj* **unable to** lacking the necessary power, ability, or authority to (do something)

unaccountable adj unable to be explained; (foll by to) not answerable to ▸ **unaccountably** adv

unadulterated adj with nothing added, pure

unanimous [yew-nan-im-uss] adj in complete agreement; agreed by all ▸ **unanimity** n

unannounced adv without warning e.g. he turned up unannounced

unarmed adj without weapons

unassailable adj unable to be attacked or disputed

unassuming adj modest or unpretentious

unaware adj not aware or conscious ▸ **unawares** adv by surprise e.g. caught unawares; without knowing

> **USAGE NOTE**
> Note the difference between the adjective unaware, usually followed by of or that, and the adverb unawares

unbalanced adj biased or one-sided; mentally deranged

unbearable adj not able to be endured ▸ **unbearably** adv

unbeknown adv **unbeknown to** without the knowledge of (a person)

unbend v (informal) become less strict or more informal in one's attitudes or behaviour ▸ **unbending** adj

unbidden adj not ordered or asked

unborn adj not yet born

unbosom v relieve (oneself) of (secrets or feelings) by telling someone

unbridled adj (of feelings or behaviour) not controlled in any way

unburden v relieve (one's mind or oneself) of a worry by confiding in someone

uncalled-for adj not fair or justified

uncanny adj weird or mysterious ▸ **uncannily** adv

unceremonious adj relaxed and informal; abrupt or rude ▸ **unceremoniously** adv

uncertain adj not able to be accurately known or predicted; not able to be depended upon; changeable ▸ **uncertainty** n

un-Christian adj not in accordance with Christian principles

uncle n brother of one's father or mother; husband of parent's sibling

unclean adj lacking moral, spiritual, or physical cleanliness

uncomfortable adj not physically relaxed; anxious or uneasy

uncommon adj not happening or encountered often; in excess of what is normal ▸ **uncommonly** adv

uncompromising adj not prepared to compromise

unconcerned adj lacking in concern or involvement ▸ **unconcernedly** adv

unconditional adj without conditions or limitations

unconscionable adj having no principles, unscrupulous; excessive in amount or degree

unconscious adj lacking normal awareness through the senses; not aware of one's actions or behaviour ▸ n part of the mind containing instincts and ideas that exist without one's awareness ▸ **unconsciously** adv ▸ **unconsciousness** n

uncooperative adj not willing to help other people with what they are doing

uncouth adj lacking in good manners, refinement, or grace

u

uncover v reveal or disclose; remove the cover, top, etc., from

unction n act of anointing with oil in sacramental ceremonies

unctuous adj pretending to be kind and concerned

undecided adj not having made up one's mind; (of an issue or problem) not agreed or decided upon

undeniable adj unquestionably true ▷ **undeniably** adv

under prep, adv indicating movement to or position beneath the underside or base ▷ prep less than; subject to

under- prefix below e.g. *underground*; insufficient or insufficiently e.g. *underrate*

underage adj below the required or standard age

underarm adj (Sport) denoting a style of throwing, bowling, or serving in which the hand is swung below shoulder level ▷ adv (Sport) in an underarm style

undercarriage n landing gear of an aircraft; framework supporting the body of a vehicle

underclass n class consisting of the most disadvantaged people, such as the long-term unemployed

undercoat n coat of paint applied before the final coat

undercover adj done or acting in secret

undercurrent n current that is not apparent at the surface; underlying opinion or emotion

undercut v charge less than (a competitor) to obtain trade

underdog n person or team in a weak or underprivileged position

underdone adj not cooked enough

underestimate v make too low an estimate of; not realize the full potential of

underfoot adv under the feet

undergarment n any piece of underwear

undergo v experience, endure, or sustain

undergraduate n person studying in a university for a first degree

underground adj occurring, situated, used, or going below ground level; secret ▷ n electric passenger railway operated in underground tunnels; movement dedicated to overthrowing a government or occupation forces

undergrowth n small trees and bushes growing beneath taller trees in a wood or forest

underhand adj sly, deceitful, and secretive

underlie v lie or be placed under; be the foundation, cause, or basis of ▷ **underlying** adj fundamental or basic

underline v draw a line under; state forcibly, emphasize

underling n subordinate

undermine v weaken gradually

underneath prep, adv under or beneath ▷ adj, n lower (part or surface)

underpants pl n man's undergarment for the lower part of the body

underpass n section of a road that passes under another road or a railway line

underpin v give strength or support to

underprivileged adj lacking the rights and advantages of other members of society

underrate v not realize the full potential of ▷ **underrated** adj

underseal v noun protective coating of tar etc. applied to the underside of a motor vehicle to prevent corrosion

underside n bottom or lower surface

understand v know and comprehend the nature or meaning of; realize or grasp (something); assume, infer, or believe ▶ **understandable** adj ▶ **understandably** adv ▶ **understanding** n ability to learn, judge, or make decisions; personal interpretation of a subject; mutual agreement, usu. an informal or private one ▶ adj kind and sympathetic

understate v describe or represent (something) in restrained terms; state that (something, such as a number) is less than it is ▶ **understatement** n

understudy n actor who studies a part in order to be able to replace the usual actor if necessary ▶ v act as an understudy for

undertake v agree or commit oneself to (something) or to do (something); promise ▶ **undertaking** n task or enterprise; agreement to do something

undertaker n person whose job is to prepare corpses for burial or cremation and organize funerals

undertone n quiet tone of voice; underlying quality or feeling

undertow n strong undercurrent flowing in a different direction from the surface current

underwater adj, adv (situated, occurring, or for use) below the surface of the sea, a lake, or a river

underwear n clothing worn under the outer garments and next to the skin

underworld n criminals and their associates; (Greek and Roman myth) regions below the earth's surface regarded as the abode of the dead

underwrite v accept financial responsibility for (a commercial project); sign and issue (an insurance policy), thus accepting liability

underwriter n person who underwrites (esp. an insurance policy)

undesirable adj not desirable or pleasant, objectionable ▶ n objectionable person

undo v open, unwrap; reverse the effects of; cause the downfall of ▶ **undone** adj ▶ **undoing** n cause of someone's downfall

undoubted adj certain or indisputable ▶ **undoubtedly** adv

undue adj greater than is reasonable, excessive ▶ **unduly** adv

undulate v move in waves ▶ **undulation** n

undying adj never ending, eternal

unearth v reveal or discover by searching; dig up out of the earth

unearthly adj ghostly or eerie; ridiculous or unreasonable

uneasy adj (of a person) anxious or apprehensive; (of a condition) precarious or uncomfortable ▶ **uneasily** adv ▶ **uneasiness** n ▶ **unease** n feeling of anxiety; state of dissatisfaction

unemployed adj out of work ▶ pl n **the unemployed** people who are out of work ▶ **unemployment** n

unequivocal adj completely clear in meaning ▶ **unequivocally** adv

unerring adj never mistaken, consistently accurate

unexceptionable adj beyond criticism or objection

unfailing adj continuous or reliable ▶ **unfailingly** adv

unfair adj not right, fair, or just ▶ **unfairly** adv ▶ **unfairness** n

unfaithful adj having sex with someone other than one's regular

u

partner; not true to a promise or vow ▸ **unfaithfulness** n

unfeeling adj without sympathy

unfit adj unqualified or unsuitable; in poor physical condition

unflappable adj (informal) not easily upset ▸ **unflappability** n

unfold v open or spread out from a folded state; reveal or be revealed

unfollow v cease to receive messages posted by (a blogger or microblogger)

unforgettable adj impossible to forget, memorable

unfortunate adj unlucky, unsuccessful, or unhappy; regrettable or unsuitable ▸ **unfortunately** adv

unfounded adj not based on facts or evidence

unfrock v deprive (a priest in holy orders) of his or her priesthood

ungainly adj -**lier**, -**liest** lacking grace when moving

ungodly adj (informal) unreasonable or outrageous e.g. *an ungodly hour*; wicked or sinful

ungrateful adj not grateful or thankful

unguarded adj not protected; incautious or careless

unguent [ung-gwent] n (lit) ointment

unhand v (old-fashioned or lit) release from one's grasp

unhappy adj sad or depressed; unfortunate or wretched ▸ **unhappily** adv ▸ **unhappiness** n

unhealthy adj likely to cause poor health; not fit or well; morbid, unnatural

unhinge v derange or unbalance (a person or his or her mind)

uni n (informal) short for **university**

uni- combining form of, consisting of, or having only one e.g. *unicellular*

unicorn n imaginary horselike creature with one horn growing from its forehead

uniform n special identifying set of clothes for the members of an organization, such as soldiers ▷ adj regular and even throughout, unvarying; alike or like ▸ **uniformly** adv ▸ **uniformity** n

unify v -**fying**, -**fied** make or become one ▸ **unification** n

unilateral adj made or done by only one person or group ▸ **unilaterally** adv

unimpeachable adj completely honest and reliable

uninterested adj having or showing no interest in someone or something

union n uniting or being united; short for **trade union**; association or confederation of individuals or groups for a common purpose ▸ **unionist** n member or supporter of a trade union ▸ **unionize** v organize (workers) into a trade union ▸ **unionization** n ▸ **Union Jack, Union Flag** national flag of the United Kingdom

unique [yoo-**neek**] adj being the only one of a particular type; without equal or like ▸ **uniquely** adv

▎**USAGE NOTE**
▎Because of its meaning, avoid
▎using *unique* with modifiers like
▎*very* and *rather*

unisex adj designed for use by both sexes

unison n complete agreement; (*Music*) singing or playing of the same notes together at the same time

unit n single undivided entity or whole; group or individual regarded as a basic element of a larger whole; fixed quantity etc., used as a standard of measurement; piece of furniture designed to be fitted with other similar pieces ► **unit trust** investment trust that issues units for public sale and invests the money in many different businesses

Unitarian n person who believes that God is one being and rejects the Trinity ► **Unitarianism** n

unitary adj consisting of a single undivided whole; of a unit or units

unite v make or become an integrated whole; (cause to) enter into an association or alliance

unity n state of being one; mutual agreement

universe n whole of all existing matter, energy, and space; the world ► **universal** adj of or typical of the whole of mankind or of nature; existing everywhere ► **universally** adv ► **universality** n

university n, pl -**ties** institution of higher education with the authority to award degrees

unkempt adj (of the hair) not combed; slovenly or untidy

unknown adj not known; not famous ▷ n unknown person, quantity, or thing

unleaded adj (of petrol) containing less tetraethyl lead, in order to reduce environmental pollution

unleash v set loose or cause (something bad)

unless conj except under the circumstances that

unlike adj dissimilar or different ▷ prep not like or typical of

unlikely adj improbable

unload v remove (cargo) from (a ship, truck, or plane); remove the ammunition from (a firearm)

unmanned adj having no personnel or crew

unmask v remove the mask or disguise from; (cause to) appear in true character

unmentionable adj unsuitable as a topic of conversation

unmistakable, unmistakeable adj not ambiguous, clear ► **unmistakably, unmistakeably** adv

unmitigated adj not reduced or lessened in severity etc.; total and complete

unmoved adj not affected by emotion, indifferent

unnatural adj strange and frightening because not usual; not in accordance with accepted standards of behaviour

unnerve v cause to lose courage, confidence, or self-control

unnumbered adj countless; not counted or given a number

unorthodox adj (of ideas, methods, etc.) unconventional and not generally accepted; (of a person) having unusual opinions or methods

unpack v remove the contents of (a suitcase, trunk, etc.); take (something) out of a packed container

unparalleled adj not equalled, supreme

unpick v undo (the stitches) of (a piece of sewing)

unpleasant adj not pleasant or agreeable ► **unpleasantly** adv ► **unpleasantness** n

unprecedented adj never having happened before, unparalleled

unprintable adj unsuitable for printing for reasons of obscenity or libel

unprofessional adj contrary to the accepted code of a profession ► **unprofessionally** adv

unqualified adj lacking the necessary qualifications; total or complete

unravel v -elling, -elled reduce (something knitted or woven) to separate strands; become unravelled; explain or solve

unreasonable adj immoderate or excessive; refusing to listen to reason ▸ **unreasonably** adv

unremitting adj never slackening or stopping

unrequited adj not returned e.g. unrequited love

unrest n rebellious state of discontent

unrivalled adj having no equal

unroll v open out or unwind (something rolled or coiled) or (of something rolled or coiled) become opened out or unwound

unruly adj -lier, -liest difficult to control or organize

unsavoury adj distasteful or objectionable

unscathed adj not harmed or injured

unscrupulous adj prepared to act dishonestly, unprincipled

unseat v throw or displace from a seat or saddle; depose from an office or position

unsettled adj lacking order or stability; disturbed and restless; constantly changing or moving from place to place

unsightly adj unpleasant to look at

unsocial adj (also **unsociable**) avoiding the company of other people; falling outside the normal working day e.g. unsocial hours

unsound adj unhealthy or unstable; not based on truth or fact

unspeakable adj indescribably bad or evil ▸ **unspeakably** adv

unstable adj lacking stability or firmness; having abrupt changes of mood or behaviour

unstinting adj generous, gladly given e.g. unstinting praise

unsuitable adj not right or appropriate for a particular purpose ▸ **unsuitably** adv

unsuited adj not appropriate for a particular task or situation

unsung adj not acclaimed or honoured e.g. unsung heroes

unswerving adj firm, constant, not changing

unsympathetic adj not feeling or showing sympathy; unpleasant, not likeable; (foll by to) opposed to

unthinkable adj out of the question, inconceivable

untidy adj messy and disordered ▸ **untidily** adv ▸ **untidiness** n

untie v open or free (something that is tied); free from constraint

until conj up to the time that ▸ prep in or throughout the period before **not until** not before (a time or event)

untimely adj occurring before the expected or normal time; inappropriate to the occasion or time

unto prep (old-fashioned) to

untold adj incapable of description; incalculably great in number or quantity

untouchable adj above reproach or suspicion; unable to be touched ▸ n (offens) member of the lowest Hindu caste in India

untoward adj causing misfortune or annoyance

untrue adj incorrect or false; disloyal or unfaithful ▸ **untruth** n statement that is not true, lie

unusual adj uncommon or extraordinary ▸ **unusually** adv

unutterable adj incapable of being expressed in words
▶ **unutterably** adv

unvarnished adj not elaborated upon e.g. *the unvarnished truth*

unveil v ceremonially remove the cover from (a new picture, plaque, etc.); make public (a secret)
▶ **unveiling** n

unwarranted adj not justified, not necessary

unwieldy adj too heavy, large, or awkward to be easily handled

unwind v relax after a busy or tense time; slacken, undo, or unravel

unwitting adj not intentional; not knowing or conscious
▶ **unwittingly** adv

unwonted adj out of the ordinary

unworthy adj not deserving or worthy; lacking merit or value **unworthy of** beneath the level considered befitting (to)

unwrap v remove the wrapping from (something)

unwritten adj not printed or in writing; operating only through custom

up prep, adv indicating movement to or position at a higher place ▷ adv indicating readiness, intensity or completeness, etc. e.g. *warm up; drink up* ▷ adj of a high or higher position; out of bed ▷ v **upping, upped** increase or raise **up against** having to cope with **up and** (informal) do something suddenly e.g. *he upped and left* **what's up?** (informal) what is wrong? ▶ **ups and downs** alternating periods of good and bad luck ▶ **upward** adj directed or moving towards a higher place or level ▷ adv (also **upwards**) from a lower to a higher place, level, or condition

upbeat adj (informal) cheerful and optimistic ▷ n (Music) unaccented beat

upbraid v scold or reproach

upbringing n education of a person during the formative years

update v bring up to date

upend v turn or set (something) on its end

upfront adj open and frank ▷ adv, adj (of money) paid out at the beginning of a business arrangement

upgrade v promote (a person or job) to a higher rank

upheaval n strong, sudden, or violent disturbance

uphill adj sloping or leading upwards; requiring a great deal of effort ▷ adv up a slope ▷ n (SAfr) difficulty

uphold v maintain or defend against opposition; give moral support to ▶ **upholder** n

upholster v fit (a chair or sofa) with padding, springs, and covering
▶ **upholsterer** n

upholstery n soft covering on a chair or sofa

upkeep n act, process, or cost of keeping something in good repair

upland adj of or in an area of high or relatively high ground ▶ **uplands** pl n area of high or relatively high ground

uplift v raise or lift up; raise morally or spiritually ▷ n act or process of improving moral, social, or cultural conditions ▶ **uplifting** adj

upload v transfer (data or a program) from one's own computer into the memory of another computer

upon prep on; up and on

upper adj higher or highest in physical position, wealth, rank, or status ▷ n part of a shoe above

u

the sole ▸ **uppermost** adj highest in position, power, or importance ▹ adv in or into the highest place or position ▸ **upper class** highest social class ▸ **upper-class** adj ▸ **upper crust** (Brit, Aust & NZ informal) upper class ▸ **upper hand** position of control

uppish, uppity adj (Brit informal) snobbish, arrogant, or presumptuous

upright adj vertical or erect; honest or just ▹ adv vertically or in an erect position ▹ n vertical support, such as a post ▸ **uprightness** n

uprising n rebellion or revolt

uproar n disturbance characterized by loud noise and confusion ▸ **uproarious** adj very funny; (of laughter) loud and boisterous ▸ **uproariously** adv

uproot v pull up by, or as if by, the roots; displace (a person or people) from their native or usual surroundings

upscale adj expensive and of superior quality

upset adj emotionally or physically disturbed or distressed ▹ v tip over; disturb the normal state or stability of; disturb mentally or emotionally; make physically ill ▹ n unexpected defeat or reversal; disturbance or disorder of the emotions, mind, or body ▸ **upsetting** adj

upshot n final result or conclusion

upside down adj turned over completely; (informal) confused or jumbled ▹ adv in an inverted fashion; in a chaotic manner

upstage adj at the back half of the stage ▹ v (informal) draw attention to oneself from (someone else)

upstairs adv to or on an upper floor of a building ▹ n upper floor ▹ adj situated on an upper floor

upstanding adj of good character

upstart n person who has risen suddenly to a position of power and behaves arrogantly

upstream adv, adj in or towards the higher part of a stream

upsurge n rapid rise or swell

uptake n **quick on the uptake, slow on the uptake** (informal) quick or slow to understand or learn

uptight adj (informal) nervously tense, irritable, or angry

up-to-date adj modern or fashionable

upturn n upward trend or improvement ▸ **upturned** adj facing upwards

uranium n (Chem) radioactive silvery-white metallic element, used chiefly as a source of nuclear energy

Uranus n (Greek myth) god of the sky; seventh planet from the sun

urban adj of or living in a city or town; denoting modern pop music of African-American origin, such as hip-hop ▸ **urbanize** v make (a rural area) more industrialized and urban ▸ **urbanization** n

urbane adj characterized by courtesy, elegance, and sophistication ▸ **urbanity** n

urchin n mischievous child

Urdu [oor-doo] n language of Pakistan

urethra [yew-reeth-ra] n canal that carries urine from the bladder out of the body

urge n strong impulse, inner drive, or yearning ▹ v plead with or press (a person to do something); advocate earnestly; force or drive onwards

urgent adj requiring speedy action or attention ▸ **urgency** n ▸ **urgently** adv

urine n pale yellow fluid excreted by the kidneys to the bladder and passed as waste from the body ▶ **urinary** adj ▶ **urinate** v discharge urine ▶ **urination** n ▶ **urinal** n sanitary fitting used by men for urination

URL uniform resource locator: a standardized address of a location on the internet

urn n vase used as a container for the ashes of the dead; large metal container with a tap, used for making and holding tea or coffee

ursine adj of or like a bear

US, USA United States (of America)

us pron objective case of **we**

USB Universal Serial Bus: standard for connecting sockets on computers ▶ **USB drive** n (Computing) small portable data storage device with a USB connection

use v put into service or action; take advantage of, exploit; consume or expend ▶ n using or being used; ability or permission to use; usefulness or advantage; purpose for which something is used ▶ **user** n ▶ **user-friendly** adj easy to familiarize oneself with, understand, and use ▶ **username** n (Computing) name entered into a computer for identification purposes ▶ **usable** adj able to be used ▶ **usage** n regular or constant use; way in which a word is used in a language ▶ **use-by date** (Aust, NZ & S Afr) date on packaged food after which it should not be sold ▶ **used** adj second-hand ▶ **used to** adj accustomed to ▶ v used as an auxiliary to express past habitual or accustomed actions e.g. I used to live there ▶ **useful** adj ▶ **usefully** adv ▶ **usefulness** n ▶ **useless** adj ▶ **uselessly** adv ▶ **uselessness** n

usher n official who shows people to their seats, as in a

church ▶ v conduct or escort ▶ **usherette** n female assistant in a cinema who shows people to their seats

USSR (formerly) Union of Soviet Socialist Republics

usual adj of the most normal, frequent, or regular type ▶ **usually** adv most often, in most cases

> **SPELLING TIP**
> The Collins Corpus shows that it's very common to write usualy, forgetting the double l of **usually**

usurp [yewz-zurp] v seize (a position or power) without authority ▶ **usurpation** n ▶ **usurper** n

usury n practice of lending money at an extremely high rate of interest ▶ **usurer** [yewz-yoor-er] n

ute [yoot] n (Aust & NZ informal) utility truck

utensil n tool or container for practical use e.g. cooking utensils

uterus [yew-ter-russ] n womb ▶ **uterine** adj

utilitarian adj useful rather than beautiful; of utilitarianism ▶ **utilitarianism** n (Ethics) doctrine that the right action is the one that brings about the greatest good for the greatest number of people

utility n usefulness; (pl **-ties**) public service, such as electricity ▶ adj designed for use rather than beauty ▶ **utility room** room used for large domestic appliances and equipment ▶ **utility truck** (Aust & NZ) small truck with an open body and low sides

utilize v make practical use of ▶ **utilization** n

utmost adj, n (of) the greatest possible degree or amount e.g. the utmost point; I was doing my utmost to comply

u

Utopia [yew-**tope**-ee-a] *n* any real or imaginary society, place, or state considered to be perfect or ideal ▶ **Utopian** *adj*

utter[1] *v* express (something) in sounds or words ▶ **utterance** *n* something uttered; act or power of uttering

utter[2] *adj* total or absolute ▶ **utterly** *adv*

uttermost *adj*, *n* same as **utmost**

U-turn *n* turn, made by a vehicle, in the shape of a U, resulting in a reversal of direction; complete change in policy e.g. *a humiliating U-turn by the Prime Minister*

UV ultraviolet

uvula [yew-**view**-la] *n* small fleshy part of the soft palate that hangs in the back of the throat ▶ **uvular** *adj*

uxorious [ux-**or**-ee-uss] *adj* excessively fond of or dependent on one's wife

Vv

V volt

v. versus; very

vacant *adj* (of a toilet, room, etc.) unoccupied; without interest or understanding ▶ **vacantly** *adv* ▶ **vacancy** *n*, *pl* -**cies** unfilled job; unoccupied room in a guesthouse; state of being unoccupied

vacate *v* cause (something) to be empty by leaving; give up (a job or position) ▶ **vacation** *n* time when universities and law courts are closed; (*chiefly US*) holiday

vaccinate *v* inject with a vaccine ▶ **vaccination** *n* ▶ **vaccine** *n* substance designed to cause a mild form of a disease to make a person immune to the disease itself

vacillate [**vass**-ill-late] *v* keep changing one's mind or opinions ▶ **vacillation** *n*

vacuous *adj* not expressing intelligent thought ▶ **vacuity** *n*

vacuum *n*, *pl* **vacuums** *or* **vacua** empty space from which all or most air or gas has been removed ▶ *v* clean with a vacuum cleaner ▶ **vacuum cleaner** electrical appliance which sucks up dust and dirt from carpets and upholstery ▶ **vacuum flask** double-walled flask with a vacuum between the walls that keeps drinks hot or cold ▶ **vacuum-packed** *adj* contained in packaging from which the air has been removed

vagabond *n* person with no fixed home, esp. a beggar

vagary [vaig-a-ree] *n, pl* **-garies** unpredictable change

vagina [vaj-jine-a] *n* (in female mammals) passage from the womb to the external genitals ▶ **vaginal** *adj*

vagrant [vaig-rant] *n* person with no settled home ▷ *adj* wandering ▶ **vagrancy** *n*

vague *adj* not clearly explained; unable to be seen or heard clearly; absent-minded ▶ **vaguely** *adv*

vain *adj* excessively proud, esp. of one's appearance; bound to fail, futile **in vain** unsuccessfully

vainglorious *adj* (*lit*) boastful

valance [val-lenss] *n* piece of drapery round the edge of a bed

vale *n* (*lit*) valley

valedictory [val-lid-**dik**-tree] *adj* (of a speech, performance, etc.) intended as a farewell ▶ **valediction** *n* farewell speech

valence [vale-ence] *n* molecular bonding between atoms

valency *n, pl* **-cies** power of an atom to make molecular bonds

valentine *n* (person to whom one sends) a romantic card on Saint Valentine's Day, 14 February

valerian *n* herb used as a sedative

valet *n* man's personal male servant

valetudinarian [val-lit-yew-din-**air**-ee-an] *n* person with a long-term illness; person overconcerned about his or her health

valiant *adj* brave or courageous

valid *adj* soundly reasoned; having legal force ▶ **validate** *v* make valid ▶ **validation** *n* ▶ **validity** *n*

valise [val-**leez**] *n* (*old-fashioned*) small suitcase

Valium® *n* drug used as a tranquillizer

valley *n* low area between hills, often with a river running through it

valour *n* (*lit*) bravery

value *n* importance, usefulness; monetary worth ▷ *pl* moral principles ▷ *v* **valuing, valued** assess the worth or desirability of; have a high regard for ▶ **valuable** *adj* having great worth ▶ **valuables** *pl n* valuable personal property ▶ **valuation** *n* assessment of worth ▶ **valueless** *adj* ▶ **valuer** *n* ▶ **value-added tax** (*Brit & S Afr*) see **VAT** ▶ **value judgment** opinion based on personal belief

valve *n* device to control the movement of fluid through a pipe; (*Anat*) flap in a part of the body allowing blood to flow in one direction only; (*Physics*) tube containing a vacuum, allowing current to flow from a cathode to an anode ▶ **valvular** *adj*

vamp¹ *n* (*informal*) sexually attractive woman who seduces men

vamp² *v* **vamp up** make (a story, piece of music, etc.) seem new by inventing additional parts

vampire *n* (in folklore) corpse that rises at night to drink the blood of the living ▶ **vampire bat** tropical bat that feeds on blood

van¹ *n* motor vehicle for transporting goods; railway carriage for goods, luggage, or mail

van² *n* short for **vanguard**

vanadium *n* (*Chem*) metallic element, used in steel

vandal *n* person who deliberately damages property ▶ **vandalism** *n* ▶ **vandalize** *v*

vane *n* flat blade on a rotary device such as a weathercock or propeller

vanguard *n* unit of soldiers leading an army; most advanced group or position in a movement or activity

vanilla *n* seed pod of a tropical climbing orchid, used for flavouring

V

vanish v disappear suddenly or mysteriously; cease to exist

vanity n, pl **-ties** (display of) excessive pride

vanquish v (lit) defeat (someone) utterly

vantage n **vantage point** position that gives one an overall view

vape v (informal) inhale vapour from an e-cigarette

vapid adj lacking character, dull

vapour n moisture suspended in air as steam or mist; gaseous form of something that is liquid or solid at room temperature ▶ **vaporize** v ▶ **vaporizer** n ▶ **vaporous** adj

variable adj not always the same, changeable ▷ n (Maths) expression with a range of values ▶ **variability** n

variant adj differing from a standard or type ▷ n something that differs from a standard or type **at variance** in disagreement

variation n something presented in a slightly different form; difference in level, amount, or quantity; (Music) repetition in different forms of a basic theme

varicose veins pl n knotted and swollen veins, esp. in the legs

variegated adj having patches or streaks of different colours ▶ **variegation** n

variety n, pl **-ties** state of being diverse or various; different things of the same kind; particular sort or kind; light entertainment composed of unrelated acts

various adj of several kinds ▶ **variously** adv

varnish n solution of oil and resin, put on a surface to make it hard and glossy ▷ v apply varnish to

varsity n (informal) university

vary v varying, varied change; cause differences in ▶ **varied** adj

vascular adj (Biol) relating to vessels

vas deferens n, pl **vasa deferentia** (Anat) sperm-carrying duct in each testicle

vase n ornamental jar, esp. for flowers

vasectomy n, pl **-mies** surgical removal of part of the vas deferens, as a contraceptive method

Vaseline® n thick oily cream made from petroleum, used in skin care

vassal n (Hist) man given land by a lord in return for military service; subordinate person or nation ▶ **vassalage** n

vast adj extremely large ▶ **vastly** adv ▶ **vastness** n

VAT (Brit & S Afr) value-added tax: tax on the difference between the cost of materials and the selling price

vat n large container for liquids

Vatican n the Pope's palace

vaudeville n variety entertainment of songs and comic turns

vault[1] n secure room for storing valuables; underground burial chamber ▶ **vaulted** adj having an arched roof

vault[2] v jump over (something) by resting one's hand(s) on it ▷ n such a jump

vaunt v describe or display (success or possessions) boastfully ▶ **vaunted** adj

VC Vice Chancellor; Victoria Cross

VCD video compact disc: an optical disc used to store computer, audio or video data

VCR video cassette recorder

VD venereal disease

VDU visual display unit

veal n calf meat

vector n (Maths) quantity that has size and direction, such as force; animal, usu. an insect, that carries disease

veer v change direction suddenly

vegan [vee-gan] n person who eats no meat, fish, eggs, or dairy products ▷ adj suitable for a vegan ▶ **veganism** n

vegetable n edible plant; (offens) severely brain-damaged person ▷ adj of or like plants or vegetables

vegetarian n person who eats no meat or fish ▷ adj suitable for a vegetarian ▶ **vegetarianism** n

vegetate v live a dull boring life with no mental stimulation

vegetation n plant life of a given place

vehement adj expressing strong feelings ▶ **vehemence** n ▶ **vehemently** adv

vehicle n machine, esp. with an engine and wheels, for carrying people or objects; something used to achieve a particular purpose or as a means of expression ▶ **vehicular** adj

veil n piece of thin cloth covering the head or face; something that masks the truth e.g. a veil of secrecy ▷ v cover with or as if with a veil **take the veil** become a nun ▶ **veiled** adj disguised

vein n tube that takes blood to the heart; line in a leaf or an insect's wing; layer of ore or mineral in rock; streak in marble, wood, or cheese; feature of someone's writing or speech e.g. a vein of humour; mood or style e.g. in a lighter vein ▶ **veined** adj

Velcro® n fastening consisting of one piece of fabric with tiny hooked threads and another with a coarse surface that sticks to it

veld, veldt n high grassland in southern Africa ▶ **veldskoen, velskoen** n (SAfr) leather ankle boot

vellum n fine calfskin parchment; type of strong good-quality paper

velocity n, pl **-ties** speed of movement in a given direction

velour, velours [vel-loor] n fabric similar to velvet

velvet n fabric with a thick soft pile ▶ **velvety** adj soft and smooth ▶ **velveteen** n cotton velvet

venal adj easily bribed; characterized by bribery

vend v sell ▶ **vendor** n ▶ **vending machine** machine that dispenses goods when coins are inserted

vendetta n prolonged quarrel between families, esp. one involving revenge killings

veneer n thin layer of wood etc. covering a cheaper material; superficial appearance e.g. a veneer of sophistication

venerable adj worthy of deep respect ▶ **venerate** v hold (a person) in deep respect ▶ **veneration** n

venereal disease [ven-eer-ee-al] n disease transmitted sexually

Venetian adj of Venice, port in NE Italy ▶ **Venetian blind** window blind made of thin horizontal slats that turn to let in more or less light

vengeance n revenge ▶ **vengeful** adj wanting revenge

venial [veen-ee-al] adj (of a sin or fault) easily forgiven

venison n deer meat

venom n malice or spite; poison produced by snakes etc. ▶ **venomous** adj

venous adj (Anat) of veins

vent¹ n outlet releasing fumes or fluid ▷ v express (an emotion) freely **give vent to** release (an emotion) in an outburst

vent² n vertical slit in a jacket

ventilate v let fresh air into; discuss (ideas or feelings) openly ▶ **ventilation** n ▶ **ventilator** n

ventral adj relating to the front of the body

ventricle n (Anat) one of the four cavities of the heart or brain

ventriloquist n entertainer who can speak without moving his or her lips, so that a voice seems to come from elsewhere ▶ **ventriloquism** n

venture n risky undertaking, esp. in business ▷ v do something risky; dare to express (an opinion); go to an unknown place ▶ **venturesome** adj daring ▶ **venture capitalist** provider of capital for new commercial enterprises

venue n place where an organized gathering is held

Venus n planet second nearest to the sun; Roman goddess of love

veracity n habitual truthfulness ▶ **veracious** adj

verandah, veranda n open porch attached to a house

verb n word that expresses the idea of action, happening, or being ▶ **verbal** adj spoken; of a verb ▶ **verbally** adv ▶ **verbalize** v express (something) in words

verbatim [verb-**bait**-im] adv, adj word for word

verbena n plant with sweet-smelling flowers

verbiage n excessive use of words

verbose [verb-**bohss**] adj speaking at tedious length ▶ **verbosity** n

verdant adj (lit) covered in green vegetation

verdict n decision of a jury; opinion formed after examining the facts

verdigris [ver-**dig**-riss] n green film on copper, brass, or bronze

verdure n (lit) flourishing green vegetation

verge n grass border along a road **on the verge of** having almost reached (a point or condition) ▶ **verge on** v be near to (a condition)

verger n (C of E) church caretaker

verify v -ifying, -ified check the truth or accuracy of ▶ **verifiable** adj ▶ **verification** n

verily adv (obs) in truth

verisimilitude n appearance of being real or true

veritable adj rightly called, without exaggeration e.g. a veritable feast ▶ **veritably** adv

verity n, pl -ties true statement or principle

vermicelli [ver-me-**chell**-ee] n fine strands of pasta

vermiform adj shaped like a worm ▶ **vermiform appendix** (Anat) same as **appendix**

vermilion adj orange-red

vermin pl n animals, esp. insects and rodents, that spread disease or cause damage ▶ **verminous** adj

vermouth [ver-**muth**] n wine flavoured with herbs

vernacular [ver-**nak**-yew-lar] n most widely spoken language of a particular people or place

vernal adj occurring in spring

vernier [ver-**nee**-er] n movable scale on a graduated measuring instrument for taking readings in fractions

veronica n plant with small blue, pink, or white flowers

verruca [ver-**roo**-ka] n wart, usu. on the foot

versatile adj having many skills or uses ▶ **versatility** n

verse n group of lines forming part of a song or poem; poetry as distinct from prose; subdivision of a chapter of the Bible ▶ **versed in** knowledgeable about ▶ **versification** n technique of writing in verse

version n form of something, such as a piece of writing, with some differences from other forms;

account of an incident from a particular point of view

verso n, pl -**sos** left-hand page of a book

versus prep in opposition to or in contrast with; (Sport, Law) against

vertebra n, pl **vertebrae** one of the bones that form the spine ▶ **vertebral** adj ▶ **vertebrate** n, adj (animal) having a spine

vertex n, pl -**texes** or -**tices** (Maths) point on a geometric figure where the sides form an angle; highest point of a triangle

vertical adj straight up and down ▶ n vertical direction

vertigo n dizziness, usu. when looking down from a high place ▶ **vertiginous** adj

vervain n plant with spikes of blue, purple, or white flowers

verve n enthusiasm or liveliness

very adv more than usually, extremely ▶ adj absolute, exact e.g. the very top; the very man

vesicle n (Biol) sac or small cavity, esp. one containing fluid

vespers pl n (RC Church) (service of) evening prayer

vessel n ship; (lit) container, esp. for liquids; (Biol) tubular structure in animals and plants that carries body fluids, such as blood or sap

vest n undergarment worn on the top half of the body; (US & Aust) waistcoat ▶ v (foll by in or with) give (authority) to (someone) ▶ **vested interest** interest someone has in a matter because he or she might benefit from it

vestibule n small entrance hall

vestige [vest-ij] n small amount or trace ▶ **vestigial** adj

vestments pl n priest's robes

vestry n, pl -**tries** room in a church used as an office or robing room by the priest or minister

vet[1] n short for **veterinary surgeon** ▶ v **vetting, vetted** check the suitability of

vet[2] n (US, Aust & NZ) military veteran

vetch n climbing plant with a beanlike fruit used as fodder

veteran n person with long experience in a particular activity, esp. military service ▶ adj long-serving

veterinarian n (US) veterinary surgeon

veterinary adj concerning animal health ▶ **veterinary surgeon** medical specialist who treats sick animals

veto n, pl -**toes** official power to cancel a proposal ▶ v -**toing, -toed** enforce a veto against

vex v frustrate, annoy ▶ **vexation** n something annoying; being annoyed ▶ **vexatious** adj ▶ **vexed question** much debated subject

VHF very high frequency: radio frequency band between 30 and 300 MHz

via prep by way of

viable adj able to be put into practice; (Biol) able to live and grow independently ▶ **viability** n

viaduct n bridge over a valley

Viagra® [vie-**ag**-ra] n drug used to treat impotence in men

vial n same as **phial**

viands pl n (obs) food

vibes pl n (informal) emotional reactions between people; atmosphere of a place; short for **vibraphone**

vibrant [vibe-rant] adj vigorous in appearance, energetic; (of a voice) resonant; (of a colour) strong and bright

vibraphone n musical instrument with metal bars that resonate electronically when hit

vibrate v move back and forth rapidly; (cause to) resonate ▸ **vibration** n ▸ **vibrator** n device that produces vibratory motion, used for massage or as a sex aid ▸ **vibratory** adj

vibrato n, pl **-tos** (Music) rapid fluctuation in the pitch of a note

Vic Victoria

vicar n (C of E) member of the clergy in charge of a parish ▸ **vicarage** n vicar's house

vicarious [vick-air-ee-uss] adj felt indirectly by imagining what another person experiences; delegated ▸ **vicariously** adv

vice[1] n immoral or evil habit or action; habit regarded as a weakness in someone's character; criminal immorality, esp. involving sex

vice[2] n tool with a pair of jaws for holding an object while working on it

vice[3] adj serving in place of

vice chancellor n chief executive of a university

vice president n officer ranking immediately below the president and serving as his or her deputy ▸ **vice-presidency** n

viceroy n governor of a colony who represents the monarch ▸ **viceregal** adj

vice versa [vie-see ver-sa] adv (Latin) conversely, the other way round

vicinity [viss-in-it-ee] n surrounding area

vicious adj cruel and violent ▸ **viciously** adv ▸ **vicious circle**, **vicious cycle** situation in which an attempt to resolve one problem creates new problems that recreate the original one

vicissitudes [viss-iss-it-yewds] pl n changes in fortune

victim n person or thing harmed or killed ▸ **victimize** v punish unfairly; discriminate against ▸ **victimization** n

victor n person who has defeated an opponent, esp. in war or in sport

Victoria Cross n (Brit) highest award for bravery in battle

Victorian adj of or in the reign of Queen Victoria (1837–1901); characterized by prudery or hypocrisy; of or relating to the Australian state of Victoria

victory n winning of a battle or contest ▸ **victorious** adj

victuals [vit-tals] pl n (old-fashioned) food and drink

vicuña [vik-koo-nya] n S American animal similar to the llama; fine cloth made from its wool

video n, pl **-os** short sequence of moving images; short for **video cassette**, **video cassette recorder** ▸ v **videoing**, **videoed** record (a TV programme or event) on video ▸ adj relating to or used in producing television images ▸ **video cassette** n cassette containing video tape ▸ **video cassette recorder** n tape recorder for recording and playing back TV programmes and films ▸ **video tape** n magnetic tape used to record video-frequency signals in TV production; magnetic tape used to record programmes when they are broadcast ▸ **videotape** v record (a TV programme) on video tape ▸ **video tape recorder** n tape recorder for vision signals, used in TV production

vie v **vying, vied** compete (with someone)

Vietnamese adj of Vietnam, in SE Asia ▷ n (pl **-ese**) native of Vietnam; language of Vietnam

view n opinion or belief; everything that can be seen from a given place; picture of this ▷ v think of (something) in a particular way **in view of** taking into consideration **on view** exhibited to the public ▶ **viewer** n person who watches television; hand-held device for looking at photographic slides ▶ **viewfinder** n window on a camera showing what will appear in a photograph

Viewdata® n videotext service linking users to a computer by telephone

vigil [vij-ill] n a night-time period of staying awake to look after a sick person, pray, etc. ▶ **vigilant** adj watchful in case of danger ▶ **vigilance** n

vigilante [vij-ill-ant-ee] n person, esp. as one of a group, who takes it upon himself or herself to enforce the law

vignette [vin-yet] n concise description of the typical features of something; small decorative illustration in a book

vigour n physical or mental energy ▶ **vigorous** adj ▶ **vigorously** adv

Viking n (Hist) seafaring raider and settler from Scandinavia

vile adj very wicked; disgusting ▶ **vilely** adv ▶ **vileness** n

vilify v **-ifying, -ified** attack the character of ▶ **vilification** n

villa n large house with gardens; holiday home, usu. in the Mediterranean

village n small group of houses in a country area; rural community ▶ **villager** n

villain n wicked person; main wicked character in a play ▶ **villainous** adj ▶ **villainy** n

villein [vill-an] n (Hist) peasant bound in service to his lord

vinaigrette n salad dressing of oil and vinegar

vindicate v clear (someone) of guilt; provide justification for ▶ **vindication** n

vindictive adj maliciously seeking revenge ▶ **vindictiveness** n ▶ **vindictively** adv

vine n climbing plant, esp. one producing grapes ▶ **vineyard** [vinn-yard] n plantation of grape vines, esp. for making wine

vinegar n acid liquid made from wine, beer, or cider ▶ **vinegary** adj

vino [vee-noh] n (informal) wine

vintage n wine from a particular harvest of grapes ▷ adj best and most typical ▶ **vintage car** car built between 1919 and 1930

vintner n dealer in wine

vinyl [vine-ill] n type of plastic, used in mock leather and records; (collectively) records made from vinyl

viol [vie-oll] n early stringed instrument preceding the violin

viola¹ [vee-oh-la] n stringed instrument lower in pitch than a violin

viola² [vie-ol-a] n variety of pansy

violate v break (a law or agreement); disturb (someone's privacy); treat (a sacred place) disrespectfully; rape ▶ **violation** n ▶ **violator** n

violence n use of physical force, usu. intended to cause injury or destruction; great force or strength in action, feeling, or expression ▶ **violent** adj ▶ **violently** adv

violet n plant with bluish-purple flowers ▷ adj bluish-purple

V

violin n small four-stringed musical instrument played with a bow.
▶ **violinist** n

VIP very important person

viper n poisonous snake

virago [vir-**rah**-go] n, pl **-goes** or **-gos** aggressive woman

viral adj of or caused by a virus ▷ adv **go viral** spread quickly and widely among internet users

virgin n person, esp. a woman, who has not had sexual intercourse ▷ adj not having had sexual intercourse; not yet exploited or explored ▶ **virginal** adj like a virgin ▷ n early keyboard instrument like a small harpsichord ▶ **virginity** n

virile adj having the traditional male characteristics of physical strength and a high sex drive ▶ **virility** n

virology n study of viruses

virtual adj having the effect but not the form of; of or relating to virtual reality ▶ **virtually** adv practically, almost ▶ **virtual reality** computer-generated environment that seems real to the user

virtue n moral goodness; positive moral quality; merit **by virtue of** by reason of ▶ **virtuous** adj morally good ▶ **virtuously** adv

virtuoso n, pl **-sos** or **-si** person with impressive esp. musical skill ▶ **virtuosity** n

virulent [vir-**yew**-lent] adj very infectious; violently harmful

virus n microorganism that causes disease in humans, animals, and plants; (Computing) program that propagates itself, via disks and electronic networks, to cause disruption

visa n permission to enter a country, granted by its government and shown by a stamp on one's passport

visage [**viz**-zij] n (lit) face

vis-à-vis [veez-ah-**vee**] prep in relation to, regarding

viscera [**viss**-er-a] pl n large abdominal organs

visceral [**viss**-er-al] adj instinctive; of or relating to the viscera

viscid [**viss**-id] adj sticky

viscose n synthetic fabric made from cellulose

viscount [**vie**-count] n British nobleman ranking between an earl and a baron

viscountess [**vie**-count-iss] n woman holding the rank of viscount in her own right; wife or widow of a viscount

viscous adj thick and sticky ▶ **viscosity** n

visible adj able to be seen; able to be perceived by the mind. ▶ **visibly** adv ▶ **visibility** n range or clarity of vision

vision n ability to see; mental image of something; foresight; hallucination ▶ **visionary** adj showing foresight; idealistic but impractical ▷ n visionary person

visit v -**iting**, -**ited** go or come to see; stay temporarily with; (foll by upon) (lit) afflict ▷ n instance of visiting; official call ▶ **visitor** n ▶ **visitation** n formal visit or inspection; catastrophe seen as divine punishment

visor [**vize**-or] n transparent part of a helmet that pulls down over the face; eyeshade, esp. in a car; peak on a cap

vista n (beautiful) extensive view

visual adj done by or used in seeing; designed to be looked at ▶ **visualize** v form a mental image of ▶ **visualization** n ▶ **visual display unit** device with a screen for displaying data held in a computer

vital adj essential or highly important; lively; necessary to maintain life ▶ **vitals** pl n bodily organs necessary to maintain life ▸ **vitally** adv ▸ **vitality** n physical or mental energy ▸ **vital statistics** statistics of births, deaths, and marriages; (informal) woman's bust, waist, and hip measurements

vitamin n one of a group of substances that are essential in the diet for specific body processes

vitiate [vish-ee-ate] v spoil the effectiveness of

viticulture n cultivation of grapevines

vitreous adj like or made from glass

vitriol n language expressing bitterness and hatred; sulphuric acid ▸ **vitriolic** adj

vituperative [vite-tyew-pra-tiv] adj bitterly abusive ▸ **vituperation** n

viva[1] [veev-a] interj long live (a person or thing)

viva[2] [vive-a] n (Brit) examination in the form of an interview

vivace [viv-vah-chee] adv (Music) in a lively manner

vivacious adj full of energy and enthusiasm ▸ **vivacity** n

viva voce [vive-a voh-chee] adv by word of mouth ▸ n same as **viva**[2]

vivid adj very bright; conveying images that are true to life ▸ **vividly** adv ▸ **vividness** n

vivisection n performing surgical experiments on living animals ▸ **vivisectionist** n

vixen n female fox; (Brit, Aust & NZ informal) spiteful woman

viz. (introducing specified items) namely

vizier [viz-zeer] n high official in certain Muslim countries

vizor n same as **visor**

v-mail n video message sent by e-mail

vocabulary n, pl **-aries** all the words that a person knows; all the words in a language; specialist terms used in a given subject; list of words in another language with their translation

vocal adj relating to the voice; outspoken ▸ **vocals** pl n singing part of a piece of pop music ▸ **vocally** adv ▸ **vocalist** n singer ▸ **vocalize** v express with or use the voice ▸ **vocalization** n ▸ **vocal cords** membranes in the larynx that vibrate to produce sound

vocation n profession or trade; occupation that someone feels called to ▸ **vocational** adj directed towards a particular profession or trade

vociferous adj shouting, noisy

VOD video on demand: TV system allowing users to watch programmes at a time of their own choosing

vodka n (Russian) spirit distilled from potatoes or grain

voetsek interj (S Afr offens) expression of rejection

vogue n popular style; period of popularity

voice n (quality of) sound made when speaking or singing; expression of opinion by a person or group; property of verbs that makes them active or passive ▸ v express verbally ▸ **voiceless** adj ▸ **voice mail** electronic system for the transfer and storage of telephone messages, which can be dealt with by the user at a later time ▸ **voice-over** n film commentary spoken by someone off-camera

void adj not legally binding; empty ▸ n empty space ▸ v make invalid; empty

voile [voyl] n light semitransparent fabric

vol. volume

volatile adj liable to sudden change, esp. in behaviour; evaporating quickly ▸ **volatility** n

V

vol-au-vent [voll-oh-von] *n* small puff-pastry case with a savoury filling

volcano *n, pl* **-noes** *or* **-nos** mountain with a vent through which lava is ejected ► **volcanic** *adj*

vole *n* small rodent

volition *n* ability to decide things for oneself **of one's own volition** through one's own choice

volley *n* simultaneous discharge of ammunition; burst of questions or critical comments; (*Sport*) stroke or kick at a moving ball before it hits the ground ► *v* discharge (ammunition) in a volley; hit or kick (a ball) in a volley ► **volleyball** *n* team game where a ball is hit with the hands over a high net

volt *n* unit of electric potential ► **voltage** *n* electric potential difference expressed in volts ► **voltmeter** *n* instrument for measuring voltage

volte-face [volt-fass] *n* reversal of opinion

voluble *adj* talking easily and at length ► **volubility** *n* ► **volubly** *adv*

volume *n* size of the space occupied by something; amount; loudness of sound; book, esp. one of a series ► **voluminous** *adj* (of clothes) large and roomy; (of writings) extensive ► **volumetric** *adj* relating to measurement by volume

voluntary *adj* done by choice; done or maintained without payment; (of muscles) controlled by the will ► *n, pl* **-taries** organ solo in a church service ► **voluntarily** *adv*

volunteer *n* person who offers voluntarily to do something; person who voluntarily undertakes military service ► *v* offer one's services; give

(information) willingly; offer the services of (another person)

voluptuous *adj* (of a woman) sexually alluring through fullness of figure; sensually pleasurable ► **voluptuary** *n* person devoted to sensual pleasures

volute *n* spiral or twisting turn, form, or object

vomit *v* **-iting, -ited** eject (the contents of the stomach) through the mouth ► *n* matter vomited

voodoo *n* religion involving ancestor worship and witchcraft, practised in the West Indies, esp. in Haiti.

voracious *adj* craving great quantities of food; insatiably eager ► **voraciously** *adv* ► **voracity** *n*

vortex *n, pl* **-texes** *or* **-tices** whirlpool

vote *n* choice made by a participant in a shared decision, esp. in electing a candidate; right to this choice; total number of votes cast; collective voting power of a given group e.g. *the Black vote* ► *v* make a choice by a vote; authorize (something) by vote ► **voter** *n*

votive *adj* done or given to fulfil a vow

vouch *v* **vouch for** give one's personal assurance about; provide evidence for

voucher *n* ticket used instead of money to buy specified goods; record of a financial transaction; receipt

vouchsafe *v* (*old-fashioned*) give, entrust

vow *n* solemn and binding promise ► *pl* formal promises made when marrying or entering a religious order ► *v* promise solemnly

vowel *n* speech sound made without obstructing the flow of breath; letter representing this

vox pop n (Brit) interviews with members of the public on TV or radio

vox populi n public opinion

voyage n long journey by sea or in space ▷ v make a voyage ▶ **voyager** n

voyeur n person who obtains pleasure from watching people undressing or having sex ▶ **voyeurism** n

vs versus

V-sign n offensive gesture made by sticking up the index and middle fingers with the palm inwards; similar gesture, with the palm outwards, meaning victory or peace

VSO (in Britain) Voluntary Service Overseas

VSOP (of brandy or port) very superior old pale

VTOL vertical takeoff and landing

VTR video tape recorder

vulcanize v strengthen (rubber) by treating it with sulphur

vulgar adj showing lack of good taste, decency, or refinement ▶ **vulgarly** adv ▶ **vulgarity** n ▶ **vulgarian** n vulgar (rich) person ▶ **vulgar fraction** simple fraction

Vulgate n fourth-century Latin version of the Bible

vulnerable adj liable to be physically or emotionally hurt; exposed to attack ▶ **vulnerability** n

vulpine adj of or like a fox

vulture n large bird that feeds on the flesh of dead animals

vulva n woman's external genitals

vuvuzela [voo-voo-**zay**-la] n (S Afr) plastic instrument that is blown to make a trumpeting sound

vying v present participle of **vie**

Ww

W watt; West(ern)

WA Western Australia

wacky adj **wackier, wackiest** (informal) eccentric or funny ▶ **wackiness** n

wad n small mass of soft material; roll or bundle, esp. of banknotes ▶ **wadding** n soft material used for padding or stuffing

waddle v walk with short swaying steps ▷ n swaying walk

waddy n, pl **-dies** heavy wooden club used by Australian Aborigines

wade v walk with difficulty through water or mud; proceed with difficulty ▶ **wader** n long-legged water bird ▷ pl angler's long waterproof boots

wadi [**wod**-dee] n, pl **-dies** (in N Africa and Arabia) river which is dry except in the wet season

wafer n thin crisp biscuit; thin disc of unleavened bread used at Communion; thin slice

waffle¹ (informal) v speak or write in a vague wordy way ▷ n vague wordy talk or writing

waffle² n square crisp pancake with a gridlike pattern

waft v drift or carry gently through the air ▷ n something wafted

Wag n (informal) wife or girlfriend of a famous sportsman

wag v **wagging, wagged** move rapidly from side to side ▷ n wagging movement; (old-fashioned) humorous witty person ▶ **wagtail** n small long-tailed bird

W

wage n (often pl) payment for work done, esp. when paid weekly ▷ v engage in (an activity)

wager n, v bet on the outcome of something

waggle v move with a rapid shaking or wobbling motion

wagon, waggon n four-wheeled vehicle for heavy loads; railway freight truck

wahoo n food and game fish of tropical seas

waif n young person who is, or seems, homeless or neglected

wail v cry out in pain or misery ▷ n mournful cry

wain n (poetic) farm wagon

wainscot, wainscoting n wooden lining of the lower part of the walls of a room

waist n part of the body between the ribs and hips; narrow middle part ▸ **waistband** n band of material sewn on to the waist of a garment to strengthen it ▸ **waistcoat** n sleeveless garment which buttons up the front, usu. worn over a shirt and under a jacket ▸ **waistline** n (size of) the waist of a person or garment

wait v remain inactive in expectation of something; be ready (for something); delay or be delayed; serve in a restaurant etc. ▷ n act or period of waiting ▸ **waiter** n man who serves in a restaurant etc. ▸ **waitress** n fem

Waitangi Day n February 6th, the national day of New Zealand commemorating the Treaty Of Waitangi in 1840

waive v refrain from enforcing (a law, right, etc.)

waiver n act or instance of voluntarily giving up a claim, right, etc.

waka n (NZ) Māori canoe

wake¹ v **waking, woke, woken** rouse from sleep or inactivity ▷ n vigil beside a corpse the night before the funeral ▸ **waken** v wake ▸ **wakeful** adj

wake² n track left by a moving ship **in the wake of** following, often as a result

walk v move on foot with at least one foot always on the ground; pass through or over on foot; escort or accompany on foot ▷ n act or instance of walking; distance walked; manner of walking; place or route for walking **walk of life** social position or profession ▸ **walker** n ▸ **walkabout** n informal walk among the public by royalty etc. ▸ **walkie-talkie** n portable radio transmitter and receiver ▸ **walking stick** stick used as a support when walking ▸ **walk into** v meet with unwittingly ▸ **Walkman®** n small portable cassette player with headphones ▸ **walkout** n strike; act of leaving as a protest ▸ **walkover** n easy victory

wall n structure of brick, stone, etc. used to enclose, divide, or support; something having the function or effect of a wall ▷ v enclose or seal with a wall or walls ▸ **wallflower** n fragrant garden plant; (at a dance) woman who remains seated because she has no partner ▸ **wallpaper** n decorative paper to cover interior walls

wallaby n, pl **-bies** marsupial like a small kangaroo

wallaroo n large stocky Australian kangaroo of rocky regions

wallet n small folding case for paper money, documents, etc.

walleye n fish with large staring eyes (also **dory**)

wallop (informal) v **-loping, -loped** hit hard ▷ n hard blow

▶ **walloping** (*informal*) *n* thrashing ▷ *adj* large or great

wallow *v* revel in an emotion; roll in liquid or mud ▷ *n* act or instance of wallowing

wally *n, pl* **-lies** (*Brit slang*) stupid person

walnut *n* edible nut with a wrinkled shell; tree it grows on; its wood, used for making furniture

walrus *n, pl* **-ruses** or **-rus** large sea mammal with long tusks

waltz *n* ballroom dance; music for this ▷ *v* dance a waltz; (*informal*) move in a relaxed confident way

wampum [wom-pum] *n* shells woven together, formerly used by Native Americans for money and ornament

WAN wide area network

wan [rhymes with **swan**] *adj* **wanner, wannest** pale and sickly looking

wand *n* thin rod, esp. one used in performing magic tricks

wander *v* move about without a definite destination or aim; go astray, deviate ▷ *n* act or instance of wandering ▶ **wanderer** *n* ▶ **wanderlust** *n* great desire to travel

wane *v* decrease gradually in size or strength; (of the moon) decrease in size **on the wane** decreasing in size, strength, or power

wangle *v* (*informal*) get by devious methods

want *v* need or long for; desire or wish ▷ *n* act or instance of wanting; thing wanted; lack or absence; state of being in need, poverty ▶ **wanted** *adj* sought by the police ▶ **wanting** *adj* lacking; not good enough

wanton *adj* without motive, provocation, or justification; (*old-fashioned*) (of a woman) sexually unrestrained or immodest

WAP Wireless Application Protocol: a system that allows mobile phone users to access the internet and other information services

wapiti [wop-pit-tee] *n, pl* **-tis** large N American deer, now also common in New Zealand

war *n* fighting between nations; conflict or contest ▷ *adj* of, like, or caused by war ▷ *v* **warring, warred** conduct a war ▶ **warring** *adj* ▶ **warlike** *adj* of or relating to war; hostile and eager to have a war ▶ **war crime** crime, such as killing, committed during a war in violation of accepted conventions ▶ **war criminal** person who has committed war crimes ▶ **warfare** *n* fighting or hostilities ▶ **warhead** *n* explosive front part of a missile ▶ **warmonger** *n* person who encourages war ▶ **warship** *n* ship designed and equipped for naval combat

waratah *n* Australian shrub with crimson flowers

warble *v* sing in a trilling voice

warbler *n* any of various small songbirds

ward *n* room in a hospital for patients needing a similar kind of care; electoral division of a town; child under the care of a guardian or court ▶ **warder** *n* prison officer ▶ **wardress** *n* fem ▶ **ward off** *v* avert or repel ▶ **wardroom** *n* officers' quarters on a warship

warden *n* person in charge of a building and its occupants; official responsible for the enforcement of regulations

wardrobe *n* cupboard for hanging clothes in; person's collection of clothes; costumes of a theatrical company

ware *n* articles of a specified type or material e.g. *silverware* ▷ *pl* goods for sale ▸ **warehouse** n building for storing goods prior to sale or distribution

warlock n man who practises black magic

warm adj moderately hot; providing warmth; (of a colour) predominantly yellow or red; affectionate; enthusiastic ▸ v make or become warm ▸ **warmly** adv ▸ **warmth** n mild heat; cordiality; intensity of emotion ▸ **warm up** v make or become warmer; do preliminary exercises before a race or more strenuous exercise; make or become more lively ▸ **warm-up** n

warn v make aware of possible danger or harm; caution or scold; inform (someone) in advance ▸ **warning** n something that warns; scolding or caution ▸ **warn off** v advise (someone) not to become involved with

warp v twist out of shape; pervert ▸ n state of being warped; lengthwise threads on a loom

warrant n (document giving) official authorization ▸ v make necessary; guarantee ▸ **warranty** n, pl **-ties** (document giving) a guarantee ▸ **warrant officer** officer in certain armed services with a rank between a commissioned and noncommissioned officer

warren n series of burrows in which rabbits live; overcrowded building or part of a town

warrigal (Aust) n dingo ▸ adj wild

warrior n person who fights in a war

wart n small hard growth on the skin ▸ **wart hog** kind of African wild pig

wary [ware-ree] adj **warier**, **wariest** watchful or cautious ▸ **warily** adv ▸ **wariness** n

was v first and third person singular past tense of **be**

wash v clean (oneself, clothes, etc.) with water and usu. soap; be washable; flow or sweep over or against; (informal) be believable or acceptable e.g. that excuse won't wash ▸ n act or process of washing; clothes washed at one time; thin coat of paint; disturbance in the water after a ship has passed by ▸ **washable** adj ▸ **washer** n ring put under a nut or bolt or in a tap as a seal ▸ **washing** n clothes to be washed ▸ **washing-up** n (washing of) dishes and cutlery needing to be cleaned after a meal ▸ **wash away** v carry or be carried off by moving water ▸ **washout** n (informal) complete failure ▸ **wash up** v wash dishes and cutlery after a meal

wasp n stinging insect with a slender black-and-yellow striped body ▸ **waspish** adj bad-tempered

waste v use pointlessly or thoughtlessly; fail to take advantage of ▸ n act of wasting or state of being wasted; anything wasted; rubbish ▸ pl desert ▸ adj rejected as worthless or surplus to requirements; not cultivated or inhabited ▸ **wastage** n loss by wear or waste; reduction in size of a workforce by not filling vacancies ▸ **wasteful** adj extravagant ▸ **wastefully** adv ▸ **waster**, **wastrel** n layabout ▸ **waste away** v (cause to) decline in health or strength ▸ **wastepaper basket** container for discarded paper

watch v look at closely; guard or supervise ▸ n portable timepiece for the wrist or pocket; (period of) watching; sailor's spell of duty ▸ **watchable** adj ▸ **watcher** n ▸ **watchful** adj vigilant or alert ▸ **watchfully** adv ▸ **watchdog**

n dog kept to guard property; person or group guarding against inefficiency or illegality ▸ **watch for** v be keenly alert to or cautious about ▸ **watchman** n man employed to guard a building or property ▸ **watchword** n word or phrase that sums up the attitude of a particular group

water n clear colourless tasteless liquid that falls as rain and forms rivers etc.; body of water, such as a sea or lake; level of the tide; urine ▷ v put water on or into; (of the eyes) fill with tears; (of the mouth) salivate ▸ **watery** adj ▸ **water buffalo** oxlike Asian animal ▸ **water closet** (old-fashioned) (room containing) a toilet flushed by water ▸ **watercolour** n paint thinned with water; painting done in this ▸ **watercourse** n bed of a stream or river ▸ **watercress** n edible plant growing in clear ponds and streams ▸ **water down** v dilute, make less strong ▸ **waterfall** n place where the waters of a river drop vertically ▸ **waterfront** n part of a town alongside a body of water ▸ **water lily** water plant with large floating leaves ▸ **watermark** n faint translucent design in a sheet of paper ▸ **watermelon** n melon with green skin and red flesh ▸ **water polo** team game played by swimmers with a ball ▸ **waterproof** adj not letting water through ▷ n waterproof garment ▷ v make waterproof ▸ **watershed** n important period or factor serving as a dividing line; line separating two river systems ▸ **watersider** n (NZ) person employed to load and unload ships ▸ **water-skiing** n sport of riding over water on skis towed

by a speedboat ▸ **watertight** adj not letting water through; with no loopholes or weak points ▸ **water wheel** large wheel which is turned by flowing water to drive machinery

watt [wott] n unit of power ▸ **wattage** n electrical power expressed in watts

wattle [wott-tl] n branches woven over sticks to make a fence; Australian acacia with flexible branches formerly used for making fences

wave v move the hand to and fro as a greeting or signal; move or flap to and fro ▷ n moving ridge on water; curve(s) in the hair; prolonged spell of something; gesture of waving; vibration carrying energy through a medium ▸ **wavy** adj ▸ **wavelength** n distance between the same points of two successive waves

waver v hesitate or be irresolute; be or become unsteady ▸ **waverer** n

wax¹ n solid shiny fatty or oily substance used for sealing, making candles, etc.; similar substance made by bees; waxy secretion of the ear ▷ v coat or polish with wax ▸ **waxen** adj made of or like wax ▸ **waxy** adj ▸ **waxwork** n lifelike wax model of a (famous) person ▷ pl place exhibiting these

wax² v increase in size or strength; (of the moon) get gradually larger

way n manner or method; characteristic manner; route or direction; track or path; distance; room for movement or activity e.g. you're in the way; passage or journey ▸ **wayfarer** n (lit) traveller ▸ **waylay** v lie in wait for and accost or attack ▸ **wayside** adj, n (situated by) the side of a road

wayward adj erratic, selfish, or stubborn ▸ **waywardness** n

W

WC water closet

we pron (used as the subject of a verb) the speaker or writer and one or more others; people in general; formal word for 'I' used by writers, editors and monarchs

weak adj lacking strength; liable to give way; unconvincing; lacking flavour ▶ **weaken** v make or become weak ▶ **weakling** n feeble person or animal ▶ **weakly** adv feebly ▶ **weakness** n being weak; failing; self-indulgent liking

weal n raised mark left on the skin by a blow

wealth n state of being rich; large amount of money and valuables; great amount or number ▶ **wealthy** adj

wean v accustom (a baby or young mammal) to food other than mother's milk; coax (someone) away from former habits

weapon n object used in fighting; anything used to get the better of an opponent ▶ **weaponry** n weapons collectively

wear v **wearing, wore, worn** have on the body as clothing or ornament; show as one's expression; (cause to) deteriorate by constant use or action; endure constant use ▷ n clothes suitable for a particular time or purpose e.g. beach wear; damage caused by use; ability to endure constant use ▶ **wearer** n ▶ **wear off** v gradually decrease in intensity ▶ **wear on** v (of time) pass slowly

weary adj **-rier, -riest** tired or exhausted; tiring ▷ v **-rying, -ried** make or become weary ▶ **wearily** adv ▶ **weariness** n ▶ **wearisome** adj tedious

weasel n small carnivorous mammal with a long body and short legs

weather n day-to-day atmospheric conditions of a place ▷ v (cause to) be affected by the weather; come safely through **under the weather** (informal) slightly ill ▶ **weather-beaten** adj worn, damaged, or (of skin) tanned by exposure to the weather ▶ **weathercock, weathervane** n device that revolves to show the direction of the wind

weave v **weaving, wove** or **weaved, woven** or **weaved** make (fabric) by interlacing (yarn) on a loom; compose (a story); move from side to side while going forwards ▶ **weaver** n

web n net spun by a spider; anything intricate or complex e.g. web of deceit; skin between the toes of a duck, frog, etc. **the Web** short for **World Wide Web** ▶ **webbed** adj ▶ **webbing** n strong fabric woven in strips ▶ **Web 2.0** the internet where interactive experience plays a more important role than simply accessing information ▶ **webcam** n camera that transmits images over the internet ▶ **webcast** n broadcast of an event over the internet ▶ **weblog** n person's online journal (also **blog**) ▶ **webmail** n system of electronic mail that allows account holders to access their mail via an internet site rather than downloading it onto their computer ▶ **website** n group of connected pages on the World Wide Web

wed v **wedding, wedded** or **wed** marry; unite closely ▶ **wedding** n act or ceremony of marriage ▶ **wedlock** n marriage

wedge n piece of material thick at one end and thin at the other ▷ v fasten or split with a wedge; squeeze into a narrow space

▶ **wedge-tailed eagle** large brown Australian eagle with a wedge-shaped tail

Wednesday n fourth day of the week

wee adj (Brit, Aust & NZ informal) small or short

weed n plant growing where undesired; (informal) thin ineffectual person ▷ v clear of weeds ▶ **weedy** adj (informal) (of a person) thin and weak ▶ **weed out** v remove or eliminate (what is unwanted)

weeds pl n (obs) widow's mourning clothes

week n period of seven days, esp. one beginning on a Sunday; hours or days of work in a week ▶ **weekly** adj, adv happening, done, etc. once a week ▷ n, pl **-lies** newspaper or magazine published once a week ▶ **weekday** n any day of the week except Saturday or Sunday ▶ **weekend** n Saturday and Sunday

weep v **weeping, wept** shed tears; ooze liquid ▶ **weepy** adj liable to cry ▶ **weeping willow** willow with drooping branches

weevil n small beetle which eats grain etc.

weft n cross threads in weaving

weigh v have a specified weight; measure the weight of; consider carefully; be influential; be burdensome ▶ **weigh anchor** raise a ship's anchor or (of a ship) have its anchor raised ▶ **weighbridge** n machine for weighing vehicles by means of a metal plate set into the road

weight n heaviness of an object; unit of measurement of weight; object of known mass used for weighing; heavy object; importance or influence ▷ v add weight to; slant (a system) so that it favours one side rather than another ▶ **weightless** adj ▶ **weightlessness** n

weighting n (Brit) extra allowance paid in special circumstances

weighty adj **weightier, weightiest** important or serious; very heavy ▶ **weightily** adv

weir n river dam

weird adj strange or bizarre; unearthly or eerie

> **SPELLING TIP**
>
> The pronunciation of **weird** possibly leads people to spell it with the vowels the wrong way round. The Collins Corpus shows that **wierd** is a common misspelling

weirdo n, pl **-dos** (informal) peculiar person

welch v same as **welsh**

welcome v **-coming, -comed** greet with pleasure; receive gladly ▷ n kindly greeting ▷ adj received gladly; freely permitted

weld v join (pieces of metal or plastic) by softening with heat; unite closely ▷ n welded joint ▶ **welder** n

welfare n wellbeing; help given to people in need ▶ **welfare state** system in which the government takes responsibility for the wellbeing of its citizens

well¹ adv **better, best** satisfactorily; skilfully; completely; intimately; considerably; very likely ▷ adj in good health ▷ interj exclamation of surprise, interrogation, etc.

well² n hole sunk into the earth to reach water, oil, or gas; deep open shaft ▷ v flow upwards or outwards

wellbeing n state of being well, happy, or prosperous

well-disposed adj inclined to be friendly or sympathetic

well-heeled adj (informal) wealthy

W

wellies *pl n* (*Brit & Aust informal*) wellingtons

wellingtons *pl n* (*Brit & Aust*) high waterproof rubber boots

well-meaning *adj* having good intentions

well-spoken *adj* speaking in a polite or articulate way

well-worn *adj* (of a word or phrase) stale from overuse; so much used as to be affected by wear

Welsh *adj* of Wales ▸ *n* language or people of Wales ▸ **Welsh rarebit, Welsh rabbit** dish of melted cheese on toast

welsh *v* fail to pay a debt or fulfil an obligation

welt *n* raised mark on the skin produced by a blow; raised or strengthened seam

welter *n* jumbled mass

welterweight *n* boxer weighing up to 147lb (professional) or 67kg (amateur)

wen *n* cyst on the scalp

wench *n* (*facetious*) young woman

wend *v* go or travel

went *v* past tense of **go**

wept *v* past of **weep**

we're we are

were *v* form of the past tense of **be** used after *we*, *you*, *they*, or a plural noun; subjunctive of **be**

weren't were not

werewolf *n* (in folklore) person who can turn into a wolf

west *n* (direction towards) the part of the horizon where the sun sets; region lying in this direction; (**W-**) western Europe and the US ▸ *adj* to or in the west; (of a wind) from the west ▸ *adv* in, to, or towards the west ▸ **westerly** *adj* ▸ **western** *adj* of or in the west ▸ *n* film or story about cowboys in the western US ▸ **westernize** *v* adapt to the customs and culture of the West ▸ **westward** *adj*, *adv* ▸ **westwards** *adv*

wet *adj* **wetter, wettest** covered or soaked with water or another liquid; not yet dry; rainy; (*Brit informal*) (of a person) feeble or foolish ▸ *n* moisture or rain; (*Brit informal*) feeble or foolish person ▷ *v* **wetting, wet** or **wetted** make wet ▸ **wet blanket** (*informal*) person who has a depressing effect on others ▸ **wetland** *n* area of marshy land ▸ **wet nurse** woman employed to breast-feed another's child ▸ **wet room** waterproofed room with a drain in the floor, often used as an open-plan shower ▸ **wet suit** close-fitting rubber suit worn by divers etc.

whack *v* strike with a resounding blow ▷ *n* such a blow; (*informal*) share; (*informal*) attempt ▸ **whacked** *adj* exhausted ▸ **whacking** *adj* (*informal*) huge

whale *n* large fish-shaped sea mammal ▸ **have a whale of a time** (*informal*) enjoy oneself very much ▸ **whaler** *n* ship or person involved in whaling ▸ **whaling** *n* hunting of whales for food and oil

wharf *n*, *pl* **wharves** or **wharfs** platform at a harbour for loading and unloading ships ▸ **wharfie** *n* (*Aust*) person employed to load and unload ships

what *pron* which thing; that which; request for a statement to be repeated ▷ *interj* exclamation of anger, surprise, etc. ▷ *adv* in which way, how much e.g. *what do you care?* **what for?** why? ▸ **whatever** *pron* everything or anything that; no matter what ▸ **whatnot** *n* (*informal*) similar unspecified things ▸ **whatsoever** *adj* at all

wheat *n* grain used in making flour, bread, and pasta; plant producing this ▶ **wheaten** adj ▶ **wheatear** n small songbird

wheedle v coax or cajole

wheel n disc that revolves on an axle; pivoting movement ▷ v push or pull (something with wheels); turn as if on an axis; turn round suddenly **wheeling and dealing** use of shrewd and sometimes unscrupulous methods to achieve success ▶ **wheeler-dealer** n ▶ **wheelbarrow** n shallow box for carrying loads, with a wheel at the front and two handles ▶ **wheelbase** n distance between a vehicle's front and back axles ▶ **wheelchair** n chair mounted on wheels for use by people who cannot walk ▶ **wheel clamp** immobilizing device fixed to one wheel of an illegally parked car

wheeze v breathe with a hoarse whistling noise ▷ n wheezing sound; (informal) trick or plan ▶ **wheezy** adj

whelk n edible snail-like shellfish

whelp n pup or cub; (offens) youth ▷ v (of an animal) give birth

when adv at what time? ▷ conj at the time that; although; considering the fact that ▷ pron at which time ▶ **whenever** adv, conj at whatever time

whence adv, conj (obs) from what place or source

where adv in, at, or to what place? ▷ pron in, at, or to which place ▷ conj in the place at which ▶ **whereabouts** n present position ▷ adv at what place ▶ **whereas** conj but on the other hand ▶ **whereby** pron by which ▶ **wherefore** (obs) adv why ▷ conj consequently ▶ **whereupon** conj at which point ▶ **wherever** conj, adv at whatever place ▶ **wherewithal** n necessary funds, resources, etc.

Where includes the idea at so avoid the use of this preposition: where was it? (not where was it at?)

whet v whetting, whetted sharpen (a tool) **whet someone's appetite** increase someone's desire ▶ **whetstone** n stone for sharpening tools

whether conj used to introduce an indirect question or a clause expressing doubt or choice

whey [way] n watery liquid that separates from the curd when milk is clotted

which adj, pron used to request or refer to a choice from different possibilities ▷ pron used to refer to a thing already mentioned ▶ **whichever** adj, pron any out of several; no matter which

whiff n puff of air or odour; trace or hint

Whig n member of a British political party of the 18th–19th centuries that sought limited reform

while conj at the same time that; whereas ▷ n period of time ▶ **whilst** conj while ▶ **while away** pass (time) idly but pleasantly

whim n sudden fancy ▶ **whimsy** n capricious idea; light or fanciful humour ▶ **whimsical** adj unusual, playful, and fanciful

whimper v cry in a soft whining way ▷ n soft plaintive whine

whin n (Brit) gorse

whine n high-pitched plaintive cry; peevish complaint ▷ v make such a sound ▶ **whining** n, adj

whinge (Brit, Aust & NZ informal) v complain ▷ n complaint

whinny v -nying, -nied neigh softly ▷ n, pl -nies soft neigh

whip n cord attached to a handle, used for beating animals or people; politician responsible for

w

organizing and disciplining fellow party or caucus members; call made on members of Parliament to attend for important votes; dessert made from beaten cream or egg whites ▷ **whipping, whipped** strike with a whip, strap, or cane; (*informal*) pull, remove, or move quickly; beat (esp. eggs or cream) to a froth; rouse into a particular condition; (*informal*) steal ▶ **whip bird** (*Aust*) bird with a whistle ending in a whipcrack note ▶ **whiplash injury** neck injury caused by a sudden jerk to the head, as in a car crash ▶ **whip-round** n (*informal*) collection of money

whippet n racing dog like a small greyhound

whirl v spin or revolve; be dizzy or confused ▷ n whirling movement; bustling activity; confusion or giddiness ▶ **whirlpool** n strong circular current of water ▶ **whirlwind** n column of air whirling violently upwards in a spiral ▷ *adj* much quicker than normal

whirr, whir n prolonged soft buzz ▷ v **whirring, whirred** (cause to) make a whirr

whisk v move or remove quickly; beat (esp. eggs or cream) to a froth ▷ n utensil for beating eggs, cream, etc.

whisker n any of the long stiff hairs on the face of a cat or other mammal ▷ *pl* hair growing on a man's face **by a whisker** (*informal*) only just

whisky n, pl **-kies** spirit distilled from fermented cereals ▶ **whiskey** n, pl **-keys** Irish or American whisky

whisper v speak softly, without vibration of the vocal cords; rustle ▷ n soft voice; (*informal*) rumour; rustling sound

whist n card game in which one pair of players tries to win more tricks than another pair

whistle v produce a shrill sound, esp. by forcing the breath through pursed lips; signal by a whistle ▷ n whistling sound; instrument blown to make a whistling sound **blow the whistle on** (*informal*) inform on or put a stop to ▶ **whistling** n, adj

whit n **not a whit** not the slightest amount

white adj of the colour of snow; pale; light in colour; (of coffee) served with milk ▷ n colour of snow; clear fluid round the yolk of an egg; white part, esp. of the eyeball; (**W-**) member of the race of people with light-coloured skin ▶ **whiten** v make or become white or whiter ▶ **whiteness** n ▶ **whitish** adj ▶ **white-collar** adj denoting professional and clerical workers ▶ **white elephant** useless or unwanted possession ▶ **white flag** signal of surrender or truce ▶ **white goods** large household appliances such as cookers and fridges ▶ **white-hot** adj very hot ▶ **white lie** minor unimportant lie ▶ **white paper** report by the government, outlining its policy on a matter

whitebait n small edible fish

whitewash n substance for whitening walls ▷ v cover with whitewash; conceal or gloss over unpleasant facts

whither adv (*obs*) to what place

whiting n edible sea fish

Whitsun, Whitsuntide n Christian festival celebrating the descent of the Holy Spirit to the apostles, Pentecost

whittle v cut or carve (wood) with a knife ▶ **whittle down, whittle**

away v reduce or wear away gradually

whizz, whiz v **whizzing, whizzed** make a loud buzzing sound; (*informal*) move quickly ▷ n, pl **whizzes** loud buzzing sound; (*informal*) person skilful at something ▶ **whizz kid, whiz kid** (*informal*) person who is outstandingly able for his or her age

who pron which person; used to refer to a person or people already mentioned ▶ **whoever** pron any person who; no matter who

whodunnit, whodunit [hoo-dun-nit] n (*informal*) detective story, play, or film

whole adj containing all the elements or parts; uninjured or undamaged ▷ n complete thing or system **on the whole** taking everything into consideration ▶ **wholly** adv ▶ **wholefood** n food that has been processed as little as possible ▶ **wholehearted** adj sincere or enthusiastic ▶ **wholemeal** adj (of flour) made from the whole wheat grain; made from wholemeal flour ▶ **whole number** number that does not contain a fraction

wholesale adj, adv dealing by selling goods in large quantities to retailers; on a large scale ▶ **wholesaler** n

wholesome adj physically or morally beneficial

whom pron objective form of **who**

whoop v, n shout or cry to express excitement

whoopee interj (*informal*) cry of joy

whooping cough n infectious disease marked by convulsive coughing and noisy breathing

whopper n (*informal*) anything unusually large; huge lie ▶ **whopping** adj

whore [hore] n prostitute

whorl n ring of leaves or petals; one turn of a spiral

whose pron of whom or of which

why adv for what reason ▷ pron because of which

Wicca n modern polytheistic religious movement, based on elements of pre-Christian paganism ▶ **Wiccan** n, adj

wick n cord through a lamp or candle which carries fuel to the flame

wicked adj morally bad; mischievous ▶ **wickedly** adv ▶ **wickedness** n

wicker adj made of woven cane ▶ **wickerwork** n

wicket n set of three cricket stumps and two bails; ground between the two wickets on a cricket pitch

wide adj large from side to side; having a specified width; spacious or extensive; far from the target; opened fully ▷ adv to the full extent; over an extensive area; far from the target ▶ **widely** adv ▶ **widen** v make or become wider ▶ **widespread** adj affecting a wide area or a large number of people

widgeon n same as **wigeon**

widget n small mechanism or device, the name of which is unknown or temporarily forgotten; (*Computing*) computer program that can be executed from a PC desktop

widow n woman whose spouse is dead and who has not remarried ▶ **widowed** adj ▶ **widowhood** n ▶ **widower** n man whose spouse is dead and who has not remarried

width n distance from side to side; quality of being wide

wield v hold and use (a weapon); have and use (power)

wife n, pl **wives** female partner in marriage

Wi-Fi n system of accessing the internet from computers with wireless connections

wig n artificial head of hair

wigeon n duck found in marshes

wiggle v move jerkily from side to side ▷ n wiggling movement

wigwam n Native American's tent

wiki (Computing) n website that can be edited by anyone ▷ adj of the software that allows this

wild adj (of animals) not tamed or domesticated; (of plants) not cultivated; lacking restraint or control; violent or stormy; (informal) excited; (informal) furious; random ▶ **wilds** pl n desolate or uninhabited place ▶ **wildly** adv **wildness** n ▶ **wild-goose chase** search that has little chance of success

wildcat n European wild animal like a large domestic cat ▶ **wildcat strike** sudden unofficial strike

wildebeest n gnu

wilderness n uninhabited uncultivated region

wildfire n ▶ **spread like wildfire** spread quickly and uncontrollably

wildlife n wild animals and plants collectively

wiles pl n tricks or ploys ▶ **wily** adj crafty or sly

wilful adj headstrong or obstinate; intentional ▶ **wilfully** adv

will¹ v, past tense **would** used as an auxiliary to form the future tense or to indicate intention, ability, or expectation

> **USAGE NOTE**
> Will is not formal for discussing the future. The use of shall with I and we is a matter of preference, not of rule

will² n strong determination; desire or wish; directions written for disposal of one's property

after death ▷ v use one's will in an attempt to do (something); wish or desire; leave (property) by a will ▶ **willing** adj ready or inclined (to do something); keen and obliging ▶ **willingly** adv **willingness** n ▶ **willpower** n ability to control oneself and one's actions

will-o'-the-wisp n elusive person or thing; light sometimes seen over marshes at night

willow n tree with thin flexible branches; its wood, used for making cricket bats ▶ **willowy** adj slender and graceful

willy-nilly adv whether desired or not

willy wagtail n (Aust) black-and-white flycatcher

willy-willy n (Aust) small tropical dust storm

wilt v (cause to) become limp or lose strength

wimp n (informal) feeble ineffectual person

wimple n garment framing the face, worn by medieval women and now by nuns

win v winning, won come first in (a competition, fight, etc.); gain (a prize) in a competition; get by effort ▷ n victory, esp. in a game ▶ **winner** n ▶ **winning** adj gaining victory; charming ▶ **winnings** pl n sum won, esp. in gambling ▶ **win over** v gain the support of (someone) ▶ **win-win** adj guaranteeing a favourable outcome for all involved e.g. a win-win situation

wince v draw back, as if in pain ▷ n wincing

winch n machine for lifting or hauling using a cable or chain wound round a drum ▷ v lift or haul using a winch

wind¹ n current of air; hint or suggestion; breath; flatulence;

w

idle talk ▷ *v* render short of breath ▶ **windy** *adj* ▶ **windward** *adj, n* (of or in) the direction from which the wind is blowing ▶ **windfall** *n* unexpected good luck; fallen fruit ▶ **wind farm** collection of wind-driven turbines for generating electricity ▶ **wind instrument** musical instrument played by blowing ▶ **windmill** *n* machine for grinding or pumping driven by sails turned by the wind ▶ **windpipe** *n* tube linking the throat and the lungs ▶ **windscreen** *n* front window of a motor vehicle ▶ **windscreen wiper** device that wipes rain etc. from a windscreen ▶ **windsock** *n* cloth cone on a mast at an airfield to indicate wind direction ▶ **windsurfing** *n* sport of riding on water using a surfboard propelled and steered by a sail

wind² *v* **winding, wound** coil or wrap around; tighten the spring of (a clock or watch); move in a twisting course ▶ **wind up** *v* bring to or reach an end; tighten the spring of (a clock or watch); (*informal*) make tense or agitated; (*slang*) tease

windlass *n* winch worked by a crank

window *n* opening in a wall to let in light or air; glass pane or panes fitted in such an opening; display area behind the window of a shop; area on a computer screen that can be manipulated separately from the rest of the display area; period of unbooked time in a diary or schedule ▶ **window-dressing** *n* arrangement of goods in a shop window; attempt to make something more attractive than it really is ▶ **window-shopping** *n* looking at goods in shop windows without intending to buy

wine *n* alcoholic drink made from fermented grapes; similar drink made from other fruits ▶ **wine and dine** entertain or be entertained with fine food and drink

wing *n* one of the limbs or organs of a bird, insect, or bat that are used for flying; one of the winglike supporting parts of an aircraft; projecting side part of a building; faction of a political party; part of a car body surrounding the wheels; (*Sport*) (player on) either side of the pitch ▷ *pl* sides of a stage ▷ *v* fly; wound slightly in the wing or arm ▶ **winged** *adj* ▶ **winger** *n* (*Sport*) player positioned on a wing

wink *v* close and open (an eye) quickly as a signal; twinkle ▷ *n* winking; smallest amount of sleep

winkle *n* shellfish with a spiral shell ▶ **winkle out** *v* (*informal*) extract or prise out

winnow *v* separate (chaff) from (grain); examine to select desirable elements

winsome *adj* charming or winning

winter *n* coldest season ▷ *v* spend the winter ▶ **wintry** *adj* of or like winter; cold or unfriendly ▶ **winter sports** open-air sports held on snow or ice

wipe *v* clean or dry by rubbing; erase (a tape) ▷ *n* wiping ▶ **wipe out** *v* destroy completely

wire *n* thin flexible strand of metal; length of this used to carry electric current; (*obs*) telegram ▷ *v* equip with wires ▶ **wiring** *n* system of wires ▶ **wiry** *adj* lean and tough; like wire ▶ **wire-haired** *adj* (of a dog) having a stiff wiry coat

wireless *adj* (of a computer network) connected by radio rather than by cables or fibre optics ▷ *n* (*old-fashioned*) same as **radio**

w

wireless application protocol
see **WAP**

wisdom n good sense and judgment; accumulated knowledge ▶ **wisdom tooth** any of the four large molar teeth that come through usu. after the age of twenty

wise[1] adj having wisdom ▶ **wisely** adv ▶ **wiseacre** n person who wishes to seem wise

wise[2] n (obs) manner

-wise adv suffix indicating direction or manner e.g. clockwise; likewise; with reference to e.g. businesswise

> **USAGE NOTE**
> Adding -wise to a noun to create the meaning 'in respect of', as in: Defencewise, the team is strong, is generally unacceptable except in very informal usage

wisecrack (informal) n clever, sometimes unkind, remark ▶ v make a wisecrack

wish v want or desire; feel or express a hope about someone's wellbeing, success, etc. ▶ n expression of a desire; thing desired ▶ **wishful** adj too optimistic ▶ **wishbone** n V-shaped bone above the breastbone of a fowl

wishy-washy adj (informal) insipid or bland

wisp n light delicate streak; twisted bundle or tuft ▶ **wispy** adj

wisteria n climbing shrub with blue or purple flowers

wistful adj sadly longing ▶ **wistfully** adv

wit n ability to use words or ideas in a clever and amusing way; person with this ability; (sometimes pl) practical intelligence ▶ **witless** adj foolish

witch n person, usu. female, who practises (black) magic; ugly or wicked woman ▶ **witchcraft** n use of magic ▶ **witch doctor** (in certain societies) a man appearing to cure or cause injury or disease by magic ▶ **witch-hunt** n campaign against people with unpopular views

witchetty grub n wood-boring edible Australian caterpillar

with prep indicating presence alongside, possession, means of performance, characteristic manner, etc. e.g. walking with his dog; a man with two cars; hit with a hammer; playing with skill ▶ **within** prep, adv in or inside ▶ **without** prep not accompanied by, using, or having

withdraw v -drawing, -drew, -drawn take or move out or away ▶ **withdrawal** n ▶ **withdrawn** adj unsociable

wither v wilt or dry up ▶ **withering** adj (of a look or remark) scornful

withers pl n ridge between a horse's shoulder blades

withhold v -holding, -held refrain from giving

withstand v -standing, -stood oppose or resist successfully

witness n person who has seen something happen; person giving evidence in court; evidence or testimony ▶ v see at first hand; sign (a document) to certify that it is genuine

witter v (chiefly Brit) chatter pointlessly or at unnecessary length

wittingly adv intentionally

witty adj **wittier, wittiest** clever and amusing ▶ **wittily** adv ▶ **witticism** n witty remark

wives n plural of **wife**

wizard n magician; person with outstanding skill in a particular field ▶ **wizardry** n

wizened [wiz-end] adj shrivelled or wrinkled

w

WMD weapon(s) of mass destruction

woad n blue dye obtained from a plant, used by the ancient Britons as a body dye

wobbegong n Australian shark with brown-and-white skin

wobble v move unsteadily; shake ▷ n wobbling movement or sound ▶ **wobbly** adj

wodge n (informal) thick chunk

woe n grief ▶ **woeful** adj extremely sad; pitiful ▶ **woefully** adv ▶ **woebegone** adj looking miserable

wok n bowl-shaped Chinese cooking pan, used for stir-frying

woke v past tense of **wake**¹ ▶ **woken** v past participle of **wake**¹

wold n high open country

wolf n, pl **wolves** wild predatory canine mammal ▷ v eat ravenously ▶ **cry wolf** raise a false alarm ▶ **wolf whistle** whistle by a man to show he thinks a woman is attractive

wolverine n carnivorous mammal of Arctic regions

woman n, pl **women** adult human female; women collectively ▶ **womanhood** n ▶ **womanish** adj effeminate ▶ **womanly** adj having qualities traditionally associated with a woman ▶ **womanizing** n practice of indulging in casual affairs with women ▶ **womanizer** n ▶ **Women's Liberation** movement for the removal of inequalities between women and men (also **women's lib**)

womb n hollow organ in female mammals where babies are conceived and develop

wombat n small heavily-built burrowing Australian marsupial

won v past of **win**

wonder v be curious about; be amazed ▷ n wonderful thing; emotion caused by an amazing or unusual thing ▷ adj spectacularly successful e.g. a wonder drug ▶ **wonderful** adj very fine; remarkable ▶ **wonderfully** adv ▶ **wonderment** n ▶ **wondrous** adj (old-fashioned) wonderful

wonky adj **-kier, -kiest** (Brit, Aust & NZ informal) shaky or unsteady

won't will not

wont [rhymes with **don't**] adj accustomed ▷ n custom

woo v try to persuade; (old-fashioned) try to gain the love of

wood n substance trees are made of, used in carpentry and as fuel; area where trees grow; long-shafted golf club, usu. with wooden head ▶ **wooded** adj covered with trees ▶ **wooden** adj made of wood; without expression ▶ **woody** adj ▶ **woodbine** n honeysuckle ▶ **woodcock** n game bird ▶ **woodcut** n (print made from) an engraved block of wood ▶ **woodland** n forest ▶ **woodlouse** n small insect-like creature with many legs ▶ **woodpecker** n bird which searches tree trunks for insects ▶ **woodwind** adj, n (of) a type of wind instrument made of wood ▶ **woodworm** n insect larva that bores into wood

woof¹ n cross threads in weaving

woof² n barking noise made by a dog ▶ **woofer** n loudspeaker reproducing low-frequency sounds

wool n soft hair of sheep, goats, etc.; yarn spun from this ▶ **woollen** adj ▶ **woolly** adj of or like wool; vague or muddled ▷ n knitted woollen garment

woomera n notched stick used by Australian Aborigines to aid the propulsion of a spear

woozy adj **woozier, wooziest** (informal) weak, dizzy, and confused

w

wop-wops pl n (NZ informal) remote rural areas

word n smallest single meaningful unit of speech or writing; chat or discussion; brief remark; message; promise; command ▷ v express in words ▶ **wordy** adj using too many words ▶ **wording** n choice and arrangement of words ▶ **word processor** keyboard, microprocessor, and VDU for electronic organization and storage of text ▶ **word processing**

wore v past tense of **wear**

work n physical or mental effort directed to making or doing something; paid employment; duty or task; something made or done ▷ pl factory; total of a writer's or artist's achievements; (informal) full treatment; mechanism of a machine ▷ adj of or for work ▷ v (cause to) do work; be employed; (cause to) operate; (of a plan etc.) be successful; cultivate (land); manipulate, shape, or process; (cause to) reach a specified condition ▶ **workable** adj ▶ **worker** n ▶ **workaholic** n person obsessed with work ▶ **workhorse** n person or thing that does a lot of dull or routine work ▶ **workhouse** n (in England, formerly) institution where the poor were given food and lodgings in return for work ▶ **working class** social class consisting of wage earners, esp. manual workers ▶ **working-class** adj ▶ **working party** committee investigating a specific problem ▶ **workman** n manual worker ▶ **workmanship** n skill with which an object is made ▶ **workshop** n room or building for a manufacturing process ▶ **workstream** n one of the areas into which a company's business is divided ▶ **worktop** n surface in a kitchen, used for food preparation ▶ **work-to-rule** n protest in which workers keep strictly to all regulations to reduce the rate of work

world n the planet earth; mankind; society of a particular area or period; sphere of existence ▷ adj of the whole world ▶ **worldly** adj not spiritual; concerned with material things; wise in the ways of the world ▶ **world-weary** adj no longer finding pleasure in life ▶ **World Wide Web** global network of linked computer sites

worm n small limbless invertebrate animal; (informal) wretched or spineless person; shaft with a spiral thread forming part of a gear system; (Computing) type of virus ▷ pl illness caused by parasitic worms in the intestines ▷ v rid of worms **worm one's way** crawl; insinuate (oneself) ▶ **wormy** adj ▶ **worm-eaten** adj eaten into by worms ▶ **worm out** v extract (information) craftily

wormwood n bitter plant

worn v past participle of **wear**

worry v -**rying**, -**ried** (cause to) be anxious or uneasy; annoy or bother; (of a dog) chase and try to bite (sheep etc.) ▷ n, pl -**ries** (cause of) anxiety or concern ▶ **worried** adj ▶ **worrying** adj

worse adj, adv comparative of **bad**, **badly** ▶ **worst** adj, adv superlative of **bad**, **badly** ▶ n worst thing ▶ **worsen** v make or grow worse

worship v -**shipping**, -**shipped** show religious devotion to; love and admire ▷ n act or instance of worshipping; (**W-**) title for a mayor or magistrate ▶ **worshipper** n ▶ **worshipful** adj worshipping

worsted [wooss-tid] n type of woollen yarn or fabric

worth prep having a value of; meriting or justifying ▷ n value or price; excellence; amount to be had for a given sum ► **worthless** adj ► **worthy** adj deserving admiration or respect ▷ n (informal) notable person ► **worthily** adv ► **worthiness** n ► **worthwhile** adj worth the time or effort involved

would v used as an auxiliary to express a request, describe a habitual past action, or form the past tense or subjunctive mood of **will**¹ ► **would-be** adj wishing or pretending to be

wouldn't would not

wound¹ [woond] n injury caused by violence; injury to the feelings ▷ v inflict a wound on

wound² [wownd] v past of **wind**²

wove v a past tense of **weave** ► **woven** v a past participle of **weave**

wow interj exclamation of astonishment ▷ n (informal) astonishing person or thing

wowser n (Aust & NZ slang) puritanical person; teetotaller

wpm words per minute

wrack n seaweed

wraith n ghost

wrangle v argue noisily ▷ n noisy argument

wrap v **wrapping, wrapped** fold (something) round (a person or thing) so as to cover ▷ n garment wrapped round the shoulders; sandwich made by wrapping a filling in a tortilla ► **wrapper** n cover for a product ► **wrapping** n material used to wrap ► **wrap up** v fold paper round; put warm clothes on; (informal) finish or settle (a matter)

wrasse n colourful sea fish

wrath [roth] n intense anger ► **wrathful** adj

wreak v **wreak havoc** cause chaos **wreak vengeance on** take revenge on

wreath n twisted ring or band of flowers or leaves used as a memorial or tribute ► **wreathed** adj surrounded or encircled

wreck v destroy ▷ n remains of something that has been destroyed or badly damaged, esp. a ship; person in very poor condition ► **wrecker** n ► **wreckage** n wrecked remains

Wren n (informal) (in Britain) member of the former Women's Royal Naval Service

wren n small brown songbird; Australian warbler

wrench v twist or pull violently; sprain (a joint) ▷ n violent twist or pull; sprain; difficult or painful parting; adjustable spanner

wrest v twist violently; take by force

wrestle v fight, esp. as a sport, by grappling with and trying to throw down an opponent; struggle hard with ► **wrestler** n ► **wrestling** n

wretch n despicable person; pitiful person

wretched [retch-id] adj miserable or unhappy; worthless ► **wretchedly** adv ► **wretchedness** n

wrier adj a comparative of **wry** ► **wriest** adj a superlative of **wry**

wriggle v move with a twisting action; manoeuvre oneself by devious means ▷ n wriggling movement

wright n maker e.g. wheelwright

wring v **wringing, wrung** twist, esp. to squeeze liquid out of; clasp and twist (the hands); obtain by forceful means

wrinkle n slight crease, esp. one in the skin due to age ▷ v make

W

or become slightly creased ▸ **wrinkly** adj

wrist n joint between the hand and the arm ▸ **wristwatch** n watch worn on the wrist

writ n written legal command

write v **writing, wrote, written** mark paper etc. with symbols or words; set down in words; communicate by letter; be the author or composer of ▸ **writing** n ▸ **writer** n author; person who has written something specified ▸ **write-off** n (informal) something damaged beyond repair ▸ **write-up** n published account of something

writhe v twist or squirm in or as if in pain

wrong adj incorrect or mistaken; immoral or bad; not intended or suitable; not working properly ▸ adv in a wrong manner ▸ n something immoral or unjust ▸ v treat unjustly; malign ▸ **wrongly** adv ▸ **wrongful** adj ▸ **wrongfully** adv ▸ **wrongdoing** n immoral or illegal behaviour ▸ **wrongdoer** n

wrote v past tense of **write**

wrought [rawt] v (lit) past of **work** ▸ adj (of metals) shaped by hammering or beating ▸ **wrought iron** pure form of iron used for decorative work

wrung v past of **wring**

wry adj **wrier, wriest** or **wryer, wryest** drily humorous; (of a facial expression) contorted ▸ **wryly** adv

wt. weight

WWW World Wide Web

wych-elm n elm with large rough leaves

Xx

X indicating an error, a choice, or a kiss; indicating an unknown, unspecified, or variable factor, number, person, or thing

xenon n (Chem) colourless odourless gas found in very small quantities in the air

xenophobia [zen-oh-fobe-ee-a] n fear or hatred of people from other countries

Xerox® [zeer-ox] n machine for copying printed material; copy made by a Xerox machine ▸ v copy (a document) using such a machine

Xmas [eks-mass] n (informal) Christmas

XML extensible markup language: a computer language used in text formatting

X-ray, x-ray n stream of radiation that can pass through some solid materials; picture made by sending X-rays through someone's body to examine internal organs ▸ v photograph, treat, or examine using X-rays

xylem [zy-lem] n plant tissue that conducts water and minerals from the roots to all other parts

xylophone [zile-oh-fone] n musical instrument made of a row of wooden bars played with hammers

Yy

ya interj (S Afr) yes

yabby n, pl -**bies** (Aust) small freshwater crayfish; marine prawn used as bait

yacht [yott] n large boat with sails or an engine, used for racing or pleasure cruising ▶ **yachting** n ▶ **yachtsman, yachtswoman** n

yak[1] n Tibetan ox with long shaggy hair

yak[2] v **yakking, yakked** (slang) talk continuously about unimportant matters

yakka n (Aust & NZ informal) work

yam n tropical root vegetable

yank v pull or jerk suddenly ▷ n sudden pull or jerk

Yankee, Yank n (slang) person from the United States

yap v **yapping, yapped** bark with a high-pitched sound; (informal) talk continuously ▷ n high-pitched bark

yard[1] n unit of length equal to 36 inches or about 91.4 centimetres ▶ **yardstick** n standard against which to judge other people or things

yard[2] n enclosed area, usu. next to a building and often used for a particular purpose e.g. builder's yard

yarmulke [yar-mull-ka] n skullcap worn by Jewish men

yarn n thread used for knitting or making cloth; (informal) long involved story

yashmak n veil worn by a Muslim woman to cover her face in public

yaw v (of an aircraft or ship) turn to one side or from side to side while moving

yawl n two-masted sailing boat

yawn v open the mouth wide and take in air deeply, often when sleepy or bored; (of an opening) be large and wide ▷ n act of yawning ▶ **yawning** adj

yd yard

ye [yee] pron (obs) you

year n time taken for the earth to make one revolution around the sun, about 365 days; twelve months from January 1 to December 31 ▶ **yearly** adj, adv (happening) every year or once a year ▶ **yearling** n animal between one and two years old

yearn v want (something) very much ▶ **yearning** n, adj

yeast n fungus used to make bread rise and to ferment alcoholic drinks ▶ **yeasty** adj

yebo interj (S Afr informal) yes

yell v shout or scream in a loud or piercing way ▷ n loud cry of pain, anger, or fear

yellow n the colour of gold, a lemon, etc. ▷ adj of this colour; (informal) cowardly ▷ v make or become yellow ▶ **yellow belly** (Aust) freshwater food fish with yellow underparts ▶ **yellow fever** serious infectious tropical disease ▶ **yellowhammer** n European songbird with a yellow head and body ▶ **Yellow Pages**® telephone directory which lists businesses under the headings of the type of service they provide

yelp v, n (give) a short sudden cry

yen[1] n, pl **yen** monetary unit of Japan

yen[2] n (informal) longing or desire

yeoman [yo-man] n, pl -**men** (Hist) farmer owning and farming his own land ▸ **yeoman of the guard** member of the ceremonial bodyguard of the British monarchy

yes *interj* expresses consent, agreement, or approval; used to answer when one is addressed ▸ **yes man** person who always agrees with their superior

yesterday *adv, n* (on) the day before today; (in) the recent past

yet *conj* nevertheless, still ▸ *adv* up until then or now; still; now

yeti *n* same as **abominable snowman**

yew *n* evergreen tree with needle-like leaves and red berries

Yiddish *adj, n* (of or in) a language of German origin spoken by many Jews in Europe and elsewhere

yield *v* produce or bear; give up control of, surrender; give in ▸ *n* amount produced ▸ **yielding** *adj* submissive; soft or flexible

YMCA Young Men's Christian Association

yob, yobbo *n* (*Brit slang*) bad-mannered aggressive youth

yodel *v* -**delling, -delled** sing with abrupt changes between a normal and a falsetto voice

yoga *n* Hindu method of exercise and discipline aiming at spiritual, mental, and physical wellbeing ▸ **yogi** *n* person who practises yoga

yogurt, yoghurt *n* slightly sour custard-like food made from milk that has had bacteria added to it, often sweetened and flavoured with fruit

yoke *n* wooden bar across the necks of two animals to hold them together; frame fitting over a person's shoulders for carrying buckets; (*lit*) oppressive force e.g. *the yoke of the tyrant*; fitted part of a garment to which a fuller part is

attached ▸ *v* put a yoke on; unite or link

yokel *n* (*offens*) person who lives in the country and is usu. simple and old-fashioned

yolk *n* yellow part of an egg that provides food for the developing embryo

Yom Kippur *n* annual Jewish religious holiday

yonder *adj, adv* (situated) over there

yonks *pl n* (*informal*) very long time

yore *n* (*lit*) **of yore** a long time ago

Yorkshire pudding *n* baked batter made from flour, milk, and eggs

you *pron* referring to the person or people addressed; referring to unspecified person or people in general

young *adj* in an early stage of life or growth ▸ *pl n* young people in general; offspring, esp. young animals ▸ **youngster** *n* young person

your *adj* of, belonging to, or associated with you; of, belonging to, or associated with an unspecified person or people in general ▸ **yours** *pron* something belonging to you ▸ **yourself** *pron*

youth *n* time of being young; boy or young man; young people as a group ▸ **youthful** *adj* ▸ **youthfulness** *n* ▸ **youth club** club that provides leisure activities for young people ▸ **youth hostel** inexpensive lodging place for young people travelling cheaply

YouTube® *n* website on which subscribers can post video files

yowl *v, n* (produce) a loud mournful cry

yo-yo *n, pl* -**yos** toy consisting of a spool attached to a string, by which it is repeatedly spun out and reeled in

yttrium [it-ree-um] n (Chem) silvery metallic element used in various alloys

yucca n tropical plant with spikes of white leaves

yucky adj **yuckier, yuckiest** (slang) disgusting, nasty

Yugoslav n person from the former Yugoslavia ▷ adj of the former Yugoslavia

Yule n (lit) Christmas (season)

yuppie n young highly-paid professional person, esp. one who has a materialistic way of life ▷ adj typical of or reflecting the values of yuppies

YWCA Young Women's Christian Association

Zz

zany [zane-ee] adj **zanier, zaniest** comical in an endearing way

zap v **zapping, zapped** (informal) kill (by shooting); change TV channels rapidly by remote control ▶ **zapper** n (informal) remote-control device for TV

zeal n great enthusiasm or eagerness ▶ **zealot** [zel-lot] n fanatic or extreme enthusiast ▶ **zealous** [zel-luss] adj extremely eager or enthusiastic ▶ **zealously** adv

zebra n black-and-white striped African animal of the horse family ▶ **zebra crossing** pedestrian crossing marked by black and white stripes on the road

zebu [zee-boo] n Asian ox with a humped back and long horns

Zen n Japanese form of Buddhism that concentrates on learning through meditation and intuition

zenith n highest point of success or power; point in the sky directly above an observer

zephyr [zef-fer] n soft gentle breeze

zeppelin n (Hist) large cylindrical airship

zero n, pl **-ros** or **-roes** (symbol representing) the number o; point on a scale of measurement from which the graduations commence; lowest point; nothing, nil ▷ adj having no measurable quantity or size ▶ **zero in on** v aim at; (informal) concentrate on

z

zest n enjoyment or excitement; interest, flavour, or charm; peel of an orange or lemon

zigzag n line or course having sharp turns in alternating directions ▷ v -zagging, -zagged move in a zigzag ▷ adj formed in or proceeding in a zigzag

zinc n (Chem) bluish-white metallic element used in alloys and to coat metal

zing n (informal) quality in something that makes it lively or interesting

Zionism n movement to found and support a Jewish homeland in Israel ▶ **Zionist** n, adj

zip n fastener with two rows of teeth that are closed or opened by a small clip pulled between them; (informal) energy, vigour; short whizzing sound ▷ v **zipping, zipped** fasten with a zip; move with a sharp whizzing sound

zircon n mineral used as a gemstone and in industry

zirconium n (Chem) greyish-white metallic element that is resistant to corrosion

zither n musical instrument consisting of strings stretched over a flat box and plucked to produce musical notes

zodiac n imaginary belt in the sky within which the sun, moon, and planets appear to move, divided into twelve equal areas, called signs of the zodiac, each named after a constellation

zombie, zombi n person who appears to be lifeless, apathetic, or totally lacking in independent judgment; corpse brought back to life by witchcraft

zone n area with particular features or properties; one of the divisions of the earth's surface according to temperature ▷ v divide into zones ▶ **zonal** adj

zoo n, pl **zoos** place where live animals are kept for show

zoology n study of animals ▶ **zoologist** n ▶ **zoological** adj ▶ **zoological garden** zoo

zoom v move or rise very rapidly; make or move with a buzzing or humming sound ▶ **zoom lens** lens that can make the details of a picture larger or smaller while keeping the picture in focus

zucchini [zoo-**keen**-ee] n, pl **-ni** or **-nis** (US & Aust) courgette

Zulu n member of a tall Black people of southern Africa; language of this people

Zumba® n system of keep-fit exercises performed to Latin American dance music

zygote n fertilized egg cell

PRACTICAL WRITING GUIDE

INTRODUCTION

A dictionary can tell you what a word means and when it can be used accurately. It cannot, though, give you guidance on how to write clearly and appropriately in a variety of situations. The *Collins Practical Writing Guide* supplement has been written to help you express yourself effectively at work and at home. It includes advice on how to structure your writing, and how to adapt tone, style, and content to different forms of communication – from letters and emails to social media.

BEFORE YOU START WRITING

It is amazing how much more effective your writing will be with a bit of thinking time beforehand. There are three questions which you should be able to answer about any piece of writing, whether it's an email, text, letter, or post:

- **Who am I writing to?** This will determine the style and tone that you use. If you are writing an email to a friend, for instance, then you are likely to use less formal language than if you are writing to apply for a job.

- **What do I want to say?** Make sure that all the information you want to communicate is included, and that it is set out as clearly as possible.

- **Why do I want to say it?** In other words, what do you want to happen as a result of your communication? Whether it's for a job application, to ask someone out on a date, or to offer your condolences for a bereavement, what you want to achieve should be clearly stated.

Once you've answered these questions, writing becomes much easier.

TIPS AND TRAPS

Here are some tips for making your writing successful and some common traps to avoid.

👍 Tips:

- **Use plain English.** Aim for concise, simple expression which will make your writing easy to read.

- **Plan.** Think about what you want to say, and how you want to say it. Planning will save you time and improve your writing. It needn't take a lot of time but, even if you're only sending a text, it will pay dividends.

- **Vary the length of your sentences.** Short sentences are powerful. Longer sentences can express more complicated thoughts, but try to keep them to a manageable length or else they become tiring to read. Try to stick to the principle of including one main idea in a sentence, and maybe one related point.

- **Use active rather than passive verbs.** It is usually better to use active verbs because it makes your writing simpler and less complicated to read:

  > *The programme was watched by an audience of one million people. (passive)*

  > *One million people watched the programme. (active)*

  The verb 'watched' is active in the second example because the subject – 'One million people' – 'does' it. The sentence is shorter and clearer as a result.

- **Read your work aloud.** If the sentences work well, it will be easy to read. Check that you have commas where there are natural pauses.

- **Think about register and tone.** Are you using the right level of formality ('register')? Does your writing accurately express your attitude to the subject and the reader ('tone')?

- **Always check your writing before sending it – and then check it again.** You'll be surprised how easy it is to overlook mistakes. Computers have introduced new errors – for instance, did you delete the original passage that you copied and pasted later in the document?

> **PROFESSIONAL TIP**
>
> *I wish people would read through what they have written before pressing 'send'. It would save me a lot of time and make their applications more successful.*
>
> (HR manager)

🖐Traps:

- **Jargon.** Specialist words which are understood by a particular group of people, or overly technical language. Don't talk about 'interfacing' with someone if you simply mean 'talking' to them.

- **Clichés.** Words or phrases that are used too often, and have little meaning. They will annoy your reader and distract from what you are trying to say. Examples include sayings like *a different kettle of fish* and *at the end of the day*.

- **Long sentences.** Avoid sentences longer than 15-20 words – they can be difficult to read, and can usually be divided into clearer statements. Try reading your work aloud. If your sentences work well, they will be easy to read.

4

- **Repeating words.** Using the same word more than once or twice in a sentence can be clumsy, and there is usually an alternative. For instance, *'The date of the CSEC exam is the same date as the CAPE exam'* is better expressed as *'The CSEC and CAPE exams are on the same date'*.
- **Long words.** Unnecessarily long or complex words can often be substituted with shorter and clearer ones.. For example, don't use 'proffer' when you mean 'give' or 'articulate' when you mean 'say'.
- **Redundancy.** Avoid using ten words where two will do: *'I am meeting Sophie later'* is clearer than *'Sophie and I are due to hook up together at some point in the day'*.
- **Ambiguity.** Many words can be understood in more than one way. Always try to put yourself in the reader's place to make sure that the meaning of your statement is clear.
- **Causing offence.** A simple rule to follow is: *treat everyone equally in your writing, regardless of sex, age, race, religion, sexual orientation, or physical difference.* Be aware of current customs and values, and also consider different cultures.
- **Exclamation marks.** The misuse or overuse of exclamation marks is distracting. Avoid ending a sentence with an exclamation mark unless it really is an exclamation.

These are only a few points to consider before you start writing, but if you refer to them regularly, they will help you express yourself clearly and consistently in all your communications.

EMAILS

Emails are an important form of written communication in many people's lives. Whether at home or at work, we spend a lot of our time sending and receiving them.

Email correspondence can feel more like a conversation than an exchange of letters. The tone is generally less formal, and the time between sending your message and getting a reply can be minutes or even seconds. The fact that it is instant can be good and bad. Good, because it can be a very efficient way of corresponding; bad, because you might write quickly and make careless mistakes.

Addressing emails

When addressing emails, the general rule of thumb is that the fewer people you email, the better. There are three address fields to consider, and each serves a different purpose:

- **'To':** This is for the main recipient, or recipients, of the information or request to do something.
- **'Cc':** If you are simply informing someone of your actions or requests, put those people in the 'Cc' ('Carbon or Courtesy Copy') field.
- **'Bcc':** If you are copying someone in but you don't want the other addressees to know, you should use the 'Bcc' ('Blind Carbon or Courtesy Copy'). This is frequently used for mailing large groups, where you don't want individuals to know who else is receiving the email.

PROFESSIONAL TIP

Bear in mind that if you send an email to one person, you are 95 per cent likely to get a reply; if you send it to 10 people, the response rate drops to 5 per cent.

(Linguistics professor)

Greeting and ending

Emails on work-related issues or personal business require a formal style. You can never go wrong with 'Dear Mr Blake' or 'Dear Peter'. If the contact is long-standing and the relationship is not a formal one, then 'Hi Peter' is acceptable.

In initial exchanges of formal email, you might sign off in the same way that you would in a formal letter, with 'Yours sincerely' or 'Yours faithfully'. As your correspondence becomes less formal, then 'Kind regards' or 'Best wishes' is acceptable.

Subject line

You can really help your correspondents by being precise in the subject line. For example, if you send out a regular set of meeting minutes by email, don't just write 'Launch Meeting Minutes', but add the date so people can quickly find which set they are looking for. If your email contains a specific question, it is good practice to add 'Q:' followed by the question in the subject line.

Layout

Use a paragraph per point you wish to make, and put headings above each paragraph if there are more than three. If your email is long and will require the reader to scroll down the screen, consider writing it as a Word document and attaching it to an email – long emails are not easy to read and respond to.

Content

- Always remember that, with email, your writing can be forwarded to anyone with a single mouse click. Be careful that what you write is not defamatory, offensive, or detrimental to you or your business.

- Restrict the email to a single subject. If you want to
  email the same person or people about other issues,
  use separate emails. It makes filing and action points
  much easier to follow.
- Don't reply straight away to an email that annoys you.
  You may not be able to hide your anger and may make
  the situation worse. Wait until you have calmed
  down enough to think through your response and compose
  a measured reply dealing with the points raised.
- Don't assume that the person you are writing to has the
  same cultural reference points, sense of humour, or values.

Attachments

Email is great for sharing information held in spreadsheets,
PDFs, and Word documents. But sending large attachments
can be challenging. Most email providers (and certainly most
companies) allocate a storage limit to each email address;
if an inbox becomes too full, you cannot receive or send
emails. So be considerate when you send anything as an
attachment.

Replying

Sometimes you get an email that has a series of questions. It is perfectly acceptable to reply to each of these questions by adding your comments (sometimes in a different colour or with your initials in square brackets before your answer) in the body of the original email. This saves you having to type the questions or write replies that incorporate the original questions. For example:

From: Asif Iqbal
To: Lataya Stewart
SUBJECT: My paintings

Dear Lataya

Thank you very much for your email about the posters I sell. I've put my answers below your questions with [AI] before each.

With thanks and all best wishes
Asif
Tel: 18696 603135
www.asifiqbal.art.gallery.org

Do you have a website?
[AI] Yes. You can see all my work at www.asifiqbal.art.gallery.org
What sizes do the posters come in?
[AI] Anything from A5 to A1. I can also frame them to order – there's a selection of frames on the website.
What range of prices are there?
[AI] All prices are shown on the website.
If you do not stock a poster I want, can you source it for me?
[AI] Of course. I'd be happy to help in any way. Have a look through the website and if you can't find what you want, just drop me a line and I'll do my best!

Formal emails

The rules for writing formal emails are similar to those for formal letters.

- If you are communicating with someone for the first time, you should adopt the structure of a formal letter. (See the WRITING FORMAL LETTERS section of the supplement.)

- It is usual and proper to use a greeting of some sort when you begin your email. 'Dear' can never be misunderstood and rarely strikes the wrong note. If you are more familiar with the person you are writing to, 'Hi' or 'Hello' is fine. If you're writing to close friends, then use whatever greeting you are accustomed to in your social circle.

- If you are emailing someone for the first time without being invited to, it is always proper to explain at the very start of the email who you are and why you are writing to them.

- Never leave the subject line blank. In most formal or professional correspondence, you should aim to keep the email to one subject. Think clearly what the email is about and be as precise as you can. Keep the subject as short as possible, as the recipient's inbox will often only display a limited amount of the line.

- If the email is going to be long, it is polite to indicate this in the opening few sentences.

- Structure your email so that each point is addressed in a separate paragraph. If you wish, it is entirely acceptable to add a heading to each paragraph. Your reader can then see at a glance the points you are covering.

- In formal emails, texting abbreviations, emojis, and emoticons should be avoided.

- When you end your email, follow the same rules as with formal letters, using 'Yours faithfully' or 'Yours sincerely' as appropriate.

   PROFESSIONAL TIP

   I avoid using multiple clauses and long sentences, and use lists or bullet points rather than block text.

                                                   (Marketing manager)

Informal emails

For emails to friends and family you can be more relaxed in your style and tone.

- 'Hi' or 'Hey' or other informal greetings are appropriate and you can close the email with 'See you' or 'Lots of love' or other phrases.

- Even when you're writing to a friend, remember that email can seem terse and abrupt if there is no greeting or close.

- It is acceptable to use texting abbreviations, emojis, and emoticons in informal emails – just make sure the person you're writing to understands them all!

Emailing at work

Emails are the default means of communication for most businesses. Whenever you write an important email at work, it needs to be concise, clear, and well thought-out. In almost all cases, the tone should be formal.

Some research conducted across a wide range of professions suggested that on average people get well over 100 emails per day. Email adds a lot of extra reading and writing to an already busy work day. So the first question you should ask is, 'Is email the best way to say what I have to say?'. A frequent complaint of people in business is that they are sent or copied in on emails that they really don't need to see.

PROFESSIONAL TIP

My tip for effective communication in an organization? Don't email. Phone or speak to someone in person.
(CEO, International Management Consultancy)

Compare the two emails on the following pages, which show the difference between good and bad practice in business email writing.

Business emails: Good vs Bad

BAD EMAIL

> Subject:
> From "Gray, Tara" <Tara.Gray@bigbooks.com>
> To: "Jay Dixon"
>
> Jay
> There are problems re the arrangements for the conference dinner. I met with David yesterday to finalize details and here's what we decided: The venue (Carmichael Hall) is booked for Dec 18th 7:00 – 12:00. Send out the invitations to all conference delegates the week starting November 3rd. Find out from the printers when the invitations are being returned to us and get them send out in a timely fashion to the delegates. You'll need to check against th otiginal delegate list. Hallidays need to know final numbers by Dec. 10th., latest. Call them to let them know Liaise with Ravi over the timing of the speeches and let the relevant people know when their slot is.
> Can't think of anything else at the moment but it's down to you now.

- The tone is abrupt, ill-tempered, and dictatorial, without being helpful.
- The email is very sloppily written: there are spelling and grammatical mistakes throughout.
- Information required to carry out the actions is not provided.
- People the recipient needs to liaise with and get information from are not included in the email.
- The email generates more work for both the sender and the addressee.
- The email is one block of text, which makes identifying what needs to be done laborious and time-consuming.

GOOD EMAIL

Subject: Conference dinner action points 25/10
From: "Gray, Tara" <Tara.Gray@bigbooks.com>
To: "Jay Dixon", "Naidoo, Ravi" <Ravi.Naidoo@bigbooks.com>

Hi Jay

Thanks again for offering to help with the arrangements for the conference dinner. I do appreciate it. David and I had a meeting yesterday to finalize the details. I've included Ravi on the email so you can liaise directly with him on a couple of the points.

Venue
The venue (Carmichael Hall) is booked for December 18th 7:00 – 12:00. The contact there is Wendy Lewis. Her number is 6603136.

Invitations
The invitations are currently with the printers. When they are returned to us (this Friday, 1st November) they will need checking against the original delegate list.
Ravi: could you send the list to Jay please?
Invitations should go to all conference delegates the week starting November 3rd.

The caterers
Williams, the caterers, will need to know final numbers by Dec. 10th. Please call them to confirm these numbers (2222276).

Speeches
Finally, please could you liaise with Ravi over the timing of the speeches and let the relevant people know when their slot is? The list of speakers is on the shared drive. (K:/conference2015/speakerlist.doc)

Any questions, give me a call. I'm here all this week apart from Thursday afternoon.

Many thanks,
Tara

- The writer includes all the people required to do something in the 'To' box.
- There is a precise 'Subject' line.
- The paragraphs are set out so that addressing each point is easy.
- All the relevant information is included (dates, contact names and numbers, and file locations).
- The tone is collaborative, helpful, and professional.

> **PROFESSIONAL TIPS**
>
> *I keep all emails as brief as possible and use short sentences to get across all important points.*
>
> (Food journalist)
>
> *I read the email back to myself as though I were reading someone else's email prior to sending.*
>
> (Course coordinator)

When not to use email

It would be unwise to think that email can be used in all situations. For example, email isn't very good for conveying tone or emotion accurately. If you want to discuss something sensitive or personal with someone, it might be better to talk with them in person or write them a note.

SOCIAL MEDIA

The development of social media has significantly changed the way we communicate with each other. They differ greatly from conventional forms of communication: letters, phone calls, emails, etc. You are not speaking to a few specific people, but with tens, hundreds, or even of thousands at once. As well as words, you use pictures, clips, and internet links to share information. The best way to describe what you are doing when you use social media, such as Facebook, Twitter, or a blog, is that you are projecting a representation of yourself to a wide audience.

This makes social media very powerful and positive communication tools – for everyone from teenage friends to large corporations. The downside is that you have to be careful about what information you share, and how you share it: how you represent yourself.

As social media are constantly developing – in terms of their reach and the purposes for which they are being used – the ground rules for successfully using them are changeable. There are, however, some fundamental 'do's' and 'don'ts' which will be considered here along with the basic mechanics.

This guide focuses on Facebook and Twitter, because they are currently the most popular forms of social media. There are many others – like Instagram and LinkedIn – which have different formats and purposes, but most of the observations made here will still be applicable.

Social media 'do's'

- **Decide why you're using them.** Is it to keep in touch with friends, be entertained, promote a business or service, or a mixture of all three? Blogging, Twitter, and Facebook are used for all the above reasons, and they have different strengths and weaknesses.

- **Try to be consistent.** If you want to attract more followers on Twitter and other media, this is more likely to happen if people grow to trust and like your opinions or tweets.

- **Be positive.** It's a good principle to keep in mind, even if you don't always follow it. Anger and negativity do not generally translate well into social media, but a positive response to a negative issue can be effective and motivating. Consider whether you might say the same thing or share the same information, if your audience were in the room with you. If the answer is 'no', then think twice about posting. This is true whether your audience is made up of personal or business contacts, and particularly true if it contains both.

- **Be clear.** Nouns and facts work better on Facebook and Twitter than adjectives and adverbs – there is less room for misinterpretation of intention or mood if the message is clear and unambiguous. Of course, if you're casually chatting with friends on Instagram or Twitter, then it's a different matter.

- **Check your privacy settings.** This will help you avoid inadvertently sharing private information on Facebook and other media, or being embarrassed by something posted on your timeline to a wider audience than you would like.

- **Reread your message before you post it.** It's very easy to make mistakes in spelling or tone, especially if you're posting from a mobile phone with a very small keyboard. Consider using emoticons if you suspect your message is ambiguous.

Social media 'don'ts'

- **'Retweet' (RT) too often.** It can be off-putting to followers. Sharing a link to a video or article you've enjoyed is often welcomed, but do this infrequently to make a greater impact.

- **Post or tag pictures of friends or acquaintances on Facebook.** Unless you're sure they won't mind, keep intimate details of shared events for private Facebook messaging.

- **Be rude.** It's very tempting to react angrily to tweets or posts that we strongly disagree with – don't. Arguments can escalate very quickly in the online environment and you will almost certainly say things from behind your computer screen that you would not in real life. A well-reasoned objection or counter-argument will have more influence than an abusive message, but serious issues are unlikely to be resolved online. If someone you're following or have 'friended' on Facebook is a continual source of annoyance, then unfollow or block them.

- **Get upset if you are 'unfollowed'.** Twitter is a more impersonal medium than Facebook, and people change whom they follow with great frequency.

- **Mix up your work life and personal life.** This applies particularly on Facebook. Your work colleagues may be amused by a picture of you at last night's party but your boss, who may be a Facebook friend of one of them, might find it less amusing – especially if you call in sick the next day.

- **Post sensitive news which might be better relayed by telephone.** It's very easy to say things online that might be difficult to express over the phone, but this doesn't necessarily make it the better option.
- **Be repetitive in your posts.** Many people repeat variations on a theme they consider important, hoping to elicit a response, while their audience gets fed up with reading the same information over and over again.
- **Ramble.** On Twitter you're restricted in the number of characters you can use, which is good practice for learning how to express yourself succinctly. Try to keep Facebook updates to one or two lines, if possible. If you have a lot to say, you might be better off sending an email, or video-calling.
- **Forget to punctuate.** Just because messages are short, it doesn't mean that they will make sense without commas, full stops, and other punctuation marks.

Glossary of social media terms

- **@.** The @ sign is used in tweets directly before a username to turn it into a link to that person's profile: for example, @janesmith.
- **# (hashtag).** Hashtags are symbols which allow tweets on a subject to be grouped together and located by a 'hashtag search'. For instance, by including the expression '#spacestation' in a tweet, the user makes it possible for

other tweeters to search and locate all tweets with this expression in them.

- **Blocking.** Preventing someone from reading your tweets or posts by denying them access to your account.
- **Follow.** Receive messages posted by a particular social media user.
- **Friend.** A contact on Facebook.
- **Links.** Web page references included in Facebook posts and tweets, which can be clicked on to take the reader to the relevant web article or image.
- **Retweet(ing) (RT).** Resending or forwarding someone else's tweet with or without a comment of your own.
- **Status update.** The space on your Facebook account where you let your 'friends' know how you are feeling or what you are thinking about. Entries should be kept short.
- **Timeline.** The personal page of your Facebook account where 'friends' can post messages, links, and other material.
- **Tweeter.** A user of Twitter.
- **Tweet(ing).** The act of posting a tweet to Twitter.
- **Tweets.** Posts on Twitter.
- **Tweetup.** A physical meeting of tweeters – a 'Twitter meet-up'.

Blogging

A blog – or 'weblog' – is a website where someone writes their thoughts and opinions in the form of a post. Blogs range in content from online diaries to a promotional tool for business. Mostly they are expressions of personal opinions and beliefs, and so tend to be informal and chatty in tone.

Regardless of why you are blogging, some basic rules apply:

- **Think about your reader.** Are you planning to inform them? Entertain them? Persuade them? Whether it is one of the above or all three, you will have to write accordingly and be consistent in your approach, or the reader's interest will wane.

- **Be sincere and engaging.** A blog is usually quite conversational in style, so it may help to imagine you are chatting with the reader as you write. Share your experiences and tips, particularly if you're writing about a hobby or interest.

- **Keep posts short and interesting.** People tend to 'scan' blogs for words and images of interest – they may not be paying full attention. For this reason, try to keep sentences short and punchy. Include images and links to other web pages that reinforce your opinions or make the page look attractive. Add headings to break up the text, and keep paragraphs short.

PROFESSIONAL TIP

I use Facebook to keep in touch with friends around the world, and Twitter to keep tabs on the news and celebrities.
(Writer)

WRITING FORMAL LETTERS

Emails and other digital media are the primary forms of communication for most of us today, but there are still occasions when a handwritten or typed letter is more appropriate.

This section lays out the basic structure and the rules which underpin most formal letter writing.

As with any written communication, your letter should be in three discernible parts:

- **An introduction.** This is where you introduce yourself, acknowledge any previous correspondence, and briefly state the reason why you are writing. Ideally, it should be no longer than a paragraph of three or four sentences.

- **The middle.** This is the section where you expand your argument, provide further details, and raise any questions you have. The middle of the letter should be a series of paragraphs set out in a logical order. Each paragraph should make a clear, separate point. If the letter is long and covers a range of subjects, it may be appropriate to divide the contents by subheadings.

- **The ending or conclusion.** The final paragraph should set out what you would like to happen as a result of the communication, whether that be a written response, a meeting to discuss the contents of the letter, or a demand for a refund.

On the next page is an example of a formal letter:

14 Julian Road
Belleville

25 March 2018

Mr D Jacobs
Belleville Building Ltd
340 Shorter Street
Belleville

Dear Mr Jacobs

Estimate for extension to living room, 14 Julian Road

I am writing to thank you for the written estimate which I received this morning. I have queries about a few details in your letter which I would like to be resolved before we proceed any further.

First, can you say exactly when you would propose to begin work on the extension? I realize this depends, to some extent, on how quickly you can finish your current project. I need to know which week work would commence, however, so that I can make arrangements to store the living room furniture.

Second, can you tell me when you propose to fit the additional plumbing, so that I can arrange to stay with friends while there is no running water? Also, are there any other times when you anticipate that I shall be without water or electricity?

Finally, there is no mention of additional costs for materials. Can I assume, therefore, that these are included in the estimate you have provided for the overall cost of the extension?

Assuming I receive satisfactory answers in writing to these queries, I shall be happy to accept your proposal and go ahead with the project as discussed.

Yours sincerely

Joseph Mokoena

Joseph Mokoena

Points to remember:

- **Your address** – should be written in the top right corner of the letter. Do not write your name here, and don't put commas after each line.

- **The date** comes next. This should come under your address, also on the right. It is common practice to write the date as 25 March 2018, instead of 25th March 2018.

- **The recipient's address** appears under the date, but on the left side of the page. Again, no punctuation is required.

- **The greeting.** If you are writing to a friend, then you will use 'Dear Luke', for instance. Otherwise, it should be 'Dear Mr Jacobs'. Note that if you are writing to a woman and do not know whether she prefers to be addressed as 'Dear Mrs Jacobs', or 'Dear Miss Jacobs', then you should use 'Dear Ms Jacobs'. If you do not know the person's name, use 'Dear Sir or Madam'.

- **Headings.** If you are using a heading, it should summarize the subject matter of the letter, and appear between the greeting and the first paragraph. Headings should be written in **bold** but not capital letters.

- **The ending.** If you have used the name of the person in the greeting, then you should end with 'Yours sincerely'. Otherwise, end the letter with 'Yours faithfully'. 'Yours...' should always begin with a capital letter.

- **Punctuation.** It is not necessary to include a comma after the greeting or after the ending.

- **Signature.** Write your signature but include your typewritten name underneath. If writing freehand, print your name in capital letters beneath your signature.
- **Further contact details.** If you are including your email address or telephone number as contact details, these should be included underneath your postal address:

> 14 Julian Road
> Belleville
> jmokoena@email.com
> 4253468

Once you have written your letter, check that:
- You have explained why you are writing in the first paragraph.
- You have made all the points that you wanted to.
- Each sentence is clear, concise, and unambiguous in meaning.
- You have not included too much information or any irrelevant details.
- Your language has been courteous and polite, even if you are writing a letter of complaint.
- There are no spelling mistakes – it is surprisingly easy to miss them!

SOCIAL COMMUNICATION

Although seemingly belonging to a different world – predating mobile phones, texts, and tablet devices, all of which convey informal messages instantly and very well – there is still a time and a place for a carefully handwritten note or card, or a printed invitation.

Here are some general points to consider when writing social communications:

Tone

Apart from formal invitations, the tone of most social communications is informal and friendly:

- Salutations can range from 'Dear' to 'Hi'.
- Language is usually quite conversational, with shortened sentences and contractions ('I'm', 'won't'). It is more emotive and less factual than in business correspondence.
- Endings are similarly warm – 'Lots of love', 'Love', 'Speak/write soon'. They may be slightly more formal – 'Best wishes', 'Kind regards', 'All the best' – if you don't know your correspondent quite so well.

Format

Just because it is handwritten, it doesn't mean that a note shouldn't have a structure. It is still usual to have:

- An introductory line or paragraph, stating the purpose of the letter ('I was sorry to hear about your loss'; 'Thank you for the birthday card').

- A middle section expounding on the subject ('She was a wonderful woman'; 'The party went really well, all things considered').

- An ending ('I shall hope to speak to you at the memorial service'; 'Let's meet up before another year goes by').

PROFESSIONAL TIP

One thing I can't stand? The computer-generated Christmas card – it's so impersonal – "Look, I can do a mail merge on my PC!". If you can't be bothered to handwrite a greetings card, don't bother!

(Publishing assistant)

Here are some examples of different kinds of social correspondence:

Letter of condolence

46 Eve Street
Whitewater

12 May

Dear Grant

I am writing to say how sorry I was to hear of your loss, and that I am thinking of you at this difficult time. Although I was aware that Mariam was ill, I was nevertheless shocked to hear of her passing.

I know she was never happier than when she had met you, and the two of you made a lovely couple. She seemed to light up the life of everyone who met her.

I shall certainly attend the memorial service next Thursday, but if there is anything I can do in the meantime, please don't hesitate to call me. I'm sure Kay and Mateo are a great comfort to you at the moment.

Thinking of you all with love and affection.

Michelle

- The writer should be acutely sensitive to the addressee's feelings, rather than trying to express their own emotions. The references made to the deceased in this letter are mainly in the context of her relationship with the bereaved partner, rather than the writer.

- A handwritten note can be more appropriate than a phone call in situations of grief and loss. Writing a letter also gives you more time to think about what you want to say, and how you want to say it.

- Note that the letter, although it is informal in address and tone, still has a discernible structure: the introductory sentence explains the purpose of the letter; the middle paragraph expands on the theme with the writer's memories of the deceased; and the final paragraph acknowledges the future by accepting an invitation and offering support.

Invitations

Invitations are frequently made by email or by text these days, but there are occasions when a written or printed invitation is still the prevalent form of communication.

A WEDDING INVITATION

Martin and Ruth Edwards
request the pleasure of the company of
Phillip and Sally Thomas
at the wedding of their daughter Naomi
to James Adams
on Saturday July 14 2018
at St Francis Church, Belmont at 2pm.

R.S.V.P.
Ruth Edwards, 121 Donald Street, Belmont
Tel: 625 5431.

- The most important feature of an invitation is that it must possess all the necessary information to allow the recipient to respond with an acceptance or a refusal. There is no point sending out wedding invitations to 400 guests without the date on them!

- Social invitations can be informal or formal. Formal invitations – to a wedding or a christening, for example – will usually be printed.

Replying to invitations

- When replying to invitations, match the style and formality of the invitation.

- If you have to decline an invitation, it is good practice to sound apologetic and regretful and explain the reason why you cannot attend.